Geometry

AN INTEGRATED APPROACH

GERVER, SGROI, HANSEN
LYNCH & MOLINA

National Textbook Company
a division of NTC/CONTEMPORARY PUBLISHING GROUP
Lincolnwood, Illinois USA

Vice President of Publishing	Peter McBride
Team Leader	Thomas A. Emrick
Project Manager	Enid Nagel
Editors	Tamara S. Jones, Charles B. Sonenshein, Edna Stroble
Assistant Editor	Darrell E. Frye, Suzanne Thomas
Marketing Manager	Colleen J. Skola
National Mathematics Consultants	Carol Ann Dana, Everett T. Draper
Manufacturing Coordinators	Jennifer D. Carles, Brian Harvey
Art and Design Coordinator	John Robb
Photo Editing	Kathryn A. Russell/Pix Inc
Editorial Assistant	Lisa M. Bell
Marketing Assistant	Laurie L. Brown
Composition/Scans/ Prepress/Imaging	Better Graphics, Inc.
Cover Design	Photonics Graphics

About the Cover The cover design is a collage of real world images from the chapter themes. How many can you find?

ISBN: 0-538-67122-X

Published by National Textbook Company,
a division of NTC/Contemporary Publishing Group, Inc.
4255 West Touhy Avenue,
Lincolnwood (Chicago), Illinois 60712-1975 U.S.A.

Manufactured in the United States of America.

6 7 8 VH 04 03 02 01 00

Experience a world of
APPLICATIONS AND CONNECTIONS
when you explore geometry

 NATURE You'll analyze many objects from nature through the symmetry of their geometric models. (*Lesson 1.1, page 5*)

NAVIGATION You'll discuss the path of a sailboat as you learn about identifying vertical angles and their relationships. (*Lesson 3.1, page 134*)

 BRIDGES You'll examine geometric properties that make bridges stable and strong. (*Chapter 6 Project, page 312*)

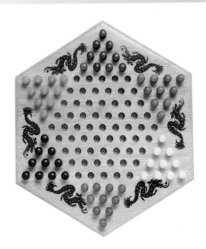

GAMES You'll discover how games use circles in their designs, as well as in strategies for winning. (*Chapter 12 Project, page 662*)

. . . and many more

Connecting the World to Geometry

To the Student

You are about to embark on an exciting geometry program!

South-Western Geometry bridges your skills and knowledge of today to what you will be doing in the future. In *South-Western Geometry*, you will connect geometry to topics such as earth science, travel, technology, art, ecology, and architecture.

Chapter themes help you see how mathematics applies to daily life, and in the **Data Activity** lessons you will use real world data in various ways, including reading and interpreting charts and graphs. **GeometryWorks** showcases a specific job or career which requires the geometry you are learning. You will see the importance geometry has in everyday life!

The **Chapter Project** helps you to develop mathematical understanding and connect geometry to the real world. You can research your projects on the Internet using the address given in the *Internet Connection*.

The **Problem Solving File** lessons provide strategies that will sharpen your critical thinking and problem solving skills!

Geometry Workshops and *Explore* activities help you understand ideas so you can become comfortable with geometry.

The **Focus on Reasoning** lessons allow you to work with a group of students to explore and practice a key reasoning skill that will be important throughout your life. **Now You See It Now You Don't** sharpens your visual skills by examining optical illusions.

Review and **Assessment** give you the opportunity to test and practice your skills and knowledge.

You will use **calculators** and **computers** as tools for learning and working with mathematics. Graphing calculators and geometry software can help you focus on geometry elements.

South-Western Geometry will make you a confident problem solver who can clearly communicate in mathematics. You will feel empowered as you realize the endless opportunities that geometry brings to you.

GEOMETRY: AN INTEGRATED APPROACH

Reviewers

Wilma R. Allen
Mathematics Teacher
Alamo Heights High School
San Antonio, Texas

Nancy Borchers
Mathematics Teacher
Taylor High School
North Bend, Ohio

Marilyn Dewoody
Mathematics Department Chair
Fort Gibson High School
Fort Gibson, Oklahoma

Alfred DiCaprio
Mathematics Teacher
Kingston High School
Kingston, New York

Kathleen Gandin
Mathematics Teacher
Clear Creek High School
League City, Texas

Mary Margaret Jones
Mathematics Teacher
Germantown High School
Memphis, Tennessee

Joan C. Lamborne
Mathematics Supervisor
Egg Harbor Township High School
Egg Habor Township, New Jersey

Geri Margin
Assistant Principal
North Brunswick Twp High School
North Brunswick, New Jersey

Donna Mechley
Mathematics Teacher
Turpin High School
Cincinnati, Ohio

Roger O'Brien
Supervisor of Mathematics
Polk County Schools
Bartow, Florida

Stephanie Ogden
Mathematics Teacher
Webb School of Knoxville
Knoxville, Tennessee

Nitza Peraza
Mathematics Teacher
Bassett High School
La Puente, California

Beth Petersen
Mathematics Teacher
State University Station
Fargo, North Dakota

Audrey C. Reineck
Mentor Teacher
Milwaukee Public Schools
Milwaukee, Wisconsin

Greta Staley Robertson
Mathematics Teacher
Independence High School
Columbus, Ohio

Steve Taylor
Mathematics Teacher
Perry Meridian High School
Indianapolis, Indiana

Mary Williams
Mathematics Department Chair
Robert E. Lee High School
Tyler, Texas

Patricia Wadecki
Mathematics Teacher
New Trier High School
Winnetka, Illinois

Mary Sue Wyss
Mathematics Teacher
Anchor Bay High School
New Baltimore, Michigan

ABOUT THE Authors

Robert Gerver, Ph.D., has been a mathematics instructor at North Shore High School in Glen Head, New York since 1977. During that time, he has also taught at several other institutions, grade 4 through graduate level. He has served on the New York Regents Competency Test Committee. He received his B.A. and M.S. degrees from Queens College of the City University of New York where he was elected to Phi Beta Kappa. He received his Ph.D. from New York University. In 1988, Dr. Gerver received the Presidential Award for Excellence in Mathematics Teaching for New York State. In 1997, he was awarded the Tandy Prize for Excellence in Mathematics Teaching and the 1997 Chevron Best Practices in Education Award.

Chicha Lynch, mathematics department head at Capuchino High School, San Bruno, California, has taught geometry during most of her career. Most recently Ms. Lynch has been assisting the mathematics department in implementing an integrated curriculum using technology at all levels. Ms. Lynch has been a Master teacher for the Stanford Teacher Education Program, and received its LaBosky Award for her contribution to its teacher education program. She was elected by her peers to serve two terms as a California State Mentor Teacher, and was a state finalist for the Presidential Award for Excellence in Mathematics Teaching. Ms. Lynch is a graduate of the University of Florida.

David Molina, Ph.D., is the Associate Director of the Charles A. Dana Center, and an Adjunct Professor of Mathematics Education at the University of Texas at Austin. He has taught secondary mathematics in Maryland, and his work includes contributions to educational technology, mathematics education policy, curriculum development, systematic changes, teacher preparation reform, and school restructuring. Dr. Molina received his B.S. degree from the University of Notre Dame, and his M.A. and Ph.D. from the University of Texas, Austin.

Richard Sgroi, Ph.D., is mathematics director of Newburgh Enlarged City School District, Newburgh, New York. He has taught mathematics at all levels for more than 20 years. While teaching at North Shore High School, Glen Head, New York, he and his colleague Robert Gerver developed both *Dollars and Sense: Problem Solving Strategies in Consumer Mathematics* and *Sound Foundations: A Mathematics Simulation.* He and Dr. Gerver received recognition from the U.S. Department of Education's Program Effectiveness Panel for *Sound Foundations* to be included in Educational Programs That Work. Dr. Sgroi received his B.A. and M.S. degrees from Queens College of the City University of New York, and his Ph.D. from New York University.

Mary Hansen, a mathematics teacher at Independence High School, Independence, Kansas, teaches algebra and geometry. While teaching in Texas and North Carolina, Ms. Hansen was active in mathematics teachers staff development. She received her B.A. and M.A.T. degrees from Trinity University, San Antonio, Texas, where she was elected to Phi Beta Kappa.

Exploring Geometry

THEME: Nature

Applications and Connections

Technology and Other Tools

2 Geometry and Logic

THEME: Computer Technology

3 Lines, Planes, and Angles

THEME: Navigation

4 Transformations in the Plane

THEME: Textiles and Designs

5 Triangles and Congruence

THEME: Art

6 A Closer Look at Triangles

THEME: Architecture

Applications and Connections

Technology and Other Tools

7 Quadrilaterals

THEME: Landmarks

8 Perimeters and Areas of Polygons

THEME: Landscaping

9 Similarity

THEME: Maps

10 Applications of Similarity and Trigonometry

THEME: Construction and Carpentry

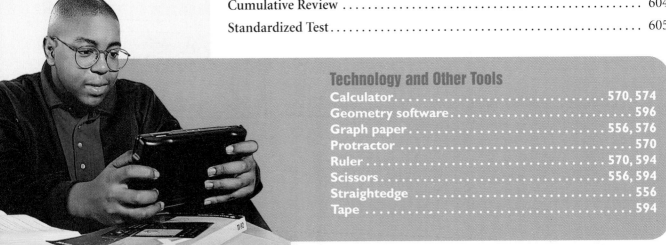

11 Geometry of the Circle

THEME: Inventions

12 Circles: Lines, Segments, and Angles

THEME: Toys and Games

13 Surface Area and Volume

THEME: Packaging

Applications and Connections

Technology and Other Tools

Also...

Take a Look AHEAD

Make notes about things that look familiar.
- What geometric shapes do you recognize? What are their names?
- How do you think coordinate planes will be involved in your study of geometry?

Make notes about things that look new.
- List the types of drawings that you will be studying in this chapter. Why do you think drawing techniques are important in geometry?
- What is the difference between a two-dimensional figure and a three-dimensional figure?

DATA Activity

Diamonds in the Rough

A diamond is the world's most precious crystal. Diamonds are made of carbon, which is the same substance that makes up the "lead" of an ordinary pencil. Diamonds are formed by pressures that exist more than 150 mi below Earth's surface at temperatures greater than 1200°C. Over the years, some remarkably large diamonds have been mined.

SKILL FOCUS

- ► Compare numbers.
- ► Convert measures.
- ► Determine the mean of a set of data.
- ► Estimate a fractional part of a number.
- ► Find a percent of a number.
- ► Construct a bar graph.

GEOMETRY WORKS

Nature

In this chapter, you will see how:

- **FORENSIC SCIENTISTS** use pattern investigation in fingerprint analysis. (Lesson 1.2, page 15)
- **CHEMICAL TECHNICIANS** use symmetry to analyze chemical compounds. (Lesson 1.3, page 22)
- **TOURIST BUREAU MANAGERS** use scale to interpret maps and to answer tourists' questions. (Lesson 1.6, page 39)

FAMOUS DIAMONDS	
Name	**Original Weight (carats)**
Blue Tavernier	112.5
Cullinan	3106
Excelsior	995
Great Mogul	817
Jacobus Jonker	726
Koh-i-nor	191
Regent	410
Vargas	726.6

Use the table to answer the following questions.

1. One carat is equal to one fifth of a gram. What is the mass in grams of the largest diamond listed? of the smallest diamond listed?

2. What is the mean weight in carats of all the diamonds listed?

3. Which diamond had an original weight that was about one eighth the original weight of the Cullinan?

4. It is believed that the famous Hope Diamond was cut from the Blue Tavernier. The cut weight of the Hope Diamond is about 40.5% the original weight of the Blue Tavernier. What is the weight in carats of the Hope Diamond?

5. **WORKING TOGETHER** Make a bar graph of the data in the table. How do you decide what intervals to use on your scale of weights? Compare your graph with those of other groups and discuss similarities and differences.

It's Crystal Clear!

A crystal is a solid substance whose molecules are arranged in a fixed geometric pattern. The word *crystal* is derived from the Greek word *krystallos*, which means "clear ice."

For centuries, people have studied both the mysteries and science of crystals. The mysteries? Many ancients felt that crystals held magical powers. The science? Metallurgists, chemists, physicists, geologists and other scientists have explored the atomic structure of the crystal.

PROJECT GOAL

To investigate and understand crystals.

Getting Started

Work with a group.

1. Gather the following materials: two wide-mouthed glass jars, two pencils, two bowls, table salt, sugar, tape, and cotton thread.

2. Tie a length of cotton thread to the middle of one pencil. Tape one end of the thread to the center of the inside bottom of one jar. Roll up the thread on the pencil until it is tight. Tape the pencil to the jar. Repeat this procedure for the other pencil and jar.

3. Fill each bowl with warm water. In the first bowl, dissolve as much salt as possible. In the second bowl, dissolve as much sugar as possible.

4. Pour the salt water into one jar and the sugar water into the other. Label the jars *salt* and *sugar*. Cover each jar with a napkin. Place the jars in a location out of direct sunlight.
You should be able to observe the jars without moving them.

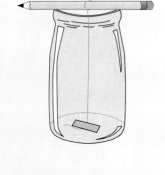

Internet Connection

www.learninggeometry.com

PROJECT Connections

Lesson 1.2, page 14:
Research facts about crystals and collect samples.

Lesson 1.4, page 28:
Identify symmetries on surfaces of crystals.

Lesson 1.6, page 38:
Model an epsomite crystal and examine its symmetries.

Lesson 1.7, page 45:
Research systems of crystals and make isometric and orthogonal drawings of their basic structures.

Chapter Assessment, page 59:
Plan a presentation to communicate final project results.

1.1 Geometry Workshop

Exploring Shapes

Think Back/Working Together

Geometry is the mathematics of shapes. Shapes can be used to understand nature.

Work with a partner. Examine these photographs.

SPOTLIGHT ON LEARNING

WHAT? In this lesson you will learn
- to work with geometric models.
- to identify figures with reflection symmetry.

Why? You can analyze many objects from nature through the symmetry of their geometric models.

1. Name any geometric shapes you recognize in the objects shown.

2. Which of the objects do you think are most alike in shape? Explain your reasoning.

3. Identify the object whose shape you think is most unlike the shape of the other objects. Explain your reasoning.

4. Name as many of these geometric figures as you can. Give the most specific name you know.

 I.   II.  III.

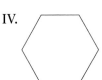 IV.   V.  VI.

5. Which of figures I – VI do you think are most alike in shape? Explain your reasoning.

6. Which of figures I – VI do you think has a shape most different from the others? Explain your reasoning.

Explore

Work with a partner. Use a mirror.

7. Consider figures I – VI above. For each figure, decide if it is possible to place a mirror so the part of the figure that is visible together with its mirror image appears to form the original figure.

Copy the rectangle at the right onto a sheet of paper. Show how to place a mirror so the visible part of the figure together with its mirror image forms the following.

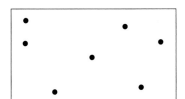

8. two dots inside a square

9. an "empty" square

10. ten dots inside a rectangle

11. fourteen dots inside a six-sided figure

12. Draw three straight lines that divide the interior of the rectangle into regions that each contain exactly one dot.

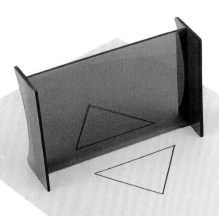

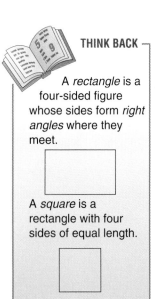

PROBLEM SOLVING TIP

If you do not have a mirror, trace the figure onto a piece of paper. If you can fold the figure so that one half fits exactly over the other half, then each half is the mirror image of the other.

THINK BACK

A *rectangle* is a four-sided figure whose sides form *right angles* where they meet.

A *square* is a rectangle with four sides of equal length.

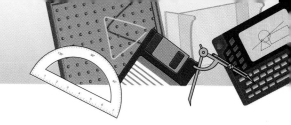

Make Connections

At the right is a *geometric model* of the leaf in the photograph on page 5. A **geometric model** is a geometric figure that represents a real life object. A good model shows all the important characteristics of the object it represents.

13. What characteristics of the actual leaf are shown in the model?

COMMUNICATING ABOUT GEOMETRY

In algebra, you used equations to model real life situations. How are *algebraic models* like geometric models? How are they different from geometric models?

Often you use a geometric model as a tool for studying an object. When you make a geometric model, you may leave out those characteristics of the object that are not important to your study.

14. What characteristics of the leaf are *not* shown in the model above?

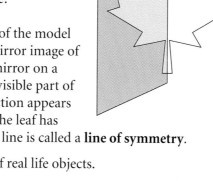

One important characteristic of the model leaf is that one half of it is a mirror image of the other half. If you place a mirror on a vertical line at the center, the visible part of the leaf together with its reflection appears to form the whole leaf. Then the leaf has **reflection symmetry**, and the line is called a **line of symmetry**.

TECHNOLOGY TIP

You can use geometry software to reflect any figure. Each drawing you make will have a line of symmetry

Here are some other models of real life objects.

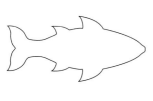   

Identify the model(s) with the given characteristic.

15. only a vertical line of symmetry

16. only a horizontal line of symmetry

17. both a vertical and a horizontal line of symmetry

18. neither a vertical nor a horizontal line of symmetry

CHECK UNDERSTANDING

What is the difference between vertical and horizontal?

Natural objects seldom have the perfect reflection symmetry of geometric models. Often small irregularities make one part of an object slightly different from its "mirror image."

19. On the left is a photograph of Noriko. How are the other two photographs related to the original photograph?

Summarize

20. WRITING MATHEMATICS Suppose you are a magazine editor. Write a caption for the photograph at the left in which you describe the reflection symmetry of the scene.

21. THINKING CRITICALLY Make a list of ten natural objects other than those described in this lesson that have reflection symmetry.

GOING FUTHER Trace and cut out ten copies of the triangle at the right. Then arrange the copies so the "dotted" corners all meet and you form a figure with the given characteristic. You may turn the shapes over but the dotted corners must still meet.

22. only a vertical line of symmetry

23. only a horizontal line of symmetry

24. neither a vertical nor a horizontal line of symmetry

25. Use only eight triangles to create a figure that has both a vertical and a horizontal line of symmetry.

26. Use the figure that you formed in Question 25. Give it a one-fourth turn around its center point. What do you notice?

1.2 Plane Figures and Symmetry

Explore

- You will need a rectangular sheet of paper and a pair of scissors.

  1. Fold and cut the paper as shown below. Discard the piece you cut off. Unfold the remaining paper. What shape is it?

  2. In how many different ways can you fold the shape so that one half fits exactly over the other half?

  3. Place the paper on a flat surface. With a pencil point at the center, turn the paper slowly. During one complete turn, how many times does the paper coincide with its original position?

> **SPOTLIGHT ON LEARNING**
>
> **WHAT?** In this lesson you will learn
> - to use the undefined terms point, line, and plane.
> - to identify the reflection symmetry and rotation symmetry of a plane figure.
>
> **WHY?** Symmetry can help you understand objects as tiny as a snowflake and as large as the rings of the planet Saturn.

Build Understanding

- The terms *point, line,* and *plane* are considered the basic *undefined terms* of geometry. The word **point** identifies a location. You use a dot to model a point, but a point actually has no size. A capital letter is usually used to name a point.

  A **line** is a set of points that extends without end in two opposite directions. When you draw a line, it appears to have thickness, but actually it has none. Two of its points are usually used to name a line. A lowercase letter can also be used. Below is line ℓ or line PQ, written $\overrightarrow{PQ}$ or $\overrightarrow{QP}$.

  A **plane** is a set of points that extends without end in all directions along a flat surface. Like a line, a plane has no thickness. Although a plane has no edges, you usually show it as a four-sided shape.

  plane *Z*

  A *figure* is any set of points A set of points that all lie in the same plane is called a **plane figure**. Much of your work in geometry will involve the study of plane figures and their properties.

COMMUNICATING ABOUT GEOMETRY

Examine the figures shown in the text at the left. Describe how points, lines, and planes are labeled and named.

A plane figure has **reflection symmetry**, or **line symmetry**, if you can divide it along a line into two parts that are mirror images of each other. The line is called a **line of symmetry**. Some figures have one line of symmetry, others have two or more, and others have none.

EXAMPLE 1

Draw all lines of symmetry in each figure.

a.

b.

c.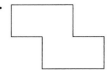

Solution

a.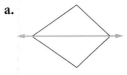

b.

c. There are no lines of symmetry.

A plane figure has **rotation symmetry** if you can turn it around a point so it coincides with its original position two or more times during a complete turn. The point is called the **center of symmetry**. The number of times the figure coincides with its original position during the complete turn is called its **order of rotation symmetry**. If n is the number of times, then the figure has *n-fold rotation symmetry*.

EXAMPLE 2

GAMES AND TOYS Determine the order of rotation symmetry for the face of the pinwheel shown at the left.

Solution

Model the face as a plane figure. Turn the figure clockwise around point C, the center of symmetry. It coincides with its original position four times during a complete turn.

    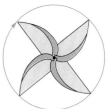

one-fourth turn   one-half turn   three-fourths turn   complete turn

So, the face of the pinwheel has 4-fold rotation symmetry.

EXAMPLE 3

Determine the order of rotation symmetry of the figure at the right.

Solution

Turn the figure clockwise around point *P*. It coincides with its original position only after one complete turn.

So, the figure has no rotation symmetry. ◄

CHECK UNDERSTANDING

In Example 3, think of a box whose lid has a shape like this.

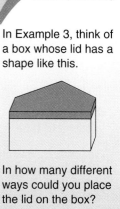

In how many different ways could you place the lid on the box?

TRY THESE

Make a drawing to illustrate each figure. Label your drawing.

1. plane $\mathcal{M}$ **2.** point *Y*   **3.** line *AB*

Trace each figure onto a sheet of paper. Then draw all its lines of symmetry. If a figure has no lines of symmetry, write *none*.

4. 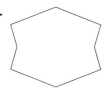   **5.**   **6.**   **7.**

Determine the order of rotation symmetry for each figure. If a figure has no rotation symmetry, write *none*.

8. 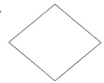   **9.**   **10.**   **11.**

MANIPULATIVES **For each exercise, carefully fold a rectangular piece of paper in fourths as shown at the right. Then cut the folded paper as indicated. Predict how the paper will look after you unfold it.**

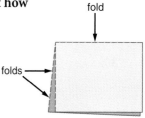

12. 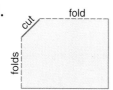   **13.** 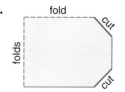   **14.**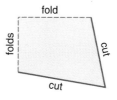

15. Describe the symmetries of the designs that result when the paper is unfolded in Exercises 12 through 14. Explain how you are able to predict the design of the paper before you unfold it.

16. CRYSTALS Describe the symmetries of the snowflake pictured at the right.

17. WRITING MATHEMATICS Explain why a circle has infinitely many lines of symmetry.

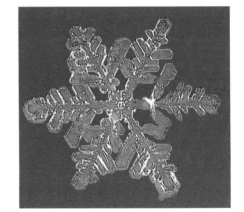

PRACTICE

Replace each ___?___ with the geometric term that makes the statement true.

18. A ___?___ has no thickness, but extends infinitely in two opposite directions.

19. A ___?___ has no thickness, but extends infinitely in all directions along a flat surface.

20. A ___?___ is a location. It has no thickness or length.

Trace each figure onto a sheet of paper. Then draw all its lines of symmetry. If a figure has no lines of symmetry, write *none*.

21.   **22.** 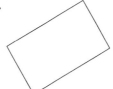   **23.**    **24.**

Determine the order of rotation symmetry for each figure. If a figure has no rotation symmetry, write *none*.

25.    **26.**    **27.**    **28.**

29. WRITING MATHEMATICS Explain how the geometric terms *point*, *line*, and *plane* are related to the types of symmetry that you studied in this lesson.

NATIVE AMERICAN BASKETRY Describe the symmetries of each Navajo basket design.

30. 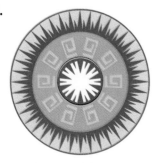   **31.**    **32.**

EXTEND

Trace each figure onto a sheet of paper. Then complete the figure so that $\overleftrightarrow{AB}$ is a line of symmetry.

33.

34.

35.

You know that a three-sided plane figure is called a triangle. **Draw a triangle that satisfies each condition.**

36. It has only one line of symmetry.

37. It has three lines of symmetry.

38. It has no lines of symmetry.

39. It has rotation symmetry.

LANGUAGE **For Exercises 40–44, use the letters of the alphabet as shown below.**

A B C D E F G H I J K L M
N O P Q R S T U V W X Y Z

40. The letters B and E each have a horizontal line of symmetry, and so the word BE has a horizontal line of symmetry.

List all the letters that have a horizontal line of symmetry. Then use the letters to create two other words that each have a horizontal line of symmetry.

41. The letters A and T each have a vertical line of symmetry. So the word AT has a vertical line of symmetry when written as shown.

List all the letters that have a vertical line of symmetry. Then create two other words that each have a vertical line of symmetry.

42. List all the letters that have both a vertical and a horizontal line of symmetry. Can you create any words that have both vertical and horizontal lines of symmetry?

43. What is special about the symmetries of the word MOM? Can you create any other words that have the same symmetries?

44. List all the letters that have rotation symmetry. Can you create any words that have rotation symmetry?

THINK CRITICALLY

45. ASTRONOMY Many objects in space travel in paths that have the shape of an *ellipse*, which is shown at the right. Since the time of their discovery, it was thought that the rings of the planet Saturn were shaped like ellipses. In 1888, Russian mathematician Sonya Kovalevsky (1850 – 1891) proved that the rings of Saturn are egg-shaped. Describe the differences in the symmetries of these two shapes.

ellipse

46. A *palindrome* is a number, word, or group of words that reads the same forward and backward. Do you think that palindromes are geometrically symmetric? Explain your reasoning.

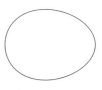

egg

47. INDUSTRIAL DESIGN The metal plates over entrances to underground utilities are popularly called "manhole covers." Often these are shaped like circles. Why do you think this shape is so practical for this purpose?

ALGEBRA AND GEOMETRY REVIEW

Evaluate each expression when $x = -2$, $y = 3$, and $z = -0.5$.

48. xyz 49. xyz^2 50. $xy + z$ 51. $\dfrac{x + y}{z}$ 52. $x^3 + 2y - z$

53. STANDARDIZED TESTS Which figure has neither reflection nor rotation symmetry?

A.    B.    C.    D.

CONSUMERISM **The following are one-way local bus fares in nine cities: 75¢, $1.00, 60¢, 85¢, 55¢, $1.00, 80¢, 75¢, 65¢. Find each statistical measure to the nearest cent for these data.**

54. range   55. mean   56. median   57. mode(s)

PROJECT *Connection*   In this activity, you will begin the process of gathering crystals and facts about crystals.

1. Research crystals from a historical perspective. What are some of the myths and mysteries associated with them?

2. Talk with a science teacher to learn how you might grow crystals other than salt and sugar crystals. Set up an experiment in which you grow some of these other types of crystals.

3. Collect five different samples of crystals that are already formed. Look for crystals of substances such as salt, sugar, brown sugar, rock salt, kosher salt, and cream of tartar. You might also find crystals on the rims of jars of honey, maple syrup, soy sauce, molasses, lemon juice, and similar foods.

A forensic scientist applies scientific methods to criminal investigations. Such investigations often involve the analysis of fingerprints.

Fingerprints are formed when oil and perspiration on the ridges of a person's fingertips transfer the pattern of the ridges to an object. Each person's fingerprints have a unique pattern.

Decision Making

Two basic fingerprint patterns are *arches* and *loops*.

   plain arch   tented arch   right-hand ulnar loop left-hand ulnar loop

1. Describe the differences you see between the two types of arches.

2. Describe the differences you see between the two types of loops.

A third basic fingerprint pattern is the *whorl*.

   plain whorl   central-pocket loop   double loop   accidental

3. Describe the differences you see among the four types of whorls.

4. Identify any symmetries you see in the patterns shown above.

5. Fingerprint yourself by gently pressing your fingertips first on an ink pad, then on a white sheet of paper. Classify your fingerprints according to the arch, loop, and whorl types.

6. Work in a group of four or five. Each student should make a copy of his or her right thumbprint on two separate index cards. Place all the index cards together and shuffle them. Work together to find the matching prints.

Three-Dimensional Figures and Symmetry

Explore

● Below are some geometric figures.

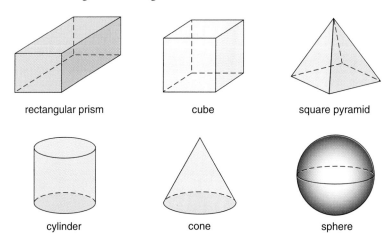

rectangular prism   cube   square pyramid

cylinder   cone   sphere

1. How are the geometric figures above different from those you worked with in Lesson 1.2?

2. What is the meaning of the dashed lines in each picture?

3. How are the figures in the top row different from those in the bottom row?

4. Give an example of a real-life object each figure might model.

Build Understanding

● In Lessons 1.1 and 1.2, you used plane figures to model real objects. Another term that describes such models is *two-dimensional*. However, real objects are not actually two-dimensional. They have an added dimension, which often is called *depth*.

Space is defined as the set of all points. Geometric figures that extend beyond a single plane into space are called *three-dimensional* figures. All the figures you examined in Explore are examples of three-dimensional figures called *solid figures*. A **solid figure**, or **solid**, is a three-dimensional figure that consists of all its surface points and all the points the surface encloses.

A three-dimensional figure has **reflection symmetry**, or **plane symmetry**, if you can divide it along a plane into two parts that are mirror images of each other. The plane is called a **plane of symmetry**.

EXAMPLE 1

Find the number of planes of symmetry
of the box at the right.

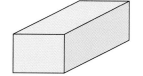

Solution

The shape of the box is a rectangular prism.
The planes of symmetry are shown below.
There is one horizontal plane, plane P.
There are two vertical planes, planes Q and R.

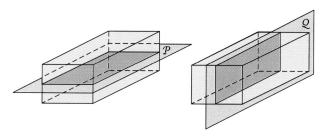

So the box has three planes of symmetry in all.

The **intersection** of two or more geometric figures is the set of points
common to all the figures. When a solid and a plane intersect, the
intersection is called a **cross section** of the solid. In Example 1, the
dark rectangular regions are cross sections of the solid.

Not all cross sections of a rectangular solid are
rectangular. For example, when plane S
intersects the same rectangular solid, as shown
at the right, the cross section is triangular.

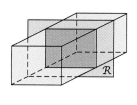

CHECK — UNDERSTANDING

Is it possible for the
intersection of a plane
and a solid to be just
one point? Can the
intersection be part of
a line? Explain.

EXAMPLE 2

Draw the cross section that results when plane N
intersects the spherical solid at the right.

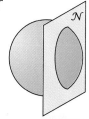

Solution

Imagine that you could slice the
solid as you would slice an orange.
The shape of the cut would be
a circular region, as shown.

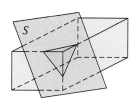

A three-dimensional figure has **rotation symmetry** if you can turn it around a line so it coincides with its original position two or more times during a complete turn. The line is called an **axis of symmetry**.

EXAMPLE 3

At the right is a game piece that is shaped like a house. Find the number of axes of symmetry.

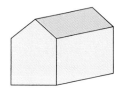

Solution

If you turn the figure clockwise around line *m*, it coincides with its original position twice during a complete turn.

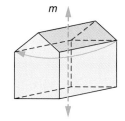

one half turn

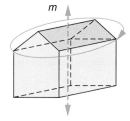

complete turn

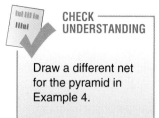

CHECK UNDERSTANDING

Draw a different net for the pyramid in Example 4.

There are no other such lines. So, the figure has exactly one axis of symmetry. ◄

A **net** for a solid is a two-dimensional figure that, when folded, forms the surface of the solid. Understanding the symmetries of a solid can help you draw a net for it.

EXAMPLE 4

SOCIAL SCIENCES Marianna wants to make a model of one of the pyramids of Giza. Draw a net for her model.

Solution

The visible surface of each pyramid consists of four identical triangular regions. The pyramid rests on a base that is a square. Below is one possible net for her model.

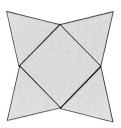

◄

18 CHAPTER 1   **Exploring Geometry**

Shown at the right is a solid shaped like a *triangular pyramid* with a base that is an equilateral triangle. You also are given a net for the solid.

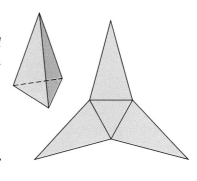

1. How many planes of symmetry does the solid have?

2. How many axes of symmetry does the solid have?

Sketch the cross section where each plane intersects the solid.

3.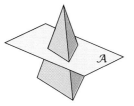

4.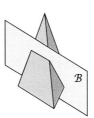

PACKAGING **Consider the soup can shown at the right. Imagine that it is resting on a horizontal surface.**

5. Name the geometric figure you can use to model the shape of the can.

6. **a.** How many horizontal planes of symmetry does the model have?
   b. How many vertical planes of symmetry does the model have?
   c. How many planes of symmetry does the model have in all?

7. How many axes of symmetry does the model have?

8. Which could be a net for the surface of the model?

   I.   **II.**   **III.**   **IV.**

MODELING **You will need a pencil, tape, paper, and scissors.**

9. Tape a rectangular piece of paper to a pencil as shown at the right. "Spin" the paper by quickly rolling the pencil back and forth between the palms of your hands. Describe the geometric figure the spinning paper appears to form.

10. Repeat Exercise 9 by spinning each figure along the dashed line.

   a.   **b.**   **c.**   **d.**

11. How are Exercises 9 and 10 related to the symmetries you studied in this lesson?

12. WRITING MATHEMATICS Compare the symmetries of plane figures and three-dimensional figures. How are they alike? How are they different?

PRACTICE

Imagine that the box shown is resting on a horizontal surface.

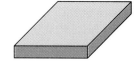

13. What geometric figure can you use to model the shape of the box?

14. WRITING MATHEMATICS Describe how the shape of this box is different from the shape of the box in Example 1 on page 17.

15. **a.** How many horizontal axes of symmetry does the box have?
   b. How many vertical axes of symmetry does it have?
   c. How many axes of symmetry does it have in all?

16. How many planes of symmetry does the box have in all?

CELEBRATIONS **Consider the party hat shown at the right.**

17. What geometric figure can you use to model the shape of the hat?

18. How many planes of symmetry does the model have?

19. How many axes of symmetry does the model have?

20. Suppose you work at a store that sells party favors. You have to stack several of these hats on a shelf. In how many ways can one hat be stacked on top of another?

21. PACKAGING The shape of the box at the right is called a *triangular prism*. Draw a net for the box.

Each figure shows the intersection of a cylinder and a plane. Match each figure with its cross section.

22.

23.

24.

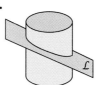

25.

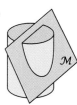

I.

II.

III.

IV.

CONSTRUCTION **The drawing shows an in-ground swimming pool. Notice that it is deep at one end and shallow at the other.**

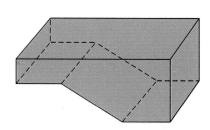

26. Describe the three-dimensional symmetries of the pool.

27. When the pool is filled with water, the surface of the still water represents a horizontal cross section of the pool. What is the shape of the cross section?

EXTEND

28. **ARCHITECTURE** The basic shape of the Transamerica Tower in San Francisco is a square pyramid. Each floor is a horizontal cross section of the pyramid. What is its shape? How is the size of each floor related to the size of the floor below it?

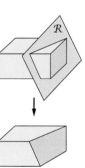

A cube is a special rectangular prism. Its surface consists of six identical square regions.

29. Describe the three-dimensional symmetries of a cube. Be sure to consider the cube from several different points of view.

30. Which can be a net for the surface of a cube?

I.   II.  III.  IV. 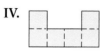 V.

31. Draw five possible nets for the surface of a cube.

THINK CRITICALLY

32. When a plane and a cube intersect, is it possible for the cross section to be a six-sided figure? Justify your answer.

33. **GENETICS** Many genes in the human body have the shape of a *helix*. You can model one type of helix using figures like that shown at the right. To visualize the figure, imagine plane $\mathcal{R}$ "cutting" a rectangular solid on a slant as shown. Now imagine several of the resulting figures with the "cut" end of one lying flat against the opposite "uncut" end of the next. What is the shape of the helix that results?

ALGEBRA AND GEOMETRY REVIEW

34. Describe the symmetries of the "star" at the right.

35. Name the three words that are considered the undefined terms of geometry.

Solve each equation.

36. $-9 = 1 + 4y$ 37. $4 - 3b = 0$ 38. $3.5k = 24.5$

39. $\dfrac{t}{6} + 7 = 5$ 40. $8x + 2 = 3x$ 41. $5 + m - 8 = 22$

42. Draw the cross section that results when the plane intersects the conical solid in the figure at the right.

43. What is 72% of 15?   44. 18 is 15% of what number?

45. 36 is what percent of 54?   46. 36 is what percent of 24?

Career
Chemical Technician

Chemical technicians assist chemists and chemical engineers in the development, testing, and manufacture of chemical products. Often these technicians work as part of an applied research team, so they must have a broad knowledge of basic chemical concepts. For example, the basic unit of any chemical compound is the *molecule*. One important property of a molecule is its symmetry.

Decision Making

The cluster of balloons at the right is a model of one molecule of *methane* CH_4. The unseen point where the balloons meet at the center represents one atom of carbon C. The four balloons represent four identical atoms of hydrogen H_4 bonded to the carbon atom and evenly spaced around it.

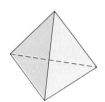

1. How many planes of symmetry does the model have? *Hint:* It may help to visualize the outer tips of the balloons as the corners of a pyramid.

2. How many axes of symmetry does the model have?

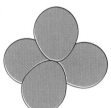

Below are models of three other carbon compounds. The different colors of the balloons represent different atoms and molecules bonded to the carbon atom at the center. Describe the symmetries of each model. Consider the different colors when determining the symmetries of the models.

3. *methanol,* CH_3OH 4. *dichloromethane,* CH_2Cl_2 5. *dichlorobromomethane,* $CHCl_2Br$

      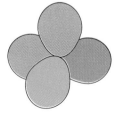

One molecule of *sulfur hexaflouride* SF_6 consists of one central sulfur atom S with six atoms of flourine F_6 bonded to it and evenly spaced around it.

6. Draw a balloon model of one molecule of sulfur hexaflouride.

7. Describe the symmetries of a sulfur hexaflouride molecule.

1.4 Perspective, Orthogonal, and Isometric Drawings

Explore/Working Together

• First work by yourself to make a drawing of each object.

    1. a rectangle   **2.** a table with a rectangular top

Now work with a partner.

    3. Compare your drawing of a rectangle to your partner's drawing. How are they alike? How are they different?

    4. Compare your drawing of a table to your partner's drawing. How are they alike? How are they different?

    5. Work together to make a drawing of a table with a circular top.

Build Understanding

• The railroad tracks appear to intersect in the distance. From your experience, you know that these tracks actually are *parallel.* That is, the tracks lie in the same plane along lines that never intersect.

The point where the tracks appear to meet is called a **vanishing point.** The vanishing point lies on a line called the **horizon line.** You can locate the vanishing point in some drawings of three-dimensional figures by extending line segments in the figure.

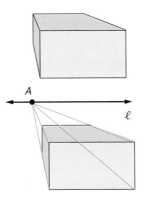

EXAMPLE 1

Locate the vanishing point and horizon line in the drawing of the rectangular prism at the right.

Solution

Draw line *segments* from the four front corners of the prism until they intersect as shown.

Point A is the vanishing point.

Line ℓ is the horizon line.

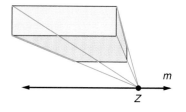

The location of the vanishing point and horizon line indicate the observer's point of view. In Example 1, the observer is looking at the prism from above and to the left. In the prism at the right, the observer is looking from below and to the right.

When you use a vanishing point to draw an object, you create a *perspective drawing*. A **perspective drawing** is made on a two-dimensional surface in such a way that three-dimensional objects appear true-to-life. When the drawing has one vanishing point, it is said to have **one-point perspective**.

EXAMPLE 2

Draw a cube in one-point perspective.

Solution

Step 1 Draw a square to show the front surface of the cube. Draw a horizon line *j* and a vanishing point *A* on line *j*.

Step 2 Lightly draw line segments that connect the vertices of the square to the vanishing point.

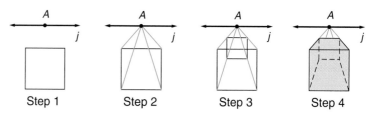

Step 3 Draw a smaller square whose vertices touch the four line segments.

Step 4 Connect the vertices of the two squares. Then use dashed segments to indicate the edges of the cube hidden from view. ◄

A drawing may have two vanishing points, such as the drawing of the cube below. A drawing like this is said to have **two-point perspective**.

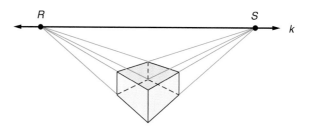

A perspective drawing shows an object realistically, but you may not be able to read certain details from it. An **orthogonal drawing**, or **orthographic drawing**, gives top, front, and side views of an object.

THINK BACK

The point where two sides of a square intersect is a *vertex* of the square. *Vertices* is the plural form of the word vertex.

COMMUNICATING ABOUT GEOMETRY

Describe how the one-point and two-point perspective drawings on this page show a cube in a true-to-life way.

EXAMPLE 3

The perspective drawing at the right shows a stack of cubes. Draw front, top, and right-side views of the figure.

Solution

Draw the front view first.

Draw the top view above it. Make sure it has the same width as the front view.

Draw the right-side view at the right. Make sure it has the same height as the front view. Make sure it has the same depth as the top view.

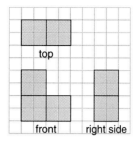

top

front   right side

◀

CHECK — UNDERSTANDING

In Example 3, do you think you would add any other important information about the figure if you included
a back view?
a bottom view?
a left-side view?

Like a perspective drawing, an **isometric drawing** shows a complete object rather than individual views of it. In an isometric drawing, though, there is no vanishing point. All parallel edges of the object are shown as parallel line segments in the drawing.

On isometric dot paper, a simple cube is drawn as shown at the right.

COMMUNICATING — ABOUT GEOMETRY

How could an isometric drawing of an object help you to make a net for the object?

EXAMPLE 4

ARCHITECTURAL DESIGN
The orthogonal drawing at the right shows part of an air conditioning duct. Make an isometric drawing of the duct.

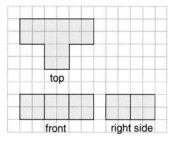

top

front   right side

Solution

Think of the duct as a combination of four cubes. The drawing on isometric dot paper is shown below.

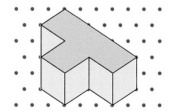

◀

Trace each figure onto a piece of paper. Locate and label the vanishing point and the horizon line of the drawing.

1.   2.   3.

Draw each figure in one-point perspective.

4. 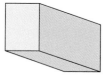   5.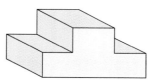

Each figure consists of a stack of cubes. No cubes are hidden from view. Draw front, top, and right-side views of each figure.

6. 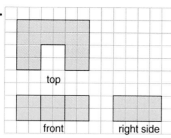   7. 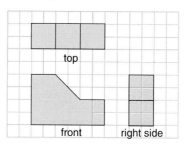   8.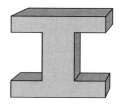

9. **WRITING MATHEMATICS** Explain the difference between a perspective drawing and an isometric drawing.

The figure at the right is an isometric drawing of a cube with a "half cube" resting on top of it. Tell which view of the figure is shown.

10.   11.   12.   13.

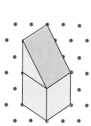

For each set of views, make an isometric drawing of the three-dimensional object.

14.
top
front   right side

15.
top
front   right side

16. **ENGINEERING** Most skyscrapers are built using steel girders shaped like the capital letter I. This shape avoids buckling and can support heavy loads. A perspective drawing of an "I-beam" is shown at the right. Make an orthogonal drawing of the I-beam.

PRACTICE

Trace each figure onto a piece of paper. Locate and label the vanishing point and the horizon line of the drawing.

17.

18.

19.

20.

21. Draw the L-shaped figure at the right in one-point perspective.

Each figure consists of a stack of cubes. No cubes are hidden from view. Draw front, top, and right-side views of each figure.

22.

23.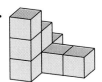

24.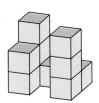

For each set of views, make an isometric drawing of the three-dimensional object.

25.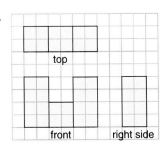

26.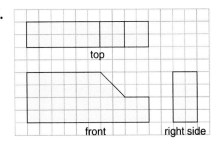

27. ENGINEERING The isometric drawing at the right shows a support for a highway overpass. Make an orthogonal drawing of the support.

28. WRITING MATHEMATICS Describe an everyday situation in which you would make a perspective drawing and a situation in which you would make an orthogonal drawing.

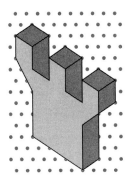

EXTEND

29. Draw a rectangular prism other than a cube in two-point perspective.

GRAPHIC ARTS **Letters of the alphabet drawn in perspective create an interesting visual effect. At the right A and Z are drawn in perspective.**

30. Trace the drawing of the letters onto a piece of paper. Locate the vanishing point(s) and the horizon line.

31. Make a one-point or two-point perspective drawing of your name.

THINK CRITICALLY

32. The drawing at the right shows top and front views of a building. Draw three possible right-side views for this building.

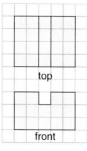

POLYCUBES **Certain arrangements of cubes are called *polycubes*.**

These are polycubes.   These are not polycubes.

33. WRITING MATHEMATICS Refer to the figures above. Write a definition of *polycube*.

34. TRANSFORMATIONS Each of the eight figures below shows a three-cube polycube. Explain why there really are only two *different* three-cube polycubes.

  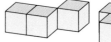

35. How many different four-cube polycubes are there? Make a drawing of each.

ALGEBRA AND GEOMETRY REVIEW

Solve each proportion for the variable.

36. $\dfrac{2}{3} = \dfrac{r}{21}$ **37.** $\dfrac{m}{18} = \dfrac{18}{27}$ **38.** $\dfrac{4.6}{y} = \dfrac{23}{7.5}$ **39.** $\dfrac{10}{1.75} = \dfrac{0.8}{a}$

The figure at the right is a stack of cubes. No cubes are hidden from sight.

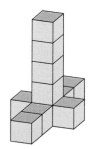

40. Draw front, right-side, and top views of the stack of cubes.

41. How many planes of symmetry does the stack of cubes have?

42. How many axes of symmetry does the stack of cubes have?

PROJECT *Connection*   You will need either a strong magnifying lens or a microscope.

1. Check the salt and sugar crystals that you are growing. If they have developed to the point where the solids are clearly visible and well formed, remove them from the solution. If not, keep them there longer.

2. Examine your crystal samples under the magnifying lens or microscope. Make a sketch of the surfaces of the crystals that you see. Are there any patterns that are common to different types of crystals? Explain.

3. Describe any symmetries that you see in the crystals.

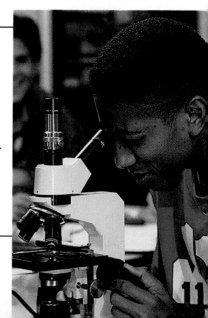

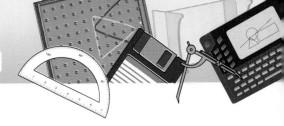

Geometry Workshop
Exploring Measurement

Think Back

- The origins of the word *geometry* are in the ancient Greek language. The prefix *geo-* means "earth," and the word *metron* means "a measure." So, it is believed that geometry originally was developed as a means of measuring land.

  Pictured here are two measurement tools that you probably have used in the past. They are a *ruler* and a *protractor*.

1. What type of measurements do you make with a ruler?

2. What type of measurements do you make with a protractor?

3. How are a ruler and a protractor alike? How are they different? Name as many likenesses and differences as you can.

4. Name two other measurement tools that you have used. What type of measurements did you make with them?

Explore

- At the right is line *RS*. Recall that you can write this as $\overleftrightarrow{RS}$.

5. Give another name for $\overleftrightarrow{RS}$.

6. Do you think you can measure $\overleftrightarrow{RS}$? Explain your reasoning.

7. Examine these figures. Describe how they are related to $\overleftrightarrow{RS}$.

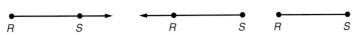

8. Which figure(s) in Question 7 do you think you can measure? Explain your reasoning.

9. How are these figures alike? How are they different?

10. Suppose you are talking to a friend on the telephone. You want to tell your friend how to draw the figure at the right using a ruler and a protractor. Write a step-by-step set of instructions that you might read to your friend.

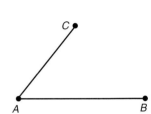

Make Connections

A **ray** is part of a line that begins at one point, called its **endpoint**, and extends without end in one direction. The endpoint and another point are used to name a ray. Below is ray *FG*, written $\overrightarrow{FG}$.

THINK BACK

Point, line, and *plane* are the basic undefined terms of geometry.

A **line segment**, more simply called a **segment**, is part of a line that begins at one endpoint and ends at another. Its endpoints are used to name a segment. Below is segment *JK*, written $\overline{JK}$. You can also call this segment *KJ*, written $\overline{KJ}$.

You use a ruler to measure the *length* of a segment. To refer to the length of a specific segment, you write the names of the endpoints but omit the symbol above them. For example, the length of $\overline{JK}$ is 2 cm. Write *JK* = 2 cm.

For 11–14, refer to the figure at the right.

11. Explain the differences among these symbols.

$\overleftrightarrow{LN}$ $\overrightarrow{LN}$ $\overline{LN}$ LN

PROBLEM SOLVING TIP

To find the length of the segments in Question 14, measure from the center of the dot that represents one endpoint to the center of the dot that represents the other endpoint.

12. Using the labeled points, list as many different names for the line as you can.

13. Identify as many different rays as you can.

14. Find as many different lengths as you can. If possible, use both a metric ruler and a customary ruler. Write the lengths using the proper notation.

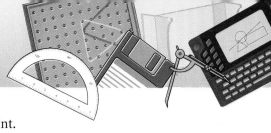

An **angle** is formed by two rays that have a common endpoint. The endpoint is the **vertex** of the angle, and each ray is a **side** of the angle. The symbol for angle is ∠. For the angle below, point *B* is the vertex and the sides of the angle are $\overrightarrow{BA}$ and $\overrightarrow{BC}$. This angle can be named ∠*ABC*, ∠*CBA*, ∠*B* or ∠1.

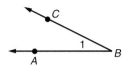

15. Refer to the figure at the right. Which is another name for ∠2?
  ∠*WXZ*   ∠*ZWX*   ∠*WZX*

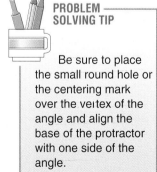

16. Why do you think it is *not* appropriate to refer to any of these angles as ∠*Z*?

17. List as many other names as possible for ∠*YZX*.

A common unit for measuring angles is the *degree*. You use a protractor to find the *degree measure* of an angle.

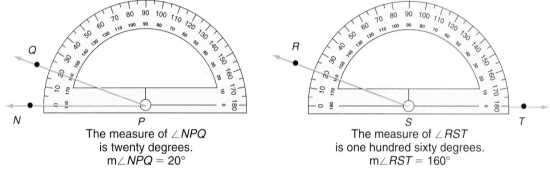

The measure of ∠*NPQ*
is twenty degrees.
m∠*NPQ* = 20°

The measure of ∠*RST*
is one hundred sixty degrees.
m∠*RST* = 160°

18. A protractor has two scales, each numbered from 0° to 180°. How do you know which scale to read for a given angle?

Use a protractor to measure each angle. Then write the measure using the proper notation.

19.

20.

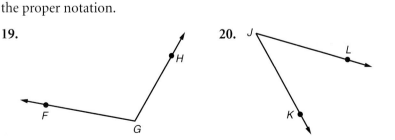

PROBLEM SOLVING TIP

Be sure to place the small round hole or the centering mark over the vertex of the angle and align the base of the protractor with one side of the angle.

1.5 **Geometry Workshop: Exploring Measurement** **31**

GEOMETRY: WHO, WHERE, WHEN

The Greek mathematician Euclid is considered the first person to organize basic geometric principles into a single work. His set of books, *Elements*, was published around 300 B.C. and was used as a text well into modern times. It is said that no single work has had greater influence on scientific thought.

Using a protractor, draw an angle of the given measure.

21. 60°   **22.** 145°   **23.** 90°   **24.** 180°

Summarize

● **25. WRITING MATHEMATICS** Write a paragraph that compares the two types of measurement you studied in this lesson.

26. THINKING CRITICALLY Explain why the following is *not* a good definition of the term *line segment.*

A line segment is a part of a line.

GOING FURTHER Draw two segments that are each $4\frac{1}{2}$ in. long, share a common endpoint, and form an angle whose measure is 60°.

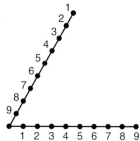

27. Divide each segment into nine segments of equal length. What is the length of each small segment?

28. Label the endpoints of each small segment with the integers from 1 to 9, as shown above.

29. Draw a segment connecting point 1 to point 1, a segment connecting point 2 to point 2, and so on until you connect point 9 to point 9. Describe the shape that has appeared.

30. How do you think the figure would be different if you used an angle with a larger measure? a smaller measure?

TECHNOLOGY TIP

You can use geometry software to measure segments and angles. You can also find out if two segments are the same length, or if two angles have the same measure.

GOING FURTHER Draw two segments that are each $7\frac{1}{2}$ in. long, intersect at their midpoints, and form four 90° angles.

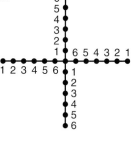

31. Divide each segment into 12 segments of equal length. What is the length of each segment?

32. Connect the points on the sides of each of the four angles as you did with the angle in Question 29.

33. The figures you drew in Questions 27–32 are called *line designs.* Use the methods you learned in this workshop to create your own original line design.

1.6 Scale Drawings and Scale Models

Explore

In the photograph at the right, the "Bike for Sale" sign is actually 24 in. high and 36 in. wide. The diameter of a wheel of the bicycle is 28 in.

1. Measure the height and width of the sign and the diameter of the wheel.

2. Calculate each ratio.

   a. $\dfrac{\text{height of sign in photograph}}{\text{actual height of sign}}$

   b. $\dfrac{\text{width of sign in photograph}}{\text{actual width of sign}}$

   c. $\dfrac{\text{diameter of wheel in photograph}}{\text{actual diameter of wheel}}$

3. What appears to be true of all your answers in Question 2?

4. How might you use the ratio to estimate the person's height?

Build Understanding

Geometric figures that have exactly the same shape and size are called **congruent figures.**

EXAMPLE 1

Choose the lettered figure that appears to be congruent to the figure on the left.

  A. **B.** **C.**

Solution

Triangle A has the same shape, but it is smaller in size.

Triangle B seems about the same size, but has a different shape.

Triangle C is in a different position, but it has exactly the same size and shape.

So, triangle C appears to be congruent to the given triangle.

PROBLEM SOLVING TIP

To determine whether two figures are congruent, carefully trace one of the figures onto a separate sheet of paper. The figures are congruent if you can position your tracing so it fits exactly over the other figure. You may flip your tracing over if necessary.

COMMUNICATING
ABOUT GEOMETRY

Explain why congruent
figures also can be
described as similar
figures.

Similar figures have the same shape, but not necessarily the same size.
In Explore the representation of the sign in the photograph is similar
to the actual sign.

EXAMPLE 2

Choose the lettered figure that appears to be similar to the figure on
the left.

 A. **B.** **C.**

Solution
The rectangle at the left seems to be about twice as long as it is wide.

Rectangle A has a different shape. It seems about four times as long as
wide. Rectangle B is larger than the given rectangle. However, it seems
to be about twice as long as it is wide. Rectangle C has a different
shape. It seems nearly square.

So, rectangle B appears to be similar to the given rectangle.   ◄

A **scale drawing** is a two-dimensional drawing that is similar to the
object it represents. The **scale** of the drawing is the ratio of the size of
the drawing to the actual size of the object. A common example of a
scale drawing is a road map.

THINK BACK

A *ratio* is a
comparison of two
quantities by division.
You can write *the ratio
of a to b* in three ways.

a to b   a : b $\frac{a}{b}$

A *proportion* is a
statement that two
ratios are equal.

$$\frac{a}{b} = \frac{c}{d}$$

You can use cross
products to solve the
proportion.

$$\frac{a}{b} = \frac{c}{d} \rightarrow ad = bc$$

EXAMPLE 3

TRAVEL The scale of a road map is 1 in. : 200 mi. Using a ruler, you
find that the map distance between Des Moines, Iowa, and Duluth,
Minnesota, is $1\frac{3}{4}$ in. What is the actual distance?

Solution
The ratio of map distance to actual distance is 1 in. to 200 mi.
The map distance is $1\frac{3}{4}$ in. Let d represent the actual distance.
Write and solve a proportion.

$$\frac{1}{200} = \frac{1\frac{3}{4}}{d} \qquad \frac{\text{map distance, in.}}{\text{actual distance, mi}}$$

$$1 \cdot d = 200\left(1\frac{3}{4}\right)$$

$$d = 200 \cdot \frac{7}{4}$$

$$d = 350$$

The actual distance between Des Moines and Duluth is about 350 mi. ◄

A **scale model** of an object is a three-dimensional figure whose surfaces are similar to the corresponding surfaces of the actual object.

TECHNOLOGY TIP

You can use geometry software to calculate ratios of lengths you have measured.

EXAMPLE 4

CIVIL ENGINEERING The world's largest road tunnel is in San Francisco. Its height is 56 ft, its width is 77 ft, and its length is 540 ft. An engineering technician plans to make a scale model of the tunnel using the scale 1 in. : 15 ft. What will be the dimensions of the model?

Solution

The ratio of model length to actual length is 1 in. to 15 ft. The actual height is 56 ft. Let h represent the height of the model. Write and solve a proportion.

$$\frac{1}{15} = \frac{h}{56} \qquad \frac{\text{model height, in.}}{\text{actual height, ft}}$$

$$56\left(\frac{1}{15}\right) = h \qquad \text{Multiply each side by 56.}$$

$$3.73 \approx h$$

The height of the scale model will be about 3.73 in.

Using similar reasoning, the width of the model will be about 5.13 in., and the length will be exactly 36 in.

As you have seen, the *linear measures* of a scale drawing or model are proportional to the actual measures of the object. However, the *angular measures* of the drawing or model are equal to the actual angular measures. You will learn more about scale in Chapter 9.

THINK BACK

In the solution of Example 4, the symbol $\approx$ means *is approximately equal to.*

CHECK UNDERSTANDING

Write and solve the proportions that you can use to find the width and length of the model in Example 4.

TRY THESE

1. Choose the lettered figure that appears to be congruent to the figure on the left.

   A.   B.   C.   D.

2. Choose the lettered figure that appears to be similar to the figure on the left.

   A.   B.   C.   D.

3. **WRITING MATHEMATICS** Explain how the concepts of congruence and similarity are related to scale drawings and scale models.

**TRAVEL Suppose the scale of a map is 1 in. : 250 mi.
For each map distance, find the actual distance.**

4. 6 in.   5. $\frac{1}{2}$ in.   6. $3\frac{3}{4}$ in.   7. $1\frac{7}{8}$ in.

8. **ARCHITECTURE** The scale of an architect's model for a house is 1 cm : 1.5 m. One room of this house will be 4.5 m wide, 6 m long, and 2.5 m high. What will be the dimensions of this room in the model to the nearest tenth?

9. **TECHNOLOGY** The picture at the right shows a microprocessor chip. The actual length of each side of the chip is $\frac{1}{4}$ in. What is the scale of the picture?

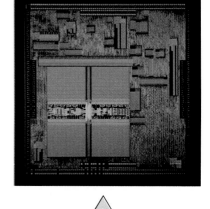

MODELING You will need grid paper, a ruler, a protractor, scissors, and tape.

10. On a sheet of grid paper, use a ruler and a protractor to draw the net shown at the right. The length of each side of the square should be 3 in. Cut out the net. Tape the sides of the triangular surfaces together to form a pyramid.

11. Use a ruler to find each measure in your model.
    a. the length of one side of the square base
    b. the length of one "slanting" edge
    c. the height

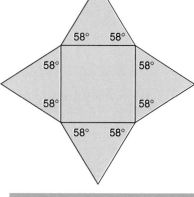

12. **SOCIAL STUDIES** The pyramid that you made is a model of the Great Pyramid of Khufu in Egypt. The table at the right lists actual measures for this pyramid. Use this table and your answers to Exercise 11 to estimate the scale of your model.

length of one side of square base	756 ft
length of one "slanting" edge	719 ft
original height	481 ft

PRACTICE

13. Which two figures appear to be congruent?

   A.   **B.**   **C.**    **D.**   **E.**

14. Which two figures appear to be similar?

   A.   **B.**   **C.**    **D.**   **E.**

Suppose you are making a floor plan with the scale 1 in. : 4 ft. For each actual length, find the length you would use in the floor plan.

15. 18 ft   **16.** 3 ft   **17.** $5\frac{1}{2}$ ft   **18.** 24 in.   **19.** 9 in.   **20.** 3 yd

SOLAR ENERGY A *solar cell* converts energy from the sun into electrical energy. The figure at the right shows the basic shape of one type of solar cell. The length of each side of the actual cell is 24 in., and the measure of each angle is 120°. Suppose you must make a scale drawing of this cell. What would be the measures of the sides and angles for each scale?

21. 1 in. : 6 in.   **22.** 1 in. : 16 in.   **23.** 1 in. : 9 in.   **24.** 1 in. : $1\frac{1}{2}$ in.

25. WRITING MATHEMATICS Describe a situation in which you might make a scale drawing or a scale model that is *larger* than the actual object.

26. LANDSCAPING A landscape architect is making a scale model of a proposed garden. Each step of a staircase in this garden is shaped like a rectangular prism that is 6 in. tall, 10 in. deep, and 3 ft wide. Each step in the model is $\frac{3}{4}$ in. tall, $1\frac{1}{4}$ in. deep, and $4\frac{1}{2}$ in. wide. What is the scale of the model?

EXTEND

Copy the rectangle at the right onto a sheet of grid paper.

27. Draw two other similar rectangles that are each larger than the given rectangle.

28. Draw two other similar rectangles that are each smaller than the given rectangle.

TRAVEL **A map of downtown San Diego is shown at the right. Its scale is 1 in. : 0.5 mi.**

29. What is the actual distance between Pacific Highway and Harbor Drive along Broadway?

30. Suppose you walk at a rate of 4 mi/h. About how long would it take to walk along Harbor Drive from Market Street to Grape Street?

31. ENTERTAINMENT All television screens must be similar because of the way in which television signals are transmitted. In the instruction manual for your television, the diagonal measure of the screen is listed as 19 in. You measure and find that the width of the screen is about 15 in. and the height is about 11.5 in. Estimate the width and height of a television screen with a 27-in. diagonal.

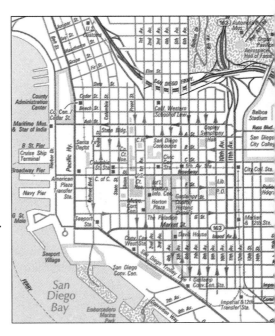

THINK CRITICALLY

32. The figure at the right shows two ways to divide a four-by-four grid square into two congruent parts. Find at least two other ways to do this. You may only divide the square along the grid lines.

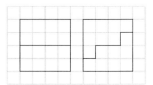

33. Trace each figure onto a sheet of paper. Then find a way to divide it into four congruent parts that are each similar to the original figure.

a.   **b.**   **c.**

ALGEBRA AND GEOMETRY REVIEW

Write an ordered pair to describe each point graphed at the right.

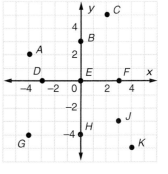

34. A **35.** B **36.** C **37.** D **38.** E

39. F **40.** G **41.** H **42.** J **43.** K

Which points graphed at the right show solutions of the equation?

44. $y = x$ **45.** $y = -x$ **46.** $y = 0$ **47.** $x = 0$

48. $y = 3$ **49.** $x = -3$ **50.** $y = -\frac{1}{2}x$ **51.** $y = x + 3$

Make a drawing to illustrate each figure.

52. $\overrightarrow{YZ}$ **53.** $\overline{GH}$ **54.** $\overleftrightarrow{CD}$ **55.** $\angle RST$ **56.** point Q **57.** plane $\mathcal{W}$

58. The scale of a blueprint is 1 in. : 3 ft. What will be the blueprint dimensions of a room whose width is 9 ft and whose length is 5 yd?

PROJECT *Connection*   You will need heavy paper, a ruler, scissors, and glue to make a model of a crystal.

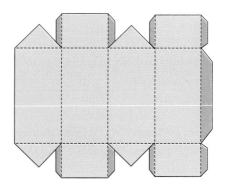

1. The figure at the right is a pattern for a model of a crystal of a mineral called *epsomite*. Choose a scale for enlarging the pattern. Calculate the new dimensions of the pattern and draw the enlarged pattern on the paper.

2. Cut out the enlarged pattern along the solid lines. Fold it along the dashed lines. Use tape or glue on the tabs to hold the model together.

3. How many planes of symmetry does the model have? How many axes of symmetry?

Career
Tourist Bureau Manager

Much of the natural beauty of the United States is preserved in a vast system of national parks. In many areas of the country, the manager of a tourist bureau is responsible for informing the public about the park and answering questions from potential visitors.

Decision Making

Great Smoky Mountains National Park is located on the border of Tennessee and North Carolina. It is shown on the map at the bottom of this page.

1. What is the scale of the map?

Use the map to estimate the distance between Great Smoky Mountains National Park and each city.

2. Nashville, Tennessee

3. Charleston, West Virginia

4. Louisville, Kentucky

5. Atlanta, Georgia

6. Birmingham, Alabama

7. What is the approximate distance from Great Smoky Mountains National Park to Cumberland Gap National Historic Park?

8. What is the approximate distance across Great Smoky Mountains National Park from east to west? from north to south?

9. The manager of a tourist bureau is preparing a brochure for visitors to Great Smoky Mountains National Park. A map of the park will be printed on a page of the brochure that is $8\frac{1}{2}$ in. wide and 11 in. long. What scale would you use for this map? Explain your reasoning.

1.6   **Scale Drawings and Scale Models**   **39**

1.7 Distances in a Coordinate System

THINK BACK

An absolute value is always a nonnegative number.

Absolute value is defined algebraically as follows.

For every real number a,

$|a| = a$ if $a \geq 0$

$|a| = -a$ if $a < 0$.

Explore

Use grid paper on which the distance between grid lines is one-fourth inch or larger. Make a "ruler" by cutting out a strip of fifteen adjacent squares, like this.

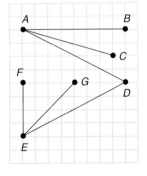

Now copy the segments at the right onto a sheet of grid paper. Then use your "ruler" to find each length. Give the length to the nearest half-unit.

1. AB 2. AC 3. AD 4. EF 5. EG 6. ED

7. Were any lengths easier to find than others? Explain.

8. Can you tell without measuring which two segments must have exactly the same length? Explain your reasoning.

Build Understanding

On a **number line**, every point corresponds to a real number. The number is called the **coordinate** of the point. The point that corresponds to zero is called the *origin of the number line.*

The **length** of a segment is the distance between its endpoints. When the endpoints are two points on a number line, you find this distance by finding the absolute value of the difference of their coordinates.

EXAMPLE 1

Find PQ on the number line.

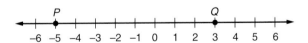

Solution

The coordinate of P is -5. The coordinate of Q is 3.
Find the absolute value of the difference of -5 and 3.
The order in which you subtract does not matter.

$$PQ = |-5 - 3| \qquad \text{or} \qquad PQ = |3 - (-5)|$$

$$= |-8| = 8 \qquad\qquad\qquad = |8| = 8$$

Using either order, you find that $PQ = 8$.

A **coordinate plane** is formed by two number lines that intersect at their origins. The number lines are *perpendicular*. That is, the measure of each angle formed at their intersection is 90°.

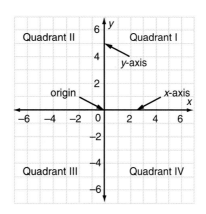

The horizontal number line is the **x-axis**, and the vertical number line is the **y-axis**. These *axes* divide the coordinate plane into four **quadrants**. The point where the axes intersect is called the **origin**.

GEOMETRY: WHO, WHERE, WHEN

The ancient Egyptians and Romans used coordinates to survey land thousands of years ago. However, the French mathematician René Descartes (1596–1650) formally developed the idea of studying geometry by analyzing coordinates of points. For this reason, the coordinate plane often is called the *Cartesian* plane.

Every point of a coordinate plane corresponds to an **ordered pair** of real numbers (x, y). The first number of the pair is the **x-coordinate** of the point. It indicates distance left or right of the y-axis. The second number is the **y-coordinate**. It indicates distance above or below the x-axis. For example, the point $K(-6, 8)$ is 6 units *left* of the y-axis and 8 units *above* the x-axis.

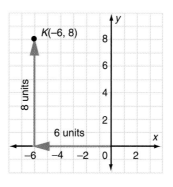

To find the distance between two points on a coordinate plane, you can use the following formula.

> **DISTANCE FORMULA ON A COORDINATE PLANE**
>
> **The distance d between two points with coordinates (x_1, y_1) and (x_2, y_2) is**
> $$d = \sqrt{(x_2 - x_1)^2 + (y_2 - y_1)^2}$$

EXAMPLE 2

Find the distance between the points $R(-4, -3)$ and $S(2, 5)$.

Solution

Let $(-4, -3)$ be (x_1, y_1) and $(2, 5)$ be (x_2, y_2).

$$d = \sqrt{(x_2 - x_1)^2 + (y_2 - y_1)^2}$$ Use the distance formula.

$$= \sqrt{(2 - (-4))^2 + (5 - (-3))^2}$$ Substitute.

$$= \sqrt{6^2 + 8^2}$$ Simplify.

$$= \sqrt{100} = 10$$

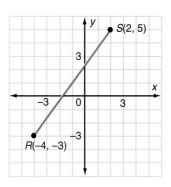

The distance between $R(-4, -3)$ and $S(2, 5)$ is 10 units. ◄

To locate points in space, you use a three-dimensional coordinate system like the one at the right. In this system, there are three axes. The *x*-axis and *y*-axis lie on a horizontal plane, and the *z*-axis is vertical. Every point in space then corresponds to an **ordered triple** of real numbers (x, y, z).

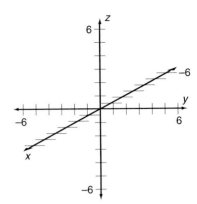

The three axes divide space into eight **octants**. In octant I, all the coordinates of a point are positive.

COMMUNICATING ABOUT GEOMETRY

How is the distance formula in a three-dimensional coordinate system like the distance formula on a coordinate plane? How are the formulas different?

EXAMPLE 3

The figure at the right shows a cube in octant I of a three-dimensional coordinate system. Find the coordinates of point *D*.

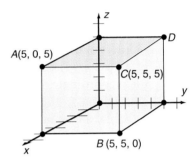

Solution

Point *D* is located 0 units along the *x*-axis, positive 5 units along the *y*-axis, and positive 5 units along the *z*-axis.

So, the ordered triple that locates point *D* is $(0, 5, 5)$.

To find the distance between two points in space, you use the following distance formula. It is an extension of the distance formula on a coordinate plane.

DISTANCE FORMULA IN THREE DIMENSIONS

The distance *d* between two points with coordinates (x_1, y_1, z_1) and (x_2, y_2, z_2) is

$$d = \sqrt{(x_2 - x_1)^2 + (y_2 - y_1)^2 + (z_2 - z_1)^2}$$

EXAMPLE 4

AVIATION An air traffic control tower is at the origin of a three-dimensional coordinate system. The basic unit of distance in this system is the mile. At a given time, the coordinates of a Boeing 757 are $(-3, 5, 2)$, and the coordinates of a commuter plane are $(9, -3, 7)$. What is the distance between the two airplanes?

Solution

Let $(-3, 5, 2)$ be (x_1, y_1, z_1) and $(9, -3, 7)$ be (x_2, y_2, z_2).

$$d = \sqrt{(x_2 - x_1)^2 + (y_2 - y_1)^2 + (z_2 - z_1)^2} \quad \text{Distance formula.}$$
$$= \sqrt{(9 - (-3))^2 + (-3 - 5)^2 + (7 - 2)^2} \quad \text{Substitute.}$$
$$= \sqrt{12^2 + (-8)^2 + 5^2} \quad \text{Simplify.}$$
$$= \sqrt{233} \approx 15.3$$

The planes are about 15.3 mi apart. ◄

TRY THESE

Find each length on the number line below.

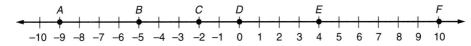

1. AB **2.** AD **3.** EF **4.** BD **5.** AE **6.** CF

Which point graphed at the right has the given coordinates?

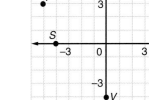

7. $(5, -3)$ **8.** $(-5, 3)$ **9.** $(5, 3)$ **10.** $(-3, -5)$

11. $(0, 4)$ **12.** $(4, 0)$ **13.** $(-4, 0)$ **14.** $(0, -4)$

Find the distance between two points on a coordinate plane with the given coordinates. Round to the nearest tenth if necessary.

15. $(3, 6)$ and $(-5, -9)$ **16.** $(-2, 4)$ and $(10, -1)$

17. $(3, -8)$ and $(-4, 6)$ **18.** $(11, -10)$ and $(5, 2)$

19. SURVEYING In the figure at the right, a surveyor has placed a coordinate grid over a map of a triangular plot of land. The basic unit of distance on this grid is the kilometer. What is the total distance around this plot of land? Round your answer to the nearest tenth of a kilometer.

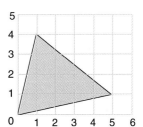

20. WRITING MATHEMATICS How is the process of locating a point on a number line similar to the process of locating a point on a coordinate plane? How is it different?

At the right is a rectangular prism in a three-dimensional coordinate system. Find the coordinates of each point.

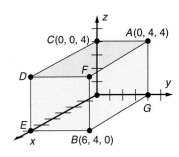

21. D **22.** E **23.** F **24.** G

Find the distance between the two points in a three-dimensional coordinate system. Round to the nearest tenth.

25. $J(-1, 5, 7)$ and $K(2, 9, 7)$ **26.** $P(-3, 6, 3)$ and $Q(-2, 5, -1)$

PRACTICE

Find each length on the number line below.

27. *KL*   **28.** *JM*   **29.** *KL*   **30.** *KN*   **31.** *JP*   **32.** *LQ*

```
      J   K           L       M   N           P       Q
  ────●───●───┬───┬───●───┬───●───●───┬───┬───●───┬───●───▶
    -10 -9 -8 -7 -6 -5 -4 -3 -2 -1  0  1  2  3  4  5  6  7  8  9  10
```

Find the distance between two points on a coordinate plane with the given coordinates. Round to the nearest tenth.

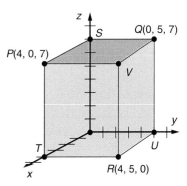

33. $(-8, 1)$ and $(-5, 5)$ **34.** $(-4, 9)$ and $(2, 1)$

35. $(-4, 2)$ and $(3, 7)$ **36.** $(2, 6)$ and $(-3, -5)$

At the right is a rectangular prism in a three-dimensional coordinate system. Find the coordinates of each point.

37. *S*   **38.** *T*   **39.** *U*   **40.** *V*

41. WRITING MATHEMATICS A classmate says that you can double the length of a line segment on a coordinate plane by doubling the coordinates of each endpoint. Do you agree or disagree? Explain your reasoning.

42. OCEANOGRAPHY A research ship is anchored at the origin of a three-dimensional coordinate system. Two probes are sent beneath the surface. The first probe stops at the point with coordinates $(-8, -4, -6)$. The second probe stops at the point with coordinates $(4, -9, -2)$. Which probe is farther from the ship? Explain.

EXTEND

At the right is a cube in a three-dimensional coordinate system. Decide whether the point with the given coordinates is *inside*, *outside*, or *on* the cube. Explain your reasoning.

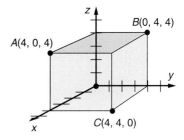

43. $(2, 2, 2)$ **44.** $(3, 3, -3)$ **45.** $(0, 3, 3)$

46. $(4, 4, 4)$ **47.** $(5, 2, 2)$ **48.** $(0, 0, 3.5)$

49. LOOKING AHEAD The midpoint of a segment is the point located at an equal distance from each endpoint. Use the distance formula to show that $M(-3, 5)$ is the midpoint of the segment with endpoints $A(-1, -1)$ and $B(-5, 11)$.

THINK CRITICALLY

50. USING ALGEBRA On a number line, how many points are located d units from the point whose coordinate is n? Write a variable expression that represents the coordinate of each such point assuming $d > 0$.

51. On a coordinate plane, how many points are located exactly six units from the origin? Explain your reasoning.

Describe the set of all points that satisfy the given condition.

52. the set of all points on a number line whose coordinate is 2

53. the set of all points on a coordinate plane for which the x-coordinate is 2

54. the set of all points in a three-dimensional coordinate system for which the x-coordinate is 2

ALGEBRA AND GEOMETRY REVIEW

55. Describe the symmetries of the figure shown at the right.

56. Evaluate $3x^2 - 2x + 1$ when $x = -1$.

Write each set of numbers in order from least to greatest.

57. $-1\frac{7}{8}, 2\frac{2}{3}, -1.5, 2.3, -\sqrt{4}, \sqrt{8}$ **58.** $-3, \sqrt{13}, -\sqrt{8}, -\frac{16}{5}, \sqrt{17}, 2.8$

59. On a map, the distance between two cities is $2\frac{3}{4}$ in. The scale of this map is 1 in. : 350 mi. What is the actual distance between the cities?

60. STANDARDIZED TESTS Which point is located less than three units from $P(-3, 2)$ on a coordinate plane?

    A. $Q(5, 2)$ **B.** $R(-3, -1)$ **C.** $S(0, 4)$ **D.** $T(-2, 1)$

PROJECT *Connection*

In this activity, you will examine the structure that is at the heart of a crystal.

1. A crystal is classified as belonging to one of seven crystal *systems* by the shape of its internal structure. Research the systems listed in the table at the right. Identify the geometric characteristics of each shape, and identify its symmetries.

2. Make an isometric drawing and an orthogonal drawing of the shape associated with each crystal system.

Crystal System	Example
cubic (isometric)	alum
tetragonal	zircon
orthorhombic	topaz
monoclinic	cane sugar
triclinic	feldspar
trigonal (rhombohedral)	quartz
hexagonal	ice

Taxicab Geometry

If you are walking in an open flat field, you can walk directly from one point to another. However, if you are a taxicab driver in an urban area like New York City, getting from one point to another might not be that simple. Frequently you must steer a course along a grid of streets.

Euclidean geometry is somewhat like the flat field. The distance between two points always is the length of the segment that joins them. You know you can use the following formula to find the Euclidean distance between two points $A(x_1, y_1)$ and $B(x_2, y_2)$ on a coordinate plane.

$$AB = \sqrt{(x_2 - x_1)^2 + (y_2 - y_1)^2}$$

In **taxicab geometry**, the distance between two points is defined as the least number of horizontal and vertical units between the points. You can use the following formula to find this *taxi distance* between $A(x_1, y_1)$ and $B(x_2, y_2)$.

$$AB = |x_2 - x_1| + |y_2 - y_1|$$

1. Copy and complete the following to find the Euclidean distance between point $A(1, 2)$ and point $B(-2, -2)$.

$$AB = \sqrt{(x_2 - x_1)^2 + (y_2 - y_1)^2}$$
$$= \sqrt{((-2) - 1)^2 + (\blacksquare - \blacksquare)^2}$$
$$= \sqrt{\blacksquare^2 + (-4)^2} = \sqrt{\blacksquare} = \blacksquare \text{ units}$$

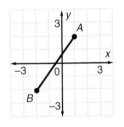

2. Copy and complete the following to find the taxi distance between point $A(1, 2)$ and point $B(-2, -2)$.

$$AB = |x_2 - x_1| + |y_2 - y_1|$$
$$= |(-2) - \blacksquare| + |\blacksquare - 2|$$
$$= |-3| + |\blacksquare| = \blacksquare + 4 = \blacksquare \text{ units}$$

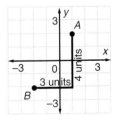

Locate point $C(-3, 1)$ and point $D(1, 3)$ on a coordinate grid.

3. Use the taxi distance formula to find CD.

4. Suppose you can only move from C to D by moving *right* and *up*. Identify three possible paths and find the length of each.

5. How does your answer to Question 4 compare to your answer to Question 3?

Find the taxi distance between each pair of points in the figure at the right.

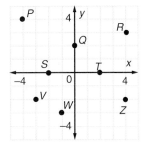

6. *P* and *Z*   7. *Q* and *V*

8. *S* and *T*   9. *Q* and *T*

10. Name all the labeled points in the figure at the right whose taxi distance from point S is four units.

11. Find a pair of labeled points in the figure above for which the taxi distance is equal to the Euclidean distance.

12. In general, when is the taxi distance between two points equal to the Euclidean distance between the points?

13. Suppose the taxi distance between two points is *not* equal to the Euclidean distance. How are the distances related?

14. Draw a set of coordinate axes. Locate all the points whose taxi distance from the origin is exactly six units.

15. Complete each statement with the name of a geometric figure.

   a. In Euclidean geometry, the set of all points in a plane that are a given distance from a given point forms a ___?___.
   b. In taxicab geometry, the set of all points in a plane that are a given distance from a given point forms a ___?___.

> **TAXICAB GEOMETRY** is an example of a *non-Euclidean geometry.* As you proceed through this book, you will have a chance to explore other such geometries.

If you extend taxicab geometry to a three-dimensional coordinate system, you might call the new geometry *staircase geometry.* The *stair distance* between two points is still the least number of horizontal and vertical units between them. But now many of the paths between two points look somewhat like staircases and hallways in a large building.

16. The figure at the right shows a staircase path between the points *A*(5, 0, 0) and *E*(0, 6, 5). What are the coordinates of point *B* and point *D*?

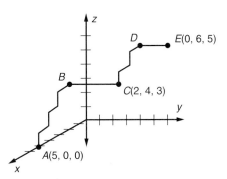

17. What is the stair distance between point *A*(5, 0, 0) and point *E*(0, 6, 5)?

18. **THINKING CRITICALLY** Devise a formula for the stair distance between two points.

19. Use your formula from Question 18 to find the stair distance between point *R*(−2, 0, 3) and *S*(3, 3, −5).

20. When will the stair distance between two points be equal to the Euclidean distance between them? When will the stair distance be equal to the taxi distance?

Making Constructions

Making precise drawings can help you explore properties of geometric figures. Two tools for drawing that have been used since the time of the ancient Greeks are a *compass* and an unmarked *straightedge*. You use the *compass* to draw circles and arcs. You use the *straightedge* to draw lines, rays, and segments. You may use a ruler as a straightedge but you must ignore its markings. Drawings made with these two tools are called **compass-and-straightedge constructions**.

Problem

Use a compass and straightedge to construct a segment congruent to a given segment.

Explore the Problem

Start by examining the figures you can draw with a compass.

1. Draw a point *O*. Use the compass to draw several circles of different sizes with center at point *O*. How is the size of each circle related to the size of the compass setting?

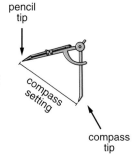
pencil tip
compass setting
compass tip

2. Draw a point *Z*. Place the compass tip at *Z* and draw an arc. Keep the compass tip at *Z*, increase the compass setting, and draw another arc. Repeat this process several times. What happens to the appearance of the arc as the compass setting increases?

Now follow these steps to solve the problem.

3. Use a straightedge to draw a segment. Label its endpoints *A* and *B*.

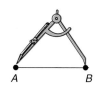
A   B

4. Use the straightedge to draw a line ℓ elsewhere on the sheet of paper. Choose a point on line ℓ and label it *C*.

5. Set the compass to the length of $\overline{AB}$.

6. Using this setting, place the compass tip at point *C* and draw an arc that intersects line ℓ. Label the point of intersection as point *D*.

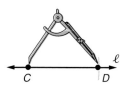
ℓ
C   D

7. What is the relationship between $\overline{AB}$ and $\overline{CD}$?

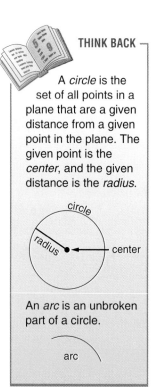

Investigate Further

The **midpoint** of a segment is the point that divides the segment into two congruent segments. Any segment, ray, line, or plane that intersects a segment at its midpoint is called a **segment bisector**. The following questions show you how to construct a segment bisector.

8. Use a straightedge to draw a segment. Label its endpoints E and F.

9. Place the compass tip at point E. Adjust the setting so it is slightly more than half EF. Draw arcs above and below $\overline{EF}$.

10. Using the same setting, place the compass tip at point F and draw arcs above and below $\overline{EF}$. These arcs should intersect the arcs you drew in Question 9. Label the points of intersection of the arcs as point G and point H.

PROBLEM
SOLVING TIP

When drawing arcs with a compass, make the arcs long enough so they eventually intersect with other arcs. If two arcs are meant to intersect but do not, you will have to extend one or both of them.

11. Use the straightedge to draw $\overleftrightarrow{GH}$. Label the point where $\overleftrightarrow{GH}$ intersects $\overline{EF}$ as point M. This is the midpoint of $\overline{EF}$.

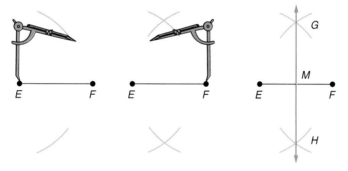

12. How can you use the compass to verify that point M is the midpoint of $\overline{EF}$?

Constructions can also be performed by paper folding. Using waxed paper or patty paper will help you see **paper-folding constructions** more easily. You may use a straightedge in paper-folding constructions, but you may not use a compass or other measuring tools.

13. The picture at the right shows a segment that was bisected by paper-folding. Describe how it was done.

14. Use a straightedge to draw a segment on a piece of paper. Label its endpoints X and Y. Construct the segment bisector of $\overline{XY}$ by paper-folding.

Apply the Strategy

15. Trace $\overline{PQ}$ onto a piece of paper. Use a compass and straightedge to divide $\overline{PQ}$ into four congruent segments.

P •————————————————————————• Q

16. Use a compass and straightedge to construct a segment whose length is twice the length of $\overline{RS}$.

R •————————————————————• S

17. Use a compass and straightedge to construct a segment whose length is one and one half times the length of $\overline{AZ}$.

A •————————————————————• Z

18. Repeat Questions 15–17, but this time perform each construction using paper-folding.

19. WRITING MATHEMATICS Explain how using a compass to make a congruent copy of a segment is different from using a ruler to make a copy. Which method do you think is easier?

20. Use a straightedge to draw a large triangle. Label the endpoints of the sides *A*, *B*, and *C*. Use compass and straightedge to bisect the sides. Label the midpoints of the sides *D*, *E*, and *F*. Draw $\overline{DE}$, $\overline{EF}$, and $\overline{DF}$. How do the lengths of these segments seem to be related to the lengths of the sides of the triangle? How can you verify this without using a ruler?

21. ART Each of these designs can be constructed using the basic methods you learned in this lesson. Choose your two favorite designs and construct them.

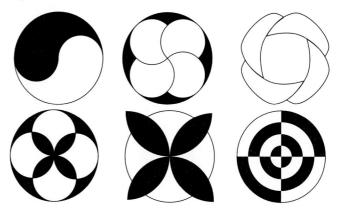

22. Create your own original design similar to those in Question 21.

REVIEW PROBLEM SOLVING STRATEGIES

Slicing the Pizza

1. Charlotte and Jon slice a pizza into two pieces with one straight cut. They find that with a second straight cut intersecting the first one they get four pieces and with three cuts they can get a maximum of seven pieces. What is the maximum number of pieces they can get with seven straight cuts?

John works the problem by drawing a circle and seven straight lines intersecting it. He then tries to count the pieces.

Charlotte makes a table recording some easier examples and looks for a pattern. Try both methods and see which one your prefer.

Oranges For $ale

2. In a special promotion, the Corner Grocery Stores stacked its oranges in a pyramid shaped pile with a triangular base. There are an equal number of oranges on each side of the base. If the display was 11 layers high, how many oranges were in the display?

Answer the following questions to guide your thinking.

How many oranges are in the top layer? second layer? third layer? Do you see a pattern? Altogether how many oranges are there in the 11 layers?

T F

3. Usually if a statement is true then the opposite of the statement is false. Do you agree that the following statement is true?

The number of n's in this sentence is five.

See if you can add just one word to this sentence to form a new statement that has the opposite meaning and yet is still true.

1.9 Focus on Reasoning

Conditional Statements

Discussion

SPOTLIGHT ON LEARNING

WHAT? In this lesson you will learn
- to determine whether a conditional statement is true or false.
- to write conditional statements.

WHY? You can use conditional statements to solve arguments and to write advertisements.

Work with a partner. Consider this incomplete statement.

> If a plane figure has line symmetry, then ____?____ .

Mark, Katrina, and Lea each completed the statement in different ways. Mark wrote the following.

> If a plane figure has line symmetry, then you can divide it along a line into two parts that are mirror images of each other.

1. Draw two figures that support Mark's statement.

2. Do you agree or disagree with Mark's statement? Explain your reasoning.

Katrina wrote this statement.

> If a plane figure has line symmetry, then it does not have rotation symmetry.

She drew all these figures to support her statement.

3. Do you agree or disagree with Katrina's statement? Explain your reasoning.

Lea wrote this statement.

> If a plane figure has line symmetry, then it has rotation symmetry.

4. Draw at least two figures that support Lea's statement.

5. Do you agree or disagree with Lea's statement? Explain your reasoning.

Build a Case

- The statements that Mark, Katrina, and Lea wrote, whether true or false, are called **conditional statements**, or simply **conditionals**. Conditional statements contain two parts that usually are introduced with the words *if* and *then*. The part that follows *if* is called the **hypothesis** of the statement, and the part that follows *then* is the **conclusion**.

6. Identify the hypothesis and the conclusion of the statements made by Mark, Katrina, and Lea.

Conditional statements often are modeled by letters. For example, you can use the letters p and q to represent the hypothesis and conclusion of this statement.

If it is 12:00 noon, then it must be daytime.
$\underbrace{\hspace{2.5cm}}_{p} \qquad \underbrace{\hspace{3cm}}_{q}$

You can write the statement as follows.

If p, then q.

You also can write the statement using a symbol.

$p \Rightarrow q$

7. Use letters and the symbol $\Rightarrow$ to model the conditional statements made by Mark, Katrina, and Lea. Identify what each letter represents.

To show that a conditional statement is false, you only need to find one **counterexample** for which the hypothesis is true and the conclusion is false. You probably decided that the statements made by Katrina and Lea are both false.

8. Describe how one of Katrina's figures could serve as a counterexample to Lea's conditional statement.

9. Give another counterexample to show Lea's statement is false.

10. Describe how a figure that supports Lea's statement could serve as a counterexample to Katrina's conditional statement.

11. Give another counterexample to show Katrina's statement is false.

To show that a conditional statement is true, you must construct a logical argument using reasons.

12. What fact from this chapter supports Mark's statement?

EXTEND AND DEFEND

Write the hypothesis and conclusion of each conditional statement.

13. If a solid has an infinite number of axes of symmetry, then it is a spherical solid.

14. If a point is in the second quadrant of a coordinate plane, then its y-coordinate is a positive number.

Decide whether each conditional statement is *true* or *false*.

15. If a plane figure has rotation symmetry, then it has line symmetry.

16. If the scale of a map is 1 cm : 4.5 km, then 1.5 cm of map distance represents 3 km of actual distance.

Give a counterexample to show that each conditional statement is false.

17. If a figure is part of a line, then it is a line segment.

18. If a plane figure has exactly one line of symmetry, then it is not a triangle.

Give a logical argument to explain why each conditional statement is true.

19. If point Z is the midpoint of $\overline{JK}$, then it divides $\overline{JK}$ into two congruent segments.

20. If point J is the common endpoint of $\overrightarrow{JK}$ and $\overrightarrow{JL}$, then it is the vertex of $\angle KJL$.

CONSUMERISM **The advertisement at the right shows one way that conditional statements are used.**

21. Write the hypothesis and the conclusion of the conditional statement.

22. The automobile manufacturer wants you to believe that this conditional statement is true. What assumption(s) must you make in order to believe it?

Replace each __?__ with a conclusion that will make a true conditional statement.

23. If a point is on the x-axis of a coordinate plane, then __?__ .

24. If 4 in. of length in a scale model represents 6 ft of length in the object it represents, then __?__ .

If you want quality and comfort at an affordable price, *then Jupiter is your only choice.*

Replace each __?__ with a hypothesis that will make a true conditional statement.

25. If __?__ , then the drawing is a geometric construction.

26. If __?__ , then the distance between point *J* and point *K* on a coordinate plane is four units.

LOOKING AHEAD **Consider these two conditional statements.**

If two plane figures are congruent, then they are similar.
If two plane figures are similar, then they are congruent.

27. Decide whether each statement is true or false.

28. These two statements are called *converses* of each other. Describe what you think is meant by the word *converse*.

29. If a statement is true, do you think its converse must also be true? Explain your reasoning.

30. Write two other conditional statements that are converses. Determine whether each statement is true or false.

Now You See It
Now You Don't

Optical Illusions

Examine the cube at the right. Where is point *A*? Different people who look at the figure see different things. These conditionals describe two different points of view.

If you are looking down at the cube from above, then point *A* is at the back of the cube.

If you are looking up at the cube from below, then point *A* is at the front of the cube.

The drawing of the cube is an example of an *optical illusion*. Optical illusions are pictures that "play tricks" on your eyes.

The figure at the right is another optical illusion similar to the cube. Write two conditional statements to describe the different points of view.

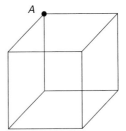

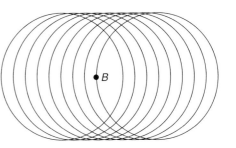

· · · CHAPTER REVIEW · · ·

Choose the word from the list that correctly completes each statement.

1. A(n) __?__ is a set of points that extends without end in all directions along a flat surface.

2. The number of times that a figure coincides with its original position during a complete turn is called its __?__ .

3. A(n) __?__ is the intersection of a solid and a plane.

4. A(n) __?__ gives the top, front, and side views of an object.

5. A(n) __?__ is part of a line that begins at one point and extends without end in one direction.

a. order of rotation symmetry

b. ray

c. plane

d. cross section

e. orthogonal drawing

Lesson 1.1 and 1.2   EXPLORING SHAPES; PLANE FIGURES AND SYMMETRY   pages 5–15

- A plane figure has **reflection symmetry**, or **line symmetry**, if you can divide it into two parts that are mirror images of each other.
- A plane figure has **rotation symmetry** if you can turn it around a point so it coincides with its original position two or more times during a complete turn.

6. Trace the figure shown at the right. Draw all lines of symmetry.

7. Draw a figure with rotation but not reflection symmetry.

8. Draw a figure that has four sides and no symmetry.

9. Draw a figure that has 5-fold rotation symmetry.

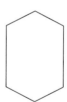

Lesson 1.3   THREE-DIMENSIONAL FIGURES AND SYMMETRY   pages 16–22

- Three-dimensional figures can have reflection and/or rotation symmetry.
- When a plane and a solid intersect, a **cross section** is formed.

10. Identify any planes of symmetry in the figure at the right.

11. Identify any axes of symmetry in the figure.

12. How would a plane intersect the figure in order to produce the following cross sections?

a.

b.

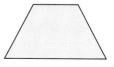

c.

- Using a *vanishing point* to draw an object results in a **perspective drawing**. An **orthogonal drawing** gives top, front, and side views of an object. An **isometric drawing** shows a complete object in which all parallel edges of the object are shown as parallel segments.

For each given figure, make the indicated drawing(s).

13. one-point and two-point perspective

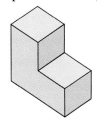

14. orthogonal drawing

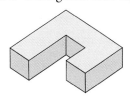

15. isometric drawing

- A **ray** and a **segment are each part of a line. An angle** is formed by two rays that have a common endpoint. A **ruler** is used to measure line segments and a **protractor** to measure angles.
- **Congruent figures** have the same size and shape. **Similar figures** have the same shape but not necessarily the same size. A **scale drawing** or **model** is similar to the object it represents.

16. Which figure is congruent to the one at the left?

17. Which figure is similar to the one at the left?

18. The scale of a road map is 1 in. : 160 mi.
The map distance between two cities is 5.25 in. What is the actual distance?

- The distance between two points with coordinates (x_1, y_1) and (x_2, y_2) is $\sqrt{(x_2 - x_1)^2 + (y_2 - y_1)^2}$.

19. Find the distance to the nearest tenth around a triangle with vertices $A(1, 2)$, $B(8, 8)$, and $C(5, -4)$.

- Constructions are made by using a compass and straightedge, by paper-folding, or by using software.

20. Construct $\overline{A'B'}$ so that $A'B' : AB = 4 : 1$, given:

- In a **conditional statement**, the part that follows *if* is the **hypothesis** and the **conclusion** follows *then*.

21. Identify the hypothesis and conclusion of the following conditional statement.

  If a drawing shows the top, front, and side views, then it is an orthogonal drawing.

22. What is the converse of the above conditional statement? Is the converse true? Explain.

CHAPTER ASSESSMENT

CHAPTER TEST

Refer to the figure at the right.

1. Name five different triangles each with a side on $\overleftrightarrow{BG}$.

2. Name five different angles each with its vertex at *L*.

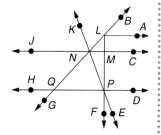

3. STANDARDIZED TEST In a model of a kitchen, the length of a countertop is 6 in. The actual length is 14 ft. What is the scale of the model?

    A. 1 in. : 14 ft   **B.** 6 in. : 14 ft

    C. 2 in. : 7 ft   **D.** 14 ft : 6 in.

Use the graph to identify the coordinates of the point.

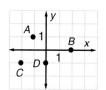

4. *A*   5. *B*

6. *C*   7. *D*

8. WRITING MATHEMATICS Explain how to use a protractor to a friend you are tutoring.

Determine the order of rotation symmetry.

9.
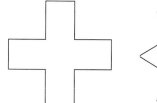

10.

Trace the figure. Draw all lines of symmetry.

11.

12.

13. The drawing shown is that of a cube. What kind of drawing is it?

14. Make a one-point perspective drawing of a cube.

15. A plane intersects a cube from its top left edge to its bottom right edge. What is the shape of the cross section?

16. Draw a triangle. Use a compass and straightedge to bisect each side. Connect the midpoints. Make an observation about the resulting figure.

17. What is the hypothesis and the conclusion in the following conditional statement?

    If *A*(3, −4) is in a coordinate plane, then it is in Quadrant IV.

18. Is the converse of the above conditional statement true or false? Explain.

Find the distance between the points. Answer to the nearest tenth.

19. *A*(−4, 1) and *B*(7, −3) are on a coordinate plane

20. *R*(−5, 0, 3) and *S*(6, −3, 1) are in a three-dimensional coordinate system

PERFORMANCE ASSESSMENT

ASTRONOMY Some star clusters have had shapes and images attributed to them. These special clusters are called *constellations*. Research any two constellations. Draw a model of each and label its stars. In each model, identify and classify line segments, angles, and shapes. Write a geometric description of each constellation.

BOTANY Go on a nature walk to collect five different specimens of leaves. Before you go, find out about dangerous plants such as poison ivy so that you can avoid them. Make a tracing or a rubbing of each leaf you have collected. In each tracing, identify any lines of symmetry, any symmetries in the veins of the leaf, and any symmetries in patterns on the leaf.

WOODWORKING A woodworker has made letters out of blocks of wood. The letters are each 5 in. tall, 2 in. thick, and no more than 4 in. wide. Make isometric and orthogonal drawings of the two wooden letters that form your initials.

3-D GRAPH From a cardboard box, cut off three sides that share a common edge. Label the three interior edges the x-, y-, and z- axes, to establish a three-dimensional coordinate system. Mark units on your axes. Determine the ordered triples that name the vertices of your box. Place the corresponding sides of smaller boxes on top of the first box. Determine the ordered triples that name the vertices of the smaller boxes.

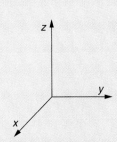

PROJECT ASSESSMENT

PROJECT *Connection*   Form four groups. Each group will prepare a presentation on one aspect of the crystal project. Each group will make its presentation to the class. Plan a schedule for the presentations.

1. One group will summarize all the information gathered about the myths and mysteries associated with crystals. The presentation should include photographs of the types of crystals to be discussed—for example: amethyst, rose quartz, or citrine.

2. A second group will summarize all the information about and experiences with the growing projects. Separate the projects into two categories: those that grew quickly and those that needed help to grow. The presentation should include a synopsis of techniques used.

3. A third group will present the epsomite crystal. This group must research and summarize additional information about this mineral—for example, where it is found, how it is mined, and how it is used. The presentation should include a display of the models made. Each model should have a label indicating its scale with respect to the original diagram in your text.

4. A fourth group will summarize all the information gathered about the seven crystal systems. The presentation should list the geometric characteristics of the shape of a crystal in each system and report on its symmetries. The presentation should include a display of the isometric and orthogonal drawings made of the shape associated with each system.

Fill in the blank.

1. A __?__ has no thickness, but extends infinitely in two opposite directions.

2. The intersection of a solid and a plane is called a __?__ .

3. A drawing that shows top, front, and side views of an object is called __?__ .

4. Two figures that have the same shape but not the same size are __?__ .

Determine the number of lines of symmetry and the order of rotation symmetry for each figure.

5.   6.

Find each length on the number line below.

7. *AC*   8. *BD*   9. *AB*

10. **BLUEPRINTS** An architect is drawing plans for a new house. The architect has chosen to use a scale of 1 in. : 3 ft.

 a. One of the bedrooms is drawn 4 in. by 5 in. What are the actual dimensions of the bedroom?

 b. The living room is to be 15 ft by 21 ft. What dimensions should the living room be drawn on the blueprint?

Describe the shape of the cross section for the triangular prism.

11. with a vertical plane

12. with a horizontal plane

Each figure consists of a stack of cubes. No cubes are hidden from view. Draw the front, top, and right-side view of each figure.

13.   14.

15. Which two figures appear to be congruent?

 A.   B.   C.   D.

16. Which two figures appear to be similar?

 A.   B.3   C.   D.

17. **WRITING MATHEMATICS** On a map, you need to show towns that are 12 miles apart, 30 miles apart, and 42 miles apart. What scale would you choose for your map? Why?

18. Determine the number of planes of symmetry and the number of axes of symmetry for the figure at the right.

Find the distance between two points on a coordinate plane with the given coordinates. Round to the nearest tenth if necessary.

19. $(4, -3)$ and $(-4, 3)$ 20. $(1, 1)$ and $(-2, -2)$

Use the rectangular prism for Exercises 21 and 22.

21. Find the coordinates of point C.

22. Find the distance between points *A* and *B*, to the nearest tenth.

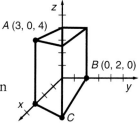

STANDARDIZED TEST

STANDARD FIVE-CHOICE Select the best choice for each question.

1. On a number line, which pair of coordinates is 6 units apart?

   A. −1 and 7   **B.** −5 and 1   **C.** 2 and 4
   D. 6 and −6   **E.** none of these

2. Which of the following is true about similar figures?

   I. They have the same shape.
   II. They are the same size.
   III. The figures have either three sides or four sides.

   A. I only   **B.** II only
   C. I and III   **D.** I and II
   E. I, II, and III

3. Which diagram shows the top view of the following stack of cubes?

   A.   **B.**   **C.**

   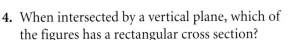

   D.   **E.**

   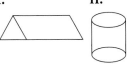

4. When intersected by a vertical plane, which of the figures has a rectangular cross section?

   I.   **II.**   **III.**

   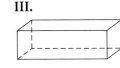

   A. I only   **B.** II only
   C. III only   **D.** II and III
   E. I, II, and III

5. On a coordinate plane, which point is 10 units from the point $(3, -1)$?

   A. $(8, 4)$ **B.** $(13, 9)$ **C.** $(9, 7)$
   D. $(6, 0)$ **E.** two of these

6. The figure shown has rotation symmetry of order

   A. 4   **B.** 3   **C.** 2
   D. 1   **E.** 0

7. A model-maker used a scale of 2 mm : 7 in. To show an actual length of 30 in., the model-maker used a length between

   A. 100 mm and 100 mm
   B. 8 mm and 9 mm
   C. 9 mm and 10 mm
   D. 8 in. and 9 in.
   E. 9 in. and 10 in.

8. Which of the following could be a net for the figure shown?

   A.   **B.**

   C.   **D.**

      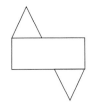

   E. none of these

2 Geometry and Logic

Take a Look AHEAD

Make notes about things that look new.
- Locate a postulate and a theorem.
- Locate two types of reasoning. What are they?

Make notes about things that look familiar.
- List the vocabulary words in this chapter that you have learned before.
- Find the lesson where you will use patterns.

DATA Activity

Everybody Online!

Not long ago, computers were large expensive machines primarily used by engineering firms, research facilities, and big businesses. With the microprocessor, it became possible to manufacture powerful computers that were considerably smaller and much less expensive. Personal computers can now be found in more than forty million American households, and that number is increasing every year. Of these households, at least one fourth have a modem and use it to access an *online service* through a telephone line.

SKILL FOCUS

- ▶ Find the percent of a number.
- ▶ Compare percents.
- ▶ Construct a graph of given data.
- ▶ Determine if it is possible to make predictions based on given data.
- ▶ Gather data.

COMPUTER TECHNOLOGY

In this chapter, you will see how:

- **KITCHEN DESIGNERS** use their knowledge of points, lines, and planes along with CAD software to plan a kitchen.
(Lesson 2.1, page 71)

- **MECHANICAL ENGINEERING TECHNICIANS** use measures of angles and segments to guide a robot.
(Lesson 2.6, page 99)

HOW AMERICANS USE THEIR TIME ONLINE			
	Percent of Users		
Task	**1993**	**1994**	**1995**
information access/retrieval	70	66	64
e-mail	65	64	62
downloading of files	60	55	55
real-time chat	20	40	50
bulletin boards/forums	40	60	60

Use the table to answer the following questions.

1. Suppose the data were gathered by surveying 1200 computer users in each year. How many of those surveyed used their computers for real-time chat in 1993? How many used their computers to download files in 1995?

2. For which task was there the greatest increase in percent of users from 1993 to 1995? the greatest decrease?

3. Create a graph of the data. Explain why you chose the type of graph you made.

4. Can you use the data to make predictions about Americans' use of online time in the year 2010? Explain your reasoning.

5. **WORKING TOGETHER** Gather your own set of data by surveying students in your class or in your school concerning their use of online time. Make a table of your data that is similar to the table on this page. Are the results of your survey similar to the results in the table on this page? Describe any differences. What do you think might be the reason for these differences?

A Bit of This, a Byte of That

When you quickly glance at a word or picture on a computer screen, you probably see straight lines and smooth curves. The longer you look at the image, though—and the more closely you look at it—you can sometimes see irregularities in the lines and curves. Graphic artists refer to these irregularities as "jaggies."

Images on a computer screen are drawn on a grid of tiny squares. Each square is called a *picture element*, or *pixel* for short. In the situation of a two-color monitor, the image is formed by simply turning the pixels "on" or "off" in an appropriate pattern.

In this project, you will investigate the way pixels form the images you see on the screen. You also will relate pixels and the images they create to *binary numbers*, the heart of all computer systems.

PROJECT GOAL

To create a bitmapped image of your name.

Getting Started

Work with a group.

1. Gather samples of several different kinds of *typefaces* and determine the meaning of these related terms: *bold, italic, roman, serif, sans serif.*

2. Research the reasons that binary numbers are so well-suited to the operation of a computer.

3. Make a list of other everyday objects in which you have seen letters, numbers, or pictures formed by a set of "dots."

Internet Connection

www.learninggeometry.com

PROJECT *Connections*

Lesson 2.1, page 70:
Draw images on a grid and write binary numbers to represent them.

Lesson 2.4 page 87:
Learn to separate information into bytes and explore the concept of resolution.

Lesson 2.8, page 109:
Explore typefaces and fonts and learn the meaning of bitmapping.

Chapter Assessment, page 123:
Plan a presentation to display images created by students and to report on insights that emerge from sharing results within the group.

2.1 Points, Lines, and Planes

Explore

- You will need three small flat objects, such as coins, for this activity.

  1. Drop two of the objects onto a flat surface. Determine whether you could place a yardstick on top of the two objects so that it touches both objects at the same time. Do this four times and record your results.

  2. In Question 1, to be able to place the yardstick on both objcts, are you restricted in how far apart the objects are dropped? If the yardstick represents a line, are you restricted? Explain.

  3. Repeat Questions 1 and 2 using three objects.

  4. Compare your results in Question 1 to your results in Question 3. Can you use the results to make a generalization?

  5. Suppose you drop only one object. How many ways could you place the yardstick so it touches the object?

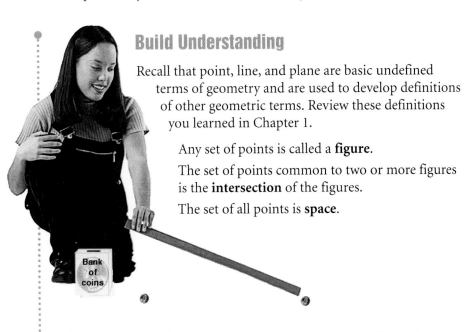

Build Understanding

Recall that point, line, and plane are basic undefined terms of geometry and are used to develop definitions of other geometric terms. Review these definitions you learned in Chapter 1.

Any set of points is called a **figure**.

The set of points common to two or more figures is the **intersection** of the figures.

The set of all points is **space**.

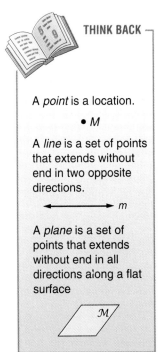

The intersection of two or more figures forms a figure. Often, this figure is a point or a line. A drawing can help you see the set of points the figures have in common. For example, consider all possible ways two lines can intersect. Drawings will show you that the lines either intersect at a single point or the lines coincide.

Collinear points are points that lie on the same line. Points that do not lie on the same line are called **noncollinear points**.

Coplanar points are points that lie in the same plane. Points that do not lie in the same plane are called **noncoplanar points**.

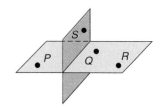

A, B, and C are collinear.
A, B, and D are noncollinear.

P, Q, and R are coplanar.
P, Q, R, and S are noncoplanar.

Just as the meanings of some words are accepted without definition, some statements are accepted as true without proof. Statements like these are called **postulates**. Several fundamental postulates of geometry involve relationships between points, lines, and planes.

POSTULATE 1   UNIQUE LINE POSTULATE   **Through two distinct points, there is exactly one line.**

POSTULATE 2   **A line contains at least two distinct points.**

POSTULATE 3   UNIQUE PLANE POSTULATE   **Through three noncollinear points, there is exactly one plane.**

POSTULATE 4   **A plane contains at least three noncollinear points.**

POSTULATE 5   **If two distinct points lie in a plane, then the line joining them lies in that plane.**

POSTULATE 6   **If two distinct planes intersect, then their intersection is a line.**

COMMUNICATING ABOUT GEOMETRY

Postulate 1 often is stated as *Two points determine a line.*

Similarly, Postulate 3 often is stated as *Three noncollinear points determine a plane.*

In these statements, what is the meaning of the word *determine*? How is it related to the use of the word *determine* in other real-world situations?

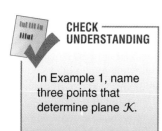

CHECK UNDERSTANDING

In Example 1, name three points that determine plane 𝒦.

EXAMPLE 1

Use the figure. Which postulate justifies the answer?

a. Name three points that determine plane 𝒥.
b. Name the intersection of planes 𝒥 and 𝒦.

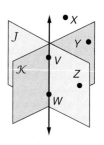

Solution

a. By Postulate 3, three noncollinear points determine a plane. Three points that determine plane 𝒥 are points V, W, and Z.

b. According to Postulate 6, the intersection of two planes is a line. Planes 𝒥 and 𝒦 intersect in $\overleftrightarrow{VW}$.

Recall that a **line segment**, or **segment**, is part of a line that begins at one point and ends at another. These points are called the **endpoints** of the segment.

S •————————• T

EXAMPLE 2

ART Many works of the artist Piet Mondrian consist only of rectangles and parts of rectangles. The painting shown is *Lozenge Compositon with Red, Gray, Blue, Yellow, and Black.* Below is a geometric model of it.

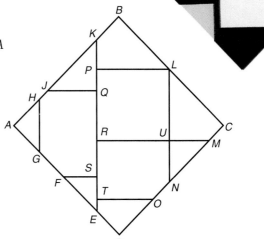

a. Name all the labeled points in the model that are collinear with point *A* and point *D*.
b. Identify the intersection of $\overline{LN}$ and $\overline{RM}$.
c. Identify the intersection of $\overline{KQ}$ and $\overline{EP}$.
d. Name all segments shown for which point *L* is one endpoint.

Solution

a. Points *E*, *F*, and *G* are collinear with *A* and *D*.
b. $\overline{LN}$ and $\overline{RM}$ intersect at point *U*.
c. The intersection of $\overline{KQ}$ and $\overline{EP}$ is $\overline{PQ}$.
d. There are five segments with one endpoint at point *L*: $\overline{LB}$, $\overline{LP}$, $\overline{LU}$, $\overline{LN}$, and $\overline{LC}$.

◄

TRY THESE

Use the figure at the right to answer each question. Which postulate justifies your answer?

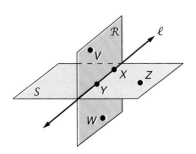

1. Name two points that determine line ℓ.

2. Name three points that determine plane $\mathcal{R}$.

3. Name the intersection of plane $\mathcal{R}$ and plane $\mathcal{S}$.

4. Name three lines that lie in plane $\mathcal{S}$.

5. **MASONRY** When a mason builds a straight wall, two sticks are placed in the ground and a rope is pulled tight between them to align the wall and make certain it is straight. Which postulate is the mason applying to this situation? Explain.

6. **WRITING MATHEMATICS** Explain why it is important to include the word *noncollinear* in the statement of Postulate 3.

CERAMICS In the figure, a geometric model has been superimposed on a pattern of ceramic tiles.

7. Identify the intersection of $\overline{EC}$ and $\overline{BD}$.

8. Identify the intersection of $\overline{AL}$ and $\overline{HD}$.

9. Name all the points that appear collinear with points J and M.

10. Name all segments shown for which point H is one endpoint.

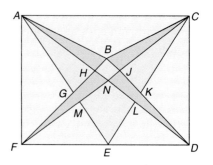

PRACTICE

Refer to the figure at the right. Tell whether each statement is true or false.

11. Points G and O determine line a.

12. Points M and T determine plane $\mathcal{U}$.

13. The intersection of lines b and c is point T.

14. The intersection of planes $\mathcal{U}$ and S is points R and T.

15. Points G, E, and O are collinear.

16. Points M, T, R, and Y are coplanar.

17. Points G and R determine a line.

18. Points M, R, and Y determine a plane.

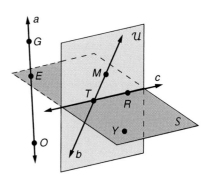

Draw a figure to illustrate each situation.

19. Points A, B, and C are noncollinear.

20. Points W, X, Y, and Z are noncoplanar.

21. Line m intersects plane Z at point Q.

22. Planes J and $\mathcal{K}$ intersect at line t.

23. **WRITING MATHEMATICS** Set aside a section of your notebook and title it *Postulates*. Copy each postulate on page 66 and draw a figure to illustrate it. Plan to update the section as more postulates are introduced.

24. **PHOTOGRAPHY** At the right you see a video camera placed on a stand called a tripod. The photographer uses the tripod to hold the camera steady. Why do you think the photographer uses a tripod instead of a stand with four legs?

TEXTILES **Refer to the fabric design shown below.**

25. Draw and label a geometric model of the design.

26. Identify four collinear points in your model.

27. Identify two segments in your model whose intersection is a segment.

28. Identify four segments in your model that share a common endpoint.

EXTEND

29. Describe an everyday situation that illustrates Postulate 6.

30. Give a counterexample to show why the following statement is false.

Two distinct lines determine a plane.

31. LOOKING AHEAD Explain why the following statement must be true and use postulates in this lesson to justify your response.

A line and a point not on the line determine a plane.

32. Is the following statement *always, sometimes,* or *never* true? Explain your answer.

If two distinct points lie on the curved surface of a cylinder, then the segment joining them lies on the curved surface of the cylinder.

33. TRANSFORMATIONS Trace $\overline{JK}$, shown at the right, onto a sheet of paper. Draw any lines of symmetry that lie in the plane of the paper.

THINK CRITICALLY

34. How many lines are determined by the corner points of the cube at the right?

35. Which postulate allows you to answer Exercise 34 without naming all the lines. Explain.

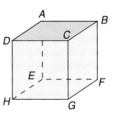

36. How many planes are determined by the corner points of the cube?

37. How many lines are determined by the corner points of a rectangular prism? How many planes?

38. How many lines are determined by the corner points of a square pyramid? How many planes?

ALGEBRA AND GEOMETRY REVIEW

Find the value of each.

39. $|2.75|$ **40.** $|-0.44|$ **41.** $|-3 + 5|$ **42.** $|-3| + 5$ **43.** $|5 - 6|$ **44.** $|5| - 6$

Find each length on the number line.

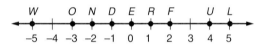

45. WD **46.** RN **47.** FL **48.** OU

Refer to the figure at the right.

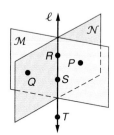

49. Name three segments. **50.** Name three collinear points.

51. Name four coplanar points. **52.** Name four noncoplanar points.

53. Identify the intersection of planes $\mathcal{M}$ and $\mathcal{N}$.

54. True or false: Point Q and line ℓ determine plane $\mathcal{N}$.

Solve each equation.

55. $6j + 8 - 2j = 10$ **56.** $-16 = t + 5 - 2t$ **57.** $8z - 3 = 5 - 2z$

58. PROBABILITY Suppose you toss a coin and then pick a card at random from a standard deck of 52 cards. What is the theoretical probability that you toss a head and pick a diamond?

PROJECT Connection

At the right the letter Y is drawn in a grid representing the *pixels* on a computer screen.

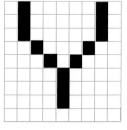

1. Below the grid are nine *binary numbers*. Each represents one row of the grid. What is the meaning of 1 in each number? of 0?

2. Translate the binary number for the first row into a *decimal number* as shown below.

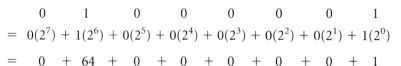

Translate the other eight binary numbers into decimal numbers.

3. Use this set of nine decimal numbers: 28, 34, 65, 1, 6, 24, 96, 127, 0. Translate them into binary numbers. Use the binary numbers as a code for shading a grid eight pixels wide and nine pixels deep. What do you see?

4. Draw the first letter of your name on a grid eight pixels wide and nine pixels deep. Keep the left column and bottom row blank. Write binary numbers to represent the image. Translate them into decimal numbers.

01000001
01000001
01000001
00100010
00010100
00001000
00001000
00001000
00000000

Career
Kitchen Designer

Kitchen designers plan new kitchens and kitchen renovations for private homes and commercial establishments. Usually a designer is expected to prepare detailed scale drawings that will give the client a realistic view of the plans. Increasingly, computer-aided design (CAD) software is used to prepare these drawings, so they can be revised easily as the designer and client experiment with different ideas. A designer must have a good sense of fundamental geometric concepts and language.

The figure below is an isometric drawing of a planned kitchen as it might appear in CAD software. Points have been labeled with letters for easy reference.

Decision Making

In CAD software, walls, floors, and other flat surfaces are considered planes. Name three points in the room that determine each plane.

1. the west wall   **2.** the floor   **3.** the ceiling

When designers speak of the "corner" of a room, they are referring to the intersection of two walls.

4. Name two points that determine the northwest "corner" of this kitchen.

5. This drawing highlights the corner determined by points *C* and *D*. In terms of compass directions which corner is this?

6. The client would like to see how the refrigerator would look if placed in this corner. Copy the given figure and draw a refrigerator in the requested location.

7. Prepare an isometric drawing that highlights the northeast corner of this kitchen. Be sure to label points *A* through *H* and the four compass directions. Draw any appliances or other objects you think should appear in this view.

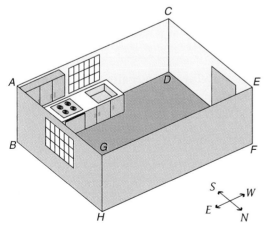

Explore

● Each figure shows a ruler placed in position to measure a segment.

- Find x, the ruler number that corresponds to the left endpoint.
- Find y, the ruler number that corresponds to the right endpoint.
- Calculate $|y - x|$ and $|x - y|$.

1.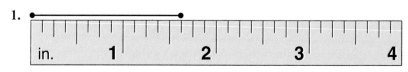

2.

3.

4. In general, how do you use a ruler to measure a length?

5. What is the relationship among the three segments?

Build Understanding

THINK BACK

The length of a segment is the distance between its endpoints.

● When you found distances between points on a number line, you were using a basic assumption about segments called the Ruler Postulate.

> ── **RULER POSTULATE** ──
>
> **POSTULATE 7**   The points on a line can be paired, one-to-one, with the real numbers so that any point is paired with 0 and any other point is paired with 1.
>
> The real number that corresponds to a point is the *coordinate* of that point.
>
> The *distance* between two points on the line is equal to the absolute value of the difference of their coordinates.

A ruler is a type of number line. Whenever you use a ruler to measure a length, you are using the ruler postulate.

Given a segment, such as $\overline{AB}$, write the length of $\overline{AB}$ as AB, without a bar above the letters.

TECHNOLOGY TIP

Using geometry software, you can make a *dynamic drawing* to help you see why the Segment Addition Postulate is assumed to be true.

• Use the *segment* tool to draw $\overline{AB}$.

• Use the *point on object* command to draw point C on $\overline{AB}$.

• Use the *measure length* command to find AB, AC, and CB.

• Use the *calculate* command to find $AC + CB$.

Now grab point C and observe the results as you move it.

EXAMPLE 1

A ruler is positioned next to $\overline{RS}$. Point R corresponds to 6.5 cm and point S corresponds to 4.8 cm. What is the length of $\overline{RS}$?

Solution

Find the absolute value of the difference of 6.5 and 4.8. The order in which you subtract does not matter.

$$RS = |6.5 - 4.8| \qquad\qquad RS = |4.8 - 6.5|$$
$$RS = |1.7| \qquad \text{or} \qquad RS = |-1.7|$$
$$RS = 1.7 \qquad\qquad\qquad RS = 1.7$$

Using either order, $RS = 1.7$ cm.   ◀

On a number line, a point C is **between** points A and B if the coordinate of C is between the coordinates of A and B. This leads to a second postulate about segments, the Segment Addition Postulate.

SEGMENT ADDITION POSTULATE

POSTULATE 8   If point C is between points A and B, then $AC + CB = AB$.

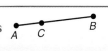

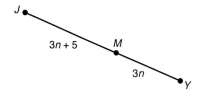
EXAMPLE 2

USING ALGEBRA In the figure below, point M is between points J and Y and $JY = 29$. Find JM.

Solution

In the figure, $JM = 3n + 5$ and $MY = 3n$. You know $JY = 29$. Use the Segment Addition Postulate to write and solve an equation.

$$\begin{array}{ll} JM + MY = JY & \text{Segment Addition Postulate} \\ (3n + 5) + 3n = 29 & \text{Substitute.} \\ 6n + 5 = 29 & \text{Combine like terms.} \\ 6n = 24 & \text{Subtract 5 from each side.} \\ n = 4 & \text{Divide each side by 6.} \end{array}$$

The value of n is 4. So $JM = 3n + 5 = 3(4) + 5 = 17$.   ◀

CHECK UNDERSTANDING

In Example 2, find MY. How can you use this measure to check the solution of Example 2?

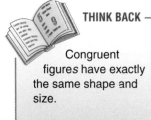
Segments that are equal in length are called **congruent segments**. In the figure below, $\overline{KL}$ is congruent to $\overline{MN}$. You write $\overline{KL} \cong \overline{MN}$. Indicate congruent segments by marking them with an equal number of "tick marks," like those shown in blue below.

You can use congruent segments to make a *linear model* for certain probability problems. An outcome is the length of a segment.

EXAMPLE 3

SCHEDULING An appliance store schedules a delivery to your house between 4:00 P.M. and 4:30 P.M. today. You cannot get home until 4:05 P.M. What is the probability that you will miss the delivery?

Solution

Draw a segment to represent the time from 4:00 P.M. to 4:30 P.M. Divide it into six congruent segments as shown below.

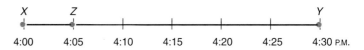

$\overline{XY}$ represents the time when the package delivery is scheduled. $\overline{XZ}$ represents the time when you might miss the delivery.

$$P(\text{miss the delivery}) = \frac{\text{length of } \overline{XZ}}{\text{length of } \overline{XY}} = \frac{1}{6}$$

The probability that you will miss the delivery is $\frac{1}{6}$. ◄

The **midpoint** of a segment is the point that divides the segment into two congruent segments. Any segment, ray, line, or plane that intersects a segment at its midpoint is called a **segment bisector**. On a number line, the coordinate of the midpoint of a segment whose endpoints have coordinates a and b is $\frac{a + b}{2}$.

EXAMPLE 4

Find the midpoint of $\overline{GH}$ on the number line at the right.

Solution

The coordinate of G is -3. The coordinate of H is 2.

$$\frac{-3 + 2}{2} = -\frac{1}{2} \qquad \frac{a + b}{2}$$

The coordinate of the midpoint of $\overline{GH}$ is $-\frac{1}{2}$. ◄

MIDPOINT FORMULA

On a coordinate plane, the midpoint of a segment whose endpoints have coordinates (x_1, y_1) and (x_2, y_2) is the point with coordinates $\left(\dfrac{x_1 + x_2}{2}, \dfrac{y_1 + y_2}{2}\right)$.

EXAMPLE 5

Find the midpoint of $\overline{ST}$ with endpoints $S(-3, 4)$ and $T(5, -3)$.

Solution

Let $(-3, 4)$ be (x_1, y_1) and $(5, -3)$ be (x_2, y_2).

$$\frac{x_1 + x_2}{2} = \frac{-3 + 5}{2} = \frac{2}{2} = 1 \qquad \frac{y_1 + y_2}{2} = \frac{4 + (-3)}{2} = \frac{1}{2} = 0.5$$

The midpoint of $\overline{ST}$ is $M(1, 0.5)$. ◄

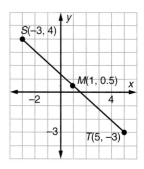

TRY THESE

Points C and D correspond to the given ruler numbers. Find CD.

1. C: 2.7 cm; D: 9.2 cm
2. C: $6\frac{1}{4}$ in.; D: $2\frac{7}{16}$ in.

3. **WRITING MATHEMATICS** Explain how a number line and a ruler are alike and how they are different.

4. Identify all the congruent segments in the figure at the right. Write a statement about each congruence using the symbol $\cong$.

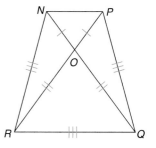

In segment JL, point K is between points J and L and $JL = 44.5$.

5. Find JK.
6. Find KL.

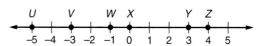

Find the midpoint of each segment on the number line.

7. $\overline{XZ}$
8. $\overline{UX}$
9. $\overline{VY}$
10. $\overline{VZ}$

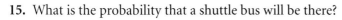

Find the midpoint of each segment in the figure at the right.

11. $\overline{AB}$
12. $\overline{CD}$
13. $\overline{EF}$
14. $\overline{GH}$

TRAVEL An airport shuttle bus arrives at the entrance to your hotel every twenty minutes. The shuttle driver waits six minutes before departing for the airport. Suppose you arrive at the hotel entrance at a random time during a one-hour period.

15. What is the probability that a shuttle bus will be there?

16. What is the probability that a shuttle bus will not be there?

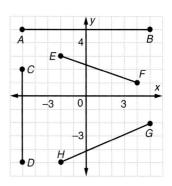

PRACTICE

Refer to the diagram below. Use the Ruler Postulate to find each length. Show your work.

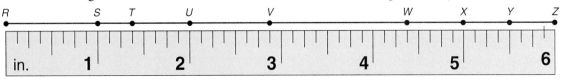

17. *UX*   **18.** *WT*   **19.** *SV*   **20.** *TX*   **21.** *VY*   **22.** *VW*

23. In the figure below, point *N* is between points *M* and *P* and *MP* = 104. Find *NP*.

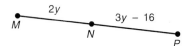

24. In the figure below, point *Y* is the midpoint of $\overline{XZ}$. Find *XY*, *YZ*, and *XZ*.

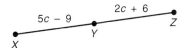

Refer to the number line below. Tell whether each statement is true or false.

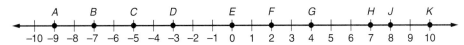

25. $\overline{BE} \cong \overline{FJ}$ **26.** $\overline{AD} \cong \overline{DG}$ **27.** $BC + CD = BD$ **28.** $AD + EG = AG$

29. Point *J* is the midpoint of $\overline{HK}$.   **30.** Point *E* is the midpoint of $\overline{BH}$.

Find the midpoint of the segment that has the given endpoints on a coordinate plane.

31. $M(-2, 15); E(8, 3)$

32. $U(4, -5); S(-3, -9.5)$

33. $P(6\frac{1}{2}, 9); Q(3, -1)$

34. $C(-7, 3\frac{1}{4}); D(0, 5\frac{1}{2})$

35. CONSTRUCTION A builder must cut a 15-ft piece from a long strip of molding. However, the builder only has a 6-ft tape measure. Explain how the builder can apply the Segment Addition Postulate to this situation.

36. BROADCASTING Radio station WGSW plays ninety seconds of commercials at the start of every ten-minute period. Suppose you tune in to WGSW at a random time during your favorite one-hour program. What is the probability that a commercial will be playing?

37. WRITING MATHEMATICS Write an original probability problem that you can solve by using a linear model. Show how to use the model to solve your problem.

EXTEND

38. The length of $\overline{AB}$ is $2\frac{7}{8}$ in. A ruler is positioned next to $\overline{AB}$ so that point A corresponds to $4\frac{3}{16}$ in. Find two ruler numbers that could correspond to point B.

39. On a coordinate plane, the midpoint of $\overline{ST}$ is the point $M(6, -2)$. One endpoint of $\overline{ST}$ is the point $S(-1, 8)$. Find the coordinates of the other endpoint.

40. Point K is between points J and L, $JL = 51$, and $JK = 2KL$. Find JK and KL.

41. TRANSFORMATIONS True or false: A line that is a bisector of a segment also is a line of symmetry for the segment. Explain.

Points G, H, J, K, L, and M are six distinct collinear points, point J is the midpoint of $\overline{GM}$, point H is the midpoint of $\overline{GJ}$, and $\overline{JK} \cong \overline{KL} \cong \overline{LM}$.

42. Draw a figure using the information given.

43. Suppose a point of $\overline{GM}$ is chosen at random. What is the probability it is on $\overline{JK}$?

44. Suppose a point of $\overline{GK}$ is chosen at random. What is the probability it is on $\overline{JK}$?

THINK CRITICALLY

45. Refer to the three-dimensional coordinate system at the right. Find the coordinates of the midpoint of $\overline{PQ}$.

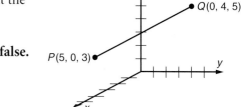

Decide whether each conditional statement is true or false. Justify your answer.

46. If point B is the midpoint of $\overline{AM}$, then $\overline{MB} \cong \overline{AB}$.

47. If $\overline{AM} \cong \overline{MB}$, then point M is the midpoint of $\overline{AB}$.

ALGEBRA AND GEOMETRY REVIEW

48. The rule for a function f is $f(x) = 3x - 4$. Find $f(0)$, $f(-5)$, $f(2.75)$, and $f\left(-\frac{1}{2}\right)$.

Write a rule for the function defined by each table.

49.

x	-2	-1	0	1	2
$f(x)$	-8	-7	-6	-5	-4

50.

x	-2	-1	0	1	2
$f(x)$	10	5	0	-5	-10

51. Find $f(0.5)$ for the function in Exercise 49 and $f\left(\frac{1}{2}\right)$ for the function in Exercise 50.

52. STANDARDIZED TESTS Point Y is between points X and Z. Which must be true?

A. $XY + YZ = XZ$
B. Point Y is the midpoint of $\overline{XZ}$.
C. $\overline{XY} \cong \overline{YZ}$
D. All of these statements are true.

53. True or false: Through three distinct points, there is exactly one plane.

NETWORKS

Children in Zaire play a game called *Shongo*.
A child sketches a figure in the sand like the
ones shown. Then the child challenges someone
unfamiliar with the game to trace around the
figure without lifting a finger from the sand
and without retracing any segment. The person
may, however, pass through a point more than
once.

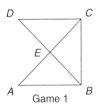

Game 1

1. Copy the diagram for Game 1. Place your
   finger at point A. Can you trace all the way
   around the figure without lifting your finger
   and without retracing any segment? Use
   arrows to show the path on your diagram.

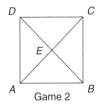

Game 2

2. Rework Question 1 choosing a different
   starting point.

3. Copy the diagram for Game 2. Can you trace
   all the way around the figure without lifting
   your finger and without retracing any
   segment? Explain your response.

Using knowledge of networks, or graphs, you can determine whether a
Shongo network is traceable before you attempt a tracing. A **network**, or
graph, is a finite set of points, called *vertices* or *nodes*, joined by line
segments or arcs, called *edges*. The study of graphs and networks is
called *graph theory*.

4. How many edges does the network for Game 1 have?

A network is *traceable* if you can begin at one vertex and travel to all
other vertices by tracing each edge exactly once. To determine whether a
network is traceable, you need to study its vertices and edges.

Each vertex of a graph has a **degree**, the number of edges that have that
vertex as an endpoint. In the graph of Game 2, vertex A has degree 3
and vertex E has degree 4. An **odd vertex** is one whose degree is odd. An
even vertex is one whose degree is even.

5. List each vertex and its degree for the network in Game 1.

6. List each vertex and its degree for the network in Game 2.

7. Can you trace around this graph? List each vertex and its degree.

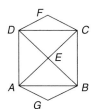

8. Can you trace around this graph? List each vertex and its degree.

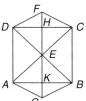

9. Use your answers for Questions 5–8 to make a conjecture based on the evenness of the vertices about the traceability of a network.

10. Can you trace around the graph shown? List each vertex and its degree for this network.

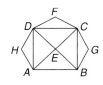

11. Use your answers for Questions 6 and 10 to make a conjecture based on the oddness of the vertices about the traceability of a graph.

Use your conjectures from Questions 9 and 11 to determine whether each graph for Questions 12–13 is traceable. If the graph is traceable, show a possible path. Explain how your conjectures are satisfied.

12.

13.

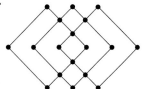

The following famous problem, posed in the early 1700s, gave rise to the mathematical study of networks.

KÖNIGSBERG BRIDGE PROBLEM Is it possible for a walker to start at a point on one of the land masses and walk across each of the seven bridges over the Pregol River in Kaliningrad, Russia, exactly once, that is, without crossing any bridge twice?

14. Try to determine a path that begins at A, B, C, or D and covers each edge exactly once. Describe your efforts.

15. List each vertex and its degree. How many odd vertices does the network have?

16. Make a conjecture about the traceability of a network that has more than two odd vertices.

17. Draw three graphs to illustrate all your conjectures.

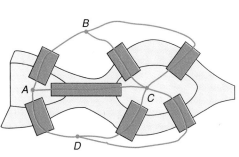

Finding Patterns

You can solve many everyday problems by looking for a pattern among given data.

Problem

A local company has a unique savings plan. After you complete one month of employment, the company deposits $1 into a special savings account for you. Then $3 more is deposited after two months, $6 more after three months, $10 more after four months, and so on. What amount is deposited into this account after twelve months?

Explore the Problem

To find a pattern among data, organize the data into a table.

month	1	2	3	4	5	6	7	8	9	10	11	12
amount, $	1	3	6	10								

1. Describe how each number in the *month* row of the table is related to the number to its left.

2. Describe how each number in the *amount* row of the table is related to the number to its left.

3. Using your answer to Question 2, copy and complete the table. What amount is deposited after twelve complete months?

THINK BACK

When one quantity depends upon another, the first quantity is said to be a **function** of the second. So the amount deposited is a function of the number of complete months of employment. You can express this relationship by using *function notation* as follows.

$a(1) = 1,$

$a(2) = 3,$

$a(3) = 6,$

$a(4) = 10, \ldots$

Questions 1–3 show you how to solve the problem by extending the pattern until both rows have an equal number of entries. Now you will see how to generalize the pattern by writing a *function rule*.

4. Let the variable n represent the number of complete months of employment. What does the expression $n + 1$ represent?

5. Evaluate $n(n + 1)$ when $n = 1$, $n = 2$, $n = 3$, and $n = 4$.

6. Compare your answers in Question 5 to the first four entries in the *amount* row of the table. What do you notice?

7. Use your answers to Questions 5 and 6. Write a rule for the function a that relates the amount deposited to the number n of complete months of employment.

8. Show how to use your function rule from Question 7 to find the amount deposited after twelve complete months. Check that the amount is the same as the amount you calculated in Question 3.

9. WRITING MATHEMATICS Suppose you want to find the amount deposited in the account after five complete years. Would you extend the table, use your function rule, or do both? Explain.

┌─────────────────────┐
│ ┌ PROBLEM ───────── │
│ SOLVING PLAN   │
│   │
│ • Understand   │
│ • Plan   │
│ • Solve   │
│ • Examine   │
│   │
└─────────────────────┘

Investigate Further

Figure numbers are numbers that can be modeled by a specific pattern of points on a plane. The amounts in the problem just explored are a type of figurate numbers called *triangular numbers*. Here are geometric models for the first four triangular numbers.

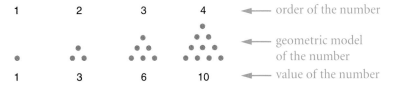

1   2   3   4   ◄—— order of the number

◄—— geometric model
of the number

1   3   6   10   ◄—— value of the number

10. Draw a geometric model for the fifth triangular number. What is the value of this number?

11. List the values of the first twelve triangular numbers.

12. Write a rule for the function *t* that relates the value of a triangular number to its order *n*. This is the *triangular number function*.

Below are geometric models for the first four *oblong numbers*.

13. Draw a geometric model for the fifth and sixth oblong numbers. What are the values of these numbers?

14. Write a rule for the oblong number function.

15. The figures below illustrate a relationship between the oblong and triangular numbers. What is the relationship?

GEOMETRY: WHO, WHERE, WHEN

More than 2500 years ago, the Greek mathematician Pythagoras and his followers searched for connections between arithmetic and geometry. They are credited with the earliest investigations of figurate numbers and their properties.

Apply the Strategy

At the right are geometric models for the first four square numbers.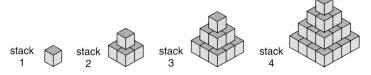

16. Draw geometric models for the fifth and sixth square numbers.

17. List the first twelve square numbers.

18. Write a rule for the square number function.

19. RETAILING A new product is packaged in boxes shaped like cubes. The manager plans to display these cubes as shown. How many cubes would be in the tenth stack of this display?

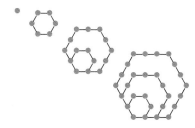

stack 1   stack 2   stack 3   stack 4

Shown below are two additional types of figurate numbers. For each type, make a similar model for the fifth number. Then list the first twelve numbers of that type.

20. pentagonal numbers

21. hexagonal numbers

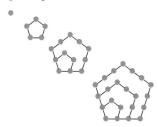

THINK BACK

Pentagons and hexagons are special types of plane figures called *polygons*. A pentagon has five sides. A hexagon has six sides.

22. A different model for the pentagonal numbers is shown below. Make a model like this for the fifth pentagonal number.

23. Use the model from Question 22. Write a rule for the pentagonal number function.

24. Adjust the pentagonal number model from Question 22 to create a different model for the hexagonal numbers. Then use this new model to write a rule for the hexagonal number function.

25. WRITING MATHEMATICS Create an original geometric model for a numerical pattern different from those in this lesson.

REVIEW PROBLEM SOLVING STRATEGIES

angles → angles →
everywhere

2. Ray AB and ray AC form right angle BAC. Juanita draws $\overrightarrow{AD}$ in the interior of ∠BAC and notices that now she can name three different angles in the drawing. If she draws a second ray in the angle's interior how many different angles can she name? If she draws a total of 11 rays in the angle's interior how many different angles can Juanita name?

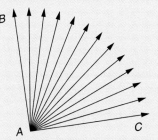

Hint: Work an easier problem and look for a pattern.

A party of 5

1. Ed, Ida, Lou, Ned, and Sally are sitting along one side of a long table. Ned is between Lou and Sally. Ed is next to Ned. Ida is between Lou and Ned. Ed is not next to Ida. What is the order of the five people at the table?

THE WANDERING ANT

3. An ant is standing at the origin of a rectangular coordinate system. It walks one unit to the right, two units up, and three units to the left. The trip takes 30 seconds. It makes the same trip during the next 30 seconds, one unit to the right, two units up, and three units to the left. If the ant continues to make this same trip every 30 seconds for a total of ten minutes, what are the coordinates of the point on which the ant finally lands?

2.4 Focus on Reasoning

Inductive Reasoning

Discussion

Work with a partner. When Mai, Malcolm, Kristen, and Shawn considered the following problem, each gave a different solution.

> How many lines are determined by eight points, no three of which are collinear?

1. Mai drew the figure at the right. She decided that eight points determine one line. What error did she make?

2. Malcolm remembered the Unique Line Postulate, which states that two points determine a line. Since $8 \div 2 = 4$, he reasoned that eight noncollinear points determine four lines. What is the flaw in his reasoning?

3. Kristen drew the figure below. She decided that eight noncollinear points determine eight lines. Do you agree or disagree? Explain.

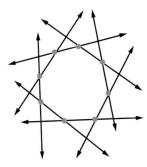

4. Shawn drew the three figures below. He said this set of figures proves that eight noncollinear points determine 28 lines. How do you think Shawn arrived at this solution?

5. Do you agree or disagree with Shawn's solution? Explain.

6. Do you think Shawn is correct to say that he has proved his solution? Explain your reasoning.

Build a Case

In Discussion, Mai, Malcolm, and Kristen only considered the number of lines determined by eight points, which is why each of their solutions is incorrect.

TECHNOLOGY TIP

Often you will be asked to make conjectures about geometric situations. Geometry software can be extremely helpful in making and testing conjectures.

Shawn realized that there are many lines determined by eight points and it could be confusing to draw and count them all. So he used a process called **inductive reasoning**. In this type of reasoning, you look for patterns among a set of data and use the patterns to make an educated guess. This educated guess is called a **conjecture**.

7. Make a table of Shawn's data. Copy the table below. Enter the data for the three figures Shawn drew.

number of noncollinear points								
number of lines determined								

8. Describe the patterns among Shawn's data in the table.

9. Complete the table by extending the patterns.

10. Using the table, what conjecture can you make about the number of lines determined by nine points, no three of which are collinear?

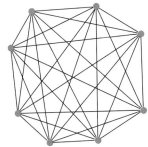

To show that Shawn's conjecture is true, you can carefully draw a figure like the one at the right and count the segments as you draw. (Segments are used instead of lines to simplify the figure.)

Often you can generalize a conjecture by writing a function rule. Shawn wrote the following conditional statement about his conjecture.

If you are given n points, no three of which are collinear, then the number of lines determined by the points is given by the function $L(n) = \dfrac{n(n-1)}{2}$.

11. Show how to use Shawn's function rule to derive each of the eight entries in the table.

12. Use Shawn's function rule to make a conjecture about the number of lines determined by twenty noncollinear points.

It is important to realize that a conjecture derived from inductive reasoning might be true, but it is not necessarily true. As you continue to study geometry you will learn how to prove if conjectures are true.

EXTEND AND DEFEND

For each pattern of points, draw the next figure of the pattern. Then write a function rule to generalize the pattern.

13.

14.

15. Study the figures below. Draw the next figure of the pattern. Make a conjecture about the number of regions formed when sixteen lines intersect at one point inside a circle.

16. Study the figures below. Draw the next figure of the pattern. Then make a conjecture about the maximum number of regions formed when sixteen lines intersect inside a circle. (Hint: Copy the fourth figure. Then look for a way to draw a line that intersects each of the lines in that figure.)

Use inductive reasoning to make a conjecture about the solution of each problem.

17. How many rays are determined by eighteen distinct coplanar lines that intersect in one point?

18. How many segments are determined by twenty collinear points?

Study the given number pattern. Then make a conjecture about the number that makes the last statement true. Use a calculator to prove or disprove your conjecture.

19.
$$1(9) + 2 = 11$$
$$12(9) + 3 = 111$$
$$123(9) + 4 = 1111$$
$$1234(9) + 5 = 11,111$$

$$12,345,678(9) + 9 = \underline{\ ?\ }$$

20.
$$9(9) + 7 = 88$$
$$98(9) + 6 = 888$$
$$987(9) + 5 = 8888$$
$$9876(9) + 4 = 88,888$$

$$9876543(9) + 1 = \underline{\ ?\ }$$

21. WRITING MATHEMATICS Design a pattern of points or a pattern of figures. Write a function rule to generalize your pattern. See if a classmate can draw the next figure in your pattern and determine your function rule.

For Questions 22–23, explain how inductive reasoning was used in making each conjecture. Do you agree or disagree with the conjecture? Explain your reasoning.

22. WEATHER Peter noticed that the outdoor temperature was above normal on each of the last ninety days. He makes a conjecture that the temperature will be above normal tomorrow.

23. CONSUMERISM In each of the past eight months, Deanna received her telephone bill in the mail on the first or second day of the month. She makes a conjecture that she will receive her next telephone bill on the first or second day of this month.

LOOKING AHEAD For Question 24, use geometry software if it is available.

24. Draw a large triangle and label the vertices *A*, *B*, and *C*.

25. Use a compass and straightedge or the point at midpoint command of your software. Locate the midpoint of each side. Label the midpoints *D*, *E*, and *F*.

26. Draw $\overline{AE}$, $\overline{BF}$, and $\overline{CD}$.

27. Examine the figure. Using a ruler and protractor or the measure menu of your software, make a list of conjectures about the measures of segments and angles in the figure.

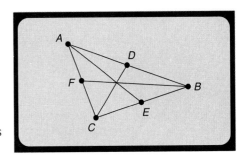

PROJECT *Connection*

The number of pixels an image is divided into is called the resolution of the image.

In the figure at the right, the letter Y is drawn on a grid eight pixels wide and nine pixels deep. Its resolution is 8 by 9. In the figure below it, the resolution has been increased to 16 by 18. At this higher resolution, the pixels are smaller. With the smaller pixels you can fill in some detail to make the curved parts of the Y appear straighter.

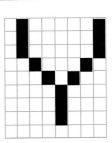

1. In a binary number, each 0 and 1 is a *bit* of information. Computers commonly process information in sets of eight bits, called *bytes*. Notice that the higher-resolution image of the letter Y has been separated into two parts by a segment. How many bytes of information have been created?

2. Draw the initials of your name on a grid so that the resolution is 16 by 18. Write a binary number of each byte of information you have created. Translate it into a decimal number.

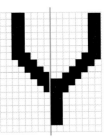

Writing Definitions

Think Back

● Refer to the figure. What geometric term
describes each?

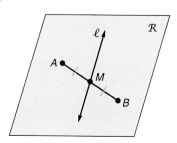

1. A, M, and B **2.** ℓ

3. $\mathcal{R}$ **4.** $\overline{AB}$

Replace each __?__ with the geometric
term that makes the statement true.

5. M is the __?__ of $\overline{AB}$.

6. ℓ is a __?__ of $\overline{AB}$.

7. Refer to your answers to Questions 1–6. Which terms are
undefined? Which are defined?

Explore

● Work with a partner. Review these definitions.

A *segment* is part of a line that begins at one point and ends at
another. The points are called the *endpoints* of the segment.

Congruent segments are segments that are equal in length.

The *midpoint of a segment* is the point that divides the segment
into two congruent segments.

A *segment bisector* is any segment, ray, line, or plane that intersects
a segment at its midpoint.

8. What geometric terms must you know in order to understand the
definition of segment?

9. Why do you think the definition of congruent segments is listed
before the definition of midpoint?

10. Explain why the following is *not* a good definition.

A segment bisector is any figure that intersects a segment at its
midpoint.

11. Assume that all the other definitions remain the same. Explain
why the following is *not* a good definition.

The midpoint of a segment is the point that is a segment
bisector.

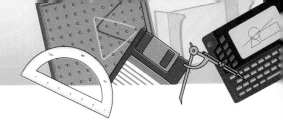

Make Connections

- A **definition** is a statement of the meaning of a word, term, or phrase. A critical part of the structure of Euclidean geometry is a carefully developed set of definitions. In writing these definitions, it is important to consider what makes a definition a good definition.

┌─────────── CHARACTERISTICS OF A GOOD DEFINITION ───────────┐

A good definition of an idea or an object must:

1. **Use only words that have been previously defined or are among the undefined terms.**

2. **Be _precise_, accurately describing the attributes of the idea or object that distinguish it from other ideas or objects of its type.**

3. **Be _concise_, giving no more information than is necessary.**

└──┘

COMMUNICATING ABOUT GEOMETRY

Good definitions are important to effective communication. Have you ever been involved in a misunderstanding that arose because two people did not have the same understanding of a term? Describe the situation.

For Questions 12–14, consider these definitions.

A _ray_ is part of a line that begins at one point and extends without end in one direction. The point is called the _endpoint_ of the ray.

12. Which words in these definitions are either undefined terms or previously defined terms?

13. Which other geometric figure is "part of a line"? Explain how these definitions distinguish a ray from that figure.

14. Explain why the following statement would _not_ be a good definition of a ray.

A ray is part of a line that begins at one point and extends without end in one direction, and it can be a segment bisector.

15. LAW Good definitions are important to maintaining a fair society. For instance, suppose a person is accused of _disturbing the peace_. Why must this term be carefully defined in the laws of the community?

THINK BACK

One common unit for measuring angles is the *degree*.

Remember that an angle is formed by two rays that have a common endpoint. The endpoint is the vertex of the angle, and each ray is a side of the angle. The figure at the right shows ∠*PQR*, which has a measure of 50°.

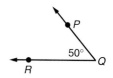

There are several special terms associated with angles. Study the examples and counterexamples given below. Then use inductive reasoning to write a definition of the term printed in blue.

16. These are right angles.   These are *not* right angles.

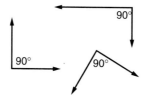   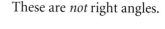

17. These are obtuse angles.   These are *not* obtuse angles.

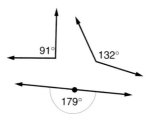   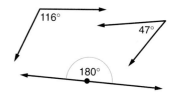

18. Angle pairs in this column are complementary angles.   Angle pairs in this column are *not* complementary angles.

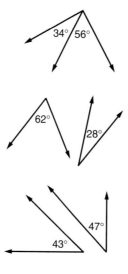   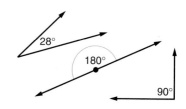

When writing definitions, you must avoid a problem called *circularity*. For example, suppose you draw the figure at the right. You then decide that you can write a definition of the term *line* as follows:

ℓ

180°

A line is an angle whose measure is 180°.

Watch what happens when you work backward from this definition.

A *line* is an (angle) whose measure is 180°.

An *angle* is the union of two (rays) that have a common endpoint.

A *ray* is part of a (line) that begins at one point and extends without end in one direction.

In trying to define the term *line*, you have "circled back" to the term *line*. So your definition is not valid.

Summarize

19. WRITING MATHEMATICS Why do you think it is important to write good definitions of geometric terms?

THINKING CRITICALLY Refer to the figure at the right.

20. Create an original geometric term that you think would be a good name for a figure of this type.

21. What do you think are the important characteristics of this figure? Use your answer to write a definition of the term you created in Question 20.

GOING FURTHER One way to verify that you have made a good definition is to check that it is "reversible." For instance, here is how you can check the definition of congruent segments.

Congruent segments are segments that are equal in length.
Segments that are equal in length are congruent segments.

Since both statements are true, the first statement is a good definition.

22. Show how to reverse this definition: Space is the set of all points.

23. Explain why the following "definition" of midpoint is not reversible: The midpoint of a segment is a point between the endpoints of the segment.

Explore/Working Together

1. Draw ∠ABC on a file folder, as shown at the right.

2. Position the folder so m∠ABC is
   a. 90°
   b. greater than 0° but less than 90°
   c. 180°
   d. greater than 90° but less than 180°

3. Work with a partner. Can you position your two file folders so the sum of the measures of the angles together is 90°? 180°?

Build Understanding

Recall that a *ray* is part of a line that begins at one point and extends without end in one direction. The point is the *endpoint* of the ray. The figure below shows ray BA, written $\overrightarrow{BA}$. Its endpoint is point B.

If point B is between points A and C, then $\overrightarrow{BA}$ and $\overrightarrow{BC}$ are **opposite rays**.

An *angle* is formed by two rays that have a common endpoint. The common endpoint is the *vertex* of the angle, and each ray is a *side* of the angle. The figure in Explore is angle ABC, written $\angle ABC$. Its vertex is point B. Its sides are $\overrightarrow{BA}$ and $\overrightarrow{BC}$.

The Protractor Postulate is a basic assumption about angles.

PROTRACTOR POSTULATE

POSTULATE 9   Let O be a point on $\overleftrightarrow{AB}$ such that O is between A and B. Consider $\overrightarrow{OA}$, $\overrightarrow{OB}$, and all the rays that can be drawn from O on one side of $\overleftrightarrow{AB}$. These rays can be paired with the real numbers from 0 to 180 so that

1. $\overrightarrow{OA}$ is paired with 0 and $\overrightarrow{OB}$ is paired with 180.

2. If $\overrightarrow{OP}$ is paired with x and $\overrightarrow{OQ}$ is paired with y, then the number paired with $\angle POQ$ is $|x - y|$. This number is called the *measure*, or the *degree measure*, of $\angle POQ$.

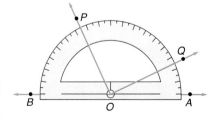

EXAMPLE 1

In the figure at the right, find the measure of $\angle POQ$.

Solution

Notice that the protractor has two scales.

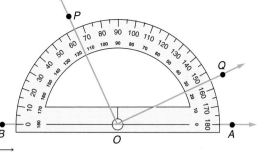

On the inner scale, $\overrightarrow{OP}$ is paired with 115 and $\overrightarrow{OQ}$ is paired with 25.

$$|115 - 25| = |90| = 90$$

On the outer scale, $\overrightarrow{OP}$ is paired with 65 and $\overrightarrow{OQ}$ is paired with 155.

$$|65 - 155| = |-90| = 90$$

By the Protractor Postulate, the m$\angle POQ$ is 90°.

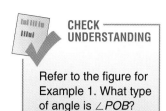

CHECK UNDERSTANDING

Refer to the figure for Example 1. What type of angle is $\angle POB$? $\angle QOB$?

Angles can be classified by their measures. This chart defines these classifications and gives an example of each from Example 1.

Type of Angle	Measure	Example
acute	greater than 0° and less than 90°	$\angle QOA$
right	equal to 90°	$\angle POQ$
obtuse	greater than 90° and less than 180°	$\angle POA$
straight	equal to 180°	$\angle BOA$

Pairs of angles may be classified by the *sum* of their measures.

Complementary angles are two angles whose measures have a sum of 90°. Each angle is the *complement* of the other.

Supplementary angles are two angles whose measures have a sum of 180°. Each angle is the *supplement* of the other.

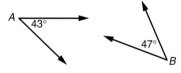
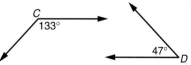

m$\angle A$ + m$\angle B$
43° + 47° = 90°
$\angle A$ and $\angle B$ are complementary.

m$\angle C$ + m$\angle D$
133° + 47° = 180°
$\angle C$ and $\angle D$ are supplementary.

An angle with measure less than 180° divides a plane into three sets of points: the angle itself, the points in the *interior* of the angle, and the points in the *exterior* of the angle.

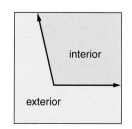

interior

exterior

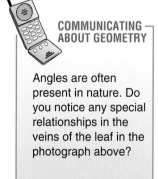

COMMUNICATING ABOUT GEOMETRY

Angles are often present in nature. Do you notice any special relationships in the veins of the leaf in the photograph above?

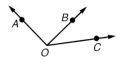

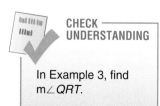
As with segments, you can add angles.

ANGLE ADDITION POSTULATE

POSTULATE 10   If point *B* is in the interior of ∠AOC, then
m∠AOB + m∠BOC = m∠AOC.

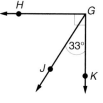

In a figure, a 90° angle is often noted by placing a small square bracket on the interior of the angle at the vertex, as shown in Example 2

EXAMPLE 2

Find m∠HGJ.

Solution

From the figure, m∠JGK = 33°. The square bracket indicates m∠HGK = 90°.

$$m\angle HGJ + 33° = 90°$$
$$m\angle HGJ = 90° - 33° \qquad \text{Angle Addition Postulate}$$
$$m\angle HGJ = 57°$$

Adjacent angles are two angles in the same plane that share a common side and a common vertex, but have no interior points in common. In Example 2, ∠HGJ and ∠JGK are adjacent angles. Two adjacent angles whose noncommon sides are opposite rays form a **linear pair** as shown in Example 3.

LINEAR PAIR POSTULATE

POSTULATE 11   If two angles form a linear pair, then they are supplementary.

EXAMPLE 3

USING ALGEBRA In the figure $\overrightarrow{RP}$ and $\overrightarrow{RT}$ are opposite rays. Find m∠PRQ.

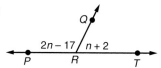

Solution

Since $\overrightarrow{RP}$ and $\overrightarrow{RT}$ are opposite rays, ∠PRQ and ∠QRT form a linear pair. Use the Linear Pair Postulate to write and solve an equation.

$$\begin{aligned} m\angle PRQ + m\angle QRT &= 180° &&\text{Linear Pair Postulate} \\ (2n - 17) + (n + 2) &= 180 &&\text{Substitute.} \\ 3n - 15 &= 180 &&\text{Combine like terms.} \\ 3n &= 195 &&\text{Add 15 to each side.} \\ n &= 65 &&\text{Divide each side by 3.} \end{aligned}$$

$$m\angle PRQ = 2n - 17 = 2(65) - 17 = 113°.$$

So, m∠PRQ = 113°.

Angles that are equal in measure are called **congruent angles**. In the figure at the right, ∠X and ∠Y are a pair of congruent angles. You write ∠X ≅ ∠Y. You indicate the angles are congruent by marking them with an equal number of arcs, as shown in blue.

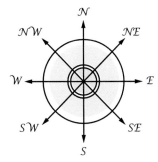

For any given angle, the **angle bisector** is the ray that divides the angle into two congruent angles.

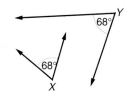

COMMUNICATING
ABOUT GEOMETRY

In the figure at the left, the drawing of ∠Y appears larger than the drawing of ∠X. Nonetheless, the angles are congruent. Explain how this is possible.

EXAMPLE 4

NAVIGATION A *compass card* is shown at the right. The four *cardinal points* of the card are North, East, South, and West. The other letters represent the four *intercardinal points*. Each of these eight directions can be considered as a ray.

The measure of the angles formed by any two successive cardinal points, like N and E, is 90°. The intercardinal points bisect these angles. What is the measure of each small angle on the card?

Solution
Each 90° angle is divided into two congruent angles. Since 45° + 45° = 90°, the measure of each small angle is 45°.   ◄

TRY THESE

Find the measure of each angle. Then classify the angle.

1. ∠CZF   2. ∠DZA   3. ∠EZB

4. ∠BZD   5. ∠AZF   6. ∠CZE

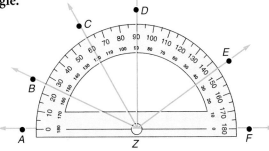

Refer to the figure at the right. Identify a pair of angles of each type. (There may be more than one correct answer.)

7. congruent angles that are not right angles

8. complementary angles that are not adjacent

9. supplementary angles that are not adjacent

10. adjacent angles that are neither complementary nor supplementary

11. a linear pair

12. WRITING MATHEMATICS Compare the Protractor Postulate and the Ruler Postulate. How are they alike? How are they different?

Find _x_ in each figure. Justify your answer.

13.
53° _x_

14.
56° _x_

15.
_x_

16.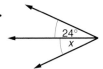
24° _x_

17. In the figure at the right, find m∠*JMK* and m∠*KML*.

18. **PHYSICS** According to the *law of reflection*, when a ray of light is reflected from a plane surface, the *angle of incidence* is congruent to the *angle of reflection*. The figure below shows a ray of light reflected from a mirror. Copy the figure on your paper. Name as many angle measures as you can.

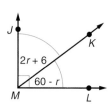
2r + 6
60 - r

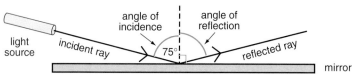
angle of incidence   angle of reflection
light source   incident ray   75°   reflected ray
mirror

PAPER FOLDING You will need unlined paper, a pencil, and a straightedge.

19. Draw an angle of any size on a sheet of paper. Fold the paper so one side of the angle lies directly on top of the other. When you unfold the paper, it should look very much like the picture at the right. What does the crease in the paper represent?

20. Work with a partner. Fold a sheet of paper so that the top edge meets the bottom edge. Then fold so that the right edge meets the left edge. Unfold the paper and use a pencil and straightedge to trace over the creases. Label the segments with the four cardinal compass directions, N, S, E, and W, as shown in Example 4 on page 95. Now devise a way to locate NE, SE, SW, and NW by paper folding.

PRACTICE

Refer to the figure at the right for Exercises 21–24.

21. Name the straight angle.

22. Name three right angles.

23. Name all the obtuse angles. Give the measure of each.

24. Name all the acute angles. Give the measure of each.

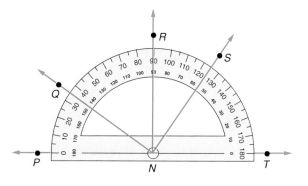

25. **WRITING MATHEMATICS** Compare the Angle Addition Postulate and the Segment Addition Postulate. How are they alike? How are they different?

In the figure ∠*BGE* **is a straight angle. Identify the following.**

26. two right angles

27. two pairs of congruent angles

28. two angle bisectors

29. a pair of opposite rays

30. four linear pairs

31. a pair of complementary angles

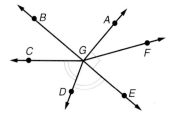

Find the measure of the complement of an angle of the given measure.

32. 77°

33. 6.5°

34. $t°$

35. $(c + 30)°$

Find the measure of the supplement of an angle of the given measure.

36. 87°

37. 161.4°

38. $y°$

39. $(n + 50)°$

AVIATION On the compass card at the right, $\overrightarrow{OA}$ shows the path of airplane *A* on course for New York City. The airplane is said to be on a *heading* that is 32° west of north.

40. Airplane *B* is on a heading that is 32° north of west. Copy the compass card and draw a ray that shows its heading.

41. What is the measure of the angle between the heading of airplane *A* and the heading of airplane *B*?

42. Airplane *C* is on a heading that is directly opposite to the heading of airplane *A*. Sketch its heading on your compass card. Then describe its heading in words.

In each figure, $\overrightarrow{OX}$ **and** $\overrightarrow{OY}$ **are opposite rays. Find m∠** *XOZ*.

43.

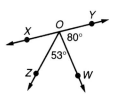

44.

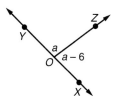

45.

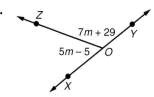

In each figure, ∠*RST* **is a right angle. Find m∠** *RSP*.

46.

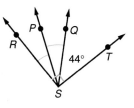

47.

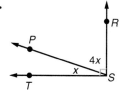

48.

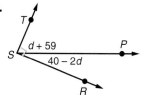

EXTEND

49. INDUCTIVE REASONING The figures below show a pattern of rays in the same plane that have a common endpoint. Draw the next figure of the pattern.

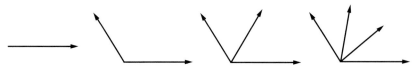

50. Copy and complete the table.

number of rays	1	2	3	4	5	6	7	8
number of angles determined	0	1	3					

51. Write a function rule to generalize the pattern in Question 50.

52. The measure of $\angle R$ is thirty-four degrees more than the measure of its complement. Find m$\angle R$.

53. The measure of $\angle S$ is fifteen degrees less than twice the measure of its supplement. Find m$\angle S$.

THINK CRITICALLY

Determine whether each conditional statement is true or false.

54. If an angle is obtuse, then its supplement is acute.

55. If an angle is acute, then its complement is obtuse.

56. If two angles form a linear pair, then they are adjacent.

57. If two angles are adjacent, then they form a linear pair.

58. TRANSFORMATIONS Describe the symmetries of the building in the photograph.

ALGEBRA AND GEOMETRY REVIEW

Find the distance between two points on a coordinate plane with the given coordinates. Round to the nearest tenth if necessary.

59. (6, 9) and (9, 5)   **60.** (−1, 0) and (0, 3)   **61.** (5, −7) and (−3, −2)

62. STANDARDIZED TESTS If m$\angle A$ + m$\angle B$ = 180°, then $\angle A$ and $\angle B$ are

A. right angles   **B.** a linear pair   **C.** supplementary   **D.** adjacent

Determine whether the ordered pair is a solution of the system of equations.

63. (1, 3) $\begin{cases} y = 4x - 1 \\ y = -x + 4 \end{cases}$ **64.** (0.5, 1) $\begin{cases} 2x + 3y = 4 \\ -8x - 2y = 6 \end{cases}$ **65.** (−3, 0) $\begin{cases} x - 3y = 9 \\ x + 2y = -6 \end{cases}$

66. Draw a four-sided figure that has reflection symmetry but not rotation symmetry.

Career
Mechanical
Engineering Technician

One of the purposes of mechanical engineering is to convert energy into motion. Mechanical engineering technicians help engineers develop and manufacture machinery, tools, and other equipment with moving parts. Some of these technicians work in the specialized field called *robotics*.

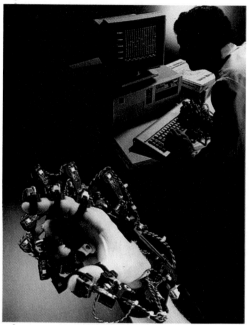

When people hear the word "robot," they usually think of a machine that moves and talks like a human being. Actually a robot is any mechanical device that works automatically. In many factories, for instance, robots are simply mechanical "arms" that repeat a set of motions over and over again. In one area of ongoing research, robotic wheelchairs are being developed for people who have lost the use of their limbs.

Decision Making

Below is the floor plan of a bedroom. The symbol → shows the position of a robotic wheelchair and the direction it is facing.

1. The red segments show a path the wheelchair can follow from the door to the desk. Find the total length of this path.

2. A technician wrote the following set of coded instructions to lead the wheelchair along the red path.

   FD 3 LF 90 FD 5.5 LF 90

   What is the meaning of this code?

3. Write similar coded instructions to lead the wheelchair on a path from a position facing the chest of drawers to a position facing the desk.

4. Choose four other paths that you think would be frequently used. Write coded instructions to guide the wheelchair along each path.

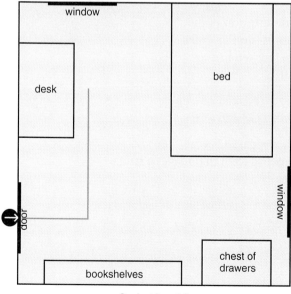

Scale 1 in. : 4 ft

Copying and Bisecting Segments and Angles

Think Back

1. The figures below show key steps in the compass-and-straightedge construction of a segment congruent to a given segment. Use these figures to complete the description given below.

 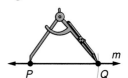

a. Start by using a straightedge to draw a segment. Label its endpoints __?__ and __?__.

b. Use the straightedge to draw a line elsewhere on the sheet of paper. Label it __?__.

c. Choose a point on line *m* and label it __?__.

d. Set the compass to the length of __?__.

e. Using this setting, place the compass tip at point __?__ and draw an arc that intersects line *m*. Label the point of intersection as point __?__.

f. It follows that __?__ ≅ __?__.

2. What construction has been performed?

Work with a partner. Refer to the figure at the right below.

3. Write a description of the construction similar to the description given in Question 1.

4. What is the relationship between $\overline{GZ}$ and $\overline{ZH}$?

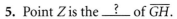

Replace each __?__ with the geometric term that makes a true statement.

5. Point *Z* is the __?__ of $\overline{GH}$.

6. $\overleftrightarrow{JK}$ is a __?__ of $\overline{GH}$.

Explore

In Think Back, you reviewed constructing a segment congruent to a given segment, or copying a segment. Now you will extend your construction skills to include copying a given angle.

7. Use a straightedge to draw an angle. Label its vertex *A*.

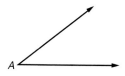

8. Use the straightedge to draw a ray. Label its endpoint *D*.

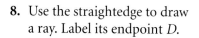

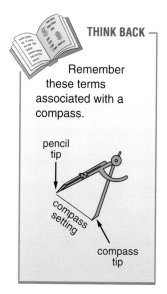
9. Place the compass tip at point *A*. Draw a large arc that intersects both sides of ∠*A*. Label the points of intersection *B* and *C*.

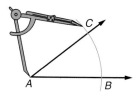

10. Using this same compass setting, place the compass tip at point *D*. Draw a large arc that intersects the ray. Label the point of intersection *E*.

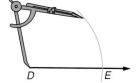

11. Place the compass tip at point *B*. Adjust the compass setting so that the pencil tip is at point *C*.

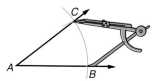

12. Using this new setting, place the compass tip at point *E*. Draw an arc that intersects the large arc. Label the point of intersection *F*.

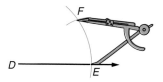

13. Use the straightedge to draw $\overrightarrow{DF}$.

14. What is the relationship between ∠*CAB* and ∠*FDE*?

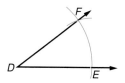

In Think Back, you also reviewed constructing a segment bisector. In Questions 15–19, you will learn how to construct an angle bisector.

15. Use a straightedge to draw an angle. Label its vertex *R*.

16. Place the compass tip at point *R*. Draw an arc that intersects both sides of ∠*R*. Label the points of intersection *S* and *T*.

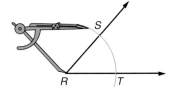

17. Choose any compass setting. Place the compass tip at point *S* and draw an arc in the interior of ∠*R*.

18. Using this same compass setting, place the compass tip at point *T*. Draw an arc that intersects the arc you drew in Question 17. Label the point of intersection *Q*.

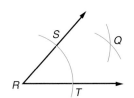

19. Draw $\overrightarrow{RQ}$. This is the bisector of ∠*SRT*.

20. How can you use the compass to verify that $\overrightarrow{RQ}$ bisects ∠*SRT*?

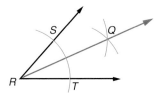

Make Connections

Copying and bisecting segments and angles are fundamental constructions. You can use these constructions to perform more complex constructions.

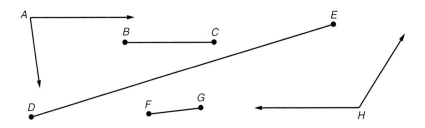

Carefully trace each of the figures below onto a sheet of paper. Then use the figures for Questions 21–25.

21. Construct a segment with length equal to $2BC + FG$.

22. Divide $\overline{DE}$ into four congruent segments.

23. Construct an angle with measure equal to $m\angle H - m\angle A$.

24. Divide $\angle H$ into four congruent angles.

25. Show that the measure of $\angle H$ is one and one half times the measure of $\angle A$.

COORDINATE GEOMETRY Refer to the figure below for Questions 26–27.

26. Copy the figure on grid paper. Using a compass and straightedge, bisect $\overline{JK}$. What are the coordinates of the midpoint?

27. Use algebra to calculate the coordinates of the midpoint of $\overline{JK}$. Compare your answer to the coordinates you identified in Question 26. Are they the same?

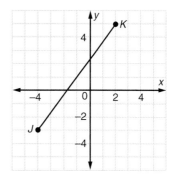

GEOMETRY: WHO, WHERE, WHEN

Origami, the Japanese art of folding paper, uses angle and segment bisectors to make animals and other decorative creations. This art form dates back at least 400 years and was once practiced by masters for ceremonial purposes. Today, it is used in speech, mental health and occupational therapy.

Summarize

28. **WRITING MATHEMATICS** Compare the construction of an angle bisector to the construction of a segment bisector. How are these constructions alike? Why does it make sense that these two constructions have common characteristics?

29. **THINKING CRITICALLY** Suppose you must bisect a very long segment. You have your compass open to the widest possible setting, but it is not wide enough to perform a standard construction. How can you use your compass and straightedge to bisect the segment?

COMMUNICATING ABOUT GEOMETRY

In Question 29, you were asked to consider a situation in which a segment is too large to be bisected using the standard construction. Is it possible for an *angle* to be too small to be bisected using the standard construction?

30. **GOING FURTHER** Carefully trace the triangle at the right onto a sheet of paper. Using the four constructions you studied in this lesson, devise a way to construct a copy of the triangle.

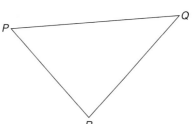

Explore

- Determine whether each conditional statement is true or false.

1. **a.** If today is July 4, then it is U.S. Independence Day.
   b. If today is U.S. Independence Day, then it is July 4.
   c. If today is not July 4, then it is not U.S. Independence Day.
   d. If today is not U.S. Independence Day, then it is not July 4.

2. **a.** If today is Saturday, then it is a weekend day.
   b. If today is a weekend day, then it is Saturday.
   c. If today is not Saturday, then it is not a weekend day.
   d. If today is not a weekend day, then it is not Saturday.

3. Look back at Question 1 and Question 2. How is the statement in part a related to the statement in part b? part c? part d?

Build Understanding

- When the hypothesis and conclusion of a conditional statement are interchanged, the result is the **converse** of the original statement.

    conditional statement: $p \Rightarrow q$ converse: $q \Rightarrow p$

The converse of a true conditional statement is not necessarily true.

EXAMPLE 1

Write the converse of the statement and tell whether it is true or false.

If three points are collinear, then they lie on the same plane.

Solution

First identify the hypothesis and conclusion.

hypothesis	conclusion

If three points are collinear, then they lie on the same plane.

Now interchange them to form the converse.

hypothesis	conclusion

If three points lie on the same plane, then they are collinear.

The converse is false. As a counterexample, consider points *A*, *B*, and *C* below. Notice that the points are not collinear.

A
● 　　 *B*
　　　 ● 　　 *C*
　　　　　　 ●

◄

If a conditional and its converse are each true, you can combine them to form one **biconditional statement**, or **biconditional**. The parts of a biconditional are connected by the phrase *if and only if*.

conditional statement	If p, then q.	$p \Rightarrow q$
converse	If q, then p.	$q \Rightarrow p$
biconditional	p if and only if q.	$p \Leftrightarrow q$

In Example 1, suppose the word "collinear" is replaced by "coplanar." Now both the statement and its converse are true.

> If three points are coplanar, then they lie on the same plane.
> If three points lie on the same plane, then they are coplanar.

So the two statements can be written as one biconditional.

> Three points are coplanar if and only if they lie on the same plane.

This biconditional is true because it is a definition. Any good definition can be written as a biconditional.

EXAMPLE 2

Write this definition as a biconditional.

> A right angle is an angle whose measure is 90°.

Solution

Conditional statement: If an angle is a right angle, then its measure is 90°.

Converse: If the measure of an angle is 90°, then it is a right angle.

Biconditional: An angle is a right angle if and only if its measure is 90°. ◄

The definition in Example 2 illustrates the fact that conditional statements may be written in some form other than if-then form.

EXAMPLE 3

Rewrite this statement in *if-then* form: Any two teachers are similar.

Solution

First identify the hypothesis and conclusion.

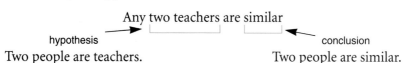

Any two teachers are similar

hypothesis   conclusion
Two people are teachers.   Two people are similar.

Now rewrite the statement. Place *if* in front of the hypothesis, and place *then* in front of the conclusion.

> If two people are teachers, then they are similar. ◄

 CHECK UNDERSTANDING

Write the converse of the statement given in Example 3. Then determine whether the statement and its converse are true or false.

Two other statements related to conditionals are the **inverse** and the **contrapositive**.

conditional statement	If p, then q.	$p \Rightarrow q$
inverse	If not p, then not q.	$\sim p \Rightarrow \sim q$
contrapositive	If not q, then not p.	$\sim q \Rightarrow \sim p$

When writing the inverse and contrapositive in symbols, notice that the symbol $\sim$ shows the *negation* of the hypothesis and conclusion.

EXAMPLE 4

Write the inverse and contrapositive of the statement. Then tell whether each is true or false.

> If two angles form a linear pair, then they are supplementary.

Solution

First identify the hypothesis and conclusion.

$$\underbrace{\text{If two angles form a linear pair,}}_{\text{hypothesis}} \text{ then } \underbrace{\text{they are supplementary.}}_{\text{conclusion}}$$

inverse: If two angles do not form a linear pair, then they are not supplementary.   False

contrapositive: If two angles are not supplementary, then they do not form a linear pair.   True

CHECK UNDERSTANDING

In Example 4, explain why the inverse is false and the contrapositive is true.

You may use a *Venn diagram* to model a given conditional statement.

EXAMPLE 5

COLLEGE APPLICATIONS In the essay he wrote for his college application, John said, "I can dance any ballet." Can the admissions director infer that John cannot do other types of dancing.

Solution

You can write John's statement as a conditional.

> If the dance is ballet, then John can dance it.

The admissions director must decide if the inverse is true.

> If the dance is not ballet, then John cannot dance it.

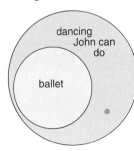

The Venn diagram models John's statement. The red point shows it is possible John can do other types of dancing, such as tap dancing. So, the admissions director cannot infer that the inverse of John's statement is true.

Write the converse of each statement. Then tell whether the statement and its converse are true or false.

1. If an angle is acute, then its measure is greater than 0° and less than 90°.

2. If the endpoints of $\overline{ST}$ are $S(-9, 2)$ and $T(5, -5)$, then its midpoint is $M(-2, -1.5)$.

3. Write the following definition as a biconditional.

   Congruent segments are segments that are equal in length.

Rewrite each statement in if-then form.

4. Congruent figures are similar.

5. A line contains at least two distinct points.

Write the inverse and contrapositive of each statement. Then tell whether each is true or false.

6. If point M is the midpoint of $\overline{XY}$, then $XM + MY = XY$.

7. All right angles are congruent.

8. **ADVERTISING** Refer to the advertisement at the right. How do you think the manufacturer of *Good Sport* shoes is trying to use the types of statements you studied in this lesson to influence your choice of a pair of shoes?

People who wear *Good Sport* shoes have all the fun.

9. **WRITING MATHEMATICS** Write an original conditional statement. Then write the converse, inverse, and contrapositive of your statement. Determine whether each statement is true or false.

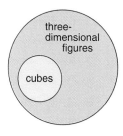

three-dimensional figures

cubes

10. **MODELING** Write a true conditional statement that can be modeled by the Venn diagram at the right. Then write the contrapositive of the statement. Show how the diagram can help you determine whether the contrapositive is true or false.

PRACTICE

Let p = two angles form a linear pair and q = two angles are adjacent. Translate each set of symbols into words. Then tell if the statement is true or false.

11. $p \Rightarrow q$

12. $q \Rightarrow p$

13. $\sim p \Rightarrow \sim q$

14. $\sim q \Rightarrow \sim p$

Write each definition as a biconditional.

15. Complementary angles are two angles whose measures have a sum of 90°.

16. The midpoint of a segment is the point that divides the segment into two congruent segments.

Rewrite each statement in if-then form.

17. The sides of a straight angle are opposite rays.

18. Two points determine a line.

Write the converse, inverse, and contrapositive of each statement. Then tell whether each of the four statements is true or false.

19. If point C is between points A and B, then $AC + CB = AB$.

20. If the x-coordinate of a point is a negative number, then the point is in the third quadrant of the coordinate plane.

21. A ray is part of a line.

22. A square has four lines of symmetry.

23. METEOROLOGY A television meteorologist said, "If the winds are blowing at speeds greater than one hundred miles per hour, then they are hurricane-force winds." Kanella heard this statement and decided that hurricane-force winds are defined as winds blowing at speeds greater than 100 mi/h. Assume that the meteorologist's statement is true. Do you think Kanella reasoned correctly? Justify your answer.

24. MODELING Draw a Venn diagram to model the statement below.

All squares are rectangles.

25. WRITING MATHEMATICS Describe a job that you think requires strong logical reasoning skills. How do you think the person who holds this job might have to work with conditional statements?

EXTEND

USING ALGEBRA **Write the converse, inverse, and contrapositive of each statement. Then tell whether each of the four statements is true or false.**

26. If $n = -3$, then $3n + 5 = -4$.

27. If $|r| = 5$, then $r = -5$.

28. If $a = 7$, then $a^2 = 49$.

29. If $y^3 = -27$, then $y = -3$.

Replace each ___?___ with a word or phrase that will make a true statement.

30. The inverse of the converse of a conditional statement is the ___?___.

31. The contrapositive of the converse of the inverse of a conditional statement is the ___?___.

THINK CRITICALLY

For each statement, copy and complete the table at the right.

32. If $JM = MK$, then point M is the midpoint of $\overline{JK}$.

33. If $m\angle W = 90°$, then $\angle W$ is a right angle.

34. If $\sqrt{s} = 64$, then $s = 8$.

35. If I live in New York City, then I live in New York State.

	True or False?
given statement	▪
converse	▪
inverse	▪
contrapositive	▪

36. LOOKING AHEAD Refer to your answers to Exercises 32–35. Which do you think is considered to be *logically equivalent* to a given conditional: its converse, its inverse, or its contrapositive? Explain your reasoning.

ALGEBRA AND GEOMETRY REVIEW

37. Draw front, top, and right-side views of the stack of cubes shown at the right. No cubes are hidden from view.

Consider this statement: Three noncollinear points determine a plane.

38. Rewrite the statement in if-then form.

39. Write the converse, inverse, and contrapositive of the statement. Then tell whether each of the four statements is true or false.

Simplify.

40. $(4m^2 + 2m - 9) + (m^3 - m + 9)$

41. $(5v^3 - 7v^2 + v) - (7v^2 - 3v + 10)$

PROJECT *Connection*

A set of letters, numbers, and other symbols with a particular design is called a typeface. For instance, at the right the letter A is printed in two traditional typefaces called *Helvetica* and *Times*.

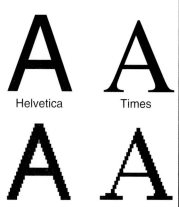

Helvetica   Times

The computer code that creates images in a certain typeface is called a computer *font*. For the majority of personal computers, the font that creates images for the monitor must describe each letter, number, and symbol as a pattern of pixels on a grid. The pattern is called a *bitmap* for the object. At the right you see bitmapped images of the letter A in the Helvetica and Times fonts.

1. From the samples of typefaces you have gathered, choose the one that is your favorite. Give its name and its characteristics, such as italic, bold, serif, sans serif, and so on.

2. Choose a resolution. Create a bitmap of each letter in your first name in the typeface you chose. Identify the resolution of your bitmapped letter.

3. Separate the bitmap for the first letter of your name into bytes of information. Write a binary number for each byte. Translate it into a decimal number.

2.9 Properties from Algebra and Justifying Conclusions

Explore

• Below is a step-by-step solution of $-2(y + 5) = 2y + 8$. Replace the __?__ at the right of each step with a reason that supports it.

$$-2(y + 5) = 2y + 8$$

1.	$-2y - 10 = 2y + 8$	__?__
2.	$-2y = 2y + 18$	__?__
3.	$-4y = 18$	__?__
4.	$y = -4.5$	__?__

Build Understanding

• In your study of algebra, you used several properties of real numbers. These properties are also important in geometry. Here is a summary of the properties of equality that you will use most often.

COMMUNICATING ABOUT GEOMETRY

In Chapter 1, you learned about symmetry in relation to geometric figures. How is the concept of symmetry related to the Symmetric Property of Equality?

PROPERTIES OF EQUALITY

For all real numbers a, b, and c,

If $a = b$, then $a + c = b + c$.	Addition Property
If $a = b$, then $a - c = b - c$.	Subtraction Property
If $a = b$, then $ac = bc$.	Multiplication Property
If $a = b$ and $c \neq 0$, then $\dfrac{a}{c} = \dfrac{b}{c}$.	Division Property
If $a = b$, then a may replace b or b may replace a in any statement.	Substitution Property
$a = a$	Reflexive Property
If $a = b$, then $b = a$.	Symmetric Property
If $a = b$ and $b = c$, then $a = c$.	Transitive Property

To show that a conditional statement is false, you must find only one specific counterexample for which the hypothesis is true and the conclusion is false. To show that a conditional statement is true, you must be able to provide a justification. To justify statements that relate to real numbers, you may use the properties of equality.

EXAMPLE 1

Name the property that justifies the conclusion of this statement.

If $AB + XY = CD + XY$, then $AB = CD$.

Solution

AB, CD, and XY are lengths of segments, so they are real numbers.

$$AB + XY = CD + XY$$
$$AB + XY - XY = CD + XY - XY \qquad \text{Subtract } XY \text{ from each side.}$$
$$AB = CD$$

So the justification is the Subtraction Property of Equality.

Name the property that justifies the conclusion of this statement.

If $2m\angle A = m\angle B$, then $m\angle A = \frac{1}{2}m\angle B$.

The reflexive, symmetric, and transitive properties of equality lead directly to these related properties of congruence.

PROPERTIES OF CONGRUENCE

	For any segments $\overline{AB}$, $\overline{CD}$, and $\overline{EF}$,	For any angles $\angle R$, $\angle S$, and $\angle T$,
Reflexive Property	$\overline{AB} \cong \overline{AB}$	$\angle R \cong \angle R$
Symmetric Property	If $\overline{AB} \cong \overline{CD}$, then $\overline{CD} \cong \overline{AB}$.	If $\angle R \cong \angle S$, then $\angle S \cong \angle R$.
Transitive Property	If $\overline{AB} \cong \overline{CD}$ and $\overline{CD} \cong \overline{EF}$, then $\overline{AB} \cong \overline{EF}$.	If $\angle R \cong \angle S$ and $\angle S \cong \angle T$, then $\angle R \cong \angle T$.

EXAMPLE 2

Replace __?__ with a conclusion that will make a true statement. Justify the conclusion.

If $\angle 1 \cong \angle 2$ and $\angle 2 \cong \angle 3$, then __?__.

Solution

A valid conclusion is $\angle 1 \cong \angle 3$. The justification is the Transitive Property.

In Example 2, notice that the justification is stated simply as the "Transitive Property," rather than the "Transitive Property of Congruence." Because the reflexive, symmetric, and transitive properties of congruence are so closely related to the corresponding properties of equality, it is not necessary to distinguish between them.

Just as you can use the properties of equality and congruence to justify conclusions, you also can use geometric definitions and postulates as justifications.

EXAMPLE 3

In the figure it is given that $\angle ABD$ and $\angle CBD$ form a linear pair. Justify each conclusion.

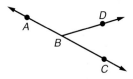

a. $\angle ABD$ and $\angle CBD$ are adjacent angles.
b. $\angle ABD$ and $\angle CBD$ are supplementary angles.

Solution

a. A linear pair is defined as a pair of adjacent angles.
So, you write the justification as Definition of linear pair.

b. Postulate 11 states that, if two angles form a linear pair, then they are supplementary. So, you write the justification as Linear Pair Postulate.

As you proceed in your study of geometry, you will find that the Substitution Property is used quite frequently. It also is a principle that is used in many real world situations.

EXAMPLE 4

COOKING Mr. Bright read this statement in a cookbook.

> When substituting cornstarch for flour as a thickener, use only half as much.

He has a recipe that calls for 3 tablespoons of flour to be used as a thickener. How much cornstarch should he substitute?

Solution

Mr. Bright should substitute half of 3 tablespoons of cornstarch, which is $1\frac{1}{2}$ tablespoons. Unfortunately, a half-tablespoon measure is not a standard measure included in his set of measuring spoons. He must make another substitution.

He knows: 1 tablespoon = 3 teaspoons

So: $1\frac{1}{2}$(1 tablespoon) $= 1\frac{1}{2}$(3 teaspoons)

$= 4\frac{1}{2}$ teaspoons

Mr. Bright should substitute $4\frac{1}{2}$ teaspoons of cornstarch.

CHECK — UNDERSTANDING

In Example 3, what conclusion can you make about $\overrightarrow{BA}$ and $\overrightarrow{BC}$? What is the justification for this conclusion?

Name the property of equality that justifies each conclusion.

1. If m∠A = m∠B, then m∠B = m∠A.   **2.** If JK = KL, then 2JK = 2KL.

Replace each ___?___ with a conclusion that will make a true statement. Justify the conclusion.

3. If GH + HJ = HJ + JK, then ___?___ .   **4.** If ∠P ≅ ∠Q, and ∠Q ≅ ∠R, then ___?___ .

In the figure at the right, it is given that point N is the midpoint of $\overline{MO}$. Justify each conclusion.

5. MN = NO   **6.** MN + NO = MO

RELATIONSHIPS Decide whether each of these human relationships has reflexive, symmetric, and transitive properties. For example, the relationship "is the cousin of" is symmetric, but it is not reflexive or transitive.

7. is the sister of   **8.** is a geometry classmate of   **9.** is the son of

10. is an ancestor of   **11.** lives next door to   **12.** lives across the street from

13. WRITING MATHEMATICS Describe the reflexive, symmetric, and transitive properties of equality in your own words.

PRACTICE

Name the property, definition, or postulate that justifies each conclusion.

14. If ∠X and ∠Y are supplementary angles, then m∠X + m∠Y = 180°.

15. If points R, S, and T are noncollinear, then they determine a plane.

16. If m∠J = m∠K and m∠K = 90°, then m∠J = 90°.

17. If m∠T = 90°, then ∠T is a right angle.

18. If $\overline{AB} \cong \overline{CD}$, then $\overline{CD} \cong \overline{AB}$.

19. If 2PR = PS, then $PR = \frac{1}{2}PS$.

In the figure at the right, $\overrightarrow{AW}$ and $\overrightarrow{AZ}$ are opposite rays, and $\overrightarrow{AX}$ is the bisector of ∠WAY. Write a conclusion about the figure that can be justified by each definition or postulate.

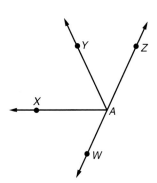

20. Angle Addition Postulate   **21.** definition of straight angle

22. Linear Pair Postulate   **23.** definition of angle bisector

24. WRITING MATHEMATICS How do you show that a conditional statement is true? How do you show that a conditional is false? Give an example of each process.

25. WOODWORKING Joshua needs to make two identical wooden brackets to support a decorative shelf he is making to hold his CDs. He made the template shown at the right from heavy cardboard, then used it to draw two outlines of the brackets on a piece of wood. What property helps justify his conclusion that the two outlines are identical in shape?

EXTEND

Refer to the figure at the right. Replace each __?__ with a hypothesis that will make a true statement. Then justify the conclusion.

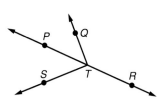

26. If __?__, then ∠PTR is a straight angle.

27. If __?__, then $\overrightarrow{TP}$ is the bisector of ∠STQ.

28. If __?__, then $\overline{PT} \cong \overline{TR}$.

29. If __?__, then m∠STQ = 90°.

30. If __?__, then ∠PTS and ∠STR form a linear pair.

THINK CRITICALLY

Tell whether each statement is true or false. Give an example.

31. If a relationship is reflexive and symmetric, then it is transitive.

32. If a relationship is transitive, then it is reflexive and symmetric.

ALGEBRA AND GEOMETRY REVIEW

33. In the figure at the right, point B is between points A and C and AC = 24. Find AB and BC.

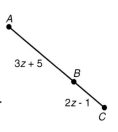

34. Refer to your answer to Exercise 33. Suppose a point of $\overline{AC}$ is chosen at random. What is the probability it is on $\overline{BC}$?

35. Name the property that justifies the conclusion of this statement.

    If MN = 4 cm and JK = MN, then JK = 4 cm.

Simplify each expression.

36. $\sqrt{500}$ **37.** $\sqrt{3} \cdot \sqrt{3}$ **38.** $\dfrac{\sqrt{98}}{\sqrt{2}}$

39. $\dfrac{\sqrt{3}}{\sqrt{5}}$ **40.** $\sqrt{y^2}$ **41.** $\sqrt{48k^2}$

Focus on Reasoning

Deductive Reasoning and Theorems

Discussion

● Work with a partner. Consider the statement and figure below.

If $AB = CD$,
then $AC = BD$.

Mike, Rita, Kim, and Sheree agree that the conclusion of this statement is true. However, they disagree about how to justify it.

Mike measured the segments in the given figure and made the list at the right.

AB = 1 cm
CD = 1 cm
AC = 5 cm
BD = 5 cm

1. Mike then reasoned as follows.

  $AC = 5$ cm and $BD = 5$ cm, so the Transitive Property of Equality justifies the conclusion $AC = BD$.

Do you agree or disagree? Explain.

Rita said that Mike only showed the statement to be true for one specific case. She said that more "evidence" is needed before you can generalize.

AB	BC	CD	AC	BD
1 cm	4 cm	1 cm	5 cm	5 cm
2 cm	4 cm	2 cm	6 cm	6 cm
3 cm	4 cm	3 cm	7 cm	7 cm
4 cm	4 cm	4 cm	8 cm	8 cm
5 cm	4 cm	5 cm	9 cm	9 cm

Rita likes the process of inductive reasoning, so she made the table at the right.

2. Rita reasoned that the pattern in her table justifies the conclusion $AC = BD$. Do you agree or disagree? Explain.

Kim claimed he could give a much more general justification. He reasoned by writing the following argument.

  It is given that $AB = CD$. You can add the same number to each side of an equation, so $AB + BC = BC + CD$. By the Segment Addition Postulate, $AB + BC = AC$ and $BC + CD = BD$. This means you can substitute $AB + BC$ for AC, and you can substitute $BC + CD$ for BD. So $AC = BD$.

3. Kim justified one of his statements by naming the Segment Addition Postulate. Name the other properties Kim has used.

Deductive reasoning is also used in scientific investigations. In some science courses, students make specific inferences based on general scientific principles. The students then construct *mental models* using deductive reasoning. They test the models, eliminating some and futher developing others, eventually discovering a conclusion for their problem. Discuss a circumstance when you have used deductive reasoning to solve a problem.

Sheree agreed with Kim. In fact, she remembered that her older brother wrote arguments like Kim's when he studied geometry last year. However, Sheree also remembered that her brother organized his arguments in a form that looked like this.

Given $AB = CD$

Prove $AC = BD$

Statements	Reasons
1. $AB = CD$	1. Given
2. $BC = BC$	2. Reflexive Property
3. $AB + BC = BC + CD$	3. Addition Property of Equality
4. $AB + BC = AC$ $BC + CD = BD$	4. Segment Addition Postulate
5. $AC = BD$	5. Substitution Property

4. How are Kim's and Sheree's arguments alike? How are they different?

Build a Case

- In Discussion, both Mike and Rita demonstrated that the statement is true for only a limited number of cases. They did not show that the conclusion must be true every time the hypothesis is true.

In contrast, Kim and Sheree used a technique called *deductive reasoning*. In **deductive reasoning**, you assume the hypothesis is true, then write a series of statements that lead to the conclusion. Each statement is accompanied by the reason that justifies it. The set of statements and reasons is called a **proof**. From Kim's and Sheree's proofs, you know that the conclusion of the statement will be true whenever the hypothesis is true, no matter what the specific lengths.

Proofs can be written in different forms. Kim wrote his statements and reasons in "running text," using a form that is called a *paragraph proof*. Sheree's proof is more formal, with the statements and reasons clearly identified and separated into columns. This is called a *two-column proof*.

5. Refer to the figure at the right. Write a paragraph proof of this statement.

If m∠*APB* = m∠*CPD*, then m∠*APC* = m∠*BPD*.

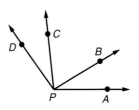

6. Write a two-column proof of the statement in Question 5.

Any statement that can be proved true is called a **theorem**. To prove a theorem, you may use definitions, postulates, properties from algebra, and any previously provided theorems to justify your statements.

EXTEND AND DEFEND

7. Copy and complete the two-column proof of this theorem.

> **THEOREM 2.1** **MIDPOINT THEOREM**
>
> If point *M* is the midpoint of $\overline{AB}$,
> then $AM = \frac{1}{2}AB$ and $MB = \frac{1}{2}AB$.

Given Point *M* is the midpoint of $\overline{AB}$.

Prove $AM = \frac{1}{2}AB$; $MB = \frac{1}{2}AB$

Proof

Statements	Reasons
1. ___?___	1. Given
2. *AM* = *MB*	2. Definition of midpoint
3. *AM* + *MB* = *AB*	3. ___?___ Postulate
4. *AM* + *AM* = *AB*, or 2*AM* = *AB*	4. ___?___ Property
5. $\frac{2AM}{2} = \frac{AB}{2}$, or $AM = \frac{1}{2}AB$	5. ___?___ Property of Equality
6. $MB = \frac{1}{2}AB$	6. Substitution Property

> **PROBLEM SOLVING TIP**
>
> In writing proofs, you might prefer to state a definition or property rather than simply naming it. So, for reason 2 in the proof of the Midpoint Theorem, write: The midpoint of a segment divides it into two congruent parts.

8. Write a two-column proof of the following theorem.

> **THEOREM 2.2** **ANGLE BISECTOR THEOREM**
>
> If $\overrightarrow{BX}$ is the bisector of ∠*ABC*, then
> $m\angle ABX = \frac{1}{2}m\angle ABC$ and $m\angle XBC = \frac{1}{2}m\angle ABC$.

Often a theorem is stated in words only, with no figure given. In such cases, the first step in writing a proof is to draw an appropriate figure.

When you prove a theorem, you probably will find it helpful to write the statement of the theorem in *if-then* form. The hypothesis of the statement is called the "Given", and the conclusion is identified as "Prove."

9. Copy and complete the paragraph proof of this theorem.

THEOREM 2.3   **CONGRUENT SUPPLEMENTS THEOREM**

Supplements of congruent angles are congruent.

Given $\angle 1$ is supplementary to $\angle 3$.
          $\angle 2$ is supplementary to $\angle 4$.
          $m\angle 3 = m\angle 4$

Prove $m\angle \underline{\ ?\ } = m\angle \underline{\ ?\ }$

Proof   It is given that $\underline{\ ?\ }$ and $\underline{\ ?\ }$. By the definition of $\underline{\ ?\ }$, $m\angle 1 + m\angle 3 = 180°$ and $m\angle 2 + m\angle 4 = 180°$. Since both quantities are equal to 180°, $m\angle 1 + m\angle 3 = \underline{\ ?\ }$. It is $\underline{\ ?\ }$ that $m\angle 3 = m\angle 4$. Since you can subtract equal quantities from each side of an equation, it follows that $m\angle 1 = m\angle 2$. ◄

10. Write a paragraph proof of the following theorem.

THEOREM 2.4   **CONGRUENT COMPLEMENTS THEOREM**

Complements of congruent angles are congruent.

---

11. WRITING MATHEMATICS Which do you like better: writing paragraph proofs or two-column proofs? Explain your answer.

12. Write a proof of this theorem in whichever form you choose.

THEOREM 2.5

All right angles are congruent.

---

13. Write a proof of this statement in whatever form you choose.

     A line and a point not on the line determine a plane.

14. LOOKING AHEAD Refer to the figure. Write a proof.

Given $\overrightarrow{ZA}$ and $\overrightarrow{ZC}$ are opposite rays.
          $\overrightarrow{ZB}$ and $\overrightarrow{ZD}$ are opposite rays.

Prove $m\angle 5 = m\angle 7$

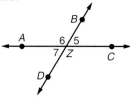

15. WRITING MATHEMATICS Explain the difference between a postulate and a theorem.

Inductive reasoning frequently is described as "particular to general." In contrast, *deductive reasoning* is described as "general to particular."

16. The figures above show a pattern of segments drawn between points on a circle. Copy and complete this table.

number of points on circle	1	2	3	4	5
number of regions formed inside circle	1	2	4		

17. Use inductive reasoning to make a conjecture about the number of regions formed when all possible segments are drawn between six points on a circle.

18. Draw six points on a circle. Then draw all possible segments between them. Now count the regions formed inside the circle. Is the result the same as your conjecture in Question 17?

Now You See It
Now You Don't

MOIRÉ PATTERNS

The figure at the left below is a group of 120 segments with a common endpoint. In the figure at its right, a second identical group of segments has been positioned so that it partially overlaps the first. The resulting illusion is called a *moiré pattern*. Describe the illusion.

There are several types of moiré patterns, and they have a wide range of practical applications. Can you think of a real life situation in which you have seen a moiré pattern?

CHAPTER REVIEW

• • • ● ●

VOCABULARY

Choose the word from the list that correctly completes each statement.

1. Points that lie on the same line are ___?___.
2. A statement accepted as true without proof is a ___?___.
3. A ___?___ is an educated guess based on patterns in data or figures.
4. Angles that have a sum of 180° are ___?___.
5. ___?___ rays form a straight angle.

a. Postulate
b. Supplementary
c. Collinear
d. Opposite
e. Conjecture

Lessons 2.1 and 2.2   POINTS, LINES, PLANES, AND SEGMENTS   pages 65–77

- Several postulates of geometry involve relationships between points, lines, and planes.
- The midpoint of a segment divides the segment into two congruent segments.

Use the figure at the right to identify each of the following. Which postulate justifies your answer?

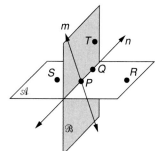

6. Name two points that determine line n.
7. Name three points that determine plane $\mathcal{A}$.
8. Name the intersection of plane $\mathcal{A}$ and plane $\mathcal{B}$.
9. Name the plane that contains $\overleftrightarrow{SR}$.

In the figure at the right, $EG = 39$. Find each of the following.

10. EF 11. FG
12. EM, if M is the midpoint of $\overline{EG}$

Lessons 2.3, 2.4, and 2.5   PATTERNS; INDUCTIVE REASONING; DEFINITIONS   pages 80–91

- Inductive reasoning involves looking for a pattern and making a conjecture.
- A good definition is precise, concise, and uses only previously defined words. It avoids circularity and is reversible.

Use inductive reasoning to make a conjecture.

13. A rectangle is alternately folded in thirds and in halves as shown. Make a conjecture about the number of regions in the tenth figure.

Explain why each statement is not a good definition.

14. A bisector divides something in two parts.

15. Collinear points are on the same line and they can be an equal distance apart.

Lessons 2.6 and 2.7   ANGLE DEFINITIONS AND POSTULATES; BISECTORS   pages 92–103

- An angle can be classified as acute, right, obtuse, or straight.
- A pair of angles can be described as complementary, supplementary, and/or congruent.
- A ray called an angle bisector divides an angle into two congruent angles.

In the figure at the right, $\angle BPE$ and $\angle DPA$ are straight angles and m$\angle BPA = 60°$. Identify the following. Give the measure of each angle.

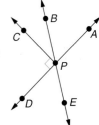

16. an acute angle   **17.** an obtuse angle   **18.** a right angle

19. a pair of complementary angles   **20.** a pair of supplementary angles

21. Use a compass and straightedge to bisect an obtuse angle.

Lesson 2.8   MORE ABOUT CONDITIONAL STATEMENTS   pages 104–109

- To form the converse of a conditional, interchange the hypothesis and the conclusion. To form the inverse, negate both parts. To form the contrapositive, negate both parts and then interchange them.
- The two parts of a biconditional statement are connected by the phrase *if and only if.*

Rewrite each statement in *if-then* form. Then write the converse, inverse, and contrapositive. Tell whether each of the four statements is true or false.

22. The midpoint of a segment divides it in two congruent parts.

23. The angles in a linear pair are adjacent.

Write each definition as a biconditional.

24. Angles that are equal in measure are called congruent angles.

25. The set of points common to two figures is their intersection.

Lessons 2.9 and 2.10   USING ALGEBRA AND DEDUCTIVE REASONING   pages 110–119

- To write a deductive proof, assume the hypothesis is true and then write a series of statements that lead to the conclusion. For each statement, include a reason that justifies it.
- Definitions, postulates, and properties can be used to justify reasons in a deductive proof.

Write a paragraph proof of this statement.

26. The complement of an acute angle is less than its supplement.

Write a two-column proof of this statement.

27. If two angles are congruent and supplementary, then they are right angles.

CHAPTER ASSESSMENT

CHAPTER TEST

For Questions 1–4, tell whether each statement is true or false.

1. Two points determine a line.

2. Two planes can intersect in a single point.

3. Three coplanar points lie on the same line.

4. Three noncollinear points lie on the same plane.

5. **STANDARDIZED TESTS** Points P, Q, and R are not on the same line. Which of the following describes points P, Q, and R?

   A. collinear   **B.** coplanar   **C.** intersecting

In the figure below, $AD = 21$.

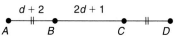

6. Find AB.   7. Find CB.

8. Write the next equation in the sequence below. Look for a pattern that relates the number of odd integers to their sum. Make a conjecture and predict the sum of the first 20 odd integers.

$$1 + 3 = 4$$
$$1 + 3 + 5 = 9$$
$$1 + 3 + 5 + 7 = 16$$

9. **WRITING MATHEMATICS** Write a paragraph explaining the characteristics of a good definition.

Use the figure below for Questions 10–13. $\overrightarrow{YX}$ and $\overrightarrow{YZ}$ are opposite rays.

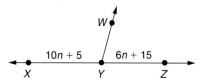

10. Find $m\angle WYX$.   11. Find $m\angle WYZ$.

12. What is the measure of an angle complementary to $\angle WYZ$?

13. If you constructed $\overrightarrow{YP}$ to bisect $\angle WYX$, what would be the measure of $\angle PYX$?

14. **WRITING MATHEMATICS** Write a true conditional statement whose converse is false. Justify your answer.

15. Rewrite this statement in *if-then* form: All acute angles are congruent.

16. Write this definition as a biconditional: The vertex of an angle is the common endpoint of the two rays.

Write the converse of each statement. Tell whether the statement and its converse are true or false.

17. If two lines are perpendicular lines, then the lines form four right angles.

18. If two rays intersect, then the rays have a common endpoint.

19. **STANDARDIZED TESTS** Which of the following represents the inverse of a conditional statement $p \Rightarrow q$?

   A. $q \Rightarrow p$ **B.** $\sim q \Rightarrow \sim p$ **C.** $\sim p \Rightarrow \sim q$

20. Write a proof of this statement in whatever form you choose.

   Points A, B, and C are collinear. M is the midpoint of $\overline{AB}$ and N is the midpoint of $\overline{BC}$. Prove that $AC = 2MN$.

21. **WRITING MATHEMATICS** Describe the difference between a paragraph proof and a two-column proof.

PERFORMANCE ASSESSMENT

DEFINE IT YOURSELF Create a name and write a definition for each of the following figures.

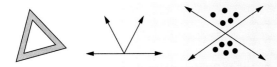

Create four more figures of your own, give each a name, and then write a definition for it.

CHOOSE YOUR FAVORITES Choose one postulate, one property, and one theorem from this chapter that you feel are particularly important or interesting. Copy each statement, provide an example of its use, and then explain why you have chosen it.

CONSTRUCTION INSTRUCTION Create a poster to show someone how to use a compass and straightedge to bisect a line segment.

GEOMETRY IN THE MIRROR Draw two large intersecting lines on a sheet of paper. Use a plane mirror and place it in different positions on the intersecting lines. Sketch and describe the figures you get.

PRACTICAL PROOF Choose an issue that is important to you such as getting your own car, staying up late on weekends, or getting a part-time job. Write a proof in two-column form to convince someone that you should get what you want.

AN IF-THEN STORY Write a humorous story using all *if-then* statements. For example, you might start, "If dogs had wings, then they could fly. If a dog could fly, then he would chase birds. If a dog chased a bird, then the bird would . . . ."

PROJECT ASSESSMENT

PROJECT *Connection* Work in a group to prepare a class presentation.

1. Make a group display of the bitmaps, binary numbers, and decimal numbers for the names of the students in the group. Point out similarities and differences in the way the students prepared their bitmaps and wrote their number "codes."

2. As a group, brainstorm to answer these questions: How are pixels and the images they form on a computer screen similar to points and the geometric figures they form? How are they different? Make a summary of your conclusions and include it as part of your class presentation.

3. As a group, make a list of related questions that were encountered in the process of research. For example: How are "outline" fonts different from bitmapped fonts? What information must be included in a binary code besides simple instructions concerning which pixels are turned on? Each member of the group should choose a different topic from the list, gather information about the topic from group members and prepare a report.

Fill in the blank.

1. Points that lie on the same line are called ___?___ .

2. The set of all points is ___?___ .

3. If two distinct planes intersect, then their intersection is a ___?___ .

4. Two angles whose measures have a sum of 90° are ___?___ .

5. In the figure below, point B is between points A and C and $AC = 73$. Find BC.

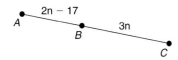

A map of a town uses a scale of 1 in. : 0.5 mi

6. On the map, it is 7 in. from Yi's house to the library. What is the actual distance?

7. Jorge knows that from his house to the movie theater is 2 mi. How far is this on the map?

For each figure, draw all the lines of symmetry and determine the order of rotational symmetry.

8.    9.

Let p represent *two angles are supplementary* and q represent *the measures of two angles have a sum of 180°*. Translate each set of symbols into words, and tell if the statement is true or false.

10. $q \Rightarrow p$ 11. $\sim p \Rightarrow \sim q$

Find the midpoint of the segment that has the given endpoints on a coordinate plane.

12. $(-1, 4)$ and $(5, 10)$ 13. $(4, -3)$ and $(-7, -4)$

14. **WRITING MATHEMATICS** Consider the conditional statement *If 1 = 2, then 3 = 3*. Is it true or false? Explain your reasoning.

Name the property that justifies each conclusion.

15. If $m\angle A + m\angle B = m\angle K$, and $m\angle K = 90°$, then $m\angle A + m\angle B = 90°$.

16. If $RT = TR$, then $TR = RT$.

17. If $m\angle A + 40° = 90°$, then $m\angle A = 50°$.

A rectangular prism is shown. Find the coordinates of each point.

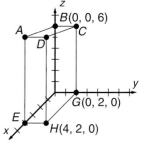

18. A 19. C

20. D 21. E

22. **STANDARDIZED TESTS** What is the angle measure of the complement of the supplement of a 150° angle?

   a. 30°   **b.** 50°   **c.** 60°   **d.** 80°

In each figure, find $m\angle JKL$.

23.   24.

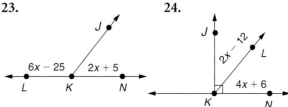

Write each definition as a conditional statement.

25. Parallel lines do not intersect.

26. Perpendicular lines form right angles.

Find the distance between two points on a coordinate plane with the given coordinates. Round to the nearest tenth if necessary.

27. $(6, -2)$ and $(3, -5)$ 28. $(-2, -4)$ and $(4, 2)$

STANDARDIZED TEST

STUDENT-PRODUCED ANSWERS Solve each question and on the answer grid write your answer at the top and fill in the ovals.

Notes: Mixed numbers such as $1\frac{1}{2}$ must be gridded as 1.5 or 3/2. Grid only one answer per question. If your answer is a decimal, enter the most accurate value the grid will accommodate.

1. Find the product of the coordinates of the midpoint of the segment with endpoints $(-2, 7)$ and $(6, -1)$.

2. What is the degree measure of an angle that is the supplement of the complement of a 40° angle?

3. Determine the order of rotational symmetry for the figure shown.

4. A map uses a scale of 1 cm : 150 mi. On the map, the distance between two cities measured 6.5 cm. How many miles is the actual distance between the cities?

5. On a number line, the coordinate of M is -2 and the coordinate of N is 5. If M is the midpoint of NP, what is the coordinate of P?

6. In the figure below, $m\angle ABE = m\angle DBC$. Find the measure of $m\angle DBE$.

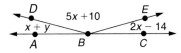

7. The top, front, and right side views of a collection of cubes is shown. How many cubes are in the collection?

8. On a map of the world, a 7 cm : 2000 mi scale is used. From San Francisco to Tokyo is approximately 5000 mi. About how many centimeters is this on the map?

9. The point $(7, -2)$ lies on a circle whose center is at the origin. To the nearest tenth, what is the radius of the circle?

10. In the figure, $JK = 90$. Express the ratio JM to MK as a fraction.

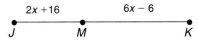

11. A model ship is built according to a scale of 150 ft : 1 in. If the ship is 1125 ft in length, how many inches long is the model?

12. A rectangular prism is 3 ft wide, 4 ft long, and 12 ft tall. How long is a piece of string that connects one corner to its opposite corner? Round to the nearest tenth.

13. On a segment, point R is at -7 and point S is at -2. If S is between R and T, and S is the midpoint of $\overline{RT}$, find the coordinate of point T.

14. Find the measure of $\angle DBC$.

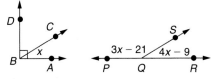

15. If $4 \star x = 35$ and $6 \star y = 4 \star x$, what is $6 \star y$?

16. On the coordinate plane, C is the midpoint of $\overline{AB}$ and D is the midpoint of $\overline{CB}$. Grid the product of the coordinates of point D.

Lines, Planes, and Angles

Take a Look
AHEAD

Make notes about things that look new.
- Find five names given to angles that have special relationships.
- Make a list of symbols used to describe relationships of angles or lines.

Make notes about things that look familiar.
- Recall the different ways two lines can intersect. Draw an example of each.
- How do you think the slope can be used to show two lines are parallel?

DATA Activity

SKILL FOCUS

▶ Organize data.

▶ Add and subtract real numbers.

▶ Draw conclusions.

Which Way is North?

A compass does not seek out and point to the *geographic* North Pole, but points to the *magnetic* pole. The angle formed by the geographic North Pole, your location, and the magnetic pole is called variation and can be large enough to cause problems in navigation. The angle measure is followed by E or W which indicates whether the longitudinal line through the magnetic North Pole is east or west of the location.

Navigation

In this chapter, you will see how:

- **RESCUE WORKERS** use intersecting lines to find locations on maps and plan rescue routes. (Lesson 3.3, page 146)

- **SHIP NAVIGATORS** use the angle between the magnetic pole and the line of sight to locate their position on a map. (Lesson 3.5, page 160)

City	Variation
Chicago	0°
New York	13°W
San Diego	13°E
Tampa	3°W
Raleigh	8°W
Seattle	19°E

The table gives different cities and their variation from the geographic North Pole.

1. Arrange the cities' variations from West to East on a horizontal line.

2. Which city has the greatest variation from the geographic North Pole?

3. Which city has the least variation from the geographic North Pole?

4. If you are in Chicago using a compass what will you know about the reading taken?

5. **WORKING TOGETHER** Discuss how to adjust a compass reading to account for variation. If you are in New York and your compass reads 170°, what compass reading points you to the geographic North Pole?

127

Over Hill Over Dale

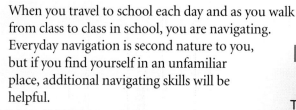

When you travel to school each day and as you walk from class to class in school, you are navigating. Everyday navigation is second nature to you, but if you find yourself in an unfamiliar place, additional navigating skills will be helpful.

A compass and map are tools for navigation. Careers such as forest rangers and rescue team members require navigational skills, but others find their way through unfamiliar territory when participating in the sport of *orienteering*. In this project you will use some orienteering skills.

PROJECT GOAL

To use angles and simple navigational skills to travel over the land.

Getting Started

Work in a group of three to four students.

1. Each person in the group should write a detailed description of how to get from school to home.

2. Share your descriptions.

3. As a group make a list of *navigational aids* included in the descriptions. This would be any tool or location references to assist someone in traveling.

4. Add to your list any other *navigational aids* that group members did not include that can be helpful in navigating from one place to another.

PROJECT *Connections*

Lesson 3.2, page 138:
Use perpendicular lines to locate the cardinal directions when a compass is not available.

Lesson 3.4, page 153:
Estimate distance by counting steps.

Lesson 3.6, page 166:
Make a counting device to assist in counting steps.

Lesson 3.9, page 181:
Determine angle measures using your hand.

Chapter Assessment, page 189:
Make a terrain sketch.

Internet Connection

www.learninggeometry.com

3.1 Relationships Among Lines

Explore/Working Together

● Work with a partner. Join two straws with a paper clip, as shown.

1. Move one straw so that ∠1 becomes smaller. What happens to ∠3? What happens to ∠2 and ∠4?

2. Move one straw so that ∠2 becomes smaller. What happens to ∠4? What happens to ∠1 and ∠3?

3. Make a conjecture: What appears to always be true about ∠1 and ∠3? What appears to always be true about ∠2 and ∠4?

4. Use a ruler to draw two intersecting lines. Measure each angle. Was your conjecture correct?

5. What do you think will happen if ∠1 is a right angle? Make a conjecture. Then devise a way to test your conjecture.

SPOTLIGHT ON LEARNING

WHAT? In this lesson you will learn
- to identify relationships between lines.
- to identify vertical angles and their measure.

WHY? Identifying vertical angles and their relationships helps you solve problems in marketing, design, and navigation.

Build Understanding

● **Vertical angles** are two angles whose sides form two pairs of opposite rays. In Explore the straws represent two intersecting lines which form vertical angles. In the figure below, ∠1 and ∠3 are a pair of vertical angles, as are ∠2 and ∠4.

THEOREM 3.1	VERTICAL ANGLES THEOREM

If two angles are vertical angles, then they are congruent.

Given In the figure, ∠1 and ∠3 are vertical angles.

Prove ∠1 ≅ ∠3

Proof Since the noncommon sides of ∠1 and ∠2 are opposite rays, they form a linear pair. For the same reason, ∠3 and ∠2 form a linear pair. By the Linear Pair Postulate,

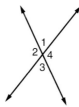

$m\angle 1 + m\angle 2 = 180°$
$m\angle 3 + m\angle 2 = 180°$

So, by the Congruent Supplements Theorem, ∠1 ≅ ∠3. ◄

TECHNOLOGY TIP

You can use geometry software to show that many different pairs of vertical angles are congruent.

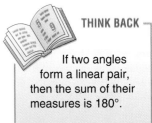

THINK BACK

If two angles form a linear pair, then the sum of their measures is 180°.

You can use the Vertical Angles Theorem to find the measures of unknown angles.

EXAMPLE 1

In the figure, m∠1 = 70°.
Find m∠2, m∠3, and m∠4.

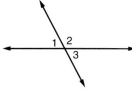

Solution

Since ∠1 and ∠3 are vertical angles,
m∠3 = m∠1 = 70°.

Since ∠1 and ∠2 form a linear pair,
m∠2 = 180° − 70° = 110°.

Since ∠4 and ∠2 are vertical angles,
m∠2 = m∠4 = 110°. ◄

Angle measurements can be given as variable expressions.

EXAMPLE 2

In the figure, m∠1 = $(9x)°$ and
m∠3 = $(5x + 12)°$. Find m∠2.

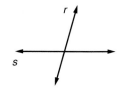

Solution

m∠1 = m∠3	∠1 and ∠3 are vertical angles.
9x = 5x + 12	Substitute for ∠1 and ∠3.
4x = 12	Solve for x.
x = 3	

Substitute for x to find m∠1 = 9(3)° = 27°.
Since ∠1 and ∠2 form a linear pair, m∠2 = 180° − 27° = 153°. ◄

CHECK UNDERSTANDING

In Example 2, find m∠3 if m∠1 = (9x)° and m∠2 = (5x + 12)°. What equation can you use to find the solution?

Two lines that intersect to form a right angle are **perpendicular lines**. Lines that intersect and are not perpendicular are **oblique lines**.

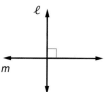

Line ℓ is perpendicular to line m.
ℓ ⊥ m

Lines r and s are a pair of oblique lines.

THINK BACK

Recall that a square bracket (⌐) in a figure indicates a right angle. The measure of a right angle is 90°.

THEOREM 3.2

If two lines are perpendicular, then they intersect to form four right angles.

EXAMPLE 3

In the figure, $\overrightarrow{AD} \perp \overrightarrow{EB}$. Find m$\angle BOC$.

Solution

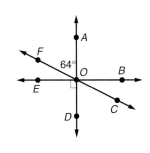

Since $\overrightarrow{AD} \perp \overrightarrow{EB}$, it follows from Theorem 3.2 that
$\angle AOE = 90°$. By the Angle Addition Postulate,

$$m\angle AOF + m\angle EOF = m\angle AOE$$

$$m\angle EOF = m\angle AOE - m\angle AOF$$

$$m\angle EOF = 90° - 64°$$

$$m\angle EOF = 26°$$

Since $\angle BOC$ and $\angle EOF$ are vertical angles, m$\angle BOC = 26°$. ◄

Coplanar lines that do *not* intersect are called **parallel lines**.
Noncoplanar lines are called **skew lines**. Rays and segments are also
described as *perpendicular, oblique, parallel,* or *skew*.

THINK BACK

Coplanar lines
are lines which lie in
the same plane.

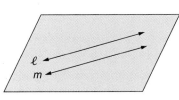 

Line ℓ is parallel to line m.   Lines p and q are skew lines.

CHECK UNDERSTANDING

Decide how to fill in
the blanks in these two
statements.

If two lines are
coplanar, then they
are either __?__ lines,
__?__ lines, or __?__
lines.

If two lines are not
coplanar, then they
are __?__ lines.

EXAMPLE 4

DESIGN A manufacturer is designing a new
tissue box. The shape is to be half of a cube as
shown. Identify each pair of edges as
perpendicular, oblique, parallel, or *skew*.

  a. $\overline{PQ}$ and $\overline{PX}$ **b.** $\overline{PQ}$ and $\overline{PR}$

  c. $\overline{PR}$ and $\overline{XZ}$ **d.** $\overline{XZ}$ and $\overline{QY}$

Solution

  a. $\overline{PQ}$ and $\overline{PX}$ form a right angle, so they are
      perpendicular.

  b. $\overline{PQ}$ and $\overline{PR}$, intersect but do *not* form a
      right angle, so they are oblique.

  c. $\overline{PR}$ and $\overline{XZ}$ lie in the same plane. If you
      could extend them indefinitely, they
      would never intersect. They are parallel.

  d. $\overline{XZ}$ and $\overline{QY}$ do not lie in the same
      plane, so they are skew. ◄

Find each unknown angle measure in each figure.

1.
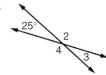

2.
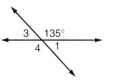

3.

4.
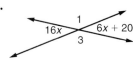

Use the figure at the right for Exercises 5–8.

5. Name the relationship between $\overrightarrow{AB}$ and $\overleftrightarrow{CD}$.

6. Name the relationship between $\overrightarrow{AB}$ and $\overrightarrow{EF}$.

7. Find m∠FOD.

8. Find m∠AOF.

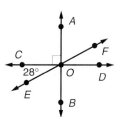

9. **MARKETING** Customer appeal is a factor when considering package design and manufacturing costs. TJ Toys is researching the design shown for a new package. Analyze the package which is the lower part of a square pyramid. Identify each pair of edges as *perpendicular*, *oblique*, *parallel*, or *skew*.

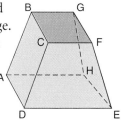

  a. $\overline{AD}, \overline{DE}$ **b.** $\overline{BC}, \overline{GF}$ **c.** $\overline{CF}, \overline{EF}$ **d.** $\overline{AH}, \overline{CD}$

10. **WRITING MATHEMATICS** Think about packages of items that you use daily. Are most of the edges that intersect perpendicular or oblique? Why do you think that most objects are packaged this way?

PRACTICE

Find each unknown angle measure in each figure.

11.
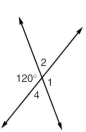

12.
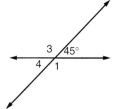

13.
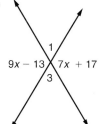

14.
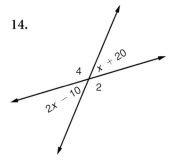

Two lines intersect so that $\angle 1$ and $\angle 3$ are vertical angles and $\angle 1$ and $\angle 2$ form a linear pair. Find the measure of each of the unknown angles.

15. $m\angle 4 = 34°$

16. $m\angle 1 = 147°$

17. $m\angle 3 = 8x - 10, m\angle 1 = 3x + 55$

18. $m\angle 4 = -2x + 18, m\angle 2 = 4x + 42$

19. $m\angle 1 = 8x + 20, m\angle 2 = 4x - 8$

20. $m\angle 3 = 5x + 43, m\angle 4 = 2x - 10$

21. INTERIOR DESIGN Two rectangular panels are used to construct the four-office cubicle at the right. They are designed to pivot at center $\overline{NS}$. If the two panels are pivoted to increase the size of one office what happens to the adjacent offices? the opposite office? In order for all four offices to be the same size, what position should one panel have to the other?

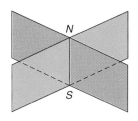

Use the figure at the right for Exercise 22–25.

22. Name an angle congruent to $\angle AOF$.

23. Name the line perpendicular to $\overleftrightarrow{CD}$.

24. Name an angle that forms a linear pair with $\angle COE$.

25. Find $m\angle AOF$.

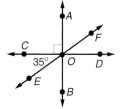

26. WRITING MATHEMATICS Describe an object or objects that model each of the following: parallel lines, perpendicular lines, oblique lines, and skew lines. Include an explanation of what part of the objects model each type of lines.

EXTEND

SAILING Sailboats cannot sail directly into the wind. Instead, sailors use a technique called *tacking*, where the boat sails a series of short paths. In the figure, the boat is tacking first 45° to the left of the oncoming wind, then 45° right of the oncoming wind, and so on until the boat reaches its destination. Tacking 45° to the left and right of the wind is considered to be the ideal path for a sailboat desiring to sail toward a location directly in the wind.

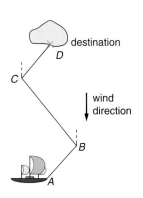

27. Choose the type of lines that best describes the relationship between $\overline{AB}$ and $\overline{BC}$. Explain your answer.

    a. parallel   **b.** perpendicular

    c. oblique   **d.** skew

28. What appears to be true about $\overline{AB}$ and $\overline{CD}$?

29. Sketch the path of a boat that is tacking at an angle larger than 45° to the left and right of the wind. Why would this path be considered not as desirable?

30. Sketch the path of a boat that is tacking at an angle smaller than 45° to the east and west of the wind. Why would this path be difficult to sail?

31. WRITING MATHEMATICS Examine the figure below. Explain why ∠1 and ∠2 are not vertical angles.

THINK CRITICALLY

Two lines intersect so that ∠1 and ∠3 are vertical angles and ∠1 and ∠2 form a linear pair. Solve for *x* to find the measure of ∠1, ∠2, ∠3, and ∠4.

32. $m\angle 2 = x^2 - 3x$, $m\angle 4 = 6x + 10$

33. $m\angle 2 = x^2 + 4x$, $m\angle 4 = -5x - 20$

34. $m\angle 1 = 3x + 3y$, $m\angle 2 = -5x + 5y$, $m\angle 3 = 8x + y$, $m\angle 4 = 5x + y$

35. $m\angle 1 = 5x + y - 6$, $m\angle 2 = x + y$, $m\angle 3 = x + 6y + 4$, $m\angle 4 = 3x - y - 14$

ALGEBRA AND GEOMETRY REVIEW

Refer to the figure for Exercises 36–39.

36. Name two angles that are complementary.

37. Name a linear pair.

38. Name two angles that are supplementary.

39. Use a protractor to measure ∠*BAE*.

40. Draw a figure that has no lines of symmetry.

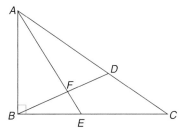

41. STANDARDIZED TESTS Two lines intersect so that ∠1 and ∠3 are vertical angles and ∠1 and ∠2 form a linear pair. Solve for *x* to find m∠1 when m∠1 = 3*x* and m∠4 = 6*x* + 9.

    A. m∠1 = 19°   **B.** m∠1 = 57°   **C.** m∠1 = 123°   **D.** m∠1 = 161°

Solve each system.

42. $2x + 3y = 7$
$\quad x - 2y = -7$

43. $-x + 4y = -2$
$\quad y = \dfrac{1}{2}$

44. $2x + 5y - 2z = 6$
$\quad -x - y - z = 1$
$\quad 4x - 4y + 3z = -2$

Think Back/Working Together

● Work with a partner.

1. Use your desktop to represent a plane and use strips of paper to represent coplanar lines. Manipulate them to determine the number of intersection points possible for each column in the table. Copy and complete the table.

number of coplanar lines	2	3	4	5
number of intersection points possible	0 or 1			

2. If some of the lines are parallel, how does that affect the number of intersection points?

3. Can four lines intersect in exactly two points? Explain.

4. Explain how can you make 20 lines intersect in 19 points?

Explore

● The intersection of two geometric figures is the set of all points common to both figures. Work with a partner to model a pair of intersecting planes. Each partner needs one index card.

5. Label the corners of the card A, B, C, and D as shown. To find the center of the card, draw $\overline{AC}$ and $\overline{BD}$. Label the point of intersection E.

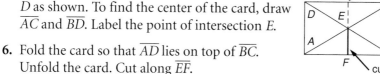

6. Fold the card so that $\overline{AD}$ lies on top of $\overline{BC}$. Unfold the card. Cut along $\overline{EF}$.

7. Join your card with your partner's card as shown at the right.

8. Recall when two lines intersect, the intersection is a point. Look at the place where the two cards intersect. When two planes intersect, the intersection is a ___?___

9. Position two index cards so they do *not* intersect. Imagine that both cards extended in all directions indefinitely. Will your cards ever intersect?

10. Position two index cards so that they model two planes that do not intersect. Describe to your partner the position of the planes and how you know they will not intersect.

11. Describe the different ways two planes can be related.

For Questions 12–15 use a pencil and one index card to model relationships between a line and a plane.

12. Is it possible that the intersection of a line and a plane is a line? Explain and sketch the positions of the line and card.

13. Is it possible that the intersection of a line and a plane is a point? Explain and sketch the positions of the line and card.

14. Is it possible that a line and a plane do not intersect at all? Explain and sketch the positions of the line and card.

15. Describe the different ways a line and a plane can be related.

Make Connections

• You use the words *parallel*, *perpendicular*, *oblique*, and *skew* to describe relationships among lines. You can apply most of these same terms to relationships among lines and planes.

> **A line and a plane (or two planes) are parallel if and only if they do not intersect.**

16. Look around your classroom. Identify a model of a line and a plane that are parallel.

17. Identify a model of two planes that are parallel.

18. How is the definition given above similar to the definition of parallel lines in Lesson 3.1? How is it different?

Use a sheet of paper to represent a plane. Draw a pair of intersecting lines that lie in the plane.

19. Use your pencil to represent a line which is perpendicular to both lines. Where must your pencil be placed?

20. Now draw another line which lies in the plane and passes through the point where the first two lines intersected. Place your pencil in the same position as you did in Question 19. Is it perpendicular to the new line also?

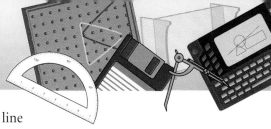

21. Will the pencil in that position be perpendicular to any line in the plane that passes through the intersection point?

> A line and a plane are perpendicular if and only if the line is perpendicular to two distinct lines in the plane at their point of intersection.

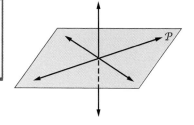

22. Look around your classroom. Identify a model of a line and a plane that are perpendicular.

Use your paper and the pencil in the position from Question 20. Visualize a plane that contains the line the pencil represents.

> Two planes are perpendicular if and only if one plane contains a line that is perpendicular to the other plane.

23. Look around your classroom. Identify a model of two planes that are perpendicular.

24. Perpendicular lines are two lines that intersect to form a right angle. Explain why you cannot define perpendicular planes as two planes that intersect to form a right angle.

25. Look around your classroom. Identify a model of a line and a plane that you think are oblique.

> A line and a plane (or two planes) are oblique if and only if they are neither parallel nor perpendicular.

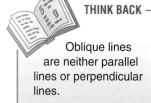

THINK BACK

Oblique lines are neither parallel lines or perpendicular lines.

26. Identify a model of two planes that are oblique.

27. Why do you think there is no definition of *skew* planes?

Summarize

28. THINKING CRITICALLY In Questions 22, 25, and 26, why were you instructed to identify models of lines and planes ?

29. MODELING Air traffic controllers assign cruising altitudes in thousands of feet above the ground. Eastbound airplanes are assigned an odd thousands of feet altitude while westbound airplanes are assigned an even thousands of feet altitude. Name the relationship of the cruising altitudes and explain why controllers use this method to avoid midair collisions.

30. **GOING FURTHER** Use the index cards from Questions 5–7 and others like them to represent planes. Manipulate the cards to determine the possible number of intersections for each column in the table. Copy and complete the table.

number of planes	2	3	4	5
number of intersections possible	0 or 1			

When two planes intersect, four **half-planes** are formed. The line where the planes intersect is the **edge** of each half-plane. Each pair of half-planes forms a **dihedral angle**. The half-planes are the **faces** of the angle, and their common edge is the **edge** of the angle.

31. If the measure of the dihedral angle shown is 45°, what is the measure of ∠*ABC*?

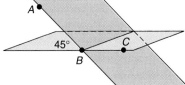

32. Use an index card to model a dihedral angle measuring about 110°. Compare with others in class.

33. **WRITING MATHEMATICS** Compare the definition of dihedral angle to the definition of angle. How are they alike? How are they different?

34. **THINKING CRITICALLY** Use the index cards from Questions 5–7 to devise a method for measuring a dihedral angle.

PROJECT Connection

If a compass is not available, the sun and a stick can be used to identify which direction is north, south, east, or west.

1. Put a stick into the ground at a level place. Mark the tip of the shadow with a stone, twig, or some other means. Label this point W.

2. Wait 10 to 15 min until the shadow has moved a few inches. Mark the new location of the tip of the shadow. Label this point E.

3. Draw a line through the two marks. This is an approximate east-west line. The first shadow mark is to the west of the second mark. This explains why you marked the tips as W and E accordingly.

4. Draw a line perpendicular to the east-west line. This is your north-south line.

5. Label the north direction and the south direction on this line. You can now use this pair of perpendicular lines to locate the direction needed.

3.3 Parallel Lines and Transversals

Explore/Working Together

1. Work with a partner. Use lined paper and a ruler. Trace over a set of lines printed on the paper to create a set of parallel lines.

2. Draw a third line which intersects the two parallel lines. Label the angles as shown.

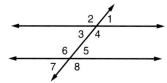

3. Name all sets of linear pairs and vertical angles.

4. Trace and cut out a copy of ∠1 and ∠2.
   a. Slide and rotate the copy of ∠1 to compare it to ∠5, ∠6, ∠7, and ∠8. What do you notice?
   b. Slide and rotate the copy of ∠2 to compare it to ∠5, ∠6, ∠7, and ∠8. What do you notice?

5. Make a new drawing and color it so that congruent angles have matching color.

Build Understanding

The third line you drew in Explore is called a **transversal**. A line is a transversal if it intersects two or more coplanar lines at different points.

Angles formed by a transversal have special names, in the figure ∠3, ∠4, ∠5, and ∠6 are called **interior angles** and ∠1, ∠2, ∠7, and ∠8 are called **exterior angles**.

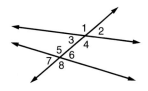

A pair of nonadjacent angles, one exterior and one interior, both on the same side of the transversal are called **corresponding angles**. In the figure above, the following are corresponding angles.

∠1 and ∠5   ∠2 and ∠6   ∠3 and ∠7   ∠4 and ∠8

> **CORRESPONDING ANGLES POSTULATE**
>
> **POSTULATE 12** If two parallel lines are cut by a transversal, then the pairs of corresponding angles are congruent.

Alternate interior angles are two nonadjacent interior angles on opposite sides of a transversal. For example, ∠3 and ∠5 are alternate interior angles.

Alternate exterior angles are two nonadjacent exterior angles on opposite sides of a transversal. For example, ∠2 and ∠8 are alternate exterior angles. In Explore, you discovered these angles have special relationships when parallel lines are cut by a transversal.

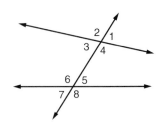

THEOREM 3.3	**ALTERNATE INTERIOR ANGLES THEOREM**

> If two parallel lines are cut by a transversal, then the pairs of alternate interior angles are congruent.

Given In the figure, $a \parallel b$

Prove ∠1 ≅ ∠3

Proof

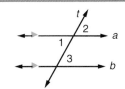

Statements	Reasons
1. $a \parallel b$	1. Given
2. ∠2 ≅ ∠3	2. Corresponding angles are congruent.
3. ∠1 ≅ ∠2	3. Vertical angles are congruent.
4. ∠1 ≅ ∠3	4. Transitive property

THEOREM 3.4	**ALTERNATE EXTERIOR ANGLES THEOREM**

> If two parallel lines are cut by a transversal, then the pairs of alternate exterior angles are congruent.

The proof of Theorem 3.4 is similar to the one shown above and will be completed in the exercises.

EXAMPLE 1

In the figure, m∠1 = 50°. Find the measures of the other angles. Justify your answers.

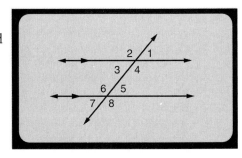

Solution

Angle	Why	Type of Angles
$m\angle 2 = 130°$	$m\angle 1 + m\angle 2 = 180°$	linear pair
$m\angle 3 = 50°$	$\angle 1 \cong \angle 3$	vertical angles
$m\angle 4 = 130°$	$m\angle 1 + m\angle 4 = 180°$	linear pair
$m\angle 5 = 50°$	$\angle 3 \cong \angle 5$	alternate interior angles
$m\angle 6 = 130°$	$\angle 2 \cong \angle 6$	corresponding angles
$m\angle 7 = 50°$	$\angle 1 \cong \angle 7$	alternate exterior angles
$m\angle 8 = 130°$	$\angle 2 \cong \angle 8$	alternate exterior angles

CHECK UNDERSTANDING

Are there other ways to justify the measures of the angles in Example 1? Explain.

When two lines are intersected by a transversal, **same-side interior angles** are two interior angles on the same side of the transversal.

> **THEOREM 3.5**   **SAME-SIDE INTERIOR ANGLES THEOREM**
>
> If two parallel lines are cut by a transversal, then the pairs of same-side interior angles are supplementary.

THINK BACK

Supplementary angles are angles whose measures have a sum of 180°

You can use the relationships of angles to find their measures when you are given an algebraic expression for the angle.

EXAMPLE 2

Find $m\angle 1$, $m\angle 2$, and $m\angle 3$.

Solution

The figure shows $a \parallel b$, so $\angle 2$ and the angle labeled $9x + 82$ are corresponding angles.

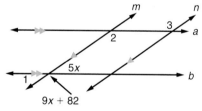

$$m\angle 2 + (5x) = 180 \quad \text{Same-side interior angles have a sum of 180°.}$$
$$(9x + 82) + 5x = 180 \quad \text{Substitute } 9x + 82 \text{ for } m\angle 2.$$
$$14x = 98 \quad \text{Solve for } x.$$
$$x = 7$$

So, $9x + 82 = 9(7) + 82 = 145$ and $m\angle 2 = 145°$.

The figure also shows $m \parallel n$, so $\angle 1$ and the angle labeled $5x$ are vertical angles. Since $x = 7$, $m\angle 1 = 35°$.

Since $m \parallel n$, $m\angle 3 = 145°$ because $\angle 2$ and $\angle 3$ are alternate interior angles.

EXAMPLE 3

DECORATING Ana is wallpapering her vaulted family room and needs to measure the angle formed by the ceiling and the wall so she can accurately precut the wallpaper. She uses a plumb bob to create 2 parallel lines on the wall. A plumb bob is a weight attached to a string and used to model a line perpendicular to the floor.

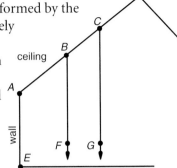

a. How is m∠*CBF* related to m∠*DCG*?

b. How can m∠*CBF* be used to find the m∠*BCG*?

c. How will Ana use the m∠*CBF* to prepare the wallpaper for hanging?

d. How can she use the m∠*CBF* to prepare the wallpaper for ∠*BAE*?

Solution

a. m∠*CBF* ≅ m∠*DCG*

b. Same-side interior angles formed by the plumb lines and the ceiling are ∠*CBF* and ∠*BCG*, so m∠*BCG* = 180 − m∠*CBF*.

c. Ana uses m∠*CBF* for the cut along the top of each strip of paper.

d. The corner should be parallel to the plumb lines which make ∠*BAE* ≅ ∠*CBF*, assuming the wall is indeed vertical.   ◄

In Example 3, if Ana does not have a vaulted ceiling, the ceiling models a perpendicular transversal, so ∠*CBF* is a right angle.

THEOREM 3.6	**PERPENDICULAR TRANSVERSAL THEOREM**

If a transversal is perpendicular to one of two parallel lines, then it is perpendicular to the other.

TRY THESE

Find the measure of each angle. Name a pair of angles for each of the six angle relationships you have learned.

1.

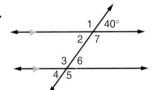

2.

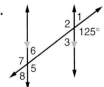

3.

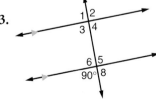

4. WRITING MATHEMATICS Write a paragraph proof explaining how you found the measures of the angles in Exercise 3. Use the angle pair relationships. Which theorem is Exercise 3 illustrating?

5. Parallel lines *r* and *s* are cut by transversal *t* so that ∠1 and ∠5 are corresponding angles. Alternate interior angles are ∠3 and ∠5, and ∠4 and ∠6. Alternate exterior angles are ∠2 and ∠8, and ∠1 and ∠7. Find the measure of each angle if m∠3 = 5*x* and m∠5 = 7*x* − 30.

Use the figure below to list all angle pairs named in Exercises 6–11.

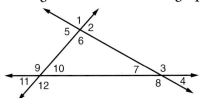

6. linear pairs

7. corresponding angles

8. alternate exterior angles

9. alternate interior angles

10. same-side interior angles

11. vertical angles

Copy each figure. Label the measure of each angle.

12.

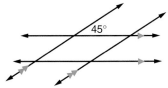

13.

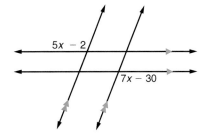

14. **CARPENTRY** A carpenter makes two parallel cuts at each end of the board at the right. The measure of one angle is given. Find the measures of the other angles.

15. **WRITING MATHEMATICS** A carpenter is going to attach another piece of wood at $\overline{AB}$ to form a *miter joint* as shown in the figure. Find the measures needed for ∠5 and ∠6 and explain your answers.

PRACTICE

Find the measure of each angle. Name a pair of angles for each of the six angle relationships you have learned.

16.

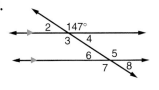

17.

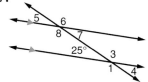

18.

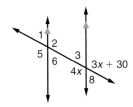

19. Parallel lines *u* and *v* are cut by transversal *t* so that ∠1 and ∠5 are corresponding angles. Alternate interior angles are ∠3 and ∠5, and ∠4 and ∠6. Alternate exterior angles are ∠2 and ∠8, and ∠1 and ∠7. Find the measure of each angle if m∠2 = 2*x* and m∠1 = 8*x* + 5.

Use the figure below to list all angle pairs named in Exercises 20–25.

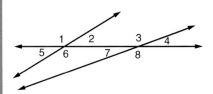

20. linear pairs

21. corresponding angles

22. alternate exterior angles

23. alternate interior angles

24. same-side interior angles

25. vertical angles

Use the figure at the right for Exercises 26–28.

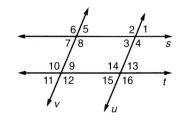

26. If $s \parallel t$, name all of the angles congruent to $\angle 1$.

27. If $u \parallel v$, name all of the angles congruent to $\angle 1$.

28. If $s \parallel t$ and $u \parallel v$, name all of the angles congruent to $\angle 1$.

Copy each figure. Label the measure of each angle.

29.

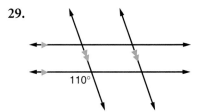

30.

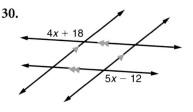

31.

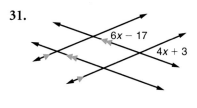

CARPENTRY Jose is a contractor reviewing the building specifications for a roofing job. The diagram he is looking at shows all of the hip jack rafters are parallel to the common rafter.

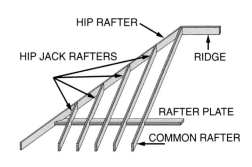

32. If each hip jack rafter is parallel to the common rafter, what must also be true about all of the hip jack rafters? Why?

33. If the first hip jack rafter meets the hip rafter in an angle of 30°, at what angle must all of the hip jack rafters meet the hip rafter? How do you know?

34. WRITING MATHEMATICS The rafter plate is parallel to the ridge and perpendicular to the common rafter. What is true about each hip jack rafter and the rafter plate? Why?

Find the values of x and y.

35.

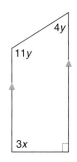

36.

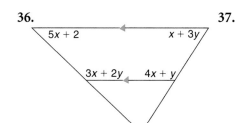

37.

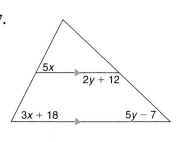

EXTEND

For Exercises 38–40, refer to the drawing at the right.

38. Find the measure of each angle if $m\angle 2 = x$, $m\angle 4 = y + 30$, $m\angle 8 = 2x - 140$.

39. Write a proof of Theorem 3.4: If two parallel lines are cut by a transversal, then the pairs of alternate exterior angles are congruent.

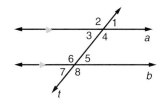

40. Write a proof of Theorem 3.5: If two parallel lines are cut by a transversal, then the pairs of same-side interior angles are supplementary.

NAVIGATION Submarines use a periscope to see above the surface of the water. The diagram shows the location of two parallel mirrors used in the periscope.

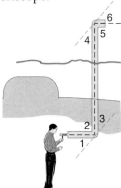

41. Which angles are alternate interior angles? What must be true about these angles?

42. If $\angle 2$ and $\angle 5$ are right angles, what must be true about $\angle 1$ and $\angle 6$? Justify your answer.

43. Given $\angle 1 = 45°$, $m\angle 2 = 90°$, and $m\angle 5 = 90°$, find $m\angle 3$, $m\angle 4$, and $m\angle 6$.

THINK CRITICALLY

Same-side exterior angles are two exterior angles on the same side of the transversal. In the figure for Exercise 38–40, $\angle 1$ and $\angle 8$ are same-side exterior angles, as are $\angle 2$ and $\angle 7$.

44. WRITING MATHEMATICS Investigate the relationship between same-side exterior angles of parallel lines. Draw several sets of parallel lines cut by a transversal and measure the same-side exterior angles. Write a sentence describing your conclusion.

45. Prove your conclusion about the relationship of same-side exterior angles.

ALGEBRA AND GEOMETRY REVIEW

46. Make a drawing that has at least one pair of parallel lines, one pair of perpendicular lines and one pair of oblique lines. Add the appropriate markings and labels. List each pair of lines and their relationship.

47. Explain the difference between complementary and supplementary angles.

48. Simplify each radical.

   a. $\sqrt{8}$ **b.** $\sqrt{27}$ **c.** $\sqrt{75}$ **d.** $\sqrt{40}$

49. Solve for x: $x^2 + 5x - 24 = 0$

HELP WANTED
SKILLED JOBS

Career
Rescue Worker

Rescue workers must often locate people who are missing or lost in national parks or forests. A compass and map are navigational tools essential to successful rescue efforts. Often the location of those in need is uncertain and the terrain or weather offer additional challenges. Working together, teams of rescue workers use the intersection of lines and angle measures to establish on a map the location of the rescue.

Two teams of rescue workers have spotted smoke from a fire. The fire most likely was started by two hikers who have been lost in the woods for several days. Before they set out to locate the hikers, the rescuers pinpoint the location of the smoke.

You can simulate how the rescue workers locate the missing hikers.

Decision Making

1. Trace the map onto another piece of paper.

2. Team 1 spots smoke at 70° east of north. Draw a north reference line at Team 1's position. Draw an angle 70° clockwise from north. The fire is located somewhere on the ray used to make a 70° with the north reference line.

3. Team 2 radios that they see the smoke at 12° west of north. Draw a north reference line at their position and draw an angle 12° counter-clockwise from north. The fire is located where the ray in Question 2 intersects the ray you just drew.

The teams radio the rescue station to send a helicopter to the area where they suspect the lost hikers are. The rescue station uses the point of intersection to determine in what direction and at what angle they must fly.

4. Draw a north reference line at the position of the station. Draw a line from the station through the point of intersection.

5. Measure the angle using the rescue station as the vertex. Explain the angle and direction the helicopter should fly.

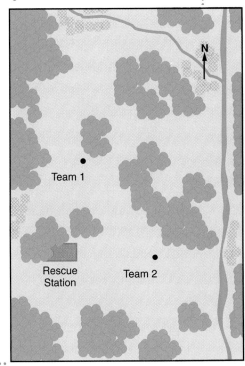

146 CHAPTER 3 **Lines, Planes, and Angles**

Explore

● You are exposed to conditional statements and their converses daily. In some circumstances these conditional statements are stated clearly, but often they are implied. For example, advertisers use conditional statements to influence the consumer's decision about purchases.

1. Write the slogan below as a conditional statement in if-then form.

2. State the hypothesis and conclusion.

3. Use a Venn diagram to model the slogan.

4. State the converse.

5. Use a Venn diagram to model the converse.

6. What message have the advertisers implied, hoping to influence your decision to buy Jiles Jeans? How do the advertisers determine if the implied message is working?

7. How can you determine if the converse is true? Explain if this proves the converse is true.

8. Write your own advertising slogan without using the words *if* and *then*. Restate the slogan as a conditional statement in if-then form. State the hypothesis and conclusion. State the converse. Explain how you can or cannot test the truth of the converse.

Build Understanding

● The postulate and theorems from Lesson 3.3 each have a converse. The conclusion of these converses is *then the lines are parallel*. The idea that parallel lines exist and are unique is a fundamental principle to Euclidean geometry. In the figure, the dashed line represents the only line that can be drawn parallel to line ℓ so that it passes through the point A.

Recall that a postulate is a statement that is accepted as true. The converses of many postulates are also postulates.

Many real world applications model two lines that must be parallel.

EXAMPLE 1

CONSTRUCTION The drawing shows the plans for the roof of a house and its garage. The garage front rafters and the house front rafters are to be parallel. Describe two ways a framer can tell if the front and garage rafters are parallel.

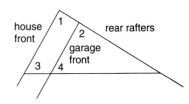

Solution

If $\angle 1 \cong \angle 2$, the rafters must be parallel because $\angle 1$ and $\angle 2$ form corresponding angles with respect to the rear rafters.

If $\angle 3 \cong \angle 4$, the rafters would also be parallel, because they form corresponding angles with respect to the horizontal. ◀

Without proof of its truth, you cannot use a converse to determine if two lines are parallel. Unlike the conditional statement in Explore, the converse to each of the theorems can be proven true.

THEOREM 3.7   CONVERSE OF ALTERNATE INTERIOR ANGLES THEOREM

If two lines are cut by a transversal so that the alternate interior angles are congruent, then the lines are parallel.

Given   In the figure, $\angle 1$ and $\angle 2$ are congruent alternate interior angles.

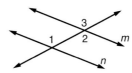

Prove $m \parallel n$

Proof

Statements	Reasons
1. $\angle 1 \cong \angle 2$	Given
2. $\angle 2 \cong \angle 3$	Vertical angles are congruent.
3. $\angle 1 \cong \angle 3$	Transitive Property
4. $m \parallel n$	Converse of Corresponding Angles Postulate

The proofs for the other converses are similar.

THEOREM 3.8   CONVERSE OF ALTERNATE EXTERIOR ANGLES THEOREM

If two lines are cut by a transversal so that alternate exterior angles are congruent, then the lines are parallel.

THEOREM 3.9   CONVERSE OF SAME-SIDE INTERIOR ANGLES THEOREM

If two lines are cut by a transversal so that same-side interior angles are supplementary, then the lines are parallel.

EXAMPLE 2

What value of x will make $a \parallel b$? What is the measure of the given angles?

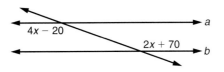

CHECK UNDERSTANDING

In Example 2 verify that substituting the value of x in the other expression, $2x + 70$, will give the same angle measure.

Solution

For the lines to be parallel, the given angles must be congruent.

$$4x - 20 = 2x + 70 \qquad \text{Alternate interior angles are congruent.}$$
$$2x = 90 \qquad\qquad\quad \text{Solve for } x.$$
$$x = 45$$

Since the angles are congruent, substitute into either expression to find the angle measure.

$$4(45) - 20 = 160°$$

◄

3.4   **Proving Lines are Parallel**   **149**

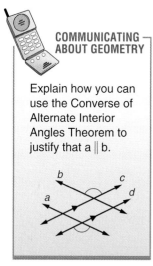
Identifying angle pairs helps you determine if two lines are parallel.

EXAMPLE 3

Name the theorem or postulate that justifies that $a \parallel b$.

a.

b.

c.

Solution

a. Converse of Same-Side Interior Angles Theorem
b. The angle which forms a linear pair with the angle measuring 102° has a measure of 78°. By the converse of Alternate Exterior Angles Theorem or Converse of Corresponding Angles Postulate, $a \parallel b$.
c. Converse of Alternate Exterior Angles Theorem ◄

In some circumstances you need to show that series of lines are parallel. The proofs for both theorems will follow in Lesson 3.8.

THEOREM 3.10

If two lines are parallel to a third line, then they are parallel to each other.

THEOREM 3.11

If two coplanar lines are perpendicular to a third line, then they are parallel to each other.

EXAMPLE 4

CIVIL ENGINEERING Tristann Jones is a civil engineer designing a parking lot for a new shopping mall. Her sketch shows she has decided to use Theorem 3.10 to guarantee uniform slanted parking spaces. What angle relationships must be true for lines a, b, and c to be parallel? How would her sketch be different if she wants straight parking spaces?

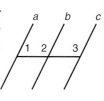

Solution

If same-side interior angles ∠1 and ∠2 are supplementary, then $a \parallel b$. If ∠2 ≅ ∠3, then ∠2 and ∠3 are corresponding angles and $b \parallel c$. Finally, if a and c are both parallel to b, then $a \parallel c$.

For straight parking spaces, the transversal is perpendicular to lines a, b, and c. So, ∠1, ∠2, and ∠3 are right angles. ◄

Name the postulate or theorem that justifies $b \parallel c$.

1.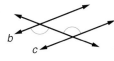

2.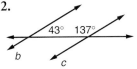

3.

4.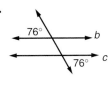

Lines p and m are cut by transversal t. Find the value of x that makes $p \parallel m$.

5. $\angle 2$ and $\angle 8$ are alternate exterior angles
$m\angle 2 = 20x - 14$ $m\angle 8 = 7x + 103$

6. $\angle 3$ and $\angle 6$ are same-side interior angles
$m\angle 3 = 2x - 6$ $m\angle 6 = 6x + 2$

Use the figure for Exercises 7–8.

7. If $\angle 4 \cong \angle 14$, name the postulate or theorem that implies $a \parallel b$.

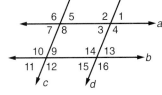

8. If $\angle 10$ and $\angle 13$ are supplementary angles, name the postulate or theorem that implies $c \parallel d$.

CARPENTRY The photograph shows the framing for a house. The vertical studs must be parallel to one another and perpendicular to the floor.

9. Explain how to tell if the studs are parallel.

10. **WRITING MATHEMATICS** Draw a model of the studs, floor, and cross brace. Label your drawing. Describe which angles and how many angles must the carpenter measure to assure that the studs are all perpendicular to the floor. Explain.

11. **WRITING MATHEMATICS** Example 4 describes a civil engineer who designs parking lots using a series of parallel lines. Name another real world application where a series of parallel lines are used. Explain why the lines must be parallel.

PRACTICE

Name the postulate or theorem that justifies $r \parallel s$.

12.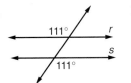

13.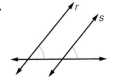

14.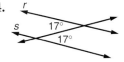

15.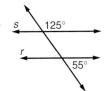

Lines *a* and *b* are cut by transversal *c*. Find the value of *x* that makes *a* ∥ b.

16. ∠4 and ∠5 are alternate exterior angles

$m\angle 4 = 3x + 10$ $m\angle 5 = 6x - 14$

17. ∠6 and ∠10 are same-side interior angles

$m\angle 6 = 9x + 32$ $m\angle 10 = 11x + 8$

Determine which segments are parallel. Justify your answer.

18.

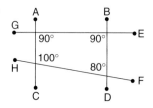

19.

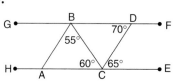

20.

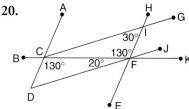

NAVIGATION The captain of a ship must be constantly aware of other ships in the area. To keep track of the paths of other ships, the captain sights another ship and measures the angle from his course to the other ship. This angle is measured at several different times. If the measure of the angles remains the same, the ships are on a *collision course*.

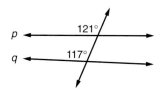

21. Explain what you know about $\overline{AB}$, $\overline{CD}$, and $\overline{EF}$. Why?

22. If the captain of ship 2 measured the angle from his course to ship 1, what would be true about each of the angles that captain measured?

23. Draw a new diagram of ship 1 and ship 2 where ship 1 sights angles of 85°, 65°, and 45°. Will ship 1 and ship 2 collide under these conditions? If not, which ship will pass in front of the other?

EXTEND

24. Write a proof for Theorem 3.8: If two lines are cut by a transversal so that the alternate exterior angles are congruent, then the lines are parallel.

25. Write a proof for Theorem 3.9: If two lines are cut by a transversal so that the same side interior angles are supplementary, then the lines are parallel.

Recall that the contrapositive results when the hypothesis and conclusion of a conditional statement are interchanged and both negated. The figure at the right illustrates the following contrapositive.

If the corresponding angles of two lines cut by a transversal are not congruent, then the lines are not parallel.

26. Write the contrapositive of each of the theorems relating parallel lines and transversals. Draw a figure for each that illustrates angle measures that produce lines that are not parallel.

THINK CRITICALLY

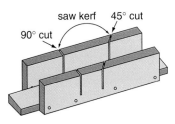

WOOD CRAFTING A miter box is used to make parallel cuts in wood. A piece of wood is placed in the box and the location of the desired cut is lined up with the slits (labeled saw kerf) in the box. A woodworker uses a saw inserted through the slits to cut the wood. To make a parallel cut, the wood is slid along the box until the locations to be cut are lined up with the slits. This miter box is designed to make 45° or 90° cuts.

27. Draw a piece of wood cut at each end using the 45° slits in the miter box. Label the angles.

28. Imagine that before the second cut in Question 27 is made, the wood is picked up, turned over and positioned back in the miter box. Will the second cut still be parallel to the first? Explain your answer and draw the piece of wood.

29. WRITING MATHEMATICS Explain how the miter box guarantees parallel cuts.

CARPENTRY A T-bevel is used by carpenters to mark the angles of cuts. To make parallel cuts the bevel is locked at a certain angle and the first cut is marked. Without changing the size of the angle on the bevel, the bevel is moved to the new location and the same angle marked for a cut.

30. Draw a diagram to show how the T-bevel will produce parallel cuts.

31. WRITING MATHEMATICS Discuss how the bevel is more versatile than the miter box.

32. Prove: If two parallel lines are cut by a transversal then the bisectors of the corresponding angles are parallel.

ALGEBRA AND GEOMETRY REVIEW

33. With a ruler draw a 5 cm segment. With a compass and straightedge, construct a segment congruent to the 5 cm segment.

34. With a protractor draw a 55° angle. With a compass and straightedge, construct an angle congruent to the 55° angle.

35. Construct the angle bisector in Exercise 34.

36. Multiply the binomials $(x + 4)(2x - 5)$.

37. Factor the trinomial $x^2 - 8x - 20$.

38. Graph: $A(-2, 4)$, $B(6, 3)$, $C(-1, -2)$, $D(0, 3)$

PROJECT *Connection*

Another important skill in land navigation is estimating distance traveled. Counting steps is a common method. Since everyone does not take the same size steps, it is important to know your personal measure.

1. Mark off a pre-determined distance, such as 100 m.

2. Count how many steps it takes you to walk the full length.

3. Repeat the process several times and average the number of steps.

4. Make a note of your average number of steps. You will use it later in the project.

Spherical Geometry

The foundations of Euclidean geometry rest in a basic understanding of three undefined terms: point, line, and plane. From a practical point of view, however, the "plane" that people encounter every day is the surface of Earth. It is not flat like a Euclidean plane, but instead curves like a sphere.

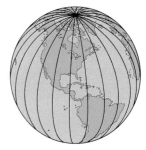

What happens if you begin with a different understanding of point, line, and plane? Mathematicians began to explore this question in earnest in the nineteenth century.

To begin, you must first understand two terms from Euclidean geometry. A **sphere** is the set of all points in space that are a given distance from a given point, called its **center**. A **great circle** is a cross section of the sphere whose center is the center of the sphere. For this lesson when you hear the words *plane* and *line*, you need to understand them to mean *sphere* and *great circle* respectively.

Thinking of lines on a curved surface probably is not a new idea to you. For example, maps and globes often are marked with *parallels of latitude* and *meridians of longitude*, as shown below.

parallels of latitude   meridians of longitude

GEOGRAPHY The shape of Earth is not a perfect sphere, but it is very close to one. In Questions 1–4, assume that Earth is a spherical solid.

1. Explain why all meridians of longitude are parts of great circles.

2. Explain why, in general, parallels of latitude are *not* great circles. Is any parallel of latitude a great circle?

3. Which do not intersect each other: meridians of longitude or parallels of latitude?

4. Which *do* intersect each other: meridians of longitude or parallels of latitude? Where do they intersect?

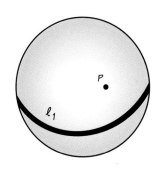

MODELING For Questions 5–8, you will need a felt-tip marker and a spherical solid that you can write on, such as a round balloon. Assume that the terms *parallel* and *perpendicular* have the same meaning as in Euclidean geometry.

5. Use the marker to draw any line on the sphere. (Remember that, on this surface, a line is a great circle.) Label the line ℓ_1. Draw a point *not* on ℓ_1. Label it P.

6. In Euclidean geometry, there are infinitely many lines through a point. Is the same true on a sphere? Specifically, are there infinitely many lines through P? Explain your answer.

7. Draw a line that passes through P and is perpendicular to ℓ_1. Label it ℓ_2. Where do ℓ_1 and ℓ_2 intersect?

8. Can you draw a line through P that is parallel to ℓ_1? Explain.

Question 8 should reveal a surprising fact: When your plane is a sphere, the Parallel Postulate of Euclidean geometry does not apply. On this plane, where a line is a great circle, *there are no parallel lines*. As it turns out, it is possible to work from this assumption to develop an entire geometry of definitions, postulates, and theorems that is logically consistent. It is called *spherical geometry*.

This leaves one other question concerning the undefined terms of spherical geometry: What is the meaning of *point*? Fundamentally, the meaning is the same as it is in Euclidean geometry. However, a slight adjustment to the definition of this term is necessary.

Each figure at the right shows two points on a sphere. In figure II, line ℓ is an axis of symmetry of the sphere.

 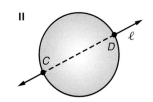

> **A GEOMETRY** in which there are no parallel lines is called an *elliptic* geometry. Spherical geometry is one type of elliptic geometry. Elliptic geometries also are called *Riemannian geometries* after Bernhard Riemann (1828–1866), a German mathematician who first understood that such geometries are possible. These are considered to be among the classical non-Euclidean geometries.

9. In Euclidean geometry, two distinct points determine a unique line. Is a unique line determined by points A and B? C and D?

10. In spherical geometry, points C and D are called a *polar pair*. What do you think is the meaning of this term?

11. Assume that the term *point* means either a single location or a polar pair. Is it now true that two distinct points determine a unique line? Draw sketches to illustrate your answer.

12. WRITING MATHEMATICS Do you think there are segments in spherical geometry? If so, what do they look like? Do you think there is a Segment Addition Postulate? Explain.

Geometry Workshop

Constructing Perpendicular and Parallel Lines

Think Back / Working Together

● **Work with a partner.**

1. On an index card, label points A and E in the lower corners. Fold point A onto point E to form $\overline{CD}$. Label points C and D.

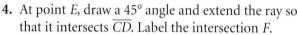

2. Describe the relationship of $\overline{CD}$ and $\overline{AE}$.

3. At point A, draw a 30° angle and extend the ray so that it intersects $\overline{CD}$. Label the intersection B.

4. At point E, draw a 45° angle and extend the ray so that it intersects $\overline{CD}$. Label the intersection F.

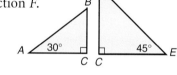

5. Cut along $\overline{CD}$, $\overline{AB}$, and $\overline{EF}$ to form two triangles.

6. Name the angle measures of each triangle.

Drafters use triangles like the ones you just created to form angles and parallel lines.

7. Use $\triangle ABC$. Set a ruler on your paper and place $\overline{AC}$ next to the ruler. Trace along $\overline{AB}$. Label this line ℓ.

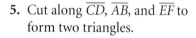

8. Slide the triangle along the ruler to a new position. Draw another line. Repeat several times. Label the lines and the angles formed by the lines as shown.

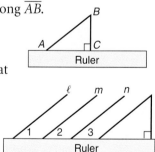

9. The ruler models a transversal. What is the relationship of $\angle 1$, $\angle 2$ and $\angle 3$? What is true about the measures of these angles?

10. What is the relationship of ℓ, m, and n? Explain your reasoning.

11. Would your answers to Questions 9 and 10 have been the same if you had used the 45° angle? Why?

Explore

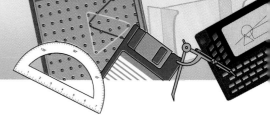

In Question 2 you may have described $\overline{CD}$ as a bisector of $\overline{AE}$. A line, ray, or segment that is perpendicular to a segment at its midpoint is a **perpendicular bisector** of the segment. The paper-folding construction of a perpendicular bisector you did in Think Back can also be completed using a compass and straightedge. Another construction where the perpendicular bisector is useful is the construction of a line perpendicular to a line through a point on the line.

TECHNOLOGY TIP

You can use geometry software to construct a perpendicular bisector. Draw a line segment, construct its midpoint, then construct a perpendicular line at the midpoint.

12. Draw line p containing point W. Choose a compass setting. Place the compass tip on W and draw an arc that intersects p to the right and left of W. Label points S and T where the arcs intersect p.

13. Adjust the compass setting so that it is slightly longer than SW. Place the compass tip on S and draw an arc above p. Move the tip to T and draw another arc above p. Label the point where the arcs intersect X.

THINK BACK

SW without the segment symbol means "the length of segment SW."

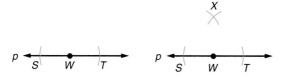

14. What is the final step in constructing the perpendicular?

You know that to draw a line you need at least two points. In the previous construction, one of the points of the perpendicular was located on the given line. In the next construction, the points which make the perpendicular are located one on each side of the given line. The construction of a line perpendicular to a line through a point not on the line is similar to the previous one.

15. Begin with line f and point A not on the line. Place the compass tip on A and draw an arc that intersects the line in two places. Label the intersection points B and C.

TECHNOLOGY TIP

You can use geometry software to construct a pair of lines that always remains perpendicular no matter how you move the original line.

16. Adjust the compass setting so it is greater than half of AB. Place the compass tip on B. Draw an arc below line f.

17. Then place the compass tip on C. Draw another arc below line f intersecting the first arc. Label the intersection D.

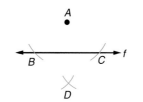

18. What is the final step in constructing the perpendicular?

3.5   **Geometry Workshop: Constructing Perpendicular and Parallel Lines   157**

Geometry Workshop

THINK BACK —

Recall the Converse of the Corresponding Angles Postulate states that if two lines are cut by a transversal so that corresponding angles are congruent, then the lines are parallel.

THINK BACK —

Refer to Lesson 2.7 to review how to copy an angle by construction.

In the Think Back you created parallel lines the way a drafter would create them, using a straightedge and drafter's triangles. Although the tools used to construct a line parallel to a line through a point not on the line are different from the drafter's, the Converse of Corresponding Angles Postulate serves as the justification for both.

19. Draw a line ℓ and a point A not on the line as shown below.

20. Draw a line n through point A and intersecting line ℓ. Label the point of intersection B.

21. Set your compass with the tip at B and the pencil at A. Swing an arc that intersects line ℓ. Label the intersection C.

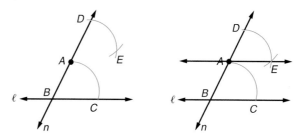

22. Using the same compass setting, set the tip at A and swing an arc that intersects line ℓ as shown above. Label the intersection D.

23. Set your compass tip at A and the pencil at C. Using this setting, set the tip of the compass at D and swing an arc that intersects the previous arc. Label the intersection E.

24. Draw $\overleftrightarrow{AE}$.

25. Which lines are parallel? Why are they parallel?

26. WRITING MATHEMATICS Compare and contrast the way drafters use triangles to create parallel lines and the way you use a straightedge and compass to construct parallel lines.

27. Describe how to do this construction using paper-folding.

Make Connections

● The lines, points, perpendicular lines, and parallel lines used in the previous constructions can be models of many real world objects.

28. Look around your classroom. Identify a model of a line and a point not on that line, a model of two perpendicular lines, a model of two parallel lines.

29. Explain how you might find the distance from a point to a line. Share your method with others.

30. Would each method described in Question 29 produce the same distance? If not, explain the problems the differences can create.

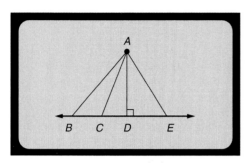

31. Measure segments AB, AC, AD, and AE in the drawing.

32. Which segment in Question 31 can you construct given a line and a point not on that line? Why?

TECHNOLOGY TIP

You can use geometry software to make this diagram. Construct line BE and any point D on that line. Construct a perpendicular to line BE at D. Construct point C anywhere between points B and D.

33. If a line parallel to $\overleftrightarrow{BE}$ passes through point A, how can you find the distance between the lines?

Summarize

● The following definitions summarize the results in Make Connections.

> The distance between a point and a line is the length of the perpendicular segment from the point to the line.
>
> The distance between two parallel lines is the distance between one line and any point on the other line.

34. GOING FURTHER Write definitions for the following.
   a. Distance between a point and a plane.
   b. Distance between two parallel planes.

35. THINKING CRITICALLY Graph line ℓ as shown. Construct a line perpendicular to line ℓ through point A. Construct a line parallel to line ℓ through point A.

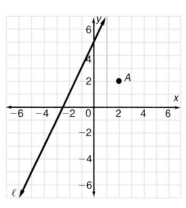

36. LOOKING AHEAD Use points $(-2, 1)$ and $(0, 5)$ to describe the slant of line ℓ. Choose two points on the line perpendicular to ℓ and describe the slant of that line. Do the same for the line parallel to line ℓ.

The use of a compass is traced back to sailors of the middle ages. Although many claim to be the inventor, no one knows who actually developed the compass. Today, ship navigators use a compass and mathematical calculations to determine an object's *bearing*. The **bearing of an object** is the measure, taken in clockwise direction, of an angle whose vertex is the observer's location, and whose sides are a ray along a *reference meridian* and a ray along the observer's line of sight to the object.

If the reference meridian is the longitudinal line through the geographic North Pole, the bearing is a *true bearing*. If the reference meridian is the longitudinal line through the magnetic pole, the bearing is a *magnetic bearing*. The magnetic poles, located approximately 1300 mi from the geographic poles, are represented by oval areas on a map.

Navigators use a technique called *resection* to locate a ship's position on a map. To use resection, select two objects which are known points on a map and find their magnetic bearing. Then change the magnetic bearing to the true bearing. By plotting these true bearings and finding the intersection, the ship's location is known.

Decision Making

1. Sketch the map on your paper.

2. The buoy is sighted at a true bearing of 80°. At the buoy, draw a north reference line and measure 80° clockwise from north. Draw a line to form the 80° angle with the reference line. Extend this line past the buoy. The ship is located somewhere on this line.

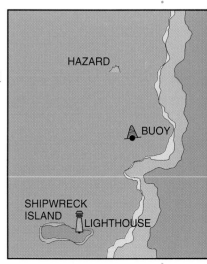

3. The lighthouse is sighted at a true bearing of 165°. Repeat the process in Question 2 for the lighthouse. Extend the line so it intersects the line drawn from the buoy. The intersection of the buoy line and the lighthouse line is the location of the ship.

4. The ship is headed on a true course of 38°. Draw the ship's course on your map.

5. What advice would you give to the captain? Explain your answer.

Explore

1. Graph the three equations on the same coordinate plane.

   a. $y = x + 3$ **b.** $y = \frac{1}{2}x + 3$ **c.** $y = 2x + 3$

2. List the equations in order of steepness beginning with the steepest.

3. Make a conjecture about the relationship of the coefficient of x and the steepness of the line.

4. Graph the three equations on the same coordinate plane.

   a. $y = -x + 3$ **b.** $y = -\frac{1}{2}x + 3$ **c.** $y = -2x + 3$

5. Describe the similarities and differences of the graphs of the lines in Questions 1 and 4.

6. Graph Questions 1a and 4a on the same coordinate plane.

7. Graph Questions 1b and 4b on the same coordinate plane.

8. Graph Questions 1c and 4c on the same coordinate plane.

9. Make a conjecture about the relationship of the coefficient of x and the direction the line slants.

10. Experiment with the conjectures you made in Questions 3 and 9. Summarize your conclusions.

Build Understanding

A line can be described by its steepness, or *slope*. For a given distance along a line, the slope of a line is the ratio of the number of units the line changes vertically (rise) to the number of units the line changes horizontally (run). On a coordinate plane, the slope of a line is the ratio of the change in the y-coordinates to the change in the x-coordinates of two points on the line.

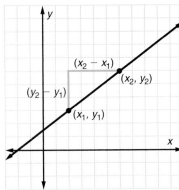

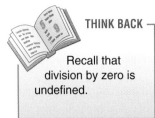

— **SLOPE OF A LINE** —

The slope m of a nonvertical line containing two points with coordinates (x_1, y_1) and (x_2, y_2) is

$$m = \frac{\text{rise}}{\text{run}} = \frac{\text{change in } y}{\text{change in } x} = \frac{y_2 - y_1}{x_2 - x_1}$$

In Explore, the lines in Question 1 increase from left to right and have a *positive slope*. The lines in Question 4 decrease from left to right and have a *negative slope*.

EXAMPLE 1

Find the slope of the line that passes through the given points. Check by graphing the points.

a. $A(4, 1)$, $B(2, -7)$ **b.** $C(-4, 3)$, $D(1, -3)$

c. $E(4, -1)$, $F(-2, -1)$ **d.** $G(-5, 3)$, $H(-5, -2)$

Solution

a. $m = \dfrac{-7 - 1}{2 - 4} = \dfrac{-8}{-2} = 4$

b. $m = \dfrac{-3 - 3}{1 - (-4)} = \dfrac{-6}{5}$

c. $m = \dfrac{-1 - (-1)}{-2 - 4} = \dfrac{0}{-6} = 0$

The line is horizontal and has a slope equal to zero.

d. $m = \dfrac{-2 - 3}{-5 - (-5)} = \dfrac{-5}{0}$

The line is vertical and has an undefined slope. ◄

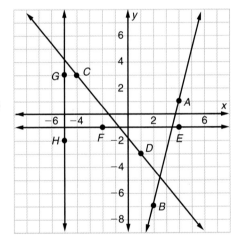

> The slope of any horizontal line is zero.
>
> The slope of any vertical line is undefined.

You can determine if two lines are parallel or perpendicular by examining their slopes. Lines that have the same slope are parallel. Lines whose slopes are *negative reciprocals* of each other are perpendicular. In Example 1 part a, every line with slope $m = 4$ is parallel to $\overleftrightarrow{AB}$. A line perpendicular to $\overleftrightarrow{AB}$ has slope $m = -\dfrac{1}{4}$.

> **POSTULATE 15** Two nonvertical lines are parallel if and only if their slopes are equal.
>
> **POSTULATE 16** Two nonvertical lines are perpendicular if and only if the product of their slopes is -1.

EXAMPLE 2

Graph each line on the same coordinate plane.

   a. line ℓ which passes through $(0, 2)$ and $m = \dfrac{3}{4}$

   b. line p is parallel to line ℓ and passes through $(1, -1)$

   c. line q is perpendicular to line ℓ and passes through $(2, 1)$

Solution

   a. Graph $(0, 2)$ then use $m = \dfrac{3}{4}$ to graph more points.

   b. The lines are parallel so $m = \dfrac{3}{4}$. Graph $(1, -1)$ and use the slope to graph more points.

   c. The lines are to be perpendicular so $m = -\dfrac{4}{3}$. Graph $(2, 1)$ and use the slope to graph more points. ◄

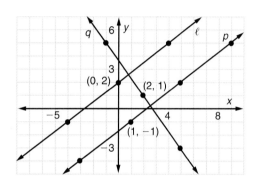

The concept of slope is used often in the real world.

EXAMPLE 3

BUILDING CODES The Uniform Federal Accessibility Standards outlines the standards for buildings to make them accessible to handicapped persons. One standard is the slope of a ramp be less than or equal to $\dfrac{1}{12}$.

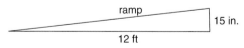

   a. Does the ramp shown meet the standard?

   b. If the ramp covers a horizontal distance of 20 ft, will the slope meet the standard?

   c. To equal a slope of $\dfrac{1}{12}$, how far must the ramp extend horizontally?

Solution

a. The rise (or vertical distance) is in inches so the horizontal distance must be converted to inches.

$$m = \frac{15}{144} = \frac{5}{48} \qquad \frac{\text{vertical distance}}{\text{horizontal distance}}$$

Since $\frac{5}{48} > \frac{1}{12}$, the ramp does not meet the standard.

b. $m = \frac{15}{240} = \frac{1}{16}$

The ramp will meet the standard because $\frac{1}{16} < \frac{1}{12}$.

c. Since the slope is $\frac{1}{12}$, use the proportion $\frac{15}{x} = \frac{1}{12}$. Solve for x to find the horizontal distance is 180 in. or 15 ft. ◄

TRY THESE

Use the graph at the right to match each line with its slope.

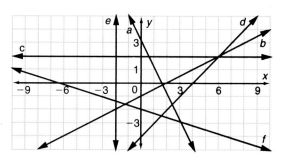

1. $m = 1$
2. $m = -\frac{1}{3}$
3. $m = 0$
4. $m = -2$
5. $m = \frac{1}{2}$
6. undefined slope

Tell whether two lines with the given slopes are parallel, perpendicular, or neither.

7. $m = 5, m = -\frac{1}{5}$
8. $m = -\frac{2}{3}, m = \frac{12}{6}$
9. $m = 1.25, m = \frac{5}{4}$
10. $m = 1$, undefined

Find the slope of the line through the given points.

11. $(1, -1), (3, -1)$
12. $(0, 1), (-2, 5)$
13. $(-2, 4), (-2, 6)$
14. $(-3, 5), (2, 3)$

15. **WRITING MATHEMATICS** Your friend needs help understanding how to find the slope of a line parallel to a line that passes through the points (2, 3) and (4, 5). Write an explanation that would help your friend.

Graph the line that passes through the given point and has the given slope.

16. $(0, -1)$ and $m = \frac{1}{2}$
17. $(-3, 1)$ and $m = -\frac{2}{3}$
18. $(4, -3)$ and $m = 0$

Visualize the line containing the points given. Describe the slope as positive, negative, zero, or undefined.

19. $(6, 4), (-2, -6)$
20. $(-2, 4), (5, -1)$
21. $(3, 2), (3, 6)$

22. CONSTRUCTION The pitch of a roof is the measure of the roof's steepness. Pitch is defined as the rise over the span as illustrated at the right. Find the pitch of the roof.

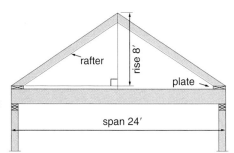

23. WRITING MATHEMATICS Write a paragraph to describe how the pitch of a roof and the geometric definition of slope are alike and how they are different.

PRACTICE

Find the slope of a line parallel to and of a line perpendicular to a line with the given slope.

24. $m = 1$

25. $m = -2$

26. $m = \dfrac{1}{3}$

Find the slope of the line passing through the given points.

27. $(5, 2), (8, 7)$ **28.** $(4, 1), (5, 0)$ **29.** $(-2, 6), (1, 2)$ **30.** $(7, -2), (2, -3)$

31. $(-2, 5), (-3, 5)$ **32.** $(-1, -3,), (5, -5)$ **33.** $(3, 2), (6, 8)$ **34.** $\left(-2, -4\right), \left(\dfrac{1}{2}, \dfrac{1}{2}\right)$

Graph the line that passes through the given point and has the given slope.

35. $(4, -2)$ and $m = \dfrac{1}{3}$ **36.** $(-5, 0)$ and slope is undefined   **37.** $(2, -1)$ and $m = -2$

38. Graph the line that passes through $(1, 1)$ and is parallel to a line with a slope of $\dfrac{2}{3}$.

39. Graph the line that passes through $(5, 0)$ and is perpendicular to a line with a slope of 0.

SAFETY ENGINEERING During the construction of a staircase, Pablo, a safety engineer, must be concerned that the *stair ratio* is within the safety standards. The stair ratio is the ratio of the riser height to the tread run (width) of the step.

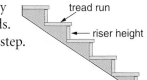

40. The maximum angle of a staircase should be 50°. A staircase with an angle of 50° is constructed with a riser height of 9.5 in. and a tread of 8 in. Pablo calculates the slope to determine if the staircase meets the safety code. What is the slope of this staircase?

41. The minimum angle of a staircase should be 20°. Does a staircase with a riser height of 6.5 in. and a tread width of 11 in. meet the safety standards? Justify your conclusions.

HIGHWAY CONSTRUCTION The grade of a highway is posted as a percent. A grade of 6% represents a slope of 0.06.

42. On a stretch of road, the starting elevation is 1200 ft and the ending elevation is 1400 ft over a horizontal distance of 1/2 mi. What is the grade of the highway to the nearest percent? (Recall: 1 mi = 5280 ft)

43. Over a horizontal distance of $\dfrac{3}{4}$ mi, a road rises to an elevation of 2500 ft. The grade of the road is 7%. What is the elevation to the nearest foot at the beginning of the section of highway?

EXTEND

LANDSCAPING The manufacturer of Green Grow lawnmowers recommends mowers not be operated on hills with a slope greater than 15° or a vertical change of no more than 2.5 ft for every 10 ft of horizontal change.

44. Write the slope of a vertical change of 2.5 ft for every 10 ft of horizontal change as a ratio. Use that ratio to write an inequality for the recommendation.

45. Make a scale drawing to determine if the slope in Exercise 44 is equivalent to a slope of 15°. An appropriate scale for the drawing is 1 cm = 1 ft.

THINK CRITICALLY

The *average rate of change* is the ratio of the change of a quantity to a period of time. Let the *x*-coordinate represent time and the *y*-coordinate represent quantity. Use the slope formula to compute the average rate of change.

46. ECOLOGY In 1966 the average person produced about 2.9 lb of trash per day. By 1988 this average had increased to 3.5 lb of trash. What was the average rate of change to the nearest hundredth?

47. POPULATION Find the average rate of change of the population of New York City if it was 7,071,639 people in 1980 and 7,322,564 people in 1990. Explain why you must round your answer to a whole number.

ALGEBRA AND GEOMETRY REVIEW

48. Identify the hypothesis and the conclusion of the statement: If the measure of an angle is 90°, then it is a right angle. Write the converse of the statement.

49. Find the slope of the line that passes through the points $(3, 7)$ and $(-2, 1)$.

50. Solve: $3(x + 4) = -24$ 51. If $m\angle A = 47°$, what is its complement? Explain.

PROJECT *Connection*   A counting device made of beads and string can be useful when you are estimating long distances by counting steps. Each bead represents a unit of measure. For example, if you know the number of steps you take in 100 m, then each bead might be equivalent to 100 m. You need some string and 20 beads that will fit on the string.

1. Cut a piece of string several inches longer than twice the length of all 20 beads.

2. String one bead and move it to the middle of the string.

3. String the second bead by feeding one end of the string through the bead from right to left and the other end through the bead from left to right. Slide the bead to the first bead.

4. Repeat Question 3 until all of the beads are on the string.

5. Tie the ends of the string in a knot, leaving about an inch of string between the last bead and the knot. The beads can be moved one at a time up and down the string.

3.7 Equations for Lines

Explore

1. What is the slope of lines a, b, and c?

2. At what point does each of the lines intersect the y-axis?

3. Line a is the graph of the equation $y = x + 1$. Line b is the graph of the equation $y = x + 4$. Line c is the graph of the equation $y = x - 2$. Look at each equation and your answers to Questions 1 and 2. Do you notice any relationships?

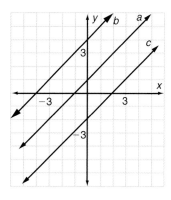

4. Use the relationship and what you know about the slopes of parallel lines to write an equation of a line that intersects the y-axis at $(0, 5)$ and is parallel to line a.

5. Use the relationship and what you know about the slopes of perpendicular lines to write an equation of a line that intersects the y-axis at $(0, -7)$ and is perpendicular to line a.

6. Test your equations by graphing the lines on the same coordinate plane as lines a, b, and c. Then summarize your conclusions about slope and y-intercept for equations that are solved for y.

SPOTLIGHT ON LEARNING

WHAT? In this lesson you will learn
- to write an equation of a line using slope and a given point.
- to write an equation of a line parallel or perpendicular to a given line.

WHY? Equations of lines can help you solve problems in chemistry, business, and quilting.

Build Understanding

When a linear equation is solved for y, it is in **slope-intercept form**.

> **SLOPE-INTERCEPT FORM**
>
> A linear equation of the form $y = mx + b$ is in slope-intercept form, where m is the slope and b is the y-intercept of the graph.

A line may also be represented by a linear equation in **standard form**.

> **STANDARD FORM**
>
> A linear equation of the form $Ax + By = C$, where A, B, and C are integers and A and B are not both zero is in standard form.

Slope-intercept form is most suitable for graphing.

THINK BACK

The y-coordinate of the point where a graph crosses the y-axis is the y-intercept. The x-coordinate of the point where a graph crosses the x-axis is the x-intercept.

EXAMPLE 1

Graph each equation using the slope-intercept form.

a. $y = -3x + 2$ **b.** $2x - 3y = 6$

Solution

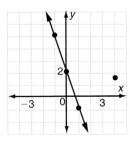

a. Determine the slope and y-intercept from the equation.

    y-intercept: $b = 2$

    slope: $m = -3$

Place a point at the y-intercept. Then use the slope to find more points on the line.

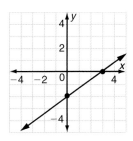

b. $2x - 3y = 6$ Solve for y.

     $-3y = -2x + 6$

       $y = \frac{2}{3}x - 2$

The y-intercept is -2 and the slope is $\frac{2}{3}$. ◄

All the points on a horizontal line have the same y-coordinate and any real number for the x-coordinate. So, the equation is in the form $y = b$ where b is the y-intercept and $m = 0$.

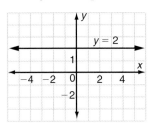
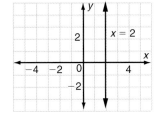

Horizontal   Vertical

All the points on a vertical line have the same x-coordinate and any real number for the y-coordinate. So, the equation of a vertical line is in the form $x = c$, where c is the x-coordinate of all the points on the line.

CHECK UNDERSTANDING

Make a list of three points that are on the horizontal line $y = 2$. What do all of the points have in common? How is this related to the equation of the line?

EXAMPLE 2

COMPUTERIZED QUILT DESIGN The *diagonal bar* is a basic design unit in quilting. To complete a quilt, the design unit is repeated until the total area of the repeated units equals the desired area of the finished quilt. Sandra uses a quilt designing software program to draw this unit. The diagonal bar design unit she is using has an area of 6 in.^{2} Find all the equations she needs to enter for the diagonal bar design at the right.

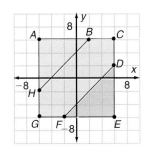

$\overleftrightarrow{AC}$	$y = 6$	Horizontal line that intersects the y-axis at 6
$\overleftrightarrow{GE}$	$y = -6$	Horizontal line that intersects the y-axis at -6
$\overleftrightarrow{AG}$	$x = -6$	Vertical line that intersects the x-axis at -6
$\overleftrightarrow{CE}$	$x = 6$	Vertical line that intersects the x-axis at 6
$\overleftrightarrow{HB}$	$y = x + 4$	$m = 1$ and $b = 4$
$\overleftrightarrow{FD}$	$y = x - 4$	$m = 1$ and $b = -4$ ◀

You can use the slope-intercept form to write the equation of a line even when you do not know the y-intercept or the slope. If you are given two points, find the slope. Then substitute the coordinates of one point and the slope into the slope-intercept form to find the y-intercept.

EXAMPLE 3

Determine the equation of the line passing through $(-4, -2)$ and $(4, 4)$.

Solution

$$m = \frac{4 - (-2)}{4 - (-4)} = \frac{6}{8} = \frac{3}{4} \qquad \text{Find the slope.}$$

Choose one point. Use the slope-intercept form to find b.

$$y = mx + b$$
$$4 = \frac{3}{4}(4) + b \qquad \text{Use } x = 4, y = 4, \text{ and } m = \frac{3}{4}.$$
$$b = 1 \qquad \text{Solve for } b.$$

The equation of the line is $y = \frac{3}{4}x + 1$. ◀

You can determine an equation of a line that is parallel or perpendicular to another line if you know the coordinates of a point on the line.

EXAMPLE 4

Given the line $y = -3x + 8$, determine an equation of a line

a. parallel to the given line and with a y-intercept of 4.

b. perpendicular to the given line and containing the point $(2, 0)$.

Solution

a. Use $m = -3$ and $b = 4$. An equation is $y = -3x + 4$.

b. Use $m = \frac{1}{3}$. Use the slope-intercept form to find b.

$$y = mx + b$$
$$0 = \frac{1}{3}(2) + b \qquad \text{Use } x = 2, y = 0, \text{ and } m = \frac{1}{3}.$$
$$-\frac{2}{3} = b$$

Since $b = -\frac{2}{3}$, an equation is $y = \frac{1}{3}x - \frac{2}{3}$. ◀

Match the equation to the line on the graph.

1. $y = x + 2$ 2. $x = -2$ 3. $y = 4$

4. $y = -\frac{2}{3}x$ 5. $y = -3x + 1$ 6. $y = \frac{1}{2}x - 1$

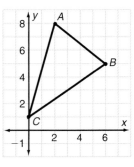

Graph each line. Name the slope and the *y*-intercept.

7. $y = 2x - 2$ 8. $x + y = 6$ 9. $2x - 3y = 6$

10. $y = -1$ 11. $2y + 4 = -x$ 12. $x = 6$

Write an equation of the line that passes through the two points.

13. $(0, 0), (1, -5)$ 14. $(1, -1), (2, 1)$ 15. $(-3, 3), (-6, 7)$

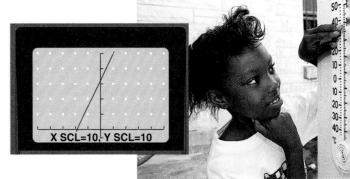

16. Write an equation of each line which forms the triangle at the right.

17. Write an equation of the line perpendicular to $y = -2x - 4$ and having the same *y*-intercept.

18. Write an equation of the line perpendicular to $y = -\frac{1}{5}x - 4$ and passing through the point $(-3, 2)$.

19. Write an equation of the line perpendicular to $y = 12$ and passing through the point $(7, 12)$.

20. WRITING MATHEMATICS The graph shows the relationship between Fahrenheit and Celsius. The equation of the line in terms of F and C is $F = \frac{9}{5}C + 32$. What is the significance of the *y*-intercept?

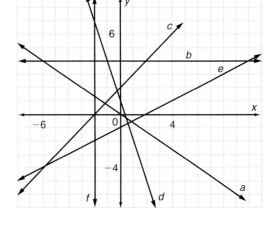

PRACTICE

Graph each line. Name the slope and the *y*-intercept of each.

21. $y = 3x - 1$ 22. $y = x$ 23. $y = -2$ 24. $x = -3$

25. $\frac{1}{4}x + y = 1$ 26. $x - y = 3$ 27. $4x + 2y = 8$ 28. $5x - 3y = 6$

Write an equation of the line that passes through the given points.

29. $(0, 0), (2, 1)$ 30. $(1, -5), (-3, 3)$ 31. $(2, 7), (-4, -14)$

32. $(2, 8), (1, 8)$ 33. $(-2, 8), (-4, 13)$ 34. $(5, 3), (5, -3)$

35. WRITING MATHEMATICS Make a list of all the combinations of information that will allow you to use the slope-intercept form to write the equation of a line. Write a description of how to write the equation for each combination you list.

Write an equation of the line that is *parallel* to the given line and passes through the given point. Then write an equation of the line that is *perpendicular* to the given line and passes through the given point.

36. $y = 4x - 3; (0, 0)$ **37.** $x + y = -5; (5, -2)$ **38.** $2x - 3y = 6; (-6, -4)$

GRAPHIC DESIGN Logos can be created on a coordinate grid to ensure symmetry and other geometric characteristics. Write the equations that could be used to form each design.

39.
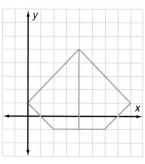

X SCL = 1, Y SCL = 1

40.
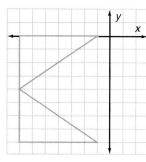

X SCL = 2, Y SCL = 2

41.
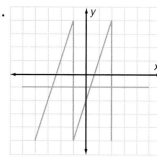

X SCL= 3, Y SCL= 3

EXTEND

LOGOS Use the equations given, any lines parallel or perpendicular to the given line, and any vertical line to design a logo. Make a list of all the equations used.

42. $y = x, \quad y = -x, \quad x = 0$ **43.** $y = 4, \quad y = -2, \quad y = 2x + 1$

THINK CRITICALLY

A **scatterplot** is a graph of ordered pairs of data used to determine the relationships among data. If the data appears to be linear, a line of best fit is drawn that best approximates the trend of the data. The equation of the line of best fit can be used to make predictions.

X SCL=20, Y SCL=10

44. SPORTS The lengths of the winning men's discus throws in the Olympics from 1900 to 1992 are shown on the scatterplot. The line of best fit is $y = 1.13x + 124.88$. If $x = 0$ corresponds to 1900, predict, to the nearest tenth of a foot, the winning throw for the year 2000.

45. WRITING MATHEMATICS What significance does the slope have in this graph?

ALGEBRA AND GEOMETRY REVIEW

46. How many lines of symmetry does a rectangle that is not a square have? a circle?

47. Find the distance between the two points: $(4, -5), (-2, 8)$

48. Find the next two numbers in the pattern: 1, 2, 4, 7, 11, 16, . . .

3.8 Focus on Reasoning

Coordinate Proofs

Discussion

SPOTLIGHT ON LEARNING

WHAT? In this lesson you will learn
• to write coordinate proofs.

WHY? Coordinate proofs can help you determine whether your conjectures are true or false.

A group of students were looking at the graph below and trying to prove that $\overline{SO} \cong \overline{TO}$.

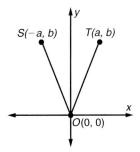

Shawn, Bill, and Lupita all thought it made sense that the segments are congruent. However, they disagreed about how to prove this fact.

Bill said that the slope of each segment is $\dfrac{b}{a}$, so the segments are congruent.

1. Explain why Bill's statement that each slope is $\dfrac{b}{a}$ is incorrect.

2. What is the slope of each segment? How are these slopes related?

3. Suppose that the slopes of the segments were equal to each other. Would this prove that $\overline{SO} \cong \overline{TO}$? Explain.

Lupita chose several values for a and b. She made a table and in each case, $SO = TO$. So Lupita decided that $\overline{SO} \cong \overline{TO}$ by inductive reasoning.

a	b	TO	SO
1	2	$\sqrt{(1-0)^2 + (2-0)^2} = \sqrt{5}$	$\sqrt{(-1-0)^2 + (2-0)^2} = \sqrt{5}$
2	3	$\sqrt{(2-0)^2 + (3-0)^2} = \sqrt{13}$	$\sqrt{(-2-0)^2 + (3-0)^2} = \sqrt{13}$
3	4	$\sqrt{(3-0)^2 + (4-0)^2} = 5$	$\sqrt{(-3-0)^2 + (4-0)^2} = 5$
4	5	$\sqrt{(4-0)^2 + (5-0)^2} = \sqrt{41}$	$\sqrt{(-4-0)^2 + (5-0)^2} = \sqrt{41}$

4. What formula did Lupita use to determine the values of TO and SO in her table?

5. Do you agree with Lupita's reasoning? Explain.

Shawn wrote the following to prove that $\overline{SO} \cong \overline{TO}$.

$$TO = \sqrt{(a-0)^2 + (b-0)^2}$$
$$= \sqrt{a^2 + b^2}$$
$$SO = \sqrt{(-a-0)^2 + (b-0)^2}$$
$$= \sqrt{(-a)^2 + b^2}$$
$$= \sqrt{a^2 + b^2}$$

Therefore, $SO = TO$ and $\overline{SO} \cong \overline{TO}$.

6. Justify each statement in Shawn's proof.

7. How is Shawn's proof similar to Lupita's "proof"? How is it different?

Build a Case

● Sometimes you must prove facts about lines or segments on a coordinate plane. In a situation like this, you justify your conclusions using concepts from coordinate geometry such as slope, equations of lines, the distance formula, and the midpoint formula. This type of proof is called a *coordinate proof*.

In Discussion, each student used coordinate geometry. Bill's use of slope was not only incorrect, but also inappropriate.

8. In what coordinate proof situations do you think the use of slope is appropriate?

Both Lupita and Shawn correctly applied the coordinate plane distance formula. The difference is that Lupita only showed the segments to be congruent in a limited number of cases. Shawn used deductive reasoning and wrote a proof that applies to all cases.

Use the graph for Questions 9–12.

9. Write a coordinate proof that $\overline{NP} \cong \overline{QO}$.

10. What is the slope of $\overline{NP}$? of $\overline{QO}$?

11. Write a coordinate proof that $\overline{NP} \perp \overline{QO}$.

12. What are the coordinates of point R? Write a convincing argument to justify your answer.

COMMUNICATING ABOUT GEOMETRY

Review with your partner the coordinate geometry concepts of slope, equation of a line, distance formula, and midpoint formula. Include any formulas associated with these concepts.

Can you think of any other concepts from algebra that can be used when doing a coordinate proof?

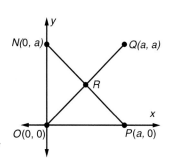

EXTEND AND DEFEND

You can use coordinate methods to prove a theorem, even though you are not given a figure on a coordinate plane. For Questions 13–15, use the following theorem.

> **THEOREM 3.10**
>
> If two lines are parallel to a third line, then they are parallel to each other.

13. Copy and complete this coordinate proof of the theorem.

    Given $\ell_1 \parallel \ell_3, \ell_2 \parallel \ell_3$

    Prove $\ell_1 \parallel \ell_2$
    Plan for Proof Let m_1, m_2, and m_3 be the slopes of ℓ_1, ℓ_2, and ℓ_3, respectively.

    Proof

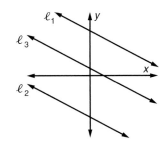

Statements	Reasons
1. ?	1. Given
2. $m_1 = m_3$, $m_2 = m_3$	2. ?
3. $m_1 = m_2$	3. ?
4. ?	4. If the slopes of two nonvertical lines are equal, then the lines are parallel.

In a *synthetic proof*, also called a *traditional proof*, you justify your conclusions using properties of algebra and geometry, without any reference to coordinates.

14. Copy and complete this synthetic proof of the theorem.

    Given $\ell_1 \parallel \ell_3, \ell_2 \parallel \ell_3$

    Prove $\ell_1 \parallel \ell_2$

    Proof

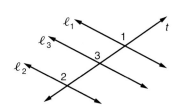

Statements	Reasons
1. $\ell_1 \parallel \ell_3$, $\ell_2 \parallel \ell_3$	1. ?
2. ? ?	2. If two parallel lines are cut by a transversal, then corresponding angles are congruent.
3. $\angle 1 \cong \angle 2$	3. ?
4. $\ell_1 \parallel \ell_2$	4. ?

15. WRITING MATHEMATICS Compare the coordinate proof in Question 13 to the synthetic proof in Question 14. How are the proofs alike? How are they different?

Use the following theorem for Questions 16 and 17.

> **THEOREM 3.11**
>
> If two coplanar lines are each perpendicular to a third line, then they are parallel to each other.

16. Write a coordinate proof.　　　　**17.** Write a synthetic proof.

Write a coordinate proof to show that $\overline{JK} \cong \overline{LM}$ and that $\overline{JK} \parallel \overline{LM}$.

18.

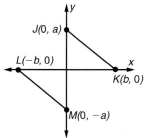

19.

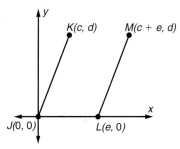

Refer to the figure at the right for Questions 20 and 21.

20. Find the slope of $\overline{PR}$ and the slope of $\overline{QS}$. Prove $\overline{PR} \perp \overline{QS}$.

21. Use the equations of $\overleftrightarrow{PR}$ and $\overleftrightarrow{QS}$ to prove $\overleftrightarrow{PR} \perp \overleftrightarrow{QS}$.

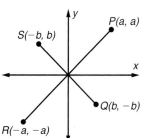

Now You See It
Now You Don't

WITHOUT PARALLEL?
Do you see the thin black segments as being parallel? Explain how to use the postulates and theorems of this chapter to demonstrate that these segments are parallel.

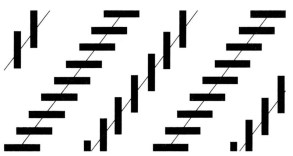

The ancient Greeks were aware of optical illusions involving parallel lines and segments. They used their knowledge in architecture to create the illusion that columns were parallel when they actually were not.

3.9 Vectors

Explore

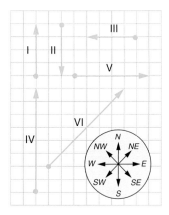

- On the grid at the right, figure I represents a wind velocity of 12 mi/h due north.

  1. Using this figure as a point of reference, what do you think figures II through VI represent?

  2. How are figures I and IV alike? How are they different?

  3. How are figures I and III alike? How are they different?

Build Understanding

- The wind velocities that you examined in Explore are examples of *vectors*. A **vector** is a quantity that has both a size and a direction. The size is called the **magnitude** of the vector.

Vectors are directed segments. The beginning point is called its **initial point**, and the ending point its **terminal point**. A vector is named using the initial and terminal points or a single letter. The figure at the right shows vector AB. In symbols, you write $\overrightarrow{AB}$ or $\vec{a}$

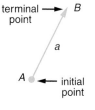

Parallel vectors are vectors that have either the same direction or opposite directions. Parallel vectors do not necessarily have the same magnitude. For example, all the vectors in the figure at the right are parallel.

For two vectors to be **equal vectors**, they must have the same magnitude and the same direction. The only equal vectors at the right are $\vec{u}$ and $\vec{v}$.

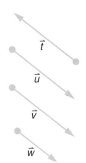

When a vector is on a coordinate plane, it can be represented by an ordered pair. For a vector with initial point $A(x_1, y_1)$ and terminal point $B(x_2, y_2)$, you determine this ordered pair as follows.

$$\overrightarrow{AB} = (\text{change in } x, \text{change in } y)$$

$$= (x_2 - x_1, y_2 - y_1)$$

COMMUNICATING ABOUT GEOMETRY

In the figure at the right, explain why $\vec{t}$ and $\vec{u}$ are not equal vectors.

To find the magnitude of $\overrightarrow{AB}$, denoted $|\overrightarrow{AB}|$, you use the coordinate plane distance formula.

$$|\overrightarrow{AB}| = \sqrt{(x_2 - x_1)^2 + (y_2 - y_1)^2}$$

EXAMPLE 1

Given $\overrightarrow{PQ}$, find its ordered pair representation and its magnitude.

Solution

The initial point (x_1, y_1) is $P(-4, 5)$.
The terminal point (x_2, y_2) is $Q(2, 1)$.

$$x_2 - x_1 = 2 - (-4) = 6$$
$$y_2 - y_1 = 1 - 5 = -4$$

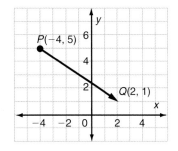

Use these values to find $\overrightarrow{PQ}$ and $|\overrightarrow{PQ}|$.

$$\overrightarrow{PQ} = (x_2 - x_1, y_2 - y_1) \qquad\qquad |\overrightarrow{PQ}| = \sqrt{(x_2 - x_1)^2 + (y_2 - y_1)^2}$$
$$= (6, -4) \qquad\qquad\qquad\qquad\qquad = \sqrt{6^2 + (-4)^2}$$
$$\qquad\qquad\qquad\qquad\qquad\qquad\qquad = \sqrt{36 + 16} = \sqrt{52} \approx 7.2$$

The ordered pair that represents $\overrightarrow{PQ}$ is $(6, -4)$.
The magnitude of $\overrightarrow{PQ}$ is $\sqrt{52}$, or approximately 7.2. ◄

CHECK UNDERSTANDING

Draw $\overrightarrow{PQ}$ from Example 1 on a set of coordinate axes. On the same axes, draw $\overrightarrow{RS}$ with initial point $R(0, 0)$ and terminal point $S(6, -4)$. What is the relationship between $\overrightarrow{PQ}$ and $\overrightarrow{RS}$?

Sometimes you can solve a problem by finding a sum of two vectors. Example 2 illustrates the *head-to-tail method of vector addition*.

EXAMPLE 2

AVIATION An airplane is flying on a northwest heading at an air speed of 200 mi/h. The wind is blowing from the east at 40 mi/h. How does the wind affect the ground speed and actual course of the airplane?

Solution

In the diagram, $\vec{u}$ represents the air speed and heading of the airplane, and $\vec{v}$ represents the wind velocity.

The vectors are drawn so the terminal point (head), of $\vec{u}$ is the initial point (tail), of $\vec{v}$. Since 200 mi/h is five times 40 mi/h, the length of $\vec{u}$ is five times the length of $\vec{v}$. The ground speed and course of the airplane are represented by $\vec{w}$.

The diagram shows that the wind increases the ground speed of the airplane and changes its actual course slightly to the west of its heading. ◄

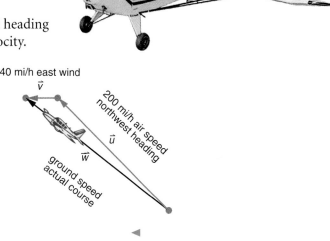

When two vectors are on a coordinate plane, you can find their sum using the *algebraic method of vector addition*. That is, the sum of $\vec{u} = (a_1, b_1)$ and $\vec{v} = (a_2, b_2)$ is defined as follows.

$$\vec{u} + \vec{v} = (a_1, b_1) + (a_2, b_2) = (a_1 + a_2, b_1 + b_2)$$

EXAMPLE 3

Given $\vec{w} = (1, 5)$ and $\vec{z} = (8, 3)$, find $\vec{w} + \vec{z}$.

Solution

$\vec{w} + \vec{z} = (1, 5) + (8, 3) = (1 + 8, 5 + 3) = (9, 8)$ ◄

You also could find the sum in Example 3 using the *parallelogram method of vector addition*. Place each vector on a set of coordinate axes with its initial point at the origin. Draw segments to complete a parallelogram. The red vector placed on the diagonal of this parallelogram is the sum.

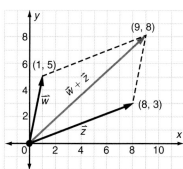

Perpendicular vectors are vectors whose directions are perpendicular. Perpendicular vectors do not necessarily intersect. In the figure at the right, for example, $\vec{s}$ and $\vec{t}$ are perpendicular.

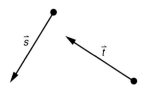

For vectors $\vec{u} = (a_1, b_1)$ and $\vec{v} = (a_2, b_2)$, the **dot product** is defined as follows.

$$\vec{u} \cdot \vec{v} = (a_1, b_1) \cdot (a_2, b_2) = a_1 a_2 + b_1 b_2$$

The dot product of two vectors is a real number, not another vector. Two nonzero vectors are perpendicular if and only if their dot product is 0.

EXAMPLE 4

Determine if each pair of vectors is perpendicular.

a. $\vec{p} = (3, 4)$; $\vec{q} = (12, -9)$ **b.** $\vec{q} = (12, -9)$; $\vec{r} = (5, 2)$

Solution

a. $\vec{p} \cdot \vec{q} = 3(12) + 4(-9) = 36 + (-36) = 0$
The dot product is 0, so $\vec{p}$ and $\vec{q}$ are perpendicular.

b. $\vec{q} \cdot \vec{r} = 12(5) + (-9)(2) = 60 + (-18) = 42$
The dot product is not 0, so $\vec{q}$ and $\vec{r}$ are not perpendicular. ◄

TRY THESE

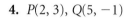

Refer to the coordinate axes at the right to answer each question. Justify each response.

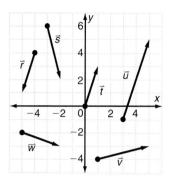

1. Name a pair of equal vectors.

2. Name a pair of vectors that are parallel but not equal.

3. Name a pair of perpendicular vectors.

A vector has initial point P and terminal point Q as given. Find the ordered pair representation of $\overrightarrow{PQ}$ and the magnitude of $\overrightarrow{PQ}$.

4. $P(2, 3)$, $Q(5, -1)$ 5. $P(-5, 3)$, $Q(2, 7)$ 6. $P(2, 6)$, $Q(-1, -1)$

7. **RECREATION** When a change in position is identified by a direction as well as a distance, it is a vector called a *displacement*. Suppose a hiker sets out from a ranger station and walks 12 mi due west to a lake. At the lake, the hiker turns and walks 8 mi southwest to a campsite. Draw a vector diagram to show the path the hiker walked. Use the diagram to describe the displacement of the hiker.

8. **WRITING MATHEMATICS** How is a vector arrow like a ray? How is it different from a ray? How is a vector arrow like a segment? How is it different from a segment?

Let $\vec{t} = (3, -2)$, $\vec{u} = (4, 1)$, $\vec{v} = (-3, -5)$, and $\vec{w} = (6, 0)$. Find each sum using the algebraic method. Then verify your answer using the parallelogram method.

9. $\vec{t} + \vec{u}$ 10. $\vec{u} + \vec{v}$ 11. $\vec{t} + \vec{v}$ 12. $\vec{u} + \vec{w}$

Let $\vec{a} = (3, 1)$, $\vec{b} = (2, -6)$, $\vec{c} = (-3, 1)$, and $\vec{d} = (0, 4)$. For each pair of vectors, find the dot product. Then tell whether the vectors are *perpendicular* or *not perpendicular*.

13. $\vec{a}, \vec{b}$ 14. $\vec{b}, \vec{c}$ 15. $\vec{a}, \vec{c}$ 16. $\vec{b}, \vec{d}$

PRACTICE

Vector JK has initial point $J(-4, 6)$ and terminal point $K(1, 0)$. Vector LM has initial point $L(2, -1)$ and terminal point $M(0, 6)$. Find each quantity.

17. $|\overrightarrow{JK}|$ 18. $|\overrightarrow{LM}|$ 19. $\overrightarrow{JK} + \overrightarrow{LM}$ 20. $\overrightarrow{JK} \cdot \overrightarrow{LM}$

A vector has initial point G and terminal point H as given. Find the ordered pair representation of $\overrightarrow{GH}$ and the magnitude of $\overrightarrow{GH}$.

21. $G(4, 1)$, $H(8, 4)$ 22. $G(-1, 4)$, $H(4, -8)$ 23. $G(3, -3)$, $H(1, -3)$ 24. $G(2, 3)$, $H(2, -4)$

25. $G(5, -3)$, $H(3, 2)$ 26. $G(-6, 2)$, $H(-4, 7)$ 27. $G(1, 1)$, $H(0, -3)$ 28. $G(0, 1)$, $H(5, 1)$

29. **BOATING** A boat is traveling due north at a speed of 10 mi/h. The water current is flowing at a speed of 4 mi/h from the southwest. Sketch a vector diagram to show the boat's actual speed and track in the water.

Let $\vec{q} = (0, 2)$, $\vec{r} = (-3, 2)$, $\vec{s} = (-4, -2)$, and $\vec{t} = (4, 0)$. **Find each sum using the algebraic method. Then verify your answer using the parallelogram method.**

30. $\vec{q} + \vec{r}$ 31. $\vec{r} + \vec{s}$ 32. $\vec{q} + \vec{s}$

33. $\vec{r} + \vec{t}$ 34. $\vec{q} + \vec{t}$ 35. $\vec{s} + \vec{t}$

36. **WRITING MATHEMATICS** Summarize the three methods of vector addition taught in this lesson. Give an example of each that is different from the example in the lesson.

Let $\vec{u} = (0, 4)$, $\vec{v} = (-1, 5)$, $\vec{w} = (10, 2)$, and $\vec{z} = (-5, 0)$. **For each pair of vectors, find the dot product. Determine if the vectors are perpendicular.**

37. $\vec{u}, \vec{v}$ 38. $\vec{u}, \vec{w}$ 39. $\vec{u}, \vec{z}$ 40. $\vec{v}, \vec{w}$ 41. $\vec{v}, \vec{z}$ 42. $\vec{w}, \vec{z}$

43. **PHYSICS** A *force* is a vector that can affect the motion of an object. For example, the diagram at the right represents a top view of two people pulling a box. Person *A* is pulling with a force of 40 lb east. At the same time, person *B* is pulling with a force of 50 lb south. On a set of coordinate axes, make a scale drawing to show the addition of the two forces. Use the drawing to describe the direction in which the box will move and the magnitude of the two forces together.

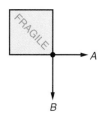

EXTEND

The *opposite* of a vector $\vec{v}$ is the vector $-\vec{v}$, which has the same magnitude as $\vec{v}$, but has the opposite direction.

44. Let $\vec{q} = (4, 3)$, $\vec{r} = (-2, -1)$, $\vec{s} = (3, -5)$, and $\vec{t} = (6, 0)$. On a set of coordinate axes, draw each of these vectors and their opposites.

45. Refer to your answers to Exercise 44. Let $\vec{v} = (x, y)$. What is the ordered pair representation of $-\vec{v}$?

46. The difference of vectors $\vec{u}$ and $\vec{v}$, denoted $\vec{u} - \vec{v}$, is defined as $\vec{u} + (-\vec{v})$. Let $\vec{z} = (3, -1)$ and $\vec{w} = (-2, -4)$. Find $\vec{z} - \vec{w}$ and $\vec{w} - \vec{z}$.

47. **WRITING MATHEMATICS** Compare vector subtraction to the subtraction of real numbers. How are the operations alike? How are they different?

THINK CRITICALLY

Explain how each situation is related to vector addition.

48. VEHICLES A wagon is drawn by two horses, each pulling with equal force in the same direction.

49. RECREATION Two teams in a tug of war each pull on a rope with equal force, but in opposite directions.

GARDENING **When you push a lawn mower like the one shown, you exert a force F in the direction of the handle. This force is the sum of two forces: *horizontal component $\vec{x}$* and *vertical component $\vec{y}$*.**

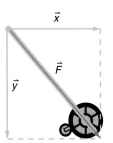

50. Which component force moves the lawn mower forward?
51. What adjustment to the position of the handle would increase the forward force?

ALGEBRA AND GEOMETRY REVIEW

52. Are the lines with equations $y = 2x - 5$ and $2x + y = 5$ perpendicular? Explain.

53. Let $\vec{r} = (5, -2)$ and $\vec{s} = (-2, 5)$. Is $\vec{r}$ perpendicular to $\vec{s}$? Explain.

54. Find the mean, median, mode, and range of these data: 45, 38, 96, 22, 58

55. Find the complement and the supplement of an angle whose measure is 27°.

PROJECT *Connection*   Estimating angle measures using common hand positions can be useful in navigation. Since hand sizes vary greatly, work in a group of four students to find the measure for each student's hand.

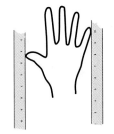

1. Step off or measure a distance of 100 m. One student holds a meter stick upright at one end of the distance and you stand at the other end. Extends arm with your hand positioned with fingers spread.

2. Sight along the edge of your hand to align the meter stick with the left side of the extended hand. Another student standing 100 m from you moves a second meter stick to align it with the other side of your hand.

3. When aligned, measure the distance between the meter sticks.

4. To find the angle A, use the following formula $A \approx \dfrac{17.8W}{D}$. Label your hand measure as W. D is the distance (100 m in this case) and W is the answer from Question 3.

5. If the distance between the objects representing the rays of the angle is the width of your hand or smaller, use a percent of W. Use the hand positions shown and their accompanying percents to adjust the value to use in the formula for W. If the distance is wider than your hand, use multiple hand widths to adjust the value of W.

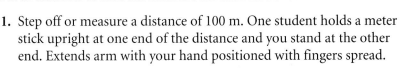

100%   60%   45%   33%   25%   10%

Using Vector Diagrams

A vector diagram can be used to obtain more detailed information about ways in which winds and water currents affect the motion of airplanes and ships.

Problem

A boat is traveling on a heading of 135°. The boat speed is 10 mi/h. The water current is flowing at a speed of 2 mi/h, and its direction is 45°. What are the ground speed and true course of the boat?

Explore the Problem

The *heading* of a boat is the direction you steer a boat. A heading is the measure of the angle the direction vector forms with north, taken in a clockwise direction. A *compass rose* like the one shown is marked with headings from 0° through 360°.

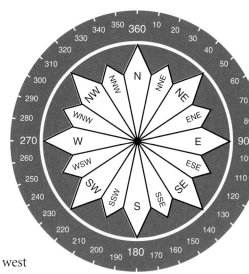

1. Give the heading of each compass direction.

   a. east
   b. west
   c. northwest
   d. south southeast

2. Name the compass direction for each heading.

   a. 45°   b. 180°   c. 315°   d. 247.5°

To draw a vector diagram for the problem above, you need a ruler and a protractor.

3. At the center of a piece of paper, draw point *A*. With point *A* as the initial point, draw a vector of any length pointing upward to represent north. Label the terminal point N_1.

4. What is the speed of the boat? of the current?.

5. To draw a vector that represents the heading of the boat, what angle should you use?

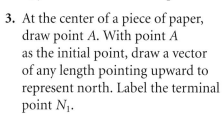

6. Let 1 cm represent 2 mi. With point A as the initial point, draw a vector to represent the velocity of the boat. Label the terminal point B. What is the length of $\overline{AB}$?

7. With point B as the initial point, draw another vector to represent north. Label the terminal point N_2.

8. What angle represents the direction of the current?

9. With point B as the initial point, draw a vector to represent the velocity of the current. Label the terminal point C. What is the length of $\overline{BC}$?

10. Draw $\overrightarrow{AC}$. Measure the angle $\overrightarrow{AC}$ forms with $\overrightarrow{AN_1}$. What does this represent?

11. The length of $\overrightarrow{AC}$ represents the ground speed of the boat, which is its speed relative to the land beneath the water. Use your ruler and the scale of your diagram to determine the ground speed?

PROBLEM SOLVING PLAN

- Understand
- Plan
- Solve
- Examine

COMMUNICATING ABOUT GEOMETRY

In Questions 3–10, you solved the boat problem using the head-to-tail method of vector addition. How could you use the parallelogram method to solve the problem?

Investigate Further

You can use a vector diagram to work backwards.

A captain wants to sail due east. The planned ship speed is 7 mi/h. The direction of the water current is 135°, and its speed is 3 mi/h. What heading should the captain set? What will be the ground speed of the ship?

12. Draw a vector diagram to represent the direction and speed of the water current. Use a scale of 1 cm = 2 mi.

13. What should be the angle of a vector that represents the true course?

14 What information do you need to draw the vector that represents the true course?

15. What vector would represent the ship's speed?

16. With your compass set at 3.5 cm, draw an arc that intersects the true course. Label the point of intersection C. What vector represents the ship's speed?

17. Determine the heading the captain should set and the resulting ground speed of the ship.

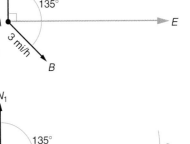

Apply The Strategy

18. WRITING MATHEMATICS What is the difference between the *heading* of a boat and its actual *course*? What is the difference between *boat speed* and *ground speed*?

Solve each problem by drawing a vector diagram.

19. AVIATION An airplane is flying on a heading of 315° at an air speed of 120 mi/h. The direction of the wind is 270°, and the wind speed is 30 mi/h. What are the actual course and ground speed of the airplane?

20. SPORTS A swimmer is heading due west across a river at a speed of 2 km/h. The water current is flowing due south at a speed of 1 km/h. What are the actual course and ground speed of the swimmer?

21. AVIATION An airplane pilot wants to fly on a course with direction 28° at an air speed of 480 km/h. A steady wind is blowing from the north at 100 km/h. What heading should the pilot set? What will be the ground speed of the airplane?

22. RECREATION Two people in a canoe are traveling across a river with a current flowing due east at 2 km/h. They want to land in a spot that is 45° from their starting point. In still water, they can paddle 2.1 km/h. What heading should they take? What will be their ground speed?

23. NAVIGATION The current of a river flows due west. To cross the river on a course due north, the pilot of a ferry must use a heading of 45°. The ferry speed is 1.8 m/s. What is the speed of the current?

24. THINK CRITICALLY Consider an airplane that is flying due east.

a. What is the effect on the airplane if the wind is blowing due east? due west?

b. In general, what wind directions have the effect of increasing the speed of the plane? of decreasing it?

REVIEW PROBLEM SOLVING STRATEGIES

May I have this dance?

1. In a dance production, the choreographer had all 20 dancers stand in a line, one behind the other. At each clang of the cymbals, the dancers in positions 10 and 20 danced forward into positions 1 and 2 respectively, while the other dancers all moved back. What is the least number of times the cymbals must clang for the dancer originally in front to return to position 1?

For some hints, answer the following questions.

a. Since two people move to the front each time, in what position will the dancer originally in front be located after the first clang? after the second clang?

b. After the fourth clang, the dancer originally in front occupies position 9. After the fifth clang, the dancer will occupy position 11. Explain why the next position the dancer will occupy is position 12.

c. When the dancer originally in front is in position 20, the position this dancer will next occupy is position 2. What position must the dancer occupy just prior to returning to position 1?

d. What is the complete sequence of position numbers occupied by the dancer originally in front? Your sequence should start and end with position 1.

CAUSE FOR ALARM

2. An old-fashioned alarm clock is shown at the left. Assuming that it is keeping time correctly, how many times each day are its hands perpendicular? How many times each day do its hands lie on opposite rays?

WHAT'S THE POINT?

3. A lattice point is a point in a rectangular coordinate plane both of whose coordinates are integers. In the first quadrant there are two lattice points on the line $29x + 33y = 2490$. One of them has coordinates (54, 28). What are the coordinates of the other lattice point? (*Hint*: Draw a picture and use the slope of the line.)

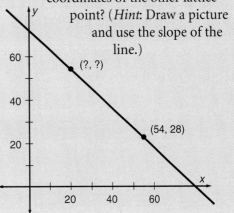

CHAPTER REVIEW

VOCABULARY

Choose the word from the list that correctly completes each statement.

1. ___?___ are two angles whose sides form two pairs of opposite rays.

2. Two lines that intersect to form a right angle are ___?___.

3. Lines that intersect and are not perpendicular are ___?___.

4. A ___?___ intersects two or more coplanar lines at different points.

a. Transversal

b. Vertical Angles

c. Perpendicular Lines

d. Oblique Lines

Lessons 3.1 and 3.2   RELATIONSHIPS AMONG LINES AND PLANES   pages 129–138

- The relationship between two lines can be classified as perpendicular, oblique, parallel, or skew.

Find the measure of each unknown.

5.

6.
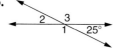

7.
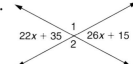

8. Draw pairs of lines that are perpendicular, parallel, and skew.

Lesson 3.3   PARALLEL LINES AND TRANSVERSALS   pages 139–146

- Given parallel lines and a transversal, you can find the measure of the remaining angles if you know the measure of one angle.

Find the measure of all angles.

9.
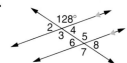

10.
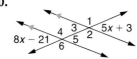

11.
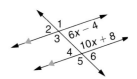

Lesson 3.4   PROVING LINES ARE PARALLEL   pages 147–153

- Use the converses of the postulates and theorems from Lesson 3.3 to prove two lines are parallel.

Name the postulate or theorem that justifies $r \parallel s$.

12.
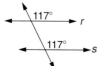

13.
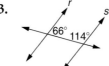

14.

- You can use compass-and-straightedge, paper-folding, or geometry software to perform constructions.

15. Copy $\overline{AB}$ at the right and construct a line p that is parallel to $\overline{AB}$.

A •――――――――• B

16. Copy $\overline{AB}$ at the right and construct a line ℓ perpendicular to $\overline{AB}$ at some point C not on $\overline{AB}$.

Lesson 3.6 THE SLOPE OF A LINE pages 161–166

- **Parallel lines** have the same slope. The product of the slopes of **perpendicular lines** is -1.
- The slope of a line containing points (x_1, y_1) and (x_2, y_2), where $x_1 \neq x_2$, is $\dfrac{y_2 - y_1}{x_2 - x_1}$.

Name the slope of a line parallel and a line perpendicular to a line with the given slope.

17. $m = 2$
18. $m = -\dfrac{3}{4}$
19. $m = -5$

Find the slope of the line passing through the given points.

20. $(3, 4), (-2, 5)$
21. $(-1, 4), (-2, 2)$
22. $(3, 0), (4, -1)$

Lessons 3.7 and 3.8 EQUATIONS FOR LINES AND COORDINATE PROOFS pages 167–175

- **The slope-intercept form of an equation,** $y = mx + b$, is useful for graphing and finding equations of lines.

Find the slope and the y-intercept of each. Then graph the lines.

23. $y = -3x + 2$
24. $y = \dfrac{1}{4}x$
25. $y = x - 5$

Write the equation of the line that passes through the given points.

26. $(2, -1), (3, 5)$
27. $(0, 1), (-5, 5)$
28. $(3, 6), (3, -1)$

29. Write a coordinate proof to prove that if $P(-b, 0)$, $Q(0, a)$, $R(0, -a)$, $S(b, 0)$ then $\overline{PQ} \parallel \overline{RS}$.

Lessons 3.9 and Lesson 3.10 VECTORS AND USING VECTOR DIAGRAMS pages 176–185

- To find the magnitude of vector AB, you can use the coordinate plane distance formula.
- The sum of vectors can be found algebraically or geometrically.

Given $A(3, -2)$, $B(-4, 1)$, $C(0, 5)$, $D(-1, 3)$, find each of the following.

30. the ordered pair representation of $\overrightarrow{AB}$

31. $|\overrightarrow{AB}|$
32. $|\overrightarrow{CD}|$
33. $\overrightarrow{AB} + \overrightarrow{CD}$
34. $\overrightarrow{AB} \cdot \overrightarrow{CD}$

35. An airplane is flying on a heading of 45° with an air speed of 120 mi/h. The direction of the wind is 90° and the wind speed is 50 mi/h. Draw a vector diagram to find the actual course and ground speed of the airplane.

CHAPTER ASSESSMENT

CHAPTER TEST

Find the measure of each unknown in the figure.

1.

2.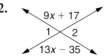

3. WRITING MATHEMATICS Write a paragraph which explains the difference between parallel, perpendicular, oblique and skew lines. Include drawings or descriptions of real life models.

4. Construct parallel lines p and q and draw transversal t. Label the angles 1–8. List all of the angle pairs below.
   a. corresponding angles
   b. alternate interior angles
   c. alternate exterior angles
   d. same side interior angles

Copy each figure. Find the measure of all angles.

5.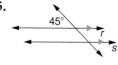

6.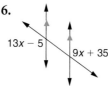

7. WRITING MATHEMATICS Write a paragraph describing at least four ways you can prove that two lines are parallel. Include sketches to illustrate your explanation.

Name the postulate or theorem that justifies $r \parallel s$.

8.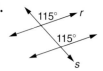

9.

Find the value that makes $a \parallel b$.

10. $\angle 1$ and $\angle 4$ are same-side interior angles.
   $m\angle 1 = 11x - 5$, $m\angle 4 = 15x + 3$.

Find the slope of the line that passes through the given points.

11. $(-3, 1), (2, 5)$ **12.** $(0, -3), (4, 3)$

13. STANDARDIZED TESTS Which statements are true about the line $3x + 2y = -6$?
   I. The slope of the line is $-\frac{3}{2}$.
   II. The y-intercept of the line is -3.
   III. The line is perpendicular to the line $y = \frac{2}{3}x$.
   IV. The line is parallel to the line $y = \frac{3}{2}x + 4$.

   a. IV only   b. I, II, and III
   c. I and IV   d. I, II, III, and IV

14. Graph the line $-x + 2y = 8$.

15. WRITING MATHEMATICS Write a paragraph which explains the slope of horizontal and vertical lines and the equations of those lines.

16. Write a coordinate proof. Given $A(a, b)$, $B(0, 0)$, $C(a, -b)$. Prove $AB = BC$.

Given $G(-3, 4)$, $E(2, -1)$, $O(0, 4)$, $M(3, -1)$, find each of the following.

17. the ordered pair representation of $\overrightarrow{GE}$

18. $|\overrightarrow{GE}|$ **19.** $\overrightarrow{GE} + \overrightarrow{OM}$ **20.** $\overrightarrow{GE} \cdot \overrightarrow{OM}$

21. Use the head-to-tail method to find $\overrightarrow{GE} + \overrightarrow{OM}$.

22. WRITING MATHEMATICS Describe how to determine if two vectors are perpendicular. Include an example of two vectors that are perpendicular.

23. A boat is traveling on a heading of 125°. The boat's speed is 15 mi/h. The water current is flowing at a speed of 2 mi/h and its direction is 30°. Make a vector drawing to find the ground speed and track of the boat.

PERFORMANCE ASSESSMENT

BOUNCING GRAPHS Set up an experiment to test how high a ball bounces when dropped from different heights. Collect and graph the data on a coordinate plane. Using two points, compute the slope of the line and then find the equation of the line that would pass through those two points. Graph the equation on the same graph as your data. Write a paragraph describing whether the equation you found is a good description of the data or not.

MODELING GEOMETRY Create a visual display with three-dimensional models to illustrate the following terms: perpendicular, parallel, oblique, and skew lines.

MATHEMATICAL ART Use your compass and straightedge to make a geometric design that incorporates parallel, perpendicular, and oblique lines. Find the equation for each of the lines in your drawing.

PARALLEL HANDBOOK Design a manual that can be used to determine if two lines are parallel. The manual should include a table of contents, written descriptions, examples, and a glosssary.

PROJECT ASSESSMENT

PROJECT *Connection* You will be using the skills you learned in the project connections to make a sketch of an area from a given point.

1. Once you are in your given location, pick out at least three permanent objects. Place an X on the paper where your group is initially located.

2. Locate the four cardinal directions and from your X, lightly draw lines to indicate each of the cardinal directions on your sketch.

3. Measure the angles formed by one of the cardinal directions and each of your objects, and draw lines from your X to represent the direction to each of your objects.

4. Estimate the distance to each of your objects by counting steps. Mark the objects' locations on the line you drew by determining a convenient scale. Label the actual distance on the sketch.

Fill in the blank.

1. The ratio of the change in the *y*-coordinates to the change in the *x*-coordinates of two points on a line is the __?__ of the line.

2. If the slopes of two lines have a product of -1, then the lines are __?__ .

3. If two lines are cut by a transversal so that the corresponding angles are congruent, then the lines are __?__ .

4. The point where a line segment is divided into two congruent segments is called the __?__ of the line segment.

Find the measure of each unknown angle.

5.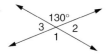

6.

State which property of equality is illustrated by each example.

7. If $4x = 20$, then $x = 5$.

8. If $12 = n$, then $n = 12$.

9. If $x + y = z$, and $z = a + b$, then $x + y = a + b$.

Sketch the cross section where each plane intersects the solid.

10.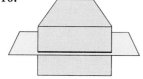

11.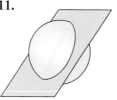

Find the complement of each angle measure.

12. $35°$ 13. $83.7°$ 14. $n°$

Find the slope of each line described.

15. passes through the points $(4, -2)$ and $(1, 4)$

16. parallel to the line that contains the points $(3, 4)$ and the origin

17. perpendicular to the line that contains the points $(-3, -2)$ and $(0, 4)$

18. perpendicular to the line that contains the points $(4, -1)$ and $(-3, -1)$

19. WRITING MATHEMATICS Describe how two figures that are similar might be different from two figures that are congruent.

Consider the statement: If point *C* is between points *A* and *B*, then $AC + CB = AB$.

20. Write the converse statement.

21. Write the inverse statement.

22. Write the contrapositive statement.

Write the equation of each line described.

23. passes through the points $(-1, -5)$ and $(2, 1)$

24. passes through the point $(4, -1)$ and is parallel to the line $y = \frac{1}{2}x + 2$

25. passes through the point $(-4, 2)$ and is perpendicular to the line $y = 2x - 5$

A vector has initial point *A* and terminal point *B* as given. Find the ordered pair representation of $\overrightarrow{AB}$ and $|\overrightarrow{AB}|$.

26. $A(1, -1)$ and $B(-2, 3)$

27. $A(-3, 7)$ and $B(5, -2)$

28. WRITING MATHEMATICS Compare the slopes of a horizontal line and a vertical line.

Find the slope and *y*-intercept of each line.

29. $6x - 3y = 9$ 30. $3x + 2y = 8$

STANDARDIZED TEST

STANDARD FIVE CHOICE Select the best choice for each question.

1. The equation of a line with a slope of $\frac{1}{3}$ and a y-intercept of -5 is

   A. $x - 3y = 15$
   B. $3x - y = 15$
   C. $x + 3y = -15$
   D. $3x + y = -15$
   E. $x + 3y = 15$

2. If $(-1, 3)$ is a point on a line, and the slope of the line is $\frac{3}{2}$, then which point is also on the line?

   I. $(2, 5)$ **II.** $(3, 9)$

   III. $(-4, 1)$ **IV.** $(-3, 0)$

   A. I only
   B. II only
   C. I and III
   D. II and IV
   E. I, II, and IV

3. In the diagram, two parallel lines are cut by a transversal. Which pair of angles is congruent because they are corresponding angles?

   A. 1 and 4
   B. 3 and 6
   C. 2 and 7
   D. 4 and 8
   E. 3 and 5

4. Point B is the midpoint of $\overline{AC}$. What is the length of $\overline{AC}$?

   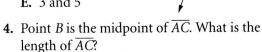
   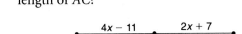

   A. 9
   B. 18
   C. 25
   D. 50
   E. cannot be determined

5. Which point is the closest to the origin?

   A. $(3, -6)$ B. $(7, 0)$

   C. $(5, 5)$ D. $(-6, 4)$

   E. $(0, -8)$

6. The supplement of the complement of an acute angle is

   A. acute
   B. right
   C. obtuse
   D. straight
   E. cannot be determined

7. The figure shown has a rotation symmetry of order

   A. 6   B. 4   C. 2
   D. 1   E. 0

8. Which of the following pairs of vectors is perpendicular?

   A. $\vec{p} = (4, -6)$ and $\vec{q} = (-3, 2)$
   B. $\vec{p} = (1, -1)$ and $\vec{q} = (-1, 1)$
   C. $\vec{p} = (3, 8)$ and $\vec{q} = (-16, -6)$
   D. $\vec{p} = (5, -2)$ and $\vec{q} = (-4, -10)$
   E. none of these

9. A circle has radius 5. Which of the following pairs of points could be the endpoints of a diameter of the circle?

   A. $(4, -6)$ and $(1, -2)$
   B. $(0, 5)$ and $(5, 0)$
   C. $(0, 0)$ and $(3, 4)$
   D. $(5, -2)$ and $(-5, 8)$
   E. $(3, -1)$ and $(-5, 5)$

10. The slope of a line perpendicular to the line $2x - 3y = 12$ is

    A. $\frac{3}{2}$

    B. $-\frac{3}{2}$

    C. $\frac{2}{3}$

    D. $-\frac{2}{3}$

    E. $\frac{1}{3}$

Take a Look AHEAD

Make notes about things that look familiar.
- What does it mean to say that one geometric figure is the image of another?
- What do you recall about flips, slides, and turns?
- Where do you see the distance formula used? In what connection is that formula applied?

Make notes about things that look new.
- What transformations will you study in this chapter? Look for reflection, translation, and rotation.
- What applications of transformations will you study? Search for and find frieze patterns.

DATA Activity

Textile Industry Inventions

Several modern inventions made a wide variety of textiles available. In 1792, Oliver Evans invented a machine to card wool. Eli Whitney invented the cotton gin which facilitated the separation of cotton fiber from seeds. In 1856, 18-year old chemistry student William H. Perkin launched the synthetic dye industry. In 1927, DuPont chemist Wallace Carothers made a fiber later called nylon. Today, patents in the textile industry often involve computers.

SKILL FOCUS

- ▶ Add, subtract, and divide real numbers.
- ▶ Determine the mean of a set of data.
- ▶ Find the percent of a number.
- ▶ Find the percent of increase or decrease.
- ▶ Analyze trends and make predictions based on a set of data.

Textiles and Design

In this chapter, you will see how:

- **PRINTERS** use the letterpress method to print patterns, words, and pictures. (Lesson 4.1, page 201)
- **APPLIQUÉ ARTISTS** use geometry to produce artistic designs. (Lesson 4.6, page 230)
- **QUILT DESIGNERS** use transformations and grids to plan a quilt patch. (Lesson 4.7, page 237)

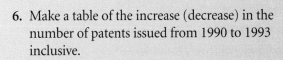

| New Patents Issued in the United States in the Textile Industry ||
Year	Number of Patents
1980	421
1985	503
1990	507
1991	498
1992	558
1993	600

1. Find the mean number of patents issued yearly in the textile industry for the years listed.

2. Assuming the mean number of patents from Question 1, how many patents would have been issued in 1981, 1982, 1983, and 1984 altogether?

3. How many more patents were issued in 1993 than in 1980? in 1990 than in 1985?

4. Of all the patents issued between 1990 and 1993 inclusive, what percent were issued in 1990?

5. Find the percent of increase in the number of patents issued from 1980 to 1993.

6. Make a table of the increase (decrease) in the number of patents issued from 1990 to 1993 inclusive.

7. Can you tell from the data how many patents were issued in 1993 for the computerization of the textile manufacturing process? Explain.

8. **WORKING TOGETHER** What overall trends do you see for the future from the patent data? Find out what textile advances and what patents issued in the textile industry are due to the progress people have made in recycling plastics? Describe how you think technology will change the textile industry in the future.

FLIPPING, SLIDING, TURNING

It was in the eighteenth century that people came upon the idea of covering walls with decorated paper. The use of the printing press made possible the manufacture of strips of decorated paper that contained pattern repeats.

In this project, you will use geometric transformations and symmetry to create your own wallpaper design.

PROJECT GOAL

To use transformations and symmetry to create a wallpaper pattern.

Getting Started

Work with a group.

1. Search through books and magazines to locate and collect wallpaper designs that reflect artistic and geometric characteristics. Make careful sketches or take photographs of those that you find attractive.

2. Talk to sales associates at stores that sell wallpaper to find out about wallpaper design and manufacture.

3. Research *frieze patterns* to find out the history and mathematics behind them.

4. Look through various mathematics textbooks to read about *transformations* such as reflections, translations, and rotations.

PROJECT *Connections*

Lesson 4.4, page 215:
Recognize reflection symmetry in wallpaper patterns and experiment with them.

Lesson 4.5, page 223:
Experiment with the placement of a shape into more wallpaper patterns.

Lesson 4.7, page 236:
Create a unique pattern integrating simple designs into more complex patterns.

Chapter Assessment, page 249:
Complete and present an attractive wallpaper pattern in full color.

Internet Connection

www.learninggeometry.com

4.1 Polygons

Explore

● You will need paper and a pair of scissors.

1. From an edge of the paper, cut a segment. Change direction. Cut another segment. Change direction again and cut a segment that brings you back to the point at which you began to cut.

2. With another sheet of paper, cut four lines, changing direction after making each cut. Describe the cutout.

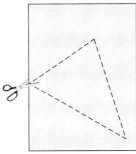

3. With another sheet of paper, cut five segments, changing direction after making each cut. Describe the cutout.

4. What happens if you want to carry out this procedure with six cuts but you cut across a cut line before you make the last cut?

Build Understanding

● A **polygon** is a plane figure formed by three or more segments such that each segment intersects exactly two others, one at each endpoint, and no segments with a common endpoint are collinear. Each segment is a **side** of the polygon. The common endpoint of two sides is a **vertex** of the polygon.

polygon   not a polygon

When its vertices are labeled, you can name a polygon by listing the vertices in order. For example, you can name the polygon at the right *ABCDE*, *CDEAB*, *CBAED*, or *EABCD*. Two sides of a polygon that share a common endpoint, such as $\overline{AB}$ and $\overline{BC}$, are *consecutive sides*. Two angles whose vertices are endpoints of the same side, such as $\angle A$ and $\angle B$, are *consecutive angles*.

A polygon completely encloses a region of the plane, called its *interior*. A polygon and its interior together form a *polygonal region*. The points of the plane not in a polygonal region are the *exterior* of the polygon.

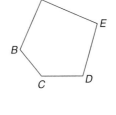

polygon

interior

exterior

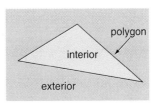

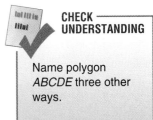

You can classify a polygon in different ways. The table at the right shows how to classify by the number of sides.

Number of Sides	Name
3	triangle
4	quadrilateral
5	pentagon
6	hexagon
7	heptagon
8	octagon
9	nonagon
10	decagon
n	n-gon

You can also classify a polygon as *convex* or *concave*. A **convex polygon** is a polygon in which no line that contains a side of the polygon contains a point in its interior. A **concave** polygon is a polygon in which a line that contains a side of the polygon also contains a point in its interior. A concave polygon is also called *nonconvex*. In a convex polygon, all points along any segment whose endpoints are inside the polygon are also inside it.

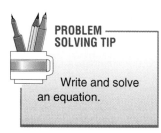

TECHNOLOGY TIP

You can use geometry software to measure the interior and exterior angles of a polygon.

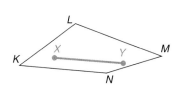

convex quadrilateral *KLMN*

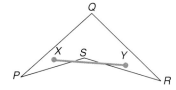

concave quadrilateral *PQRS*

EXAMPLE 1

TEXTILE DESIGN Classify the polygon that bounds the cloth patch at the right.

Solution

If you imagine points X and Y inside polygon *DEFGHJKLMNOP*, you can see that $\overline{XY}$ lies entirely inside it. Thus, *DEFGHJKLMNOP* is a convex 12-gon. ◄

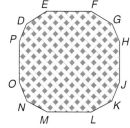

An angle that forms a linear pair with an angle of a polygon is called an **exterior angle** of the polygon. A polygon has two exterior angles at each vertex. To distinguish them from exterior angles, the polygon's own angles are called its *interior angles*.

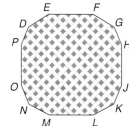

exterior angle

EXAMPLE 2

Find the measure of each exterior angle at vertex B of quadrilateral *ABCD*.

Solution

$$x + 137 = 180 \quad \text{m}\angle CBA + \text{m}\angle CBE = 180°$$
$$x = 43 \quad \text{Subtract 137 from each side.}$$

The exterior angles at B measure $43°$. ◄

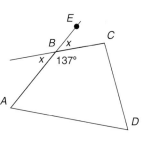

PROBLEM SOLVING TIP

Write and solve an equation.

An **equilateral polygon** is one whose sides are all congruent. An **equiangular polygon** is one whose angles are all congruent. A **regular polygon** is a convex polygon that is both equilateral and equiangular.

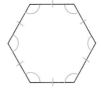

equilateral hexagon   equiangular hexagon   regular hexagon

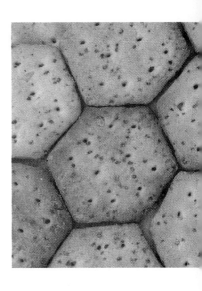

You can use the symbol △ when referring to a triangle. So, in the following examples, you read △*FAO* as "triangle *FAO*."

EXAMPLE 3

QUILT PIECES Angela cut six triangular pieces of cloth so that each side of each piece is 2 in. long and the measure of each angle is 60°. Classify △*FAO* and hexagon *ABCDEF* as equilateral, equiangular, and regular.

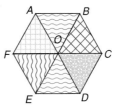

Solution

Since each side of △*FAO* is 2 in. long, △*FAO* is equilateral. Since the measure of each angle in △*FAO* is 60°, it also is equiangular. This means that △*FAO* is a regular triangle.

Since each side of hexagon *ABCDEF* is 2 in. long, it is equilateral. By the Angle Addition Postulate, the measure of each angle is 60° + 60°, or 120°. So hexagon *ABCDEF* is equiangular. This means that *ABCDEF* is a regular hexagon. ◄

A **diagonal** of a polygon is a segment whose endpoints are nonconsecutive vertices. The number $d(n)$ of diagonals in a polygon with n sides is given by .

$$d(n) = \frac{n(n-3)}{2}$$

CHECK UNDERSTANDING

Find the number of diagonals in a convex hexagon by drawing the hexagon and by evaluating the formula for $n = 6$.

EXAMPLE 4

Find the number of diagonals in pentagon *VWXYZ*. Verify your count by applying the formula above.

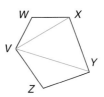

Solution

List each of the diagonals from each vertex.

V: *W:*   *X:*   *Y:*   *Z:*
$\overline{VX}$ and $\overline{VY}$ $\overline{WZ}$ and $\overline{WY}$ $\overline{XV}$ and $\overline{XZ}$ $\overline{YV}$ and $\overline{YW}$ $\overline{ZW}$ and $\overline{ZX}$.

There are ten segments in the list and each is counted twice. In all there are five diagonals in a convex pentagon. If $n = 5$,

$$d(5) = \frac{n(n-3)}{2} = \frac{5(2)}{2} = 5$$ ◄

Classify each polygon by number of sides and as convex or concave.

1.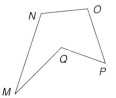

2.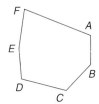

3.

4.

5. In quadrilateral *DEFG*, find the measure of each exterior angle at vertex *D* and at vertex *E*.

6. **WRITING MATHEMATICS** Sketch one example of a polygon that has seven sides and is convex and another that has seven sides and is concave. Describe in your own words the difference between a convex and a concave polygon.

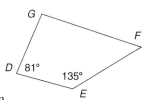

7. **TEXTILE PATCHES** In the figure at the right, m∠*S* = m∠*U* = m∠*W* = m∠*Y* = 135°, and m∠*SRT* = m∠*STR* = m∠*UTV* = m∠*UVT* = m∠*WVX* = m∠*WXV* = m∠*YXR* = m∠*YRX* = 22.5°. Classify quadrilateral *RTVX* and octagon *RSTUVWXY* as equilateral, equiangular, or regular.

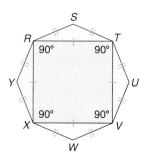

8. Find the number of diagonals of a regular decagon.

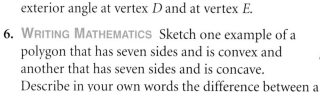

PRACTICE

Classify each polygon by number of sides and as convex or concave.

9.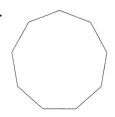

10.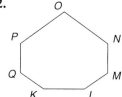

11.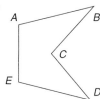

12.

13. In pentagon *PQRST*, find the measure of each exterior angle at vertex *P* and at vertex *Q*.

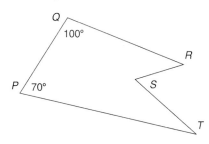

14. **TEXTILE PATCHES** Each angle in the shaded triangles measures 60°. Classify quadrilateral *BCDE* as equilateral, equiangular, or regular.

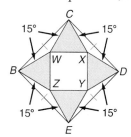

15. Find the number of diagonals of a regular 12-gon.

16. Writing Mathematics Refer to the dart board. Explain why no section of it is bounded by a polygon. In your own words, list the essential characteristics of any polygon.

17. a. Which of the following symbols are bounded by polygons? For those that are not, explain why. For those that are, classify the bounding polygons.

b. Explain why signmakers might choose different bounding shapes for different messages.

EXTEND

Sketch the polygon described by each set of characteristics.

18. The polygon is a quadrilateral with both pairs of opposite sides lying along parallel lines.

19. The polygon is a concave hexagon with at least three sides the same length.

20. The polygon is a convex pentagon with exactly two right angles.

Sketch each path *ABCDEA*. Determine whether the path determines a polygon.

21. $A(-1, 3)$; $B(0, 2)$, $C(3, 1)$; $D(3, -2)$; $E(0, -1)$

22. $A(-4, 3)$; $B(0, 2)$, $C(3, 1)$; $D(3, -2)$; $E(-4, 4)$

23. Find an equation for each diagonal of quadrilateral *KLMN*. Are the diagonals perpendicular to one another? Justify your response. Then find the coordinates of the point at which the diagonals intersect.

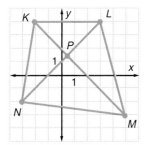

24. Points $R(0, 2)$ and $T(6, 5)$ are in the interior of polygon *ABCDE*. Find an equation for the line containing *R* and *T*. Verify $S(2, 3)$ is on the line. Since $S(2, 3)$ is in the exterior of *ABCDE* does this suggest a different way to define concave polygon? Write a different definition of concave polygon.

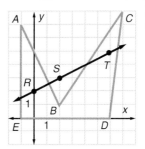

25. In quadrilateral *ABCD* at the right, $m\angle A = m\angle D$ and $m\angle B = m\angle C$. Find *x* and *y*.

26. In quadrilateral *ABCD* at the right, $AB = DC$. Find *r*.

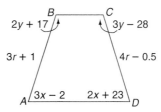

In the diagram at the right, polygon *DEFGH* is a regular pentagon, point *X* is the midpoint of $\overline{DE}$, and $\overline{GX}$ bisects $\angle HGF$. Find each length or angle measure.

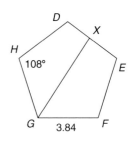

27. m$\angle$ *GFE*   **28.** *EF*   **29.** *DX*

30. m$\angle$ *DEF*   **31.** m$\angle$ *HGF*   **32.** m$\angle$ *HGX*

33. Using ruler, protractor, and compass, construct a quadrilateral that has two opposite sides congruent and two pairs of angles that are congruent.

34. WOODWORKING Katie has built the birdhouse shown. Using the language of polygons, describe its component parts.

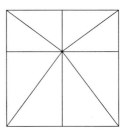

THINK CRITICALLY

35. Mikhail claimed that when he scattered *n* points, no three of which are collinear, on a sheet of paper, drew a polygon with *n* sides, and drew all its diagonals, he had drawn 50 segments in all. Do you believe he actually formed an *n*-sided polygon? Explain why. If so, what is *n*?

36. Refer to the figure at the right. How many triangles, quadrilaterals, pentagons, and hexagons does the figure contain? Include both convex and concave polygons in your counts.

37. Write a convincing argument to show that the number $d(n)$ of diagonals in a convex polygon with *n* sides is given by the equation $d(n) = \dfrac{n(n-3)}{2}$.

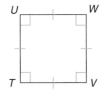

ALGEBRA AND GEOMETRY REVIEW

Solve each linear inequality.

38. $2x - 5 < 10$ **39.** $-3x - 4 \geq 23$ **40.** $7x + 3 \geq 31$

41. $-2x \leq 100$ **42.** $4x + 5 > 12$ **43.** $-3x - 5 \leq -14$

44. The length of $\overline{AB}$ is 12.5 units, point *X* is on $\overline{AB}$, and $AX = 4.5$ units. Find *XB*.

45. Classify polygon *UTVW* at the right. Justify your response.

Try reading the copy at the left, which a printer will place into a newspaper. As you can see, it is very difficult to decipher the true message. Nonetheless, a printer using the *letterpress method* of printing must reverse the material to be printed.

Lamorville—Three fire companies were dispatched late last night in response to a call reporting a motor vehicle accident along the curve of Appledate St. Fire officials reported that they rescued two occupants from the vehicle which hit a tree at a high rate of speed.

The printer does this so that the image will be readable when it is printed on paper. If you place a mirror along the right side of the paragraph, the message becomes readable in the mirror.

Decision Making

In the questions below, use a mirror.

1. If the roman letter **A** carved on the block of wood is printed on paper, what would it look like? If the italic letter **A** were printed on paper, what would it look like? Suppose you want the italic **A** to appear properly on paper. What would it look like on the wood?

2. Suppose you want to send a message to a friend by printing it on paper using the letterpress method. The message is **HI PAL**. What would it look like on the wood block?

3. Explain how you can use a mirror to figure out how to place the letters in a message on a block of wood so that the message prints properly in the letterpress method.

You can also use the letterpress method of printing to print art.

4. Refer to the wood block below. Sketch how this wallpaper pattern would look when printed on paper. The grid is meant as a guide for you.

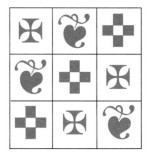

5. Stack and glue together several sheets of construction paper. Cut out an attractive shape using scissors. Glue the complete shape to cardboard or wood. Print your shape on paper using ink and the letterpress method. Do you agree that the letterpress method requires the image to be backwards so that it prints forward? Repeat the exercise with other shapes, letters, words, and pictures.

Drawing a Diagram

When you read a mathematics problem, you may find an illustrative drawing next to the prose. That drawing is provided as an aid to your understanding. Many times, however, you may find yourself in a situation where you have only a verbal description of something. Your understanding of what someone is saying will depend in part on your ability to visualize what you are given in words.

Problem

Jennifer has decided to display some of her favorite photographs on a bulletin board in her bedroom. She plans to use pushpins to attach the photographs to the board. How many pushpins will she need?

Explore the Problem

Use diagrams to help you answer each question.

1. Suppose Jennifer uses two pushpins for each photograph and no two photographs touch. How many pushpins will she need to mount 5 photographs? 12 photographs? n photographs?

2. Suppose that she decides to mount n photographs that do not touch one another and wants to use one, two, three, or four pushpins to mount each photograph. Write formulas for the number of pushpins she will need.

3. Jennifer is thinking of arranging 24 photographs in a rectangular array consisting of 6 photographs across and 4 photographs down. She will also allow each photograph to touch or overlap others. She has decided to mount each photograph using four pushpins. Draw a diagram that illustrates the arrangement. How many pushpins will she need?

4. Refer to Question 3. Suppose that Jennifer considers an array of 8 photographs across and 3 photographs down. Draw a diagram to illustrate the arrangement. How many pushpins will she need?

5. Repeat Question 3 for an array of 2 photographs across and 12 photographs down.

6. Write a formula for the number of pushpins Jennifer will need to make a rectangular arrangement that is n photographs across and m photographs down.

Investigate Further

Jennifer was able to mount all her photographs and still had many pushpins left over. So she decided to create a circular and polygonal design for her wall. She followed these steps.

- Space pushpins equally around the circle.

- Connect each pushpin to the other pushpins using a string.

PROBLEM
SOLVING PLAN

- Understand
- Plan
- Solve
- Examine

7. How many strings will Jennifer need if she uses 5 pushpins? 6 pushpins?

8. Copy and complete the table at the right that shows the number of strings needed for each number of pushpins.

Number of Pushpins	Number of Strings
3	
4	
5	
6	
7	
8	
9	
10	

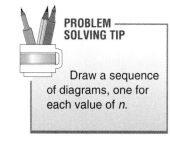

PROBLEM
SOLVING TIP

Draw a sequence of diagrams, one for each value of n.

9. Write a formula for the number S of strings needed to make a design with n pushpins.

10. Jennifer noticed in her string diagrams that certain strings had the same length. One of her practice diagrams is shown at the right. It contains eight pushpins. How many strings have the same length as the red one? the yellow one?

11. Suppose Jennifer uses n pushpins, creates a design as described, and connects every other pushpin with tan string. How many tan strings are there? Are they the same length?

12. **WRITING MATHEMATICS** State a few reasons why you might include a sketch, photograph, or diagram in a letter that tells a friend how you spent your summer.

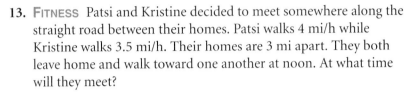

Apply the Strategy

● **Sketch and use a diagram to solve each problem.**

13. **FITNESS** Patsi and Kristine decided to meet somewhere along the straight road between their homes. Patsi walks 4 mi/h while Kristine walks 3.5 mi/h. Their homes are 3 mi apart. They both leave home and walk toward one another at noon. At what time will they meet?

14. **PEST PROTECTION** Mr. and Mrs. Omsden decided to fence in their 30-yd by 20-yd rectangular vegetable and herb garden to keep pests out. They expect to place one fence post at each corner of the garden and fence posts along the sides at one-yard intervals. How many fence posts will they need?

15. **ROOM REARRANGEMENTS** Tensie wants to rearrange her four stuffed animals on a shelf in her room. They are Rocky Raccoon, Joe Bear, Jerri Giraffe, and Tabbie Cat. If she places two on each side of her globe, how many arrangements are possible?

16. **SPORTS PRACTICE** Enid timed her friend Eric, who is racing around a circular track at school. The track is 800 yd around and has a marker every 100 yd. Enid found that Eric passed three consecutive 100-yd markers in 60 s. At this rate, how long will it take Eric to run once around the track?

17. **PARKING LOT PLANNING** The managers of Magic Moment Video Store plan to build a rectangular parking lot 102 ft wide and 120 ft long. Each rectangular parking space will be 8 ft wide and 12 ft long, and three driveways will run along the length of the lot. How many cars will the lot accommodate?

18. **COMMUNICATIONS** To notify each player on the 31-member football team, the coach will call the captain. Then each person who receives a call will call two other players. How many people will make phone calls?

19. **FARMING** Mr. Ramirez has a rectangular plot of land 800 ft long and 600 ft wide. He wants to divide it into four smaller rectangular plots with areas 120,000 ft^2, 120,000 ft^2, 80,000 ft^2, and 160,000^2. Is this possible? Draw a diagram to explain.

20. **CRIME PREVENTION** In Centerville, east/west streets are numbered consecutively starting at First, and north/south streets are lettered alphabetically starting at A. Police officers who are at the corner of Second and E Streets must respond to a call at Fourth and G Streets. They must take the shortest route possible along the streets. How many possible routes are there?

REVIEW PROBLEM SOLVING STRATEGIES

The Ups and Downs of Being a Worm

2. A worm is at the bottom of a well and wants to climb to the top. The well is 60 ft deep. The worm climbs for 12 h, at the rate of 1 ft/h. It rests for 12 h during which time it slides down the well at the rate of $\frac{1}{2}$ ft/h. If it alternates climbing and resting every 12-hour period, how long will it take the worm to reach the top of the well?

Easy as 1, 2, 3

1. Draw the next three items in the following picture sequence.

$$\underline{M}, \heartsuit, 8, \not{M},$$

Toothpick Puzzler

3. You have six toothpicks of equal length. Show how you can arrange them to form four congruent triangles. You may not break or discard any of the toothpicks and they may only touch at their endpoints.

Exploring Transformations

Think Back

1. Explain why quadrilateral *GHJK* has reflection symmetry.

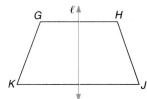

2. Explain why quadrilateral *VWXY* has rotation symmetry.

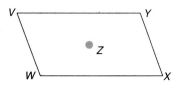

Explore

Work with a partner. You will need tracing paper, pencil, ruler, compass, and protractor.

3. On a sheet of tracing paper, sketch a triangle like △ *ABC* at the right. Draw a line *m* that does not intersect the △ *ABC*.

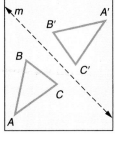

4. Measure each side of your triangle. Write the measurements along the sides.

5. Measure each angle of the triangle. Write the measurements at each angle.

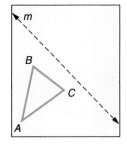

Fold the paper along line *m*. Poke holes through the paper at the vertices of the triangle. Unfold the paper. Connect the holes with segments.

6. Label the new triangle as shown at the right. (You read *A′*, *B′*, and *C′* as "*A* prime," "*B* prime," and "*C* prime.")

7. Measure each side and angle of △ *A′B′C′*.

8. How do the two triangles compare?

On a second sheet of paper, sketch a line and quadrilateral like *PQRS*. Measure each side and angle in your quadrilateral. Fold, poke holes, connect them, and label the new quadrilateral *P′Q′R′S′*.

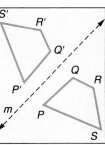

9. Measure the sides and angles of quadrilateral *P′Q′R′S′*.

10. How do the two quadrilaterals compare?

Draw a triangle something like $\triangle ABC$ shown at the right in red. Place a sheet of tracing paper over the sheet containing your triangle. Copy your triangle onto the tracing paper. Carefully slide the tracing paper right and up. (Keep the edges of the tracing paper parallel to the edges of the sheet under it.) Poke holes in the tracing paper at each vertex of the triangle. On the original sheet of paper, connect the holes to form new triangle $A'B'C'$.

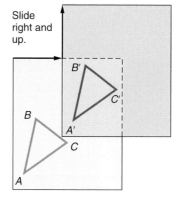

Slide right and up.

11. Measure each side and angle of the $\triangle ABC$ and $\triangle A'B'C'$.

12. How do the two triangles compare?

Place the sheet of tracing paper with the triangle you copied directly on top of the sheet containing $\triangle ABC$. Push a pin into both sheets of paper at one vertex of the triangle. Rotate the tracing paper about one third of a turn while holding the original sheet steady. Poke holes through the tracing paper at the other two vertices of the triangle on the tracing paper. Remove the tracing paper and pin. Connect the three pin holes to form a new triangle $A'B'C'$.

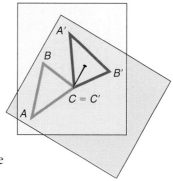

13. Measure each side and angle of $\triangle ABC$ and the new $\triangle A'B'C'$.

14. How do the two triangles compare?

Make Connections

● In Questions 3–14, you investigated geometric motions that are informally called *flips, slides,* and *turns.* In each case, moving the original figure resulted in a new figure. Each point of the new figure corresponded to exactly one point of the original figure, and each point of the original figure corresponded to exactly one point of the new figure. Motions like these are called **transformations.**

You can also perform transformations on a coordinate plane.

15. The figure below shows two polygons in the coordinate plane. After studying them, describe how each is obtained from the other by some transformation.

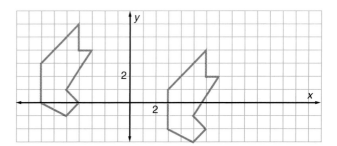

16. Copy both polygons from Question 15 onto grid paper. What would be the coordinates of the vertices of the polygon that results when you flip the polygon on the left down over the *x*-axis? (*Hint:* Think of the flip as spinning the polygon on the *x*-axis.)

17. **TEXTILE PATTERNS** Nancy noticed that the two polygons on the coordinate plane in Question 15 are congruent. She cut out copies of them and rearranged them as shown. Which polygon did she move and how did she move it?

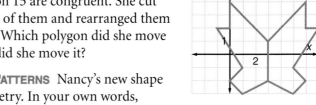

18. **TEXTILE PATTERNS** Nancy's new shape has symmetry. In your own words, describe that symmetry. What object does the symmetry suggest to you?

19. The figure at the right shows another arrangement formed by transformations of the polygons in Question 15. How is this figure similar to, yet different from, the one just above it?

20. Do you think the reflection of a concave figure is also concave? Sketch a concave polygon and its reflection about a line on grid paper to confirm your conjecture.

21. On the computer screen at the right, you see a polygon that has been flipped across a line. Classify the original polygon.

22. Sketch a quadrilateral. Locate a line so that, when you flip the quadrilateral across it, the boundary of the figure formed by the original figure and the new figure together is a hexagon.

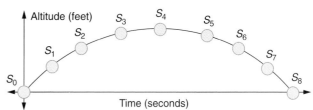

23. **PHYSICS** The diagram at the right shows the path of a soccer ball when it is kicked into the air.
   a. The path is its own reflection in some line. Describe that line.
   b. Interpret S_4.

Summarize

24. **WRITING MATHEMATICS** Compare the transformations you studied in this workshop: flips, slides, and turns. How are they alike? How are they different?

25. **THINKING CRITICALLY** Suppose you sketch a nonagon and flip it across a line. The total number of distinct vertices of the original nonagon and the new nonagon is nine. What can you say about the line?

26. **GOING FURTHER** On a coordinate plane, sketch square $OABC$ whose vertices have coordinates $O(0, 0)$, $A(0, 3)$, $B(3, 3)$, and $C(3, 0)$. Then, on the same coordinate plane, sketch the points you get when you apply the following rules.

   Multiply each x-coordinate by 3.
   Multiply each y-coordinate by 2.

Describe the effect of the transformation. Do the multiplication rules stated here suggest how to make a rectangle out of a square? How would you make a rectangle that extends vertically more than it extends horizontally?

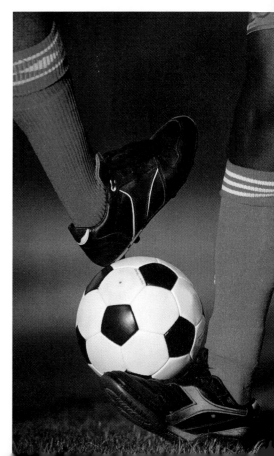

4.4 Reflections

Explore/Working Together

The picture illustrates a plastic reflector you can use to explore reflections. Place the plastic reflector along the line of reflection. Reach around the back of the reflector and trace along the outline.

Work with a partner. Each member should draw a triangle on a sheet of paper and a line that does not intersect the triangle. Use the plastic reflector to sketch the reflection image of the triangle.

1. Exchange papers. How does your partner's original triangle compare with its reflection image?

On another sheet of paper, sketch a triangle and a line that intersects the triangle. Then sketch the reflection image.

2. Exchange papers. How does your partner's second triangle compare with its reflection image?

3. Do you think that what you discovered is still true if you begin with a quadrilateral? Explain your response.

Build Understanding

Whenever you move a figure, you create a new figure. The new figure is called an **image**, and the original figure is its **preimage**. The motion creates a correspondence, or pairing, between the image and preimage. If each point of the preimage is paired with exactly one point of the image, and if each point of the image is paired with exactly one point of the preimage, then the correspondence is a **transformation**. Sometimes a transformation is called a **mapping,** and you say that it *maps* the preimage onto its image.

A **reflection** in a line ℓ is a transformation that maps every point P to a point P' such that:

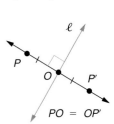

- if P is not on ℓ, then ℓ is the perpendicular bisector of $\overline{PP'}$;

- if P is on ℓ, then P' is P.

Line ℓ is the **line of reflection**, and P' is the **reflection image** of P.

A transformation in which a figure and its image are congruent is called a **rigid transformation**, or an **isometry**.

THEOREM 4.1

A reflection is an isometry.

EXAMPLE 1

Pentagon *LRPZT* is the reflection image of pentagon *ABCDE* across line *m*. Identify a segment congruent to $\overline{ED}$ and an angle congruent to $\angle TLR$.

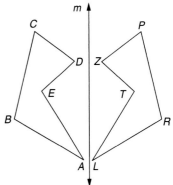

Solution

Identify corresponding vertices of *ABCDE* and *LRPZT*.

$A \leftrightarrow L \quad B \leftrightarrow R \quad C \leftrightarrow P \quad D \leftrightarrow Z \quad E \leftrightarrow T$

So $\overline{ED} \cong \overline{TZ}$, and $\angle TLR \cong \angle EAB$. ◄

If you further examined the pentagons in Example 1, you would conclude that each side and angle of *LRPZT* is congruent to a *corresponding angle* or *corresponding side* of *ABCDE*. So you can develop a precise definition of **congruent polygons**: Two polygons are congruent if their sides and angles can be placed in a correspondence such that corresponding sides are congruent and corresponding angles are congruent.

EXAMPLE 2

The vertices of $\triangle KLM$ are $K(-5, 5)$, $L(2, 0)$, and $M(-2, -4)$. Find the coordinates of the vertices of its reflection image across the *y*-axis.

Solution

For each vertex, find the point that is the same distance from the *y*-axis, but in the opposite direction.

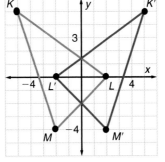

$$K(-5, 5) \rightarrow K'(5, 5)$$

$$L(2, 0) \rightarrow L'(-2, 0)$$

$$M(-2, -4) \rightarrow M'(2, -4)$$

Thus, the vertices of the reflection image of $\triangle KLM$ across the *y*-axis are the points $K'(5, 5)$, $L'(-2, 0)$, and $M'(2, -4)$. ◄

CHECK UNDERSTANDING

In Example 2, sketch $\triangle KLM$ and its reflection image across the *x*-axis. Give the coordinates of the vertices of the image.

Notice that a reflection does not preserve *orientation*. In Example 2, when you list the vertices of the preimage as *K*, *L*, *M*, you go around the figure *clockwise*. When you list the corresponding vertices of its image as *K'*, *L'*, *M'*, you go around the image *counterclockwise*.

THINK BACK

A plane figure has *reflection symmetry*, or *line symmetry*, if you can divide it along a line into two parts that are mirror images of each other. The line is called a *line of symmetry*.

THINK BACK

If the product of the slopes of two lines is −1, then the lines are perpendicular.

CHECK UNDERSTANDING

In Example 3, show that the line $y = -x$ is another line of symmetry for polygon *ABCDEFGH*.

PROBLEM SOLVING TIP

In Example 4, the segment $\overline{CC'}$ is an *auxiliary figure*. That is, it helps solve the problem, but it is not itself the solution. Drawing an auxiliary figure is often a helpful problem solving tool.

Sometimes a reflection image coincides with a preimage. When this happens, the original figure has reflection summetry, and the line of reflection is a line of symmetry for the figure.

EXAMPLE 3

The polygon *ABCDEFGH* has line symmetry. Write an equation for the line of symmetry with the following correspondence of vertices.

$A \longleftrightarrow H \quad B \longleftrightarrow G \quad C \longleftrightarrow F \quad D \longleftrightarrow E$

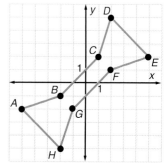

Solution

Find the midpoints *W*, *X*, *Y*, and *Z* of $\overline{AH}$, $\overline{BG}$, $\overline{CF}$, and $\overline{DE}$, respectively.

$$W\left(\frac{-5 + (-2)}{2}, \frac{-2 + (-5)}{2}\right) = W(-3.5, -3.5)$$

$$X\left(\frac{-2 + (-1)}{2}, \frac{-1 + (-2)}{2}\right) = X(-1.5, -1.5)$$

$$Y\left(\frac{1 + 2}{2}, \frac{2 + 1}{2}\right) = Y(1.5, 1.5)$$

$$Z\left(\frac{2 + 5}{2}, \frac{5 + 2}{2}\right) = Z(3.5, 3.5)$$

The midpoints *W*, *X*, *Y*, and *Z* lie on the line $y = x$, whose slope is 1. Using $\frac{\text{rise}}{\text{run}}$, you can see that $\overline{AH}$, $\overline{BG}$, $\overline{CF}$, and $\overline{DE}$ all have slope −1. So the line $y = x$ is the perpendicular bisector of $\overline{AH}$, $\overline{BG}$, $\overline{CF}$, and $\overline{DE}$. By definition, then, the line $y = x$ is a line of reflection for each half of *ABCDEFGH*. This means it is a line of symmetry for the figure. ◀

Properties of reflections can help you solve real world problems.

EXAMPLE 4

PUBLIC UTILITIES A power company plans to locate a transformer at a point *P* along the road between points *A* and *B* so that the length of cable from the home at point *C* to the transformer and then to the home at point *D* is minimized. Locate point *P*.

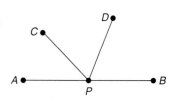

Solution

Reflect point *C* across $\overline{AB}$. Label its reflection image C'.

Draw $\overline{C'D}$.

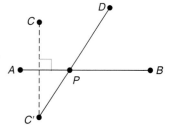

Label the intersection $\overline{AB}$ and $\overline{C'D}$ as point *P*. This is the location of the transformer. ◀

Hexagon *SCBTZA* is the reflection image of hexagon *NKJHDL* across line ℓ. Find the segment or angle that is congruent to the given segment or angle.

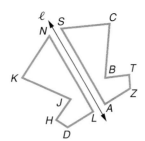
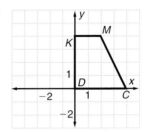

1. $\overline{KJ}$ 2. $\angle DLN$ 3. $\overline{NL}$

4. $\angle NKJ$ 5. $\overline{HD}$ 6. $\angle HDL$

Copy each figure onto grid paper. Sketch its reflection image across the *y*-axis. State the coordinates of the vertices of the image.

7.
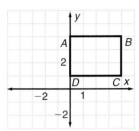

8.
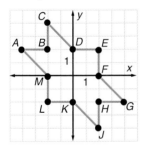

9.
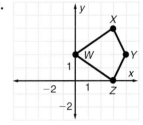

10. Determine whether the polygon shown below has line symmetry. If it does, write an equation for the lines of symmetry.

11. **WAREHOUSING** The Duffy Company plans to build a warehouse along the road linking points *X* and *Y* at a location *P* such that the distance from *T* to *P* to *W* is minimized. Copy the figure below and locate point *P*.

12. **WRITING MATHEMATICS** Suppose you are given △*ABC* on the left side of vertical line *m*. Explain how to find the reflection of △*ABC* across line *m*. Include a diagram with your explanation.

PRACTICE

In the diagram at the right, hexagon *DLAZRF* is the reflection image of hexagon *HUYQTC* under a reflection across line ℓ. Find the segment or angle that is congruent to the given segment or angle.

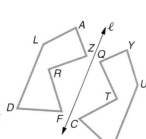

13. $\angle LDF$ 14. $\overline{AZ}$ 15. $\angle LAZ$

16. $\overline{DL}$ 17. $\angle DFR$ 18. RF

**Copy each figure onto grid paper. Sketch its image across the *x*-axis.
State the coordinates of the vertices of the image.**

19. the figure in Exercise 7   **20.** the figure in Exercise 8   **21.** the figure in Exercise 9

22. Determine whether the polygon shown below has line symmetry. If it does, write an equation for the lines of symmetry.

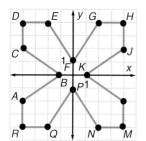

23. FARMING A farmer plans to build a roadside farm stand along the road linking points *A* and *B* at a location *P* such that the distance from *R* to *P* to *M* is minimized. Copy the figure below and locate point *P*.

24. WRITING MATHEMATICS Suppose that $\triangle ABC$ is placed on a coordinate plane and you know the coordinates of its vertices *A*, *B*, and *C*. Explain how to find the reflection image of $\triangle ABC$ across the *x*-axis. Include a diagram with your explanation.

EXTEND

Copy each diagram onto grid paper. Then sketch the image of each set of squares under a reflection across line ℓ.

25.

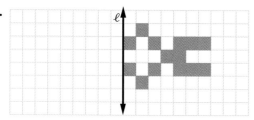

26.

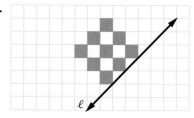

Graph each equation over the interval $-3 \le x \le 3$. Then tell whether the *y*-axis is or is not a line of symmetry of the graph.

27. $y = |x|$ **28.** $y = 3$ **29.** $y = x^2$ **30.** $y = |x + 1|$

31. $y = |x| - 1$ **32.** $y = x^2 - 1$ **33.** $y = x$ **34.** $y = (x + 1)^2$

35. Points *P* and *Q* have coordinates $P(0, 3)$ and $Q(8, 9)$. Find the coordinates of the point *R* on the *x*-axis that minimizes $PR + QR$. Then find that minimum length.

36. Sketch the symbol that completes the square at the right which consists of digits from 1 to 9.

37. Refer to Example 4 on page 212. Write a convincing argument that justifies the given method for locating point *P*.

THINK CRITICALLY

38. Use coordinates to show that $x = 1$ is a line of symmetry for nonagon *ABCDEFGHJ*, shown at the right.

39. Write a convincing argument to show that when you reflect a triangle across the *x*-axis and reflect its image across the *x*-axis, the result is the original triangle.

40. What is an equation of the line that is the reflection image of the graph of $y = 2x + 3$ in the *y*-axis? the *x*-axis?

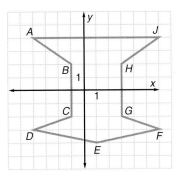

ALGEBRA AND GEOMETRY REVIEW

Simplify each expression.

41. $3^2 4^3$

42. $(-1)^5 (-3)^2$

43. $(2 + 3)^2 (3 - 2)^3$

44. $3^{4-2} + 5^{5-2}$

Find *PQ* and the coordinates of the midpoint of $\overline{PQ}$.

45. $P(1, 1)$ and $Q(7, 7)$

46. $P(-1, 0)$ and $Q(6, 0)$

47. $P(2, -3)$ and $Q(5, -2)$

48. The vertices of $\triangle ABC$ are $A(2, 3)$, $B(3, 5)$, and $C(2, -2)$. Find the coordinates of its reflection image across the *y*-axis.

PROJECT *Connection* Recognize reflection symmetry in wallpaper patterns you have collected.

1. Five simple diagrams that show wallpaper patterns are shown below. Which, if any, have reflection symmetry?

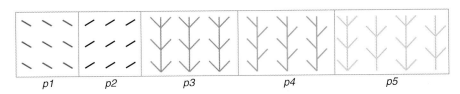

p1   p2   p3   p4   p5

2. How might you modify pattern *p1* so that it has reflection symmetry in a vertical line?

3. Sketch a simple shape. Show how it would look when positioned as suggested by each wallpaper pattern.

4. Do any of the wallpaper patterns you located in your research display reflection symmetry? Show the line of symmetry.

POLYOMINOES

A *unit square* is a square that is 1 unit on a side. When you arrange two unit squares so that they share a common side, you have a *domino*. When you arrange three unit squares so they share a common side, you have a *tromino*. The figures at the right illustrate all possible dominoes and trominoes. Notice that there is exactly one type of domino, and exactly two types of trominoes.

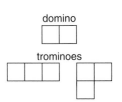

domino

trominoes

The figure at the right shows a checkerboard. It has 64 unit squares, half of which are white and half of which are green.

1. Is it possible to cover the checkerboard with dominoes made of congruent unit squares so there is no gap or overlap? Explain.

2. Is it possible to cover the checkerboard with trominoes made of congruent unit squares so there is no gap or overlap? Explain.

A *tetromino* is a set of four unit squares arranged so that each square shares a common side with at least one other square. Two tetrominoes are shown.

L-tetromino   skew tetromino

3. There are more than two tetrominoes. Draw diagrams on grid paper to find all the genuinely distinct tetrominoes. How many are there in all?

4. Do all the tetrominoes you discovered in Question 3 have the same area? Which of them has the least perimeter?

5. Using one copy of each tetromino you found in Question 3, is it possible to arrange them so as to form a rectangle?

6. The number 4 is a factor of 64. Do you think it is possible to cover the 64-square checkerboard without gap or overlap by using tetrominoes made of congruent unit squares. Draw a sketch to justify your response.

7. Sketch a diagram that shows how to cover the 64-square checkerboard without gap or overlap using L-tetrominoes made of congruent unit squares.

8. Confirm or deny the following statement. Justify your response.
   It is impossible to cover the 64-square checkerboard without gap or overlap using the skew tetromino made of congruent unit squares.

As you might expect, you can also define a pentomino. The notion of the pentomino is very old, and pentominoes have been extremely popular. A *pentomino* is a set of five unit squares arranged in such a way that each square shares a common side with at least one other square.

9. Draw figures to illustrate four pentominoes other than those shown at the right.

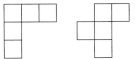

10. Conjecture how many genuinely distinct pentominoes there are. Then give a justification of your conjecture. Create one diagram for each pentomino you find.

11. **WRITING MATHEMATICS** Describe the strategy or strategies that you used to discover all the pentominoes you found.

12. Illustrate those pentominoes that could be folded along common sides to make an open-top box.

13. Explain why you cannot cover a 64-square checkerboard without gap or overlap using pentominoes made of congruent unit squares.

14. Suppose you cut off the square at each corner of the 64-square checkerboard. You now have 60 squares and 5 is a factor of 60. Is it possible to cover, without overlap, the 60-square checkerboard with pentominoes? Justify your response by using a diagram.

15. Using one copy of each pentomino you discovered, arrange them so there is as little space between them as possible. Find the area of the smallest rectangle that will contain your arrangement.

GOING FURTHER Experience with dominoes, trominoes, tetrominoes, and pentominoes suggests that you can keep going in your pursuit of the question of how unit squares can be connected or joined. The study of *polyominoes* is the study of the ways that *n* unit squares can be arranged so that each square shares a common side with at least one other square.

16. A *hexomino* is formed from six unit squares so that various sides coincide. How many hexominoes can you make that are five units long?

17. How many hexominoes can you make that are four units long?

18. Using your experience in Question 17, devise a strategy for finding out how many hexominoes there are altogether. Show how your strategy works by using tables and diagrams.

4.5 Translations

Explore / Working Together

You will need a plastic reflector or tracing paper. You also will need a ruler. Draw a triangle on a sheet of paper. Label it $\triangle ABC$. Draw a line m that does not intersect $\triangle ABC$. Reflect $\triangle ABC$ across line m. Label the image $\triangle A'B'C'$. Draw a line n parallel to line m. Reflect $\triangle A'B'C'$ across line n. Label the image $\triangle A''B''C''$.

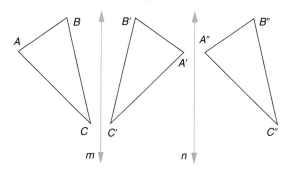

1. Exchange papers with a partner. How do the sides and angles of your partner's original triangle compare with those parts of the double-reflection image?

2. Suppose you list the vertices of $\triangle ABC$ in counterclockwise order. Would you list the corresponding vertices of $\triangle A''B''C''$ in counterclockwise fashion also?

3. Explain how you can transform $\triangle ABC$ into $\triangle A''B''C''$ by performing one operation or transformation.

4. Measure the distance between lines m and n. Then measure the distance between points A and A'', points B and B'', and points C and C''. Use your observations to make a conjecture.

CHECK UNDERSTANDING

For every point P in the figure at the right, $\overline{PP'} \parallel \overline{AA'}$. Sketch a different translation in which there is at least one point P for which $\overline{PP'}$ is collinear with $\overline{AA'}$.

Build Understanding

A **translation** by a vector $\overrightarrow{AA'}$ is a transformation that maps every point P to a point P' such that

• $\overrightarrow{PP'} = \overrightarrow{AA'}$

• $\overline{PP'} \parallel \overline{AA'}$ or $\overline{PP'}$ is collinear with $\overline{AA'}$

The vector $\overrightarrow{AA'}$ is the **translation vector**, and P' is the **translation image** of P.

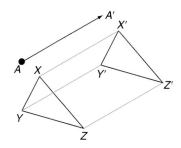

EXAMPLE 1

Sketch the translation image of $\triangle KLM$ under a translation $\vec{v}$.

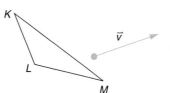

Solution

Trace the triangle and the vector. Using points K, L, and M as the initial points, draw three vectors that are equal to $\vec{v}$. Label the terminal points of these vectors K', L', and M', respectively. Then $\triangle K'L'M'$ is the translation image of $\triangle KLM$ by the vector $\vec{v}$.

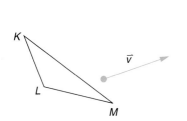

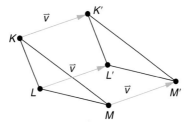

One of the important characteristics of a translation is that the image is always congruent to the preimage. This fact is stated as a theorem.

THEOREM 4.2

A translation is an isometry.

When you perform a translation on a coordinate plane, you can use the ordered pair representation of the translation vector.

EXAMPLE 2

The endpoints of $\overline{AB}$ are $A(-1, 4)$ and $B(3, -2)$. Find the endpoints of its translation image $\overline{A'B'}$ by the vector $\vec{v} = (-4, -1)$.

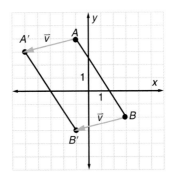

Solution

The vector $\vec{v} = (-4, -1)$ defines a translation of each point 4 units to the left and 1 unit down. So add -4 to the x-coordinate of each endpoint, and add -1 to the y-coordinate.

$$A'(-1 + (-4), 4 + (-1)) = A'(-5, 3)$$
$$B'(3 + (-4), -2 + (-1)) = B'(-1, -3)$$

The endpoints of $\overline{A'B'}$ are $A'(-5, 3)$ and $B'(-1, -3)$.

$$AB = \sqrt{(-1-3)^2 + (4-(-2))^2} = \sqrt{52}$$
$$A'B' = \sqrt{(-5-(-1))^2 + (3-(-3))^2} = \sqrt{52}$$

Thus, $A'B' = AB$.

You can apply translations to many everyday situations.

EXAMPLE 3

WALLPAPER HANGING

Annette, her brother Jon, and their parents are preparing a nursery for a new arrival. The pattern of the paper they have chosen is shown at the right. The top row of the sheet they just hung is shown below.

top of sheet hung

How should they position the next sheet to the right of this sheet?

Solution

Align the left edge of the next sheet along the right edge of the sheet just hung. Translate the new sheet upward until the right part of the butterfly matches its left half on the sheet just hung. ◄

There is a special relationship between translations and reflections.

THEOREM 4.3

If $\ell \parallel m$, then a reflection over line ℓ followed by a reflection over line m is a translation. If P'' is the image of P after the two reflections, then $\overline{PP''}$ is perpendicular to ℓ and $PP'' = 2d$, where d is the distance between ℓ and m.

EXAMPLE 4

In the figure at the right, lines ℓ and m are parallel, and the distance between them is 1.7 cm. Draw the image of $\triangle XYZ$ after a reflection across line m, followed by a reflection of the image across line ℓ.

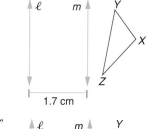

Solution

Apply Theorem 4.3. The distance between lines ℓ and m is 1.7 cm, so a reflection across line m followed by a reflection of the image across line ℓ is equivalent to a translation 3.4 cm to the left. ◄

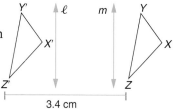

CHECK UNDERSTANDING

How many rows of animals up should they move the paper if the first row on the remaining paper is the back of the bunny?

COMMUNICATING ABOUT GEOMETRY

Discuss the fact that you cannot perform a reflection by means of translations.

Copy each figure. Sketch the translation image of the triangle under a translation along $\vec{v}$.

1.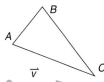

2.

3.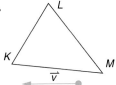

4.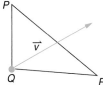

5. Find the coordinates of the translation image of pentagon *ABCDE* along the vector $\vec{v} = (3, -2)$. Use the distance formula to verify that the length of $\overline{AE}$ and its image are equal.

6. WALLPAPER HANGING Refer to the diagram that accompanies Example 3. The diagram below shows the top of a sheet just hung.

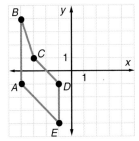

How should Annette and her brother hang the next sheet to the right of this sheet?

7. WRITING MATHEMATICS Suppose a friend does not believe that Theorem 4.3 is true. How would you convince your friend of its truth?

8. Copy the figure at the right. Draw the image of $\triangle$ *YOU* after a reflection across line ℓ, followed by a reflection of the image across line *m*.

PRACTICE

Copy each figure. Sketch the translation image of the triangle under a translation along $\vec{v}$.

9.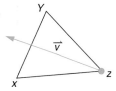

10.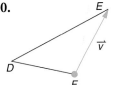

11.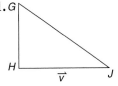

12.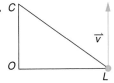

13. Refer to the figure for Exercise 5. Find the coordinates of the translation image of pentagon *ABCDE* along the vector $\vec{v} = (-2, 5)$. Use the distance formula to verify that the length of $\overline{CD}$ and its image are equal.

14. Refer to the figure for Exercise 5. Find the coordinates of the translation image of pentagon *ABCDE* along the vector $\vec{v} = (-4, 0)$. Use the distance formula to verify that the length of $\overline{BC}$ and its translation image are equal.

15. WRITING MATHEMATICS Compare and contrast the reflection and translation transformations.

16. WALLPAPER HANGING Refer to the diagram for Example 3. The diagram below shows the top of sheet just hung.

How should Annette and her brother hang the next sheet to the right of this sheet?

17. Copy the figure at the right. Draw the image of △*CST* after a reflection across line ℓ, followed by a reflection across line *m*.

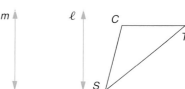

EXTEND

18. Describe a translation by which △*ABC* is transformed into △*A′B′C′*. Can you describe the movement in more than one way? Give a second description.

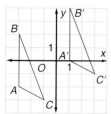

19. Copy square *ABCD* and vector $\vec{u}$ onto grid paper. Translate *ABCD* along vector $\vec{u}$. Join each point of *ABCD* to its image. Classify the three-dimensional figure suggested.

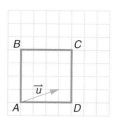

20. HEALTH The diagram at the right represents a typical heartbeat over one cycle of pumping. Copy the diagram. Then sketch the next heartbeat.

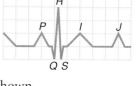

21. MUSIC The diagram below shows the musical scale that Kilea uses when she practices each day. Copy the diagram. Then sketch the next sixteen notes she would play if she repeats what is shown.

22. INTERIOR DECORATING Describe how the repeating pattern shown below is made from the upper and lower stripes and the wave shown in red.

23. Graph $y = |x|$ over the interval $-1 \le x \le 1$. Suppose the graph of a new function consists of this graph copied and translated to the right and left 2 units repeatedly. Sketch the graph of the new function over $-5 \le x \le 5$.

THINK CRITICALLY

24. Copy the figure at the right. Translate $\triangle ABC$ along $\vec{v}$. Translate the image along $\vec{u}$. Can you perform one translation to accomplish the same task? Explain your response and give an example.

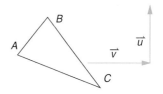

25. In the figure at the right, $\ell \parallel m$, P' is the reflection image of P across line ℓ, and P'' is the reflection image of P' across line m. The distance between lines ℓ and m is d. Show $PP'' = 2d$.

26. Suppose $\triangle ABC$ whose vertices have coordinates $A(-3, -2)$, $B(0, -1)$, and $C(2, 2)$ is reflected across the line $x = 3$, the image is reflected over the line $x = 5$, and that image is reflected across the line $x = 7$. Find the coordinates of the final image and tell whether the transformation is a reflection or a translation.

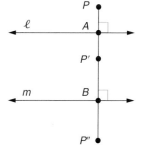

ALGEBRA AND GEOMETRY REVIEW

Solve each absolute value equation.

27. $|x - 5| = 10$ 28. $2|t + 3| - 1 = 7$ 29. $3 - |a + 1| = 11$

30. Write the converse of: If at least two sides of a triangle are congruent, then at least two angles of the triangle are congruent. Tell whether the converse is true or false.

31. Find the coordinates of the translation image of $\triangle KLM$ if the vertices have coordinates $K(-2, 1)$, $L(0, 5)$, and $M(3, 3)$ and the translation vector is $\vec{v} = (2, 1)$.

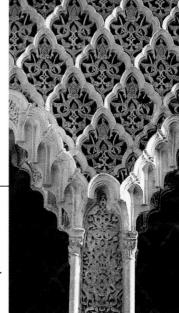

PROJECT *Connection*   A pattern has *translation symmetry* if it can be translated along a nonzero vector so that the image coincides with the preimage. Strictly speaking, only infinite patterns can have translation symmetry. So, to decide if a wallpaper pattern has translation symmetry, you must imagine that it could be extended infinitely.

1. The diagrams below show five wallpaper patterns different from those you saw in Lesson 4.4. Which of these patterns have reflection or translation symmetry?

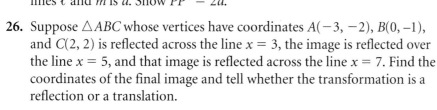

2. Place the shape you used in the Project Connection in Lesson 4.4 into the five patterns shown here.

3. Research wall decorations made by the Egyptians or the Moors. Locate and sketch at least one wall decoration that contains reflection or translation symmetry.

4.6 Rotations

Explore/Working Together

• Work with a partner. Sketch △ABC, a vertical line ℓ, and a line m that makes a 45°-angle with the vertical as shown. Then reflect △ABC across ℓ. Reflect the image of △ABC across line m.

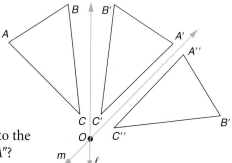

1. Using a compass, open it to the length of $\overline{OA}$. Is $OA = OA''$?

2. Using a compass, open it to the length of $\overline{OB}$. Is $OB = OB''$?

3. Using a compass, open it to the length of $\overline{OC}$. Is $OC = OC''$?

4. Find m∠AOA″, m∠BOB″, and m∠COC″. What does this suggest about the movement of A to A″, B to B″, and C to C″?

Build Understanding

• A **rotation** through $a°$ about a point O is a transformation that maps every point P to a point P' such that

- If point P is different from point O, then $OP' = OP$ and m∠$POP' = a°$.

- If point P is point O, then P' is P.

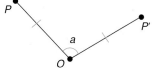

Point O is the **center of rotation**, $a°$ is the **angle of rotation**, and P' is the **rotation image** of P.

If the direction of a rotation is counterclockwise, the measure of the angle of rotation is given as a positive number. If the direction is clockwise, the measure is given as negative.

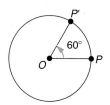

counterclockwise rotation
positive angle

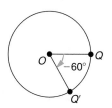

clockwise rotation
negative angle

A rotation of a point through 360° or −360° is a *complete turn*. Thus, the degree measures 180° and −180° correspond to a *half turn*. The degree measures 90° and −90° correspond to a *quarter turn*.

Sometimes when you rotate a figure about a point, the rotation image coincides with the preimage. When this happens, the given figure has rotation symmetry, and the center of rotation is the figure's center of symmetry.

THINK BACK

A plane figure has *rotation symmetry* if you can turn it around a point so it coincides with its original position two or more times during a complete turn. The point is called its *center of symmetry*.

EXAMPLE 1

TEXTILE DESIGN The textile design at the right has rotation symmetry. Find the angle(s) of rotation symmetry. (Neglect color.)

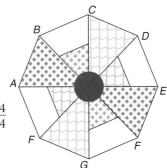

Solution

Rotating the design through $\frac{1}{4}, \frac{2}{4}, \frac{3}{4}$, and $\frac{4}{4}$ of a turn will bring the figure back into coincidence with itself. These fractions of a complete turn correspond to 90°, 180°, 270°, and 360°. Thus, the design has 4-fold rotation symmetry defined by rotations of 90°, 180°, 270°, and 360° about the center of the design. ◀

CHECK UNDERSTANDING

In Example 1, explain how you know that a rotation through 45° will *not* make the textile design coincide with itself.

As with reflections and translations, one of the characteristics of a rotation is that it preserves distance and angle measure.

THEOREM 4.4

> A rotation is an isometry.

COMMUNICATING ABOUT GEOMETRY

What do you think happens when you rotate a point *P* about a point *O* through an angle of 720°?

EXAMPLE 2

Draw the rotation image of $\triangle ABC$ after a −120° turn about point *O*.

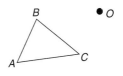

Solution

A rotation of −120° indicates a turn of 120° clockwise.

Using a protractor, draw $\angle AOA'$, $\angle BOB'$, and $\angle COC'$, each with a measure of 120°.

Using a ruler or compass, locate points A', B' and C' so that $OA = OA'$, $OB = OB'$, and $OC = OC'$.

Draw $\overline{A'B'}$, $\overline{B'C'}$, and $\overline{A'C'}$.

$\triangle A'B'C'$ is the rotation image. ◀

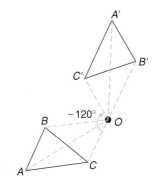

THINK BACK

If you studied rotations in a previous course, you may have simply called them *turns*.

In Example 3, another
portion of the
gathering area comes
from the rotation of
quadrilateral *OABC*
−90° about the origin.
Find the coordinates of
the image of *OABC*
under that rotation
about the origin.

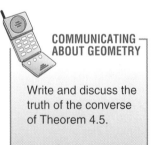

COMMUNICATING
ABOUT GEOMETRY

Write and discuss the
truth of the converse
of Theorem 4.5.

Often you may need to perform a rotation about a point and through
a special angle such as 90°, 180°, or 270°. In such a case, you may
find a coordinate approach useful.

EXAMPLE 3

PUBLIC GATHERING PLACES In
preparation for a large sporting event,
planners have decided to lay out a
public gathering area which forms a
4-fold rotation symmetry pattern.
Paving bricks will fill the four parts of
the design. Quadrilateral *OABC* is one
fourth of the finished area. Find the
vertices of the rotation image of
quadrilateral *OABC* after a 90° turn
about the origin. Each grid block is
100 ft long.

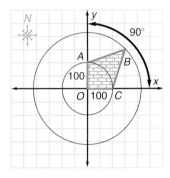

Solution

Under the rotation of 90° about the origin, map the positive *x*-axis
to the positive *y*-axis and the positive *y*-axis to the negative *x*-axis.

$$O(0, 0) \rightarrow O(0, 0) \qquad\qquad A(0, 200) \rightarrow A'(-200, 0)$$
$$B(300, 300) \rightarrow B'(-300, 300) \qquad C(200, 0) \rightarrow C'(0, 200)$$ ◄

The following theorem can help you sketch the rotation of a figure
through a specified angle around a specified point by using reflections.

THEOREM 4.5

If two lines, ℓ and m, intersect at point O, then a
reflection over line ℓ followed by a reflection over line m
is a rotation about point O. The measure of the angle of
rotation is $(2x)°$, where $x°$ is the measure of the acute or
right angle between ℓ and m.

EXAMPLE 4

Joanna wants to rotate a triangle through a 150° angle about a point
by using reflections across lines ℓ and m. At what angles should she
make lines ℓ and m intersect?

Solution

To find the angle at which ℓ and m intersect, solve $2x = 150$.

$\quad 2x = 150 \qquad$ Divide each side by 2.

$\quad x = 75$

Joanna should make lines ℓ and m intersect at an angle of 75°. ◄

1. The glass pattern below has rotation symmetry. Find the angle(s) of rotation symmetry. (Neglect color.)

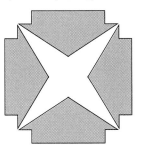

2. Copy the figure below onto a sheet of paper. Draw the rotation image of △DEF after a 65° turn about point O.

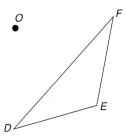

3. **ART FESTIVALS** Pentagon *OLCEK* represents one fourth of a finished display space for an art festival. Find the vertices of the rotation image of pentagon *OLCEK* after a 90° turn about the origin. (Each grid block is 100 ft long.)

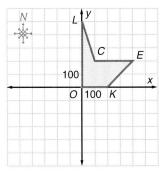

4. **WRITING MATHEMATICS** Describe how to make a 4-fold rotation symmetric design by beginning with a figure that is one fourth the final design.

5. Misha wants to rotate a triangle through a 55° angle about a point by using reflections across lines ℓ and *n*. At what angles should he make lines ℓ and *n* intersect?

PRACTICE

6. **TEXTILES** The cloth pattern below has rotation symmetry. Find the angle(s) of rotation symmetry. (Neglect color.)

7. Copy the figure below onto a sheet of paper. Draw the rotation image of quadrilateral *JKLM* after a −100° turn about point O.

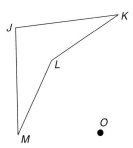

8. **WRITING MATHEMATICS** Suppose you want to rotate △ABC about point O.
   - You are given △ABC and point O on plain paper.
   - You are also given △ABC on grid paper with point O being the origin.
   Explain how you would rotate the figure 90° in each context.

9. Pat wants to rotate a triangle through a 137° angle about a point by using reflections across lines ℓ and *n*. At what angle should he make lines ℓ and *n* intersect?

10. **ART FESTIVALS** Pentagon *ZQPAO* represents one fourth of a finished display space for an art festival. Find the vertices of the rotation image of pentagon *ZQPAO* after a 90° turn about the origin. (Each grid block is 100 ft long.)

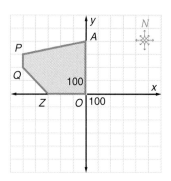

11. Jesse drew lines ℓ and *n* in such a way that they intersected at a 45° angle. He used the point of intersection as the center of rotation to make a design by rotation. Through how many degrees can he rotate a figure he begins with.

EXTEND

12. **RECREATION** Determine whether the crossword puzzle shown has rotation symmetry. If it does, find the angle(s) of rotation symmetry. Where is the center of the rotation that shows the symmetry?

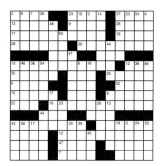

13. **AMUSEMENT RIDES** The gondola in the ride shown begins its swing at *P* and takes 6 s to swing from *P* to *Q* and back to *P*. Is the gondola swinging clockwise or counterclockwise after swinging for 43 s?

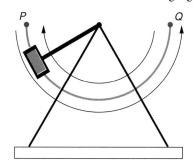

14. **ARCHITECTURE** The diagram at the right shows a planned recreational city park. The circular regions are fountains. The space consists of a regular pentagon with circles the same size centered at each vertex of the pentagon. The circles touch one another at the midpoints of the sides of the pentagon and nest inside another regular pentagon. Explain how to get each circle as a

    a. reflection of the circle centered at *A*

    b. rotation of the circle centered at *A*

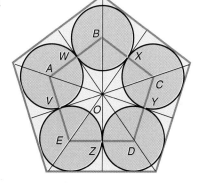

15. **GRAPHIC ARTS** A graphic specialist created the diagram shown at the right. Estimate the number of degrees and state the direction of rotation through which one part was rotated to give the second part of the creation.

16. The diagram below is an encoded message made by rotations of letters of the alphabet about the points in blue. Decode the message. Describe your rotations.

17. CHEMISTRY Magdelana and Isidore needed to time a chemical reaction. The experiment started with the second hand on their watch on 60. Through how many degrees did the tip of the second hand travel when it pointed to 24?

18. CHEMISTRY In the experiment in Exercise 17, the second hand was initially pointing to 60. The experiment lasted for 5.4 min. Through how many degrees did the second hand travel?

THINK CRITICALLY

19. You are given $\vec{v} = (3, 2)$. Find a and b such that $\vec{u} = (a, b)$ is perpendicular to $\vec{v}$. Are the values of a and b unique? How are they related?

20. Find an equation for the line whose graph is the rotation image of the graph of $y = 3.5x$ after a turn of 90° about the origin. Generalize your result to find an equation of the line whose graph is the rotation image of the graph of $y = mx$ after a 90° turn about the origin.

21. Suppose n is an integer, P has coordinates $P((-1)^n, (-1)^n)$, and O is the origin. Describe what happens to $\overline{OP}$ when $n = 1, 2, 3, \ldots$ and when $n = -1, -2, -3, \ldots$.

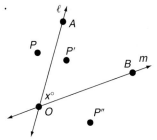

In the figure at the right, point P' is the reflection image of point P across line ℓ, point P'' is the reflection image of point P' across line m, and m$\angle AOB = x°$.

22. Show that the transformation that maps point P onto point P'' is a rotation.

23. Show that m$\angle POP'' = (2x)°$.

ALGEBRA AND GEOMETRY REVIEW

Graph each linear inequality.

24. $y \geq 2x - 1$ **25.** $y \leq 3x + 2$ **26.** $y < -2x - 2$

In the diagram at the right, $\ell \parallel m$. Find the measure of each angle.

27. m$\angle RPQ$ **28.** m$\angle BQA$ **29.** m$\angle FPQ$ **30.** m$\angle SQA$

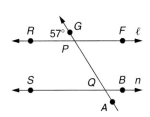

31. The coordinates of the vertices of $\triangle ABC$ are $A(0, 0)$, $B(0, -2)$, and $C(2, 1)$. Find the vertices of the rotation image of $\triangle ABC$ after a 90° turn about the origin.

Career
Appliqué Artist

Jan is an appliqué artist. She uses her talent as an expert seamstress, her creativity, and her knowledge of geometric transformations to create patterns.

Decision Making

In the questions below, use construction paper and grid paper.

1. Jan sketched a circle and its center. She marked the locations of eight points equally spaced around the circle. She then drew the segments shown. How can she use the red quadrilateral, copies of it, and transformations to complete the eight-part appliqué?

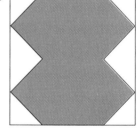

2. Identify the polygon determined by the eight vertices of the quadrilaterals that are closest to the center of the circle.

3. After laying out her appliqué pattern in Question 2, she decided to alternate the quadrilaterals by flipping every other one. When she flipped a quadrilateral, she used the line passing through the vertices separating noncongruent consecutive sides as the line of reflection. Sketch her new pattern.

4. Jan folded the square piece of fabric shown at the right. Then she cut part(s) of it off. Explain how she made the design and identify the polygon that the cuts determine when the fabric is unfolded.

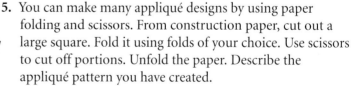

5. You can make many appliqué designs by using paper folding and scissors. From construction paper, cut out a large square. Fold it using folds of your choice. Use scissors to cut off portions. Unfold the paper. Describe the appliqué pattern you have created.

6. Using polygons and transformations, design an attractive appliqué design that could go on the front or back of the T-shirt shown at the right. Your design is meant to compliment the appliqué bird that is shown on the front of the shirt. The red lines are provided as a guide if you wish to use them.

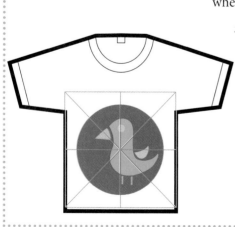

4.7 Compositions of Transformations

Explore

Work with a partner. You will need a plastic reflector or tracing paper. You also will need a ruler. Fold and crease a rectangular sheet of paper horizontally in half to make line *m*, vertically in half to make line *n*, vertically again to make line *p*, and at a slant to make line *q*. Sketch $\triangle ABC$.

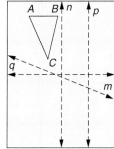

1. Describe the overall effect on $\triangle ABC$ by reflecting it across *m* and its image across *p*.

2. Describe the overall effect on $\triangle ABC$ by reflecting it across *m* and its image across *p*.

3. Describe the overall effect on $\triangle ABC$ by reflecting it across *n* and its image across *m*.

4. Describe the overall effect on $\triangle ABC$ by reflecting it across *m*, its image across *p*, and the resulting image across *q*.

Build Understanding

When you perform a transformation on a figure, then perform a second transformation on the figure's image, you are performing a **composition of transformations**. If the transformations you use are reflections, translations, and rotations, the final image you get will be congruent to the original. This fact is stated in the following theorem.

> **THEOREM 4.6**
>
> A composition of two or more isometries is an isometry.

Suppose you reflect $\triangle ABC$ across line *m* to get $\triangle A'B'C'$ and reflect $\triangle A'B'C'$ across line *n* to get $\triangle A''B''C''$. Notice from the diagram that each image of $\triangle ABC$ you get during the composition process is congruent to $\triangle ABC$.

$$\triangle ABC \cong \triangle A'B'C' \cong \triangle A''B''C''$$

You can verify Theorem 4.6 with the sketches you made in Explore by using a ruler and a protractor.

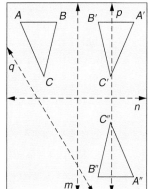

Often it is possible to identify a single reflection, translation, or rotation that is *equivalent* to a given composition.

EXAMPLE 1

The vertices of $\triangle ABC$ are $A(-5, 2)$, $B(-1, -1)$, and $C(-2, 5)$. Find its image under a reflection in the y-axis followed by a reflection in the x-axis. What single transformation is equivalent to this composition?

Solution

For each vertex of $\triangle ABC$, find the point that is the same distance from the y-axis, but in the opposite direction.

For each vertex of $\triangle A'B'C'$, find the point that is the same distance from the x-axis, but in the opposite direction.

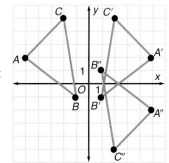

$$A(-5, 2) \quad \to A'(5, 2) \quad \to A''(5, -2)$$

$$B(-1, -1) \to B'(1, -1) \to B''(1, 1)$$

$$C(-2, 5) \quad \to C'(2, 5) \quad \to C''(2, -5)$$

The vertices of the image after the composition are $A''(5, -2)$, $B''(1, 1)$, and $C''(2, -5)$. This composition is equivalent to a single rotation of $\triangle ABC$ through a $-180°$ turn about the origin. ◀

A composition may be equivalent to a transformation other than a reflection, translation, or rotation. For example, reflecting a figure across a line and translating the image in the direction of the line results in a **glide reflection**, which is shown at the right.

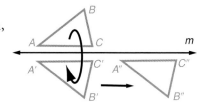

EXAMPLE 2

The vertices of $\triangle PQR$ are $P(1, 2)$, $Q(1, 5)$, and $R(5, 1)$. Find its image under a glide reflection across the x-axis and 6 units left.

Solution

For each vertex of $\triangle PQR$, find the point that is the same distance from the x-axis, but in the opposite direction. Then move each of these points 6 units left.

$$P(1, 2) \quad \to P'(1, -2) \to P''(-5, -2)$$

$$Q'(1, 5) \to Q'(1, -5) \to Q''(-5, -5)$$

$$R(5, 1) \quad \to R'(5, -1) \to R''(-1, -1)$$

The vertices of the image are $P''(-5, -2)$, $Q''(-5, -5)$, and $R''(-1, -1)$. ◀

Composition of transformations enables you to generate complex patterns from a single geometric design.

EXAMPLE 3

BANNER DESIGN At the right is part of a template for a school banner. Describe how star 3 is generated from star 1 by a composition of transformations and by a single transformation. (Consider the shapes only, not the numbers.)

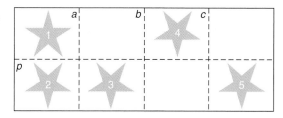

Solution

Assume that lines *a*, *b*, and *c* are each perpendicular to line *p*.

If you reflect star 1 across line *p*, you get star 2. If you then reflect star 2 across line *a*, you get star 3. So star 3 results from a composition of reflections of star 1, first across line *p*, then line *a*.

By Theorem 4.5, successive reflections of star 1 across lines *p* and *a* are equivalent to a single rotation through an angle with measure 2(90)°, or 180°, about the point where lines *p* and *a* intersect. The direction of the turn may be either 180° or −180°. ◄

CHECK UNDERSTANDING

Describe how star 5 is generated from star 3 by a composition of transformations and by a single transformation.

The table below summarizes some common compositions of transformations and the single transformations equivalent to them.

Composition	Equivalent
reflection in ℓ followed by reflection in *m*.	translation if ℓ ∥ *m* rotation if ℓ intersects *m*
translation followed by translation	translation
rotation followed by rotation	rotation
reflection across ℓ followed by a translation in the direction of ℓ	glide reflection

When you work with compositions of transformations, you must make sure that you perform the transformations in the correct order.

EXAMPLE 4

Find the coordinates of P'', the image of point $P(4, 3)$ under each composition.

a. 90° rotation about the origin, then reflection across the *x*-axis

b. reflection across the *x*-axis, then 90° rotation about the origin

Solution

a. $P(4, 3) \rightarrow P'(-3, 4) \rightarrow P''(-3, -4)$

b. $P(4, 3) \rightarrow P'(4, -3) \rightarrow P''(3, 4)$ ◄

Find the coordinates of the vertices of the image of △*DTP* under each composition. What single transformation is equivalent to this composition?

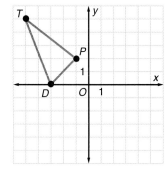

1. reflection across the *x*-axis, then translation left 2 units

2. 90° rotation about the origin, then reflection across the *y*-axis

3. translation 3 units right and 2 units down, then reflection across the *y*-axis

4. 90° rotation about the origin, then 90° rotation about the origin

Find the coordinates of the image of *P* (−4, 3) under each composition.

5. **a.** reflection in the *y*-axis, then translation down 2 units
   b. translation down 2 units, then reflection across the *y*-axis

6. **a.** 90° rotation about the origin, then reflection across the *y*-axis
   b. reflection in the *y*-axis then, 90° rotation about the origin

7. **WRITING MATHEMATICS** What does it mean to say the composition of two isometries is an isometry? Illustrate your response.

8. Describe how starburst 3 is generated from starburst 1 by a composition of transformations and by a single transformation.

PRACTICE

Refer to the diagram for Exercises 1–4. Find the coordinates of the vertices of the image of △*DTP* under each composition. What single transformation is equivalent to this composition?

9. reflection across the *x*-axis, then reflection across the *y*-axis

10. reflection across the *y*-axis, then 90° rotation about the origin

11. translation 3 units right and 2 units down, then reflection across the *x*-axis

12. −90° rotation about the origin, then −90° rotation about the origin

Find the coordinates of the image of *P* (−4, 3) under each composition.

13. **a.** reflection across the *y*-axis, then translation left 4 units
    b. translation left 4 units, then reflection across the *y*-axis

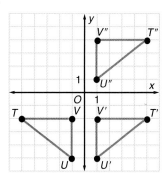

14. **a.** 90° rotation about the origin, then reflection across the *y*-axis
    b. reflection across the *y*-axis, then 90° rotation about the origin

15. **WRITING MATHEMATICS** Using the diagram at the right as an aid, explain how a composition of reflections preserves distance.

16. Refer to the diagram for Exercise 8. Describe how starburst 5 is generated from starburst 3 by a composition of transformations and by a single transformation.

EXTEND

Refer to the design below. Describe a pair of transformations which, when applied to the red pentagon, will yield.

17. the yellow pentagon   **18.** the green pentagon

19. In the diagram, $\overline{BC} \perp \overline{BE}$ and $\overline{ED} \perp \overline{BE}$.
Also $\overline{BD}$ is the reflection of $\overline{AB}$ across $\overline{BC}$ and $\overline{DF}$ is
the reflection of $\overline{BD}$ across $\overline{ED}$.

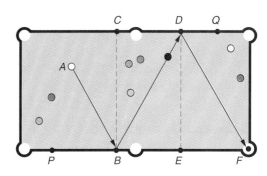

  a. Explain why m$\angle ABC$ = m$\angle CBD$.
  b. Explain why m$\angle BDE$ = m$\angle EDF$.
  c. Explain why m$\angle CBD$ = m$\angle BDE$.
  d. Justify the equation below.

  m$\angle ABP$ = m$\angle DBE$ = m$\angle CDB$ = m$\angle QDF$

20. Identify the transformation shown below.

21. In the figure at the right, points A, B, C, D, E, F, G, and H are spaced equally
around the outer edge of a circular dial. An arrow points toward A. Suppose
you turn the dial clockwise 135°, counterclockwise 225°, clockwise 450°, and
then counterclockwise 315°. Toward what letter does the arrow now point?
What single rotation is equivalent to all the turns together?

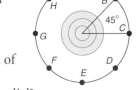

22. Suppose you turn the dial so that A moves clockwise to the original position of
F, then turn the dial so that F moves clockwise from its new position to the
original position of D. Through how many degrees in all have you turned the dial?

Identify the pair of transformations performed on P to give P' then P''.

23. $P(2, 3) \longrightarrow P'(-2, 3) \longrightarrow P''(-2, -3)$ **24.** $P(2, 3) \longrightarrow P'(-3, 2) \longrightarrow P''(-3, -2)$

25. Copy this pattern. Then draw the next four triangles in the pattern. Describe two different ways to
get the leftmost red triangle from the original blue one.

26. Let $\overline{A'B'}$ be the image of $\overline{AB}$ under an isometry, and let $\overline{A''B''}$
be the image of $\overline{A'B'}$ under an isometry. Show that the
transformation which maps $\overline{AB}$ to $\overline{A''B''}$ is an isometry.

THINK CRITICALLY

27. **a.** MOVING A mover picks up a chair and carries it overhead to the curb. He turns right, lowers the chair to the bed of the truck and pushes it to the front of the truck. Describe the transformations the mover has applied to the chair.

    b. When the mover gets to the new home, he wants to move the chair from the truck to the front room on the first floor of the house. The front sidewalk is perpendicular to the street on which the truck is parked. Describe the transformations he would apply to the chair.

28. Find the least number of transformations needed to make the design shown from one red oval and images of compositions of transformations on it.

29. Identify the polygon determined by the graph of $y = x + 1$, its reflection across the y-axis, the image's reflection across the x-axis, and that image's reflection across the y-axis. Justify your response.

ALGEBRA AND GEOMETRY REVIEW

Evaluate each formula given the values of its variables.

30. $P = 2(l + w)$
    $l = 2.2, w = 1.5$

31. $A = P + PRT$
    $P = 850, R = 0.05, T = 2$

32. $d = rt$
    $r = 52.5, t = 7.3$

33. Write the contrapositive of the following. Then tell if the contrapositive is true.

    If $\angle 1$ and $\angle 2$ form a linear pair, then m$\angle 1$ + m$\angle 2$ = 180°.

34. Find the coordinates of the vertices of the image of $\triangle XYZ$ whose vertices have coordinates $X(-3, 1)$, $Y(1, -2)$, and $Z(4, 2)$ under a reflection in the x-axis followed by a translation 2 units right and 2 units down.

PROJECT *Connection*

1. Place the shape you used in the Project Connection in Lesson 4.4 into the seven patterns shown below.

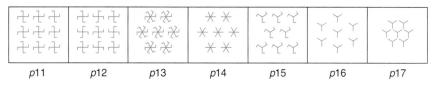

| p11 | p12 | p13 | p14 | p15 | p16 | p17 |

2. Choose one of the seventeen patterns you made from your shape in the Project Connection in Lesson 4.4. Or, if you prefer, change your original shape and create a new pattern using one of the seventeen patterns. Place three copies of the completed pattern on a long strip of paper wide enough to fit one pattern across.

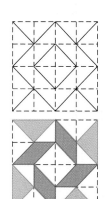

The creation of a quilt depends on the sewing ability and design sense of the artist. The designer might begin with a square *block*, perhaps measuring 12 in. on each side, and divide it into four, nine, or sixteen square *patches*. The figure at the left shows how a quilt design called *Windblown Square* arises from a sixteen-patch block.

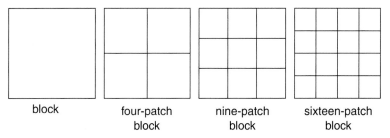

| block | four-patch block | nine-patch block | sixteen-patch block |

Decision Making

Refer to the five blocks at the bottom of the page.

1. Copy and complete the *Variable Star* block by successively rotating the shaded region 90° about the center of the block.

2. Copy and complete the *Rob Peter to Pay Paul* block by successively reflecting the shaded region across the lines through the midpoints of the sides of the block.

3. Copy and complete the *Variable Star Variation* block by successively reflecting the shaded region across the lines through the vertices of the block.

4. Copy and complete the *Waterwheel* block by successively rotating the shaded region 90° about the center of the block.

5. Copy and complete the block called *Your Choice* by using any transformations you choose. Be sure the completed block displays symmetry of some type. Describe the transformations you used to create your design.

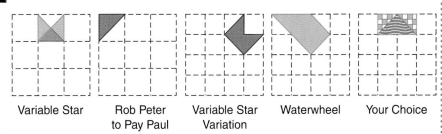

| Variable Star | Rob Peter to Pay Paul | Variable Star Variation | Waterwheel | Your Choice |

Focus on Reasoning

Algebraic Transformations

Discussion

• Work in a group of three or four students. First study these figures.

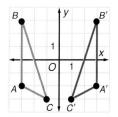 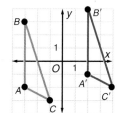

Allison, Israel, Maria, and David wondered whether there were any algebraic "formulas" for reflections. They made these conjectures about finding the reflection image of a point across the y-axis.

• Allison said you just take the opposite of each coordinate.

• Israel said that, since you are reflecting across the y-axis, you take the opposite of the y-coordinate. The x-coordinate remains the same.

• Maria said you take the opposite of the x-coordinate, while the y-coordinate remains the same.

1. Examine the figure at the left above. Which conjecture does it support? Explain your reasoning.

2. Draw $\triangle ABC$ above on a new set of coordinate axes. Then draw $\triangle A'B'C'$, its reflection image across the x-axis. How are the coordinates of the image and preimage related?

3. Make a conjecture about an algebraic method for finding the reflection image of a point across the x-axis.

David made this conjecture: To translate a point along the vector $\vec{v} = (5, 1)$, you add 5 to the x-coordinate and 1 to the y-coordinate.

4. Use the figure at the right above to verify David's conjecture.

5. Draw $\triangle ABC$ above on a set of coordinate axes. Draw $\triangle A'B'C'$, its image after a translation 2 units right and 2 units down. How are the coordinates of the image and preimage related?

6. Write an ordered pair representation of the translation vector.

7. Make a conjecture about an algebraic method for finding the translation image of a point on the coordinate plane.

Build a Case

In the Discussion, you probably discovered that Maria's conjecture about reflection across the y-axis is correct. You probably also discovered how to make a similar conjecture about reflection across the x-axis. Algebraically, these conjectures are stated as follows.

> ### REFLECTIONS ACROSS THE COORDINATE AXES
>
> The reflection image of a point $P(x, y)$ across the y-axis is the point $P'(-x, y)$.
>
> $$r_y: (x, y) \rightarrow (-x, y)$$
>
> The reflection image of a point $P(x, y)$ across the x-axis is the point $P'(x, -y)$.
>
> $$r_x: (x, y) \rightarrow (x, -y)$$

You read $r_y: (x, y) \rightarrow (-x, y)$ as "reflection across the y-axis maps (x, y) onto $(-x, y)$."

David's conjecture about translations also is correct. It can be generalized as follows.

> ### TRANSLATIONS ON THE COORDINATE PLANE
>
> The translation image of a point $P(x, y)$ moved h units horizontally and k units vertically is the point $P'(x + h, y + k)$.
>
> $$T_{h,k}: (x, y) \rightarrow (x + h, y + k)$$

TECHNOLOGY TIP

You can use geometry software to check your answers to Questions 8–10. First locate the vertices of the preimage triangle. Then use the appropriate commands from the menus to determine whether the transformation you chose results in the given image vertices.

In the questions that follow, you will investigate algebraic generalizations that can be associated with other transformations.

Consider $\triangle ABC$ with vertices $A(-2, -4)$, $B(-3, 1)$ and $C(-1, 0)$. For each set of image points A', B', and C':

a. determine the transformation that maps $\triangle ABC$ to $\triangle A'B'C'$.

b. describe the relationship between the coordinates of the vertices of $\triangle ABC$ and the vertices of $\triangle A'B'C'$.

8. $A'(2, 4)$, $B'(3, -1)$, $C'(1, 0)$

9. $A'(-4, 2)$, $B'(1, 3)$, $C'(0, 1)$

10. $A'(4, -2)$, $B'(-1, -3)$, $C'(0, -1)$

In Questions 8–10, you probably discovered that each of the transformations is a rotation about the origin. They can be generalized algebraically as follows.

ROTATIONS ABOUT THE ORIGIN

The rotation image of a point $P(x, y)$ through 90° about the origin is the point $P'(-y, x)$.

$$R_{90}: (x, y) \rightarrow (-y, x)$$

The rotation image of a point $P(x, y)$ through 180° about the origin is the point $P'(-x, -y)$.

$$R_{180}: (x, y) \rightarrow (-x, -y)$$

The rotation image of a point $P(x, y)$ through 270° about the origin is the point $P'(y, -x)$.

$$R_{270}: (x, y) \rightarrow (y, -x)$$

EXTEND AND DEFEND

Use the algebraic representations of transformations to find the coordinates of each point under the given transformation.

11. $P(5, -2)$; reflection across the y-axis

12. $P(-5, -2)$; reflection across the x-axis

13. $P(-2, 2)$; 180° rotation about the origin

14. $P(0, -2)$; 90° rotation about the origin

15. $P(3.5, 3.5)$; translation 4 units up and 7 units left

16. $P(-2, 2)$; translation 2 units right and 2 units down

The figure shows points A, B, and C and their reflection images across the line $y = x$.

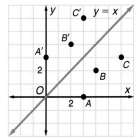

17. Write the coordinates of points A and A', points B and B', and points C and C'.

18. Copy and complete the following algebraic representation for reflection across the line $y = x$: $r_{y\,=\,x}: (x, y) \rightarrow (?, ?)$

19. Make and test a conjecture about reflection across the line $y = -x$. Then write an algebraic representation for this transformation.

Identify the two transformations that make up each composition of transformations. Then identify the single transformation that is equivalent to the composition.

20. $P(x, y) \rightarrow P'(x, -y) \rightarrow P''(-x, -y)$

21. $P(x, y) \rightarrow P'(-x, -y) \rightarrow P''(-x, y)$

22. $P(x, y) \rightarrow P'(y, x) \rightarrow P''(-y, x)$

23. $P(x, y) \rightarrow P'(y, -x) \rightarrow P''(-x, y)$

24. ANIMATION Use the notation of Questions 20–23. Write a composition of transformations that moves each point $P(x, y)$ of the simple cartoon character at the left below to each position across the coordinate plane.

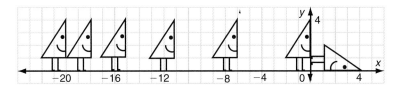

LOOKING AHEAD Using algebraic representations, you can investigate transformations that are not isometries. In Questions 25 and 26, consider square $ABCD$ with vertices $A(0, 3)$, $B(3, 3)$, $C(3, 0,$ and $O(0, 0)$.

25. Draw the image of the square under the transformation $P(x\ y) \rightarrow P'(2x, 3y)$.

26. Make a conjecture about the image under the transformation $P(x, y) \rightarrow P'(3x, 2y)$. Then verify your conjecture by drawing the image.

The vertices of square $ABCD$ are $A(-2, -2)$, $B(-2, 0)$, $C(0, 0)$, and $D(0, -2)$. Draw the image of $ABCD$ under each composition.

27. $P(x, y) \rightarrow P'(x - 1, y - 1) \rightarrow P''(2x, 3y) \rightarrow P'''(x - 2, y - 2) \rightarrow P''''(x, -y)$

28. $P(x, y) \rightarrow P'(x - 2, y - 2) \rightarrow P''(-x, -y) \rightarrow P'''(x - 2, y - 2) \rightarrow P''''(2x, -y)$

29. $P(x, y) \rightarrow P'(2x, y - 2) \rightarrow P''(x, -y) \rightarrow P'''(x - 2, y + 1) \rightarrow P''''(2x, -2y)$

30. $P(x, y) \rightarrow P'(-x, y) \rightarrow P''(x, -y) \rightarrow P'''(x + 1, y + 2) \rightarrow P''''(-x, -y)$

31. For each composition in Questions 27–30, write an algebraic representation for the single transformation that is equivalent to it.

Now You See It
Now You Don't

VERDICT OR VERTICAL

- Are the dark bars straight? Are they perfectly vertical?

- What do you think contributes to the illusion that arises from looking at the picture?

- How might you use a transformation to disprove the illusion?

Geometry Workshop
Frieze Patterns

Think Back/Working Together

• Draw the next figure of each pattern.

1.

2.

3.

Explore

• Work in a group of three or four students. Shown below are three strips of fabric, each with a distinct pattern. Imagine that each is a small part of a fabric strip that is infinitely long. Make a list of *all* the transformations that could map the pattern onto itself.

4.

5.

6.
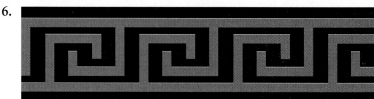

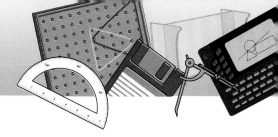

Make Connections

A **frieze pattern**, or **strip pattern**, extends infinitely in such a way that the pattern can be mapped onto itself by a horizontal translation in either direction. Some frieze patterns can be mapped onto themselves by other transformations. Using transformations as a basis for classification, there are exactly seven types of frieze patterns, shown below. Every frieze pattern involves a repeating block of shapes.

GEOMETRY: WHO, WHERE, WHEN

The punched card has been key to technological growth in many areas.

1805 Joseph Jacquard (1752–1834) builds an automatic loom. The patterns are controlled by punch cards.

1833 Charles Babbage designs the first all-purpose programmable computer, the Analytical Engine.

Charles Babbage's chief assistant explained "The Analytical Engine weaves algebraical patterns just as the Jacquard loom weaves flowers and leaves."

1890 Hermann Hollerith (1860–1929) uses a punched card system to make the U.S. Census calculations hundreds of times faster.

1946 Engineers in the United States build the first electronic digital computer. Punched cards contained computer instructions.

Types of Frieze Patterns	
Code	**Example**
T	▶ ▶ ▶ ▶ ▶ ▶ ▶ ▶
TV	▶ ◀ ▶ ◀ ▶ ◀ ▶ ◀
TR	pattern
TG	pattern
THG	pattern
TVRG	pattern
THVRG	pattern

T = translation R = 180° rotation
H = reflection across a horizontal line G = glide reflection
V = reflection across a vertical line

7. Use the codes above to classify each fabric strip in Questions 4–6.

Write the code that describes each frieze pattern below. Then copy the pattern and draw the next shape at the right and at the left.

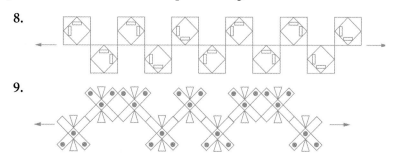

8.

9.

10. Create an original frieze pattern that would be classified TVRG.

Geometry Workshop

THINK BACK

The *circumference* of a cylinder is the distance around the curved surface.

MANUFACTURING In a rotary printing press like the one shown in the photograph, the paper unwinds from the rolls and passes by a rotating cylindrical *impression drum* that contains a pattern to be printed on the paper.

11. Suppose the cylindrical impression drum turns at 1200 revolutions per minute and has a circumference of 18 in. How many copies of the pattern will be printed in 1 min? 40 min? 3 h?

12. Suppose the length of the blank paper on one roll is 2,000 ft. How many copies of the pattern from the drum can fit on the entire roll of paper?

13. For the letter shown below to print properly on the paper, it must be placed "improperly" on the impression drum. Sketch the letter to be placed on the drum.

R

14. **WRITING MATHEMATICS** Explain how the printing process discussed in Questions 11–13 is related to frieze patterns.

There is a connection you can make between frieze patterns and repeating decimals.

15. How is the repeating decimal $\frac{7}{13} = 0.53846153846153846\ldots$ similar to and different from the frieze pattern in Question 8?

16. Write a rational number in fraction form and as a repeating decimal that has a repeating block of three digits. Then make a simple frieze pattern that consists of repetitions of a block of three figures. Explain the connections between the rational number and the frieze pattern.

You may be surprised to learn that there is a connection between frieze patterns and *periodic functions*. An example of part of the graph of a periodic function $y = f(x)$ is shown below.

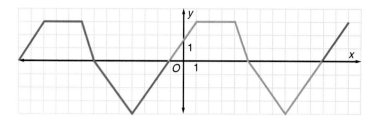

Find the value of y for each value of x.

17. -1 **18.** 0 **19.** 1 **20.** 11 **21.** 13 **22.** 2

23. The graph suggests that $f(x) = f(x + 12)$ for all values of x. Verify that $f(-3) = f(-3 + 12) = f(9)$ and $f(3) = f(3 + 12) = f(15)$.

24. Use the repeating pattern in the graph to predict $f(20)$, $f(32)$, $f(44)$, and $f(56)$. Verify your prediction.

25. Find five values of x for which $y = -4$. Justify your answers.

26. The graph shown extends infinitely both right and left. Describe a transformation that you can use to map the graph onto itself.

Summarize

27. WRITING MATHEMATICS Explain how frieze patterns are made and tell something about how you can tell whether patterns on a strip are frieze patterns.

28. THINKING CRITICALLY Suppose that a frieze pattern has horizontal line symmetry. Explain why it must be true that there is a glide reflection that you can use to map the frieze pattern onto itself.

29. GOING FURTHER Another way to show how frieze patterns are classified is to use a tree diagram like the one at the right. In this type of diagram, you represent the possible types of transformations as decision points. The branch of the tree that takes you to the TV classification is shown. Copy and complete the tree to show how you arrive at the other six classifications.

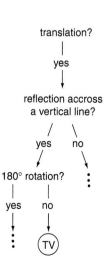

CHAPTER REVIEW

VOCABULARY

Choose the word from the list that correctly completes each statement.

1. A(n) ___?___ of a polygon forms a linear pair with an angle of the polygon.

2. One example of a composition of transformations is a(n) ___?___ .

3. In a(n) ___?___ , the mapping of preimage to image is defined by a vector.

4. Any transformation in which a figure and its image are congruent is a(n) ___?___ .

   a. isometry

   b. exterior angle

   c. translation

   d. glide reflection

Lesson 4.1   POLYGONS   pages 195–200

- If no line that contains a side of a polygon contains a point in its interior, the polygon is convex. If such a line exists, the polygon is concave.

Classify each polygon by number of sides and as convex or concave. Then find the measure of each exterior angle at vertex P.

5.

6.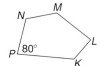

7.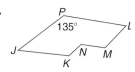

8.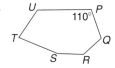

Lesson 4.2   MAKE A DIAGRAM   pages 202–205

- A useful strategy in solving some problems is drawing a diagram that helps you visualize the conditions of the problem.

9. A small regional airline plans to establish service among six cities so that each city is connected to each other by a direct route. How many direct routes must they establish?

Lessons 4.3 and 4.4   TRANSFORMATIONS AND REFLECTIONS   pages 206–215

- In a transformation, each point of the preimage corresponds to exactly one point of the image, and each point of the image corresponds to exactly one point of the preimage.
- A figure is reflected across a line. In a reflection, the preimage and image are congruent, but they do not have the same orientation.

Draw the reflection image of each figure across the y-axis. Give coordinates of the vertices of the image.

10.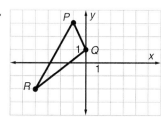

11.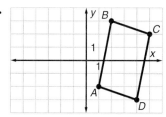

12.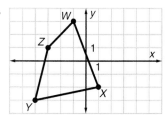

- A figure is translated along a vector (x, y). The x-coordinate indicates a move right (positive) or left (negative). The y-coordinate indicates a move up (positive) or down (negative).
- A figure is rotated about a point. A counterclockwise rotation is assigned a positive angle measure. A clockwise rotation is assigned a negative angle measure.

Draw the specified translation image. Give the coordinates of the vertices of the image.

13. along $\vec{v} = (-3, 1)$
Use the figure for Exercise 10.

14. along $\vec{v} = (2, -2)$
Use the figure for Exercise 11.

15. along $\vec{v} = (0, -4)$
Use the figure for Exercise 12.

Draw the specified rotation image. Give the coordinates of the vertices of the image.

16. 90° about the origin
Use the figure for Exercise 10.

17. −90° about the origin
Use the figure for Exercise 11.

18. 180° about the origin
Use the figure for Exercise 12.

- In a composition of transformations, you perform two or more transformations in sequence.
- A glide reflection is a reflection across a line, then a translation in the direction of the line.

Find the coordinates of the image of the figure under each composition. What single transformation is equivalent to the composition?

19. reflection across the y-axis, translation right 2 units
Use the figure for Exercise 10.

20. 90° rotation about the origin, reflection across the x-axis
Use the figure for Exercise 11.

21. reflection across the x-axis, reflection across the y-axis
Use the figure for Exercise 12.

- Several transformations on the coordinate plane can be represented algebraically as follows.

translations	reflections across the axes	rotations about the origin
$T_{h,k}: (x, y) \rightarrow (x + h, y + k)$	$r_y: (x, y) \rightarrow (-x, y)$ $r_x: (x, y) \rightarrow (x, -y)$	$R_{90}: (x, y) \rightarrow (-y, x)$ $R_{180}: (x, y) \rightarrow (-x, -y)$ $R_{270}: (x, y) \rightarrow (y, -x)$

Find the coordinates of the image of each point under the given transformation.

22. $P(-4, 1); r_x$ **23.** $P(2, -3); r_y$ **24.** $P(5, -2); R_{180}$ **25.** $P(3, 8); T_{-5,2}$ **26.** $P(5, -1); R_{270}$

- A frieze pattern extends infinitely in two opposite directions. It can be mapped onto itself by a horizontal translation in either direction.

27. Write the code that describes this frieze pattern.

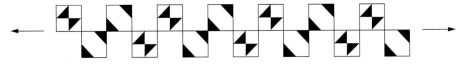

CHAPTER ASSESSMENT

CHAPTER TEST

Classify each polygon as concave or convex.

1.   2.   3.

4. **WRITING MATHEMATICS** Describe the difference between a convex and a concave polygon.

Find the measure of each exterior angle at vertex Y.

5. 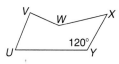 6.

7. **WRITING MATHEMATICS** Write a paragraph describing how to find a function for the number of diagonals of a polygon.

8. What is the least number of pushpins needed to display twelve congruent rectangular photos if a pushpin must go through each corner?

For Exercises 9–14, draw the image of this figure under each transformation or composition. Give the coordinates of the vertices of the image.

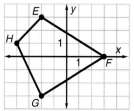

9. reflection across the x-axis

10. reflection across the y-axis

11. translation along the vector $\vec{v} = (-3, 2)$

12. $-90°$ rotation about the origin

13. reflection across the y-axis, then 90° rotation about the origin

14. translation along the vector $\vec{v} = (-1, 0)$, then reflection across the x-axis

15. In Exercises 13–14, what single transformation is equivalent to the given composition?

16. **STANDARDIZED TESTS** Which of these isometries do not necessarily preserve orientation? Choose A, B, C, or D.

   I. rotation   **II.** reflection

   III. translation   **IV.** composition

   A. I, II, and III   **B.** I, III, and IV

   C. I and IV   **D.** II and IV

17. **STANDARDIZED TESTS** Lines n and m are parallel. What is the result when a figure is reflected across n, then its image is reflected across m?

   A. a translation   **B.** a reflection

   C. a rotation   **D.** a glide reflection

18. **WRITING MATHEMATICS** Explain why it is important to pay attention to order when performing a composition of transformations.

Find the coordinates of the image of each point under the given transformation.

19. $P(-1, 2)$; r_x 20. $P(-3, -1)$; r_y

21. $P(3, -4)$; R_{180} 22. $P(-2, 5)$; $T_{3, -4}$

23. **STANDARDIZED TESTS** Which of these is a reflection of $P(-2, 1)$ across the x-axis?

   A. $P'(2, -1)$ **B.** $P'(2, 1)$

   C. $P'(1, -2)$ **D.** $P'(-2, -1)$

24. Write the code that describes this frieze pattern.

PERFORMANCE ASSESSMENT

PUZZLING POLYGONS Design a jigsaw puzzle composed of all concave polygons. Include at least ten polygons. The entire puzzle must be a rectangle, but it can be any size you wish.

RANDOM QUADS Use two number cubes with faces numbered from 1 through 6 to randomly generate the coordinates for the vertices of a quadrilateral. Repeat the activity several times, graphing each quadrilateral on a set of coordinate axes. Which is more likely, a concave or a convex quadrilateral? Give reasons for your conjecture.

MODELING GEOMETRY Create a visual display showing how a rotation is determined by both the center and the angle of rotation. Use the same figure in all your examples so the comparisons will be obvious to your viewers.

A ROTATION EXPERIMENT Start with a polygon that does not have reflection symmetry. Rotate it 60° about one vertex. Repeat this five times to make a design. Repeat the activity starting with a polygon that has one line of symmetry. Compare and contrast the results.

 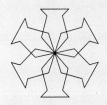

TRANSFORMATION ART Create a wallpaper border that uses a combination of two or more transformations. Explain in writing how you created your border.

PROJECT ASSESSMENT

PROJECT *Connection* Work with your group to finalize your wallpaper pattern. Prepare a presentation that addresses the following.

1. Decide on how you will organize the wallpaper patterns the team has collected and present descriptions of geometric transformations reflected in them.

2. Consider how you will display your wallpaper pattern and describe the process you went through to create it.

3. Describe a room in which your wallpaper pattern would be suitable and attractive.

4. Decide on how you can gather all the finalized wallpaper patterns and display them together in a showroom setting.

• • • CUMULATIVE REVIEW • • •

Fill in the blank.

1. A transformation of the plane in which every point is moved the same distance in the same direction is called a(n) __?__ .

2. A transformation that preserves length and angle measurement is called a(n) __?__ .

3. A transformation that reflects a figure across a line and then translates the image in the direction of the line is called a(n) __?__ .

4. The __?__ of a line is the ratio of the difference of the y-coordinates to the difference of the x-coordinates of two points on the line.

Sketch the image of each figure under a reflection across the y-axis. State the coordinates of the image.

5.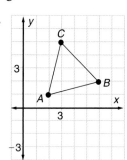
6.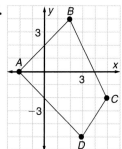

Write the equation of each line described in slope-intercept form.

7. $m = 5, b = -3$

8. passes through $(4, -2)$ and $(-1, 8)$

9. passes through $(3, 0)$ and perpendicular to $y = 3x - 1$

Find the midpoint of the segment with the given endpoints.

10. $A(6, -3)$ and $B(-5, 11)$

11. $M(1.5, -10)$ and $N(-8, 1)$

12. $W(12, -7)$ and $Z(-12, 7)$

Make an isometric drawing from each orthogonal drawing.

13.
14.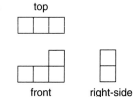

15. Find the image of the triangle under a translation 3 units right and 2 units down. State the coordinates of the image.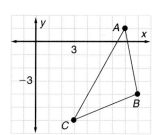

Give the justification for each conclusion.

16. If $6x = 84$, then $x = 14$.

17. If an angle has a measure of 90°, then it is a right angle.

18. If $x + y = z$, then $z = x + y$.

19. **WRITING MATHEMATICS** Describe how reflections over two parallel lines result in a single translation.

20. Find the images of the triangle under rotation of 90° and −90° about the origin. State the coordinates of each image.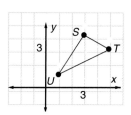

21. Find the measure of each angle in the figure where $m\angle 4 = (13x + 5)°$ and $m\angle 2 = (17x - 23)°$.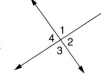

· · · STANDARDIZED TEST · · ·

QUANTITATIVE COMPARISON In each question, compare the quantity in Column 1 with the quantity in Column 2. Select the letter of the correct answer from these choices:

A. The quantity in Column 1 is greater.
B. The quantity in Column 2 is greater.
C. The two quantities are equal.
D. The relationship cannot be determined by the information given.

Notes: In some questions, information which refers to one or both columns is centered over both columns. A symbol used in both columns has the same meaning in each column. All variables represent real numbers. Most figures are not drawn to scale.

	Column 1	Column 2
1.	distance from the origin to the point $(-10, 0)$	distance from the origin to the point $(-6, 8)$
2.	the number of diagonals in an octagon	the number of diagonals in two hexagons
3.	the complement of a supplement	the supplement of a complement
4.	the dot product of two perpendicular vectors	the product of the slopes of two perpendicular lines

5.

x	y

	Column 1	Column 2
6.	the number of points needed to determine a line	the number of points needed to determine a plane

	Column 1	Column 2
7.	the complement of an acute angle	the supplement of an obtuse angle

8. Point C is between points A and B

	AC	CB

9.	the magnitude of the vector PQ with $P(2, -1)$ and $Q(-3, 7)$	the length of the segment PQ with $P(2, -1)$ and $Q(-3, 7)$

10. $3x + y = 12$

	the slope of a line parallel to the given line	the slope of a line perpendicular to the given line

11. A scale drawing is made using the scale 1 cm : 20 ft.

	the actual distance represented by a 3.5 cm segment	the actual distance represented by a 20 cm segment

12.

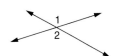

measure of $\angle 1$	measure of $\angle 2$

13. A triangle in Quadrant I is reflected over the y-axis.

	the y-coordinate of any point on the original triangle	the y-coordinate of any point on the image triangle

14. The triangle in Question 13 is reflected over the y-axis.

	the x-coordinate of any point on the original triangle	the x-coordinate of any point on the image triangle

Take a Look
AHEAD

Make notes about things that look new.
- What criteria for triangle congruence will you study? Look for the abbreviations SSS, SAS, ASA, and AAS.
- What information can you get from a triangle congruence? Find the abbreviation CPCTC.

Make notes about things that look familiar.
- What polygons do you see?
- What does it mean to say that two geometric figures are congruent?

DATA Activity

The Artist Population

You can see artistic expression almost anywhere you look, whether in a landscape design, an advertisement, stained glass windows, or sculpture. Many people with artistic ability are employed in fields such as graphic design and photography.

To see something about the change in the artist population in the United States, examine and study the table of data on the next page.

SKILL FOCUS

▶ Determine ratios.

▶ Use ratios to compare.

▶ Write numbers as fractions, decimals, and percents.

▶ Determine percent of increase (decrease).

▶ Analyze trends and make predictions based on a set of data.

In this chapter, you will see how:

- **SCULPTORS** use angle measurements to assess the layout for a monument. (Lesson 5.2, page 266)
- **GRAPHIC ARTISTS** use computer graphics programs to produce geometric designs. (Lesson 5.5, page 281)
- **ILLUSTRATORS** use geometry to create technical illustrations for textbooks. (Lesson 5.9, page 303)

OCCUPATIONAL DATA								
	1983				1994			
	Total	Percent of total			Total	Percent of total		
	(1000s)	Female	Black	Hispanic	(1000s)	Female	Black	Hispanic
Designers	393	52.7	3.1	2.7	548	55.3	3.4	5.8
Painters, sculptors, craft-artists, artist printmakers	186	47.4	2.1	2.3	225	50.5	4.6	5.2
Photographers	113	20.7	4.0	3.4	148	28.4	4.6	5.1

Use the table to answer the following questions.

1. By what percent did the number of Female designers increase from 1983 to 1994?

2. Find the ratio of Black photographers to Hispanic photographers in 1994.

3. For painters, sculptors, craft-artists, and artist printmakers, was there greater increase for Blacks or Hispanics from 1983 to 1994?

4. How many Female artists were in the work force in 1994?

5. How many more Female artists were in the work force in 1994 than in 1983?

6. **WORKING TOGETHER** Work in groups.

   a. What overall trends do you see for the future from the data?

   b. If you were to predict the percent of females in the arts occupation in the year 2004, what would your prediction be?

   c. Briefly describe how you think technology will change the artist work force.

   d. What job opportunities do you think may be created?

   e. What mathematical and technological skills do you think artists will need?

CREATING ART

Have you ever wondered how complex, beautiful, and symmetric designs like the stained glass window shown here are made? You may be surprised to learn that they are engineered by using geometric tools like the ruler and compass.

In this project, you will use geometric constructions and facts about congruence and symmetry to create your own artistic designs.

PROJECT GOAL

To use constructions to create artistic designs.

Getting Started

Work with a group.

1. Search through books and magazines to locate and collect designs and logos that reflect artistic and geometric characteristics. Make careful sketches or take photographs of art with geometric characteristics that you see on walls and structures in your locality.

2. Talk to artists to find out how they use geometry in creating their art. Find out if the art begins as a rough sketch and evolves into something more precise or elaborate.

3. List the different types of art with underlying geometry that you have found.

4. Set aside one or more pictures of geometric art for futher study.

PROJECT Connections

Lesson 5.2, page 265:
Construct a geometric design as line art and look for constructions in a picture you chose.

Lesson 5.5, page 280:
Use constructions to revise or introduce greater detail to the design you started to create.

Lesson 5.7, page 293:
Identify symmetries and congruences in the design.

Chapter Assessment, page 307:
Complete and present an attractive geometric design in full color.

✉ **Internet Connection**

www.learninggeometry.com

5.1 Triangles

Explore

- Draw a large triangle on a sheet of paper. Place the numbers 1, 2, and 3 inside each of the angles as shown in the diagram at the left below.

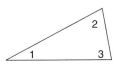 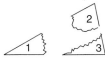

1. Carefully tear the three angles from the triangle as shown in the middle diagram above. Arrange the three pieces of paper as shown in the right diagram above.

2. What can you conclude about m∠1 + m∠2 + m∠3?

3. Draw a triangle different from the one you already drew. Tear off the angles and rearrange them as shown in the diagram at the right above. Does your conclusion in Question 2 remain the same? Explain your response.

Build Understanding

- Triangles can be classified in two ways. One classification depends on the number of sides that are congruent. The diagrams below illustrate this classification scheme.

scalene   isosceles   equilateral

The second classification scheme depends on the angles of the triangle. The diagrams below illustrate this classification scheme.

 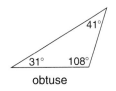

acute   right   obtuse

The table on the next page gives the definitions of different types of triangles based on these two classification schemes.

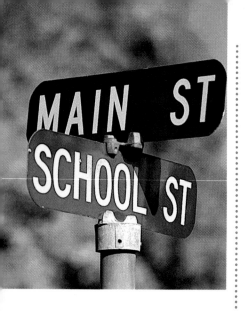

Classification of a Triangle by Its Sides	
scalene	a triangle with no congruent sides
isosceles	a triangle with at least two congruent angles
equilateral	a triangle with three congruent sides
Classification of a Triangle by Its Angles	
acute	a triangle with three acute angles
right	a triangle with one right angle
obtuse	a triangle with one obtuse angle

EXAMPLE 1

Classify $\triangle ABC$ shown at the right by lengths of sides and by angle measures.

Solution

Since the sides have different lengths and the angles are acute, $\triangle ABC$ is scalene and acute.   ◄

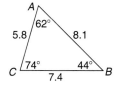

In Explore, you investigated an important fact about triangles.

THEOREM 5.1	ANGLE-SUM THEOREM FOR TRIANGLES

The sum of the measures of the angles of a triangle is 180°.

EXAMPLE 2

STREET MAPS What is the measure of the smaller angle formed by Main Street and School Street?

Solution

The three streets shown determine a triangle. To find the angle at which Main Street and School Street intersect, apply the Angle-Sum Theorem for Triangles.

$$m\angle A + m\angle B + m\angle C = 180°$$

$$90° + m\angle B + 39° = 180°$$

$$m\angle B + 129° = 180°$$

$$m\angle B = 51°$$

Since Main Street and School Street intersect at point B, the measure of the smaller angle they form is 51°.   ◄

PROBLEM SOLVING TIP

In Example 2, you are given two of three numbers whose sum is known. Write and solve an equation to find the third number.

CHECK UNDERSTANDING

In Example 2, suppose that the measure of the angle formed by Main Street and Richard Street were 91°. How would the solution be different?

The Angle-Sum Theorem for Triangles is proved below.

Theorem 5.1 The sum of the measures of the angles of a triangle is 180°.

Given $\triangle ABC$

Prove $m\angle A + m\angle B + m\angle C = 180°$

Plan for Proof Draw $\overleftrightarrow{XY}$ through B and parallel to $\overline{AC}$. Use alternate interior angles.

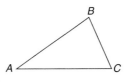

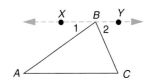

Proof Write a paragraph proof.

Through point B, draw $\overleftrightarrow{XY}$ parallel to $\overline{AC}$. You know this line exists by the Parallel Postulate. When parallel lines are cut by a transversal, alternate interior angles are congruent. Therefore, $\angle A \cong \angle 1$ and $\angle C \cong \angle 2$. By the Angle Addition Postulate, $m\angle 1 + m\angle B + m\angle 2 = m\angle XBY$. Since $\angle XBY$ is a straight angle, $m\angle 1 + m\angle B + m\angle 2 = 180°$. Using the Substitution Property, you can conclude $m\angle A + m\angle B + m\angle C = 180°$. ◄

In the above proof, notice that it was necessary to add an *auxiliary line* $\overleftrightarrow{XY}$ to the given triangle. An **auxiliary figure** is a point, line, ray, or segment added to a figure to aid in a proof. You may use an auxiliary figure in a proof as long as you can justify that it exists.

Theorem 5.1 concerns the interior angles of a triangle. In the figure at the right, you see the two *exterior angles* at vertex A of $\triangle ABC$. In relation to these exterior angles, $\angle B$ and $\angle C$ are called *remote interior angles* of the triangle.

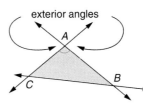

exterior angles

THEOREM 5.2 **EXTERIOR-ANGLE THEOREM FOR TRIANGLES**

The measure of an exterior angle of a triangle is equal to the sum of the measures of the two remote interior angles.

EXAMPLE 3

Find $m\angle 2$, $m\angle 3$, and $m\angle 4$, given $m\angle 1 = 33°$ and $m\angle 5 = 24°$.

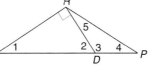

Solution

$$m\angle 2 + 33° + 90° = 180°$$
$$m\angle 2 = 57°$$

$$m\angle 3 = m\angle 1 + m\angle FRD$$
$$= m\angle 1 + 90°$$
$$m\angle 3 = 123°$$

$$m\angle 4 + m\angle 3 + m\angle 5 = 180°$$
$$m\angle 4 + 123° + 24° = 180°$$
$$m\angle 4 = 33°$$

Therefore, $m\angle 2 = 57°$, $m\angle 3 = 123°$, and $m\angle 4 = 33°$. ◄

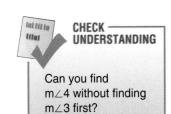

Classify each triangle by its sides and angles. (Lengths are rounded.)

1.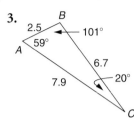
K
50° 6.7
4.0
L 40° M
5.0

2.
D
44°
5.4 5.4
E 68° 68° F
4.0

3.
B
2.5 ← 101°
59°
A
6.7
7.9 ↘ 20°
C

4.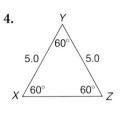
Y
60°
5.0 5.0
X 60° 60° Z

5. STREET MAPS At what angle do Jasmine Ave. and Rogers Rd. intersect?

6. WRITING MATHEMATICS Briefly describe the different types of triangles in this lesson. Draw a right isosceles triangle and explain how you know it is both isosceles and right.

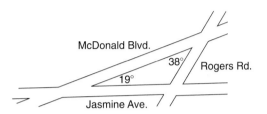
McDonald Blvd.
38°
19° Rogers Rd.
Jasmine Ave.

Find m∠1, m∠2, m∠3, and m∠4. Explain your method.

7.
N
2 3
K 34° 52° 4
W 1 O

8.
Q
3
37°
D 1 2 25° 4
M J

Classify each triangle by its sides and angles.

9.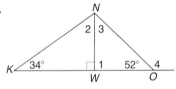
E
74°
5.2 5.2
F 53° 53° D
6.2

10.
E
60°
536 536
U 60° 60° Q
536

11.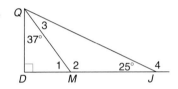
R
4.0 45°
5.7
T
4.0 45°
S

12.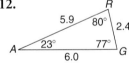
R
5.9 80° 2.4
A 23° 77° G
6.0

13. PICTURE FRAMING At what angle do $\overline{AB}$ and $\overline{AC}$ in the picture frame meet?

14. WRITING MATHEMATICS Explain how to find the measure of one angle of a triangle given the measures of the others.

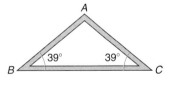
A
39° 39°
B C

Find m∠1, m∠2, and m∠3. Explain your method.

15.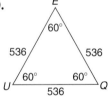
Z
28° 1 2
68°
Y 3
E

16.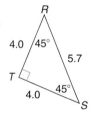
G
67°
M
1
67° 2
L 3 25° N

Find x.

17.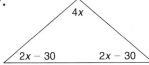

18.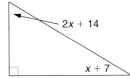

19.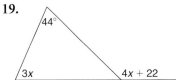

In each diagram m ∥ n. Find x.

20.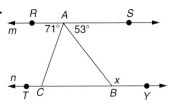

21.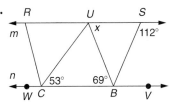

22.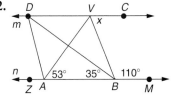

23. In $\triangle KLM$, there is a right angle at K. Show that $\angle L$ and $\angle M$ are complementary.

24. Refer to the diagram at the right. Find $m\angle 1 + m\angle 2 + m\angle 3$.

25. Refer to the result of Exercise 24. Make and prove a conjecture about the sum of the exterior angles, one at each vertex, of a triangle.

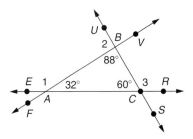

26. Prove that if two angles of one triangle are congruent to two angles of another triangle, then the third angles are congruent.

THINK CRITICALLY

27. Show that no triangle can have two obtuse angles.

28. Using the diagram at the right, find the sum expressed below.

$$m\angle ABC + m\angle BCD + m\angle CDE +$$
$$m\angle DEF + m\angle EFA + m\angle FAB$$

ALGEBRA AND GEOMETRY REVIEW

Evaluate each formula given the specified values of the variables.

29. $P = 2(l + w)$
$l = 2.5, w = 3.5$

30. $A = lw$
$l = 10.2, w = 4.5$

31. $V = lwh$
$l = 3, w = 3.3, h = 10.5$

Solve each equation for the specified variable.

32. $d = rt; t$

33. $I = prt; p$

34. $V = lwh; h$

35. $ax + b = c; x$

36. $y = mx + b, b$

37. $ax + by = c; y$

38. In $\triangle KLM$, $m\angle KLM = 42°$ and $m\angle LMK = 65°$, find $m\angle MKL$.

5.2   Angles of Polygons

Explore

SPOTLIGHT ON LEARNING

WHAT? In this lesson you will learn
• to find measures of angles related to polygons.

WHY? Precise angle measurements are necessary to construct an attractive and stable structure such as a gazebo.

1. Make a table that lists the number of sides in each convex polygon below and the number of triangles formed by drawing diagonals from point A.

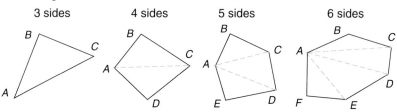

| 3 sides | 4 sides | 5 sides | 6 sides |

2. To the table in Question 1, add a column for the number of triangles formed by drawing diagonals only from point B.

3. Make a conjecture about a formula that gives the number of triangles formed in a convex *n*-gon when diagonals are drawn from one point.

4. Into how many triangles can you subdivide a convex polygon having forty sides? Justify your response.

TECHNOLOGY TIP

You can use geometry software to find the angle sums of convex polygons with more than six sides.

Build Understanding

From the Explore questions, you can derive the following theorem.

> **THEOREM 5.3**   **ANGLE-SUM THEOREM FOR POLYGONS**
>
> The sum of the measures of the angles of a convex polygon with *n* sides is $(n - 2)180°$.

You can apply the theorem to find unknown measures of angles.

CHECK UNDERSTANDING

In the first equation at the right, why is 180° multiplied by 2?

EXAMPLE 1

Find m∠*U* in quadrilateral *RSTU*.

Solution

Use Theorem 5.3.

$$m\angle R + m\angle S + m\angle T + m\angle U = 2(180°)$$
$$217° + m\angle U = 360°$$
$$m\angle U = 143°$$

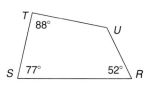

Thus, m∠*U* = 143°.

A **corollary** to a theorem is a theorem that follows easily from a previously proved theorem. The corollary below tells you about angles in regular polygons as shown in the diagram at the right.

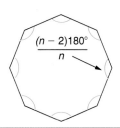

$$\frac{(n-2)180°}{n}$$

> **COROLLARY 5.4** The measure of each angle of a regular n-gon is $\dfrac{(n-2)180°}{n}$.

EXAMPLE 2

HOME IMPROVEMENTS Joshua Nester is planning to build a gazebo whose base is a regular octagon. At what angle x should he cut the lumber to frame the base?

Solution

$$\frac{(n-2)180°}{n} \qquad \text{Replace } n \text{ by 8.}$$

$$\frac{(8-2)180°}{8} = 135°$$

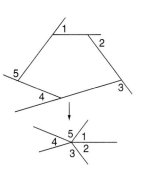

The measure of the angle the cut forms with the edge of the lumber should be 135°. ◄

In the figure at the right, one exterior angle has been drawn at each vertex of a convex pentagon. Below the pentagon, these exterior angles have been rearranged so they share a common vertex and do not overlap. As you can see, the sum of their measures is 360°. If you were to experiment, you would arrive at a similar result with the exterior angles of any convex polygon.

> **CHECK UNDERSTANDING**
>
> How would the solution of Example 2 change if the base of the gazebo is a dodecagon?

> **THEOREM 5.5** **EXTERIOR-ANGLE THEOREM FOR POLYGONS**
>
> The sum of the measures of the exterior angles of a convex polygon, one angle at each vertex of the polygon, is 360°.

A corollary to Theorem 5.5 tells you about the measures of the exterior angles of a regular polygon.

> **COROLLARY 5.6** The measure of each exterior angle of a regular n-gon is $\dfrac{360°}{n}$.

> **COMMUNICATING ABOUT GEOMETRY**
>
> Compile a list of facts about regular polygons.

You will need to use your equation-solving skills when you approach a geometry problem involving angle measurements.

EXAMPLE 3

In quadrilateral *ABCD*, you are given that

$$m\angle A = m\angle C = (3x + 8)°,$$
$$m\angle B = (7x - 22)°, \text{ and } m\angle D = (8x - 12)°.$$

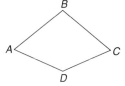

Find the measure of each angle of *ABCD*.

Solution

Apply the Angle-Sum Theorem for Polygons.

$$m\angle A + m\angle B + m\angle C + m\angle D = 360°$$

Substitute the given value for each angle.

$(3x + 8) + (7x - 22) + (3x + 8) + (8x - 12) = 360$

$(3x + 7x + 3x + 8x) + (8 - 22 + 8 - 12) = 360$ Combine like terms.

$21x - 18 = 360$ Add 18 to each side.

$21x = 378$ Divide each side by 21.

$x = 18$

Thus, with $x = 18$

$$m\angle A = m\angle C = 3(18) + 8 = 62°$$

$$m\angle B = 7(18) - 22 = 104°$$

$$m\angle D = 8(18) - 12 = 132°$$
◄

To test a conjecture about angles, you can use a theorem.

EXAMPLE 4

STUDENTS Marianna claimed there is a regular polygon whose angles each measure 100°. Determine whether her claim is true.

Solution

$\dfrac{(n - 2)180}{n} = 100$ Solve for *n*.

$(n - 2)180 = 100n$ Multiply each side by *n*.

$180n - 360 = 100n$ Apply the distributive property.

$80n = 360$

$n = 4.5$

Since *n* is not a whole number, there is no regular polygon whose angles each measure 100°. Marianna's claim is not true. ◄

Find the unknown angle measure in each quadrilateral.

1.

2.

3.

4.

5. **CONCRETE WORK** Employees of the Bedrock Foundation Works are readying an excavation site for a regular pentagonal foundation wall. At what angle will the concrete forms meet?

6. **WRITING MATHEMATICS** Write a plausible argument that shows that every convex polygon can be divided into $(n - 2)$ triangles and, therefore, the Angle-Sum Theorem for Polygons is sensible.

Find the measure of each angle in the given polygon.

7.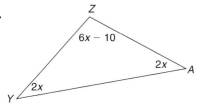

8.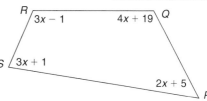

9. Is it possible for a regular polygon to have angles whose measures are each 150°? Justify your answer.

PRACTICE

Find the unknown angle measure in each quadrilateral.

10.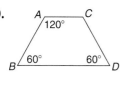

11.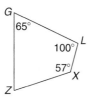

12.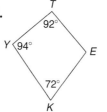

13.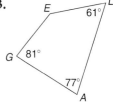

14. **HOBBIES** Emile wants to cut a piece of stained glass in the form of a regular polygon from a rectangular pane of glass as represented at the right. At what angle must she cut along adjacent edges of one piece?

15. **WRITING MATHEMATICS** Summarize the facts you learned about the measures of angles related to polygons.

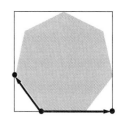

Find the measure of each angle in the given polygon.

16.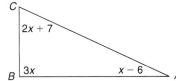

17.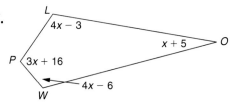

18. Is it possible for a regular polygon to have 240° as the measure of each of its angles? Justify your answer.

EXTEND

Jewelia and some of her friends began to make the design at the right by drawing the dashed lines as shown. Refer to this diagram for Exercises 19-21.

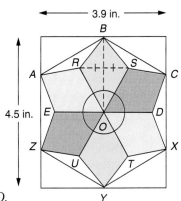

19. Find the sum of the measures of the angles of polygon *ABCDE*.

20. Each of the shaded quadrilaterals are congruent. Find m∠*ROS*, m∠*SOD*, m∠*DOT*, m∠*TOU*, m∠*UOE*, and m∠*EOR*.

21. You are given that m∠*RBS* = 80° and ∠*BRO* ≅ ∠*BSO*. Using this information and the result of Exercise 20, find m∠*BRO* and m∠*BSO*.

22. Refer to the diagrams at the right.

   a. Use the top figure to show that the sum of the measures of the angles of a convex pentagon is 540°.

   b. Repeat Exercise 22a using the bottom figure.

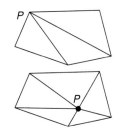

23. In Exercise 22, does it make any difference whether point *P* is located in the interior of the pentagon or on a side of the pentagon?

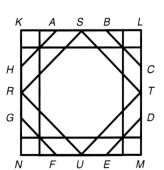

For Exercises 24–27, refer to the diagram at the right. Quadrilaterals *KLMN* and *RSTU* are squares and polygon *ABCDEFGH* is a regular octagon. Also m∠*GRU* = 45°.

24. Show that ∠*KHA* ≅ ∠*KAH*.

25. Show that $\overline{HA} \parallel \overline{RS}$.

26. Show that $\overline{GF} \parallel \overline{RU}$.

27. Show that $\overline{UT} \parallel \overline{ED}$.

28. Prove Corollary 5.4.

The measure of each angle in a regular *n*-gon is $\frac{(n-2)180°}{n}$.

29. Prove Corollary 5.6.

The measure of each exterior angle of a regular *n*-gon is $\frac{360°}{n}$.

THINK CRITICALLY

30. For what regular n-gons is the measure of each interior angle between 100° and 200°?

31. When you solve $a = \dfrac{(n-2)180}{n}$ for n, you get $n = \dfrac{360}{180-a}$. For what values of a will n be a whole number?

32. Use algebra to simplify $\dfrac{((n+1)-2)180}{n+1} - \dfrac{(n-2)180}{n}$. What does the simplified difference, $\dfrac{360}{n(n+1)}$, tell you about the change in the angle measure as n becomes larger and larger?

ALGEBRA AND GEOMETRY REVIEW

Find x. Give irrational real solutions to the nearest tenth. If an equation has no real solutions, write "no real solution."

33. $x^2 = 49$ **34.** $x^2 = 200$ **35.** $x^2 = 400$

36. $x^2 = 150$ **37.** $x^2 = 160$ **38.** $x^2 = -100$

Find the slope of each line.

39. $y = -2$ **40.** $y = -2.3x + 10$ **41.** $x = 5$ **42.** $2x - 3y = 1$

43. Find the measures of the interior angles and exterior angles of a regular decagon.

44. STANDARDIZED TESTS Given $\overline{AB} \perp \overline{XY}$ and $\overline{AB}$ intersects $\overline{XY}$ in O, which statement is false?

     A. $\angle AOX$ and $\angle YOB$ are supplementary.
     B. $\angle AOX$ and $\angle YOB$ are complementary.
     C. $\angle AOX$ and $\angle YOB$ are right angles.
     D. $m\angle AOX = m\angle YOB = 90°$.

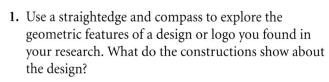

The design at the right is taken from a collection of tiling designs. The white arcs and lines show the results of using a straightedge and compass to examine the geometry of the design.

1. Use a straightedge and compass to explore the geometric features of a design or logo you found in your research. What do the constructions show about the design?

2. If you were to create a geometric design, how would you begin a rough sketch using a straightedge and compass?

3. Use a straightedge and compass, paper-folding, or geometry software to begin a design involving segment and angle bisectors. Keep a final design in mind when you begin to create.

Career
Sculptor

The beauty of a sculpture lies within the artistry and precision with which it was created. Many geometric concepts are involved in obtaining the precise details of a design and the artistry is the unique talent that the sculptor brings to the craft. Before a sculptor begins a design, whether it be chiseling marble, modeling clay, or casting in metal, the sculptor considers the layout of the location in which the sculpture will be displayed.

Decision Making

The diagram at the right below shows a small park bounded by existing streets where a monument is being constructed.

1. Suppose that sidewalks $\overline{DO}$ and $\overline{CO}$ are laid out so that m∠ODC = 42° and m∠OCD = 39°. Find m∠DOC.

2. Surveying measurements indicate that m∠OAB = 31° and m∠OBA = 56°. Find m∠AOB.

Use geometry software. Sketch a polygon like DCBAJ. Locate and mark point O in the interior of the polygon. Connect O to each of the vertices of polygon DCBAJ. Measure each of the angles in the triangles formed by your sketch.

3. Suppose that m∠ODC = 20° and m∠OCD = 20°. Find m∠DOC. What would be the location of point O relative to the pond?

4. If the decision was made to locate the monument at point O, where would the monument be located if △ODC were equiangular, that is, m∠ODC = m∠OCD = m∠DOC?

5. Why is the location of the monument independent of the sum of the measures of the angles of DCBAJ?

6. Suppose there is a small hill just east of the pond. Where would you locate the monument? Explain your response. Then find the measures of the angles of △DOC.

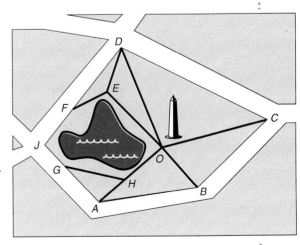

5.3 Geometry Workshop
Tessellations

Think Back/ Working Together

Work with a partner. Refer to the diagram below.

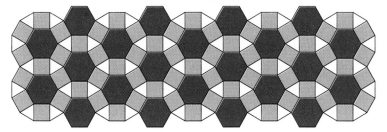

SPOTLIGHT ON LEARNING

WHAT? In this lesson you will learn
- to analyze a tessellation of the plane.
- to create a tessellation of the plane.

Why? You can use tessellation to create designs for floors, wallpaper, or fabric.

1. Identify all the different types of regular polygons that appear to make up the diagram.

2. Find the measure of each angle in each type of regular polygon you identified in Question 1.

3. Suppose the diagram above represents tiles placed on a floor. What would happen if the tiles were rearranged in such a way that some tiles overlapped and other tiles had gaps between them?

Use a compass to draw a circle. Using the same compass setting as for the radius, draw six arcs as shown at the right. Join, in order, the intersections of the circle and arcs. The polygon formed is a regular hexagon.

4. Can you arrange congruent regular hexagons to make a tile diagram with no gaps or overlap? Share your conclusions with others.

Explore

A **tessellation of the plane** is a repeating pattern of shapes that completely cover the plane without gaps or overlaps.

Work with a partner. Refer to the figures at the right.

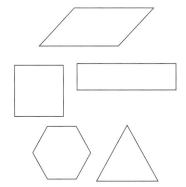

5. Choose one of the shapes shown at the right. Can you place copies of it together without gaps and having no pieces overlap and completely cover the plane? Explain your response.

6. Repeat Question 5 using one of the other shapes shown.

7. Can you make a tessellation of the plane using the parallelogram and square together? Explain and illustrate your response.

8. Can you tessellate the plane using the square and hexagon together? Explain and illustrate your response.

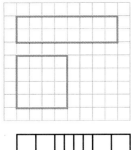

In Questions 9 and 10, use the fact that the rectangle shown at the left is twice as long and half as tall as the square. Use graph paper to make squares 4 units on a side and rectangles 8 units by 2 units.

9. A tessellation of the plane using copies of the square and rectangle is shown at the left. Explain and illustrate a different tessellation using the square and rectangle.

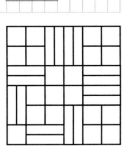

10. Is there a different arrangement than the one you explored in Question 9 that will make a tessellation of the plane? Explain and illustrate your response.

11. Explain how to generate many different tessellations of the plane from the square and the rectangle.

Make Connections

In the tile diagram on the preceding page, you can see the regular hexagon, square, and equilateral triangle meet at a point. The table below shows the angle measure of each type of polygon.

Polygon	Angle
regular hexagon	120°
square	90°
equilateral triangle	60°

Notice the tessellation pattern consists of one regular hexagon, two squares, and one equilateral triangle. Notice also the sum below.

$$120° + 2(90°) + 60° = 360°$$

12. Can you place regular pentagons, squares, and equilateral triangles together so that they meet at a point, do not overlap, and leave no gaps? Justify your response.

In a **regular tessellation**, each shape is a regular polygon, and all the shapes are congruent. The figure at the right shows a regular tessellation of equilateral triangles.

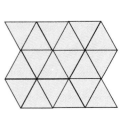

13. Identify two other shapes that can be used to form a regular tessellation.

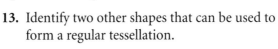
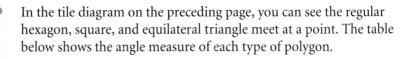

One tessellation of the plane using a scalene triangle is shown at the right. Trace the triangle and make several copies of it.

14. The dot indicates a point in which six copies of the triangle meet. Find the sum of the measures of the angles that have the red dot as a common vertex.

15. Using the triangle and copies of it, make a different tessellation of the plane.

16. Locate a point that is the common vertex of polygons in the tessellation you made in Question 15. Find the sum of the measures of the angles that have that point as a common vertex.

Frequently, an interesting tessellation arises from a transformation of one basic shape. Follow the steps below to make a puzzle piece out of a square. Trace and make several copies of the puzzle piece.

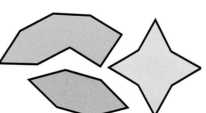

Section the square using the midpoints of the sides.

Draw four circles the same size. Position as shown.

Cut two of the circles along the dotted lines.

Paste the cut half circles along the top and right side.

17. Can you make a tessellation of the plane by rotating, flipping, and sliding copies of the puzzle piece? Explain your response.

You cannot make a tessellation of the plane using a regular pentagon. However, you can make a tessellation of it using the pentagon at the right. Trace the pentagon and make copies of it.

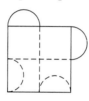

18. Using copies of the pentagon, make a tessellation of the plane.

Trace the three puzzle pieces shown at the right. Make copies of each of them.

19. Using all three pieces, make a tessellation of the plane.

20. Can you make a tessellation of the plane using only two puzzle pieces? Explain your response.

Geometry Workshop

TECHNOLOGY TIP

You can use geometry software or any computer drawing program to make your own tessellations.

USING ALGEBRA Suppose that you have e equilateral triangles, s squares, p regular pentagons, and h regular hexagons. Suppose that all of these polygons are one unit on a side.

21. Does the data $e = 1$, $s = 2$, $p = 0$, and $h = 1$ satisfy the equation $60e + 90s + 108p + 120h = 360$?

22. Find another set of values for e, s, p, and h that satisfy the equation $60e + 90s + 108p + 120h = 360$. What does your solution tell you about polygons that will make a tessellation of the plane?

Summarize

23. WRITING MATHEMATICS Briefly describe the characteristics of a tessellation of the plane.

24. THINKING CRITICALLY Write an expression you can use to test whether an equilateral triangle, square, regular octagon, or regular decagon will make a tessellation of the plane.

25. Using the regular polygons in Question 24 and your equation from that question, is there a pair of regular polygons that will make a tessellation of the plane? Justify your response.

26. GOING FURTHER The illustration at the lower left is a representation of the artwork of the Dutch graphic artist M. C. Escher (1898–1972). To see how to make it, examine the diagrams at the right below. Copy the diagram in Step 3. Draw the diagrams that come next in the sequence of diagrams to make the tessellating figure. Cut and paste to show the tessellating figure.

Step 1
Draw an equilateral triangle.

Step 2
Mark out the head and the forward edges of the wings along the bottom edge of the triangle.

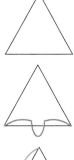

Step 3
Trace the curve from Step 2. Rotate it 60° counterclockwise. Place the rotated copy along the left edge of the triangle.

5.4 Geometry Workshop
Exploring Congruent Triangles

Think Back/Working Together

● Work with a partner. Refer to the set of figures below.

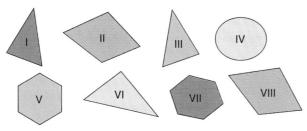

1. Describe each pair of figures as congruent or not congruent.

   a. I and III   **b.** V and VII   **c.** VII and VIII

   d. II and VIII   **e.** II and VI   **f.** I and VI

2. Explain why you cannot apply the term congruent to figures IV and V.

3. Explain why you cannot apply the term congruent to figures III and VI.

4. Write your own definition of polygon congruence using the terms, translate, flip, and turn.

Explore

● Work with a partner. You will need paper, pencil, ruler, compass, and protractor.

5. Each person should construct a triangle whose sides have length 5 cm, 7 cm, and 10 cm.

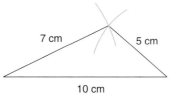

7 cm   5 cm

10 cm

6. Exchange papers with your partner. How is the triangle of your partner the same as and different from the triangle you drew? Summarize the similarities and differences.

7. Place the papers on top of one another and try to position one triangle on the other. Then hold the papers up to the light. What can you and your partner conclude about two triangles made from the same three line segments?

Continue to work with a partner.

8. Each person should construct a triangle whose sides have length 6 cm, 7 cm, and a 55° angle between them.

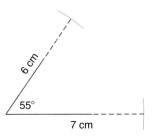

9. Exchange papers with your partner. How is the triangle of your partner the same as and different from the triangle you drew? Summarize the similarities and differences.

10. Place the papers on top of one another and try to position one triangle on top of the other. Then hold the papers up to the light. What can you and your partner conclude about two triangles made from the same two line segments and the given angle between them?

11. What do you think will happen when you and your partner try to draw triangles given segments 6 cm and 7 cm long and a 55° angle that is not between those segments?

Continue to work with a partner.

12. Now, each person should construct another triangle with one side measuring 10 cm, and angles measuring 55° and 30° each having one side lying along the segment and one vertex at each end of the segment.

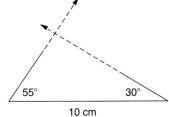

13. Exchange papers with your partner. How is the triangle of your partner the same as and different from the triangle you drew? Summarize the similarities and differences.

14. Place the papers on top of one another and try to position one triangle on top of the other. Then hold the papers up to the light. What can you and your partner conclude about two triangles made from the same line segment and pair of angles lying along that segment?

Make Connections

● Two triangles are **congruent** if there is a correspondence between them such that corresponding sides are congruent and corresponding angles are congruent.

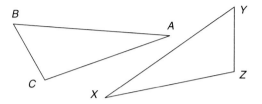

$$A \longrightarrow X \qquad B \longrightarrow Y \qquad C \longrightarrow Z$$
$$\overline{AB} \cong \overline{XY} \qquad \overline{BC} \cong \overline{YZ} \qquad \overline{CA} \cong \overline{ZX}$$
$$\angle A \cong \angle X \qquad \angle B \cong \angle Y \qquad \angle C \cong \angle Z$$
$$\triangle ABC \cong \triangle XYC$$

Decide if you think the pairs of triangles are congruent. If the triangles are congruent, state the correspondence between the vertices. Justify your reasoning.

15. $\triangle STU$ and $\triangle VWX$

$\angle U \cong \angle X$

$\angle S \cong \angle V$

$\overline{SU} \cong \overline{VX}$

16. $\triangle KLM$ and $\triangle PQR$

$\overline{KL} \cong \overline{PQ}$

$\overline{KM} \cong \overline{PR}$

$\overline{LM} \cong \overline{QR}$

17. $\triangle DEF$ and $\triangle RST$

$\angle D \cong \angle R$

$\angle E \cong \angle S$

$\overline{ED} \cong \overline{SR}$

18. $\triangle JKL$ and $\triangle TUV$

$\angle J \cong \angle T$

$\overline{JL} \cong \overline{TV}$

$\overline{JK} \cong \overline{TU}$

TRANSFORMATIONS **In the diagram at the right, $\triangle LMN$ is the reflection of $\triangle ABC$ in line ℓ.**

19. What can you say about the relationship between $\triangle LMN$ and $\triangle ABC$? Justify your response by using what you know about transformations.

20. Based on your answer to Question 19, write relationships between the sides of $\triangle LMN$ and $\triangle ABC$.

21. Based on your answer to Question 19, write relationships between the angles of $\triangle LMN$ and $\triangle ABC$.

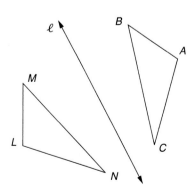

COORDINATE GEOMETRY **The diagram at the right shows** $\triangle LMN$ **and** $\triangle ABC$ **on a coordinate plane.**

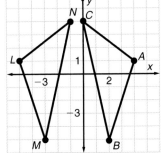

22. What appears to be the relationship between $\triangle LMN$ and $\triangle ABC$?

23. Based on your answer to Question 22, write relationships between the sides of $\triangle LMN$ and those of $\triangle ABC$.

24. Based on your answer to Question 22, write relationships between the angles of $\triangle LMN$ and of $\triangle ABC$.

25. Use the distance formula to find the lengths of the sides of $\triangle LMN$ and those of $\triangle ABC$. Which Explore activity suggests that you can conclude that $\triangle LMN \cong \triangle ABC$?

Summarize

26. **WRITING MATHEMATICS** Summarize what you have learned about how to tell whether two triangles are congruent.

27. **WRITING MATHEMATICS** Describe how you would test to see if two right triangles are congruent.

28. **WRITING MATHEMATICS** Describe how you would test to see if two regular n-gons are congruent.

29. **GOING FURTHER** Devise a test for the congruence of two triangles other than any of those you explored already. Your test should involve congruence of sides and angles.

You might find geometry software helpful for the following questions.

30. **THINKING CRITICALLY** Confirm or deny the following proposed test for the congruence of two triangles.

    If the sum of the lengths of the sides of one triangle equals the sum of the lengths of the sides of a second triangle, then those triangles are congruent.

31. **THINKING CRITICALLY** Confirm or deny the following statements about polygons and congruence.

    All squares are congruent. All regular pentagons are congruent. In general, all regular polygons with the same number of sides are congruent.

Proving Triangles Congruent

Explore

● Fold an 8.5 in.-by-11 in. sheet of paper left to right and top to bottom as shown. Fold and cut out triangles 1, 2, 3, and 4.

5.5 in.

5.5 in.

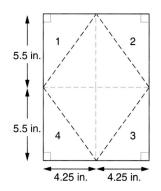

4.25 in. 4.25 in.

1. Are triangles 1, 2, 3, and 4 congruent? How does placing them on top of one another show you they are congruent?

2. What facts tell you triangles 1, 2, 3, and 4 are congruent?

Fold a second sheet of paper in half from left to right and fold it from top to bottom less than halfway down the paper. Label four triangles 5, 6, 7, and 8 as above. Fold and cut out the four triangles.

3. What mathematical facts about triangles 5, 6, 7, and 8 tell you which triangles are congruent and which are not?

Build Understanding

● Stated below are triangle congruence postulates you can use to test a pair of triangles to see if they are congruent.

SIDE-SIDE-SIDE (SSS) CONGRUENCE POSTULATE

POSTULATE 17 If three sides of one triangle are congruent to three sides of another triangle, then the triangles are congruent.

SIDE-ANGLE-SIDE (SAS) CONGRUENCE POSTULATE

POSTULATE 18 If two sides and the included angle of one triangle are congruent to two sides and the included angle of another triangle, then the triangles are congruent.

ANGLE-SIDE-ANGLE (ASA) CONGRUENCE POSTULATE

POSTULATE 19 If two angles and the included side of one triangle are congruent to two angles and the included side of another triangle, then the triangles are congruent.

COMMUNICATING
ABOUT GEOMETRY

Mass produced
automobile parts make
it possible to replace a
defective part. How
is this related to
congruence?

The SSS, SAS, and ASA congruence postulates are very useful when
you need to draw conclusions about pairs of triangles. In Example 1,
you use a postulate in conjunction with a theorem you learned about
vertical angles to show a pair of triangles are congruent.

EXAMPLE 1

In the diagram at the right, △ABC and
△EDC appear to be congruent. Use
the congruences marked to show that
the triangles are congruent.

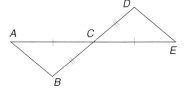

Solution

You are given in the figure that $\overline{AC} \cong \overline{EC}$ and $\overline{BC} \cong \overline{DC}$.

Since vertical angles are congruent, $\angle ACB \cong \angle ECD$.

Since the conditions of the SAS Congruence Postulate are satisfied, you
can conclude that △ABC ≅ △EDC.   ◄

The reflexive property of congruence is very useful when you
encounter two triangles that have a side in common.

EXAMPLE 2

Show that △ABC ≅ △ZCB, given the
congruences marked in the figure.

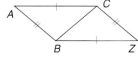

Solution

You are given in the figure that $\overline{AB} \cong \overline{ZC}$ and $\overline{AC} \cong \overline{ZB}$.

By the reflexive and symmetric properties of congruence, you can
reason that $\overline{BC} \cong \overline{CB}$.

Since the conditions of the SSS Congruence Postulate are satisfied, you
can conclude that △ABC ≅ △ZCB.   ◄

The diagrams below show there is only one triangle you can draw
given three line segments. Notice that $\overline{AB} \cong \overline{DE}$, $\overline{BC} \cong \overline{EF}$, and
$\overline{AC} \cong \overline{DF}$. By the SSS Congruence Postulate, △ABC ≅ △DEF. The
triangles are congruent and differ only in their position on the page.

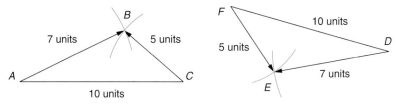

You can apply the congruence postulates and the distance formula to
the problem of determining whether two triangles on a coordinate
plane are congruent.

EXAMPLE 3

Determine whether $\triangle ABC$ and $\triangle XYZ$ are congruent.

Solution

Use the distance formula.

$$AB = \sqrt{(-4 - (-1))^2 + (5 - 1)^2} = 5$$
$$XY = \sqrt{(4 - 1)^2 + (-1 - 3)^2} = 5$$
$$AC = \sqrt{(-4 - (-5))^2 + (5 - 2)^2}$$
$$= \sqrt{10}$$
$$XZ = \sqrt{(4 - 5)^2 + (-1 - 2)^2}$$
$$= \sqrt{10}$$
$$BC = \sqrt{(-1 - (-5))^2 + (1 - 2)^2} = \sqrt{17}$$
$$YZ = \sqrt{(1 - 5)^2 + (3 - 2)^2} = \sqrt{17}$$

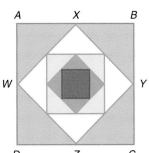

THINK BACK

The distance d between points with coordinates (x_1, x_2) and (y_1, y_2) is

$$d = \sqrt{(x_2 - x_2)^2 + (y_2 - y_1)^2}$$

The calculations of the distances show that the conditions of the SSS Congruence Postulate are satisfied. Thus, $\triangle ABC \cong \triangle XYZ$. ◄

You can use the congruence postulates to solve problems in designing patterns for quilts.

EXAMPLE 4

QUILTING In the quilt pattern at the right, quadrilateral $ABCD$ is a square and points W, X, Y, and Z are the midpoints of the sides. Show that $\triangle WAX \cong \triangle YCZ$.

CHECK UNDERSTANDING

What pieces of information in the solution of Example 4 allow you to reason that $\triangle XBY$ and $\triangle WDZ$ in the quilt pattern are congruent?

Solution

Since you are given that quadrilateral $ABCD$ is a square,

$$\overline{AB} \cong \overline{BC} \cong \overline{CD} \cong \overline{DA}$$
$$m\angle A = m\angle B = m\angle C = m\angle D = 90°$$

Since you are given W, X, Y, and Z as midpoints of the sides of the square,

$$\overline{AX} \cong \overline{XB} \cong \overline{BY} \cong \overline{YC} \cong \overline{CZ} \cong \overline{ZD} \cong \overline{DW} \cong \overline{WA}$$

Thus, $\overline{WA} \cong \overline{YC}$, $\overline{AX} \cong \overline{CZ}$ and $\angle A \cong \angle C$.

You have now satisfied the hypothesis of the SAS Congruence Postulate. Therefore, you can conclude that $\triangle WAX \cong \triangle YCZ$. ◄

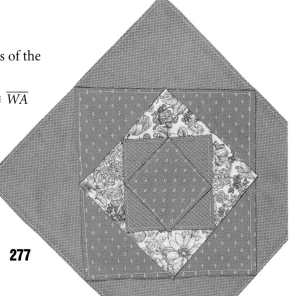

Write a logical argument to show that each pair of triangles is congruent.

1. $\triangle ABC \cong \triangle CDA$

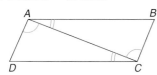

2. $\triangle KON \cong \triangle MOL$

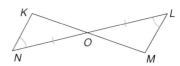

3. Given $\triangle XYZ$ with vertices $X(-5, 5)$, $Y(0, 4)$, and $Z(-4, 1)$ and $\triangle RST$ with vertices $R(5, -4)$, $S(0, -3)$, and $T(4, 0)$, show that $\triangle XYZ \cong \triangle RST$.

4. ARCHITECTURE The diagram at the right shows a decoration on the door of a building. In the diagram, D, E, and F are midpoints of $\overline{AB}$, $\overline{BC}$, and $\overline{AC}$, respectively. Also $\overline{DE} \parallel \overline{AC}$ and $\overline{DE} \cong \overline{AF}$. Show that $\triangle DBE \cong \triangle ADF$.

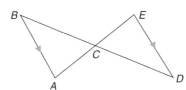

5. WRITING MATHEMATICS Compare and contrast the SSS, SAS, and ASA Congruence Postulates.

6. Give a reason for each step in the proof.

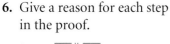

Given $\overline{AB} \parallel \overline{DE}$
C is the midpoint of $\overline{BD}$

Prove $\triangle ABC \cong \triangle EDC$

Statements	Reasons
1. $\overline{AB} \parallel \overline{DE}$	**1.** Given
2. $\angle B \cong \angle D$	**2.**
3. $\angle BCA \cong \angle ECD$	**3.**
4. C is the midpoint of $\overline{BD}$	**4.**
5. $\overline{BC} \cong \overline{CD}$	**5.**
6. $\triangle ABC \cong \triangle EDC$	**6.**

PRACTICE

Write a logical argument to show that each pair of triangles is congruent.

7. $\triangle PQR \cong \triangle NOM$

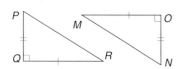

8. $\triangle YDE \cong \triangle FLE$

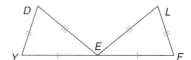

9. WRITING MATHEMATICS To apply the ASA Congruence Postulate, what information do you need to know?

10. **Floor Tiling** Square $ACEG$ at the right represents a custom flooring pattern made of wood pieces. In the diagram, $\angle C$ and $\angle G$ are right angles, $HG = \frac{1}{2} AG$, $CD = \frac{1}{2} CE$, $BC = \frac{3}{4} AC$, and $GF = \frac{3}{4} GE$. Show that $\triangle BCD \cong \triangle FGH$.

11. Given $\triangle PQR$ with vertices $P(-5, 4)$, $Q(2, 3)$, and $R(-4, 0)$ and $\triangle RST$ with vertices $R(5, -2)$, $S(-2, -1)$, and $T(4, 2)$, show that $\triangle PQR \cong \triangle RST$.

For Exercises 12 and 13, refer to the bridge diagram below.

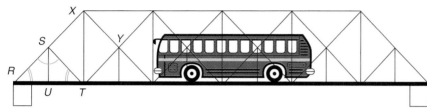

12. Given: $\overline{XS} \parallel \overline{YT}$ and $\overline{XY} \parallel \overline{ST}$
    Prove: $\triangle SXT \cong \triangle YTX$

13. Given: $\overline{SR} \cong \overline{ST}$
    Prove: $\triangle RSU \cong \triangle TSU$

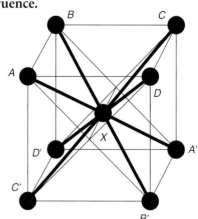

EXTEND

Exercises 14 and 15 refer to the diagram at the right below. It is the diagram you used in Example 1. In that example, you saw that $\triangle ABC \cong \triangle EDC$.

14. Is $\triangle EDC$ the image of $\triangle ABC$ under some single transformation? Explain your response.

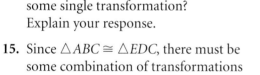

15. Since $\triangle ABC \cong \triangle EDC$, there must be some combination of transformations for which $\triangle EDC$ is the image of $\triangle ABC$. What are they?

Refer to the diagram at the right. Prove each triangle congruence.

16. Given: $\overline{AB} \parallel \overline{A'B'}$, $\overline{AB'} \parallel \overline{BA'}$, and $\overline{AB} \cong \overline{A'B'}$

    Prove: $\triangle ABX \cong \triangle A'B'X$

17. Given: $\overline{A'B} \parallel \overline{B'A}$, $\overline{AB} \parallel \overline{A'B'}$, and $\overline{AB'} \cong \overline{A'B}$

    Prove: $\triangle AXB' \cong \triangle A'XB$

18. Given: $\overline{C'D'} \parallel \overline{CD}$, $\overline{C'D} \parallel \overline{D'C}$ and $\overline{C'D'} \cong \overline{CD}$

    Prove: $\triangle D'XC' \cong \triangle DXC$

19. Given: $\overline{C'D'} \parallel \overline{CD}$, $\overline{C'D} \parallel \overline{D'C}$, and $\overline{C'D} \cong \overline{CD'}$

    Prove: $\triangle D'CX \cong \triangle DC'X$

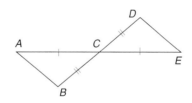

In the figure at the right, points *X*, *Y*, and *Z* lie on a line perpendicular to the plane containing *A*, *B*, *C*, and *D*. Also ∠*ZYA* ≅ ∠*ZYB* ≅ ∠*ZYC* ≅ ∠*ZYD* and ∠*YAX* ≅ ∠*YBX* ≅ ∠*YCX* ≅ ∠*YDX*.

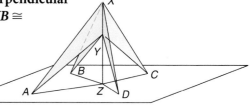

20. Prove: △*YAX* ≅ △*YBX* ≅ △*YCX* ≅ △*YDX*

21. Prove: △*YZA* ≅ △*YZB* ≅ △*YZC* ≅ △*YZD*

THINK CRITICALLY

22. In △*ABC* and △*DEF*, m∠*A* = m∠*D* = 30°, m∠*B* = m∠*E* = 60°, and m∠*C* = m∠*F* = 90°. Use examples to show that △*ABC* may not be congruent to △*DEF*. What additional information do you need to show a congruence?

23. The vertices of △*ABC* have coordinates *A*(−4, 0), *B*(−3, 5), and *C*(2, 0). The vertices *P* and *Q* of △*PQR* have coordinates *P*(−2, 0) and *Q*(−1, −5). Find the coordinates of *R* so that △*ABC* ≅ △*PQR*.

ALGEBRA AND GEOMETRY REVIEW

Solve each equation.

24. $x^2 + 5x + 6 = 0$

25. $x^2 - x - 12 = 0$

26. $x^2 + 11x + 18 = 0$

27. $2x^2 + 7x - 15 = 0$

28. $x^2 + 15x + 8 = 0$

29. $9x^2 + 3x - 2 = 0$

Find the length of each segment. Leave answers in simplest radical form.

30. *A*(−4, 1) and B(−3, 1)

31. *X*(3, 1) and *Y*(5, 5)

32. *D*(3, −1) and *K*(−2, 0)

33. Given $\overline{AB} \perp \overline{BC}$, $\overline{DC} \perp \overline{BC}$, *E* is the midpoint of $\overline{BC}$, and ∠*AEB* ≅ ∠*DEC*. Show that △*ABE* ≅ △*DCE*.

34. Standardized Tests In △*ABC* and △*XYZ*, ∠*B* ≅ ∠*Y*. Which information allows you to conclude △*ABC* ≅ △*XYZ*?

A. $\overline{AB} \cong \overline{XY}$
$\overline{AC} \cong \overline{XZ}$

B. ∠*A* ≅ ∠*X*

C. $\overline{AB} \cong \overline{XY}$
$\overline{BC} \cong \overline{YZ}$

D. ∠*C* ≅ ∠*Z*

PROJECT *Connection*   Introduce greater detail in your design.

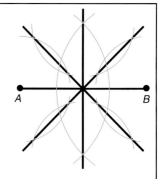

1. Using the art you have been working with, continue to look for any geometry resulting from circles, bisectors, and regular polygons.

2. The diagram at the right shows a perpendicular bisector of a segment and the angle bisectors of the right angles formed. Trace the diagram. From your copy, form a regular octagon.

3. Continue the creation of your design using constructions.

Today's graphic artists use computers to help them draw complex and beautiful designs. For example, the five-point star shown at the right was easily drawn using a star tool in a computer graphics program. The artist selects the number of points the star is to have, and the computer does the rest.

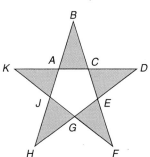

How can you determine that a five-point star consists of five congruent triangles lying along the sides of a regular pentagon? If you know that the pentagon is regular, you can use your congruence theorems and postulates to prove the triangles are congruent.

Decision Making

Suppose that polygon *ACEGJ* above is a regular polygon.

1. Classify polygon *ACEGJ*.

2. Give the reason why $\overline{AC} \cong \overline{CE} \cong \overline{EG} \cong \overline{GJ} \cong \overline{JA}$.

3. Find the measure of each of its angles. Then find the measure of each exterior angle.

4. Explain why $\angle BAC \cong \angle BCA \cong \angle DCE \cong \angle DEC \cong \angle FEG \cong \angle FGE \cong \angle HGJ \cong \angle HJG \cong \angle KJA \cong \angle KAJ$.

5. What postulate or theorem can you use to conclude that $\triangle ABC \cong \triangle CDE \cong \triangle EFG \cong \triangle GHJ \cong \triangle JKA$?

At the left is a seven-point star drawn using a computer.

6. Write a brief argument that shows the following congruences involving the blue triangles.

$\triangle XAR \cong \triangle RCS \cong \triangle SET \cong \triangle TGU \cong \triangle UJV \cong \triangle VLW \cong \triangle WNX$

7. Show that $\triangle XRS \cong \triangle UVW$.

8. Do you think the triangles formed by three consecutive vertices of polygon *XRSTUVW* are congruent? Explain your response.

9. Make a conjecture about the geometric figure you get when you draw a regular hexagon and extend all the sides until the extensions of them intersect.

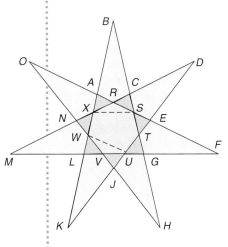

Overlapping Triangles

Tessellations use geometric figures that do not overlap. However, there are many times when overlapping geometric figures are precisely the figures you need.

Problem

Kyle and his father own a print shop. Often customers come in with the request to have their company logo created and reproduced for their stationery. The customer depends on Kyle and his father to make a technically correct rendering of it.

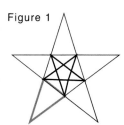

Suppose a customer wants a logo to look like the star pattern above. In what ways can Kyle use a simpler geometric shape to create the logo?

Explore the Problem

Make a copy of the star logo shown above.

Figure 1

1. Can you use the red triangle in Figure 1 and copies of it transformed in various ways to reproduce the star logo? Explain your answer.

2. Can you use the red triangle in Figure 2 and copies of it transformed in various ways to reproduce the star logo? Explain your answer.

Figure 2

3. Can you use the red triangle in Figure 3 and copies of it transformed in various ways to reproduce the star logo? Explain your answer.

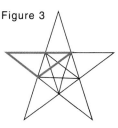

4. How are the triangles assembled from the triangle in Figure 1 different from the triangles assembled in Figures 2 and 3?

Figure 3

5. Kyle began to wonder whether other shapes could be used for the creation of the star logo. Will copies of the red quadrilateral in Figure 4 serve the purpose of creating the star logo? Explain and illustrate your answer.

Figure 4

6. What are some other polygons Kyle can use to make the logo?

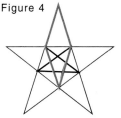

Investigate Further

● Shown at the right is a quick sketch one of Kyle's customers gave him. Kyle noticed that it was rough but could also see a pattern in the sketch.

7. Where do you think a line of symmetry is? What does this say about the locations of the two bottom points and the two points along the left and right sides of it?

The red drawing at the right shows one possibility for a representation of the customer's request.

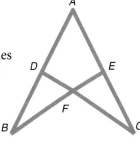

8. Name two overlapping congruent triangles that make up Kyle's representation of the logo.

9. WRITING MATHEMATICS Write a convincing argument that shows $\triangle BAE \cong \triangle CAD$.

10. Make a different representation of the customer's request.

Apply the Strategy

● Each design shown below contains one or more pairs of overlapping congruent triangles. Identify the congruent triangles.

11.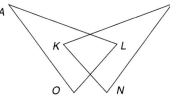

12.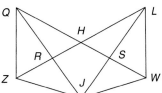

13.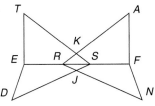

14.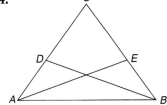

15. **Logo Design** Kyle decided to explore the creation of the rough sketch of the logo further. He drew the red triangle shown at the right. Use your knowledge of transformations to describe how to complete an accurate sketch of the logo the customer wants.

16. **Logo Design** The rough sketch at the right shows a proposed design for a logo. The design is to have a line of symmetry that contains points *B* and *F*. Write the triangle congruences for the two pairs of overlapping congruent triangles in the design.

17. **Printing** Refer to the star diagram at the right.
    a. Is it possible to create the star pattern by using transformed copies of the red triangle?
    b. How many copies of the triangle would be needed?
    c. Can you recreate the star pattern by using copies of the blue triangle? If so, how many copies would be needed?

18. **Real Estate** The diagram at the right shows a plot of land. The Masters family owns the plot marked by △*ABE* and the Gallo family owns the land marked by △*GBE*. What can you say about △*JBE*?

19. **Real Estate** The Gallo family owns the land marked by △*GBE* and the Wesson family owns the plot marked by △*AGE*. What property do they own jointly?

20. **School Logos** The students at Starsville High School want to create the logo at the right. How can they use the least number of overlapping triangles to create it?

REVIEW PROBLEM SOLVING STRATEGIES

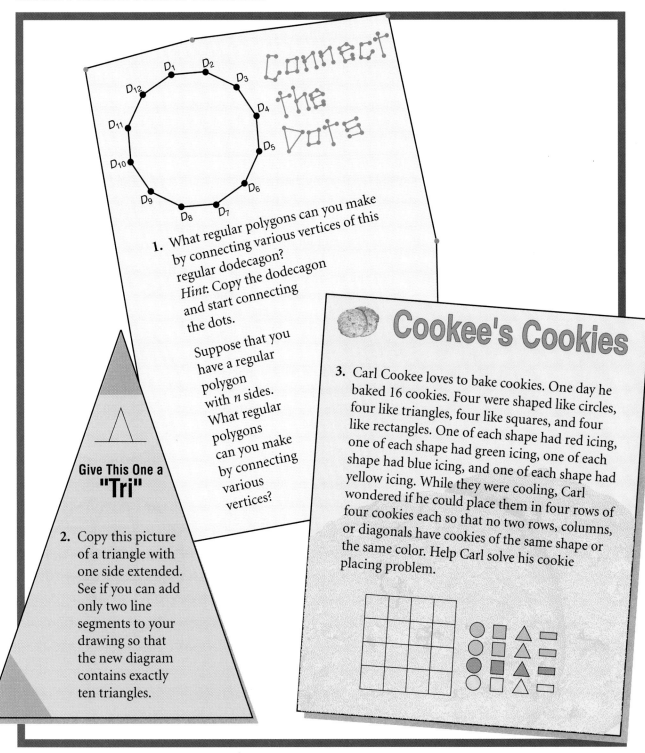

1. What regular polygons can you make by connecting various vertices of this regular dodecagon? *Hint.* Copy the dodecagon and start connecting the dots.

Suppose that you have a regular polygon with n sides. What regular polygons can you make by connecting various vertices?

Give This One a "Tri"

2. Copy this picture of a triangle with one side extended. See if you can add only two line segments to your drawing so that the new diagram contains exactly ten triangles.

Cookee's Cookies

3. Carl Cookee loves to bake cookies. One day he baked 16 cookies. Four were shaped like circles, four like triangles, four like squares, and four like rectangles. One of each shape had red icing, one of each shape had green icing, one of each shape had blue icing, and one of each shape had yellow icing. While they were cooling, Carl wondered if he could place them in four rows of four cookies each so that no two rows, columns, or diagonals have cookies of the same shape or the same color. Help Carl solve his cookie placing problem.

Geometry Excursion

Spherical Angles and Triangles

If you travel a short distance on Earth, you can think of your travel as moving in a plane. If, however, you make a long journey on Earth, such as a trip from location *A* in Canada to location *B* in Australia, you cannot really think of that journey as travel in the plane.

But you might think of your travel as moving on the surface of a sphere, perhaps on a great circle. Recall that a great circle is a cross section of a sphere whose center is the center of the sphere.

1. Refer to the top diagram at the right. Justify this statement.
   Two points *A* and *B* on a sphere lie along exactly one great circle.

2. Refer to the middle diagram at the right. Justify this statement.
   The distance between *A* and *B* along a great circle is greater than the distance between *A* and *B* along a line straight through the sphere.

Two great circles determine a **spherical angle**. The bottom diagram at the right illustrates the spherical angle formed by points *L*, *N*, and *M*.

3. Points *L* and *M* lie along the equator and the distance between them is one quarter of the length of the equator. Find m∠ *LOM*. What do you think is a reasonable measure to assign to the spherical angle *LNM*?

4. Repeat Question 3 for points *L* and *M* if the distance between them is one third of the length of the equator.

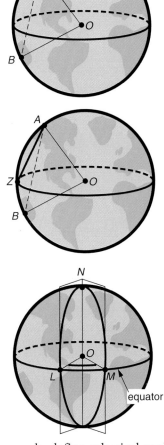

5. WRITING MATHEMATICS In your own words, define spherical angle and describe how to find its measure.

6. Justify this statement.
   Three points on a sphere and not lying on the same great circle determine a triangle on the sphere.

The diagram at the right shows the spherical triangle formed by points A and B on the equator and point N on the sphere and directly above the center O of the sphere. Plane $\mathcal{M}$ containing A, N, and O and plane $\mathcal{N}$ containing B, N, and O are perpendicular to plane $\mathcal{S}$ containing A, B, and O. Refer to this diagram for Questions 7–10.

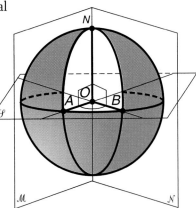

7. Would you call the spherical triangle formed by points A, N, and B a right triangle? How many right angles does it have?

8. Suppose that you move points A and B along the equator in such a way that m$\angle AOB = 90°$.
   a. What is the measure of spherical angle ANB?
   b. What is the sum of the measures of spherical angles NAB, NBA, and ANB?

9. Keep A where it is on the equator and move point B along the equator so that it is almost opposite A. What is the sum of the measures of spherical angles NAB, NBA, and ANB in spherical triangle ANB?

10. Confirm or deny each statement.
   a. If A and B are points along the equator and less than half the circumference of the equator apart and N is on the sphere and directly above its center O, then the triangle formed by A, N, and B is always a right triangle.
   b. In the spherical triangle described in part a, the sum S of the measures of the angles satisfies $180° < S < 360°$.

11. In the diagram below, a "square" is shown as if you were looking straight down on it. How do you think you might make the "square"?

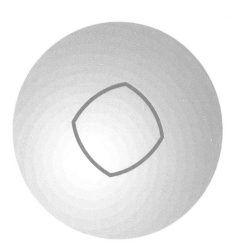

WHEN INTERNATIONAL and commercial trade caused people to view the Earth as a spherical surface, Scottish mathematician John Napier (1550–1617) developed rules by which navigators could find the length of a journey along a great circle.

In the nineteenth century, mathematicians began to look at the sphere as a geometric space in which they could make new and non-Euclidean geometries. The German mathematician Bernhard Riemann developed the geometry of the sphere, where lines are circles, triangles are curved figures, and some theorems from Euclidean geometry are no longer true.

5.7 Other Ways to Prove Triangles Congruent

Explore

● Jamie claimed that she knew of triangle congruence criteria other than the SSS, SAS, and ASA criteria presented in Lesson 5-5. She claimed that the statement below is true.

> If two sides and a nonincluded angle of one triangle are congruent to two sides and the corresponding nonincluded angle of another triangle, then the triangles are congruent.

You will need a compass and straightedge.

1. Draw a 55° angle. Along one side of it, mark point *P* at a distance 8 cm from the vertex. Set your compass at 7 cm. At point *P*, draw a circle.

2. What does your drawing tell you about Jaime's claim? Does your drawing validate Jamie's claim or does it provide a counterexample to it?

3. Repeat Questions 1 and 2 using a 3-cm segment instead of a 7-cm segment.

4. Do you think that there is a side-side-nonincluded angle triangle congruence postulate? Justify your response.

Build Understanding

● The SSS, SAS, and ASA congruence postulates along with other geometry facts provides another test for congruence.

THEOREM 5.7   ANGLE-ANGLE-SIDE (AAS) CONGRUENCE THEOREM

If two angles and a nonincluded side in one triangle are congruent to two angles and the corresponding nonincluded side in another triangle, then the triangles are congruent.

Notice that the hypothesis of the theorem above is different from the hypothesis of the ASA Congruence Postulate.

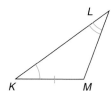
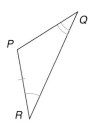

EXAMPLE 1

In the diagram at the right, $\angle F \cong \angle G$, $\angle EKF \cong \angle HKG$, and quadrilateral *EKHJ* is a square. Prove that $\triangle EFK \cong \triangle HGK$.

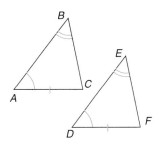

Solution

The angle congruences below are given.

$\angle EFK \cong \angle HGK$ and $\angle EKF \cong \angle HKG$

Since *EKHJ* is a square, $\overline{EK} \cong \overline{HK}$.

The hypothesis of the Angle-Angle-Side Congruence Theorem is satisfied. So, you can conclude that $\triangle EFK \cong \triangle HGK$. ◄

The proof of Theorem 5.7, the Angle-Angle-Side Congruence Theorem, utilizes the Angle-Side-Angle Congruence Postulate.

Given $\angle A \cong \angle D$, $\angle B \cong \angle E$, and $\overline{AC} \cong \overline{DF}$

Prove $\triangle ABC \cong \triangle DEF$

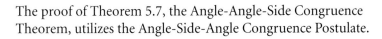

Plan for Proof   Show $\angle C \cong \angle F$. Then apply the Angle-Side-Angle Congruence Postulate.

Proof   Write a two-column proof.

Statements	Reasons
1. $\angle A \cong \angle D$, $\angle B \cong \angle E$, $\overline{AC} \cong \overline{DF}$	**1.** Given
2. $m\angle A = m\angle D$, $m\angle B = m\angle E$	**2.** Definition of congruent angles
3. $m\angle A + m\angle B + m\angle C = 180°$ $m\angle D + m\angle E + m\angle F = 180°$	**3.** Angle-Sum Theorem for Triangles
4. $m\angle A + m\angle B + m\angle C =$ $m\angle D + m\angle E + m\angle F$	**4.** Substitution
5. $m\angle C = m\angle F$	**5.** Subtraction Property
6. $\angle C \cong \angle F$	**6.** Definition of congruent angles
7. $\triangle ABC \cong \triangle DEF$	**7.** ASA Congruence Postulate

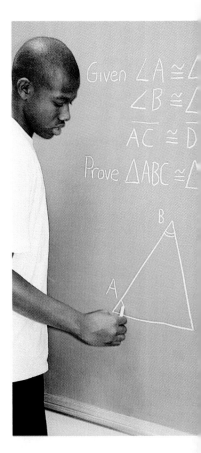

In Explore, you examined the "Side-Side-Angle" situation, in which two sides and a nonincluded angle of one triangle were congruent to two sides and a nonincluded angle of another. You probably discovered that this does not necessarily result in congruent triangles. However, a special situation does occur when the nonincluded angle is a right angle.

Recall that a triangle with one right angle is a right triangle. In a right triangle, the side opposite the right angle is the **hypotenuse**. The sides opposite the acute angles are the **legs**. The following is a congruence theorem that applies only to right triangles.

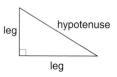

| THEOREM 5.8 | HYPOTENUSE-LEG (HL) CONGRUENCE THEOREM |

If the hypotenuse and one leg of a right triangle are congruent to the hypotenuse and one leg of another right triangle, then the triangles are congruent.

EXAMPLE 2

In the figure at the right, $\overline{AB} \cong \overline{EF}$ and $\overline{BC} \cong \overline{FB}$. Prove $\triangle ABC \cong \triangle EFB$.

Solution
Both $\triangle ABC$ and $\triangle EFB$ are right triangles.

Since $\overline{AB} \cong \overline{EF}$, the hypotenuses of $\triangle ABC$ and $\triangle EFB$ are congruent. Since $\overline{BC} \cong \overline{FB}$, a pair of legs in $\triangle ABC$ and $\triangle EFB$ are congruent.

By the Hypotenuse-Leg Congruence Theorem, $\triangle ABC \cong \triangle EFB$. ◄

You can apply congruence postulates and theorems to triangles that do not lie in the same plane.

EXAMPLE 3

RADIO AND TELEVISION The tower at the right is secured by four strong cables all the same length. Show that the tower, the support cables, and the ground determine four congruent right triangles.

Solution
The tower is perpendicular to the ground. So $\triangle AXP$, $\triangle BXP$, $\triangle CXP$, and $\triangle DXP$ are right triangles and share leg $\overline{XP}$.

The four support cables represented by $\overline{AP}$, $\overline{BP}$, $\overline{CP}$, and $\overline{DP}$ are the hypotenuses of the right triangles and are congruent.

By the Hypotenuse-Leg Congruence Theorem, you can conclude that $\triangle AXP \cong \triangle BXP \cong \triangle CXP \cong \triangle DXP$. ◄

CHECK UNDERSTANDING

In Example 3, suppose that AP = 792 ft. How long are each of the other support cables? How much cable is used in all four support cables?

TRY THESE

Use the drawing and given information to name the postulate or theorem (SSS, SAS, ASA, AAS, HL) that can be used to prove the triangles congruent.

1. △PQS and △RSQ

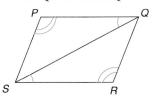

2. △XYZ and △WXL

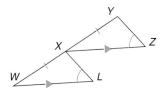

3. △PRY and △SQX

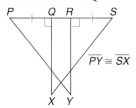

$\overline{PY} \cong \overline{SX}$

4. Quadrilateral *ACEG* is a square. △BCD and △HGF

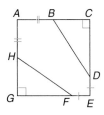

5. Quadrilateral *ABCD* is a square. △ABE and △BAF

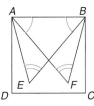

6. Point *M* is the midpoint of $\overline{LN}$. △KLM and △OMN

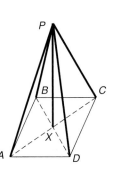

7. RADIO AND TELEVISION The angles the cables make with the ground at points *A* and *D* are congruent. That is, ∠XAP ≅ ∠XDP. The tower is perpendicular to the ground. Show that △XAP ≅ △XDP.

8. WRITING MATHEMATICS Summarize the various methods available to you to show that two given triangles are congruent. Rephrase the SAS Congruence Postulate for right triangles.

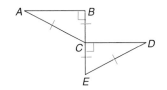

PRACTICE

Use the drawing and given information to name the postulate or theorem (SSS, SAS, ASA, AAS, HL) that can be used to prove the triangles congruent.

9. △GJK and △HKJ

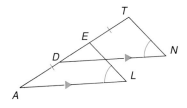

10. △OAL and △FTL

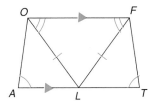

11. △AEL and △DTN

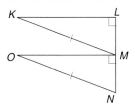

12. △KGH and △JHK

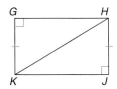

13. △JNQ and △YKF

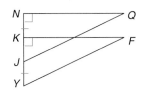

14. △ABC and △DCE

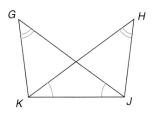

15. ART Quadrilateral *ABCD* at the right is a square in a quilt pattern, and $\overline{WC} \cong \overline{AY} \cong \overline{DX} \cong \overline{BZ}$. Show that $\triangle WDC \cong \triangle YBA$.

16. WRITING MATHEMATICS Whenever you use AAS to prove two triangles, explain why you could have used ASA instead.

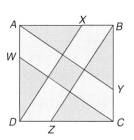

ART Show that each pair of triangles is congruent.

17. Given $\overline{MP} \parallel \overline{OR}$; $\overline{AB} \cong \overline{BC}$
Prove $\triangle APB \cong \triangle COB$.

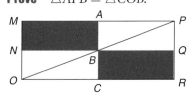

18. Given $\overline{AC} \perp \overline{NQ}$; $\overline{AB} \cong \overline{BC}$; $\overline{NA} \cong \overline{QC}$
Prove $\triangle NAB \cong \triangle QCB$.

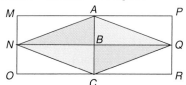

EXTEND

In Exercises 19 and 20, refer to the diagram at the right. In the diagram, $\overline{FC} \perp \overline{AE}$, $\overline{BF} \cong \overline{DF}$, and $\angle FAB \cong \angle FED$.

19. Prove $\triangle BFC \cong \triangle DFC$.

20. Prove $\triangle FAC \cong \triangle FEC$.

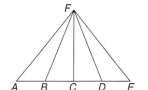

Each theorem below correlates with a previously shown congruence method.

HA THEOREM If the hypotenuse and one acute angle of one right triangle are congruent to the hypotenuse and one acute angle of a second right triangle, then the triangles are congruent.

LA THEOREM If one leg and one acute angle of one right triangle are congruent to the corresponding one leg and one acute angle of a second right triangle, then the triangles are congruent.

Use the HA Theorem or the LA Theorem to do the following proofs. Refer to the drawing at the right.

21. Given $\overline{CD} \perp \overline{AB}$
    $\overline{CD}$ bisects $\angle ACB$
Prove $\triangle ADC \cong \triangle BDC$

22. Given $\overline{AE} \perp \overline{BC}$; $\overline{BF} \perp \overline{AC}$
    $\angle EAB \cong \angle FBA$
Prove $\triangle EAB \cong \triangle FBA$

23. Given $\angle AEC$ and $\angle BFC$
    are right angles.
    $\overline{AE} \cong \overline{BF}$
Prove $\triangle AEC \cong \triangle BFC$

24. Given $\angle AFG$ and $\angle BEG$
    are right angles.
    $\overline{AG} \cong \overline{BG}$
Prove $\triangle AFG \cong \triangle BEG$

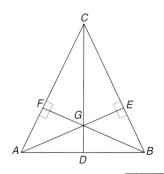

Use the information given to prove each triangle congruence.

25. In the drawing at the right $\overline{AB}$ represents a ladder leaning against a wall at $\overline{BC}$. The top of the ladder slides down the wall to the position represented by $\overline{EF}$. If $\angle ABC \cong \angle CEF$, prove that $\triangle ABC \cong \triangle FEC$.

26. In addition to the information in Question 25, what additional facts would you need to prove $\triangle EAG \cong \triangle BFG$, using ASA?

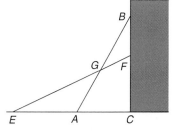

THINK CRITICALLY

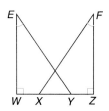

27. In the diagram at the right, what must be true of *WX* and *YZ* to assure that △*EYW* ≅ △*FXZ*? Using your answer and the given information, prove that △*EYW* ≅ △*FXZ*.

28. What happens to △*EYW* ≅ △*FXZ* when $WX = YZ = \frac{1}{2}WZ$? Prove your conjecture.

ALGEBRA AND GEOMETRY REVIEW

Evaluate each expression.

29. $2(3 + 2)^2 - 1$

30. $(2 - 3)^2(5 + 2)^3$

31. $2^3 4^2$

32. $\dfrac{(5 - 4)^3}{(2 + 3)^3}$

33. $\dfrac{(10 - 8)^2}{(3 + 1)^2}$

34. $\dfrac{3^3}{2^2} + \dfrac{(3 + 2)^2}{(3 - 4)^2}$

Write the converse of each statement. Tell whether the converse is true or false.

35. If two distinct points lie in a plane, the line joining them lies in that plane.

36. If two angles form a linear pair, then they are supplementary.

37. In the diagram at the right, $\overline{ML}$ lies along the bisector of ∠*XLY*. Show that △*XLM* ≅ △*YLM*.

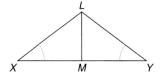

38. The supplement of an angle is five times as large as the complement of the angle. Find the measure of the angle.

For Question 39 use the drawing at the right.

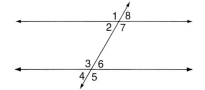

39. Name a pair of corresponding angles.

40. Name a pair of interior angles on the same side of the transversal.

PROJECT *Connection*　　Continue to add detail to your design.

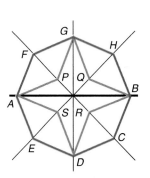

1. The diagram at the right is made by using the perpendicular bisector of $\overline{AB}$ and the angle bisectors of the right angles formed. The sides of the star inside regular octagon *AFGHBCDE* are congruent. Also $\overline{AP} \cong \overline{AF}$. Show that △*FAP* ≅ △*CBR*.

2. Look for and prove triangle congruences in one of the designs you gathered from books, magazines, or the community.

3. Look for and prove triangle congruences in the design you are making.

Flow Proofs

Discussion

● Work with a partner. Consider the diagram that Malcolm and Lupita use to help them organize different types of triangles.

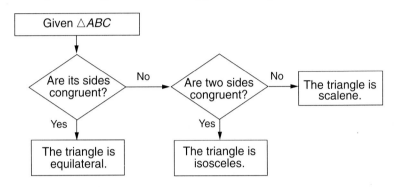

1. Do you think their scheme is logical? Explain your response.

2. Malcolm and Lupita decided they could modify the existing diagram so as to classify triangles by angles. Do you think this is possible? What would the revised diagram look like?

3. Malcolm and Lupita then decided to make a chart that would help them prove whether a pair of triangles is congruent. They decided on questions that would help them choose the appropriate congruence postulate or theorem. Assemble such a chart.

Build a Case

● Malcolm and Lupita wondered if they could organize their proof work in a *flow proof*. Their first attempt at a flow proof is shown below.

Given $\overline{BC} \parallel \overline{AD}$; $\overline{AB} \parallel \overline{CD}$

Prove $\triangle ABC \cong \triangle CDA$

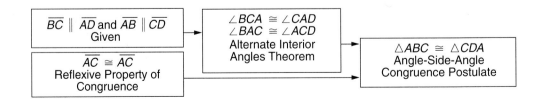

4. Did Malcolm and Lupita prove what they set out to prove? Explain how they represented their steps.

5. To verify that Malcolm and Lupita reasoned validly, you can rewrite their flow proof as a two-column proof. Rewrite their work in a two-column format.

Malcolm and Lupita thought they could rewrite a two-column proof in flow-proof format. Below is the proof they wanted to rewrite.

Given $\overline{AE} \cong \overline{BE}$;
Point E is the midpoint of $\overline{DC}$;
$\overline{AD} \perp \overline{DC}$ and $\overline{BC} \perp \overline{DC}$

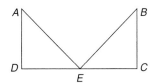

Prove $\triangle ADE \cong \triangle BCE$

Proof

Statements	Reasons
1. $\overline{AE} \cong \overline{BE}$; Point E is the midpoint of $\overline{DC}$; $\overline{AD} \perp \overline{DC}$ and $\overline{BC} \perp \overline{DC}$	1. Given
2. $\overline{DE} \cong \overline{CE}$	2. Definition of midpoint
3. $\angle ADE$ and $\angle BCE$ are right angles.	3. If two lines are perpendicular, then they form right angles.
4. $\triangle ADE$ and $\triangle BCE$ are right triangles.	4. Definition of right triangle
5. $\triangle ADE \cong \triangle BCE$	5. Hypotenuse-Leg Theorem

6. Complete the flow proof for the two-column proof above.

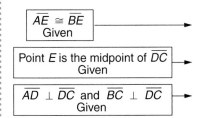

$\triangle ADE \cong \triangle BCE$
Hypotenuse-Leg Theorem

7. First write either a flow proof or a two-column proof. Then rewrite your proof in the other format.

Given $\overline{GH} \parallel \overline{LK}$ and $\overline{GJ} \cong \overline{KJ}$

Prove $\triangle GJH \cong \triangle KJL$

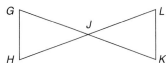

EXTEND AND DEFEND

Use the given information and the information in each diagram to write a flow proof.

8. Given S is the midpoint of $\overline{TV}$; $\overline{TR} \cong \overline{VR}$
   Prove $\triangle TSR \cong \triangle VSR$

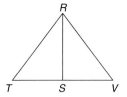

9. Given $\overline{WY}$ bisects $\angle ZWX$ and $\angle ZYX$
   Prove $\triangle WZY \cong \triangle YXW$

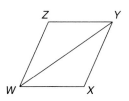

10. Given $\angle A \cong \angle B$; $\angle 1 \cong \angle 2$
   Prove $\triangle XAY \cong \triangle YBX$

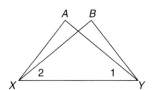

11. Given $\overline{EB} \perp \overline{AC}$; $\overline{EB} \cong \overline{CB}$; $\angle E \cong \angle C$
   Prove $\triangle AEB \cong \triangle DCB$

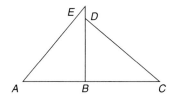

12. In $\triangle QST$ and $\triangle USR$, $\overline{QS} \cong \overline{US}$, $\angle SQT \cong \angle SUR$, and $\angle STQ \cong \angle SRU$. Prove that $\triangle QST \cong \triangle USR$.

13. In $\triangle XKY$, L is on $\overline{XY}$, $\overline{KL}$ bisects $\angle K$ and $\overline{KL} \perp \overline{XY}$. Prove that $\triangle XKL \cong \triangle YKL$.

14. Given $\angle O$ and $\angle E$ are right angles and $\overline{RO} \cong \overline{ES}$. Prove that $\triangle RES \cong \triangle SOR$.

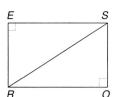

15. Given $\overline{AC} \cong \overline{BC}$ and $\overline{CD}$ bisects $\angle ACB$. Prove that $\triangle ACD \cong \triangle BCD$.

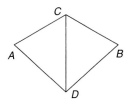

16. WRITING MATHEMATICS List some advantages of writing a proof in flow-proof format.

Now You See It
Now You Don't

STAR SEARCH

Find the four-point star.

5.9 Using Congruent Triangles

Explore

Follow the instructions below to produce a figure like the one at the right.

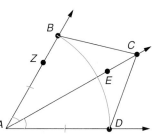

- Draw $\angle DAE$.

- Copy $\angle DAE$ so that the new vertex is A, one side is along $\overrightarrow{AE}$, and the angle is outside $\angle DAE$.

- Draw an arc with center A, passing through D, and intersecting $\overrightarrow{AZ}$. Mark the intersection of the arc and $\overrightarrow{AZ}$ as point B.

- Choose a point C on $\overrightarrow{AE}$, and draw $\overline{BC}$ and $\overline{DC}$.

1. What can you say about $\triangle DAC$ and $\triangle BAC$? Justify your answer.

2. Measure $\angle ABC$ and $\angle ADC$. What appears to be the relationship between their measures?

3. Do you expect the conclusion in Question 2 to remain the same when you measure $\angle ACB$ and $\angle ACD$? $\overline{BC}$ and $\overline{DC}$? Explain your answer.

Build Understanding

By the definition of congruent polygons, if two triangles are congruent, then the corresponding angles and sides are congruent. This is often abbreviated as **CPCTC** which stands for **C**orresponding **P**arts of **C**ongruent **T**riangles are **C**ongruent.

EXAMPLE 1

Justify the fact that $\triangle KLC \cong \triangle RFW$. Then list the side and angle congruences.

Solution

By the SSS Congruence Postulate, you can conclude that

$$\triangle KLC \cong \triangle RFW$$

By CPCTC,

$\overline{KL} \cong \overline{RF}$ $\overline{LC} \cong \overline{FW}$ $\overline{KC} \cong \overline{RW}$

$\angle C \cong \angle W$ $\angle K \cong \angle R$ $\angle L \cong \angle F$ ◄

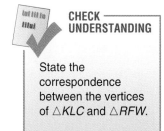

EXAMPLE 2

BRIDGE CONSTRUCTION In the diagram at the right, G is the midpoint of the bridge bed, $\overline{AE}$. Show that bridge supports $\overline{BG}$ and $\overline{DG}$ are congruent.

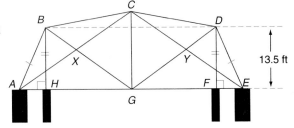

13.5 ft

Solution

You are given that $\triangle ABH$ and $\triangle EDF$ are right triangles, $\overline{AB} \cong \overline{ED}$, and $\overline{BH} \cong \overline{DF}$. By the Hypotenuse-Leg Theorem, $\triangle ABH \cong \triangle EDF$. By CPCTC, you can conclude that $\angle BAH \cong \angle DEF$.

Since G is the midpoint of $\overline{AE}$, $\overline{AG} \cong \overline{EG}$. You already know that $\angle BAH \cong \angle DEF$ and that $\overline{AB} \cong \overline{ED}$. By the SAS Congruence Postulate, $\triangle ABG \cong \triangle EDG$. By CPCTC, you can conclude that $\overline{BG} \cong \overline{DG}$.

Therefore, bridge supports $\overline{BG}$ and $\overline{DG}$ are congruent. ◀

COMMUNICATING ABOUT GEOMETRY

Discuss how you would rewrite the proof in Example 3 in flow-proof format.

Once you conclude that two angles are congruent using CPCTC, you may be able to conclude that two lines are parallel.

EXAMPLE 3

In the diagram at the right, L is the midpoint of both $\overline{JR}$ and $\overline{KT}$. Show that $\overline{JK} \parallel \overline{TR}$.

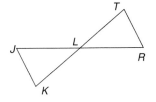

Solution

Organize your work in a two-column format.

Statements	Reasons
1. L is the midpoint of both $\overline{JR}$ and $\overline{KT}$.	1. Given
2. $\overline{JL} \cong \overline{RL}$; $\overline{KL} \cong \overline{TL}$	2. Definition of midpoint
3. $\angle JLK \cong \angle RLT$	3. Vertical angles are congruent.
4. $\triangle KJL \cong \triangle TRL$	4. SAS Congruence Postulate
5. $\angle KJL \cong \angle TRL$	5. CPCTC
6. $\overrightarrow{JK} \parallel \overrightarrow{TR}$	6. Converse of Alternate Interior Angles Theorem.
7. $\overline{JK} \parallel \overline{TR}$	7. Segments of parallel lines are parallel.

TECHNOLOGY TIP

You can use geometry software to make congruent triangles. Construct one triangle and then translate it, reflect it or rotate it. The result will be a pair of congruent triangles.

To prove that two lines or segments are perpendicular, you can use CPCTC to show that two supplementary angles are congruent.

EXAMPLE 4

Logo Design Members of the Art Club at Lockwood High School plan to design a new logo. To make the design, they want the following congruences.

$\overline{QR} \cong \overline{SR}$, $\overline{QA} \cong \overline{SA}$, $\overline{QB} \cong \overline{SB}$, $\overline{QC} \cong \overline{SC}$, and $\overline{QT} \cong \overline{ST}$.
Prove that $\overline{QS} \perp \overline{RT}$.

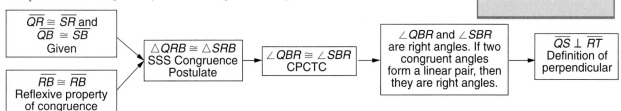

PROBLEM SOLVING TIP

When a problem contains a great deal of information, you must sort out those pieces of information that will help you solve the problem from information that is unnecessary.

Solution

You can organize your thinking in a flow proof.

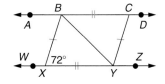

$\overline{QR} \cong \overline{SR}$ and $\overline{QB} \cong \overline{SB}$ Given		
$\overline{RB} \cong \overline{RB}$ Reflexive property of congruence		

$\triangle QRB \cong \triangle SRB$ SSS Congruence Postulate → $\angle QBR \cong \angle SBR$ CPCTC → $\angle QBR$ and $\angle SBR$ are right angles. If two congruent angles form a linear pair, then they are right angles. → $\overline{QS} \perp \overline{RT}$ Definition of perpendicular

Using CPCTC, you can find angle measures and lengths of segments.

EXAMPLE 5

From the information given in the figure at the right, find m$\angle BCY$.

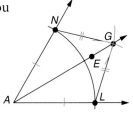

Solution

Because $\overline{BX} \cong \overline{YC}$, $\overline{BC} \cong \overline{YX}$, and $\overline{BY} \cong \overline{YB}$, $\triangle BXY \cong \triangle YCB$ by the SSS Congruence Postulate. By CPCTC, $\angle BXY \cong \angle YCB$. Therefore, m$\angle BCY = 72°$. ◄

In earlier lessons, you learned to construct an angle bisector. According to the construction, you draw segments $\overline{AL}$, $\overline{AN}$, $\overline{NG}$, and $\overline{LG}$ such that $\overline{AL} \cong \overline{AN}$ and $\overline{NG} \cong \overline{LG}$.

Outlined below is a justification of the construction of the angle bisector of a given angle.

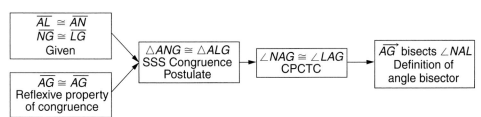

$\overline{AL} \cong \overline{AN}$ $\overline{NG} \cong \overline{LG}$ Given		
$\overline{AG} \cong \overline{AG}$ Reflexive property of congruence		

$\triangle ANG \cong \triangle ALG$ SSS Congruence Postulate → $\angle NAG \cong \angle LAG$ CPCTC → $\overrightarrow{AG}$ bisects $\angle NAL$ Definition of angle bisector

Identify the postulate or theorem you would use to prove the triangles congruent. List all the corresponding side and angle congruences.

1. $\triangle PQR \cong \triangle NTM$

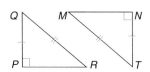

2. $\triangle ADE \cong \triangle XYZ$

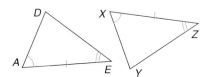

3. HYDRAULICS In the diagram of the water tower at the right, $\overline{HL} \parallel \overline{DO}$. Show that $\overline{HD} \cong \overline{LO}$.

4. WRITING MATHEMATICS Briefly explain the concept of CPCTC in a letter to a friend.

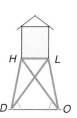

Use the given information and the diagram, prove each statement.

5. Given $\overline{JK} \parallel \overline{TR}$
Prove $\overline{JL} \cong \overline{RL}$

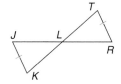

6. Prove $\angle B \cong \angle E$

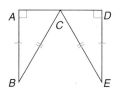

Refer to the diagram at the right. Find the measure of each angle or the length of each segment.

7. $\overline{MG}$ **8.** $\angle KLF$ **9.** $\angle MGL$ **10.** $\overline{LF}$

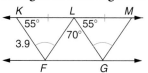

Justify each triangle congruence. List the side and angle congruences.

11. $\triangle DEC \cong \triangle STW$

12. $\triangle ABC \cong \triangle PML$

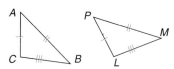

13. WRITING MATHEMATICS Write a brief justification with illustrations to justify the perpendicular bisector construction.

14. CARPENTRY Show that the stool pictured at the right makes congruent angles with the floor. That is, show that $\angle EAB \cong \angle FDC$.

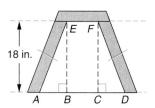

18 in.

Use the given information and the diagram to prove each statement.

15. Prove $\overline{PQ} \parallel \overline{ST}$

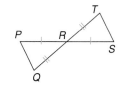

16. Given $\overline{LM} \parallel \overline{OP}$
Prove $\overline{LN} \parallel \overline{OM}$

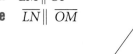

For Exercises 17–20, refer to the diagram at the right.

17. Find m∠*BYA*.   **18.** Find *CY*.

19. Find m∠*YCB*. Use the result to find m∠*YAB*.

20. Suppose $\overline{AC} \parallel \overline{XY}$. Find m∠*AYX*.

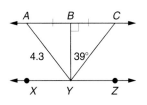

EXTEND

GARDENING **Nancy and Quan planted a young tree straight up on level ground. To support it, they made a circle on the ground with the tree at the center. That is, *OA* = *OB* = *OC*. They attached three wires to the tree at *P*, 5 ft above the ground, and to the circle on the ground at *A*, *B*, and *C*.**

21. Show the wire supports are the same length: $\overline{AP} \cong \overline{BP} \cong \overline{CP}$.

22. Show that the supports make congruent angles with the ground, that is, ∠*PAO* ≅ ∠*PBO* ≅ ∠*PCO*.

23. Suppose Nancy and Quan want to attach more support wires to the circle and the tree 5 ft above the ground. Will those be as long as the others they attached? Explain your answer.

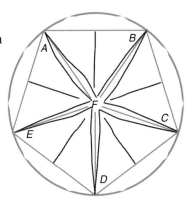

MECHANICS **In the diagram below, the three pulleys are equally spaced apart. Points *B* and *D* are midpoints of $\overline{AC}$ and $\overline{CE}$. The vertical deflections $\overline{BF}$ and $\overline{DG}$ are congruent.**

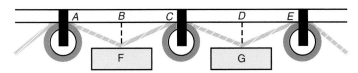

24. Show that the rope from *A* to *F* to *C* is evenly divided by *F*.

25. Show the angles of deflection, ∠*FAB* and ∠*FCB*, are congruent.

26. In the flower pattern at the right, *ABCDE* is a regular pentagon and *AF* = *BF* = *CF* = *DF* = *EF*. Write a proof showing ∠*AFB* ≅ ∠*BFC* ≅ ∠*CFD* ≅ ∠*DFE* ≅ ∠*EFA*. Then find the measure of each of those angles.

Toni decided to make an alphabet for a class project. Her letter **W** is shown at the right with *PQ = QR = XY* and *PX = QX = QY = RY.*

27. Draw any segments needed and prove that $\angle PXQ \cong \angle QYR \cong \angle XQY.$

28. Toni observed that the result of Exercise 27 implies that $\overline{PX} \parallel \overline{QY}$ and that $\overline{QX} \parallel \overline{RY}$. How do you know her observation is correct?

THINK CRITICALLY

Refer to the diagram at the right.

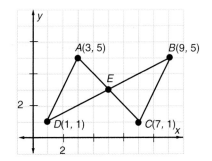

29. Use the distance formula to find *AE, DE, AD, BC, EC,* and *BE.*

30. Use the result of Exercise 29 to show that $\triangle ADE \cong \triangle CBE.$

31. What can you conclude about $\angle D$ and $\angle B$? Justify your answer.

In the diagram at the right, the figure is a cube. Also, $\overline{AW} \perp \overline{WK}$, $\overline{AW} \perp \overline{WL}$ and $\overline{KY} \cong \overline{LY}$.

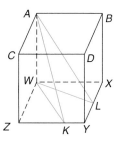

32. Show that $\overline{WK} \cong \overline{WL}.$

33. Use Exercise 32 to show that $\overline{AK} \cong \overline{AL}.$

ALGEBRA AND GEOMETRY REVIEW

Solve each equation.

34. $3(x + 2) = 7$

35. $2x - 5 = 5x + 2$

36. $2(x - 1) = 5x$

37. $3x + 6 = 2x$

38. $10x + 5 - 2x = 21$

39. $3x - 2 = 2x - 3$

Determine whether each pair of lines is parallel, perpendicular, or neither.

40. $2x - 3y = 1; 3x + 2y = 4$

41. $x = 3; y = 5$

42. $3x + 7y = 1; 3x + 7y = 4$

43. $4x - 5y = 1; 3x + 5y = 4$

44. STANDARDIZED TESTS Which of the conditionals is the contrapositive of $p \Rightarrow q$?

    A. $q \Rightarrow p$ **B.** $p \Rightarrow \sim q$ **C.** $\sim p \Rightarrow \sim q$ **D.** $\sim q \Rightarrow \sim p$

An illustrator is a person who draws visual representations of a writer's ideas. This textbook is filled with technical illustrations rendered by a technical illustrator. Many times grid paper is used for illustrations so that measurements can be made quickly and accurately. Then the drawing is traced onto a plain sheet of paper where details and color may be added.

Decision Making

In the questions below, use grid paper, straightedge, and compass.

1. Make a logo using squares and right triangles. Precisely illustrate the logo. Do you see any congruent triangles?

2. Trace your logo onto a plain sheet of paper and add detail and color.

3. Make another logo using squares and circles.

4. Trace this new logo onto a sheet of plain paper and add detail and color.

5. You and some classmates want to propose the shape for a new park near your school. The park must be a pentagon and have some symmetry. Illustrate a park that meets those demands.

6. A logo consists of $\triangle ABC$ with right angle at C, a copy of it rotated 90° about a vertex, a rotation of the copy 90° about the image of that vertex, and so on. Instructions to the artist are ambiguous if different illustrations can be obtained from the same set of instructions. Are these instructions ambiguous? Use technical illustrations to justify your response.

· · · CHAPTER REVIEW · · ·

Choose the word from the list that correctly completes each statement.

1. The statement "If two triangles are congruent, then parts that correspond are congruent" is known as __?__ .

2. A triangle in which one angle measures more than 90° is __?__ .

3. A proof using diagrams to show steps is __?__ .

4. A triangle in which no two sides are congruent is __?__ .

5. A way to cover the plane without gaps or overlap is __?__ .

a. a scalene triangle

b. a tessellation

c. an obtuse triangle

d. a flow proof

e. CPCTC

Lesson 5.1 TRIANGLES	pages 255–259

- The sum of the measures of the angles of a triangle is 180°. The measure of an exterior angle of a triangle equals the sum of the measures of the two remote interior angles.

6. Find m∠ABC.

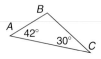

7. Find m∠KLM.

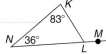

8. Find m∠PQR.

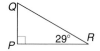

Lesson 5.2 ANGLES OF POLYGONS	pages 260–266

- The sum of the measures of the angles of a convex polygon with n sides is $(n - 2)180°$.

- The measure of each angle in a regular n-gon is $\dfrac{(n - 2)180°}{n}$.

- The sum of the measures of the exterior angles of a convex polygon, one angle at each vertex of the polygon, is 360°. The measure of each exterior angle of a regular n-gon is $\dfrac{360°}{n}$.

9. Find m∠ABC.

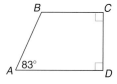

10. Find x.

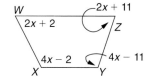

11. Find m∠JKL.

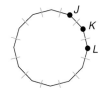

Lesson 5.3 TESSELLATIONS	pages 267–270

- A tessellation of the plane is a covering of it by nonoverlapping figures leaving no gaps.

12. Can you tessellate the plane using only equilateral triangles and squares? Justify your response.

- Two triangles are congruent if three sides of one triangle are congruent to three sides of another triangle (SSS); if two sides and the included angle of one triangle are congruent to two sides and the included angle of another triangle (SAS); if two angles and the included side of one triangle are congruent to two angles and the included side of another triangle (ASA); if two angles and a nonincluded side in one triangle are congruent to two angles and the corresponding nonincluded side in another triangle (AAS).

- Two right triangles are congruent if the hypotenuse and one leg of a right triangle are congruent to the hypotenuse and the corresponding leg of another right triangle. (HL)

13. In $\triangle DEF$ and $\triangle RST$, $\overline{DF} \cong \overline{RT}$, $\angle FDE \cong \angle TRS$, and $\angle DFE \cong \angle RTS$. Can you conclude that $\triangle DEF \cong \triangle RST$? If so, give a justification.

14. **Prove** $\triangle RST \cong \triangle VUT$

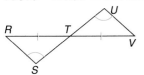

15. **Prove** $\triangle RST \cong \triangle VWU$

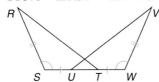

16. **Prove** $\triangle KEF \cong \triangle FHK$

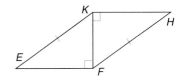

- Two triangles overlap if they share parts of their interiors.

17. Identify the two congruent triangles in the diagram at the right. Justify your answer.

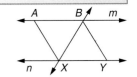

- To write a flow proof, you link information that satisfies a theorem or postulate to a conclusion.

18. In the diagram at the right, $m \parallel n$ and $\overline{AB} \cong \overline{XY}$. Write a flow proof to show that $\triangle ABX \cong \triangle YXB$.

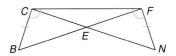

- If two triangles are congruent, then the corresponding sides are congruent and the corresponding angles are congruent (CPCTC).

19. **Prove** $\angle WNO \cong \angle ETO$

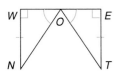

20. **Prove** $\overline{FP} \cong \overline{YS}$

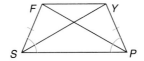

21. **Prove** $\overline{HE} \cong \overline{MO}$

CHAPTER TEST

Classify each triangle by its sides and by its angles.

1.

2.

Find m∠ABC in each triangle.

3.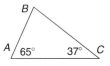

4.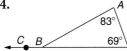

Refer to the regular polygon at the right below.

5. Find the measure of each interior angle of the polygon.

6. Find the measure of each exterior angle of the polygon.

7. Find the measure of each angle in the quadrilateral shown below.

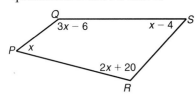

8. **STANDARDIZED TESTS** Which of the following degree measures cannot be the measure of an angle of a regular polygon?

   A. 45°   **B.** 60°   **C.** 90°   **D.** 108°

9. Illustrate a tessellation of the plane using equilateral triangles.

10. **WRITING MATHEMATICS** State and illustrate one of the triangle congruence postulates.

11. In △LMN and △PQR, $\overline{LN} \cong \overline{PR}$, $\overline{LM} \cong \overline{PQ}$, and ∠MLN ≅ ∠QPR. How can you conclude △LMN and △PQR?

12. **WRITING MATHEMATICS** Is it possible to tessellate the plane using only regular pentagons? Explain your response.

Prove each triangle congruence.

13. △FSP ≅ △SFE

14. △TAR ≅ △NRA

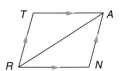

15. **WRITING MATHEMATICS** Briefly describe how the AAS Congruence Theorem follows from the ASA Congruence Postulate.

16. A mason is readying a site for a concrete patio, which will be in the shape of a regular hexagon. At what angle will the concrete forms meet?

Prove each statement.

17. $\overline{DO} \cong \overline{PL}$

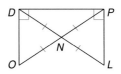

18. $\overline{TW} \cong \overline{EW}$

19. Refer to the diagram below. Write a flow proof to show that △QMC ≅ △ONA.

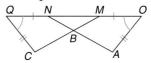

20. **STANDARDIZED TESTS** Which of the following can you use to show that △ABC ≅ △CDA?

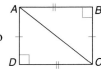

   I. SAS   **II.** SSS   **III.** ASA   **IV.** HL

   A. I and II only   **B.** I and IV only

   C. I, II, and IV only   **D.** I, II, III, and IV

PERFORMANCE ASSESSMENT

CUTTING INTO SQUARES You know that you can make a tessellation by modifying a basic shape. Beginning with a square, create a shape that you can use to tessellate the plane. One possible beginning variation on the square is shown. When you have finished making the tessellating piece, add detail to it. Then, using copies of the tessellating piece, illustrate the tessellation.

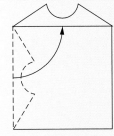

CONGRUENCE DATABASE Write a report titled *Triangles and Congruence: Putting It All Together.* In your report:

- state each congruence postulate and theorem you can use to show that two triangles are congruent,

- indicate the information about the triangles you need to know in order to establish the triangle congruence,

- write an example with an illustration of the use of each congruence postulate and theorem, and

- give an instance in real life where triangle congruence is significant and useful.

GENERATING A LOGO Draw a reasonably sized right triangle on a sheet of paper. Create two or more different artistic logos using overlapping copies of that triangle. Your logos should

- have some type of symmetry and

- involve at least five overlapping triangles.

GO WITH THE FLOW Choose a few of the exercises you were asked to prove during your study of this chapter. The exercises you choose should involve either two-column proof or paragraph proof. Recast each of those proofs in flow-proof format. Make a presentation that compares and contrasts the two types of proof.

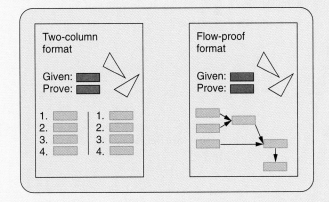

PROJECT ASSESSMENT

PROJECT *Connection* Work together with your group. Review and organize the designs and logos that reflect artistic and geometric characteristics. Gather together your notes and findings about the geometry of the design(s) you decided to study in depth. Finally include the design that you created using your knowledge of constructions and congruence. Prepare a presentation that addresses the following.

1. Explain the geometry you found appealing and informative in the picture(s) you chose.

2. Explain how you used geometry to build your design from a simple beginning to a finished product.

3. Explain how you can use what you learned about design creation to make more designs.

Fill in the blank.

1. A(n) ___?___ triangle has two sides that are the same length.

2. A(n) ___?___ triangle has no sides that are the same length.

3. If N lies between M and P, and $MN = NP$, then N is the ___?___ of segment MP.

4. A plane figure that can be turned about a point so that it coincides with its original position two or more times has ___?___ .

Classify each triangle by its lengths of sides and its measures of angles.

5.
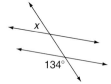

6.
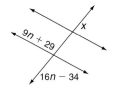

In each figure, two parallel lines are cut by a transversal. Find x.

7.
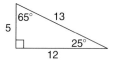

8.
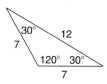

The vertices of a triangle are $A(2, 1)$, $B(5, 0)$, and $C(3, 6)$. Find the image under each transformation.

9. Reflection in the x-axis

10. Translation along $v(5, -4)$

11. Rotation of 180° about the origin

12. Glide reflection in the y-axis followed by a translation of 4 units up

13. Find the midpoint of segment GH.

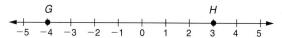

Classify each polygon by number of sides and as convex or concave.

14.   15.
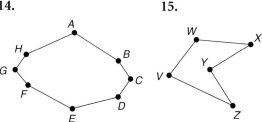

Find the measure of an interior angle and an exterior angle for each figure described.

16. regular hexagon   17. regular decagon

18. WRITING MATHEMATICS Explain why there is not a theorem or postulate for Angle-Angle-Angle congruence for triangles.

A map was drawn showing some of the parks in the town of Parkville. A scale of 2 cm = 1.5 mi was used.

19. The measured distance from Elm Park to Oak Park is 6 cm. What is the actual distance?

20. The actual distance from Rock Park to Evergreen Park is 9 mi. What is the distance between them on the map?

21. The measured distance from Rock Park to Elm Park is 3 cm. What is the actual distance?

22. STANDARDIZED TESTS Which of the following is not a theorem or postulate that allows triangles to be proven congruent?

    A. Side-Angle-Side

    B. Hypotenuse-Leg

    C. Side-Angle-Angle

    D. Angle-Side-Angle

    E. All may be used.

• • • STANDARDIZED TEST • • •

1. Consider the statement: If all three sides of a triangle have different lengths, then the triangle is scalene. Which of the following is the converse statement?

    A. A scalene triangle has different lengths for all three sides.
    B. If a triangle is scalene, then all three sides have different lengths.
    C. If all three sides of a triangle do not have different lengths, then the triangle is not scalene.
    D. If a triangle is not scalene, then all three sides are not different lengths.
    E. none of these

2. Which transformation is not an isometry?

    A. Reflection
    B. Translation
    C. Rotation
    D. Glide reflection
    E. All are isometries.

3. In the diagram, two parallel lines are cut by a transversal. Which pair of angles is supplementary because they are same-side interior angles?

    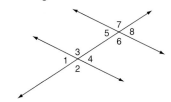

    A. 5 and 6
    B. 3 and 6
    C. 1 and 2
    D. 4 and 8
    E. 3 and 5

4. Which pair of slopes can be the slopes of two perpendicular lines?

    A. $\frac{2}{3}$ and $\frac{2}{3}$ D. $-\frac{3}{4}$ and $\frac{4}{3}$

    B. 3 and -3 E. $\frac{2}{5}$ and $-\frac{2}{5}$

    C. 4 and $\frac{1}{4}$

5. Which of the following triangles cannot exist?

    A. obtuse and equilateral
    B. acute and scalene
    C. isosceles and right
    D. obtuse and isosceles
    E. acute and equilateral

6. Which of the following equations describes a horizontal line?

    A. $x + y = 0$
    B. $2y = 10$
    C. $3x = -12$
    D. $x - y = 0$
    E. $x - 10y = 10$

7. A triangle in the first quadrant is reflected in the y-axis and then reflected in the x-axis. Which of the following transformations describes the resulting image compared to the original triangle?

    A. Translation
    B. Reflection
    C. Rotation
    D. Glide reflection
    E. none of these

8. In the figure, A is an endpoint of a segment and B is the midpoint. What is the coordinate of the other endpoint?

    A. 8   B. 2   C. -4
    D. -2 E. cannot be determined

6 A Closer Look at Triangles

Take a Look AHEAD

Make notes about things that look familiar.
- Find an example where the distance formula is used. How is it used to verify a point on a perpendicular bisector?
- What problem solving strategy is the focus of the Problem Solving File?

Make notes about things that look new.
- You have seen how air traffic control models parallel lines. Locate another geometric concept that air traffic controllers use.
- How does an indirect proof differ from proof styles used in earlier chapters?

DATA *Activity*

Tall Tales

Tall structures are architectural marvels. From early mounds of earth, through the Egyptian pyramids, to present day skyscrapers, height has been fascinating. The form and the purpose of these structures have dramatically changed throughout the years to meet the needs of society. On the next page are some fascinating statistics about the Sears Tower in Chicago, Illinois, which is presently the tallest building in the United States.

SKILL FOCUS

- ▶ Divide numbers.
- ▶ Make estimates.
- ▶ Determine approximate length of time.
- ▶ Find a percent of a number.

GEOMETRY WORKS

HELP WANTED

ARCHITECTURE

In this chapter, you will see how:

- **MASONS** use patterns in creating floor designs of brick or stone.
  (Lesson 6.2, page 323)

- **MODEL BUILDERS** use right angle geometry to build scale models.
  (Lesson 6.9, page 357)

- **ARCHITECTS** solve problems involving staircases. (Lesson 6.10, page 363)

SEARS TOWER: Facts and Figures

Height: 1,454 ft (443 m)	70 miles of water pipes
Weight: 223,000 tons	3.7 million sq ft of floor space
16,100 windows	6 automatic window washing robots
28 acres of black aluminum panels (11 hectares)	145,000 lights
76,000 tons of steel used in the construction	796 washroom taps
Enough concrete to build a 5-mile, eight-lane highway	28 double-decker elevators
Enough steel to make 52,000 cars	7 miles of elevator shafts
25,000 miles of plumbing (40,000 km)	2,232 steps from Lobby to Skydeck

Use the table to answer the following questions.

1. The distance between New York City and San Francisco is 2920 miles. If the plumbing in the Sears Tower were laid end to end, how many one way trips would it cover?

2. The observation Skydeck is located 1353 feet above ground. Double decker elevators travel at 1600 feet per minute. About how long might an elevator ride take from ground level to the Skydeck?

3. The total area of floor space in the Tower is equivalent to 65 football fields. How many square feet are in one football field?

4. The Empire State Building is shorter than the Sears Tower, but weighs 65% more. What is the weight of the Empire State Building?

5. **WORKING TOGETHER** Research two other famous architectural structures. Compile statistics about them and make up questions similar to these to challenge your classmates.

BRIDGING THE GAP

Have you ever traveled across the span of a long bridge and wondered how it was able to remain stable and support all of the traffic that was moving over it? Although bridges are complex architectural structures there is some very simple geometry that plays a role in their design. Bridges come in all shapes and sizes. Some span a mere walkway while others traverse wide waterways. In this project, you will examine some of the structural components of bridges and learn some of the terminology necessary to report on your findings.

PROJECT GOAL

To examine geometric properties that make bridges stable and strong.

Getting Started

Work in a group of four students.

1. Each group member should research four different bridges. Find the location, the type, the bridge statistics and a reproducible photograph of each bridge. Make sure that all 16 of the bridges researched are different so that some comparisons and classifications can be made in a later project connection.

2. As part of this project, you will be constructing a bridge component known as a *truss*. Research the meaning and usage of trusses in architecture.

3. In preparation for the Project Connections gather the following items: at least 20 wooden popsicle sticks, craft sticks, or tongue depressors with holes punched at each end, string, small weights, and small paper fasteners.

PROJECT *Connections*

Lesson 6.1, page 315:
Examine cable bridges to determine how they support heavy loads.

Lesson 6.2, page 322:
Examine various shapes and determine their stability.

Lesson 6.10, page 362:
Build a truss bridge.

Chapter Assessment, page 367:
Prepare a presentation showing the importance of geometric shape in bridge design.

Internet Connection

www.learninggeometry.com

Exploring Loci in the Plane

Think Back

- Recall that geometric construction with a compass and straightedge is a technique for drawing figures without using numerical measurements. Do the following constructions.

   1. Draw a line segment of any length. Construct its perpendicular bisector. What is a perpendicular bisector?

   2. Draw an angle of any measure. Construct its bisector. What is an angle bisector?

SPOTLIGHT ON LEARNING

WHAT? In this lesson you will learn
- to identify a set of points that meet a given condition.
- to identify points as a locus.

Why?
Understanding locus can be useful in solving problems in air traffic control.

Explore

- Paper folding and geometry software are other construction tools you have used. In the thirteenth century the Italian mathematician Mascheroni attempted to do all the classical constructions with only a compass. Constructions have also been done using only a double-edged straightedge.

   3. Cut a strip of uniform width from one edge of a piece of paper and fold it in half lengthwise. Use it for a double-edged straightedge.

   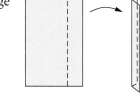

   4. On a second sheet of paper draw ∠PQR.

   5. Place one edge of your double-edged straightedge along $\overleftrightarrow{QR}$ with the other edge in the angle's interior as shown in the diagram at the right. Draw a line along this second edge and call it $\overleftrightarrow{AB}$. What is true about the distance from any point on $\overleftrightarrow{AB}$ to $\overleftrightarrow{QR}$?

   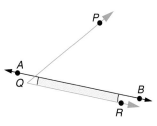

   6. This time put the straightedge along $\overleftrightarrow{QP}$ and draw a second line in the interior of ∠PQR as shown in the diagram at the right below. Call it $\overleftrightarrow{CD}$.

   7. Call the intersection of $\overleftrightarrow{AB}$ and $\overleftrightarrow{CD}$ point T. What is true about the distance from T to $\overleftrightarrow{QR}$ and from T to $\overleftrightarrow{QP}$?

   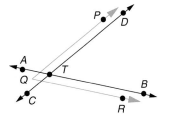

8. Repeat Questions 3–7 with the same ∠PQR, but this time use a wider double-edged straightedge. Locate another point similar to point T. Name it U. What is true about the distance from U to $\overrightarrow{QR}$ and from U to $\overrightarrow{QP}$?

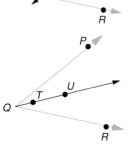

9. Draw $\overrightarrow{QU}$. Are points Q, T and U collinear?

10. Is each point of $\overrightarrow{QU}$ the same distance from $\overrightarrow{QR}$ as it is from $\overrightarrow{QP}$?

11. Measure ∠PQU and ∠RQU. Is $\overrightarrow{QU}$ an angle bisector?

Make Connections

COMMUNICATING ABOUT GEOMETRY

What is a *condition*? Give an example from everyday life in which a condition must be met in order for something to be true or to happen.

• The set of all points that satisfy a given condition or a set of given conditions is a **locus**. The plural of locus is *loci*. In Explore you located points that were equidistant from the sides of ∠PQR. All of the points that satisfy this condition form the bisector of ∠PQR. So, an angle bisector may be thought of as a locus of points.

12. Complete the following statement: An angle bisector is the locus of points that _?_ .

13. Draw a diagram to represent the locus of points that lie in plane P such that each point lies exactly 5 inches from point M in plane P.

14. How many points are in this locus?

15. What geometric figure is formed by this locus?

16. Write the definition of a circle as a locus of points.

17. Recall the perpendicular bisector you constructed in Question 1. Locate a point on the bisector and measure the distance from this point to the endpoints of the segment being bisected. What do you find true about these distances? Make a conjecture.

18. Repeat Question 17 with another point on the perpendicular bisector. Is your conjecture true?

19. Complete the following statement: The perpendicular bisector of a segment is the locus of points that _?_ .

20. Draw a line with a straightedge and find several points that are two units from this line. What is the locus of points in a plane two units from a line in that plane?

21. Lines m and n are parallel. What is the locus of points in the same plane as m and n that are equidistant from m and n?

Summarize

22. MODELING Describe how an airport radar system can model the concept of locus.

23. WRITING MATHEMATICS A particular locus of points is illustrated in the figure at the right using dashes. Describe this locus.

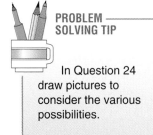

24. GOING FURTHER Sometimes, a locus depends upon more than one condition. Determine five different possibilities for the following locus.

> The locus of points in a plane two units from a given point and one unit from a given line in that plane.

PROBLEM SOLVING TIP

In Question 24 draw pictures to consider the various possibilities.

25. THINKING CRITICALLY Graphs in a coordinate plane can be described as loci. Write the equation of the locus of points equidistant from the endpoints of $\overline{AB}$ if A and B have coordinates $(2, 3)$ and $(8, 5)$, respectively.

PROJECT *Connection*

Cables carry tension and can support heavy loads. To examine how a cable bridge works, you will need about 4 ft of string, and three equal weights (fill plastic sandwich bags with pennies or stones).

1. Measuring from one end of the string, tie a knot at the 1 ft, 2 ft, and 3 ft points.

2. Hang one weight from a large paper clip and fasten the end of the clip to the center knot. Hold the ends of the string and lift. Describe the movements you must make to lift the weight. What shape does the string take?

3. Hang a weight in a similar manner from the 1 ft and 3 ft markings (but not at the center). Lift the string. Describe the motion and the shape of the string.

4. Repeat Question 3 with all three weights, one at each knot.

5. How does a real cable bridge support its weight?

6.2 Segment and Angle Bisectors

Explore/Working Together

- Work with a partner. You will need a geoboard and 2 rubber bands.

1. Stretch a rubber band across a middle vertical column of pegs from one edge of the geoboard to the opposite edge as shown.

2. Then stretch a rubber band across a horizontal row between two pegs the same distance away from a peg enclosed by the rubber band. What is the relationship of the two rubber bands?

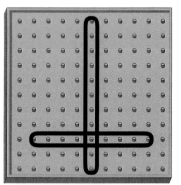

3. Stretch one strand of the horizontal rubber band over any peg not enclosed by the vertical rubber band as shown. Do the lengths determined by the two parts of the stretched side of the rubber band appear to be equal?

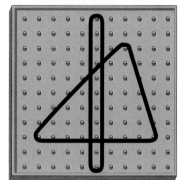

4. Stretch the same strand over a different peg. Do the lengths appear to be equal this time?

5. Repeat the procedure in Question 4 three times. What is the result?

6. Now stretch the same strand over a peg enclosed by the vertical rubber band. What is the result? Repeat this three times.

7. What can you conclude about the lengths of the stretched side of the rubber band relative to the position of the peg chosen?

Build Understanding

- In Explore you discovered that a point on the perpendicular bisector of a segment is related to the endpoints of the segment in a special way.

THEOREM 6.1	**PERPENDICULAR BISECTOR THEOREM**

If a point lies on the perpendicular bisector of a segment, then the point is equidistant from the endpoints of the segment.

To verify this theorem use a coordinate system. Segment AB has endpoints $A(-a, 0)$ and $B(a, 0)$. Since the origin is the midpoint of $\overline{AB}$ and $\overline{AB}$ lies on the x-axis, the y-axis is the perpendicular bisector of $\overline{AB}$. Select any point $C(0, b)$ on the y-axis and determine AC and BC. Use the distance formula.

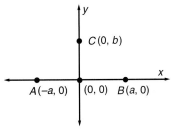

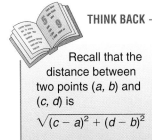

THINK BACK

Recall that the distance between two points (a, b) and (c, d) is

$$\sqrt{(c - a)^2 + (d - b)^2}$$

$$AC = \sqrt{(0 - (-a))^2 + (b - 0)^2} \qquad BC = \sqrt{(0 - a)^2 + (b - 0)^2}$$
$$= \sqrt{a^2 + b^2} \qquad\qquad\qquad = \sqrt{a^2 + b^2}$$

Therefore, $AC = BC$ and C is equidistant from A and B.

Points A and B do not have to be on an axis, nor does the bisector have to be one of the axes.

EXAMPLE 1

Segment RS has endpoints $R(1, 7)$ and $S(9, 3)$. Point $T(7, 9)$ lies on the perpendicular bisector $\overline{TM}$ of $\overline{RS}$. Verify that T is equidistant from R and S.

Solution
Show that $RT = ST$, using the distance formula.

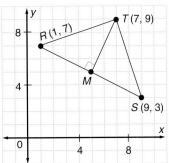

CHECK UNDERSTANDING

Draw a horizontal line segment on a coordinate grid not on the x-axis. Draw a vertical line through the midpoint of the horizontal segment and perpendicular to it. Verify that points on the vertical line are equidistant from the endpoints of the horizontal segment by following the procedure used in Example 1.

$$RT = \sqrt{(1 - 7)^2 + (7 - 9)^2} \qquad ST = \sqrt{(9 - 7)^2 + (3 - 9)^2}$$
$$= \sqrt{(-6)^2 + (-2)^2} \qquad\qquad = \sqrt{2^2 + (-6)^2}$$
$$= \sqrt{36 + 4} \qquad\qquad\qquad = \sqrt{4 + 36}$$
$$= \sqrt{40}, \text{ or } 2\sqrt{10} \qquad\qquad = \sqrt{40}, \text{ or } 2\sqrt{10}$$

Since $RT = ST$, T is equidistant from R and S.

The converse of the perpendicular bisector theorem enables you to prove that a point lies on the perpendicular bisector of a line segment.

THEOREM 6.2	CONVERSE OF PERPENDICULAR BISECTOR THEOREM

If a point is equidistant from the endpoints of a segment, then the point lies on the perpendicular bisector of the segment.

EXAMPLE 2

Determine whether $V(4, 2)$ and $W(3, 1)$ lie on line that is the perpendicular bisector of $\overline{RS}$ from Example 1.

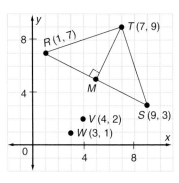

Solution

By Theorem 6.2, if point V is equidistant from the endpoints R and S, then it must lie on the perpendicular bisector of $\overline{RS}$. Use the distance formula to check point V.

$$RV = \sqrt{(1-4)^2 + (7-2)^2} \qquad SV = \sqrt{(9-4)^2 + (3-2)^2}$$
$$= \sqrt{(-3)^2 + 5^2} \qquad\qquad = \sqrt{5^2 + 1^2}$$
$$= \sqrt{9 + 25} \qquad\qquad = \sqrt{25 + 1}$$
$$= \sqrt{34} \qquad\qquad = \sqrt{26}$$

Since $RV \neq SV$, V does not lie on the perpendicular bisector of $\overline{RS}$.

Next, check point W.

$$RW = \sqrt{(1-3)^2 + (7-(1))^2} \quad SW = \sqrt{(9-3)^2 + (3-(1))^2}$$
$$= \sqrt{(-2)^2 + 6^2} \qquad\qquad = \sqrt{6^2 + 2^2}$$
$$= \sqrt{40}, \text{ or } 2\sqrt{10} \qquad\qquad = \sqrt{40}, \text{ or } 2\sqrt{10}$$

Since $RW = SW$, W does lie on the perpendicular bisector of $\overline{RS}$. ◄

Just as points on the perpendicular bisector of a segment are related to the endpoints of the segment so are the points on an angle bisector related to the sides of the angle.

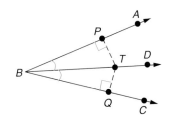

Assume $\overrightarrow{BD}$ is the bisector of $\angle ABC$. Choose any point T on $\overrightarrow{BD}$ and draw segments from T perpendicular to the sides of $\angle ABC$ and call the points where they intersect the sides P and Q. By the definition of angle bisector, $\angle ABT \cong \angle CBT$. Triangles PBT and QBT are two right triangles that share a common hypotenuse. Therefore, $\triangle PBT \cong \triangle QBT$ because of the Angle-Angle-Side Congruence Theorem. Since corresponding parts of congruent triangles are congruent, $\overline{PT} \cong \overline{QT}$. So, T is equidistant from $\overrightarrow{BA}$ and $\overrightarrow{BC}$.

This leads to the Angle Bisector Theorem and its converse given on the next page. Congruent triangles may be used to prove the converse.

THEOREM 6.3   **ANGLE BISECTOR THEOREM**

If a point lies on the bisector of an angle, then the point is equidistant from the sides of the angle.

THEOREM 6.4   **CONVERSE OF ANGLE BISECTOR THEOREM**

If a point is equidistant from the sides of an angle, then the point lies on the bisector of the angle.

EXAMPLE 3

CONSTRUCTION Trusses are used to support the roof of a storage shed. A coordinate grid is superimposed on one at the right. Coordinates of points A, B, C, and D are given and $\angle DBA$ and $\angle DCA$ are right angles. Show that beam AD bisects $\angle BAC$.

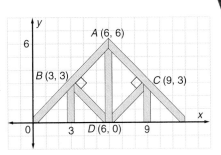

Solution

Show that $D(6, 0)$ is equidistant from $B(3, 3)$ and $C(9, 3)$.

$$BD = \sqrt{(3-6)^2 + (3-0)^2} \qquad CD = \sqrt{(9-6)^2 + (3-0)^2}$$
$$= \sqrt{(-3)^2 + 3^2} \qquad\qquad\quad = \sqrt{3^2 + 3^2}$$
$$= \sqrt{18} \qquad\qquad\qquad\quad\ = \sqrt{18}$$

Since $BD = CD$, D is equidistant from the sides of $\angle BAC$. Therefore, by Theorem 6.4, $\overrightarrow{AD}$ is the bisector of $\angle BAC$. ◄

TECHNOLOGY TIP

Sometimes, when using geometry software, it is necessary to "square" the viewing window when perpendicular lines do not appear to meet at right angles or circles do not appear circular. Read the manual that comes with your software so you will understand when such a procedure would be necessary.

TRY THESE

In the triangle at the right, $\overline{AC}$ is the perpendicular bisector of $\overline{BD}$.

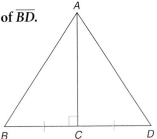

1. What kind of angle is $\angle ACB$?

2. State why $\overline{DC} \cong \overline{BC}$.

3. State why $\overline{AB} \cong \overline{AD}$.

4. Does $\overrightarrow{AC}$ bisect $\angle DAB$? Why?

In Exercises 5–10, you are given the endpoints of $\overline{AB}$ and a third point C. Determine whether or not C lies on the perpendicular bisector of $\overline{AB}$. Justify your answer.

5. $A(1, 3)$, $B(7, 3)$, $C(4, 7)$ 6. $A(-3, -4)$, $B(9, -4)$, $C(3, -8)$ 7. $A(-6, 3)$, $B(0, 3)$, $C(-4, 2)$

8. $A(0, 4)$, $B(4, 0)$, $C(6, 6)$ 9. $A(-5, 1)$, $B(7, 5)$, $C(-1, -7)$ 10. $A(-4, 5)$, $B(6, 1)$, $C(3, 8)$

11. **MODELING** On the geoboard at the right, explain how you can show that the colored band bisects the angle of the triangle.

12. **WRITING MATHEMATICS** Explain how you can use the converse of the angle bisector theorem to prove that the diagonals of a square bisect the angles of that square.

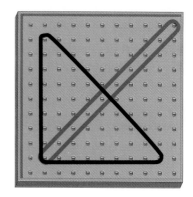

PRACTICE

Use the drawing of regular hexagon *ABCDEF*.

13. Identify two midpoints. Justify your choices.

14. Identify a perpendicular bisector. Justify your choice.

15. Identify an angle bisector. Justify your choice.

16. Identify a pair of parallel line segments. Justify your choice.

17. Copy hexagon *ABCDEF*. Draw a line segment that bisects $\overline{AC}$ and passes through *D*.

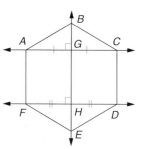

In Exercises 18–23, you are given the two endpoints of $\overline{AS}$ and a third point *M*. Determine whether or not *M* lies on the perpendicular bisector of $\overline{AS}$. Justify your answer.

18. $A(0, 7)$, $S(7, 0)$, $M(5, 5)$

19. $A(0, -6)$, $S(8, 0)$, $M(6, -6)$

20. $A(1, 1)$, $S(9, 3)$, $M(3, 10)$

21. $A(0, -2)$, $S(6, -8)$, $M(0, -8)$

22. $A(-3, 1)$, $S(-1, 7)$, $M(2, 3)$

23. $A(-3, -5)$, $S(3, -5)$, $M(0, 0)$

24. **HANG GLIDING** Many hang gliders are constructed with the following design as shown in the diagram below. Two triangular regions are joined at a keel. The keel meets the nose of the hang glider at a right angle, $AB = AD = AC$, and $BC = CD$.

   a. State why $\triangle BAC \cong \triangle DAC$.

   b. Why is $\overrightarrow{AC}$ the bisector of the angle formed at the nose of the glider?

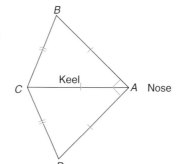

25. **CONSTRUCTION** Draw an angle and construct its bisector. Describe how you can use a construction to verify the angle bisector theorem.

26. NAVIGATION An azimuth circle is used in air and sea navigation. It assigns 0° to the direction north, 90° to the direction east, 180° to the direction south, and 270° to the direction west as shown at the right. An air traffic controller spots a 747 airplane at point P and a private jet at point J as shown on the azimuth circle.

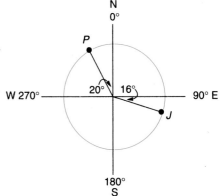

  a. What is the measure of the angle formed by the 747 airplane and the private jet with the control tower at the vertex?

  b. Suppose that a "blip" appeared on the radar screen at a point which, when joined with the center of the circle, would form the bisector of the angle found above. What would be the azimuth circle location in degrees for this "blip"?

EXTEND

Line segment AB, with endpoints A(0, 6) and B(12, 2), lies on line l.

27. Determine the equation of line l.

28. Determine the midpoint M of $\overline{AB}$.

29. What is the slope of any line that is perpendicular to line l?

30. What is the equation of the perpendicular bisector of $\overline{AB}$?

Line segment CD lies on the line whose equation is $y = \dfrac{3}{2}x - 12$. The endpoints C and D are at the x- and y-intercepts of the line.

31. What are the coordinates of points C and D?

32. What is the midpoint of $\overline{CD}$?

33. What is the equation of the perpendicular bisector of $\overline{CD}$?

34. Point $R(-5, 0)$ is on the perpendicular bisector of $\overline{CD}$. Prove this by showing that it satisfies the equation that you found in Exercise 33.

35. Point $S(-2, -2)$ is on the perpendicular bisector of $\overline{CD}$. Prove this by showing that it is equidistant from the endpoints of $\overline{CD}$.

36. The angle bisector $\overrightarrow{BD}$ of $\angle ABC$ divides the angle into $\angle ABD$ and $\angle CBD$. Find the measure of each angle given the following.

  $$m\angle ABD = (3x - 30)° \quad \text{and} \quad m\angle CBD = (25 - 2x)°$$

37. WRITING MATHEMATICS Two half-planes that have a common edge form a *dihedral angle*. Suppose that a third half-plane bisects the dihedral angle. What might you expect to be true about each point in the bisecting plane? Explain your reasoning.

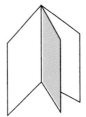

THINK CRITICALLY

Determine the endpoint $B(x, y)$ of $\overline{AB}$, given A and the midpoint M of $\overline{AB}$.

38. $A(0, -4)$, $M(2, -2)$ **39.** $A(4, 0)$, $M(0, -2)$ **40.** $A(5, -3)$, $M(3\frac{1}{2}, -2)$

41. CONSTRUCTION Construct an equilateral triangle. Is the perpendicular bisector of a side collinear with the bisector of the angle opposite that side? Explain.

42. CONSTRUCTION Draw two congruent overlapping circles such that the center of one is in the exterior of the other. Draw the line segment connecting the centers of these circles. Bisect this line segment and construct its perpendicular bisector. What is true about this perpendicular bisector? Is this true for all such pairs of circles? Explain

43. CONSTRUCTION Draw $\overline{WX}$. Construct a segment that is the perpendicular bisector of $\overline{WX}$. Label that segment $\overline{YZ}$. Construct a segment that is the perpendicular bisector of $\overline{YZ}$. Label that segment $\overline{AB}$. What is true about $\overline{AB}$ and $\overline{WX}$? Explain.

ALGEBRA AND GEOMETRY REVIEW

44. In the diagram at the right, give an example of each.

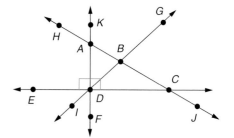

    a. an acute angle   **b.** pair of complementary angles
    c. right angle   **d.** pair of adjacent angles
    e. straight angle   **f.** pair of supplementary angles
    g. an obtuse angle

Factor each of the following.

45. $x^2 - 7x + 12$ **46.** $2x^2 + x - 15$ **47.** $5 - 20x^2$

48. STANDARDIZED TEST Complements of congruent angles are

    A. complementary   **B.** supplementary   **C.** congruent   **D.** adjacent

---

PROJECT *Connection*   In this Connection, you will examine how geometry is used to stabilize structures. You will need sticks and fasteners.

1. Connect four sticks as shown at the right. Is this shape stable?

2. Experiment by building each of the following shapes. Which shape is most stable? Why?

3. Make two equilateral triangles out of sticks and fasteners. Join them at a vertex. Is the new shape stable?

4. Keep the vertex joined and use another stick to link a vertex of one to a vertex of another as shown here. Is the shape stable? Where might you have seen such a shape before?

A mason is a skilled worker who builds with bricks or stone. Often the mason is responsible for construction of the foundation of a building, its facade, and many of the exterior pathways of the building. When laying brick or stone, masons use a variety of geometric patterns to fill a plane region. Two common patterns are the herringbone and basketweave.

Decision Making

A diagonal herringbone pattern is accomplished by setting the bricks at a 45° angle to the sides of a rectangular border. An isosceles right triangle is used to form the first stage of this pattern as shown.

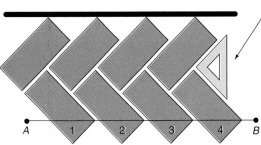

1. Small triangular pieces of brick are cut to fill the region to the edges. What are the measures of the angles of these pieces? Justify your answer.

2. In the second stage of laying the pattern, the isosceles right triangle is moved as shown above. Suppose that the path ends along segment *AB*. Identify the types of triangles that are labeled 1–4 in the illustration. Justify your response.

A basketweave pattern uses bricks laid parallel to one another to form small squares. In a diagonal basketweave pattern, the first brick is laid on a diagonal using an isosceles right triangle to identify the correct slope of the diagonal.

3. What are the measures of angles 1, 2, and 3 in this plan? Justify your answer.

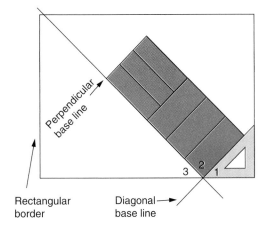

6.2   **Segment and Angle Bisectors**   **323**

6.3 Isosceles Triangles

Explore

1. Draw a large scalene triangle on a sheet of paper, cut it out with scissors, and label it $\triangle ABC$.

2. Fold side AB in half. Is $\angle A$ congruent to $\angle B$?

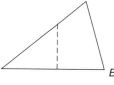

3. Unfold the triangle and repeat Question 2 for each of the other two sides of $\triangle ABC$. Are any of the angles of $\triangle ABC$ congruent?

4. Draw a large isosceles triangle. Cut it out and label it $\triangle DEF$ such that $DE = DF$ and $DE > EF$.

5. Fold side EF in half. What seems to be true about $\angle E$ and $\angle F$?

6. Let G be the point where the fold intersects side $\overline{EF}$. Examine the triangle carefully. What seems to be true about $\overline{DG}$?

7. Repeat the procedure decribed above in Questions 1–6 with two new triangles.

8. What conclusion can you draw if the lengths of two sides of a triangle are equal?

Build Understanding

In Chapter 5 you learned that an isosceles triangle has at least two congruent sides. In this lesson you will study some special properties of isosceles triangles.

Before you proceed, you will need to learn a few additonal terms associated with isosceles triangles. The two congruent sides of an isosceles triangle are called **legs**. The third side is the **base**. The angles opposite the legs are the **base angles**. The angle opposite the base is the **vertex angle**.

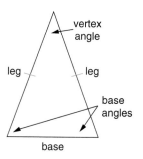
Isosceles triangle

In Explore, you probably discovered this important fact: The base angles of an isosceles triangle are congruent. This is stated formally as the following theorem.

THEOREM 6.5	BASE ANGLES THEOREM

If two sides of a triangle are congruent, then the angles opposite those sides are congruent.

Given $\overline{DE} \cong \overline{DF}$

Prove $\angle E \cong \angle F$

Plan for Proof Draw a segment from point D to the midpoint G of $\overline{EF}$, dividing $\triangle DEF$ into two congruent triangles. Then $\angle E$ and $\angle F$ are congruent corresponding parts.

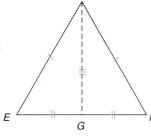

Proof Write a two-column proof.

Statements	Reasons
1. $\overline{DE} \cong \overline{DF}$	1. Given
2. G is the midpoint of $\overline{EF}$.	2. Ruler Postulate
3. $\overline{EG} \cong \overline{FG}$	3. Definition of midpoint
4. Draw $\overline{DG}$.	4. Through two distinct points, there is exactly one line.
5. $\overline{DG} \cong \overline{DG}$	5. Reflexive Property
6. $\triangle DEG \cong \triangle DFG$	6. SSS Congruence Postulate
7. $\angle E \cong \angle F$	7. CPCTC

When a figure is positioned on a coordinate plane, you can use the distance formula and the Base Angles Theorem to identify congruent angles.

EXAMPLE 1

The vertices of $\triangle ABC$ are $A(1, 1)$, $B(4, 5)$, and $C(7, 1)$. Show that this triangle is isosceles and name a pair of congruent angles.

Solution
Use the distance formula to find the length of each side.

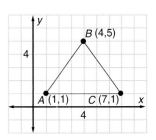

$$BA = \sqrt{(4-1)^2 + (5-1)^2} = \sqrt{3^2 + 4^2} = 5$$

$$BC = \sqrt{(4-7)^2 + (5-1)^2} = \sqrt{(-3)^2 + 4^2} = 5$$

$$AC = \sqrt{(1-7)^2 + (1-1)^2} = \sqrt{(-6)^2 + 0} = 6$$

Therefore, $AB = BC$, and $\triangle ABC$ is isosceles with base $\overline{AC}$. So, by the Base Angles Theorem, $\angle A \cong \angle C$. ◄

In an isosceles triangle, the line that contains the angle bisector of the vertex angle is a line of symmetry.

EXAMPLE 2

In the figure at the right, $\triangle RST$ is isosceles with base $\overline{RT}$. Find the coordinates of point R.

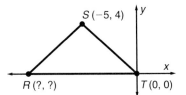
S (−5, 4)

Solution

Draw the line of symmetry, $\overleftrightarrow{SZ}$.

The line of symmetry must be the perpendicular bisector of the base. So the coordinates of point Z are $(−5, 0)$, and $\overline{RZ} \cong \overline{TZ}$.

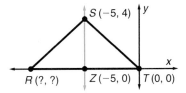
S (−5, 4)
R (?, ?)   Z (−5, 0)   T (0, 0)

The length of $\overline{TZ}$ is 5 units, so the length of $\overline{RZ}$ also must be 5 units.

So the point is $R(−10, 0)$.   ◄

Since an equilateral triangle has *three* congruent sides, it is a special type of isosceles triangle. The following corollaries about equilateral triangles follow directly from the Isosceles Triangle Theorem.

> **COROLLARY 6.6** If a triangle is equilateral, then it is equiangular.
>
> **COROLLARY 6.7** The measure of each angle of an equilateral triangle is 60°.

EXAMPLE 3

COMMUNICATION WIRING A communication line consists of three congruent cables. Each cable contains wires twisted into a cylindrical shape about a central copper core. A cross section of the line is shown at the right. The dashed lines form a triangle whose vertices are at the three copper cores. What is the measure of each angle of this triangle?

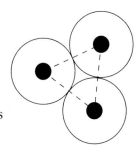

Solution

Since the cables are congruent, the radii of their circular cross sections must also be congruent. Let r represent this common radius. Then the length of each side of the triangle is $2r$. So the triangle is equilateral. By Corollary 6.7, the measure of each angle is 60°.   ◄

Congruent triangles may also be used to prove the converse of the Base Angles Theorem.

THEOREM 6.8	CONVERSE OF BASE ANGLES THEOREM

If two angles of a triangle are congruent, then the sides opposite those angles are congruent.

A corollary for equiangular triangles follows from this theorem.

COROLLARY 6.9   If a triangle is equiangular, then it is equilateral.

When you know that two angles of a triangle are congruent, you may be able to use the converse of the Base Angles Theorem to find unknown lengths of sides.

EXAMPLE 4

USING ALGEBRA Find the length of each side of $\triangle JKL$ at the right.

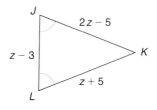

Solution

From the figure, you know $\angle J \cong \angle L$. So, by the converse of the Base Angles Theorem, it follows that $\overline{JK} \cong \overline{LK}$.

From the figure, $JK = 2z - 5$ and $LK = z + 5$. Use the fact that $JK = LK$ to write and solve an equation.

$$JK = LK$$
$$2z - 5 = z + 5 \qquad \text{Subtract } z \text{ from each side.}$$
$$z - 5 = 5 \qquad \text{Add 5 to each side.}$$
$$z = 10$$

Now replace z with 10 in the expression that represents each side.

$JK = 2z - 5$	$LK = z + 5$	$JL = z - 3$
$JK = 2(10) - 5$	$LK = 10 + 5$	$JL = 10 - 3$
$JK = 15$	$LK = 15$	$JL = 7$

TRY THESE

Determine whether $\triangle ABC$ with the given vertices is isosceles. If it is isosceles, name a pair of congruent angles. Justify your answer.

1. $A(0, 2)$, $B(5, 0)$, $C(0, -2)$ 2. $A(0, 0)$, $B(3, 2)$, $C(6, 0)$ 3. $A(-1, 0)$, $B(-3, -3)$, $C(-4, 0)$

4. $A(2, 1)$, $B(2, 4)$, $C(5, 4)$ 5. $A(-2, 1)$, $B(2, 3)$, $C(4, -2)$ 6. $A(-2, -1)$, $B(0, 3)$, $C(4, 1)$

In Exercises 7–11, $\triangle RST$ is isosceles with base $\overline{RT}$. The measure of one angle is given. Find the measures of the other two angles.

   7. $m\angle R = 40°$ **8.** $m\angle T = 80°$ **9.** $m\angle S = 60°$ **10.** $m\angle R = 35°$ **11.** $m\angle S = 101°$

Complete each set of coordinates so that $\triangle ABC$ is isosceles with base $\overline{AC}$.

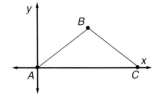

12. $A(0, 0)$, $B(4, 5)$, $C(?, ?)$ **13.** $A(0, ?)$, $B(?, 7)$, $C(11, 0)$

14. $A(0, 0)$, $B(m, n)$, $C(?, ?)$ **15.** $A(0, ?)$, $B(?, k)$, $C(j, 0)$

16. $A(0, 0)$, $B(?, h)$, $C(2g, ?)$ **17.** $A(0, 0)$, $B(3r, 2s)$, $C(?, ?)$

18. In $\triangle PQR$, $\angle P \cong \angle R$, $PQ = 3x$, $QR = 8x - 10$, and $PR = 5x$. Find PQ, QR, and PR.

19. $\triangle STU$ is equiangular with $ST = 3x + 5$, $TU = 9x - 13$, and $US = 17 - x$. Find the length of each side of this triangle.

20. WRITING MATHEMATICS Explain how knowing the measure of a base angle of an isosceles triangle assures you of being able to find the measures of the other two angles.

21. WRITING MATHEMATICS Explain how knowing the measure of the vertex angle of an isosceles triangle, assures you of being able to find the measures of the base angles.

PRACTICE

Determine whether $\triangle XYZ$ with the given vertices is isosceles. If it is isosceles, name a pair of congruent angles. Justify your answer.

22. $X(0, 0)$, $Y(-4, 0)$, $Z(-2, -3)$ **23.** $X(-1, -1)$, $Y(0, 3)$, $Z(2, -1)$

24. $X(-3, 3)$, $Y(11, -2)$, $Z(6, 3)$ **25.** $X(-2, 2)$, $Y(-2, -2)$, $Z(2, -2)$

26. $X(-2, -1)$, $Y(4, 2)$, $Z(7, -4)$ **27.** $X(4, 2)$, $Y(5, -4)$, $Z(-4, -4)$

Complete each set of coordinates so that $\triangle PQR$ is isosceles with base $\overline{PQ}$ and point R on the y-axis.

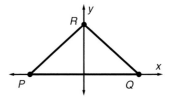

28. $P(-6, 0)$, $Q(?, 0)$, $R(?, 5)$ **29.** $P(?, ?)$, $Q(4, 0)$, $R(0, 8.5)$

30. $P(?, 0)$, $Q(a, 0)$, $R(?, b)$ **31.** $P(-j, 0)$, $Q(?, ?)$, $R(0, k)$

32. $P(-2c, ?)$, $Q(?, 0)$, $R(0, 2d)$ **33.** $P(?,?)$, $Q(8s, 0)$, $R(0, 5t)$

In Exercise 34–38, $\triangle UVW$ is isosceles with legs $\overline{UV}$ and $\overline{VW}$. The measure of one angle is given. Find the measures of the other two angles.

34. $m\angle U = 56°$ **35.** $m\angle W = 45°$ **36.** $m\angle U = 89°$ **37.** $m\angle V = 90°$ **38.** $m\angle V = 100°$

39. In $\triangle PQR$, $\overline{PQ} \cong \overline{QR}$, $PQ = 8m - 2$, $QR = 6m + 12$, and $PR = 50 - 2m$. Find PQ, QR, and PR. Name a pair of congruent angles.

40. Triangle *STU* is equilateral with $ST = 7x - 10$, $TU = 2x + 25$, and $US = x^2 - 10$. Find the length of each side of this triangle.

41. Writing Mathematics Triangle *JKL* is isosceles with base $\overline{KL}$. State as many facts as you can about this triangle.

42. Telephone Wiring On the telephone pole shown at the right two slanting braces of equal length support the lowest horizontal crosspiece. Suppose the braces are perpendicular to each other. What is the measure of the angle formed where each brace meets the crosspiece? Justify your answer.

43. Manipulatives The figure at the right shows the ancient Chinese puzzle called a *tangram*. Trace the figure onto a sheet of paper. Cut your copy along the solid lines. Rearrange the pieces so they form an isosceles triangle. Explain how you know it is isosceles.

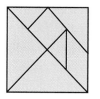

EXTEND

For Exercises 44–46, write a proof in two-column form.

44. Given $\overline{AB} \cong \overline{CB}$

$\overline{AE} \cong \overline{CD}$

Prove $\triangle ABE \cong \triangle CBD$

45. Given $\triangle ABC$ is isosceles. with base $\overline{AC}$.

$\angle ABD \cong \angle CBE$

Prove $\triangle DBE$ is isosceles.

46. Given $\overline{XY} \cong \overline{ZY}$

$\overrightarrow{XW}$ is the bisector of $\angle YXZ$.

$\overrightarrow{ZW}$ is the bisector of $\angle YZX$

Prove $\overline{XW} \cong \overline{ZW}$

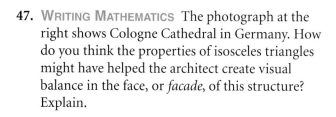

47. Writing Mathematics The photograph at the right shows Cologne Cathedral in Germany. How do you think the properties of isosceles triangles might have helped the architect create visual balance in the face, or *facade*, of this structure? Explain.

48. Prove Corollary 6.6: If a triangle is equilateral, then it is equiangular. Write the proof in paragraph form.

49. Prove Corollary 6.7: The measure of each angle of an equilateral triangle is 60°. Write the proof in two-column form.

50. Prove Theore 6.8: If two angles of a triangle are congruent, then the sides opposite those angles are congruent. Write the proof in two-column form.

51. Prove Corollary 6.9: If a triangle is equiangular, then it is equilateral. Write the proof in paragraph form.

A *median of a triangle* is a segment whose endpoints are a vertex of the triangle and the midpoint of the opposite side. In the figure at the right, △ABC is isosceles, and $\overline{AE}$ and $\overline{CD}$ are medians.

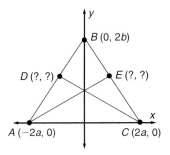

52. Give the coordinates of points D and E.

53. Write expressions that represent the lengths of $\overline{AE}$ and $\overline{CD}$.

54. Write a coordinate proof of this theorem: The medians to the legs of an isosceles triangle are congruent.

THINK CRITICALLY

55. In the figure at the right, $\overline{AG}$ and $\overline{BF}$ intersect a point C, $\overline{EG}$ and $\overline{BF}$ intersect at point D, △ABC is an equilateral triangle, and △EDF is an isosceles right triangle with legs $\overline{DE}$ and $\overline{DF}$. Find m∠CGD. Explain your reasoning.

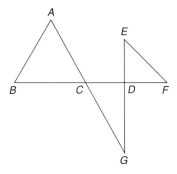

56. Points X, Y, and Z are the midpoints of the sides of △JKL. If △JKL is isosceles, show that △XYZ must be isosceles.

57. Show how it is possible to form a cross section of a cube that is shaped like an equilateral triangle.

ALGEBRA AND GEOMETRY REVIEW

In the figure at the right, m∠GDC = m∠EAB = 65°, $t \parallel u$, and $r \parallel s$.

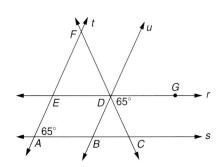

58. Find m∠DCB. Justify your answer.

59. Find m∠EFD. Justify your answer.

60. Explain why $\overline{EF} \cong \overline{FD}$.

61. The perimeter of an equilateral triangle can be represented by $5x - 6y$. Write an expression for the length of one of its sides.

62. Determine the mean, median, and mode for the following set of data: 85, 90, 75, 85, 80, 93, 82, 85, 93, 80.

Bisectors, Medians, and Altitudes

Think Back/Working Together

- Recall a perpendicular bisector of a segment is a line that intersects a segment at right angles, dividing it into two congruent parts. An angle bisector is a ray that divides an angle into two congruent angles.

  Work with a partner. You need a straightedge, a compass, and paper.

  1. Each partner selects three different triangles from the six options.

     equilateral triangle   isosceles triangle   right triangle
     obtuse triangle   scalene triangle   acute triangle

  2. Construct each of your choices on a separate sheet of paper.

  3. Construct the perpendicular bisector of each side of each triangle.

  4. Construct the angle bisector of each angle of each triangle.

  5. Compare your results. Identify any similarities and differences.

Explore

- 6. Continue to work with a partner, One person should complete Part 1 and the other Part 2.

  Part 1

  a. Draw a triangle.
  b. Construct the bisector of each angle of the triangle.
  c. Label the point of intersection of these bisectors *B*.
  d. Construct a line through *B* perpendicular to one side of the triangle. Name the intersection of the side and this line *C*.
  e. Place the compass tip on *B* and the pencil tip on *C*. Draw the circle.
  f. Describe the relationship of the circle to the triangle.

  Part 2

  a. Draw a triangle.
  b. Construct the perpendicular bisector of each side of the triangle.
  c. Label the point of intersection of these bisectors *A*.
  d. Place the compass tip on *A* and the pencil tip on any vertex. Draw the circle.
  e. Describe the relationship of the circle to the triangle.

Geometry Workshop

Make Connections

TECHNOLOGY TIP

When doing constructions on geometry software, it may be necessary to adjust the window to see where concurrent lines intersect.

- A circle is **circumscribed** about a triangle when each vertex of the triangle is on the circle. The triangle is said to be *inscribed in the circle.* A circle is **inscribed** in a triangle when each side of the triangle touches the circle at exactly one point (not a vertex). The triangle is said to be *circumscribed about the circle.*

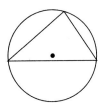

Circumscribed circle
(inscribed triangle)

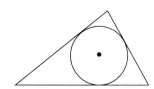

Inscribed circle
(circumscribed triangle)

7. In question 6, which circle is inscribed in the triangle? Which circle is circumscribed about the triangle?

Two or more lines that intersect at a single point are **concurrent lines**. The point of intersection is called the **point of concurrency**.

8. The point of concurrency of the perpendicular bisectors of the sides of a triangle is called the **circumcenter** of the triangle. The point of concurrency of the angle bisectors is called the **incenter**. Explain why these definitions make sense.

9. Draw a large triangle on a piece of cardboard. Cut it out. Find the point on the cardboard where the triangle will balance on the point of a pencil. Puncture the cardboard at that point.

10. Trace the triangle on a sheet of paper. Through the hole in the cardboard locate the balancing point on your tracing and name it *B*.

11. Bisect each side of the triangle. Draw a line segment from each vertex of the triangle to the midpoint of the side opposite that vertex.

12. What do you notice about the point of intersection of the line segments in Question 11?

A **median** of a triangle is a segment joining a vertex of the triangle to the midpoint of the opposite side. The point of concurrency of the medians is called the **centroid** of the triangle, or its **center of gravity**.

13. Why is the name center of gravity appropriate?

14. The centroid divides each median into two parts. Determine the ratio of the smaller part to the larger part for each median and make a conjecture about this ratio.

An **altitude** of a triangle is a perpendicular segment from a vertex of the triangle to the line containing the opposite side.

THINK BACK

An acute triangle is one in which all the angles are acute.

An obtuse triangle is one in which one angle is obtuse.

15. Draw an acute triangle and construct an altitude to each side.

16. Draw an obtuse triangle and construct an altitude to each side.

17. Compare the locations of the altitudes in Questions 15 and 16.

18. WRITING MATHEMATICS Is there a triangle in which at least one of the altitudes is a side of the triangle? Is there a triangle in which exactly one of the altitudes is a side of the triangle? Explain.

19. WRITING MATHEMATICS Is there a triangle in which the perpendicular bisector of a side, an altitude, a median, and a bisector of one of the angles are on the same line? Explain.

MUSIC A metronome is a measuring device that is used to mark time or beats when playing music.

20. Examine the picture of the metronome. Of what type of triangle are you reminded? Justify your answer.

21. The arm of the metronome swings like an upside down pendulum. It can be adjusted to change the timing of the ticks. In the illustration at the right, what types of segments in a triangle does the arm model at the position shown?

Summarize

22. WRITING MATHEMATICS Suppose $\overrightarrow{AB}$ is fixed at A and can swing freely like a pendulum. Beginning with $\overrightarrow{AB}$ on $\overrightarrow{AC}$, let $\overrightarrow{AB}$ swing toward and eventually coincide with $\overrightarrow{AD}$. Examine $\triangle ABC$ and describe the position of the altitude from each vertex. Explain your reasoning.

GEOMETRY: WHO, WHERE, WHEN

The line that you drew in Question 24 is the *Euler Line* named after the Swiss mathematician Leonard Euler (1707–1783). Euler proved that the centroid, the circumcenter, and the orthocenter are collinear.

23. GOING FURTHER In Question 15 were the altitudes concurrent? Were they concurrent in Question 16?

The point of concurrency of the lines that contain the altitudes of a triangle is called the **orthocenter** of the triangle.

24. THINKING CRITICALLY Construct a large scalene triangle. Locate and label the incenter I, the center of gravity (centroid) G, the circumcenter C, and the orthocenter O. Three of these points should be collinear. Draw the line that contains the three points.

Geometry Excursion

Dissections

A *dissection* problem involves cutting a geometric shape into pieces and rearranging them to form a new shape. For example, if equilateral triangle *ABC* is divided as shown, then the four pieces can be rearranged to form a square.

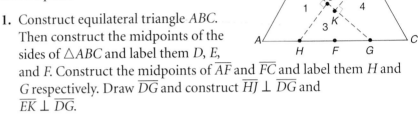

1. Construct equilateral triangle *ABC*. Then construct the midpoints of the sides of △*ABC* and label them *D, E,* and *F.* Construct the midpoints of $\overline{AF}$ and $\overline{FC}$ and label them *H* and *G* respectively. Draw $\overline{DG}$ and construct $\overline{HJ} \perp \overline{DG}$ and $\overline{EK} \perp \overline{DG}$.

2. Cut out △*ABC* and cut along the dashed line segments to divide the triangle into four pieces. Rearrange them to form a square.

3. **CARPENTRY** Maria is making a table top with four wooden pieces shaped like those in Question 2. She is going to link the pieces together with three hinges attached at certain vertices. When the pieces are folded together in one direction they will form an equilateral triangle and when folded in the opposite direction they will form a square. Where should Maria put the hinges?

The German mathematician David Hilbert proved that you can dissect any polygon to form any other polygon with equal area by using a finite number of cuts.

4. Dissect the 4 × 9 rectangle using one cut to make a square.

5. Dissect the 3 × 12 rectangle using four cuts to make two squares of equal area.

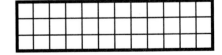

Copy and cut out each regular shape. Rearrange the pieces to make another regular shape. Describe the shape that results.

6.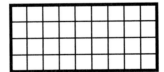

7.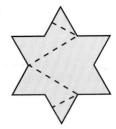

Construct a regular hexagon by drawing a circle. Then use the radius to mark off six equal arcs on the circle. These points are the vertices of a regular hexagon as shown.

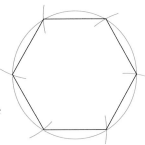

8. Cut out the hexagon that you constructed. Is it possible to dissect the hexagon into two pieces and use them to form a parallelogram?

9. Construct another hexagon. Is it possible to dissect the hexagon into three pieces and use them to form a rectangle?

10. SEWING Nancy bought a one yard by one yard square piece of fabric. She wants to make a tablecloth for a regular octagonal table top. She does not want to cut the corners off the fabric, she wants to use the entire piece. Explain how she can use Question 6 to make the tablecloth.

Refer to the pair of dissections of a mitered square shown below.

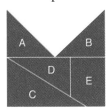

Dissection 1

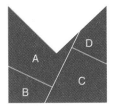

Dissection 2

11. Copy and cut out all five pieces of Dissection 1. Use the pieces to make a complete square.

12. Copy Dissection 2 and cut out two of its pieces. Make a copy of each piece. Use the four pieces to make a complete square.

13. Repeat Question 12 using the two remaining pieces of Dissection 2.

14. PUZZLE MAKING George wants to make an octagonal jigsaw puzzle. He began by drawing the diagram at the right. His goal is to subdivide the square and make it consist of a regular octagon surrounded by four congruent isosceles triangles. Explain how he can use constructions to make this happen.

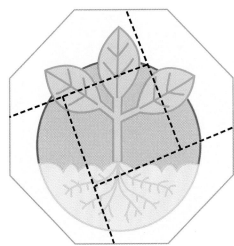

6.5 Midsegments

Explore

1. Draw a large triangle and label it $\triangle ABC$. Locate the midpoints of $\overline{AB}$, $\overline{BC}$, and $\overline{AC}$. Name them D, E, and F, respectively.

2. Draw $\overline{DE}$. Find the measures of $\angle A$, $\angle B$, $\angle C$, $\angle BDE$, and $\angle BED$. Which angles are congruent?

3. Is $\overline{DE}$ parallel to $\overline{AC}$? Justify your answer.

4. Compare the lengths of $\overline{DE}$ and $\overline{AC}$.

5. Draw $\overline{EF}$. Make a conjecture about $\overline{EF}$ and $\overline{AB}$ and their lengths. Find $m\angle EFC$, EF, and AB. Is your conjecture verified?

6. Draw $\overline{FD}$. How is it related to $\overline{BC}$? Explain.

Build Understanding

A **midsegment** of a triangle is a segment whose endpoints are the midpoints of two sides of the triangle. The following theorem states an important fact about midsegments.

THEOREM 6.10	TRIANGLE MIDSEGMENT THEOREM

> The segment joining the midpoints of two sides of a triangle is parallel to the third side, and its length is half the length of the third side.

EXAMPLE 1

Find each measure in $\triangle JKL$ at the right.

a. XZ **b.** KL **c.** YZ

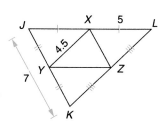

Solution

By definition, $\overline{XY}$, $\overline{XZ}$, and $\overline{YZ}$ are midsegments of $\triangle JKL$. By Theorem 6.10, each is parallel to one side of $\triangle JKL$, and its length is half the length of that side.

a. $XZ = \frac{1}{2}(JK) = \frac{1}{2}(7) = 3.5$

b. $XY = \frac{1}{2}(KL)$, so $KL = 2(XY) = 2(4.5) = 9$

c. Since point X is the midpoint of $\overline{JL}$, $JL = 2(XL) = 2(5) = 10$.

Then $YZ = \frac{1}{2}(JL) = \frac{1}{2}(10) = 5$.

Sometimes you can use the Triangle Midsegment Theorem to determine unknown angle measures.

EXAMPLE 2

In △PQR at the right, points S and T are the midpoints of $\overline{PQ}$ and $\overline{PR}$, respectively. Find:

a. m∠PTS   **b.** m∠PST   **c.** m∠QST

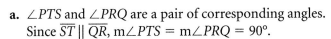

Solution

By definition, $\overline{ST}$ is a midsegment of △PQR.
By Theorem 6.10, $\overline{ST}$ is parallel to $\overline{QR}$.
Consider $\overline{PR}$ and $\overline{PQ}$ as transversals intersecting $\overline{ST}$ and $\overline{QR}$.

a. ∠PTS and ∠PRQ are a pair of corresponding angles.
Since $\overline{ST} \parallel \overline{QR}$, m∠PTS = m∠PRQ = 90°.

b. The sum of the measures of the angles of a triangle is 180°, so

m∠Q = 180° − (m∠P + m∠R) = 180°− (49° + 90°) = 41°

Since ∠PST and ∠Q are congruent corresponding angles,

m∠PST = m∠Q = 41°

c. Since ∠QST and ∠PST are a linear pair,

m∠QST = 180° − m∠PST = 180° − 41° = 139°   ◀

Sometimes you must use algebra to find unknown lengths.

EXAMPLE 3

USING ALGEBRA In △FGH at the right, find MN.

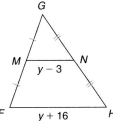

Solution

From the figure, $\overline{MN}$ is a midsegment of △FGH. By Theorem 6.10, MN is half of FH. Use this fact to write and solve an equation.

$$MN = \frac{1}{2}FH$$

$$y - 3 = \frac{1}{2}(y + 16) \qquad \text{Substitute.}$$

$$2(y - 3) = 2\left(\frac{1}{2}(y + 16)\right) \qquad \text{Multiply each side by 2.}$$

$$2y - 6 = y + 16 \qquad \text{Simplify.}$$

$$y - 6 = 16 \qquad \text{Subtract } y \text{ from each side.}$$

$$y = 22 \qquad \text{Add 6 to each side.}$$

Since MN = y − 3, it follows that MN = 22 − 3 = 19.   ◀

CHECK UNDERSTANDING

In the solution of Example 3, describe a different way that you might solve the equation.

You can use the Triangle Midsegment Theorem to solve real world problems.

EXAMPLE 4

Maps At the left below is a map of part of Washington D.C. Connecticut Avenue, S Street, and Vermont Avenue form an isosceles triangle with vertices at points C, W, and D. M Street is midway between the White House and S Street. Farragut Square and McPherson Square are midway between the White House and M Street.

a. Determine the distance along S Street from Connecticut Avenue to Vermont Avenue.

b. Determine the distance from Farragut Square to McPherson Square.

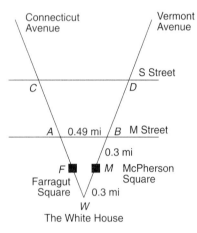

Solution

a. You need to find the length of $\overline{CD}$. In $\triangle CWD$, $\overline{AB}$ is the midsegment parallel to $\overline{CD}$, and so $AB = \frac{1}{2}CD$. Thus

$$CD = 2(AB) = 2(0.49) = 0.98 \text{ mi}$$

So the distance along S Street from Connecticut Avenue to Vermont Avenue is 0.98 mi.

b. The squares are midpoints of sides of $\triangle ABW$, so FM is a midsegment, and

$$FM = \frac{1}{2}AB = \frac{1}{2}(0.49) = 0.245 \text{ mi}$$

So the distance from Farragut Square to McPherson Square is 0.245 mi.

TRY THESE

For Exercises 1–7, refer to $\triangle RST$ at the right.

1. Name three midsegments.

Find the measure of each segment. Explain how you arrived at your answer.

2. $\overline{ST}$ 3. $\overline{XZ}$ 4. $\overline{XT}$

5. $\overline{ZY}$ 6. $\overline{RT}$ 7. $\overline{XY}$

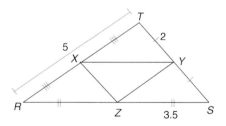

In $\triangle LMN$ at the right, Q is the midpoint of $\overline{MN}$, point P is the midpoint of $\overline{LM}$, m$\angle L = 45°$, and m$\angle N = 35°$. Find the measure of each angle.

8. $\angle M$ 9. $\angle LPQ$ 10. $\angle NQP$

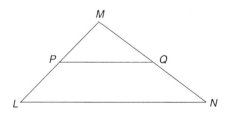

11. In $\triangle PRY$, point A is the midpoint of $\overline{PR}$, point T is the midpoint of $\overline{RY}$, $AT = 3j$, and $PY = j + 10$. Find AT and PY.

COORDINATE GEOMETRY **Refer to $\triangle ABC$ on the coordinate axes at the right.**

12. Verify that points D, E, and F are the midpoints of $\overline{AC}$, $\overline{AB}$, and $\overline{BC}$, respectively.

13. Show that $\overline{EF} \parallel \overline{AC}$ and $EF = \frac{1}{2} AC$.

14. Show that $\overline{ED} \parallel \overline{BC}$ and $ED = \frac{1}{2} BC$.

15. Show that $\overline{FD} \parallel \overline{AB}$ and $FD = \frac{1}{2} AB$.

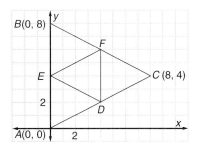

16. WRITING MATHEMATICS Suppose you are asked to find the length of $\overline{MN}$ in each triangle at the right. How would your work for triangle I be similar to your work for triangle II? How would it be different? What is the length of $\overline{MN}$ in each triangle?

BOARD GAMES **The game of Alquerque dates back to 1400 B.C. It is attributed to the early Egyptians. Alquerque is a logic game of strategy. Twenty-four pieces are equally spaced on a game board as shown below. Players take turns moving a piece, each trying to capture opponent's pieces.**

17. Make a sketch of the game board. Label each game piece as a point.

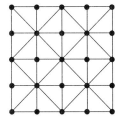

18. Name six midsegments of triangles in your sketch. Explain how you know that the segments you identified are midsegments.

PRACTICE

For Exercises 19–24, refer to $\triangle EFG$ at the right. Find the measure of each segment. Explain how you arrived at your answer.

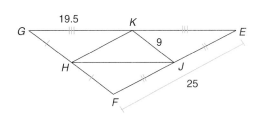

19. $\overline{FJ}$ 20. $\overline{EG}$ 21. $\overline{FG}$

22. $\overline{FH}$ 23. $\overline{HK}$ 24. $\overline{HJ}$

In △*ABC* at the right, point *X* is the midpoint of $\overline{AB}$, point *Y* is the midpoint of $\overline{BC}$, m∠*A* = 22°, and m∠*XYC* = 46°. Find the measure of each angle.

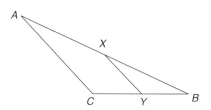

25. ∠*B*   **26.** ∠*C*   **27.** ∠*BXY*

28. ∠*BYX*   **29.** ∠*AXY*   **30.** ∠*CYX*

Find the value of *n* in each figure. Use that value to find as many unknown measures in the figure as you can.

31.
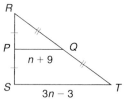

32.
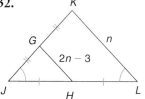

33.
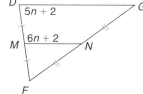

ARCHITECTURAL PLANS The figure at the right shows modified plans for the front of a medieval structure. The base has been divided into six congruent segments, each 5 m in length. Early architects used triangles to assist in determining measurements and maintaining symmetry.

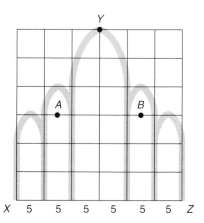

34. Show that $\overline{AB}$ is a midsegment of △*XYZ*. (*Hint:* Assign coordinates to the grid.)

35. Suppose that stained glass windows were to be positioned at points *A* and *B*. What is the distance between the centers of the windows? Explain your reasoning.

36. WRITING MATHEMATICS Write a real life problem that can be solved using the Triangle Midsegment Theorem. Show the solution of your problem.

EXTEND

In the figure at the right, point *D* is the midpoint of $\overline{AB}$ and point *E* is the midpoint of $\overline{BC}$.

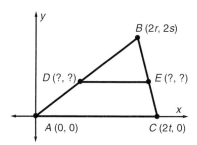

37. Give the coordinates of points *D* and *E*.

38. Find the slope of $\overline{DE}$ and the slope of $\overline{AC}$. What do the slopes tell you about the relationship between $\overline{DE}$ and $\overline{AB}$?

39. Find *DE* and *AC*. How are these lengths related?

40. Write a coordinate proof of Theorem 6.10: The segment joining the midpoints of two sides of a triangle is parallel to the third side, and its length is half the length of the third side.

THINK CRITICALLY

41. In the figure at the right, triangle ABC is isosceles with $AB = AC$. The midpoints of $\overline{AB}$ and $\overline{AC}$ are points D and E, respectively. The midpoints of $\overline{AD}$, $\overline{AE}$, and $\overline{DE}$ are points F, G, and H, respectively. Explain why triangle FGH must be an isosceles triangle.

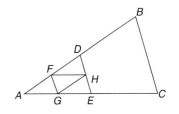

42. **LANDSCAPING** A statue is to be erected in the middle of a square. A landscaper will surround the statue on four sides with a low fence and create a grassy walkway outside the fence. The plan is shown at the right. Express the total amount of fencing needed in terms of y. Explain your reasoning.

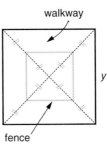

walkway

y

fence

43. **WRITING MATHEMATICS** The sides of $\triangle XYZ$ are midsegments of $\triangle ABC$, as shown at the right. Write one or two sentences to explain why the perimeter of $\triangle XYZ$ must be half the perimeter of $\triangle ABC$.

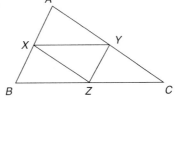

44. **USING ALGEBRA** Write an algebraic proof to support your reasoning in Exercise 43.

ALGEBRA AND GEOMETRY REVIEW

Solve by factoring.

45. $x^2 - 6x - 16 = 0$

46. $y^2 + 2y = 80$

47. $2w^2 - 6w = 20$

48. In the figure at the right, line m intersects lines ℓ and n. Determine whether $\ell \parallel n$. Justify your answer.

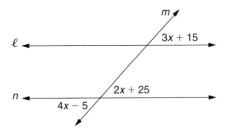

m

ℓ

$3x + 15$

n

$4x - 5$

$2x + 25$

49. In $\triangle ABC$, $m\angle A = (x^2 + 60)°$, $m\angle B = (x + 80)°$, and $m\angle C = (x + 16)°$. Find the measure of each angle of $\triangle ABC$. What kind of triangle is $\triangle ABC$?

50. **STANDARDIZED TESTS** Name the polygon that does not have an angle with a measure that is a multiple of 30.

   A. equilateral triangle   **B.** square   **C.** regular pentagon   **D.** regular hexagon

51. What is the slope of a line parallel to the line $2x + 3y = 4$ and containing point $(-2, 1)$?

52. In $\triangle HMR$, point O is the midpoint of $\overline{HM}$, point E is the midpoint of $\overline{RM}$, $\overline{HR} = 4z$ and $OE = z + 9$. Find HR and OE.

Eliminate Possibilities

One of the first problem solving strategies that you may have used is *trial* and *error*. Usually, after a number of tries, and many errors, you would find a path to a solution. Many times, your trials would be haphazard, while other times after reflecting on what you had done, reviewing the outcomes, and trying again, you would narrow in on a strategy that would work. This is known as *systematized trial and error*. One way of systematizing trial and error is the *method of elimination*. With this strategy you list all possibilities and prove all but one of them is false. To prove a possibility false you assume that it is true and then show that this assumption leads to a contradiction.

Problem

HOUSE BUILDING The Shannons are planning to put an addition onto their house. They found a sketch of a truss in a home improvement book. They wondered if there are any limitations to the size of the angles that are formed at the vertices and at the supports. What are possible measures of ∠FBC, ∠BCF, and ∠BFC that make it possible for the truss shown to be built?

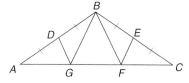

Explore the Problem

1. The truss shown above has an identifiable line of symmetry. Make a sketch of the truss and its line of symmetry.

2. What type of triangle is △ABC? Justify your answer.

3. For △AGB and △CFB what are $\overline{DG}$ and $\overline{EF}$ called?

4. List all of the possibilities for the type of triangles that △BFC and △BGA could be.

5. Assume △BFC and △BGA are equilateral. Examine the measures of the angles in the drawing under this possibility. What contradiction, if any, arises?

6. Assume $\triangle BFC$ and $\triangle BGA$ are isosceles. Examine the possible measures of the angles under this possibility. What contradiction, if any, arises?

7. Repeat Question 6 for a right triangle, an obtuse triangle, and a scalene triangle. Which triangle types can be eliminated because of contradictions? Explain.

8. **WRITING MATHEMATICS** The slope or pitch of the roof will have a direct influence on the measures of the base angles of the truss. Copy and complete the following two statements by referring to the measures of these angles.

The steeper the pitch the ___?___.
The flatter the pitch the ___?___.

> **PROBLEM SOLVING PLAN**
>
> • Understand
> • Plan
> • Solve
> • Examine

Investigate Further

● Eliminating possibilities helps you to gain more information about the data in a problem by narrowing down your choices. Consider the following problem. You are given four shapes that have something in common. This characteristic is not present in the second set of shapes. You must determine if a particular shape is or is not a *geoshape*.

These are geoshapes.

These are not geoshapes.

> **COMMUNICATING ABOUT GEOMETRY**
>
> Explain exactly what is meant by a steep roof pitch and a flat roof pitch.

9. Make a list of at least four features that geoshapes share.

10. Examine the shapes that are not geoshapes. Do any of them have any of the features you listed in Question 9? If so, this feature does not define a geoshape. Which of the features listed can be eliminated?

11. Is this a geoshape?

12. Can any of the following characteristics of geoshapes be eliminated as defining characteristics? Explain your reasoning.

Geoshapes are closed figures.
Geoshapes all contain shapes within shapes.
Geoshapes have a horizontal base.
Geoshapes are straight-sided.

13. WRITING MATHEMATICS Notice that the pentagon has a triangle enclosed within it. The hexagon has a rectangle enclosed within it. The square has a rectangle. Could this be the defining feature of a geoshape? Discuss how the triangle fits this characteristic.

14. Now that you have eliminated possibilities, what is a geoshape?

Apply the Strategy

15. These are out-a-shapes.

These are not out-a-shapes.

Is this an "out-a-shape"? Explain how you can use eliminating possibilities to determine your answer?

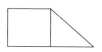

16. A shape is gradually revealed in four steps. Identify all of the possible figures that the shape could be at each stage shown. Describe how subsequent stages assist in eliminating possibilities.

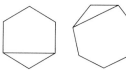
(1)   (2)

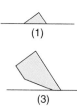 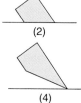
(3)   (4)

17. Examine the following conjecture: A triangle can have only one right angle. Describe how you can use the strategy of eliminating possibilities to confirm or reject this conjecture.

18. Examine the following conjecture: A triangle can have only one obtuse angle. Describe how you can use the strategy of eliminating possibilities to confirm or reject this conjecture.

19. WRITING MATHEMATICS Explain how eliminating possibilities includes looking for contradictions.

Review Problem Solving Strategies

TWINKLE, TWINKLE

1. A star is formed by extending the sides of a regular pentagon. Find the sum of the measures of the five angles determined by the points of the star.

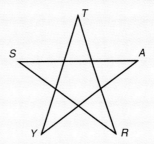

That's the Spirit!

2. Jason and Joe are two excellent geometry students who invent a game with numbers. The rules are simple. One person starts by stating any integer a such that $0 < a \leq 8$. The second person calls out an integer b such that $0 < b \leq 8$ and this is added to the first number. The game continues with each person alternately calling out a positive integer less than or equal to 8 until the sum of the integers named is equal to 76. The person who ends up at 76 is the winner. What strategy will give a person the best chance of winning?

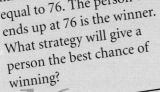

3. Place your finger on the top *I* in the figure below and move it down to an adjacent letter. Continue moving your finger down spelling the word *ISOSCELES*. How many different paths can you take tracing out this word? Remember you always start at the top *I* and move down to an adjacent letter.

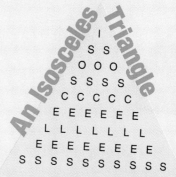

An Isosceles Triangle

```
        I
       S S
      O O O
     S S S S
    C C C C C
   E E E E E E
  L L L L L L L
 E E E E E E E E
S S S S S S S S S
```

Focus on Reasoning

Indirect Proof

Discussion

• Work with a partner. Consider this conditional statement.

   If two lines intersect, then they intersect at exactly one point.

Malcolm and Allison are discussing this statement. Malcolm believes it is true. He drew these figures as a proof.

1. Has Malcolm drawn every possible case of two lines intersecting? Explain.

2. Was Malcolm able to draw each line entirely? Explain.

3. Do you think Malcolm's figures make up a proof? Explain.

Allison says if the hypothesis of the given conditional statement is true, there are only two possible conclusions.

   If two lines intersect, then they intersect at exactly one point.
   If two lines intersect, then they intersect at more than one point.

4. Allison believes the given conclusion is true. She drew the figure at the right, then wrote the following argument to justify her belief. Copy and complete her argument.

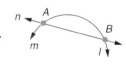

Step 1 Assume that two lines can intersect at _?_ points. In particular, assume that there are lines *n* and *m* that intersect at points _?_ and _?_ .

Step 2 By the _?_ Postulate, there is exactly one line through points *A* and *B*.

Step 3 The statements in steps 1 and 2 are contradictory. Therefore, the statement "if two lines intersect, then they intersect at more than one point" must be false. It follows that the statement "if two lines intersect, then they intersect at exactly one point" must be _?_ .

5. What problem solving strategy did Allison use in step 3 of her proof?

Build a Case

- You probably recognize that Malcolm's "proof" is not a proof at all. Allison's argument is a proof, using a process called *indirect reasoning*. In **indirect reasoning**, you eliminate all possible conclusions but one, with the result that the one remaining conclusion must be true. A proof based on indirect reasoning is called an *indirect proof*. Use the following steps when you write an indirect proof.

Writing an Indirect Proof

Step 1   Assume temporarily that the conclusion of the given statement is false.

Step 2   Reason logically until you arrive at a contradiction of the hypothesis or some other known fact.

Step 3   State that the temporary assumption must be false, and that the given statement must be true.

Consider the following statement.

> A triangle cannot have two obtuse angles.

6. State the assumption you would make to begin an indirect proof.

7. What do you know about the measure of an obtuse angle?

8. What must be true about the sum of the measures of two obtuse angles?

9. How does your answer to Question 8 contradict a known fact about triangles?

10. Using your answers to Questions 6–9, write an indirect proof of the given statement.

> **COMMUNICATING ABOUT GEOMETRY**
>
> The proofs you have written until now are *direct proofs*. What do you think distinguishes a direct proof from an indirect proof?

> **GEOMETRY: WHO, WHERE, WHEN**
>
> The process of indirect reasoning has been in use for centuries. Its classical name is *reductio ad absurdum*, or "reduction to absurdity." One of the earliest known indirect proofs is Euclid's proof that the number of primes is infinite.

┌ EXTEND AND DEFEND

Write an indirect proof of each statement.

11. If a triangle is not isosceles, then it is not equilateral.

12. If two sides of a triangle are not congruent, then the angles opposite those sides are not congruent.

13. Through a point outside a line, there is exactly one line perpendicular to the given line. (*Hint*: Consider the triangle formed in the figure at the right.)

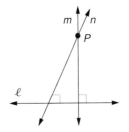

In Questions 14–16 use indirect proof.

14. Prove that in scalene △*ABC*, the bisector of ∠*ABC* is not perpendicular to $\overline{AC}$, the opposite side.

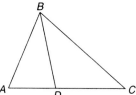

15. Prove that in △*XYZ*, if $\overline{YA}$ is a median that does not bisect ∠*XYZ*, then $\overline{XY}$ is not congruent to $\overline{YZ}$.

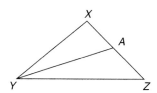

16. In △*LMN*, $\overline{MP} \perp \overline{LN}$ and *LM* ≠ *MN*. Prove $\overline{LP}$ is not congruent to $\overline{PN}$.

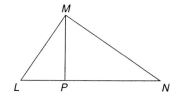

17. FINGERPRINT IDENTIFICATION A suspect claims never to have been at the scene of a crime. However, Detective Garcia matched one of this suspect's fingerprints to a fingerprint on a wall at the crime scene. Prove by indirect reasoning that the suspect once was at the scene of the crime.

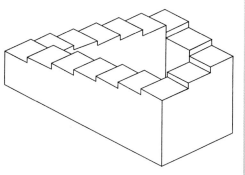

18. WRITING MATHEMATICS Describe an everyday situation in which you might use indirect reasoning. Explain how indirect reasoning would be applied to the situation.

19. THINKING CRITICALLY A student wants to write an indirect proof that a certain angle is an acute angle. The student begins as follows: Assume that the angle is an obtuse angle. Do you agree with this assumption? Explain.

20. ZIP CODES Anna remembers the first three digits of her bank's ZIP code are 125 and that the ZIP code is an odd number. She also knows that the main branch is in either Rhinebeck, Pawling, Newburgh, or Milton whose ZIP codes are 12572, 12564, 12550, and 12547, respectively. Prove indirectly that the main branch is in Milton.

Now You See It
Now You Don't

GOING UP
Use the method of indirect reasoning that you learned in this lesson to write a proof of this statement.

Every staircase has a top step.

How does the figure at the right appear to contradict the conclusion of your proof?

Exploring Triangle Inequalities

Think Back

Copy each of the following statements. Fill in the blank with either $<$ or $>$. Assume that a, b, and c are real numbers.

1. By the Trichotomy Property, exactly one of the following relationships must be true for a and b.

$$a = b \qquad a \underline{\quad?\quad} b \qquad a \underline{\quad?\quad} b$$

2. By the Transitive Property of Inequality,

If $a > b$ and $b > c$ then $a \underline{\quad?\quad} c$.

If $a < b$ and $b < c$ then $a \underline{\quad?\quad} c$.

3. If $a > b$ then $a + c \underline{\quad?\quad} b + c$ and $a - c \underline{\quad?\quad} b - c$.

4. If $a < b$, then $a + c \underline{\quad?\quad} b + c$ and $a - c \underline{\quad?\quad} b - c$.

5. If $a > b$ and $c > 0$ then $ac \underline{\quad?\quad} bc$ and $\dfrac{a}{c} \underline{\quad?\quad} \dfrac{b}{c}$.

6. If $a > b$ and $c < 0$ then $ac \underline{\quad?\quad} bc$ and $\dfrac{a}{c} \underline{\quad?\quad} \dfrac{b}{c}$.

7. By the comparison property, $a > b$ if and only if there exists a c where $c \underline{\quad?\quad} 0$, such that $a = b + c$.

Explore

8. Draw four different scalene triangles. Label each as $\triangle ABC$ and number them 1–4. Copy and complete the table below.

Triangle	AB	AC	BC	m∠C	m∠B	m∠A
1	▦	▦	▦	▦	▦	▦
2	▦	▦	▦	▦	▦	▦
3	▦	▦	▦	▦	▦	▦
4	▦	▦	▦	▦	▦	▦

9. In your table put a circle around the measure of the longest side of each triangle and another circle around the angle that lies opposite the longest side.

10. Next put a square around the measure of the shortest side of each triangle and another square around the angle that lies opposite it.

11. Make a conjecture about the measures of the angles based upon the measures of the sides opposite them.

12. Make a conjecture about the measures of the sides based upon the measures of the angles opposite them.

13. Set up another table with the following headings.

Triangle	AB + BC	CA	BC + CA	AB	CA + AB	BC

Using the data from your answer to Question 8, compute the sums and fill in the table.

14. Make a conjecture about the relationship between the sum of the lengths of two sides of a triangle and the length of the third side.

Make Connections

In a triangle, the angle with the greatest measure lies opposite the longest side and the angle with the least measure lies opposite the shortest side. Furthermore, the sum of the lengths of any two sides of a triangle is greater than the length of the third side.

15. In $\triangle RST$, m$\angle R = 62°$, m$\angle S = 68°$, m$\angle T = 50°$. List the sides from shortest to longest. Justify your answer.

16. In $\triangle UVW$, $UV = 10$ cm, $VW = 8$ cm, and $UW = 12$ cm. List the angles from least measure to greatest measure. Justify your answer.

Use the figure to answer Questions 17 and 18.

17. In $\triangle EFG$, list the sides in order from shortest to longest.

18. In $\triangle HFG$, list the sides in order from shortest to longest.

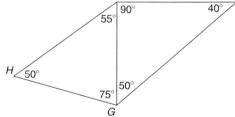

19. **USING ALGEBRA** The sides of $\triangle STU$ are represented in terms of x by $US = x + 2$, $ST = x$, and $TU = x - 2$. Which angle of $\triangle STU$ has the least measure? Justify your answer.

20. In $\triangle XYZ$, m$\angle X = 42°$ and m$\angle Y = 63°$. Which is the shortest side? Justify your answer.

21. The coordinates of the vertices of $\triangle DEF$ are $D(1, 3)$, $E(1, 8)$, and $F(13, 8)$. List the angles from least measure to greatest measure. Justify your answer.

22. CARPENTRY A shelf is to be supported with a bracket as shown at the right. Is the support longer than the shelf depth? Explain.

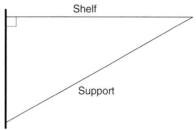

Shelf

Support

23. The coordinates of the vertices of △*GHI* are *G*(−3, −4), *H*(−1, −9) and *I*(3, −3). List the angles from least measure to greatest measure. Justify your answer.

24. Identify each of the following sets of three numbers as measures that can or cannot form triangles. Justify your answers.

    a. 9, 15, 22   **b.** 16, 45, 25

    c. 18.7, 13.5, 10.2   **d.** $19\frac{3}{4}$, $10\frac{1}{2}$, $9\frac{1}{8}$

Summarize

25. MANIPULATIVES Work with a partner. Each partner should create a scalene triangle on a geoboard using 3 different colored rubber bands. The sides should not be obviously different in length. Angles are identified as opposite a colored side. Switch geoboards. Challenge your partner to identify the smallest and largest angle.

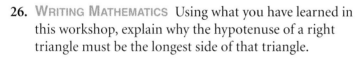

26. WRITING MATHEMATICS Using what you have learned in this workshop, explain why the hypotenuse of a right triangle must be the longest side of that triangle.

27. THINKING CRITICALLY If two sides of a triangle have lengths *p* and *r*, what is the range of possible values for the third side?

28. THINKING CRITICALLY A stick is 8 in. long. If it is broken into three pieces, which can be made into a triangle, such that the length of each piece is an integer number of inches, what is the length of the shortest piece?

29. GOING FURTHER Use the figure at the right to explain why the perimeter of a quadrilateral is greater than the sum of the length of its diagonals. Let the lengths of the diagonals be *f* and *g*.

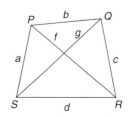

6.9 Inequalities in One Triangle

Explore

• Construct two different scalene triangles and measure their sides and angles. Copy and complete the following statements.

1. The angle opposite the longest side is __?__ than the angle opposite the shortest side.
2. The side opposite the largest angle is __?__ than the side opposite the smallest angle.
3. The sum of the lengths of any two sides of a triangle is __?__ than the length of the third side.

Build Understanding

• The triangle relationship in Explore Question 1 is a theorem.

THEOREM 6.11	UNEQUAL SIDES THEOREM

If two sides of a triangle are unequal in length, then the measure of the angle opposite the longer side is greater than the measure of the angle opposite the shorter side.

Given $\triangle RST$ with $RT > RS$

Prove $m\angle RST > m\angle T$

Plan for Proof Draw an auxiliary line segment. Then use the Base Angles Theorem, the Exterior-Angles Theorem for Triangles, and properties of algebra.

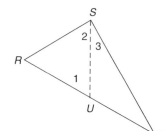

Proof Use the Ruler Postulate to locate point U on $\overline{RT}$ so $RU = RS$. Draw $\overline{SU}$. By the Angle Addition Postulate $m\angle RST = m\angle 2 + m\angle 3$. So, $m\angle RST > m\angle 2$. But $m\angle 1 = m\angle 2$ by the Base Angles Theorem. So by substitution $m\angle RST > m\angle 1$. Since $\angle 1$ is an exterior angle of $\triangle UTS$, $m\angle 1 = m\angle 3 + m\angle T$ by the Exterior-Angle Theorem for Triangles. So, $m\angle 1 > m\angle T$ because the whole is greater than its parts. Finally, $m\angle RST > m\angle T$ by the Transitive Property for Inequaliy. ◄

EXAMPLE 1

In $\triangle ABC$, $AB = 15$, $BC = 10$ and $CA = 8$. Identify the largest angle and the smallest angle in the triangle.

Solution

By the Unequal Sides Theorem, since $\overline{AB}$ is the longest side, $\angle C$, the angle opposite it, is the largest angle. Also, since $\overline{CA}$ is the shortest side, $\angle B$, the angle opposite it, is the smallest angle. ◄

SPOTLIGHT ON LEARNING

WHAT? In this lesson you will learn
• to apply the Unequal Sides Theorem.
• to apply the Unequal Angles Theorem.
• to apply the Triangle Inequality Theorem.

WHY? The three inequality theorems can help you find distances on a map and in designing objects such as bicycles.

THINK BACK

The Exterior Angles Theorem for Triangles states that the measure of an exterior angle of a triangle is equal to the sum of the measures of the two remote interior angles.

The converse of Theorem 6.12 and its proof follow.

CHECK
UNDERSTANDING

Explain how the
Trichotomy Property
assures you that all of
the possibilities have
been accounted for
in this situation.

THEOREM 6.12 **UNEQUAL ANGLES THEOREM**

If two angles of a triangle are unequal in measure, then
the side opposite the larger angle is longer than the side
opposite the smaller angle.

Given $\triangle ABC$ with $m\angle A > m\angle C$

Prove $BC > AB$

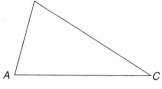

Proof Write an indirect proof. Assume
$BC \not> AB$. By the Trichotomy Property,
either $BC = AB$ or $BC < AB$. If $BC = AB$,
then $\triangle ABC$ is isosceles. Since the base angles of an isosceles triangle are
congruent, $m\angle A = m\angle C$. This contradicts the given information that
$m\angle A > m\angle C$. If $BC < AB$, then, from the Unequal Sides Theorem,
$m\angle A < m\angle C$. This too contradicts the given. All possibilities have been
eliminated except $BC > AB$. Therefore, the theorem is proved. ◄

EXAMPLE 2

In $\triangle KMS$, $m\angle K = 47°$ and $m\angle M = 51°$. Identify the longest and
shortest sides of this triangle.

Solution
Since $m\angle K + m\angle M + m\angle S = 180°$, $m\angle S = 82°$. By the Unequal
Angles Theorem, side $\overline{KM}$ is the longest side since it is opposite the largest
angle, and $\overline{MS}$ is the shortest side since it is opposite the smallest angle. ◄

The next inequality theorem can help you determine whether three
segments will form a triangle.

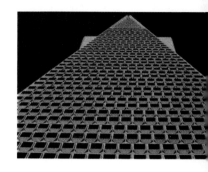

THEOREM 6.13 **TRIANGLE INEQUALITY THEOREM**

The sum of the lengths of any two sides of a triangle is
greater than the length of the third side.

EXAMPLE 3

LANDSCAPING Jill has three landscaping timbers that have lengths of
8.6 ft, 4.4 ft, and 3.8 ft. Is it possible to use the timbers to border a
triangular flower bed without cutting any of them shorter? Explain.

Solution
Check each possibility with the Triangle Inequality Theorem.

$$8.6 + 3.8 > 4.4 \qquad 8.6 + 4.4 > 3.8 \qquad 4.4 + 3.8 \not> 8.6$$

Since the last two sides are not greater than the first side, the triangle is
impossible. The 8.6 timber must be cut shorter. ◄

TECHNOLOGY TIP

Use geometry software to try to construct the triangle in Example 4. Explain what happens as you attempt to fit the sides of these lengths together.

When two sides of a triangle are known, Theorem 6.13 can be used to determine the range of possible values for the third side.

EXAMPLE 4

In $\triangle STU$, $ST = 10$, $TU = 14$, and $US = x$. What is the range of possible values for US?

Solution

Use the Triangle Inequality Theorem to write three inequalities that express the relationships among the sides in terms of x. Then solve for x.

$$10 + 14 > x \qquad 10 + x > 14 \qquad 14 + x > 10$$
$$24 > x \qquad\qquad x > 4 \qquad\qquad x > -4$$

Therefore, $4 < x < 24$. Since $US = x$, $4 < US < 24$. ◄

TRY THESE

Given the sides, identify the largest and smallest angle in each triangle.

1. $\triangle ABC$: $AB = 18$, $BC = 15$, $CA = 12$

2. $\triangle DEF$: $DE = 3$, $EF = 9$, $FD = 10$

3. $\triangle GHI$: $GH = 5\frac{1}{4}$, $HI = 7\frac{1}{8}$, $IG = 6\frac{1}{2}$

4. $\triangle JKL$: $JK = 18.8$, $KL = 15.4$, $LJ = 9.2$

Given two angles, identify the longest and shortest side in each triangle.

5. $\triangle MNO$: $m\angle M = 40°$, $m\angle N = 80°$

6. $\triangle PQR$: $m\angle Q = 90°$, $m\angle R = 42°$

7. $\triangle STU$: $m\angle S = 110°$, $m\angle U = 34°$

8. $\triangle VWX$: $m\angle V = 15°$, $m\angle W = 85°$

Determine whether or not the three measures given could be the lengths of the sides of a triangle. Explain your reasoning.

9. 7, 6, 5   **10.** 2, 9, 4   **11.** 15, 15, 15   **12.** $6\frac{1}{4}$, $10\frac{3}{4}$, 17   **13.** 7.5, 3.5, 6.5   **14.** 56, 56, 70

Given two sides of $\triangle ABC$, write an algebraic inequality that identifies the range of possible values for the third side.

15. $AB = 12$, $BC = 9$

16. $BC = 8.5$, $CA = 3.2$

17. $CA = 3\frac{3}{4}$, $AB = 5\frac{1}{2}$

18. $AB = 1$, $BC = \sqrt{2}$

19. $BC = \sqrt{3}$, $CA = 9.5$

20. $CA = \sqrt{5}$, $AB = y$

21. BICYCLES Two triangles are formed by the tubing on a particular bike as shown. Arrange the angles in each triangle from greatest to smallest measure. Justify your reasoning.

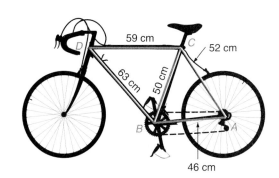

PRACTICE

Given the sides, identify the largest and smallest angle in each triangle below.

22. $\triangle ABC$: $AB = 13$, $BC = 6$, and $CA = 9$

23. $\triangle DEF$: $DE = 6$, $EF = 8$, $FD = 12$

24. $\triangle GHI$: $GH = 5\frac{1}{2}$, $HI = 3\frac{1}{16}$, $IG = 2\frac{3}{4}$

25. $\triangle JKL$: $JK = 1.9$, $KL = 1.4$, $LJ = 2.2$

Given two angles, identify the longest and shortest side in each triangle below.

26. $\triangle MNO$: m$\angle M = 10°$, m$\angle N = 100°$

27. $\triangle PQR$: m$\angle Q = 30°$, m$\angle R = 60°$

28. $\triangle STU$: m$\angle S = 11°$, m$\angle U = 45°$

29. $\triangle VWX$: m$\angle V = 120°$, m$\angle W = 35°$

Determine whether or not the three measures given could be the lengths of the sides of a triangle. Explain your reasoning.

30. 24, 27, 25

31. 1, 2, 3

32. 19, 5, 9

33. $5\frac{5}{8}$, 5, $3\frac{7}{8}$

34. 9.3, 2.7, 6.6

35. $5\sqrt{2}$, $10\sqrt{2}$, $18\sqrt{2}$

You are given two sides of $\triangle ABC$. Write an algebraic inequality that identifies the range of possible values for the third side.

36. $AB = 6.6$, $BC = 7.5$

37. $BC = 13$, $CA = 18$

38. $CA = 9\frac{1}{4}$, $AB = 5\frac{1}{8}$

39. $AB = 5\sqrt{13}$, $BC = 5$

40. $BC = \sqrt{7}$, $CA = 0.5$

41. $CA = r$, $AB = s$

42. MAPS Examine the following distance chart. Copy the map and draw the triangular route that joins these three cities. List the angles formed by these routes from greatest to smallest measure. Justify your answer.

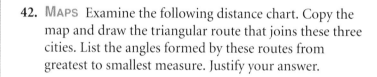

Mileage Chart			
	New York	Chicago	Atlanta
New York		807	885
Chicago	807		717
Atlanta	885	717	

43. Triangle QRS has vertices $Q(0, 0)$, $R(7, 0)$, and $S(0, 10)$. List the angles from least to greatest measure. Justify your reasoning.

44. WRITING MATHEMATICS A scalene triangle has a 60° angle. Does this angle lie opposite the longest, shortest, or other side? Explain your answer.

EXTEND

45. The shortest segment that joins a point to a line is a segment from the point that is perpendicular to the line.

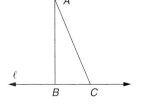

   a. Given line ℓ with point A not on ℓ, $\overline{AB} \perp \ell$, and C is any point on ℓ. What kind of angle is $\angle ABC$? Why?

   b. Justify why m$\angle ACB <$ m$\angle ABC$.

   c. Therefore, why does it follow that $AC > AB$?

46. Give the reason for each statement in the proof of the Triangle Inequality Theorem.

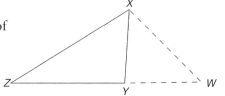

   Given $\triangle XYZ$

   Prove $XY + YZ > ZX$

Statements	Reasons
1. Draw $\overleftrightarrow{ZY}$.	
2. On $\overrightarrow{ZY}$ locate point W such that $YW = YX$.	
3. Draw $\overline{WX}$.	
4. $WZ = WY + YZ$	
5. $WZ = YX + YZ$	
6. m$\angle WXZ =$ m$\angle WXY +$ m$\angle YXZ$	
7. m$\angle WXZ >$ m$\angle WXY$	
8. m$\angle W =$ m$\angle WXY$	
9. m$\angle WXZ >$ m$\angle W$	
10. $WZ > XZ$	
11. $XY + YZ > XZ$	

THINK CRITICALLY

47. Writing Mathematics Given $\triangle ABC$ with point Q in its interior, explain why $BA + AC > BQ + QC$.

ALGEBRA AND GEOMETRY REVIEW

Use the figure at the right for Exercises 48–51.

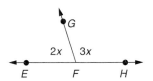

48. Find m$\angle GFH$.

49. What kind of angle is $\angle EFG$?

50. What kind of angle is $\angle GFH$?

51. What kind of angle is $\angle EFH$?

52. The larger of two complementary angles is 4° more than twice the smaller angle. Find the measure of the larger angle.

53. Simplify: $-7r + 6(-3r + 2s) - 2s(1 + r)$

54. Solve for y: $y^2 + y - 6 > 0$

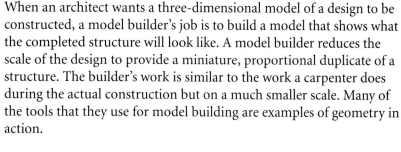

When an architect wants a three-dimensional model of a design to be constructed, a model builder's job is to build a model that shows what the completed structure will look like. A model builder reduces the scale of the design to provide a miniature, proportional duplicate of a structure. The builder's work is similar to the work a carpenter does during the actual construction but on a much smaller scale. Many of the tools that they use for model building are examples of geometry in action.

Decision Making

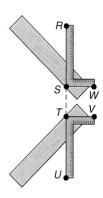

Models are made of sturdy lightweight material such as formboard or balsa wood. The size and material of the model determines the tools used, but often a protractor, drafter's triangle, or carpenter's square is used. The carpenter's square shown consists of two rulers that meet at a right angle. The square can be used in a variety of ways to *lay off* right angles, draw right triangles, and draw a variety of other geometric shapes.

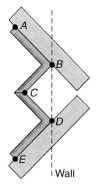

A model builder must be able to join two pieces of wood at an angle. The illustration at the right above shows two carpenter's squares lined up along imaginary $\overline{ST}$. Lines will be drawn along $\overline{SW}$ and $\overline{TV}$.

1. Explain why this assures the builder that $\overline{SW}$ is parallel to $\overline{TV}$. Why must the cuts be made along $\overline{SW}$ and $\overline{TV}$?

A builder may need to have two pieces of wood meet at congruent angles along the same wall. Two carpenter's squares can be used as shown at the left.

2. If the carpenter's squares are placed so that $\overline{BC} \cong \overline{DC}$, explain how the builder can be certain that the angles at which the wood meet the wall are the same for both pieces.

6.10 Inequalities in Two Triangles

Explore/Working Together

1. Break a strand of uncooked spaghetti into two unequal pieces.

2. Lay them on a sheet of paper so they share a common endpoint. They determine two sides and an included angle of a triangle. Call this angle the "spaghetti angle."

3. Find the length of the side of the triangle opposite the "spaghetti angle."

4. Think of the vertex of the "spaghetti angle" as a hinge and increase the angle's measure. Did the side opposite increase or decrease in length?

5. Next decrease the measure of the "spaghetti angle." What happened to the opposite side?

6. Break a new piece of spaghetti at a different place and repeat Questions 1–5.

7. Make a conjecture about the relationship between the two sides opposite the two "spaghetti angles."

TECHNOLOGY TIP

You can use geometry software to perform the activity in Explore.

COMMUNICATING ABOUT GEOMETRY

Why do you think that this theorem is known as the Hinge Theorem? Explain how the word "hinge" gives a visual image of the theorem.

Build Understanding

The opening and closing of the "spaghetti angle" in Explore suggests the following theorem.

THEOREM 6.14 **HINGE THEOREM**

If two sides of one triangle are congruent to two sides of another triangle, and the included angle of the first triangle is greater than the included angle of the second triangle, then the third side of the first triangle is longer than the third side of the second triangle.

EXAMPLE 1

In the figure $\overline{EF} \parallel \overline{HG}$, $\overline{EH} \cong \overline{HG}$, m$\angle EFH = 60°$, and m$\angle EHF = 70°$.
Which is longer $\overline{EF}$ or $\overline{FG}$?

Solution

Since $\overline{EF} \parallel \overline{HG}$, $\angle GHF \cong \angle EFH$
because of the Alternate Interior
Angles Theorem. Therefore,
m$\angle GHF = 60°$. By the Reflexive
Property, $\overline{HF} \cong \overline{HF}$. The
conditions of the Hinge
Theorem are satisfied.
Therefore, since
m$\angle EHF > $ m$\angle GHF$,
$\overline{EF}$ is longer than $\overline{FG}$.

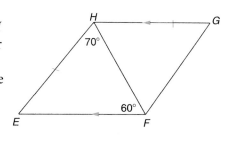

◀

You can use constructions with spaghetti or geometry software to test
the converse of the Hinge Theorem.

THEOREM 6.15 **CONVERSE OF HINGE THEOREM**

If two sides of one triangle are congruent to two sides of
another triangle, and the third side of the first triangle is
longer than the third side of the second, then the
included angle of the first triangle is larger than the
included angle of the second.

Theorem 6.15 is useful in discovering relationships between angles in
triangles when you only know the lengths of sides.

EXAMPLE 2

HOME DESIGN A sideview plan of a home indicates that the rear roof
line is raised to accommodate a vertical skylight window. The lengths
of the front and rear roofs, AC and DE, are equal and $\overline{AB} \cong \overline{BE}$. What
is the relationship between m$\angle CAB$
and m$\angle DEB$?

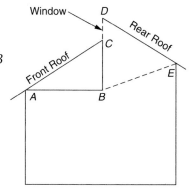

Solution

Two triangles can be identified, $\triangle CAB$
and $\triangle DEB$, in which $\overline{AC} \cong \overline{DE}$ and
$\overline{AB} \cong \overline{BE}$. From the diagram, point
C is between point D and point B,
so $DB > CB$. Therefore, by the
converse of the Hinge Theorem,
m$\angle DEB > $ m$\angle CAB$.

Given $\triangle ABC$ and $\triangle DEF$ with $\overline{AB} \cong \overline{DE}$ and $\overline{AC} \cong \overline{DF}$, determine the relationship between sides BC and EF in each case.

1. $m\angle A = 35°$ and $m\angle D = 45°$

2. $m\angle A = 62°$ and $m\angle D = 70°$

3. $m\angle A = 50°$ and $m\angle D = 50°$

4. $m\angle A = 110°$ and $m\angle D = 90°$

5. $m\angle A = (x + 2)°$ and $m\angle D = x°$

6. $m\angle A = y°$ and $m\angle D = (y - 20)°$

7. In $\triangle LMN$ and $\triangle PQR$, $\overline{MN} \cong \overline{QR}$, $\overline{LN} \cong \overline{PR}$, $m\angle M = 60°$, $m\angle L = 70°$, $m\angle Q = 62°$, and $m\angle P = 67°$. Determine the relationship between LM and PQ.

8. In $\triangle JKL$ and $\triangle TVW$, $\overline{JK} \cong \overline{TV}$, $\overline{JL} \cong \overline{TW}$, $m\angle K = 64°$, $m\angle J = 55°$, $m\angle V = 60°$, and $m\angle W = 61°$. Determine the relationship between KL and VW.

9. In the drawing at the right, $\overline{BA} \cong \overline{BD}$, $\overline{BC} \cong \overline{BE}$, $m\angle A = 90°$, $m\angle C = 50°$, $m\angle E = 50°$, and $m\angle DBE = 60°$. Is the perimeter of $\triangle ABC$ greater than or less than the perimeter of $\triangle DBE$? Explain your reasoning.

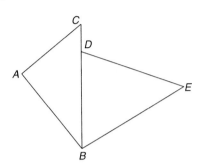

Given $\triangle ABC$ and $\triangle DEF$ with $\overline{AB} \cong \overline{DE}$ and $\overline{AC} \cong \overline{DF}$, determine the relationship between $m\angle A$ and $m\angle D$ in each case.

10. $BC = 10$ in., $EF = 12$ in.

11. $BC = 9$ cm, $EF = 5$ cm

12. $BC = 3$ in., $EF = 3$ in.

13. $BC = \sqrt{2}$ ft, $EF = \sqrt{3}$ ft

14. $BC = x$ in., $EF = 2x$ in.

15. $BC = \frac{x}{2}$ cm, $EF = x$ cm

16. **SAILBOATS** In the picture of a sailboat at the right, $\overline{AB} \cong \overline{BC}$ and $\overline{AE} \cong \overline{DC}$. Which angle is larger, $\angle EAB$ or $\angle DCB$? What theorem did you use to get your answer?

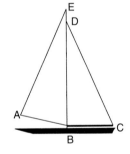

PRACTICE

Given $\triangle ABC$ and $\triangle DEF$ with $\overline{AB} \cong \overline{DE}$ and $\overline{AC} \cong \overline{DF}$, determine the relationship between BC and EF in each case.

17. $m\angle A = 135°$ and $m\angle D = 85°$

18. $m\angle A = 85°$ and $m\angle D = 95°$

19. $m\angle A = 30°$ and $m\angle D = 30°$

20. $m\angle A = 10°$ and $m\angle D = 15°$

21. $m\angle A = (2x + 3)°$ and $m\angle D = x°$

22. $m\angle A = (y - 15)°$ and $m\angle D = (y - 30)°$

23. In $\triangle LMN$ and $\triangle PQR$, $\overline{MN} \cong \overline{QR}$, $\overline{LN} \cong \overline{PR}$, $m\angle M = 85°$, $m\angle L = 45°$, $m\angle Q = 120°$, and $m\angle P = 20°$. Determine the relationship between LM and PQ.

24. In $\triangle ABC$ and $\triangle JKL$, $\overline{BC} \cong \overline{KL}$, $\overline{AC} \cong \overline{JL}$, $m\angle A = 60°$, $m\angle B = 90°$, $m\angle J = 40°$, and $m\angle L = 25°$. Determine the relationship between AB and JK.

Given △ABC and △DEF with $\overline{AB} \cong \overline{DE}$ and $\overline{AC} \cong \overline{DF}$. Determine the relationship between m∠A and m∠D in each case.

25. $BC = 9$ in., $EF = 15$ in.

26. $BC = 2$ cm, $EF = 1$ cm

27. $BC = 15$ in., $EF = 15$ in.

28. $BC = \sqrt{7}$ in., $EF = \sqrt{5}$ in.

29. $BC = (x + 5)$ in., $EF = x$ in.

30. $BC = (x - 1)$ cm, $EF = (x + 2)$ cm

In Exercises 31–33 state which segment or angle is larger, *x* or *y*. Justify your answer.

31.

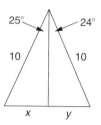

32.

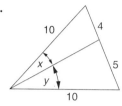

33.

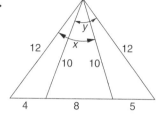

EXTEND

MAPS **Copy this map of the west coast of the United States and use it for Exercises 34–37.**

34. Draw straight-line paths that join each of the following cities: San Francisco, Los Angeles, Salt Lake City, Phoenix, and Denver.

35. The distance between San Francisco and Los Angeles is approximately the same as the distance between Los Angeles and Phoenix. Mark these congruent segments on your sketch.

36. The distance between Los Angeles and Salt Lake City is approximately the same as the distance between Phoenix and Denver. Mark these congruent segments on your sketch.

37. The angle made by the paths from Los Angeles to Phoenix to Denver is approximately 127°. The angle made by the paths from San Francisco to Los Angeles to Salt Lake City is approximately 78°. Use the Hinge Theorem to determine which distance is greater, from San Francisco to Salt Lake City or from Los Angeles to Denver.

THINK CRITICALLY

38. Triangle *LMN* and triangle *RST* are isosceles with $\overline{LM} \cong \overline{MN} \cong \overline{RS} \cong \overline{ST}$ and m∠*L* = m∠*S* = 56°. Which triangle has the longer base? Explain.

39. In the diagram below *AE* = *CD*, *AB* > *BC*, and *BE* = *BD*. Which is greater m∠*AEB* or m∠*CDB*? Explain.

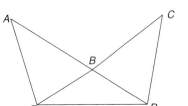

40. In the diagram below $\overline{BC} \cong \overline{EF}$ and *FC* > *BE*. Show that *AF* > *AB*.

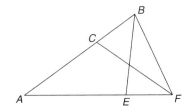

ALGEBRA AND GEOMETRY REVIEW

41. In △*RST*, m∠*R* = 2*x*, m∠*S* = 3*x* − 10, m∠*T* = 5*x* + 20. Determine the measure of each angle of the triangle.

42. ∠*A* and ∠*B* are complementary angles. If m∠*A* is 35° more than m∠*B*, find m∠*A*.

43. ∠*C* and ∠*D* are supplementary angles. If m∠*C* is 3 times m∠*D*, find m∠*D*.

44. ∠*E* and ∠*F* are supplementary angles and ∠*F* and ∠*G* are complementary angles. If m∠*G* = 15°, find the measures of the other two angles.

PROJECT *Connection*   You will need sticks and fasteners for this project. A *truss* is a rigid framework designed to support a structure. Many bridges are built with steel trusses. A truss works on the principles of tension and compression in supporting the roadway.

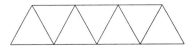

Select one of the following three truss types. Use the sticks and fasteners to build a truss support bridge of the type you have chosen. Research this type of truss and determine if there are any such bridges where you live.

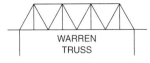

WARREN TRUSS

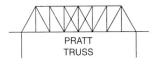

PRATT TRUSS

QUADRANGULAR WARREN TRUSS

Architects are excellent problem solvers. They collect and analyze data, devise plans, carry out the plan, and then look back at what they have done in order to make adjustments and improvements. One of the simplest and most commonplace problems that an architect faces in planning for construction is how to get people from one level to another. A staircase is something that we all take for granted and yet an architect must carefully plan its design.

Decision Making

The *total rise* of a stairway is the vertical distance from the lower level to the upper level. The *total run* of a stairway is the horizontal distance from the edge of the top landing to the edge of the bottom step. The vertical part of a step is called the *riser* and the horizontal part of the step is known as the *tread*.

1. Identify some triangles in the illustration of the staircase.

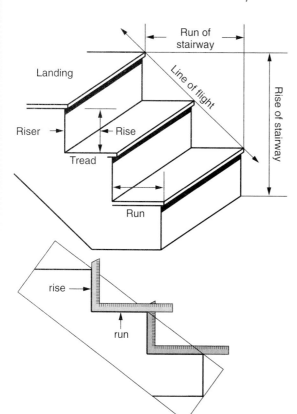

2. A right triangle can be formed whose hypotenuse is known as the *line of flight* along the edges of the treads. The base angle of this triangle is commonly between 30° and 35°. Suppose that the line of flight angle is 32°. What are the measures of the other angles in this triangle?

3. The rise and run of a staircase, along with the tread and riser lengths of each step can vary. The drawing at the left shows how a carpenters square is used to build a support for the treads. Explain how the two triangles are related.

4. An architect must plan for a stairway that is safe and easy to use. The general rule of thumb about the dimensions is that the tread T and riser R measurements should fit the formula $2R + T = 25$ in. as closely as possible. How well does a rise of $6\frac{1}{2}$ in. and a run of 12 in. fit the formula?

CHAPTER REVIEW

VOCABULARY

Choose the word from the list that correctly completes each statement.

1. A(n) __?__ of a triangle is a segment drawn from one vertex of the triangle to the midpoint of the side opposite that vertex.

2. An __?__ of a triangle is a segment drawn from a vertex of the triangle perpendicular to the line that contains the side opposite that vertex.

3. A(n) __?__ divides a segment into two congruent parts and intersects the segment at an angle whose measure is 90°.

4. A(n) __?__ divides an angle into two congruent angles.

a. perpendicular bisector

b. angle bisector

c. altitude

d. median

Lesson 6.1 EXPLORING LOCI IN THE PLANE   pages 313–315

- A **locus** of points is the set of all points that satisfy a given condition.

Identify the figure determined by the locus of points described.

5. The set of all points in a plane equidistant from a given point in the plane.

6. The set of all points in a plane equidistant from the sides of an angle.

7. The set of all points in a plane equidistant from the endpoints of a segment.

Lesson 6.2 SEGMENT AND ANGLE BISECTORS   pages 316–323

- If a point lies on the perpendicular bisector of a segment, then the point is equidistant from the endpoints of the segment.
- If a point lies on the bisector of an angle, then the point is equidistant from the sides of the angle.

Use the graph at the right to answer Questions 8 and 9.

8. Show that point R lies on the bisector of the angle formed by the positive x- and y-axes.

9. Show that point S lies on the perpendicular bisector of DE.

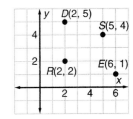

Lesson 6.3 ISOSCELES TRIANGLES   pages 324–330

- If two sides of a triangle are congruent, then the angles opposite them are congruent.
- If a triangle is equilateral, it is equiangular with each angle measuring 60°.

10. Find the coordinates of point C so that $\triangle ABC$ is isosceles with base AC, $A(-2, 2)$, and $B(3, 2)$.

11. Triangle XYZ is isosceles with $\overline{XY}$ congruent to $\overline{YZ}$. If m$\angle Y = 37°$, what are m$\angle X$ and m$\angle Z$?

- A **median** of a triangle is a line segment joining a vertex of the triangle to the midpoint of the opposite side.
- An **altitude** of a triangle is a segment drawn from a vertex of the triangle to a line that contains the side opposite that vertex so that it is perpendicular to that line.

Triangle JKL has coordinates J(1, 3), K(7, −1), L(3, −1).

12. What are the coordinates of the endpoints of the median from vertex J?

13. What are the coordinates of the endpoints of the altitude from vertex J?

- A **midsegment** is a line segment whose endpoints are the midpoints of two sides of a triangle.
- The midsegment of two sides of a triangle is parallel to the third side and half its length.

Triangle ASD has vertices A(−2, 0), S(2, 0), and D(4, −4).

14. Determine the coordinates of the midsegment joining $\overline{AS}$ and $\overline{SD}$. Then show that the midsegment is parallel to a side of the triangle.

15. Show that the midsegment that is parallel to $\overline{AS}$ is half the length of $\overline{AS}$.

- To use the strategy of elimination, you list all possibilities and prove all but one of them is false. In indirect proof, you show how an assumption leads to a contradiction.

16. Prove that if a line intersects $\triangle ABC$ at point D such that D is between A and C, then the line intersects $\overline{AB}$ or $\overline{BC}$.

- If two sides of a triangle are unequal in length, then the measure of the angle opposite the longer side is greater than the measure of the angle opposite the shorter side. If two angles of a triangle are unequal in measure, then the side opposite the larger angle is longer than the side opposite the smaller angle.
- The sum of the lengths of any two sides of a triangle is greater than the length of the third side.
- If two sides of one triangle are congruent to two sides of another triangle, and the included angle of the first triangle is greater than the included angle of the second triangle, then the third side of the first triangle is longer than the third side of the second triangle.

17. In $\triangle LMN$, $m\angle L = 82°$ and $m\angle N = 37°$. Identify the longest and shortest sides of the triangle.

18. In $\triangle ABC$, $AB = 12$, $BC = 16$, and $CA = y$. What is the range of possible values for CA?

You are given $\triangle ABC$ and $\triangle DEF$ with $\overline{AB}$ congruent to $\overline{DE}$ and $\overline{AC}$ congruent to $\overline{DF}$.

19. Determine the relationship between $m\angle A$ and $m\angle D$ if $BC = 5$ in. and $EF = 4$ in.

20. Determine the relationship between BC and EF if $m\angle A = 18°$ and $m\angle D = 14°$.

CHAPTER ASSESSMENT

CHAPTER TEST

1. The endpoints of $\overline{AB}$ are $A(-1, 5)$ and $B(9, 1)$. Does point $R(6, 8)$ lie on the perpendicular bisector of $\overline{AB}$? Explain.

2. The vertices of $\triangle XYZ$ are $X(-3, 1)$, $Y(0, 6)$, and $Z(5, 3)$. Show that this triangle is isosceles and name a pair of congruent angles.

3. **STANDARDIZED TESTS** In isosceles $\triangle JKL$ the coordinates of J and L are $(-2, 2)$ and $L(6, 4)$. What could be the coordinates of K?

   A. $(5, -9)$ B. $(-1, 15)$ C. $(4, -5)$

   D. $(0, 11)$ E. all of these

4. **WRITING MATHEMATICS** Explain the difference between an equilateral triangle and an isosceles triangle in terms of their angle bisectors and the perpendicular bisectors of the sides.

Vertices of $\triangle EFG$ are $E(-1, 5)$, $F(7, 9)$, and $G(5, 1)$. Midsegment $\overline{RS}$ joins $\overline{EF}$ and $\overline{FG}$.

5. Determine the endpoints of $\overline{RS}$.

6. Show that midsegment RS is parallel to side EG of $\triangle EFG$.

7. What is the length of $\overline{EG}$?

8. What is the length of midsegment $\overline{RS}$?

9. What is the relationship between the lengths of $\overline{EG}$ and $\overline{RS}$?

10. The sides of $\triangle UVW$ are represented in terms of x, $x > 0$, by $UV = 5x$, $VW = 3x$, and $WU = 5x + 3$. Which angle of $\triangle UVW$ has the least measure? Explain.

11. In $\triangle AMS$ $m\angle A = 45°$ and $m\angle M = 81°$. Identify the longest and shortest side in this triangle. Justify your answers.

12. **STANDARDIZED TESTS** Which of the following could not represent the sides of a triangle?

    A. $5, 9, 6$ B. $2.5, 1.7, 3.1$

    C. $2\frac{1}{2}, 3\frac{3}{4}, 1\frac{1}{4}$ D. $100, 150, 200$

    E. $1, 2, \sqrt{5}$

13. **Given** $\angle 1 \cong \angle 2$

    $\overline{VW} \cong \overline{XW}$

    $\overline{UW} \perp \overline{XV}$

    Prove $\triangle UXW \cong \triangle UVW$

14. **LANDSCAPING** The drawing below is an aerial view of a house; rectangle $ABCD$, and a light pole, point P. Point P is 30 ft from BC. A landscaper is going to plant a tree at a location 15 ft from BC and 20 ft from point P. Draw a diagram showing all possible locations for the tree.

15. Draw a scalene acute triangle and construct its altitudes.

16. The perimeter of a triangle with integral length sides is 7 in. Eliminate all other possibilities, to show that the sides of the triangle are of length 3 in., 3 in., and 1 in. or 2 in., 2 in., and 3 in.

17. Use the drawing to the right to write an inequality relating $m\angle 1$ with $m\angle 2$. Justify your answer.

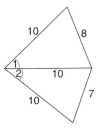

18. **WRITING MATHEMATICS** Describe the steps involved in writing an indirect proof.

PERFORMANCE ASSESSMENT

A TRIANGULAR NET Construct four large equilateral triangles as shown. Cut out the large parallelogram and crease along the sides of the triangles. How can this net be folded to form a three-dimensional figure? Describe the characteristics of this figure.

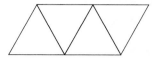

NAVIGATION The following procedure can be used to estimate the distance from a ship to a point on shore. A person aboard the ship at point Q measures $\angle TQS$, where T is the location of a landmark on shore and $\overrightarrow{QS}$ is the direction the ship is sailing. When the ship gets to point R, the point at which the measure of $\angle TRS$ is twice the measure of $\angle TQS$, $RQ = RT$. Explain why this is so.

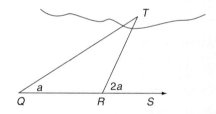

A PERPENDICULAR PERPLEXITY Construct a large equilateral triangle. Select a point in the interior of the triangle and construct three segments from this point to each of the three sides so that each segment is perpendicular to a side of the triangle.

a. Find the sum of the lengths of the three segments you constructed.

b. Select a different point in the interior of the triangle. Repeat the construction. Then find the sum of the lengths of the segments you constructed.

c. What is true about these two sums? Make a conjecture.

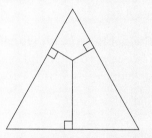

PROJECT ASSESSMENT

PROJECT *Connection* — When each group member has researched four different bridges, arrange a day for group presentations.

1. Before making a presentation, each group should sort and classify their bridges in terms of how they support the roadway. Tell whether they are cable bridges or truss bridges.

2. Determine if there are any other ways to classify the bridges.

3. Discuss how triangles play an important role in bridge design and construction. Identify any special kinds of triangles in the bridges.

CUMULATIVE REVIEW

Fill in the blank.

1. A point that lies on the __?__ of a segment is equidistant from the endpoints of the segment.

2. Two angles whose sides form two pairs of opposite rays are congruent and are called __?__ angles.

3. A polygon in which no line that contains a side of the polygon contains a point in its interior may be described as __?__ .

4. A line that cuts across two or more parallel lines is called a(n) __?__ .

Determine whether the triangle with the given vertices is an isosceles triangle and if so, identify a pair of congruent angles. Justify your answer.

5. $A(1, 2)$, $B(4, -3)$, $C(9, 0)$

6. $J(-2, 4)$, $K(5, 2)$, $L(1, -3)$

7. **STANDARDIZED TEST** Which of the following pairs of triangles is not necessarily congruent?

A.

B.

C.

D.

E. All are congruent.

Find the slope of each line described.

8. passes through $(4, -2)$ and $(1, 3)$

9. parallel to the line $3x - 2y = 8$

10. perpendicular to the line $y = 0$

Consider the statement: In an isosceles triangle, the base angles are congruent.

11. Write the statement in if-then form, then tell whether the statement is true or false.

12. Write the converse of the statement, then tell whether the converse is true or false.

13. Write the inverse of the statement, then tell whether the inverse is true or false.

14. Write the contrapositive of the statement, then tell whether the contrapositive is true or false.

15. How many planes of symmetry does the figure shown at the right have?

Justify each triangle congruence. List the side and angle congruences.

16. 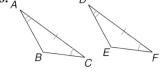   17.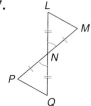

Let $\vec{a} = (4, -3)$, $\vec{b} = (-1, 2)$, $\vec{c} = (3, 0)$, and $\vec{d} = (6, 3)$. Find each of the following.

18. $\vec{a} + \vec{c}$ 19. $\vec{b} + \vec{c} + \vec{d}$

20. The dot product of $\vec{a}$ and $\vec{b}$.

21. $|\vec{a} - \vec{c}|$ 22. $|\vec{b} + \vec{d}|$

23. **WRITING MATHEMATICS** Which of the vectors used in Exercises 18–19 are perpendicular? How do you know?

STUDENT-PRODUCED ANSWERS Solve each question and on the answer grid write your answer at the top and fill in the ovals.

Note: Mixed numbers such as $1\frac{1}{2}$ must be gridded as 1.5 or 3/2. Grid only one answer per question. If your answer is a decimal, enter the most accurate value the grid will accommodate.

1. In the triangle shown, grid the greatest integer value for *x*.

2. In the diagram, two parallel lines are cut by a transversal. Find the value of *x*.

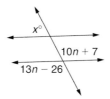

3. Grid the degree measure of an exterior angle of a regular octagon.

4. The vertex angle of an isosceles triangle measures 80°. Grid the number of degrees in the supplement of a base angle.

5. Write the equation of the line that passes through the points $(2, 1)$ and $(-2, 4)$ in standard form. Grid the coefficient of *y*.

6. To draw a particular map, a scale of 2 cm : 15 mi was used. The distance from Town A to Town B was measured to be 7 cm on the map. What is the actual distance, in miles, from Town A to Town B?

7. If $m\angle A < m\angle D$, grid the least possible integer value for *x*.

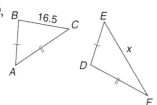

8. If a translation of 6 units up and 3 units left is applied to the point $(1, -2)$, grid the product of the coordinates of the image.

9. Find the supplement of the complement of a 70° angle.

10. If the lines $2x - 3y = 9$ and $9x + ny = 4$ are perpendicular, what is the value of *n*?

11. If the product of the coordinates of a point is 15, what is the product of the coordinates of its image under a reflection in the *x*-axis?

12. If the product of the coordinates of a point is 18, what is the product of the coordinates of its image under a rotation of 180° about the origin?

13. Find the difference between the greatest possible integer value for *x* and the least possible integer value for *x*.

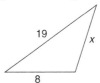

14. If *N* is the midpoint of segment *MP*, find *MP*.

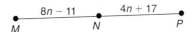

15. Given $J(2, -3)$ and $K(7, 1)$, find the magnitude of vector *JK* to the nearest tenth, if necessary.

16. On the coordinate plane, $\overline{DE}$ is a midsegment of $\triangle ABC$. Find *DE* to the nearest tenth, if necessary.

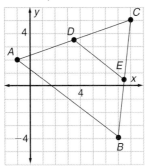

7 Quadrilaterals

Take a Look AHEAD

Make notes about things that look new.

- What special types of quadrilaterals will you study in this chapter? Look for the "tree" that shows how they are related.
- Which three-dimensional figures will you study in this chapter? How do you think they are related to quadrilaterals?

Make notes about things that look familiar.

- In what other situation did you use the word isosceles to describe a figure? What did it mean?
- What does midsegment mean in relation to a triangle? What do you think it means in relation to a trapezoid?

DATA Activity

The Long and the Tall of It

Landmarks come in many shapes and sizes. Many are well-known, while others are hardly known at all. Of those that are widely known, some are of historical, cultural, or political significance. Others are known simply because they are the largest, shortest, tallest, or widest, of their kind. The table on the next page gives data about some worldwide landmarks that are known as "tallest" or "longest."

SKILL FOCUS

- ▶ Compare numbers.
- ▶ Convert customary measurements.
- ▶ Find the percent one number is of another.
- ▶ Find a percent of a number.

GEOMETRY WORKS

HELP WANTED

Landmarks

In this chapter, you will see how:

* **RESTORATIONS EXPERTS** use drawing tools to recreate old designs. (Lesson 7.3, page 390)

* **GLAZIERS** use quadrilaterals and symmetry to repair stained-glass windows. (Lesson 7.5, page 403)

* **ARCHITECTURAL TECHNICIANS** use three-dimensional visualization skills to create accurate drawings of structures that have been lost in a disaster. (Lesson 7.9, page 425)

Major World Structures			
Category	**Name**	**Location**	**Dimensions**
tallest office building in the United States	Sears Tower	Chicago, Illinois	1454 ft
longest stairway	service staircase for Niesenbahn Funicular	Spiez, Switzerland	7759 ft
tallest self-supporting tower	CN Tower	Toronto, Ontario	1815 ft
longest roller coaster	The Ultimate	Ripon, England	1.42 mi
tallest structure	Television Tower	North Dakota	2063 ft
tallest apartment building	Metropolitan Tower	New York City	716 ft
tallest monument	Gateway to the West Arch	St. Louis, Missouri	630 ft

Use the table to answer the following questions.

1. How much longer is the longest stairway than the longest roller coaster?

2. What percent of the height of the Sears Tower is the height of the Gateway to the West Arch?

3. The world's largest Ferris wheel is in Japan. Its height is 55% of the height of the Gateway to the West Arch. How tall is this Ferris wheel?

4. How many Sears Towers laid end-to-end would you need to cover the full length of the longest stairway?

5. **WORKING TOGETHER** There are many other "longest" and "tallest" landmarks. Find data about at least three more. (Some ideas: What is the longest pier? the tallest pyramid? the tallest statue?) Use your new data to make up three questions similar to Questions 1 through 4.

Face Facts

When people go away on a vacation, they try to take in all the sights of the area they are visiting, Many times, though, people overlook the sights, or *landmarks*, right in their own communities. A landmark need not be as tall as the Empire State Building or as wide as the Grand Canyon. A landmark can take on many simpler forms.

- a *historic district*
- a *scenic landmark,* such as a park or bridge
- an *interior landmark,* such as a room of architectural significance
- an *individual landmark*, such as a monument or historic building

In this project, you will examine the exterior surface, or *facade*, of an individual landmark or other structure in your community.

PROJECT GOAL

To use grids and properties of quadrilaterals to analyze building facades.

Getting Started

Work with a group.

1. List the characteristics that you think make a structure important enough to be a landmark. Identify any structures in your community that you think are landmarks.

2. Go to a library or historical society. Find out if any structures on your list are officially listed as landmarks.

3. If a structure in your community is an official landmark, research some historical information about it. Why was it given landmark status?

4. If none of the structures in your community is an official landmark, select one that you think has the most interesting facade. Research some historical information about it.

5. Obtain a photograph of the facade of the building you selected.

PROJECT Connections

Lesson 7.2, page 383:
Make a rectified photograph of a facade.
Lesson 7.5, page 402:
Create a geometric representation of a facade on a coordinate grid.
Lesson 7.6, page 409:
Identify symmetries and transformations in a facade.
Chapter Assessment, page 429:
Plan a presentation to communicate final project results.

Internet Connection

www.learninggeometry.com

7.1 Geometry Workshop
Exploring Quadrilaterals

Think Back/Working Together

Work with a partner. Recall that a triangle is a polygon with three sides. If a polygon has four sides, it is called a *quadrilateral*.

1. Remember that you can classify a triangle by the number of its sides that are congruent. Draw a quadrilateral that has the given number of congruent sides.
   a. none
   b. exactly two
   c. exactly three
   d. four

2. You also can classify a triangle by the number of its angles that are congruent. Draw a quadrilateral that has the given number of congruent angles.
   a. none
   b. exactly two
   c. exactly three
   d. four

3. If possible, draw a quadrilateral that satisfies the given condition.
   a. It has four acute angles.
   b. It has more than one obtuse angle.

4. You probably remember that it is impossible to draw a concave triangle. Is it possible to draw a concave quadrilateral? If you think it is possible, give an example. If you think it is not possible, justify your answer.

Explore

On a sheet of graph paper, draw a set of coordinate axes. Then follow these steps to create a set of tangrams like the ones shown at the right.

5. Graph the points $A(-6, 6)$, $B(6, -6)$, $C(6, 6)$, $D(-6, -6)$, $E(0, -6)$, $F(6, 0)$, $G(3, -3)$, $H(0,0)$, $J(-3, -3)$, and $K(3, 3)$.

6. Draw $\overline{AC}$, $\overline{AD}$, $\overline{AG}$, $\overline{CB}$, $\overline{CD}$, $\overline{DB}$, $\overline{EF}$, $\overline{EJ}$, and $\overline{GK}$.

7. The polygons formed by the segments are the tangram pieces. Number them from 1 through 7, as shown. Use scissors to cut apart the tangram pieces.

8. Identify the shape of each tangram piece. For each piece, give the most accurate name that you know.

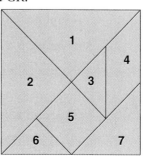

SPOTLIGHT ON LEARNING

WHAT? In this lesson you will learn
- to form special quadrilaterals using tangrams.
- to analyze relationships among special quadrilaterals.

Why? Relationships among quadrilaterals can help in solving many types of practical problems.

THINK BACK

A polygon is convex if no line that contains a side of the polygon contains a point in its interior. A polygon that is not convex is called concave.

9. **MODELING**   Arrange all seven tangram pieces to make the swan that is outlined at the right. On a sheet of paper, make a sketch that shows how you arranged the pieces.

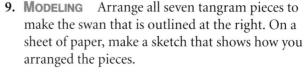

10. Copy the table below onto a sheet of paper. Use your tangrams to make each shape in the table, using the exact number of tangram pieces given in the column at the left. Record your answers by sketching the pieces. There may be more than one way to build a figure with a given number of pieces. You also may find it impossible to build a figure with a given number of pieces.

Number of Tangram Pieces	Shape to Create				
1	▨	▨	▨	▨	▨
2	▨	▨	▨	▨	▨
3	▨	▨	▨	▨	▨
4	▨	▨	▨	▨	▨
5	▨	▨	▨	▨	▨
6	▨	▨	▨	▨	▨
7	▨	▨	▨	▨	▨

Make Connections

● The following definitions describe some special quadrilaterals. You may be familiar with these from your work in a previous course.

A **parallelogram** is a quadrilateral with two pairs of parallel sides.

A **rhombus** is a quadrilateral with four congruent sides.

A **rectangle** is a quadrilateral with four right angles.

A **square** is a quadrilateral with four right angles and four congruent sides.

A **trapezoid** is a quadrilateral with exactly one pair of parallel sides.

An **isosceles trapezoid** is a trapezoid in which the nonparallel sides are congruent.

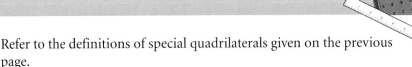

Refer to the definitions of special quadrilaterals given on the previous page.

11. Which term *or terms* do you think you can use to describe the shapes in each column of your tangram table?

12. Which do you think is the *best term* to describe the shape in each column of your tangram table?

13. Compare and contrast each pair of special quadrilaterals. How are the figures in each pair alike? How are they different? List as many likenesses and differences as you can.
    a. parallelogram, rectangle
    b. parallelogram, rhombus
    c. parallelogram, square
    d. parallelogram, trapezoid
    e. rectangle, square
    f. rhombus, square

14. List each special quadrilateral that satisfies the given set of conditions.
    a. It has four sides.
    b. It has at least one set of parallel sides.
    c. It has two sets of parallel sides.
    d. It has four congruent sides.
    e. It has four right angles.
    f. It has four right angles and four congruent sides.

COORDINATE GEOMETRY Refer to quadrilateral *RHOM* on the coordinate axes at the right.

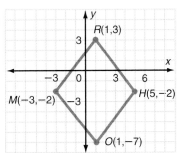

15. Use the distance formula and the definition of rhombus to demonstrate that *RHOM* is a rhombus.

16. Use slope and the definition of square to demonstrate that *RHOM* is *not* a square.

TRANSFORMATIONS Consider each type of special quadrilateral that is defined on the preceding page.

17. Does the figure have line symmetry? If it does, draw the figure and all its lines of symmetry.

18. Does the figure have rotation symmetry? If it does, draw the figure and its center of symmetry. Then identify all the angles of rotation between 0° and 360° that cause the image and preimage to coincide.

CHECK UNDERSTANDING

Is a square a rectangle? Is a rectangle a square?

TECHNOLOGY TIP

You can use geometry software to construct a quadrilateral of each type. Remember that two perpendicular lines form right angles. For the rhombus, use two radii of the same circle for the first two adjacent and congruent sides.

THINK BACK

For a nonvertical line containing points (x_1, y_1) and (x_2, y_2) the slope m is
$$m = \frac{y_2 - y_1}{x_2 - x_1} = \frac{\text{rise}}{\text{run}}$$

Geometry Workshop

Summarize

19. WRITING MATHEMATICS The diagram below is called a *quadrilateral tree*. Explain how it represents the relationships among the special quadrilaterals that you studied in this workshop.

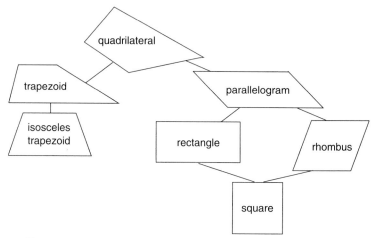

GOING FURTHER Another type of special quadrilateral is a *kite*.

These are kites.   These are not kites.

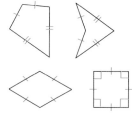 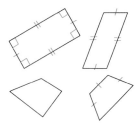

20. Use the examples and counterexamples shown above to write a definition of *kite*.

21. Copy the quadrilateral tree from Question 19 onto a sheet of paper. What is an appropriate place to insert *kite* into the tree? Explain your reasoning. Then adjust the diagram on your paper.

22. GOING FURTHER Use geometry software, if available. Draw an example of each special quadrilateral in this lesson. In each figure, join the midpoints of the sides in clockwise order. What type of quadrilateral appears to be inside the original quadrilateral?

23. THINKING CRITICALLY In Chapter 6, you learned that the perpendicular bisectors of the sides of a triangle intersect at a point that is equidistant from the vertices of the triangle. Do you think the same is true for the perpendicular bisectors of the sides of a quadrilateral? Explain your reasoning.

Explore

- Use a pencil, a straightedge, and a piece of lined paper to draw a quadrilateral as shown below. Cut out the quadrilateral.

1. The quadrilateral you cut out is called a *parallelogram*. Why do you think it is given that name?

2. Study the parallelogram and look for relationships among its angles and sides. What conjectures can you make?

Now cut the parallelogram along a diagonal to form two triangles. Lay the triangles on top of each other as shown.

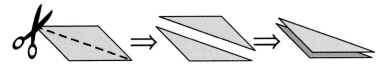

3. What is the relationship between the two triangles?

4. Does the relationship help to justify any of the conjectures you made in Question 2? Explain.

Use the method described above to draw and cut out a second parallelogram. Fold this parallelogram along a diagonal. Unfold it, then fold along the second diagonal, as shown at the right.

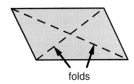

folds

5. Make a conjecture about the relationship between the diagonals.

Build Understanding

- A **parallelogram** is a quadrilateral with two pairs of parallel sides. Parallelograms are important to study because they have several properties that always are true. In Explore you may have discovered some of these properties, which are listed at the top of the following page. Each property is stated as a theorem because it can be proved.

If a quadrilateral is a parallelogram, then its opposite sides are congruent.

If a quadrilateral is a parallelogram, then its opposite angles are congruent.

If a quadrilateral is a parallelogram, then its consecutive angles are supplementary.

If a quadrilateral is a parallelogram, then its diagonals bisect each other.

Here is one way you might prove Theorem 7.1.

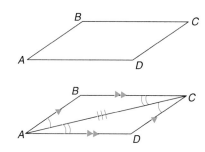

Given $ABCD$ is a parallelogram.

Prove $\overline{AB} \cong \overline{CD}$; $\overline{BC} \cong \overline{DA}$.

Plan for Proof Draw a diagonal. Prove the triangles formed are congruent. Then show the sides are corresponding parts of congruent triangles.

Proof Write a two-column proof.

Statements	Reasons
1. Draw $\overline{AC}$.	1. Through two distinct points, there is exactly one line.
2. $ABCD$ is a parallelogram.	2. Given
3. $\overline{AB} \parallel \overline{DC}$; $\overline{BC} \parallel \overline{AD}$	3. Definition of parallelogram
4. $\angle BAC \cong \angle DCA$; $\angle BCA \cong \angle DAC$	4. If two parallel lines are cut by a transversal, then alternate interior angles are congruent.
5. $\overline{AC} \cong \overline{CA}$	5. Reflexive property
6. $\triangle BAC \cong \triangle DCA$	6. ASA Congruence Postulate
7. $\overline{AB} \cong \overline{CD}$; $\overline{BC} \cong \overline{DA}$	7. CPCTC

TECHNOLOGY TIP

To make a parallelogram with geometry software, start with any two adjacent sides, $\overline{AB}$ and $\overline{BC}$. Construct a line parallel to $\overline{AB}$ through point C. Then construct a line parallel to $\overline{BC}$ through point A.

You often can use the properties of parallelograms to find unknown measures in a geometric figure.

EXAMPLE 1

In the figure at the right, *JKLM* is a parallelogram, *KL* = 20, *JL* = 26.4, and m∠*KLM* = 66°. Find each measure.

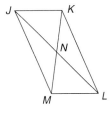

  a. *JM*   **b.** *JN*
  c. m∠*KJM*   **d.** m∠*JML*

Solution

  a. Opposite sides of a parallelogram are congruent.
      $\overline{KL}$ and $\overline{JM}$ are opposite sides.
      KL = 20, so *JM* = 20.

  b. The diagonals of a parallelogram bisect each other.
      $\overline{JL}$ and $\overline{KM}$ are diagonals.

      JL = 26.4, so $JN = \frac{1}{2}(26.4) = 13.2$.

  c. Opposite angles of a parallelogram are congruent.
      ∠*KLM* and ∠*KJM* are opposite angles.
      m∠*KLM* = 66°, so m∠*KJM* = 66°.

  d. Consecutive angles of a parallelogram are supplementary.
      ∠*KLM* and ∠*JML* are consecutive angles.
      m∠*KLM* = 66°, so m∠*JML* = (180 − 66)° = 114°.   ◀

CHECK UNDERSTANDING

Which other segment or angle measures can you find using the given information in Example 1? For which measures is there not enough information?

You also can use the properties of parallelograms to solve many real world problems.

EXAMPLE 2

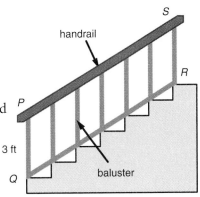

handrail

3 ft

baluster

BUILDING   A builder is adding a handrail to a staircase, as shown. The handrail must be constructed so that $\overline{PS} \parallel \overline{QR}$. It will be supported by vertical posts called *balusters*. The baluster attached to the bottom step is to be 3 ft high. What should be the height of the other balusters?

Solution

It is given that $\overline{PS} \parallel \overline{QR}$. Since the balusters must all be vertical, $\overline{PQ} \parallel \overline{SR}$. Thus, by definition, *PQRS* is a parallelogram. Since opposite sides of a parallelogram are congruent, $\overline{PQ} \cong \overline{SR}$. So, if the length of $\overline{PQ}$ is 3 ft, the length of $\overline{SR}$ must be 3 ft. Then, by similar reasoning, the length of each other baluster also must be 3 ft.   ◀

How would the
solution of Example 3
be different if the
coordinates of point A
were (3, 14)? (5, 14)?
(r, s)?

Sometimes you need to show a parallelogram on a coordinate plane.

EXAMPLE 3

Three vertices of parallelogram $ABCD$ are $A(3, 8)$, $B(0, 0)$, and $C(7, 0)$.
Find the coordinates of point D.

Solution

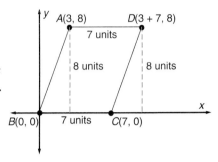

By Theorem 7.1, if $ABCD$ is a
parallelogram, then $\overline{AD} \cong \overline{BC}$.
The length of $\overline{BC}$ is 7 units, so the
length of $\overline{AD}$ also must be 7 units.
Since $\overline{BC}$ lies on the x-axis. $\overline{AD}$
must be horizontal. So the
x-coordinate of point D is
$3 + 7 = 10$

If $ABCD$ is a parallelogram, then $\overline{AD} \parallel \overline{BC}$. Point A is 8 units from the
x-axis, so point D also must be 8 units from the x-axis. So the
y-coordinate of point D is 8.

Thus, the coordinates of point D are $(10, 8)$. ◀

TRY THESE

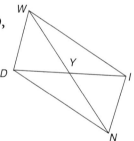

In the figure at the right, *WIND* is a parallelogram, *WI* = 5.9, *YN* = 2.9,
and m∠ *WDN* = 107°. Find each measure.

1. m∠*DNI*   **2.** m∠*WIN*   **3.** m∠*IWD*

4. *WN*   **5.** *WY*   **6.** *DN*

7. WRITING MATHEMATICS   Given the quadrilateral *RBYG* is a
parallelogram, list as many facts about *RBYG* as you can.

**Complete each set of coordinates so that *PQRS* is a parallelogram. Justify
your answer.**

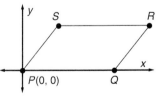

8. $Q(9, 0)$, $R(?, ?)$, $S(2, 4)$ **9** $Q(11, 0)$, $R(15, 7)$, $S(?, ?)$

10. $Q(a, 0)$, $R(?, ?)$, $S(b, c)$ **11.** $Q(2h, 0)$, $R(2j, 2k)$, $S(?, ?)$

12. URBAN PLANNING   In Center City, all lettered
streets are parallel to each other, and all numbered
streets are parallel to each other. The shaded region of
the figure at the right shows a block in Center City. The
corner of the block at F Street and 6th Street has an
angle measure of 121°. What are the angle measures at
the other three corners?

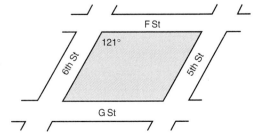

PRACTICE

Refer to parallelogram *ABCD* at the right.

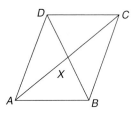

13. If $AD = 10$, then $BC =$ __?__ .

14. If $AC = 15$, then $AX =$ __?__ .

15. If $m\angle CDA = 111°$, then $m\angle ABC =$ __?__ .

16. If $m\angle DAB = 69°$, then $m\angle ABC =$ __?__ .

In the figure at the right, *PARK* is a parallelogram.
Tell whether each statement is true or false.

17. $\overline{PS} \cong \overline{KS}$ **18.** $\overline{PK} \cong \overline{RA}$

19. $\overline{PA} \perp \overline{AR}$ **20.** $\overline{PK} \parallel \overline{AR}$

21. $SR = \frac{1}{2}PR$ **22.** $m\angle ARP = \frac{1}{2}m\angle ARK$

23. $\angle KPA \cong \angle RAP$ **24.** $\angle PAR \cong \angle RKP$

25. Point *S* is the midpoint of $\overline{AK}$.

26. $\angle PKR$ and $\angle KRA$ are a pair of supplementary angles.

WOODWORKING A woodworker made parallel cuts $\overline{XY}$ and $\overline{ZW}$ in a board, as shown at the right. The edges of the board, $\overline{XZ}$ and $\overline{YW}$, also are parallel to each other. Find each measure if possible. If it is not possible to give the measure, write *not enough information*.

27. XY **28.** XZ **29.** ZW

30. $m\angle YXZ$ **31.** $m\angle XYW$ **32.** $m\angle YWZ$

Refer to the figure at the right below. Complete the coordinates so that *EFGH* is a parallelogram. Justify your answer.

33 $E(0, 0)$, $F(-3, 6)$, $G(11, 6)$, $H(?, ?)$

34. $E(0, ?)$, $F(?, 2)$, $G(4, 2)$, $H(5, 0)$

35. $E(0, 0)$, $F(?, ?)$, $G(b, 2c)$, $H(2a, 0)$

36. $E(0, 0)$, $F(s, t)$, $G(r, ?)$, $H(?, 0)$

37. **MACHINERY** The construction equipment shown at the right is a small *loader*. The shape of window frame *ABCD* is a parallelogram, $m\angle ABC = 74°$, $AB = 1.35$ m, and $BC = 0.96$ m. Find the measures of the other angles and sides of the frame.

38. WRITING MATHEMATICS A rectangle is defined as a quadrilateral with four right angles. Your friend says this means that a rectangle is not a parallelogram. Do you agree or disagree? Explain your reasoning.

In the figure at the right, *PLAN* is a parallelogram. State the theorem or definition that justifies each conclusion.

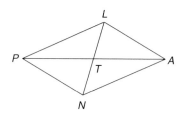

39. $\overline{PN} \cong \overline{AL}$ **40.** $\overline{PN} \parallel \overline{LA}$

41. $\angle PNA \cong \angle ALP$ **42.** $\overline{NT} \cong \overline{LT}$

43. $\angle PNA$ and $\angle NAL$ are a pair of supplementary angles.

EXTEND

In each figure, the quadrilateral is a parallelogram. Find *x*, *y*, and *z*.

44.

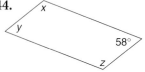

45.

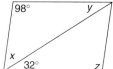

46.

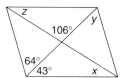

USING ALGEBRA **Refer to parallelogram *FGHJ* at the right.**

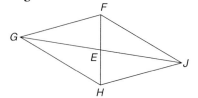

47. If $FE = 2r$ and $EH = 3r - 5$, then $r =$ ___?___.

48. If $GJ = 4b$ and $EJ = b + 9$, then $b =$ ___?___.

49. If $m\angle FJH = (4n)°$ and $m\angle GHJ = (12n + 4)°$, then $n =$ ___?___.

50. Prove Theorem 7.2: If a quadrilateral is a parallelogram, then its opposite angles are congruent. Write the proof in two-column form.

51. Prove Theorem 7.3: If a quadrilateral is a parallelogram, then its consecutive angles are supplementary. Write the proof in paragraph form.

Exercises 52–55 refer to the figure at the right. In the figure, *ABCD* is a parallelogram.

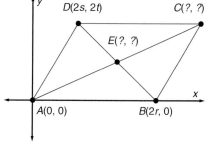

52. Give the coordinates of points C and E.

53. Write an expression that represents the length of $\overline{AE}$ and an expression that represents the length of $\overline{CE}$.

54. Use your answers to Exercise 53 to show that $\overline{AE} \cong \overline{CE}$.

55. Write a coordinate proof of Theorem 7.4: If a quadrilateral is a parallelogram, then its diagonals bisect each other.

THINK CRITICALLY

56. Three vertices of a parallelogram are $C(-4, 2)$, $H(3, 5)$, and $L(5, -4)$. Give the coordinates of all possible locations of the fourth vertex.

57. Is one diagonal of a parallelogram always longer than the other? Draw a sketch to illustrate your answer. Use geometry software if it is available.

58. Refer to the figure at the right.

Given $PQRS$ is a parallelogram.
$\overline{PA} \cong \overline{RB}$

Prove $\overline{QA} \parallel \overline{SB}$

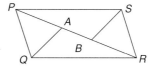

ALGEBRA AND GEOMETRY REVIEW

Find AB when A and B have the given coordinates.

59. $A(9, 7)$ and $B(-6, 15)$ **60.** $A(-2, -3)$ and $B(0, -7)$ **61.** $A(4, 5.5)$ and $B(7, 3.25)$

Simplify.

62. $(k^3)^4$ **63.** $3a^4(-2a^3)$ **64.** $\dfrac{12a^4}{-3a^3}$ **65.** $\left(\dfrac{3g^2}{2h^4}\right)^2$ **66.** $(z + 5)^2$

67. What is the measure of each angle of a regular decagon?

68. How many sides does a polygon have if the sum of the measures of its angles is 2340°?

PROJECT *Connection*   At the right is a *rectified photograph* of the facade of a building. Historians, restoration experts, and others use photographs like these to document the size, shape, and details of a building.

1. Notice that there is a coordinate grid drawn over the photograph. Explain how you could use this grid to describe the location of each window.

2. Use the photograph you obtained in Getting Started on page 372. Rectify it by drawing a coordinate grid over it. Your grid should be similar to the one above.

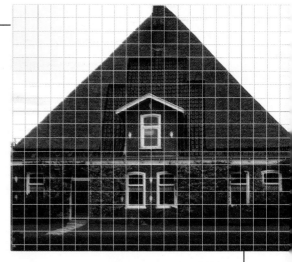

3. How do you think a rectified photograph like the one you made might be used in the process of restoring or preserving a landmark?

7.3 Proving Quadrilaterals Are Parallelograms

Explore

SPOTLIGHT ON LEARNING

WHAT? In this lesson you will learn
• to determine whether a quadrilateral is a parallelogram.

WHY? Showing that a quadrilateral is a parallelogram can help you solve problems in design, crafts, and social studies.

1. Use four strands of uncooked thick spaghetti and four miniature marshmallows.
   a. Cut two strands of spaghetti so that each has the same length as the other.
   b. Cut the two other strands so that each has the same length as the other, but a different length from part a.
   c. Use the marshmallows as vertices. Join the strands of spaghetti to form a quadrilateral with opposite sides congruent.

2. What special type of quadrilateral did you form in Question 1? Justify your answer.

3. Write the converse of Theorem 7.1.

4. Explain how the converse of Theorem 7.1 is related to Question 1.

Build Understanding

TECHNOLOGY TIP

To investigate Theorem 7.5, use a geometry software to create a quadrilateral. Measure all its sides and angles. Now adjust the lengths until opposite sides are congruent. How can you verify that the quadrilateral is a parallelogram? Drag a side of the quadrilateral. Is it still a parallelogram?

You can make similar dynamic drawings to investigate other theorems.

How can you determine whether a figure is a parallelogram? By definition, a quadrilateral is a parallelogram if it has two pairs of parallel sides. In Explore you saw that a quadrilateral is a parallelogram if it has two pairs of *congruent* sides. Each of the following theorems lists a different way to show that a figure is a parallelogram.

THEOREM 7.5

If both pairs of opposite sides of a quadrilateral are congruent, then the quadrilateral is a parallelogram.

THEOREM 7.6

If both pairs of opposite angles of a quadrilateral are congruent, then the quadrilateral is a parallelogram.

THEOREM 7.7

If an angle of a quadrilateral is supplementary to both consecutive angles, then the quadrilateral is a parallelogram.

The following is a proof of Theorem 7.5.

Given $\overline{AB} \cong \overline{CD}$; $\overline{BC} \cong \overline{DA}$

Prove *ABCD* is a parallelogram.

Plan for Proof Draw a diagonal to form two triangles. Show that the numbered angles are congruent by CPCTC. Then show that opposite sides of the quadrilateral are parallel because alternate interior angles are congruent.

Proof Write a two-column proof.

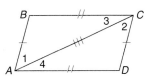

Statements	Reasons
1. Draw $\overline{AC}$.	1. Through two distinct points, there is exactly one line.
2. $\overline{AB} \cong \overline{CD}$; $\overline{BC} \cong \overline{DA}$	2. Given
3. $\overline{AC} \cong \overline{CA}$	3. Reflexive property
4. $\triangle BAC \cong \triangle DCA$	4. SSS Congruence Postulate
5. $\angle 1 \cong \angle 2$; $\angle 3 \cong \angle 4$	5. CPCTC
6. $\overline{AB} \parallel \overline{DC}$; $\overline{BC} \parallel \overline{AD}$	6. If two lines are cut by a transversal so that alternate interior angles are congruent, then the lines are parallel.
7. *ABCD* is a parallelogram.	7. Definition of parallelogram

Example 1 shows how you can use Theorem 7.7 to prove that a specific quadrilateral is a parallelogram.

EXAMPLE 1

Refer to the figure at the right. Decide if enough information is given to determine that *HOME* is a parallelogram.

Solution

HOME is a parallelogram if one angle is supplementary to both consecutive angles.

$m\angle O + m\angle H = 107° + 73° = 180°$

$m\angle O + m\angle M = 107° + 73° = 180°$

So, by Theorem 7-7, *HOME* is a parallelogram. ◀

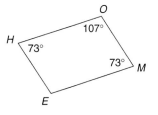

Here are two more ways to prove a quadrilateral is a parallelogram.

THEOREM 7.8

If the diagonals of a quadrilateral bisect each other, then the quadrilateral is a parallelogram.

THEOREM 7.9

If one pair of opposite sides of a quadrilateral are congruent and parallel, then the quadrilateral is a parallelogram.

You can solve many real-world problems by determining that a quadrilateral is a parallelogram.

EXAMPLE 2

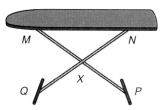

DESIGN   The ironing board shown at the right is designed so that $\overline{MX} \cong \overline{PX}$ and $\overline{QX} \cong \overline{NX}$. Explain why segment MN is parallel to segment QP on the floor.

Solution

$\overline{MP}$ and $\overline{NQ}$ are the diagonals of quadrilateral $MNPQ$. $\overline{MX} \cong \overline{PX}$ and $\overline{QX} \cong \overline{NX}$, so the diagonals bisect each other. Therefore, by Theorem 7-8, $MNPQ$ is a parallelogram. Opposite sides of a parallelogram are parallel, so $\overline{MN} \parallel \overline{QP}$. Thus, the ironing surface always will be parallel to the floor.   ◀

You can analyze a quadrilateral on the coordinate plane.

EXAMPLE 3

The vertices of quadrilateral $DECK$ are $D(-2, 1)$, $E(8, 3)$, $C(6, -2)$, and $K(-4, -4)$. Determine whether $DECK$ is a parallelogram.

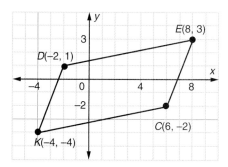

Solution

$\overline{DE}$ and $\overline{KC}$ are a pair of opposite sides. Find DE and KC to determine if they are congruent.

$DE = \sqrt{(8 - (-2))^2 + (3 - 1)^2} = \sqrt{104}$ Use the distance formula.

$KC = \sqrt{(6 - (-4))^2 + (-2 - (-4))^2} = \sqrt{104}$

$DE = KC = \sqrt{104}$, so $\overline{DE}$ and $\overline{KC}$ are congruent.

Find the slopes of $\overline{DE}$ and $\overline{KC}$ to determine if they are parallel.

slope of $\overline{DE} = \dfrac{3-1}{8-(-2)} = \dfrac{1}{5}$ slope of $\overline{KC} = \dfrac{-2-(-4)}{6-(-4)} = \dfrac{1}{5}$

The slope of $\overline{DE}$ equals the slope of $\overline{KC}$, so $\overline{DE} \parallel \overline{KC}$.

$\overline{DE}$ and $\overline{KC}$ are both congruent and parallel. Therefore, by Theorem 7.9, *DECK* is a parallelogram. ◀

TRY THESE

State the theorem or definition that proves that each quadrilateral is a parallelogram.

1.   2. 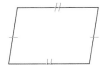   3.   4.

Determine whether each quadrilateral must be a parallelogram. Justify your answer.

5.    6. 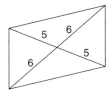   7.    8.

CRAFTS   A craftsperson measured equal lengths *AB* and *CD* along two sides of a piece of stained glass, as shown at the right. Edges $\overline{AB}$ and $\overline{CD}$ are parallel to each other. The craftsperson then cut the glass carefully along $\overline{AC}$ and $\overline{BD}$. Tell whether each statement is true or false.

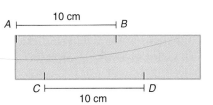

9. $\overline{AC} \parallel \overline{BD}$ 10. $\angle BAC \cong \angle CDB$

11. $AC = AB$ 12. $\mathrm{m}\angle ACD = 2(\mathrm{m}\angle BDC)$

Is the quadrilateral with the given vertices a parallelogram? Justify your answer.

13. $W(-6, 5)$, $X(1, 3)$, $Y(3, -4)$, $Z(-4, -2)$ 14. $P(6, 2)$, $O(0, 0)$, $E(0, -4)$, $M(6, -6)$

15. WRITING MATHEMATICS   Given quadrilateral *PQRS*, make a summary of all the ways you know to test whether *PQRS* is a parallelogram.

PRACTICE

Refer to quadrilateral *JKLM* at the right. For each set of conditions, determine whether *JKLM* must be a parallelogram. Justify your answer.

16. $\overline{JK} \cong \overline{LM}$ and $\overline{JM} \parallel \overline{KL}$ 17. $\overline{JK} \cong \overline{LM}$ and $\overline{JM} \cong \overline{LK}$

18. $\overline{JZ} \cong \overline{LZ}$ and $\overline{JL} \cong \overline{MK}$ 19 $\angle JML \cong \angle LKJ$ and $\angle MJK \cong \angle KLM$

20. $\overline{JK} \parallel \overline{ML}$ and $\overline{JM} \parallel \overline{KL}$ 21. $\angle JZK \cong \angle LZM$ and $\angle JZM \cong \angle LZK$

22. **WRITING MATHEMATICS** Explain how your work with parallelograms in this lesson is different from your work with parallelograms in Lesson 7.2.

23. **DESIGN** On the bookcase shown at the right, $\overline{RS}$ and $\overline{TU}$ are braces placed at the back in such a way that $\overline{RY} \cong \overline{SY}$ and $\overline{UY} \cong \overline{TY}$. Explain how this guarantees that the top and bottom shelves will be parallel.

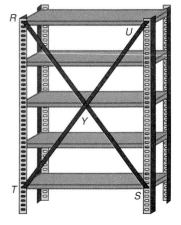

Explain why the quadrilateral with the given vertices is a parallelogram.

24. $W(-1, 10)$, $O(0, 0)$, $R(5, -2)$, $K(4, 8)$

25. $P(-6, 4)$, $L(2, 6)$, $A(2, -6)$, $Y(-6, -8)$

26. $A(5, -1)$, $B(3, 5)$, $C(-3, 3)$, $D(-1, -3)$

27. $W(-3, -3)$, $X(2, 6)$, $Y(6, 6)$, $Z(1, -3)$

Determine whether each quadrilateral must be a parallelogram. Justify your answer.

28.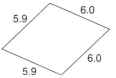

29.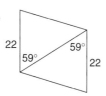

30.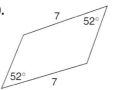

31.

32. **SOCIAL STUDIES** The flag of Congo is shown at the right. Suppose you must recreate the design of the flag on a blank rectangular sheet of paper. You have a pencil and a ruler. Explain how you could use one of the theorems presented in this lesson to assure that the yellow stripe you draw is a parallelogram.

EXTEND

Find values of x and y that guarantee that the quadrilateral is a parallelogram.

33.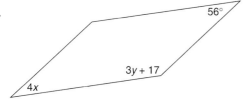

34.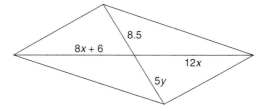

Determine whether a quadrilateral that satisfies each condition must be a parallelogram. Justify your answer.

35. The diagonals are congruent.

36. A diagonal forms two congruent triangles.

37. All pairs of consecutive angles are congruent.

38. All pairs of opposite angles are supplementary.

39. Two pairs of consecutive sides are congruent.

40. One diagonal bisects the other and one pair of opposite sides is parallel.

41. Prove Theorem 7.6: If both pairs of opposite angles of a quadrilateral are congruent, then the quadrilateral is a parallelogram.

42. Prove Theorem 7.7: If an angle of a quadrilateral is supplementary to both consecutive angles, then the quadrilateral is a parallelogram.

43. Prove Theorem 7.8: If the diagonals of a quadrilateral bisect each other, then the quadrilateral is a parallelogram.

44. Prove Theorem 7.9: If one pair of opposite sides of a quadrilateral are both congruent and parallel, then the quadrilateral is a parallelogram.

45. The vertices of parallelogram *CARE* are $C(-2, 5)$, $A(4, 8)$, $R(5, -2)$, and $E(-1, -5)$. The midpoints of the sides of *CARE* are joined in order. Show that the quadrilateral formed also is a parallelogram.

THINK CRITICALLY

46. Draw a quadrilateral that has two pairs of congruent sides, but that is not a parallelogram.

47. Draw a quadrilateral that has one pair of congruent sides and one pair of parallel sides, but that is not a parallelogram.

48. Refer to the figure at the right.

    Given   *JKLM* is a parallelogram.

    $\overline{QM} \cong \overline{PK}$

    Prove   *JPLQ* is a parallelogram.

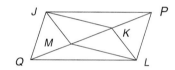

49. Determine whether the following statement is *always, sometimes,* or *never* true.

    A diagonal of a parallelogram lies on a symmetry line of the parallelogram.

ALGEBRA AND GEOMETRY REVIEW

In the figure at the right, $\overline{CF} \parallel \overline{BG}$ and $\overline{AE} \perp \overline{DH}$. Find the measure of each angle.

50. $\angle ELF$ 51. $\angle LKJ$ 52. $\angle DLC$

53. $\angle LJK$ 54. $\angle KJH$ 55. $\angle CLJ$

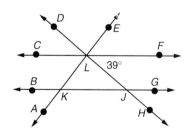

The image of $A(-4, 1)$ under a translation is $A'(2, -3)$.

56. Find the image of $\triangle PQR$ with vertices $P(-1, 4)$, $Q(2, 3)$, and $R(0, -1)$ under this same translation.

57. STANDARDIZED TESTS   Which is the equation of a line perpendicular to the line with equation $3x - 4y = 12$?

    A. $-4x + 3y = 12$ **B.** $4x + 3y = 12$ **C.** $3x - 4y = -12$ **D.** $3x + 4y = -12$

Career
Restorations Expert

Have you ever visited a historic site such as Williamsburg, the White House, Independence Hall, or Ellis Island? If so, did you feel as if you had traveled back in time? These and many other national landmarks have undergone major restoration. Teams of restoration experts recreated portions of buildings that were damaged by water, heat, cold, or decay. Due to the efforts of these experts, visitors can see these sites much as they appeared when they were first built.

Decision Making

One part of the restoration process involves copying, or *replicating*, a figure that has been damaged. This might occur in a wall hanging, a piece of art, a furnishing, or a border trim. In Questions 1 through 4, you will see how a restorations expert might replicate a quadrilateral design.

1. Trace *ABCD* onto a sheet of paper. Draw $\overline{AC}$.

2. Draw $\overrightarrow{EZ}$. Copy $\overline{AB}$ onto $\overrightarrow{EZ}$ to form $\overline{EF}$.

3. Construct an angle congruent to $\angle CAB$ with vertex *E*. Construct an angle congruent to $\angle CBA$ with vertex *F*. Extend the sides of these angles, if necessary, so they intersect at point *G*.

4. Construct an angle congruent to $\angle DAC$ with vertex *E*. Construct an angle congruent to $\angle DCA$ with vertex *G*. Extend the sides of these angles, if necessary, so they intersect at point *H*.

Quadrilateral *EFGH* is congruent to quadrilateral *ABCD*, so you have replicated the given quadrilateral.

5. Shown at the right is a section of a decorative border. Suppose you must replicate the region outlined in red in order to restore another portion of the border that has been damaged. Decide how to use constructions to replicate this region. Then perform the constructions.

Explore

- In the diagrams below, classify the quadrilateral determined by *ABCD* in Figure I. Notice $\overline{AC}$ and $\overline{BD}$ intersect in different ways.

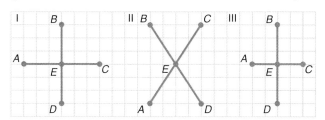

1. If you draw quadrilateral *ABCD* in Figure I, what type of quadrilateral is it?

2. If you draw quadrilateral *ABCD* in Figures II and III, what can you say about each of them?

3. Suppose that $\overline{AC}$ and $\overline{BD}$ are congruent, intersect at their midpoints, and are not perpendicular, do you think you can always conclude what you did in Question 2? Justify your answer.

Refer to the diagram at the right.

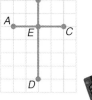

4. What can you say about $\overline{AC}$ and $\overline{BD}$?

5. If you draw quadrilateral *ABCD*, what can you say about it?

Build Understanding

- In Lesson 7.1, you learned about special quadrilaterals.

   A **rhombus** is a quadrilateral with four congruent sides.

   A **rectangle** is a quadrilateral with four right angles.

   A **square** is a quadrilateral with four right angles and four congruent sides.

Each of these figures is illustrated below. Notice that every rhombus, rectangle, and square is a parallelogram.

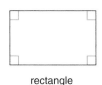

| rhombus | rectangle | square |

TECHNOLOGY TIP

You may wish to draw $\overline{AC}$ and $\overline{BD}$ in a software program and adjust them as you explore each figure.

CHECK UNDERSTANDING

Explain how you know that every rectangle is a parallelogram.

Label the three regions in the diagram below with the words *rhombus*, *rectangle*, and *square* to illustrate Theorems 7.10, 7.12, and 7.13.

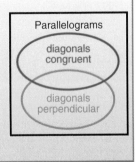

CHECK UNDERSTANDING

In Example 1, Jamie looked at the diagram and concluded that the diagonals appear to be perpendicular. She applied Theorem 7.10 to reason that quadrilateral *KLMN* was a rhombus. What error did she make?

The theorems below can help you recognize a rhombus or a rectangle.

THEOREM 7.10

A parallelogram is a rhombus if and only if its diagonals are perpendicular.

THEOREM 7.11

A parallelogram is a rhombus if and only if each diagonal bisects a pair of opposite angles.

THEOREM 7.12

A parallelogram is a rectangle if and only if its diagonals are congruent.

EXAMPLE 1

From the information about the park shown at the right below, find how many feet of fencing are needed to enclose all four sides.

Solution

PARK PLANNING The diagram indicates that the diagonals of quadrilateral *KLMN* bisect opposites angles. Thus, you can conclude by Theorem 7.11 that the park is in the shape of a rhombus. By definition of a rhombus, the four sides of the park are congruent. Planners will need 4(76) ft, or 304 ft, of fencing to enclose the park. ◄

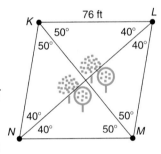

You can get new information about a figure given its classification.

EXAMPLE 2

Find each length and angle measure in rectangle *RSTU*. Give a reason for your answer.

a. m∠*STR*   **b.** *RW*   **c.** *SU*

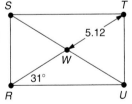

Solution
a. Since rectangle *RSTU* is a parallelogram, $\overline{ST} \parallel \overline{RU}$. Thus, m∠*STR* = m∠*TRU* = 31°.
b. Since the diagonals of a parallelogram bisect each other, *RW* = *WT* = 5.12.
c. Since the diagonals of a rectangle are congruent, *SU* = *RT*. Thus *SU* = 2(5.12) = 10.24 ◄

You can use triangle congruence to prove part of Theorem 7.11.

Given $\overline{BD}$ bisects $\angle ABC$ and $\angle ADC$.
$\overline{AC}$ bisects $\angle DAB$ and $\angle DCB$.

Prove Quadrilateral $ABCD$ is a rhombus.

Plan for Proof Use the ASA congruence Postulate
to show that $\triangle ABC \cong \triangle ADC$ and $\triangle ADB \cong \triangle CDB$.
Use CPCTC to show that $\overline{AB} \cong \overline{BC} \cong \overline{CD} \cong \overline{DA}$.

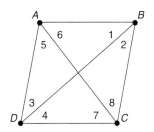

Proof Write a two-column proof.

Statements	Reasons
1. $\angle 1 \cong \angle 2$, $\angle 3 \cong \angle 4$ $\angle 5 \cong \angle 6$, $\angle 7 \cong \angle 8$	1. Given and definition of angle bisector
2. $\overline{DB} \cong \overline{DB}$ and $\overline{AC} \cong \overline{AC}$	2. Reflexive property
3. $\triangle ABC \cong \triangle ADC$ and $\triangle ADB \cong \triangle CDB$	3. ASA Congruence Postulate
4. $\overline{AB} \cong \overline{AD}$ and $\overline{BC} \cong \overline{DC}$ $\overline{AD} \cong \overline{CD}$	4. CPCTC in $\triangle ABC$ and $\triangle ADC$ CPCTC in $\triangle ADB$ and $\triangle CDB$
5. $\overline{AB} \cong \overline{BC} \cong \overline{CD} \cong \overline{DA}$	5. Transitive property
6. Quadrilateral $ABCD$ is a rhombus.	6. Definition of rhombus

TECHNOLOGY TIP

To use geometry
software to explore the
ideas in this lesson,
you will need to
construct a
parallelogram, and
then construct and
measure its diagonals.

You can use an equation with theorems from geometry to find
measures.

EXAMPLE 3

In the diagram at the right, $PR = 8.73 = SQ$
and $m\angle PSQ = 50°$. Find $m\angle QSR$.

Solution
The diagonals of quadrilateral $PQRS$ are
congruent. By Theorem 7.12, the
quadrilateral is a rectangle.

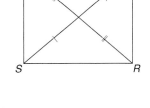

PROBLEM SOLVING TIP

To find the
measure of one angle
given the measure of
the other angle in a
pair of complementary
angles, write and
solve an equation.

$$m\angle PSR = 90°$$ Angle of a rectangle.
$$m\angle QSR + 50° = 90°$$ $\angle PSQ$ and $\angle QSR$ are complementary.
$$m\angle QSR = 40°$$ Solve for $m\angle QSR$.

Thus, $m\angle QSR = 40°$. ◀

The theorem below can help you recognize a square.

THEOREM 7.13

A parallelogram is a square if and only if its diagonals
are both perpendicular and congruent.

You can classify a parallelogram given the coordinates of its vertices.

EXAMPLE 4

Classify parallelogram $RSTU$ with vertices $R(-1, 5)$, $S(4, 3)$, $T(2, -2)$, and $U(-3, 0)$.

Solution

slope of $\overline{US} = \dfrac{3 - 0}{4 - (-3)} = \dfrac{3}{7}$

slope of $\overline{RT} = \dfrac{5 - (-2)}{-1 - 2} = \dfrac{7}{-3}$

$US = \sqrt{(4 - (-3))^2 + (3 - 0)^2} = \sqrt{58}$

$RT = \sqrt{(-1 - 2)^2 + (5 - (-2))^2} = \sqrt{58}$

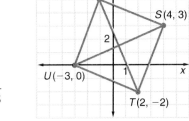

Since the diagonals of parallelogram $RSTU$ are perpendicular and congruent, parallelogram $RSTU$ is a square.

TRY THESE

1. **RECREATION** Sidewalks joining A with C and B with D in quadrilateral $ABCD$ are each 1000 ft long and bisect each other. Classify $ABCD$. If $AB = 800$ ft, find the amount of fencing to enclose $ABCD$.

2. **RECREATION** Sidewalks $\overline{PR}$ and $\overline{SQ}$ in rectangle $PQRS$ are 1225 ft long and intersect at right angles. The park is 866.2 ft on one side. Find the amount of fencing to enclose $PQRS$.

Find each length and angle measure in the given quadrilateral.

3. rectangle $WXYZ$
   $m\angle ZWT$ and WY

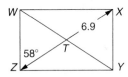

4. rhombus $TRUE$
   $m\angle UTE$ and TE

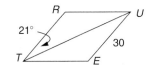

5. square $MATH$
   MB and $m\angle HMA$

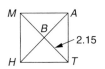

Use the given information to find each length and angle measure.

6. DE and $m\angle RLE$

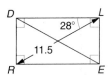

7. $m\angle YFA$, $m\angle CFA$, and YF

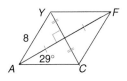

8. $m\angle MLJ$ and LJ

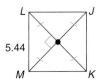

Classify quadrilateral $ABCD$.

9. $A(-3, -2)$, $B(-1, 3)$, $C(1, -2)$, $D(-1, -7)$

10. $A(0, 0)$, $B(3, 7)$, $C(10, 4)$, $D(7, -3)$

11. $A(-1, 3)$, $B(4, 0)$, $C(1, -5)$, $D(-4, -2)$

12. $A(-4, -2)$, $B(-1, 3)$, $C(5, 2)$, $D(3, -3)$

13. **WRITING MATHEMATICS** Explain how to distinguish a rectangle from a square by using information about diagonals of a quadrilateral.

PRACTICE

14. LANDSCAPING Garden restorers are using the diagram at the right to reconstruct a herb garden at a historical residence. Classify quadrilateral *EKXZ*. Justify your response. Find the amount of fencing needed to enclose the garden.

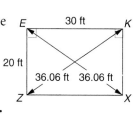

Find each length and angle measure in the given quadrilateral.

15. rhombus *PHAC*
m∠*HCA* and *PH*

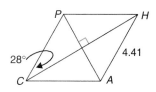

16. square *UDLM*
MD and *UD*

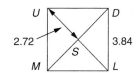

17. rectangle *WOHT*
m∠*HTO* and *JO*

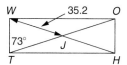

Use the given information in each diagram to find each length or angle measure.

18. m∠*DAB* and *AB*

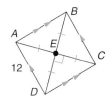

19. m∠*LTR* and m∠*TRF*

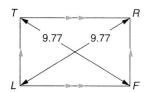

20. m∠*AZB*, m∠*ABZ*, and *ZY*

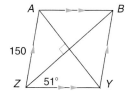

Classify quadrilateral *ABCD*.

21. $A(-1, 3), B(3, -1), C(-1, -5), D(-5, -1)$

22. $A(1, -3), B(-4, 2), C(-1, 5), D(4, 0)$

23. $A(-12, 3), B(-6, 4), C(0, 3), D(-6, 2)$

24. $A(1, 1), B(4, -4), C(-1, -1), D(-4, 4)$

25. WRITING MATHEMATICS Write Theorem 7.11 as two separate theorems.

EXTEND

Fill in the coordinates so that the quadrilateral is the specified figure. Justify your answers.

26. square

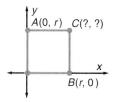

27. parallelogram

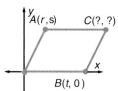

28. rectangle

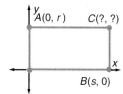

29. square

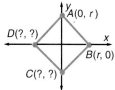

30. WOODWORKING Martha and Jon nailed two 20 in. sticks together at their midpoints. What else must they do so that they can make a square?

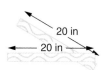

31. WOODWORKING If in Question 30 Martha and Jon nailed the sticks at their midpoints, and they made angles of 60° and 120°, what quadrilateral do the endpoints of the sticks determine?

32. In square $ABCD$ shown at the right, $\overline{WY} \parallel \overline{AB}$ and $\overline{XZ} \parallel \overline{AD}$. Explain how you know the diagonals of quadrilateral $WXYZ$ are perpendicular.

33. Refer to the diagram at the right. How would you relocate any of W, X, Y, Z, and R to make quadrilateral $WXYZ$ a rhombus?

In the diagram at the right, quadrilateral $ABCD$ is a rectangle, quadrilateral $AEFD$ is a square, and quadrilateral $URST$ is a rhombus. Find each length or angle measure. Give a reason for each answer.

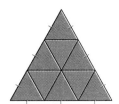

34. EB　　　　**35.** m$\angle URZ$　　　　**36.** US

37. RT　　　　**38.** m$\angle ZRS$　　　　**39.** m$\angle UER$

Prove the following parts of Theorems 7.10 and 7.13

40. If the diagonals of a parallelogram are perpendicular, it is a rhombus.

41. A parallelogram with congruent and perpendicular diagonals is a square.

THINK CRITICALLY

42. Refer to the diagram at the right in which congruences are shown. How many rhombuses can you find? How many parallelograms that are not rhombuses can you find?

Write a coordinate proof for each statement.

43. The quadrilateral $WXYZ$ formed by the midpoints of a convex quadrilateral $ABCD$ is a parallelogram.

44. The quadrilateral $WXYZ$ formed by the midpoints of rectangle $ABCD$ is a rhombus.

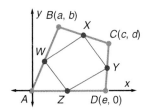

ALGEBRA AND GEOMETRY REVIEW

Find the slope of $\overline{PQ}$.

45. $P(-3, 2)$; $Q(5, 1)$　　**46.** $P(3, 2)$; $Q(7, 9)$　　**47.** $P(-5, 2)$; $Q(7, 2)$　　**48.** $P(-3, -3)$; $Q(3, 3)$

Graph each function.

49. $y = 2x + 1.5$　　**50.** $y = x^2 - 3$　　**51.** $y = 1.5x - 3$　　**52.** $y = 2x^2 + 1$

53. Quadrilateral $EFGH$ has vertices $E(-4, -4)$, $F(-2, 2)$, $G(4, 2)$, and $H(2, -4)$. Classify the quadrilateral.

54. The measure of $\angle W$ is 30° more than twice the measure of its complement. Find m$\angle W$.

55. If $\angle T$ is acute and m$\angle T = 2x + 8$, write an inequality for the range of x.

7.5 Properties of Trapezoids

Explore

1. The diagram at the right shows a rectangle whose vertices have coordinates $O(0, 0)$, $A(0, 6)$, $B(8, 6)$, and $C(8, 0)$.

   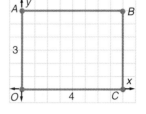

   a. Describe quadrilateral $OABC$ if you move point A to $A(2, 6)$ and leave points O, B, and C where they are.

   b. Describe $OABC$ if you move A to $A(2, 6)$, B to $B(6, 6)$, and leave O and C where they are.

2. Given quadrilateral $OABC$ with vertices $O(0, 0)$, $A(2, 6)$, $B(6, 6)$, and $C(8, 0)$, show that $\overline{AB} \parallel \overline{OC}$ but that $\overline{OA} \nparallel \overline{CB}$.

Build Understanding

A **trapezoid** is a quadrilateral with exactly one pair of parallel sides. The two parallel sides are the **bases** and the nonparallel sides are the **legs**. Two consecutive angles whose vertices are endpoints of a single base form a pair of **base angles**. The **midsegment of a trapezoid** is the line segment joining the midpoints of the legs.

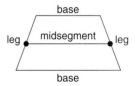

THEOREM 7.14   Midsegment Theorem for Trapezoids

The midsegment of a trapezoid is parallel to each base, and its length is equal to half the sum of the lengths of the bases.

EXAMPLE 1

The vertices of trapezoid $KLMN$ are $K(-2, 4)$, $L(1, 4)$, $M(3, -2)$, and $N(-1, -2)$. Represent trapezoid $KLMN$ on a coordinate plane and find the length of the midsegment $\overline{YZ}$.

Solution

The bases of trapezoid $KLMN$ are $\overline{KL}$ and $\overline{NM}$.

$$KL = |1 - (-2)| = 3$$
$$NM = |3 - (-1)| = 4$$
$$YZ = \frac{1}{2}(3 + 4) = 3.5$$

Thus, $\overline{YZ}$ is 3.5 units long. ◄

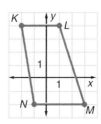

An **isosceles trapezoid** is a trapezoid whose legs are congruent.

THEOREM 7.15

If a quadrilateral is an isosceles trapezoid, then its base angles are congruent.

THEOREM 7.16

If a quadrilateral is an isosceles trapezoid, then its diagonals are congruent.

Shown below is a paragraph proof of Theorem 7.15.

Given Trapezoid $ABCD$ is isosceles with $\overline{BC} \parallel \overline{AD}$ and $\overline{AB} \cong \overline{DC}$.

Prove $\angle A \cong \angle D$ and $\angle ABC \cong \angle DCB$

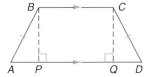

Proof Draw $\overline{BP} \perp \overline{AD}$ and $\overline{CQ} \perp \overline{AD}$ to form rectangle $PBCQ$. Since $PBCQ$ is a rectangle, opposite sides are congruent: $\overline{BP} \cong \overline{CQ}$. Since $PBCQ$ is a rectangle, m$\angle BPQ =$ m$\angle CQP = 90°$. So, m$\angle BPA =$ m$\angle CQD = 90°$. Thus, $\triangle BPA$ and $\triangle CQD$ are right triangles.

Since quadrilateral $ABCD$ is an isosceles trapezoid, $\overline{AB} \cong \overline{CD}$. Since $\triangle BPA$ and $\triangle CQD$ are right triangles, $\overline{BP} \cong \overline{CQ}$, and $\overline{AB} \cong \overline{DC}$, $\triangle BPA \cong \triangle CQD$ by the Hypotenuse-Leg Theorem. So, $\angle A \cong \angle D$ and $\angle ABP \cong \angle DCQ$ by CPCTC.

Since $\angle A \cong \angle D$, the lower base angles of $ABCD$ are congruent. Since $\angle ABP \cong \angle DCQ$ and $\angle PBC$ and $\angle QCB$ are right angles, $\angle ABC \cong \angle DCB$. Thus, the upper base angles are congruent. ◄

EXAMPLE 2

STONE CUTTING Four faces of a monument base are isosceles trapezoids. Find the measures of m$\angle SRU$, m$\angle TUR$, m$\angle RST$, m$\angle STU$, and the length of $\overline{EF}$.

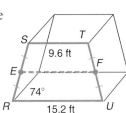

Solution

By Theorem 7.15, m$\angle SRU =$ m$\angle TUR$ and m$\angle RST =$ m$\angle STU$. Thus, m$\angle SRU =$ m$\angle TUR = 74°$ and m$\angle RST =$ m$\angle STU = 106°$. Apply the midsegment theorem for trapezoids.

$$EF = \frac{1}{2}(ST + RU) = \frac{1}{2}(9.6 + 15.2) \text{ ft, or } 12.4 \text{ ft}$$ ◄

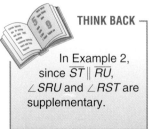

THINK BACK

In Example 2, since $\overline{ST} \parallel \overline{RU}$, $\angle SRU$ and $\angle RST$ are supplementary.

The following two theorems provide you with criteria that help you tell if a trapezoid is isosceles.

THEOREM 7.17

If the diagonals of a trapezoid are congruent, then the trapezoid is isosceles.

THEOREM 7.18

If one pair of base angles of a trapezoid are congruent, then the trapezoid is isosceles.

EXAMPLE 3

In trapezoid USTV in the diagram at the right $\overline{ST} \parallel \overline{UV}$. Determine whether trapezoid USTV is isosceles.

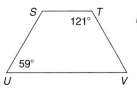

Solution
Since $\angle STV$ and $\angle TVU$ are same-side interior angles and $\overline{ST} \parallel \overline{UV}$, $m\angle TVU = 180° - 121° = 59°$. Thus, $\angle SUV \cong \angle TVU$. By Theorem 7.18, trapezoid USTV is isosceles.

CHECK UNDERSTANDING

In Example 3, show that trapezoid *USTV* is isosceles by analyzing $\angle UST$ and $\angle STV$.

Many theorems are easily proved using a coordinate proof.

EXAMPLE 4

COORDINATE PROOF Prove Theorem 7.14

Given Trapezoid *PQRS* with bases $\overline{PQ}$ and $\overline{RS}$; *M* and *N* are midpoints of $\overline{PS}$ and $\overline{QR}$, respectively,

Prove $\overline{MN} \parallel \overline{PQ} \parallel \overline{RS}$ and
$MN = \frac{1}{2}(PQ + RS)$

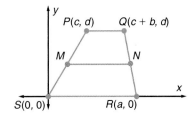

Proof By the midpoint formula, the coordinates of *M* and *N* are $M\left(\frac{c}{2}, \frac{d}{2}\right)$ and $N\left(\frac{c+b+a}{2}, \frac{d}{2}\right)$. The y-coordinates are equal, so the slope of $\overline{MN}$ is zero. Therefore, $\overline{MN}$ is a horizontal segment. So, $\overline{MN} \parallel \overline{PQ} \parallel \overline{RS}$. Since $\overline{MN}$ is horizontal, its length is the difference in the x-coordinates, $\frac{c+b+a}{2} - \frac{c}{2} = \frac{b+a}{2}$. Find *PQ* in the same way. So, $PQ = c + b - c = b$. Since $RS = a$,
$\overline{MN} = \frac{1}{2}(PQ + RS)$ by substitution.

For each trapezoid whose vertices have the given coordinates, represent it on the coordinate plane. Then find the length of the midsegment.

1. $P(0, 0)$, $Q(0, 4)$,
$R(7, 4)$, $S(5, 0)$

2. $E(0, 0)$, $A(5, 0)$,
$C(7, -4)$, $H(0, -4)$

3. $P(1, 2)$, $L(7, 2)$,
$O(6, 9)$, $T(1, 9)$

4. $A(-3, 9)$, $B(1, 5)$,
$C(1, -1)$, $D(-3, -9)$

5. POWER TRANSMISSION The diagram at the right shows a power transmission tower whose frame is made of isosceles trapezoids. Find the measure of each angle in isosceles trapezoid $ABCD$ and find the length of its midsegment, $\overline{EF}$.

6. POWER TRANSMISSION Quadrilateral $ABFE$ in the diagram at the right is an isosceles trapezoid. Refer to the results of Exercise 5. Find the measure of each angle in isosceles trapezoid $ABHG$ and find the length of its midsegment, $\overline{EF}$.

7. WRITING MATHEMATICS Explain the difference between an isosceles trapezoid and a non-isosceles trapezoid. Make a list of both types of trapezoids you can find in your classroom.

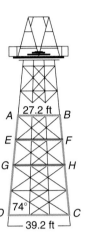

Determine whether each trapezoid is isosceles. Justify your response.

8.

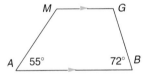

9.

10.

PRACTICE

For each trapezoid whose vertices have the given coordinates, represent it on the coordinate plane. Then find the length of the midsegment.

11. $E(-7, 5)$, $F(7, 5)$,
$G(1, -1)$, $H(-1, -1)$

12. $T(1, -5)$, $U(7, 5)$,
$V(-7, 5)$, $W(-1, -5)$

13. $M(-3, 9)$, $N(3, 3)$,
$A(3, -2)$, $L(-3, -2)$

14. $A(-2, 5)$, $B(-2, 3)$,
$C(-7, 0)$, $D(-7, 8)$

15. GARDENING Refer to the wheelbarrow at the right. Find the measures of the angles in isosceles trapezoid $WHEL$ and the length of its midsegment.

16. WRITING MATHEMATICS Describe how to make an isosceles trapezoid from a rectangle.

Determine whether each trapezoid is isosceles. Justify your response.

17.

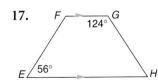

18.

19.

Determine whether each trapezoid is isosceles. Justify your response.

20.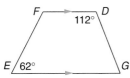

21.

22.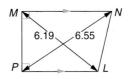

EXTEND

In the diagram at the right, trapezoid *ABCD* is an isosceles trapezoid and *X* and *Y* are the midpoints of the legs. Find each length or angle measure.

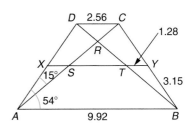

23. m∠*DCB*   24. *AD*   25. m∠*CYX*   26. *XY*

27. m∠*DCA*   28. *XS*   29. m∠*XSA*   30. *ST*

In each trapezoid, $\overline{AB}$ is the midsegment. Find *x*.

31.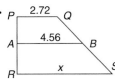

32.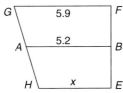

33.

34.

Trapezoid *PQRS* at the right is isosceles. Find each measure. Justify your response.

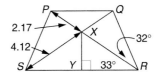

35. *XQ*   36. m∠*QSP*   37. *SQ*

38. m∠*PXQ*   39. *XR*   40. m∠*PXS*

Exercies 46–50 refer to trapezoid *ABCD*.

41. Write coordinates of *C* and *D* so that trapezoid *ABCD* is isosceles.

42. Verify that your coordinates for points *C* and *D* assure that quadrilateral *ABCD* is a trapezoid.

43. Show that $\overline{AB} \cong \overline{CD}$ and, therefore, trapezoid *ABCD* is isosceles.

44. Use the distance formula to find *CA* and *DB*.

45. Write a coordinate proof of Theorem 7.16: If a quadrilateral is an isosceles trapezoid, then its diagonals are congruent.

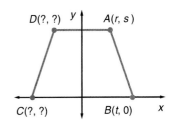

In isosceles trapezoid *PQRS* at the right, *Y* is the midpoint of $\overline{SR}$. Prove each statement.

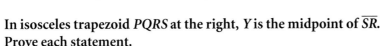

46. △*SPY* ≅ △*RQY*   47. ∠*PQY* ≅ ∠*QPY*

48. △*SPQ* ≅ △*RQP*   49. △*PQX* is isosceles.

THINK CRITICALLY

50. Prove Theorem 7.17: If the diagonals of a trapezoid are congruent, then the trapezoid is isosceles. Write the proof in two-column form.

51. Prove Theorem 7.18: If one pair of base angles of a trapezoid are congruent, then the trapezoid is isosceles. Write the proof in paragraph form.

52. In isosceles trapezoid *KLMN*, points *B* and *E* are midpoints of $\overline{KN}$ and $\overline{LM}$, *A* and *D* are midpoints of $\overline{KB}$, and $\overline{LE}$, and *C* and *F* are midpoints of $\overline{BN}$ and $\overline{EM}$. Find *AD* and *CF* in terms of *KL* and *NM*.

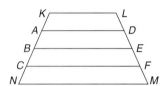

53. In the diagram below, quadrilateral *PQRS*, is an isosceles trapezoid with midsegment $\overline{XY}$. Prove that $\overline{PY} \cong \overline{QX}$ and that $\overline{XR} \cong \overline{YS}$.

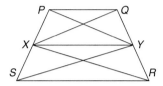

ALGEBRA AND GEOMETRY REVIEW

Find the sum of the angle measures of each convex polygon.

54. hexagon   **55.** octagon   **56.** 12-gon   **57.** 20-gon

Solve each system of equations. If the system has no solution, so state.

58. $\begin{cases} x + 3y = 13 \\ 2x - 3y = -19 \end{cases}$ **59.** $\begin{cases} -x + 5y = -10 \\ 2x - 3y = 6 \end{cases}$ **60.** $\begin{cases} 5x - y = 2 \\ 5x - y = 3 \end{cases}$

61. In trapezoid *PCTB*, m$\angle P$ = 35° and m$\angle T$ = 145°. Determine whether the trapezoid is isosceles. Justify your answer.

62. STANDARDIZED TESTS What is the maximum number of right angles a trapezoid can have?

     A. 1   **B.** 2   **C.** 3   **D.** 4

PROJECT *Connection*

In the questions below, use your rectified photograph.

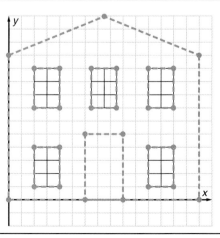

1. Identify the coordinates of defining vertices of your facade. Include information about windows, doors, roof lines, and any details important to the facade.

2. Transpose your coordinates to a clean sheet of graph paper. Connect the points to outline the details. The facade now appears as a collection of geometric shapes.

3. Describe how to use coordinate geometry to identify the shapes of items that appear on the facade of your building. How can you prove or disprove a window is rectangular?

Landmark restoration efforts range from structural work to highly detailed work, like that involved in repairing a stained glass window. Restoration of such a panel is a highly skilled craft carried out by glaziers trained in the use of soldering lead caning to hold the glass panels in place.

At the right is a representation of a glass panel made of triangles and quadrilaterals. Notice that the panel shows a symmetric pattern formed by special quadrilaterals many of which are congruent.

Decision Making

In the stained glass diagram, quadrilateral *ABCD* is a rhombus and points *W*, *X*, *Y*, and *Z* are the midpoints of the four sides. Segments $\overline{WY}$ and $\overline{ZX}$ intersect at *J*.

1. Make a conjecture regarding rhombus *ABCD* and $\overline{WY}$.

2. Prove the conjecture you made in Question 1.

3. What can you say about quadrilaterals *AXJW*, *XBYJ*, *JYCZ*, and *WJZD*? Justify your response.

4. Suppose that a glazier has a single pane of glass in the shape of a rhombus. Explain how the glazier can make four congruent pieces of glass from that one piece.

The shapes shown at the left represent pieces of stained glass. Trace each of them and reproduce them on a sheet of paper. On that sheet of paper, make many copies of each shape. Then cut them out.

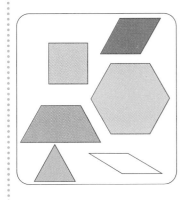

5. Classify each of the quadrilaterals in the set of shapes. Your classification may include two different names for a given shape.

6. Verify using shapes that the hexagon consists of two congruent isosceles trapezoids.

7. Create a stained glass design using all of the shapes shown at the left. Explain the geometry you used to decide on your design.

7.6 Properties of Kites

Explore

1. Does a toy kite like the one shown have special mathematical features? Explain your response.

2. Use six strands of uncooked spaghetti and four marshmallows.

   a. Cut two strands so that they have the same length and cut the two other strands so that their lengths are equal but unequal to the first two strands.

   b. Using these four stands and four marshmallows, create what you think a kite is.

   c. Using the other two strands, insert the diagonals of the quadrilateral. Do the diagonals have any special relationship with one another? Explain your response.

Build Understanding

COMMUNICATING ABOUT GEOMETRY

Under what conditions is a quadrilateral a kite, a rhombus, and a square *at the same time*?

A **kite** is a convex quadrilateral $ABCD$ in which $AB = BC = m$ and $AD = DC = n$. In other words, there are two pairs of congruent adjacent sides. If $m = n$, then all four sides of the kite are congruent and the kite is also a rhombus. In Explore you may have discovered the following.

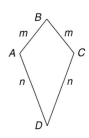

THEOREM 7.19

If a quadrilateral is a kite, then its diagonals are perpendicular.

CHECK UNDERSTANDING

Find the measures of ∠*WAZ*, ∠*ZAY*, and ∠*YAX*. Justify your answers.

EXAMPLE 1

Quadrilateral *WXYZ* is a kite labeled as shown.

a. Find *XY*.   b. Find m∠*WXZ*.

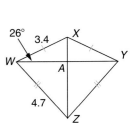

Solution

a. Since $XY = WX$, $XY = 3.4$.

b. By Theorem 7.19, m∠*XAW* = 90°.
   Thus, m∠*WXZ* + 26° = 90°. Therefore m∠*WXZ* = 64°.

The following theorem tells you that every kite has line symmetry. The points where the line of symmetry intersects the kite are called the **endpoints of the kite**.

GEOMETRY: WHO, WHERE, WHEN

> ### THEOREM 7.20
>
> Every kite has a diagonal that bisects the angles at its endpoints.

To prove the theorem, use Theorem 7.19 and congruent triangles.

Given $\overline{AB} \cong \overline{CB}$ and $\overline{AD} \cong \overline{CD}$

Prove $\overline{BD}$ bisects $\angle ABC$ and $\angle ADC$.

Plan for Proof Show that $\triangle ABT \cong \triangle CBT$ and $\triangle ATD \cong \triangle CTD$. Then by CPCTC, $\angle ABT \cong \angle CBT$ and $\angle ADT \cong \angle CDT$.

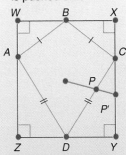

Proof Write a two-column proof.

Statements	Reasons
1. $\overline{AB} \cong \overline{CB}$ and $\overline{AD} \cong \overline{CD}$	1. Given
2. $\angle ATB, \angle CTB, \angle ATD,$ and $\angle CTD$ are right angles.	2. If a quadrilateral is a kite, its diagonals are perpendicular.
3. $\overline{BT} \cong \overline{BT}$ and $\overline{TD} \cong \overline{TD}$	3. Reflexive Property
4. $\triangle ABT \cong \triangle CBT$ $\triangle ATD \cong \triangle CTD$	4. Hypotenuse-Leg Congruence Theorem
5. $\angle ABT \cong \angle CBT$ $\angle ADT \cong \angle CDT$	5. CPCTC in $\triangle ABT$ and $\triangle CBT$ CPCTC in $\triangle ATD$ and $\triangle CTD$
6. $\overline{BD}$ bisects $\angle ABC$ and $\angle ADC$.	6. Definition of angle bisector

In Euclidean geometry kites and rectangles have different properties.

Nicolai Ivanovitch Lobachevsky (1793–1856) suggested that a kite and a rectangle are essentially the same. He claimed this because he could push outward on kite *ABCD* until its sides became parts of rectangle *WXYZ*. Point *P*, for example is pushed onto *P'*.

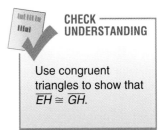

You can verify this fact by using graph paper, pushpins, and a rubber band.

You can prove a quadrilateral is a kite by using information about its diagonals.

EXAMPLE 2

KITE FLYING Ngyhen tacked two strips of wood together as shown. Prove *EFGH* is a geometric kite.

Solution
Since $\overline{FJ} \cong \overline{FJ}$, $\overline{EJ} \cong \overline{JG}$, and $\angle EJF \cong \angle GJF$, $\triangle EFJ \cong \triangle GFJ$ by the SAS Congruence Postulate. Thus, $\overline{EF} \cong \overline{GF}$ by CPCTC.

By similar reasoning, $\overline{EH} \cong \overline{GH}$.

Therefore, quadrilateral *EFGH* is a kite. ◄

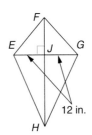

CHECK UNDERSTANDING

Use congruent triangles to show that $\overline{EH} \cong \overline{GH}$.

When you are given the coordinates of the vertices of a quadrilateral, you can use coordinate geometry to gather information about the special features of the quadrilateral.

EXAMPLE 3

Show that quadrilateral *JKLM* whose vertices have coordinates $J(-4, 4)$, $K(3, 1)$, $L(4, -4)$, and $M(-1, -3)$ is a kite.

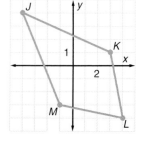

Solution

By showing that $JK = JM$ and $KL = ML$, you can show that quadrilateral *JKLM* satisfies the definition of kite.

$$JK = \sqrt{(-4 - 3)^2 + (4 - 1)^2} = \sqrt{58}$$
$$JM = \sqrt{(-4 - (-1))^2 + (4 - (-3))^2} = \sqrt{58}$$
$$KL = \sqrt{(3 - 4)^2 + (1 - (-4))^2} = \sqrt{26}$$
$$ML = \sqrt{(-1 - 4)^2 + (-3 - (-4))^2} = \sqrt{26}$$

Since $JK = JM$ and $KL = ML$, you can conclude that quadrilateral *JKLM* is a kite. ◄

You can create a kite by paper folding. Notice that the first fold midway across the sheet acts as a line of symmetry for the kite.

Fold the paper in half lengthwise. Fold the paper down from top to bottom. (Try not to fold the paper in half.)

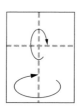

Mark where the fold lines meet the edges of the paper. Connect these points in order. Fold inward.

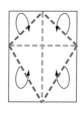

Answer each question about kite *CHOW* shown at the right.

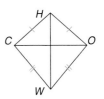

1. What can you say about $\overline{CO}$ and $\overline{HW}$? Justify your answer.

2. What can you say about $\angle CHW$ and $\angle OHW$? What can you say about $\angle CWH$ and $\angle OWH$? Justify your answers.

For each kite, find the specified length and angle measure.

3. *AH* and m$\angle AXH$
4. *TB* and m$\angle UNA$
5. *AT* and m$\angle OAL$

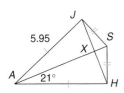
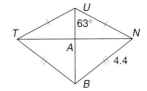
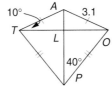

6. **SEWING** Julia drew a horizontal line intersecting a vertical line on a sheet of fabric using tailor's chalk. Then she drew rays $\overrightarrow{FG}$, $\overrightarrow{FL}$, $\overrightarrow{AG}$, and $\overrightarrow{AL}$ as shown at the right. Show that quadrilateral *FLAG* is a mathematical kite.

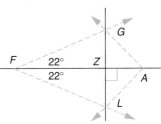

7. **WRITING MATHEMATICS** Briefly describe how to determine the vertices of a kite using two sticks. Illustrate your explanation.

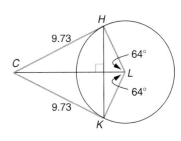

Show that each quadrilateral is a kite.

8. *WXYZ*; $W(-1, 3)$, $X(2, 4)$, $Y(3, 1)$, $Z(-2, -4)$
9. *ABCD*; $A(-3, 3)$, $B(7, 3)$, $C(-1, -3)$, $D(-5, -1)$

For each kite, find the specified length and angle measure.

10. *PT* and m$\angle PAT$
11. *DC* and m$\angle DBC$
12. *AQ* and m$\angle ADQ$

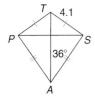
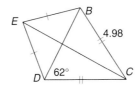
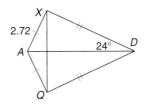

13. **MODELING** Miguel drew $\overline{CL}$. Using ruler and compass, he drew quadrilateral *CHLK*. In the diagram $\overline{CL} \perp \overline{HK}$. Show that quadrilateral *CHLK* is a mathematical kite.

14. **WRITING MATHEMATICS** Explain how to use symmetry to make a kite. Given a quadrilateral known to be a kite, explain how to find the line of symmetry for a kite.

Show that each quadrilateral is a kite.

15. *RASE*; $R(4, 4)$, $A(5, -4)$, $S(0, -4)$, $E(-3, 0)$ **16.** *BULD*; $B(-4, 3)$, $U(-1, 3)$, $L(2, -3)$, $D(-4, 0)$

17. *WOLD*; $W(0, 7)$, $O(0, 0)$, $L(7, 0)$, $D(9, 9)$ **18.** *NAQW*; $N(-4, -6)$. $A(-4, 1)$, $Q(1, -1)$, $W(3, -6)$

Exercises 19–22 refer to quadrilateral *ABCD* at the right.

19. Fill in the coordinates of *C* and *D* so that *ABCD* is a kite with $\overline{AB} \cong \overline{BC}$.

20. Use the distance formula to verify that *ABCD* is a kite.

21. Find the midpoint of $\overline{CA}$.

22. Use coordinate geometry to prove that in a kite the midpoint of one diagonal lies along the other diagonal.

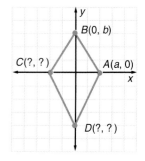

EXTEND

In the diagram at the right, *XAFY* is an isosceles trapezoid, *ACDF* is a rectangle, and *HBJE* is a kite. Find the length of each path.

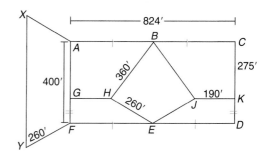

23. *X* to *A* to *B* to *J* to *E* to *D*

24. *X* to *A* to *B* to *H* to *E* to *D*

25. *Y* to *F* to *G* to *H* to *B* to *C*

26. *Y* to *F* to *A* to *X* to *A* to *B* to *J* to *K* to *C*

27. *E* to *H* to *B* to *J* to *E* to *D* to *K* to *C* to *B*

Exercises 28–30 refer to the diagram at the right.

28. In rectangle *WXYZ*, *B* and *D* are the midpoints of $\overline{XY}$ and $\overline{WZ}$, respectively, $XA = \frac{1}{4}XW$, and $YC = \frac{1}{4}YZ$. Show that quadrilateral *ABCD* is a kite. Use the two-column form.

29. Suppose that $XA = \frac{1}{n}XW$, and $YC = \frac{1}{n}YZ$, where $n > 0$. Show that quadrilateral *ABCD* is a kite. Use the two-column form.

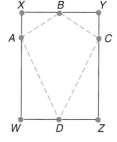

30. How does Exercise 29 justify the creation of a kite from a rectangular sheet of paper by paper folding?

31. In the diagram at the right, $\overline{UR} \cong \overline{SR}$ and $\overline{UT} \cong \overline{ST}$. Show that if *T* is reflected across $\overline{US}$ to *T'*, then *URST'* is a kite.

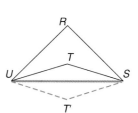

32. Illustrate a tessellation of the plane using copies of a given kite.

33. Kite *ABCD* has sides $AB = 2x + 30$, $BC = 15 - x$, and $CD = 3x + 18$. Identify the pairs of congruent sides in the kite.

THINK CRITICALLY

34. Make a conjecture about the quadrilateral you get when you join in order the midpoints of the sides of a kite. Use a coordinate proof to prove your conjecture.

35. What quadrilateral is formed by joining the midpoints of the sides of a kite? Explain.

ALGEBRA AND GEOMETRY REVIEW

Solve each equation. If an equation has no solution, write "no solution."

36. $|3x + 5| = 13$

37. $|-2x + 1| = 7$

38. $|5x - 4| + 1 = 9$

39. $|3x + 7| - 5 = 16$

40. $3|5x + 1| = 12$

41. $-2|x - 5| = 20$

Prove each statement given that $\ell \parallel m$.

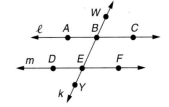

42. $\angle ABW$ and $\angle DEY$ are supplementary.

43. $\angle WBC \cong \angle DEY$

In kite $RSTU$, $\overline{RT}$ and $\overline{SU}$ intersect at Q, $\overline{RT}$ is the axis of symmetry, and $UT = 14.5$ and m$\angle SRQ = 42°$. Find each measure.

44. Find ST.

45. Find m$\angle URQ$.

46. Standardized Tests   Which set of information determines a kite?

    A. The diagonals are congruent but not perpendicular.

    B. The diagonals are perpendicular but not congruent.

    C. The diagonals are perpendicular and one bisects the other.

    D. The diagonals are not perpendicular and not congruent.

47. The measure of the vertex angle of an isosceles triangle is half the measure of a base angle. Find the measure of a base angle.

48. Write the equation for a line with slope 5 that contains the point $(2, 0)$.

PROJECT *Connection*   Use the graph paper with coordinates you made in Lesson 7–5.

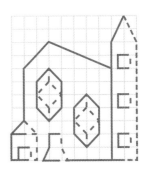

1. Inspect the rectified photograph and the grid drawing of your facade for symmetry. Identify lines of symmetry in various details on the facade.

2. Describe transformations that produce the symmetries.

3. Imagine you are an architect planning the construction of a new building. Describe the building and state its function. On graph paper, sketch the facade with windows, doors, roof lines, and so on that can be defined by using various transformations.

The Saccheri Quadrilateral

Through the centuries, many people have tried to show that Euclid's parallel postulate actually is a theorem. In 1733, an Italian priest named Girolamo Saccheri believed he had proved the postulate by indirect reasoning. That is, he assumed a different postulate and thought he had shown a contradiction. He never realized that he had, in fact, proved several fundamental theorems of non-Euclidean geometry!

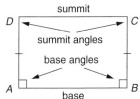

Saccheri based his work on the figure shown at the right, the *Saccheri quadrilateral ABCD* in which angles A and B are right angles and $\overline{AD} \cong \overline{BC}$. Notice that there are no right angles marks at points D and C.

1. Copy the Saccheri quadrilateral and prove that it is a parallelogram.

2. What must be the measure of each summit angle of a Saccheri quadrilateral? Justify your answer.

3. What type of parallelogram is a Saccheri quadrilateral?

Your answers to Questions 1–3 were based on the Parallel Postulate of Euclidean geometry, which states that through a point not on a line, there is exactly one line parallel to the given line. But what if you made a different assumption? Suppose that the Parallel Postulate states that there is no line parallel to the given line through a point not on a line or that there is more than one line parallel to the given line.

The mathematician Georg Friedrich Riemann (1826–1866) developed his non-Euclidean geometry by assuming that there is no line through the point parallel to the given line. A model for *Riemannian geometry* is the spherical geometry introduced in the Geometry Excursions presented in Chapters 3 and 5. Since all of the great circles of a sphere intersect, there are no lines parallel to a line through a point not on the given line.

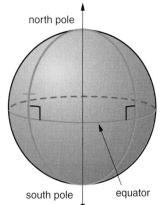

In Riemannian, or spherical geometry, lines perpendicular to the equator are great circles that intersect at the north and south poles as shown in the figure at the right.

MODELING For Questions 4–10, you will need a piece of chalk and a spherical solid that you can write on, like a basketball.

4. Use the chalk to draw any line on the sphere. Remember that, on this surface, a line is a great circle. Label it line ℓ_1.

5. Draw two more lines on the sphere perpendicular to line ℓ_1 and label the points of intersection A and B as shown in the drawing at the right. Recall that if you think of line ℓ_1 as the equator of the Earth, then lines perpendicular to line ℓ_1 will contain the poles of the Earth.

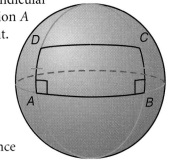
spherical Saccheri quadrilateral

6. Locate points D and C as shown so that the distance from A to D equals the distance from B to C.

7. What kind of angles do $\angle D$ and $\angle C$ appear to be?

8. Make a conjecture about the length of the summit as compared to the length of the base.

9. Is quadrilateral $ABCD$ a rectangle? Justify your answer.

THROUGH THE EFFORTS of Nicolai Ivanovitch Lobachevsky (1793–1856), geometry took on a life of its own. Not only was geometry a tool to explore the physical universe, it also became an axiomatic subject in which mathematics beyond actuality is created.

Although the Saccheri quadrilateral is a rectangle in Euclidean geometry it is not a rectangle in Riemannian geometry. In another non-Euclidean geometry, *Lobachevskian geometry*, the assumption is made that there is more than one line parallel to a given line through a point not on the line.

When you consider the Saccheri quadrilateral on a curved surface such as the one shown at the right, you will see that the summit angles, the angles at D and C, appear not to be right angles. The surface is called a *pseudosphere* and is a model of the surface that is the space in Lobachevskian geometry.

10. What type of angles do the summit angles of the Saccheri quadrilateral on the pseudosphere appear to be?

11. Make a conjecture about the length of the summit as compared to the length of the base.

12. Is quadrilateral $ABCD$ a rectangle? Justify your answer.

Problem Solving File

Make an Appropriate Drawing

An important step in solving many problems is making a drawing that accurately represents the conditions of the problem. If there is exactly one geometric figure that satisfies the conditions shown, the drawing uniquely *determines* the figure. If the drawing is labeled with too little information, it *underdetermines* the figure.

Problem

The Simon family is hiring a landscaper to dig a rectangular garden 15 m wide and 20 m long. The Simons' daughter Lisa made the drawing of the planned garden shown at the right. Does this drawing give the landscaper precise instructions about the size and shape of the garden?

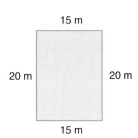

Explore the Problem

1. Does Lisa's drawing show correct lengths for the sides of the planned garden? for the angles at the corners? Explain.

2. Suppose the landscaper dug the garden as shown at the right. Does this shape satisfy the conditions shown in Lisa's drawing? Explain.

3. Explain why Lisa's drawing underdetermines the garden.

4. Copy Lisa's drawing. Then adjust the drawing so that it uniquely determines the garden as the Simons planned it.

The Simons' son Bart proposed a different shape for the garden. He made the drawing at the right to illustrate his idea.

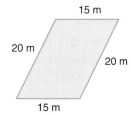

5. Explain why Bart's drawing underdetermines his proposed garden.

6. Mr. Simon says that Bart can uniquely determine his garden by adding the length of just one diagonal to his drawing. Justify this conclusion.

7. Mrs. Simon says that Bart can uniquely determine his garden by adding the measure of just one angle to his drawing. Do you agree or disagree? Justify your answer.

Investigate Further

• When drawing a figure, you also must be careful that the information you label is not *contradictory*. For example, consider the pattern for a triangular pennant that is shown at the right.

28 in.
78°
CENTERVILLE
68°
28 in.

PROBLEM
SOLVING PLAN

• Understand
• Plan
• Solve
• Examine

8. If the lengths of the *sides* of the triangle are labeled correctly, what type of triangle is the shape of the pennant?

9. If the measures of the *angles* of the triangle are labeled correctly, what type of triangle is the shape of the pennant?

10. Use your answers to Questions 8 and 9 to explain why the sketch of the pennant contains contradictory information.

11. WRITING MATHEMATICS Your friend says that you can always determine the shape of a triangle by labeling exactly three of its measures. Do you agree or disagree? Justify your response.

Apply the Strategy

• Decide whether each figure is *uniquely determined* or *underdetermined*. Explain your reasoning.

12.

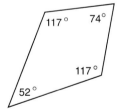

117°   74°
117°
52°

13.

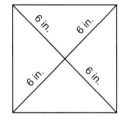

6 in.   6 in.
6 in.   6 in.

14.

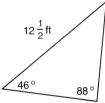

$12\frac{1}{2}$ ft
46°   88°

15.

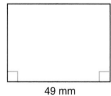

49 mm

Explain why each drawing shows information that is contradictory.

16.

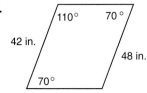

110°   70°
42 in.
48 in.
70°

17.

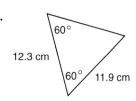

60°
12.3 cm
60° 11.9 cm

For each situation, tell whether the given drawing *determines* the figure, *underdetermines* the figure, or shows *contradictory information*. Explain your reasoning.

18. HOME IMPROVEMENT Mr. Redbird needs to replace a broken pane of glass in an old window. He measured the pane of glass, then took the sketch at the right to the glass cutter.

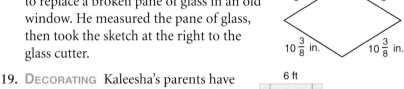

19. DECORATING Kaleesha's parents have promised to buy new carpeting for her bedroom as a birthday present. She measured the room, then took the sketch at the right to the carpet store.

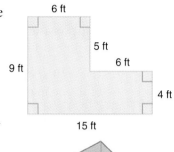

20. HOBBIES Jerry wants to order a new kite that will be built to his specifications. He made the pattern at the right, then took it to the hobby store.

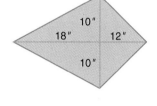

21. DESIGN The Bayview High School booster club designed a new banner to be used at team rallies. They gave the manufacturer the pattern at the right.

22. CRAFTS A quiltmaker designed a quilt in which the basic pattern piece is shaped like a rhombus with sides of length six inches and smaller angle measuring 30°. Make a drawing that determines this pattern piece.

23. CITY PLANNING An assistant to an city planner has designed a rectangular park with diagonal walkways that are each 120 ft long. Given this information, explain why it is not possible to make a drawing that uniquely determines the park.

24. JEWELRY A jeweler instructed an assistant to cut a pair of copper earrings in the shape of isosceles triangles with a base that is 1.8 cm long and legs that are each 8.5 mm long. Explain why these instructions contain contradictory information.

25. WRITING MATHEMATICS Write a paragraph explaining the importance of making drawings in which figures are uniquely determined.

REVIEW PROBLEM SOLVING STRATEGIES

TWINKLE, TWINKLE REVISITED

1. In Review Problem Solving Strategies Chapter 6, page 345 you found the sum of the measures of the five angles formed at the points of a star. The star was formed by extending the side of a regular pentagon. Now, try to find the sum of the measures of the same five angles of a star. However, this time the pentagon used to determine the star is not a regular polygon.

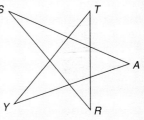

THE LOST SQUARE

3. A magic square is a square array of numbers whose rows, columns, and diagonals have the same sum. Natalie is trying to complete a 3 by 3 magic square with the integers 1 through 9. Natalie decided that a good strategy is to write each integer on a small piece of paper and shift them around. In her haste she loses the paper with the 9 on it. She does manage to arrange the remaining eight pieces in the square array (leaving one space blank) so that each of the three rows, three coluns and two diagonals have the same sum. How does she do this?

Triangles that Count

2. Count the number of triangles that satisfy the following conditions. The length of each side of the triangle is an integer and the perimeter of the triangle is 13. How many do you get?

7.8 Focus on Reasoning

Quadrilaterals and Converses

Discussion

• Work with a partner. Consider this conditional statement.

> The diagonals of a square are congruent and perpendicular.

John, Cynthia, George, and Ravi agree that this statement is true. They disagree, though, about whether the converse is true.

1. John says that you must identify the hypothesis and conclusion of a statement in order to write the converse. However, he has trouble remembering how to do this.

  a. Rewrite the statement in if-then form.
  b. What is the hypothesis? What is the conclusion?
  c. Write the converse of the statement.

2. Cynthia says, "The converse is true because the converse of any true conditional is true." Do you agree or disagree? Explain your reasoning.

3. George says the converse is true and draws this figure. Do you think George's argument is valid?

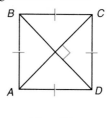

AC = BD

4. Ravi says the converse is false and draws this figure. Do you think Ravi's argument is valid?

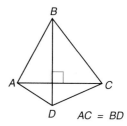

AC = BD

5. Do you think the converse is true or false? Explain your reasoning.

Build a Case

• John's observation is correct. That is, to write the converse of a conditional, you must identify the hypothesis and the conclusion.

6. Explain how you can identify the hypothesis and conclusion of any conditional statement.

7. How do you form the converse of any conditional statement?

Cynthia's observation is incorrect. That is, the converse of a true conditional statement might be true, but it is not necessarily true. For example, consider this statement about integers.

If an integer is divisible by 8, then it is divisible by 2 and by 4.

8. Give an example which shows that this conditional is true.

9. Give an example which shows that the converse is false.

In the Discussion, George used a diagram of a square to decide if the converse of the given statement is true. Most likely, his mind was not open to other possibilities because he focused on the square.

10. In general, how do you demonstrate that a conditional is true?

Ravi read the hypothesis and the conclusion of the converse carefully and tried to find a counterexample to the converse. He followed the premise and drew a quadrilateral *ABCD* with congruent and perpendicular diagonals. Notice that his quadrilateral is not a square.

11. In general, how do you demonstrate that a conditional is false?

A quadrilateral whose diagonals are perpendicular is called an *orthodiagonal quadrilateral.* A quadrilateral whose diagonals are both perpendicular and congruent is called a *skewsquare.* Ravi demonstrated that the converse of the given statement is false by drawing a skewsquare as a counterexample.

EXTEND AND DEFEND

Give the converse of each statement.

12. If both pairs of opposite sides of a quadrilateral are parallel, then the quadrilateral is a parallelogram.

13. If a quadrilateral is a kite, then it is not a trapezoid.

14. If a quadrilateral is a rhombus, then its diagonals are perpendicular and bisect each other.

15. If the diagonals of a parallelogram are congruent, then the parallelogram is a rectangle.

Write each statement in *if-then* form. Then write the converse.

16. Rectangles have congruent diagonals.

17. Base angles of an isosceles trapezoid are congruent.

18. A square is a kite.

19. Consecutive angles of a parallelogram are supplementary.

Write the converse of each statement. Then give a counterexample to show why the converse is false. Explain your counterexample.

20. All rectangles are parallelograms.

21. The diagonals of a rhombus bisect each other.

22. The diagonals of a square are perpendicular.

23. If a quadrilateral is an isosceles trapezoid, then its diagonals do not bisect each other.

Determine whether the converse of each statement is true or false. Explain your reasoning.

24. The diagonals of an isosceles trapezoid are congruent.

25. The diagonals of a convex quadrilateral are contained in the interior of the quadrilateral.

26. If four angles of one quadrilateral are congruent to four angles of another quadrilateral, then the quadrilaterals are congruent.

27. All rectangles are polygons.

28. A kite is not a trapezoid.

29. All orthodiagonal quadrilaterals are squares.

30. All squares are skewsquares.

Now You See It
Now You Don't

QUADRILATERAL QUANDRY

Which is longer, $\overline{AC}$ or $\overline{EC}$?

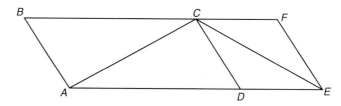

$\overline{AC}$ is the shorter diagonal of parallelogram *ABCD*, and $\overline{CE}$ is the longer diagonal of parallelogram *CDEF*. Measure each and you will see that they are, in fact, congruent!

Now decide whether the converse is true or false.

   If the shorter diagonal of one parallelogram is congruent to the longer diagonal of another parallelogram, then the parallelograms are not congruent.

7.9 Prisms and Pyramids

Explore

1. Refer to the diagram at the right.

   a. On graph paper, make a representation of the diagram.

   b. Cut out the figure along the solid segments. Then fold along the dashed segments to make a three-dimensional shape.

   c. What does it mean to stand the shape upright or on a side?

2. Describe in writing, as thoroughly as you can, all of the characteristics of the three-dimensional figure you created.

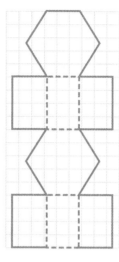

Build Understanding

A **prism** is a three-dimensional figure that consists of two parallel **bases** that are congruent polygonal regions and **lateral faces** that are bounded by parallelograms connecting corresponding sides of the bases. A segment that is the intersection of two lateral faces is called a **lateral edge** of the prism. In a **right prism**, the lateral edges are perpendicular to the bases. A prism that is not a right prism is called an **oblique prism**.

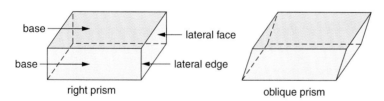

right prism   oblique prism

Notice that, in a right prism, the parallelograms that bound the lateral faces are rectangles. If the lateral faces of a prism are all congruent rectangles and the bases are both regular polygons, the figure is a **regular prism**.

regular prisms

7.9 **Prisms and Pyramids**   **419**

A **pyramid** is a three-dimensional figure that consists of a polygonal face, called a **base**, a point called the **vertex** that is not in the plane of the base, and triangular **lateral faces** that connect the vertex to each side of the base. As with a prism, a segment that is the intersection of two lateral faces is called a **lateral edge** of the pyramid. A pyramid is a **right pyramid** if its base has rotation symmetry and the segment from the vertex to the center of symmetry is perpendicular to the base. A pyramid that is not a right pyramid is called an **oblique pyramid**.

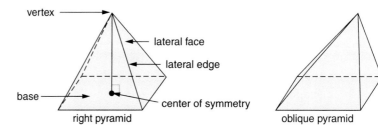

Notice that, in a right pyramid, the triangles that bound the lateral faces are isosceles triangles. If these triangles are all congruent, the pyramid is a *regular pyramid*. A **regular pyramid** is a right pyramid whose base is a regular polygonal region.

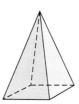

regular pyramids

To classify a prism or pyramid, you must determine whether it is best described as oblique, right, or regular. You must also identify the shape of its base or bases.

EXAMPLE 1

Classify each prism or pyramid.

a.

b.

c.

Solution

a. right triangular prism

b. regular square pyramid

c. oblique rectangular prism ◀

To draw a net for a prism or pyramid, you must know the shapes of its base or bases and of its lateral faces.

EXAMPLE 2

ARCHITECTURE Draw a net for the regular square pyramid that models the steeple of the schoolhouse shown at the right.

Solution

Draw a dashed square. Then draw four congruent isosceles triangles, with the base of each lying along one side of the square.

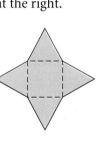

THINK BACK

A *net* is a two-dimensional figure that, when folded, forms the surface of a three-dimensional figure.

EXAMPLE 3

The net at the right consists of three regions bounded by rectangles and two regions bounded by equilateral triangles. Sketch the three-dimensional figure formed when this net is folded along the dashed lines.

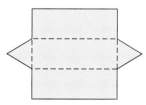

Solution

The net forms a regular triangular prism. One possible view is shown at the right.

Every regular prism or pyramid has reflection symmetry.

CHECK UNDERSTANDING

How would the solution of Example 4 be different if the figure were a regular pentagonal pyramid?

EXAMPLE 4

The figure at the right is a regular pentagonal prism. Find the number of planes of symmetry.

Solution

The base is a regular pentagon. It has five lines of symmetry, as shown at the right.

A plane perpendicular to the base that contains one of these lines of symmetry is a plane of symmetry for the prism.

One other plane of symmetry is a plane parallel to the bases and halfway between them.

So the prism has six planes of symmetry in all.

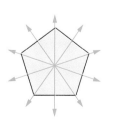

Classify each prism or pyramid. Give the name that best describes the figure.

1.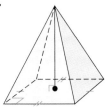

2.

3.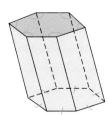

4.

Sketch a net for each.

5. right rectangular prism

6. regular pentagonal prism

7. regular triangular pyramid

8. regular pentagonal pyramid

Explain how you know that each three-dimensional figure is not a prism or pyramid.

9.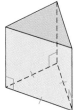

10.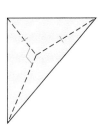

11.

12.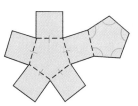

Sketch the three-dimensional figure formed if each net is folded along the dashed lines. You may assume that any angles which appear to be right angles are right angles.

13.

14.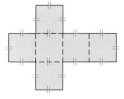

15.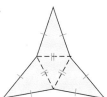

Find the number of planes of symmetry for each regular prism or pyramid.

16. hexagonal prism

17. square prism

18. hexagonal pyramid

19. square pyramid

20. triangular prism

21. triangular pyramid

22. LANDMARKS Classify the prisms and pyramids suggested by the red outlines.

23. WRITING MATHEMATICS Explain how prisms and pyramids are related to the triangles and quadrilaterals you have studied in Chapters 6 and 7.

PRACTICE

Classify each prism or pyramid. Give the name that best describes the figure.

24.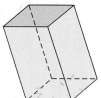

25.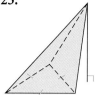

26.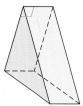

Sketch a net for each regular prism or regular pyramid.

27. right square prism

28. regular square pyramid

29. regular hexagonal prism

Sketch the three-dimensional figure formed if each net is folded along the dashed lines. You may assume that any angles which appear to be right angles are right angles.

30.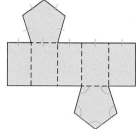

31.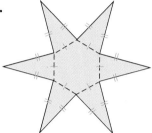

32.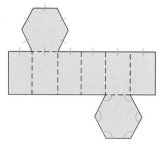

Find the number of planes of symmetry for each regular prism or pyramid.

33. octagonal prism

34. octagonal pyramid

35. heptagonal prism

36. heptagonal pyramid

37. decagonal prism

38. decagonal pyramid

39. WRITING MATHEMATICS How is a pyramid like a prism? How is it different? Explain.

EXTEND

TRANSFORMATIONS **The figure at the right is a regular hexagonal pyramid.**

40. Explain why $\overleftrightarrow{XZ}$ is an axis of symmetry for the pyramid.

41. How many times during a complete turn about the axis does the pyramid coincide with its original position?

42. Describe the rotation that takes point C to point F.

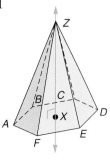

Sketch a net for each composite three-dimensional figure.

43.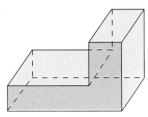

44.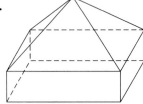

45.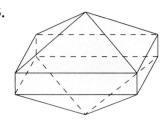

46. Refer to the figure at the right.

    Given $ABCDE$ is a regular square pyramid.
           $\overline{EM} \perp \overline{AC}$
           $\overline{EM} \perp \overline{BD}$

    Prove $\overline{EA} \cong \overline{EB} \cong \overline{EC} \cong \overline{ED}$

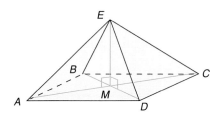

THINK CRITICALLY

47. Copy and complete the table at the right.

48. Use your table from Exercise 47. Write a rule for the function S that relates the number of planes of symmetry of a regular prism to the number n of sides of the base.

49. Repeat Exercise 48 for a regular pyramid.

50. Make generalizations about the number of axes of symmetry of a regular prism and of a regular pyramid with base of n sides.

Reflection Symmetry of Regular Figures		
Number of Sides of Base	Number of Planes of Symmetry	
	Prism	Pyramid
3	▨	▨
4	▨	▨
5	▨	▨
6	▨	▨

ALGEBRA AND GEOMETRY REVIEW

Find the real solutions of each equation by using the quadratic formula. Leave real solutions in simplest radical form. If the equation has no real solutions, write none.

51. $x^2 - 5x + 10 = 0$ **52.** $2x^2 + 3x - 5 = 0$ **53.** $x^2 + 2x + 7 = 0$

54. $5x^2 - 245 = 0$ **55.** $4x^2 + 28 = 0$ **56.** $3x^2 + 7x - 1 = 0$

In the diagram at the right, $\ell \parallel m$.

57. Find m$\angle BCS$.   **58.** Find m$\angle CBY$.

59. Find m$\angle BCA$.   **60.** Find m$\angle XBA$.

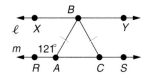

61. STANDARDIZED TESTS Which has exactly four planes of symmetry?

    A. regular triangular prism   **B.** regular square prism

    C. right rectangular prism   **D.** cube

Career
Architectural Technician

A restoration team may include an architect and an architectural technician. From old photographs, sketches, or written descriptions, they may create drawings that represent a part of a structure lost by fire or natural disaster.

Architectural technicians often may make three views of a figure in space, the *front view*, a *side view*, and a *top view*.

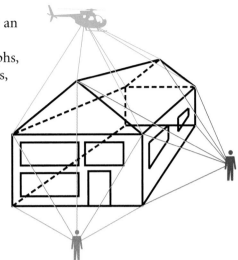

Decision Making

1. Sketch a front view of the building shown at left.

2. Sketch a left-side view of the building.

3. When you look directly down at the building, how many blocks do you see? Sketch a top view.

In addition to sketching views of a given three-dimensional figure, you can sketch a three-dimensional figure given three views of it.

Shown at the left is a set of views of a building.

4. Describe the height, left-right extent, and front-back extent of the building in terms of blocks.

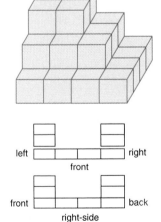

5. What can you tell and not tell from the top view?

6. Sketch the three-dimensional figure represented by these views.

7. Describe the building so that others can picture it realistically.

8. Make several views of a structure. Challenge others to identify it.

CHAPTER REVIEW

• • • •

VOCABULARY

Choose the word from the list that correctly completes each statement.

1. A ___?___ is a quadrilateral with four congruent sides.

2. A ___?___ is a quadrilateral with four right angles.

3. A ___?___ is a quadrilateral with exactly one pair of parallel sides.

a. rectangle

b. rhombus

c. trapezoid

Lessons 7.1   EXPLORING QUADRILATERALS   pages 373–376

- You can use tangrams to form special quadrilaterals.

Name the special quadrilateral formed by each set of tangram pieces.

4.   5.   6.   7.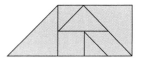

Lessons 7.2 and 7.4   PARALLELOGRAMS; SPECIAL PARALLELOGRAMS   pages 377–383, 391–396

- If a quadrilateral is a parallelogram, then its opposite sides are congruent, its opposite angles are congruent, its consecutive angles are supplementary, and its diagonals bisect each other.
- A parallelogram is: a rhombus if and only if its diagonals are perpendicular or if each diagonal bisects a pair of opposite angles; a rectangle if and only if its diagonals are congruent; a square if and only if its diagonals are both perpendicular and congruent.

In parallelogram *TWIN* at the right, *IN* = 7.6, *WN* = 8.4, and m∠*WIN* = 74°.

8. Find m∠*NTW*.   9. Find m∠*TWI*.   10. Find *WE*.   11. Find *WT*.

Classify special parallelogram *PARL* with the given vertices. Justify your answer.

12. $P(1, 10)$, $A(9, 9)$, $R(7, -7)$, $L(-1, -6)$ 13. $P(-5, 8)$, $A(-3, 0)$, $R(5, -2)$, $L(3, 6)$

14. $P(2, 0)$, $A(8, 2)$, $R(6, 8)$, $L(0, 6)$ 15. $P(1, 4)$, $A(-3, 10)$, $R(-7, 4)$, $L(-3, -2)$

Lesson 7.3   PROVING QUADRILATERALS ARE PARALLELOGRAMS   pages 384–390

- A quadrilateral is a parallelogram if both pairs of opposite sides are congruent, if both pairs of opposite angles are congruent, if an angle is supplementary to both consecutive angles, if the diagonals bisect each other, or if one pair of opposite sides are congruent and parallel.

Given each set of conditions, determine if quadrilateral *PQRS* is a parallelogram.

16. $\overline{PT} \cong \overline{RT}$ and $\overline{ST} \cong \overline{QT}$ 17. $\angle PSQ \cong \angle RQS$ and $\angle SPR \cong \angle QRP$

18. $\overline{PS} \cong \overline{RQ}$ and $\overline{PQ} \parallel \overline{SR}$ 19. $\angle PSR \cong \angle RQP$ and $\angle SPQ \cong \angle QRS$

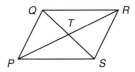

- The midsegment of a trapezoid is parallel to each base, and its length is equal to half the sum of the lengths of the bases.
- If a quadrilateral is an isosceles trapezoid, then its base angles and its diagonals are congruent.
- If the diagonals or base angles of a trapezoid are congruent, then the trapezoid is isosceles.

The vertices of trapezoid _TRAP_ are _T_(2, 7), _R_(7, 5), _A_(7, -3) and _P_(-3, 1).

20. Find the length of the midsegment.   **21.** Determine whether _TRAP_ is an isosceles trapezoid.

- If a quadrilateral is a kite, then its diagonals are perpendicular, and one diagonal bisects the angles at its endpoints.

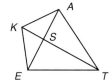

In kite _KATE_ at the right, _AT_ = 31, m∠_ETS_ = 28°, and m∠_KAS_ = 37°.

22. Find m∠_EAT_.   **23.** Find m∠_KET_.   **24.** Find m∠_EKA_.   **25.** Find _ET_.

- You can use conditional statements to analyze properties of quadrilaterals.

Determine whether the converse of each statement is true or false. Explain your reasoning.

26. If a quadrilateral is a rectangle, then its diagonals bisect each other.   **27.** The opposite sides of a parallelogram are parallel.

- A drawing _determines_ a figure if exactly one figure satisfies the conditions shown. If too little information is shown, the drawing _underdetermines_ the figure. You also must be careful that a drawing does not show _contradictory_ information.

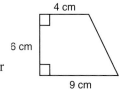

28. Decide whether the drawing at the right determines the figure, underdetermines it, or shows contradictory information. Explain your reasoning.

- A **polyhedron** is formed by polygons that share common sides and completely enclose a single region of space. The polygons are called **faces** of the polyhedron. A **prism** is a polyhedron with two congruent parallel faces and whose other faces are parallelograms. A **pyramid** is a polyhedron with one polygonal face and whose other faces are triangles.

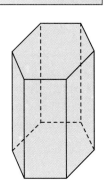

The figure at the right is a regular prism.

29. What is the shape of each base of the prism? What is the shape of each lateral face?

30. How many planes of symmetry does the prism have?

31. Draw a net for the prism.

CHAPTER ASSESSMENT

CHAPTER TEST

Name the special quadrilateral formed by each set of tangram pieces.

1.    2.

In the figure at the right, *HOME* is a parallelogram. Tell whether each statement is true or false.

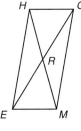

3. $\overline{HE} \cong \overline{MO}$ 4. $\overline{EO} \cong \overline{HM}$

5. $\overline{ER} \cong \overline{OR}$ 6. $\overline{HO} \parallel \overline{ME}$

7. $\angle HEM \cong \angle MOH$

8. **WRITING MATHEMATICS** List as many different ways as you can to complete the following to make a true statement.
A quadrilateral is a parallelogram if __?__.

Is the quadrilateral with the given vertices a parallelogram? Justify your answer.

9. $P(-2, 7)$, $O(0, 0)$, $E(4, 5)$, $M(6, -2)$

10. $A(-3, 4)$, $B(2, 4)$, $C(4, -3)$, $D(-2, -3)$

11. In the figure below, $m\angle QSR = 32°$. Find $m\angle QPR$.

12. In the figure below, $KL = 17.4$. Find GJ.

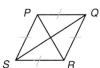

In trapezoid *TRAP*, $TR = 8$, $RY = 4.5$, $YA = 4.5$, $PA = 17$, and $m\angle R = 120°$. Find each measure.

13. XY 14. TP

15. $m\angle T$ 16. $m\angle A$

In the figure at the right, *ABCD* is a kite. Fill in each blank to make a true statement.

17. $\overline{BC} \cong$ __?__

18. $\overline{BE} \cong$ __?__

19. $\overline{BE} \perp$ __?__

20. $\angle BCA \cong$ __?__

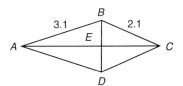

Determine whether the converse of each statement is true or false. Explain your reasoning.

21. If a quadrilateral is trapezoid, then it is not a parallelogram.

22. The diagonals of a square are congruent.

23. A tile is to be cut in the shape of a rhombus with diagonals of lengths 8 in. and 5 in. Make a drawing that uniquely determines the tile.

24. **STANDARDIZED TESTS** Which of the following names applies to the figure shown at the right?

   I. parallelogram

   II. rhombus

   III. rectangle

   IV. square

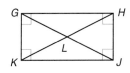

  A. I and II only
B. I and III only
  C. II, III, and IV only   **D.** I, II, III, and IV

Shown at the right is a net for a three-dimensional figure.

25. Identify the three-dimensional figure.

26. How many planes of symmetry does the three-dimensional figure have?

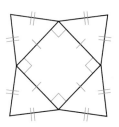

PERFORMANCE ASSESSMENT

CONSTRUCTION AHEAD   Work individually or in pairs. Use a compass and straightedge, paper folding, or use geometry software. Devise a method for constructing each of the following figures: parallelogram, rhombus, rectangle, square, isosceles trapezoid, kite. Organize all your constructions into a construction "scrapbook."

NET WORK   Shown at the right is a rectangular piece of cardboard on which has been drawn a net for a regular square prism. Clearly, much of the cardboard would be wasted if you cut out this net. Work in groups of four or five. Create a different net for the prism so that the amount of waste is reduced. Then create nets for each of the following figures: regular triangular prism, regular triangular pyramid, square pyramid, regular hexagonal prism, and regular hexagonal pyramid. Plan each net in order to minimize the amount of waste generated when it is cut out.

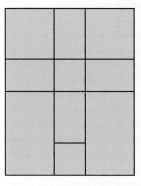

QUADRILATERAL DATABASE   Use index cards, or use database software if it is available. Make a database of facts that are always true about the special quadrilaterals you studied in this chapter. For example, this might be the *beginning* of a car showing facts about a rhombus.

rhombus			
definition: a quadrilateral with four congruent sides			
parallel sides	one	one	two
	none	pair	pair
opposite sides congruent		yes	no
opposite angles congruent		yes	no
consecutive angles supplementary		yes	no
four right angles		yes	no
four congruent sides		yes	no

Be sure to include facts about the diagonals of each figure. Then show how you can sort through your database to identify all types of quadrilaterals with a certain characteristic, such as congruent diagonals.

PUTTING IT ALL TOGETHER   Write a report titled *Triangles and Quadrilaterals: Putting It All Together.* In your report, describe how the facts about triangles that you learned in earlier chapters are related to the facts about quadrilaterals that you learned in this chapter.

PROJECT ASSESSMENT

PROJECT *Connection*   Work together with your group. Organize the historical and architectural data you gathered and the photographs, pictures, and coordinate grids you made. Also gather the work on your proposed building project. Prepare a presentation that addresses the following.

1. Explain why the landmark you examined is of historical or architectural importance.

2. Explain how you used both coordinate and transformational geometry to represent and study the facade of that landmark.

3. Describe how information about an old building and knowledge of transformations and coordinate geometry helped you design a new building.

CUMULATIVE REVIEW

Fill in the blank.

1. A quadrilateral that has exactly one pair of parallel sides is called a(n) __?__.

2. A quadrilateral that has exactly two pairs of congruent adjacent sides is called a(n) __?__.

3. A quadrilateral that has all four sides congruent is called a(n) __?__.

4. A triangle with a 110° angle is called a(n) __?__ triangle.

Classify each figure.

5.

6.

7.

8.

9. **STANDARDIZED TEST** The endpoints of a segment are $A(-2, 5)$ and $B(4, 1)$. Which of the following points lie on the perpendicular bisector of AB?

    I. $(1, 3)$ **II.** $(7, 12)$

   III. $(-3, -3)$ **IV.** $(0, 1\frac{1}{2})$

   A. I only

   B. II only

   C. I and III only

   D. I, II, and IV only

   E. I, II, III, and IV

Find the distance between each pair of points in the coordinate plane. Give answers to the nearest tenth, if necessary.

10. $(8, -3)$ and $(-1, -2)$

11. $(-4, -1)$ and $(4, 14)$

In each figure, a midsegment is drawn. Find the length of each midsegment.

12.
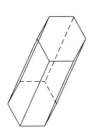

13.
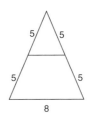

14. **WRITING MATHEMATICS** Describe the relationships between a parallelogram, a rhombus, and a square.

15. In the figure, G is between F and H. If $FH = 80$, find FG.

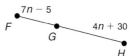

16. In the figure, Q is the midpoint of PR. Find PR.

Write a logical argument to show that each pair of triangles is congruent.

17.

Prove that $\triangle ABD \cong \triangle CDB$.

18.
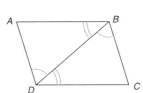

C is the midpoint of $\overline{AE}$ and $\overline{DB}$. Prove that $\triangle ABC \cong \triangle EDC$.

19. Reflected in the x-axis, what are the endpoints of the image of the segment whose endpoints are $(-2, 3)$ and $(4, -1)$?

• • • STANDARDIZED TEST • • •

STANDARD FIVE-CHOICE Select the best choice for each question.

1. Which of the following is not true as it pertains to parallelograms?

   A. The opposite sides are congruent.
   B. The opposite angles are congruent.
   C. Consecutive angles are supplementary.
   D. The diagonals bisect each other.
   E. All are true.

2. Which of the following statements is not true?

   A. All squares are rhombuses.
   B. All rectangles are parallelograms.
   C. Some rectangles are squares.
   D. Some trapezoids are rhombuses.
   E. All are true.

3. Which of the following pairs of vectors is perpendicular?

   A. $v(-4, 2)$ and $w(4, -2)$
   B. $v(3, -6)$ and $w(18, 9)$
   C. $v(5, -4)$ and $w(5, 6)$
   D. $v(7, -4)$ and $w(-3, -5)$
   E. None of these.

4. In the diagram, two parallel lines are cut by a transversal. Which pair of angles is congruent because they are alternate exterior angles?

   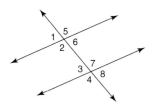

   A. 2 and 7
   B. 8 and 6
   C. 4 and 5
   D. 4 and 7
   E. 2 and 3

5. If the image of a point reflected in the y-axis is $(-3, 4)$, which of the following are the coordinates of the original point?

   A. $(-3, -4)$
   B. $(3, 4)$
   C. $(3, -4)$
   D. $(4, -3)$
   E. $(-4, 3)$

6. Which of the following cannot be used to prove that two lines that are cut by a transversal are parallel?

   A. corresponding angles that are congruent
   B. alternate interior angles that are congruent
   C. vertical angles that are congruent
   D. same-side interior angles that are supplementary
   E. All prove that the lines are parallel.

7. Which of the following statements is not true about an isosceles triangle?

   I. The base angles are acute.
   II. The sides opposite the base angles are congruent.
   III. The vertex angle is always acute.

   A. I only
   B. II only
   C. III only
   D. I and II
   E. all are true

8. A circle has its center at $(2, -1)$. The radius of the circle is 5. Which of the following points are on the circle?

   I. $(5, 3)$ II. $(-1, -5)$
   III. $(-3, -1)$ IV. $(7, -6)$

   A. I only
   B. III only
   C. IV only
   D. II and IV only
   E. I, II, and III only

Perimeters and Areas of Polygons

Take a Look AHEAD

Make notes about things that look familiar.

- What area formulas do you recognize? How are the areas of these figures related?
- Find an example of a right triangle, an acute triangle, and an obtuse triangle. State a characteristic each type of triangle has.

Make notes about things that look new.

- What new area formulas will you study in this chapter?
- List the types of figures you will use to estimate area. Why do you think estimating area is important?

DATA Activity

Green Acres

America was once a vast wilderness, landscaped by nature. In modern times, much of our green areas have disappeared. Recognizing the importance of preserving our heritage and protecting our environment, there is now a serious effort to conserve and renew our forested areas.

A commonly used unit of area for large regions such as a forest is the *acre*. The table gives the areas of the five largest and five smallest states in the United States and the number of acres of forest in each of these states.

SKILL FOCUS

- ▶ Compare numbers.
- ▶ Round numbers.
- ▶ Divide numbers.
- ▶ Convert measures.
- ▶ Write numbers as percents.
- ▶ Find a percent of a number.
- ▶ Solve percent problems.

GEOMETRYWORKS

Landscaping

In this chapter, you will see how:

- **LANDSCAPERS** use patterns to determine efficient mowing methods. (Lesson 8.2, page 447)

- **POOL DESIGNERS** use perimeter and area to incorporate walkways into a pool layout. (Lesson 8.5, page 461)

- **DECK BUILDERS** use the Pythagorean Theorem and its converse to obtain right angles during construction. (Lesson 8.8, page 480)

Rank	State	Area mi²	Acres of Forest
1	Alaska	591,000	129,045,000
2	Texas	266,807	13,656,000
3	California	158,706	39,381,000
4	Montana	147,046	21,910,000
5	New Mexico	121,593	18,526,000
46	New Jersey	7,787	1,985,000
47	Hawaii	6,471	1,748,000
48	Connecticut	5,018	1,815,000
49	Delaware	2,045	398,000
50	Rhode Island	1,212	399,000

1. There are 43,560 ft² in an acre and 5,280 ft equal 1 mi. How many acres equal 1 mi²?

2. How many square miles of Texas is forest? Round to the nearest square mile.

3. What percent of Texas is forest? Round to the nearest percent.

4. What percent of Rhode Island is forest?

5. With fewer acres of forest, how can Rhode Island have a higher percentage of forest land than Texas?

6. Which of the 50 states are smaller than the forested part of Alaska?

7. The area of New Mexico is about 100 times that of Rhode Island. The forest area of New Mexico is about 45 times greater than that of Rhode Island. Which state has a higher percentage of forest area? Explain.

8. **WORKING TOGETHER** Rank the 10 states in order of their percentages of forest land. Draw a conclusion. Create and display a poster to summarize your group's findings.

Creating a Garden

Think of some places where you have seen beautiful gardens—around homes, parks, banks, schools, and many more. A garden that commemorates a famous person or event often has a statue in it. Some gardens are areas for greeting and meeting people.

Designing a garden involves an integration of creativity, art, science, and geometry. In this project, you will plan, design, estimate cost, and give a presentation about building a commemorative garden on your school grounds.

PROJECT GOAL

To plan and design a garden.

Getting Started

Work as a class to gather ideas about the nature of the garden. At times, the class will work in groups.

1. Locate books, magazines, and articles about landscape architecture and famous gardens.

2. Visit some special gardens in your area.

3. Obtain photographs of some special gardens. You may want to take a photograph yourself.

4. Create a list of themes appropriate for a garden on your school grounds. Find out about famous graduates of your school that you may wish to honor.

5. Arrange to speak to your school administrators about your plans for the project.

PROJECT *Connections*

Lesson 8.1, page 440:
Research possible designs, materials, and plants.

Lesson 8.3, page 453:
Create four different designs, each with a scale drawing and fact sheet. Conduct a poll to choose a design.

Lesson 8.8, page 479:
Prepare a detailed estimate of costs and plan fund raisers.

Chapter Assessment, page 487:
Prepare a presentation to communicate final project plans.

Internet Connection

www.learninggeometry.com

8.1 Perimeter and Area

Explore

An environmental club is planning a garden for use by students. A store donated 100 ft of fencing for the project. Using this fencing, the club would like the garden to have the largest area possible.

1. What will the perimeter of the garden be?

2. Copy and complete the table. Recall that you can use the formula $P = 2\ell + 2w$ to find the perimeter of a rectangle, and the formula $A = \ell w$ to find its area.

Length	Width	Perimeter	Area
49	1		
40	10		
35	15		
25	25		
20	30		
15	35		

3. Make a table like the one above, using a perimeter of 64 ft. Have a partner make a similar table using a perimeter of 36 ft. Make a conjecture about the dimensions of the rectangle that gives the maximum area for a given perimeter.

Build Understanding

A plane figure has two important measures of size. The **perimeter** of a plane figure is the sum of the lengths of its sides. The **area** of a plane figure is the number of nonoverlapping square units contained in the interior of the figure. The square at the right has a perimeter of twelve *units*, while its area is nine *square units*.

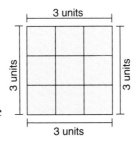

3 units

3 units · 3 units

3 units

The formulas for finding the areas of some of the special types of polygons you have studied are based on the following assumptions.

POSTULATE 20   AREA OF A SQUARE POSTULATE
The area A of a square is the square of the length s of one side, or $A = s^2$.

POSTULATE 21   AREA CONGRUENCE POSTULATE
If two polygons are congruent, then they have the same area.

POSTULATE 22   AREA ADDITION POSTULATE
The area of a region is the sum of the areas of all its nonoverlapping parts.

In Explore you used perimeter and area formulas to investigate size.

> A rectangle with length ℓ and width w.
> - has perimeter P such that $P = 2\ell + 2w$
> - has area A such that $A = \ell w$

EXAMPLE 1

Find the perimeter and area of the rectangle.

11 in.

5 in.

Solution

The length ℓ is 11 in. The width w is 5 in.

perimeter: $P = 2\ell + 2w$ area: $A = \ell w$
$P = 2(11) + 2(5)$ $A = 11(5)$
$P = 22 + 10 = 32$ $A = 55$

So the perimeter is 32 in. The area is 55 in.^{2} ◄

Sometimes a figure does not look like a rectangle, but you are able to divide it into smaller rectangular regions.

EXAMPLE 2

THINK BACK

The perimeter of a polygon is the sum of the lengths of all its sides.

DECORATING The Reids plan to carpet the L-shaped room shown. Before the carpet is laid, they must lay wood strips on the floor along the walls.

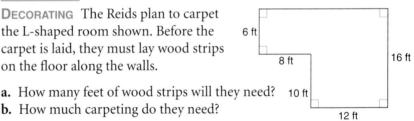

6 ft

8 ft

16 ft

10 ft

12 ft

a. How many feet of wood strips will they need?
b. How much carpeting do they need?

COMMUNICATING ABOUT GEOMETRY

In Example 2 carpeting usually is sold by the square *yard*. How can you calculate the number of square yards of carpeting the Reids will need?

Solution

a. Since the strips are placed along the perimeter walls, find the perimeter of the floor. The length of one side is not known but the figure shows the length equals 8 ft + 12 ft.

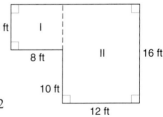

6 ft   I

8 ft   II   16 ft

10 ft

12 ft

$$P = 20 + 16 + 12 + 10 + 8 + 6 = 72$$

b. The carpeting covers the floor so find the area. Divide the figure into two rectangles labeled I and II.

area I: $A = \ell w$ area II: $A = \ell w$
$A = 8(6)$ $A = 16(12)$
$A = 48$ $A = 192$

Add the two areas to find total area equals 240 ft^2.

The Reids need 72 ft of wood strips and 240 ft^2 of carpeting.   ◄

Sometimes you can use the area of a rectangle as a geometric model to help you understand an algebraic operation.

EXAMPLE 3

MODELING Use a model to find the product $(x + 3)(x + 4)$.

Solution

Draw a rectangle with length labeled $(x + 4)$ and width labeled $(x + 3)$. The area of this rectangle is the product $(x + 3)(x + 4)$.

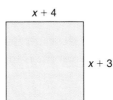

$x + 4$

$x + 3$

Divide the rectangle into four smaller rectangles, as shown at the right below.

Find the area of each small rectangle. Then apply the area addition postulate.

$$(x + 3)(x + 4) = x^2 + 4x + 3x + 12$$
$$= x^2 + 7x + 12$$

PROBLEM SOLVING TIP

Algeblocks can be used to model problems like Example 3.

TRY THESE

Determine whether the conditional statement is true or false. Justify your answer.

1. If the areas of two rectangles are equal, then their perimeters are equal.

Find the perimeter and area of each rectangle.

2.
4 ft
7 ft

3.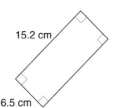
15.2 cm
6.5 cm

4.
30 in.

5.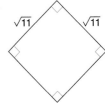
$\sqrt{11}$ $\sqrt{11}$

6. **WRITING MATHEMATICS** A rectangle is 9 ft long and 3 yd wide. A classmate says that its area is 27. Do you agree or disagree? Explain your reasoning.

Find the perimeter and area of each polygon.

7.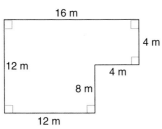
16 m
12 m
4 m
4 m
8 m
12 m

8.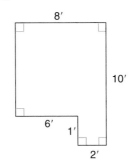
8'
10'
6'
1'
2'

9.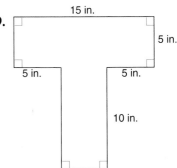
15 in.
5 in.
5 in.   5 in.
10 in.

MODELING Write the expression that represents each dimension. Find the product.

10.

11.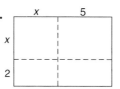

12.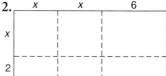

INTERIOR DECORATING A decorator has been hired to redecorate the room shown at the right.

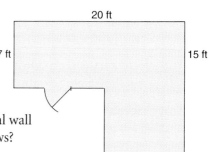

13. The decorator plans to install a decorative molding around the ceiling. How many feet of molding are needed?

14. The height of each wall is 8 ft. What is the total wall area of the room, including doors and windows?

15. The decorator wants to install carpeting that costs $29/\text{yd}^2$, including installation. What will be the cost of carpeting the room?

PRACTICE

Determine whether the conditional statement is true or false. Justify your answer.

16. If the perimeters of two rectangles are equal, then their areas are equal.

Find the perimeter and area of each rectangle.

17.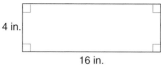

18.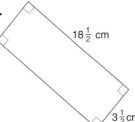

19.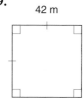

20.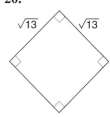

Find the perimeter and area of each polygon.

21.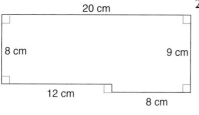

22.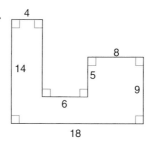

23.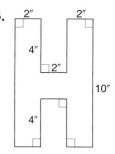

24. **WRITING MATHEMATICS** When converting between units of measure, Kevin gets confused about whether to multiply or divide by the appropriate constant. Explain how to get all measurements for a problem in the same units.

Write the expression that represents each dimension. Find the product.

25.

26.

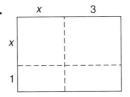

27.

28. HOME IMPROVEMENT Rose Construction is finishing the basement shown. How many square yards of carpeting are needed for all but the area taken by the staircase. Round up to the nearest ten square yards.

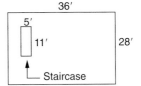

29. MUSIC The walls, ceiling, and floor of a recording studio are often carpeted to help absorb sound. There is a window and a door on a wall that separates the musicians from the recording engineers. If the entire wall (except for the window and door) is to be carpeted at a cost of $28/yd^2, find the cost of carpeting this wall.

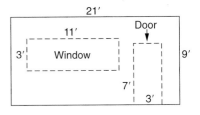

Use a numerical example to verify your conjecture.

30. If you know the area of a square, how can you find the length of its side?

31. If you know the area and one dimension of a rectangle, how can you find the other?

EXTEND

32. The base of a rectangle is 7 ft more than its height. The area is 44 ft^2. Find the perimeter.

33. A square has area represented by $x^2 + 6x + 9$. Express its perimeter in terms of x.

34. CONSTRUCTION The sides of an 8 m $\times$ 10 m patio are increased an equal amount, and the area of the new patio is 255 m^2. By how much was each side increased?

BIOLOGY A fish hatchery is planning an experiment along a river. A rectangular area will be fenced in the river itself, to keep out certain types of fish. One side of this rectangle will not require fence, since the river bank will be used as a natural border.

35. Draw a diagram that models this situation.

36. What are the dimensions of the rectangle with the maximum area that can be enclosed with 60 ft of fence?

37. GEOMETRIC CONSTRUCTION Construct a rectangle with length a and width equal to one-half its length.

THINK CRITICALLY

38. A rectangle has an area of 100 ft². What is the minimum perimeter the rectangle could have?

39. WRITING MATHEMATICS What is the maximum perimeter for a rectangle whose area is 100 in.² ? Explain using specific numerical examples.

40. Find the dimensions of a square for which the numerical values of the area and perimeter are equal.

41. WRITING MATHEMATICS Using technology, diagrams, examples, and sentences, explain what happens to the area of a rectangle when the sides are doubled.

42. TRANSFORMATIONAL GEOMETRY Joan found the perimeter of the L-shaped polygon labeled *SHAPED* by finding the perimeter of rectangle *SLED*. Use transformation to show why both figures have the same perimeter.

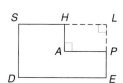

43. TRANSFORMATIONAL GEOMETRY Find the perimeter of the polygon shown. Explain the procedure you used.

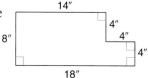

ALGEBRA AND GEOMETRY REVIEW

44. STANDARDIZED TESTS The slope of the line that passes through $(2, -3)$ and $(-8, -13)$ is

   A. $\dfrac{3}{5}$ **B.** $-\dfrac{3}{8}$ **C.** 0.625   **D.** 1

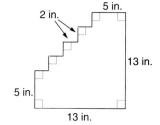

45. Find the perimeter and area of the polygon at the right. Each "step up" has the same width and height.

46. Find the additive inverse, multiplicative inverse, and negative reciprocal of $\dfrac{2}{5}$.

47. Arrange these numbers in descending order:

$$0.41, \frac{2}{5}, \frac{5}{2}, -\frac{2}{5}, -\frac{5}{2}, 0.3$$

PROJECT *Connection*   **Divide the class into four groups to research the different possible designs, materials, and plants that can be used in the garden.**

1. One group will consult with the teacher about the space that could be allotted for the garden. They should determine the maximum possible area.

2. The second group will research the different plants that thrive in your climate.

3. The third group will contact a landscaper by mail, phone, E-mail, or the Internet to get information about making a scale drawing of the garden.

4. The fourth group will research the incorporation of a sign, decks, and benches.

5. As a class, come together and make a list of all of the possible different materials you will consider using in your designs.

8.2 Areas of Parallelograms and Triangles

Explore

- On graph paper, plot the points $P(4, 6)$, $O(0, 0)$, $R(12, 0)$, $T(16, 6)$.
Connect the points to form parallelogram *PORT*.

Plot point $S(4, 0)$ and draw $\overline{PS}$, an altitude of parallelogram *PORT*.

  1. Cut out right triangle *OPS*. Translate it 12 units to the right. Tape
  it to the paper. What type of quadrilateral have you formed?

  2. Determine the area of the new quadrilateral. How do you think
  the area of the original parallelogram is related to this new area?

On a new sheet of graph paper, plot another parallelogram *PORT* that
includes altitude $\overline{PS}$.

  3. Draw $\overline{PR}$ to create triangle *OPR*. In this triangle, identify the
  altitude to base $\overline{OR}$.

  4. Cut out triangle *OPR* and describe a transformation that shows
  how triangle *OPR* is related to triangle *PRT*.

  5. How do you think the area of triangle *OPR* is related to the area of
  parallelogram *PORT*?

SPOTLIGHT
ON LEARNING

WHAT? In this
lesson you will learn
- to find the area of
  a parallelogram.
- to find the area of
  a triangle.

WHY? Knowing how
to find the areas of
parallelograms can
help you solve
everyday problems
in sports, business,
and gardening.

Build Understanding

- In Explore, Questions 1–2, you used the area of a rectangle to help
find the area of a parallelogram. You may have noticed that you
can find the area of a parallelogram from the measures of a side
and the altitude drawn to it.

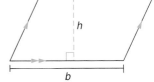

You can consider any side to be a **base of a parallelogram**. The
term *base* refers to either the line segment or to its length b. The
altitude of a parallelogram is any segment perpendicular to the line
containing the base from any point on the opposite side of the
parallelogram. The length of an altitude is the **height** h of the figure.
All the altitudes to a particular base have the same length.

THEOREM 8.1	AREA OF A PARALLELOGRAM

The area A of a parallelogram equals the product of a
base b and the height h to that base, or $A = bh$.

You may use the formula for the area of a parallelogram to find the
area of any type of parallelogram, such as a rectangle.

EXAMPLE 1

Find the area of the parallelogram.

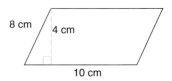

Solution

Use the side of the parallelogram to which the altitude is drawn as the
base. In the given figure, the base b is 10 cm. The height h is 4 cm.

Apply Theorem 8.1. $A = bh$
$A = 10(4)$
$A = 40$

So the area of the parallelogram is 40 cm^2. ◄

In Explore, Questions 3–5, you used the area of a
parallelogram to help find the area of a triangle.
An **altitude of a triangle** is any segment from a
vertex perpendicular to the line containing the
opposite side. The length of the altitude is the
height h of the figure. The **base of a triangle** b is
the side to which an altitude is drawn.

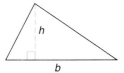

| THEOREM 8.2 | AREA OF A TRIANGLE |

The area A of a triangle equals half the product of a base
b and the height h to that base, or $A = \frac{1}{2}bh$.

You may use the formula for the area of a triangle to find the area of
any type of triangle.

EXAMPLE 2

Find the area of the triangle.

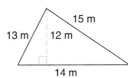

Solution

Be careful in selecting which lengths to use.
The base b is 14 m. The height h is 12 m.

$A = \frac{1}{2}bh$

$A = \frac{1}{2}(14)(12)$

$A = 84$

The area of the triangle is 84 m^2. ◄

You can use the formula for the area of a triangle to derive a formula for the area of a rhombus.

THEOREM 8.3 **AREA OF A RHOMBUS**

The area A of a rhombus equals half the product of the lengths of its diagonals d_1 and d_2, or $A = \frac{1}{2}d_1d_2$.

EXAMPLE 3

RECREATION Joelle is building a box kite that consists of panels shaped like rhombuses, one of which is shown below. Each rhombus is strengthened by the inclusion of the diagonals in the framing. She will use fabric to cover the panels. Find the amount of fabric needed to cover one panel of the box kite.

TECHNOLOGY TIP

You can use geometry software to verify the formula for the area of a parallelogram, triangle, or rhombus.

Solution

Find the lengths of the diagonals.

$d_1 = 30 + 30 = 60$ cm

$d_2 = 40 + 40 = 80$ cm

To find the area, apply Theorem 8.3.

$A = \frac{1}{2}d_1d_2$

$A = \frac{1}{2}(60)(80)$

$A = 2400$ cm^2

Jo needs 2400 cm^2 of fabric to cover each panel of her box kite. ◄

TRY THESE

Find the area of each figure.

1.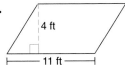
4 ft

11 ft

2.
4.5 cm

8 cm

3.
12.2 in.

20 in.

4.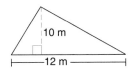
10 m

12 m

5.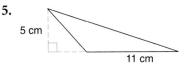
5 cm

11 cm

6.
$d_1 = 12$ cm

$d_2 = 16$ cm

7. WRITING MATHEMATICS Detail three different ways in which you could find the area of a rhombus that has a right angle. Explain what else you would need to know about the rhombus to use each formula.

8. **SPORTS** A diving board handrail is shaped like a parallelogram, as shown at the right. It is going to be filled with a colorful canvas that features the logo of the swim club. Find the area of the parallelogram.

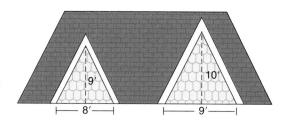

9. **BUILDING** Two triangular gables are to be built on a house, with the dimensions shown at the right. To cover the gables with shingles, the builder needs to know the total area. Find the sum of the areas of the two triangles.

PRACTICE

Find the area of each figure.

10.
6 in.
—14 in.—

11.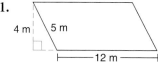
4 m   5 m
—12 m—

12.
8 cm
3.5 cm

13.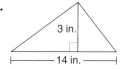
3 in.
—14 in.—

14.
10 m
3 m   3 m

15.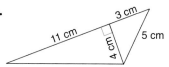
3 cm
11 cm   4 cm   5 cm

16.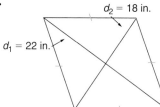
$d_2 = 18$ in.
$d_1 = 22$ in.

17.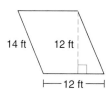
14 ft   12 ft
—12 ft—

18.
5 in.
12 in.

Find the unknown measure.

19. If the area of a parallelogram is 456 cm² and the base is 24 cm, find the height.

20. If the area of a triangle is 325 in.² and the height is 15 in., find the base.

21. If the area of a rhombus is 186 m² and the length of the shorter diagonal is 12 m, find the length of the longer diagonal.

22. **WRITING MATHEMATICS** Explain how you would find the area of the kite figure shown at the right.

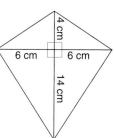
4 cm
6 cm   6 cm
14 cm

23. Draw one figure that is made of a parallelogram, a triangle, and a rhombus. Label the dimensions. Find the area of your figure. Have a classmate verify your answers.

24. LANDSCAPING A parks commissioner is reviewing plans for the center of a new park. It will consist of a circular fountain surrounded by four flowerbeds shaped like parallelograms, as shown. The commissioner needs to know the area and perimeter so the correct amount of topsoil and fence can be purchased. Find the area and perimeter of each flowerbed.

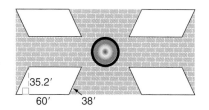

BUSINESS **Quintet's Quick Park is building a new parking lot near a post office. A diagram of the parking lot is shown.**

25. Find the area of each parking space.

26. Find the area of the handicap space.

27. Find the total area of the shaded triangles that are not used for parking.

28. Find the area of the driving lane that runs down the middle of the lot.

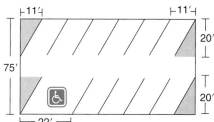

29. OPTICAL ILLUSIONS In the illusion shown at the right, you may view rhombus *TRIK* as the top of one rectangular solid or as the bottom of another. If one diagonal of the rhombus is twice the length of the other and the area is 128 square units, find the length of each diagonal.

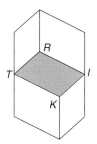

EXTEND

30. FLAGS The flag of the Congo features a yellow parallelogram and two right triangles. The flag of Trinadad and Tobago has a black parallelogram and two right triangles. Describe a transformation that maps one flag to the other. Do not consider color. Assume that the parallelograms, including the stripes, of the two flags are congruent.

CONGO

TRINIDAD & TOBAGO

31. SPORTS The baseball field shown below is to be covered with artificial turf. The entire field will be carpeted except for the *sliding boxes*—the dirt area around the bases. Divide the pentagonal-shaped sliding area into two rectangles and a triangle and find its area.

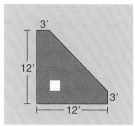

32. A triangle's height is 5 cm less than its base. Its area is 36 cm². Find the measure of the base and height.

33. One diagonal of a rhombus is 12 in. longer than the other. The area of the rhombus is 32 in.² Find the length of each diagonal.

34. On graph paper, draw line segment AB with $A(1, 1)$ and $B(8, 1)$. Plot $C(3, 5)$. Draw triangle ABC. Sketch five other triangles with base $\overline{AB}$ that have the same area as triangle ABC but are not congruent to it. Explain why all of your triangles have the same area. State the area.

35. HERON'S FORMULA Heron was a first-century Greek mathematician who discovered a formula for the area of a triangle using the lengths of the three sides.

$$A = \sqrt{s(s - a)(s - b)(s - c)}$$ where a, b, c are the lengths of the sides of the triangle and s is the semiperimeter (one-half the perimeter)

Use Heron's formula to find the area of the triangle whose sides have measures of 14, 13, and 15 in. Compare your results with that of Example 2.

THINK CRITICALLY

36. Parallelogram $EASY$ has vertices $E(1, 2)$, $A(10, 2)$, $S(13, 7)$, and $Y(4, 7)$. Parallelogram $TUFR$ has vertices $T(1, 1)$, $U(5, 4)$, $F(6, 12)$, and $R(2, 9)$. Explain why it is easier to find the area of parallelogram $EASY$ as compared to parallelogram $TUFR$.

37. Explain how you would use the formula for the area of a triangle to derive the formula for the area of a rhombus. *Hint:* Begin by focusing on one of the right triangles formed by the diagonals.

ALGEBRA AND GEOMETRY REVIEW

38. What is the equation of the line with slope -12 and y-intercept 6?

39. The measures of opposite angles of a parallelogram are represented by $(x + 12)°$ and $(2x - 4)°$. What is the measure of these angles?

40. Find the supplement of an angle of 120°.

Use *always*, *sometimes*, or *never* to make each sentence a true statement.

41. The diagonals of a rectangle are __?__ perpendicular.

42. A quadrilateral with perpendicular diagonals is __?__ a trapezoid.

Find the area of each figure.

43.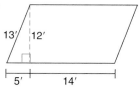

44.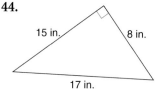

45.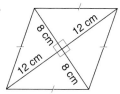

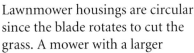

Career
Landscaper

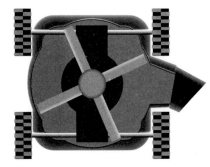

A part of a landscaper's business is mowing lawns. When choosing a new mower, landscapers must decide if it is appropriate for their particular business—if it will save time and money.

Lawnmower housings are circular since the blade rotates to cut the grass. A mower with a larger housing cuts a wider path of grass than a mower with a smaller housing. A wider path means less cutting time, but the larger mowers cost more to buy and more to operate. As lawns are cut faster, labor costs are reduced and more customers can be serviced.

Decision Making

A lawnmower with a 24-in. housing cuts a 24-in. path.

1. Draw a diagram of the area cut by the mower if it is walked 100 ft.

2. Describe the path of the cut from Question 1 in geometric terms. Explain how you could use transformations to efficiently find the area of the geometric figure.

The Ramos' lawn is a 120 ft by 60 ft rectangle that is bordered by cement walks on which the lawnmower is turned around.

3. A pattern of mowing this lawn uses 120-ft-long parallel paths. Compare cutting this lawn using a 24-in. mower to cutting it using a 36-in. mower.

4. If the landscaper can push the larger mower as fast as the smaller mower, what percent of time is saved by using the larger mower?

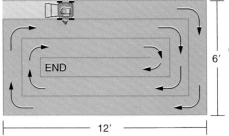

5. Suppose the pattern is changed to 60-ft-long parallel paths. How many more such paths would each mower need to make?

6. Vic uses the pattern shown at the left on a scale model of a 6 ft by 12 ft lawn with a 1-ft mower path to test a new cutting pattern. Find the length of each path Vic cuts. Then find the average length of each cut between turns. Decide if his pattern is less time consuming than six parallel 12-ft cuts. Explain your decision.

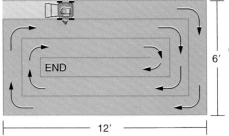

8.2 **Areas of Parallelograms and Triangles** **447**

Areas of Trapezoids

Explore

• You need two index cards, scissors, and tape.

Cut the two short sides of one index card to create a trapezoid, as shown. Label the vertices *TRAP* on the inside of the trapezoid. Label the height *h*, the shorter base b_1, and the longer base b_2.

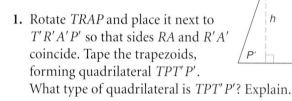

Trace *TRAP* onto a second card to create $T'R'A'P'$.

1. Rotate *TRAP* and place it next to $T'R'A'P'$ so that sides *RA* and $R'A'$ coincide. Tape the trapezoids, forming quadrilateral $TPT'P'$. What type of quadrilateral is $TPT'P'$? Explain.

2. What is the height of $TPT'P'$?

3. How long is a base of $TPT'P'$?

4. Express the area of $TPT'P'$.

5. How is the area of *TRAP* related to the area of $TPT'P'$? Write an expression for the area of *TRAP*.

Build Understanding

• In Explore, you used the area of a parallelogram to help find the area of a trapezoid. You may have noticed that you can find the area of a trapezoid if you know its height and the lengths of its two bases.

As with other polygons, the height of a trapezoid is the length of an altitude. An **altitude of a trapezoid** is any segment perpendicular to the line containing one base from any point on the opposite base.

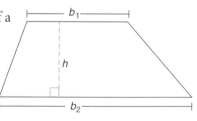

THEOREM 8.4   AREA OF A TRAPEZOID

The area *A* of a trapezoid equals half the product of its height *h* and the sum of its bases b_1 and b_2, or

$$A = \frac{1}{2}h(b_1 + b_2).$$

EXAMPLE 1

Find the area of the trapezoid.

Solution

The height h is 6 cm.
The bases are 10 cm and 16 cm.

$$A = \frac{1}{2} h(b_1 + b_2)$$

$$A = \frac{1}{2}(6)(10 + 16)$$

$$A = 3(26)$$

$$A = 78$$

The area of the trapezoid is 78 cm^2.

Carpenters often use a trapezoidal figure when working with moldings.

EXAMPLE 2

INTERIOR DESIGN Karen is a carpenter. She is going to cut wood moldings to frame the 30 in. by 80 in. door shown. The wood for the molding is 5 in. wide. Karen needs to cut it into three trapezoids. Since the face of each trapezoid will be painted, she needs to know the total area. Find the sum of the areas of the three trapezoids.

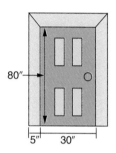

Solution

Find the areas of the two congruent trapezoids at the sides. For each of these trapezoids, the height is 5 in. and the shorter base is 80 in. The other base is 5 in. longer than the shorter base, or 85 in.

$$A = \frac{1}{2} h(b_1 + b_2)$$

$$A = \frac{1}{2}(5)(80 + 85) = (2.5)(165) = 412.5$$

The total area of the two side trapezoids is 825 in.^{2}

The trapezoid at the top has a height of 5 in. Its shorter base is 30 in. Its longer base extends 5 in. beyond the width of the door on each side, for a length of 40 in.

$$A = \frac{1}{2} h(b_1 + b_2)$$

$$A = \frac{1}{2}(5)(30 + 40) = (2.5)(70) = 175$$

The area of the trapezoid at the top is 175 in.^{2}

The sum of the areas of the three trapezoids is 1000 in.^{2}

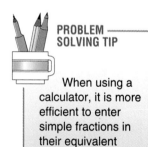

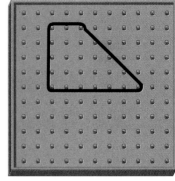

A *geoboard* has pegs at the vertices of congruent squares. The pattern formed by these vertices is called a **lattice**. You can find the area of a *geoboard polygon* or a **lattice polygon** using the number of **boundary points** and the number of **interior points**.

The lattice polygon shown has 16 boundary points and 9 interior points. You can count that there are 16 square units of area. The following theorem can also be used to find the area of a lattice polygon.

THEOREM 8.5   **PICK'S THEOREM**

If a polygon on a lattice has *B* boundary points and *I* interior points, then its area *A* is

$$A = \frac{1}{2}B + I - 1$$

Pick's Theorem is especially useful when you cannot conveniently count the number of square units in the area of a lattice polygon. You can use dot paper to simulate a lattice. In order to name the points, superimpose coordinate axes on the dot paper.

EXAMPLE 3

Find the area of trapezoid *PICK* formed by the points $P(1, 1)$, $I(2, 4)$, $C(4, 4)$, and $K(7, 1)$.

Solution

Use dot paper to plot points. Count the number of lattice points. Trapezoid *PICK* has 12 boundary points and 7 interior points.

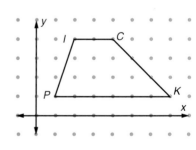

$A = \frac{1}{2}B + I - 1$ Pick's Theorem

$A = \frac{1}{2}(12) + 7 - 1$

$A = 12$ Substitue for *B* and *I*.

To verify this value for area, use Theorem 8.1. The height is 3 units, the shorter base is 2 units, and the longer base is 6 units.

$A = \frac{1}{2}h(b_1 + b_2)$

$A = \frac{1}{2}(3)(2 + 6) = (1.5)(8) = 12$

The area of trapezoid *PICK* is 12 square units.   ◄

Find the area of each trapezoid.

1.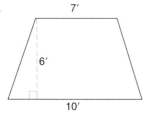

2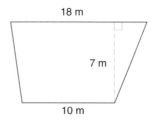

3.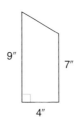

4. **AUTOMOBILE MAINTENANCE** Maple Car Supply manufactures car ramps that do-it-yourselfers can use to make car repairs easier. A trapezoidal plate is welded to the sides of a ramp so it can support heavier loads. Find the area of a plate that has the dimensions shown.

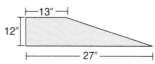

5. **PACKAGING** The Gauss Candy Company needs to design packaging for its health-food candy bar, the Algebar. The bar has four trapezoidal faces, each having a height of 1 cm, and two rectangular faces. Find the sum of the areas of the four trapezoidal faces of the Algebar.

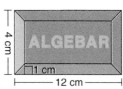

Plot each set of points on dot paper. Connect them in the given order to form a lattice polygon. Use Pick's Theorem to find the area.

6. $Z(2, 4)$, $O(0, 0)$, $I(9, 0)$, $D(8, 4)$

7. $A(2, 6)$, $X(4, 6)$, $I(6, 0)$, $O(0, 0)$, $M(0, 5)$

8. **WRITING MATHEMATICS** Describe another way to find the area of each lattice polygon of Questions 6 and 7. Use your method to find the areas and compare your results with those you obtained by using Pick's Theorem.

Find the indicated measure in each trapezoid.

9. The area of a trapezoid is 100 cm². The sum of the lengths of the bases is 50 cm. Find the height.

10. The longer base of a trapezoid is twice the length of the shorter base. The longer base measures 14 in. and the height is 4 in. Find the area.

PRACTICE

Find the area of each trapezoid.

11.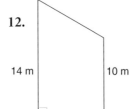

12.

13.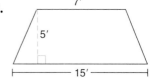

14. **CARPENTRY** A carpenter is using wood to frame a window. The framing is 4″ wide all around the window. The frame will be made up of four trapezoids. Find the sum of the areas of the four trapezoids.

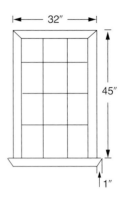

15. **MUSIC** Fibonacci's Dance Club has a stage in the shape of an isosceles trapezoid. The front measures 8 yd and the back measures 11 yd. The front and back are parallel. The depth of the stage from front to back is 4.5 yd. The stage is to be covered with carpet that costs \$34/yd² plus 6% sales tax. Find the total cost of the carpeting.

16. **WRITING MATHEMATICS** In trapezoid *ABCD*, the height is 8 cm. Base *AB* is 12 cm and base *CD* is 20 cm. Describe how you could find the area of *ABCD* by drawing diagonal $\overline{AC}$. State the area.

Plot each set of points on dot paper. Connect them in the given order to form a lattice polygon. Use Pick's Theorem to find the area.

17. $B(2, 0)$, $E(3, 2)$, $S(6, -2)$, $T(8, 0)$ 18. $F(2, 3)$, $A(4, 5)$, $C(4, -1)$, $E(2, -1)$

19. $T(2, 1)$, $R(6, 5)$, $Y(6, 1)$ 20. $E(-1, 1)$, $F(2, 5)$, $G(8, 5)$, $H(4, 1)$

Find the indicated measure in each trapezoid.

21. The area of a trapezoid is 420 m². The height is 12 m. One base is 20 m. Find the length of the other base.

22. The area of a trapezoid is 102 in.² The bases measure 14 in. and 20 in. Find the height.

EXTEND

23. One base of a trapezoid is 9 m longer than the other. The height is 8 m. The area is 88 m². Find the lengths of the bases.

24. Quadrilateral *RSTU* has vertices $R(2, 5)$, $S(5, 8)$, $T(9, 8)$ and $U(2, 1)$. Use coordinate geometry to prove that *RSTU* is an isosceles trapezoid.

PAPER FOLDING Cut off the two shorter sides of an index card to create a trapezoid. Fold the trapezoid such that the two bases lie on the same line as shown. This fold line is called the midsegment of the trapezoid. Label it *m*. Cut along the fold to create two trapezoids, I and II.

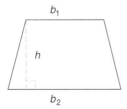
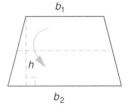
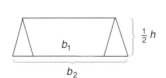

25. Write an expression for the area of the original trapezoid.

26. Write an expression the area of trapezoid I.

27. Write an expression the area of trapezoid II.

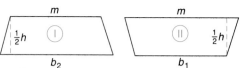

28. Use the area addition postulate to write an equation that relates the area of the original trapezoid to the areas of trapezoids I and II.

29. Use your equation to show that the length of the midsegment of a trapezoid equals half the sum of the bases.

30. Write a formula for the area of a trapezoid in terms of its height and midsegment.

THINK CRITICALLY

31. WRITING MATHEMATICS Carlos wondered if he could find the area of a triangle by viewing it as a trapezoid in which the shorter base measured 0. Explain whether or not he would get the correct area of the triangle. Use an example to illustrate your case.

32. GEOMETRIC CONSTRUCTION Given base b_1, construct an isosceles trapezoid with short base b_1, height $\frac{1}{2} b_1$, and longer base $2b_1$.

b_1 _____

ALGEBRA AND GEOMETRY REVIEW

33. In parallelogram $ABCD$, $m\angle B$ is 42° more than $m\angle C$. Find $m\angle D$.

34. Find the solutions of the equation $x^2 + 7x - 18 = 0$.

35. Is $\sqrt{36}$ rational or irrational?

36. Find the slope of a line that is perpendicular to the graph of $2y = 4x - 9$.

37. Find the perimeter of a square whose area is 49 cm^2.

38. Find the area of a trapezoid with height 9 cm and bases 17 cm and 19 cm.

PROJECT *Connection*   Divide the class into four groups.

1. Each group should design a commemorative garden for the school. Include a list of materials and a scale drawing. Make the gardens geometrically interesting. List the different polygons used in the design.

2. Each group should write a fact sheet about the garden. Include the area and outside perimeter. Also include the dimensions and areas of all the features of the garden, such as walkways, benches, decks, gazebos, signs, flowerbeds, and lawns.

3. As a class, design a flier announcing a school-wide poll in which students will pick their favorite garden from the four designs. Arrange the details of conducting the poll. Include the opportunity for students to decide what the garden should commemorate.

The Area Under A Curve

Think Back/Working Together

• Recall that the graph of the equation $y = x^2 + 1$ is a parabola.

Recall also that the coordinates of any point on a curve satisfy the equation of the curve. If you know one coordinate of a point on a curve, substitute in the equation to find the other coordinate.

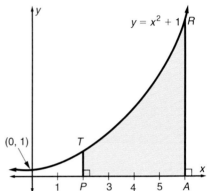

1. If T is on the parabola and $\overline{TP} \perp x$-axis with $P(2, 0)$, find the coordinates of T.

2. If R is on the parabola and $\overline{RA} \perp x$-axis with $A(6, 0)$, find the coordinates of R.

Explore

• 3. Explain why the outline of the shaded region shown in the graph at the right below is not a polygon.

You can approximate the area of the shaded region by trapezoid *TRAP*.

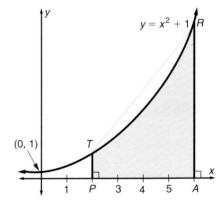

4. Which two sides of the trapezoid are the bases? What are their lengths?

5. Which segment in the graph is the altitude of the trapezoid? What is its length?

6. What is the area of trapezoid *TRAP*?

7. Is the area of trapezoid *TRAP* an over approximation or an under approximation of the area of the shaded region? Explain.

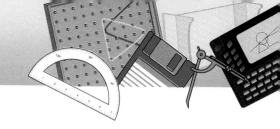

Make Connections

• You can improve the approximation of the area of the shaded region by dividing the area under the curve into several trapezoids.

8. Find the area of each of the four trapezoids I, II, III, IV.

9. Find the sum of the areas of the four trapezoids.

10. Does the sum of the areas of the four trapezoids represent an over approximation or an under approximation of the area of the shaded region?

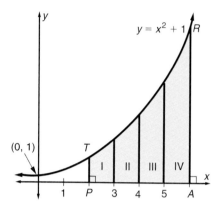

Suppose the area of the shaded region were divided into eight trapezoids, as shown at the right below.

11. Make a conjecture about the accuracy of the approximation as compared to when you used four trapezoids. Explain your reasoning.

12. Find the area of each of the eight trapezoids.

13. Find the sum of the areas of the eight trapezoids.

14. What could you do to improve the area approximation achieved by using the eight trapezoids?

15. What is a disadvantage of using too few trapezoids?

16. What is a disadvantage of using too many trapezoids?

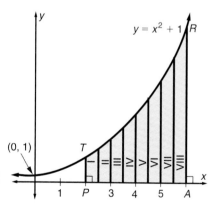

Geometry Workshop

- MODELING The algebraic steps that follow create a new formula for approximating the area under a curve. Refer to the graph below.

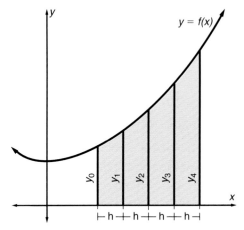

17. This step models the work you did in the case of four trapezoids. Explain what the equation represents.

$$A = \frac{1}{2}h(y_0 + y_1) + \frac{1}{2}h(y_1 + y_2) + \frac{1}{2}h(y_2 + y_3) + \frac{1}{2}h(y_3 + y_4)$$

18. Explain how the following equation was derived from the equation in Question 17.

$$A = h\left(\frac{1}{2}(y_0 + y_1) + \frac{1}{2}(y_1 + y_2) + \frac{1}{2}(y_2 + y_3) + \frac{1}{2}(y_3 + y_4)\right)$$

19. Explain how the following equation was derived from the equation in Question 18.

$$A = h\left(\frac{1}{2}y_0 + \frac{1}{2}y_1 + \frac{1}{2}y_1 + \frac{1}{2}y_2 + \frac{1}{2}y_2 + \frac{1}{2}y_3 + \frac{1}{2}y_3 + \frac{1}{2}y_4\right)$$

20. Explain how the following equation was derived from the equation in Question 19.

$$A = h\left(\frac{1}{2}y_0 + y_1 + y_2 + y_3 + \frac{1}{2}y_4\right)$$

The equation above leads to a general formula for finding the area under any curve in a given interval. Called the *Trapezoidal Rule*, the formula can be applied when you divide the given interval into any number of trapezoids of equal height h.

21. Write out the Trapezoidal Rule for eight trapezoids.

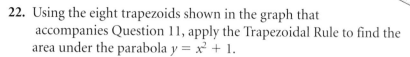

22. Using the eight trapezoids shown in the graph that accompanies Question 11, apply the Trapezoidal Rule to find the area under the parabola $y = x^2 + 1$.

23. Check that your answer agrees with the answer in Question 13.

24. If you were to use eight trapezoids to find the area under the parabola $y = x^2 + 1$ from $x = 3$ to $x = 19$, what would be the height of each trapezoid?

TECHNOLOGY TIP

Use the substitution feature of your graphing utility to compute the lengths of the bases of the trapezoids in Question 22.

25. Use the Trapezoidal Rule to approximate the area above the y-axis that is under the curve $y = 2x^2 - 4$ from $x = 2$ to $x = 5$. Use three trapezoids.

26. Use the Trapezoidal Rule to approximate the area above the y-axis that is under the curve $y = 2x^2 - 4$ from $x = 2$ to $x = 5$. Use six trapezoids.

27. Use the Trapezoidal Rule to approximate the area above the y-axis that is under the curve $y = x^2 - 7x + 12$ from $x = 5$ to $x = 7$. Use four trapezoids.

Summarize

28. THINKING CRITICALLY Why do you need to know the equation of the function to find the area under a curve?

29. MODELING Sketch a curve and shade a region under the curve for which the Trapezoidal Rule gives an under approximation of the area. Explain why the Trapezoidal Rule under approximates the area of your curve.

30. WRITING MATHEMATICS Write a paragraph explaining each step you should take when using the Trapezoidal Rule to find the area under a curve. Why is it easier than finding the area of each trapezoid separately?

31. GOING FURTHER Suppose you need to use the Trapezoidal Rule to compute an area from $x = a$ to $x = b$ using n trapezoids. Create a formula that can be used to find the height h of each trapezoid.

Geometry Workshop
Areas of Irregular Polygons

Think Back

● When a triangle, rectangle, or trapezoid is drawn in the coordinate plane, you may be able to find the lengths of horizontal or vertical line segments to determine the dimensions of the figure. Then you can apply the appropriate area formula.

1. On graph paper, plot the points $A(-3, 1)$, $B(8, 1)$, $C(4, 5)$, and $D(-1, 5)$. Connect the points in order to create trapezoid $ABCD$. Draw the altitude from D. What are the coordinates of the point at which this altitude intersects base $\overline{AB}$?

2. Find the height of trapezoid $ABCD$.

3. Find the length of base $\overline{AB}$.

4. Find the length of base $\overline{CD}$.

5. Find the area of trapezoid $ABCD$.

Explore

● 6. On the same set of axes on which you graphed trapezoid $ABCD$, plot points $E(-3, 5)$ and $F(8, 5)$. Draw $\overline{AE}$, $\overline{ED}$, $\overline{CF}$, and $\overline{FB}$.

7. Find the area of $\triangle AED$.

8. Find the area of $\triangle BCF$.

9. Find the area of rectangle $ABFE$.

10. Describe how you can use the areas of the two right triangles and the rectangle to find the area of trapezoid $ABCD$. Find the area.

11. Call the method in Question 10 the *subtraction of areas method*. Verify that your result is the same as your result in Question 5.

THINK BACK

Recall that the legs of a right triangle can represent its base and its height.

Make Connections

● 12. Draw a new set of coordinate axes. Plot the points $P(2, 2)$, $Q(7, 4)$, $R(10, 10)$, and $S(4, 12)$. Connect these points in order to form quadrilateral $PQRS$. On the same axes, plot $T(2, 12)$, $U(7, 2)$, $V(10, 2)$, $W(10, 4)$, and $Z(10, 12)$. Draw rectangle $PVZT$. Draw $\overline{QW}$ and $\overline{QU}$ to form right triangles QRW and PQU.

13. Find the areas of the four right triangles PTS, SZR, QRW, and PQU and the area of rectangle $QUVW$.

14. Find the area of rectangle *PTZV*.

15. Use the information from Questions 13 and 14 to find the area of irregular quadrilateral *PQRS*.

16. On a new graph, draw non-convex quadrilateral *JKLM* with vertices *J*(−1, 5), *K*(11, 5), *L*(2, 9), and *M*(−1, 16). On the same axes, plot *P*(11, 16). Draw $\overline{MP}$ and $\overline{KP}$. Divide the area that is outside quadrilateral *JKLM* but inside rectangle *JKPM* into two right triangles and a rectangle.

17. Find the area of the polygon that is outside quadrilateral *JKLM* but inside rectangle *JKPM*.

18. Find the area of quadrilateral *JKLM*.

19. On a new sheet of graph paper, draw parallelogram *ABCD* with vertices *A*(−3, 2), *B*(4, 2), *C*(6, 6), and *D*(−1, 6). Find the area of parallelogram *ABCD* by using the area formula for a parallelogram.

20. Find the area of parallelogram *ABCD* by using the subtraction of areas method. Explain your plan. Check your result with the answer to Question 19.

21. On a graph, draw △*PQR* with vertices *P*(2, −2), *Q*(5, 3), and *R*(11, −2). Find the area of △*PQR* by using the area formula for a triangle.

22. Find the area of △*PQR* by using the subtraction of areas method. Explain your plan. Check your result with the answer to Question 21.

23. On a new graph, draw △*RST* with vertices *R*(1, 3), *S*(9, 7), and *T*(4,10). Explain why you will not be able to easily use the formula for the area of a triangle to find the area of △*RST*.

24. Use the subtraction of areas method to find the area of △*RST*.

25. On a new graph, draw irregular pentagon *DEFGH* with vertices *D*(3, 3), *E*(8, 3), *F*(10, 6), *G*(5, 10), and *H*(2, 6). Use the subtraction of areas method to find the area of pentagon *DEFGH*. Explain your plan.

26. On a new graph, draw the non-convex hexagon *ABCDEF* with vertices *A*(−2, −4), *B*(2, −4), *C*(4, −2), *D*(5, 4),*E*(1, 1), and *F*(−2, 3). Use the subtraction of areas method to find the area of hexagon *ABCDEF*. Explain your plan.

THINK BACK

A polygon is non-convex when none of its diagonals lies in the interior of the polygon.

Summarize

27. **WRITING MATHEMATICS** Explain why you cannot use the subtraction of areas method to find the exact areas of figures with curved sides.

28. **THINKING CRITICALLY** How could you use the subtraction of areas method to find the area of an irregular polygon that was not plotted in the coordinate plane?

29. **GOING FURTHER** Graph the irregular quadrilateral *ABCD* with vertices *A*(2, 2), *B*(6, 1), *C*(8, 6), and *D*(2, 12).

30. Divide the inside of the quadrilateral into rectangles and right triangles. Use addition to find the area of *ABCD*. Explain your plan.

31. Verify the area of *ABCD* by the subtraction of areas method. Explain your plan.

32. Verify the area of *ABCD* by Pick's Theorem. Use a lattice diagram.

MODELING Jose is planning a vegetable garden in a curved "free form" design, as shown. To find the area of his garden, Jose created a polygon by using line segments to approximate the curved edges. Then he could use a grid to approximate the area.

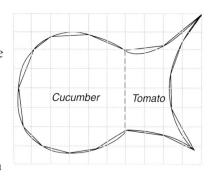

Cucumber   Tomato

33. By using the polygon, in which part of the garden will Jose under estimate the area?

34. By using the polygon, in which part of the garden will Jose overestimate the area?

35. **GOING FURTHER** Graph the following system of inequalities and find the area of the quadrilateral formed by the solution.

$$\begin{cases} x \geq 0 \\ y \geq 0 \\ y \leq -2x + 10 \\ y \geq -x + 3 \end{cases}$$

Career
Pool Designer

When a pool is included in landscape plans, the landscaper must work in conjunction with a pool designer. Their construction schedules must be coordinated. Trees and shrubs must be selected carefully since some do not thrive near pools and others deposit leaves and debris in the pool. Most of the landscaping cannot be started until the construction of the pool is completed.

Strict government regulations on swimming pools must be followed. Some of these regulations involve requirements for a fence around the pool, specific pool dimensions when a diving board is installed, and dimensions of walkways surrounding the pool.

Decision Making

The diagram below shows the minimum dimensions for a pool with a 6-ft diving board that is 20 in. above the surface of the water.

1. What type of polygon describes the shape of this pool?

2. Find the area of the surface of the pool to the nearest square foot.

3. A rope is required to separate the swimming region from the pentagonal diving region. Find the area of the diving region at water level. Round to the nearest square foot.

4. A regulation requires a walkway to have a minimum width of 4 ft. The landscaper is to put a rectangular fence around this pool. Draw and label a diagram that shows the smallest rectangle that could include the walkway and the pool.

5. Find the perimeter and area of the rectangle in your diagram that encloses the pool and the walkway.

6. Find the area of the walkway if it covers all of the area between the pool and the fence.

7. Discuss why you think there are regulations about fences, walkways, and pool depths.

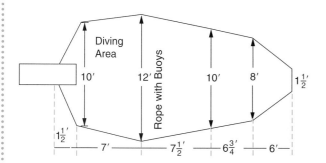

Using Probability to Find Areas

You have learned methods for finding areas and approximating the area under a curve. Now you will explore a way to find the area of a plane figure that has curved borders.

Problem

From his front lawn, Hector is planning to sculpt out a region for a rock garden, as shown.

What is the area of the region, which is of free-form shape?

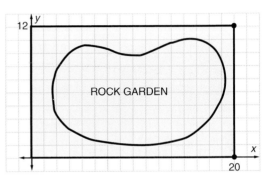

Explore the Problem

Focus first on a simpler problem with smaller dimensions, where the shape inside the rectangle is a polygon.

The graph shows a rectangle with a triangle in its interior.

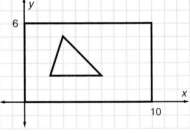

1. Use the graph and the area formulas for a rectangle and a triangle to find the area of each of the two figures.

2. What fractional part of the area of the rectangle is represented by the triangle?

3. Describe the range of the coordinates of all points (x, y) that are inside the rectangle.

Suppose you were to randomly select one of the points in the rectangle.

4. **WRITING MATHEMATICS** Which is greater, the probability that the point is in the triangle or not in the triangle? Why?

5. What is the probability that the point is inside the triangle?

Suppose a new triangle were drawn inside the rectangle. You randomly select one of the the points in the rectangle. You know that the probability that this point is inside the new triangle is $\frac{1}{9}$.

6. **WRITING MATHEMATICS** Explain how you could find the area of the new triangle. Find this area.

Investigate Further

The technique of using probabilities to compute areas is often called the *Monte Carlo method.*

7. Hector used a calculator to generate random points (x, y) that are inside the rectangle of his lawn. Use inequalities to describe the range of the coordinates of these points. Explain the meaning of using closed intervals.

Hector had generated 400 random points that were in the rectangle of the lawn. He found that 174 of these points were inside the free-form region for the rock garden and 226 were outside of it.

8. Use these results to state the fraction of the entire rectangular lawn that is inside the free-form region.

9. Use the graph to find the area of the rectangular lawn.

10. Use your answers from Questions 8 and 9 to compute the area of the free-form region to the nearest square foot.

11. **WRITING MATHEMATICS** Would your value for the area of the free-form region be more accurate or less accurate if the data had been based on 1000 points instead of 400? What if there had been 10 points instead of 400? Explain.

In the diagram, $\triangle ABC$ is framed by square *OPQR*.

12. Find the area of $\triangle ABC$.

13. Apply the Monte Carlo method to find the area of $\triangle ABC$.

    a. Use a graphing utility to generate 200 random points (x, y) with $0 \le x \le 15$ and $0 \le y \le 15$.

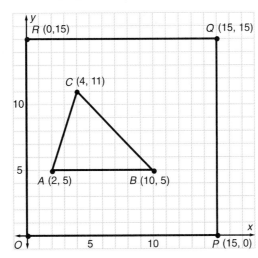

    b. Examine each of the 200 points to determine where they lie with respect to the square and the triangle. Keep a tally chart.

    c. Use your data to compute the area of $\triangle ABC$.

    d. Compare this result with your answer to Question 12.

PROBLEM SOLVING PLAN

- Understand
- Plan
- Solve
- Examine

GEOMETRY: WHO, WHERE, WHEN

The first means of generating random numbers used equipment similar to that in gambling casinos. So methods that use random numbers to simulate events are named after the Monte Carlo casino in the European principality of Monaco.

TECHNOLOGY TIP

Check the probability menu of your calculator to find the random number generator.

PROBLEM SOLVING TIP

Group your tally marks in sets of 5 so that you can count more efficiently.

Apply the Strategy

14. **ECOLOGY** An oil tanker involved in an accident leaked oil. To estimate the area of the spill, a technician drew a scale diagram, framing the region of the spill by a 40 km by 17 km rectangle. The technician then used a computer to generate 2000 random points. Of these, she found that 1234 points landed inside the spill region. What is the area of the oil spill to the nearest square yard?

15. **LANDSCAPING** Using a scale model of a golf course, landscape designer Al Koren intended to find the area of the golf greens. Al began by framing the scale drawing of the green for the first hole with a rectangle. Then he used a computer to generate 360 random points. Of these points, Al found that 225 landed in the green. Using the Monte Carlo method, he found the area of the first green to be 1800 square feet. What was the area of the framing rectangle he used?

16. **DESIGN** Part of a design to be screen printed on a t-shirt consists of a quarter circle situated in a rectangle.

    a. Work with a partner. On graph paper draw a quarter circle in Quadrant I with radius 10 units and center at $(0, 0)$.

    b. Frame the quarter circle by a 10 by 10 rectangle $ABCD$ with vertices $A(0, 0)$, $B(10, 0)$, $C(10, 10)$, and $D(0, 10)$.

    c. Use a graphing utility to generate random points to fall inside $ABCD$. Use the Monte Carlo method to find the area of the quarter circle.

    d. Verify your result by using the formula for the area of a circle, $A = \pi r^2$, to find the area of the quarter circle to the nearest square unit.

17. **CALCULUS** If you study calculus, you will learn to find the exact area between the graph of a function and the x-axis.

    a. Work with a partner. On graph paper, draw $y = x^2 - 10x + 9$ by plotting points for all x-values in the interval from 1 through 9. Frame the parabola with a 20 by 20 rectangle $ABCD$ with vertices $A(0, 0)$, $B(0, -20)$, $C(20, -20)$, and $D(20, 0)$.

    b. Use a graphing utility to generate 200 random points to fall in $ABCD$. Test these points to determine whether they fall inside or outside of the parabola in the interval $[1, 9]$.

    c. Use the Monte Carlo method to find the area between the parabola and the x-axis in the interval $[1, 9]$.

    d. Verify your result by applying the Trapezoidal Rule to find the area.

GEOMETRY: WHO, WHERE, WHEN

The early foundations of calculus began with the ancient Greek Archimedes. In the 1600's, mathematicians throughout Europe such as Newton and Leibnitz expanded these early ideas and developed fundamental concepts of calculus.

REVIEW PROBLEM SOLVING STRATEGIES

THE CASE OF THE MISSING AREA

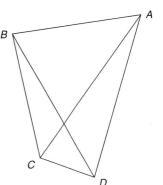

1. Sherlock Bones has a mystery to solve involving a missing area. He knows the following areas in the quadrilateral below: the area of $\triangle ACD$ is 7, the area of $\triangle BCD$ is 11, and the area of $\triangle ABD$ is 12. Sherlock needs to find the missing area of $\triangle ABC$. See if you can solve the mystery of the missing area.

2. Rose MacDonald has a trapezoid-shaped rose garden. The dimensions of the garden are shown in the diagram. Find the area of the trapezoid.

Rose's Roses

The Spider's Journey

3. The rectangular-shaped room shown here is 24 ft long, 8 ft wide, and 8 ft high. There is a spider at point *A* which is on an end wall, halfway between the side walls and 1 ft up from the floor. There is a fly at point *B* which is on the opposite end wall, halfway between the side walls and 1 ft from the ceiling. If the spider walks to the fly what is the shortest possible path? Note: the answer is *not* 32 ft.

8 ft

8 ft

24 ft

Explore

1. Create a right triangle by cutting an index card along a diagonal. On both faces of the triangle, label the legs a and b, and the hypotenuse c. Express the area of the triangle in terms of a and b.

2. On a sheet of paper, trace and transform your triangle four times to create the picture shown at the right. Express the area of $EGIK$ in terms of a and b.

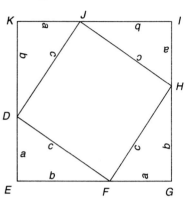

3. Find the total area of the four right triangles.

4. Use subtraction to express the area of square $DFHJ$ in terms of a and b.

5. Express the area of square $DFHJ$ in terms of c.

6. Write an equation using the two expressions for the area of $DFHJ$.

Build Understanding

The activity in Explore is one of many proofs of the Pythagorean Theorem. The area diagram you created dates back to ancient China.

> **THEOREM 8.6   PYTHAGOREAN THEOREM**
>
> If a triangle is a right triangle with legs of length a and b and hypotenuse of length c, then $a^2 + b^2 = c^2$.

EXAMPLE 1

Find the length of the hypotenuse of the right triangle whose legs measure 5 m and 12 m.

Solution

$$a^2 + b^2 = c^2 \quad \text{Pythagorean Theorem}$$
$$5^2 + 12^2 = c^2 \quad \text{Substitute for } a \text{ and } b.$$
$$169 = c^2$$
$$13 = c$$

The hypotenuse measures 13 m. ◄

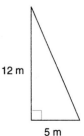

12 m

5 m

The Pythagorean Theorem can be used to help you solve many different problems.

EXAMPLE 2

HOME IMPROVEMENT Oscar is planning to paint his house. For stability, the foot of his 17-ft-long ladder should be placed 4 ft from the house. How far up the side of the house will the ladder reach?

Solution

Draw a diagram. Label the length of the ladder and the ground distance between the ladder's foot and the house.

The unknown side of the triangle is one of the legs. Use the Pythagorean Theorem to find its length.

$$a^2 + b^2 = c^2$$
$$4^2 + b^2 = 17^2$$
$$16 + b^2 = 289$$
$$b^2 = 273$$
$$b = \sqrt{273}$$
$$b \approx 16.5$$

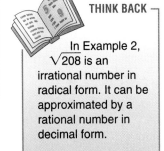

17 ft   b

4 ft

The ladder will reach about 16.5 ft up the side of the house.

◄

THINK BACK

In Example 2, $\sqrt{208}$ is an irrational number in radical form. It can be approximated by a rational number in decimal form.

You can use the Pythagorean Theorem to derive the formula for finding the distance d between any two points $A(x_1, y_1)$ and $B(x_2, y_2)$.

In the diagram, A and B are used to create right triangle ABC with $C(x_2, y_1)$.

The length of $\overline{AC}$ is $|x_2 - x_1|$.

The length of $\overline{BC}$ is $|y_2 - y_1|$.

Apply the Pythagorean Theorem.

$$AC^2 + BC^2 = AB^2$$
$$(|x_2 - x_1|)^2 + (|y_2 - y_1|)^2 = d^2$$
$$(x_2 - x_1)^2 + (y_2 - y_1)^2 = d^2$$

To find d, take the square root of both sides of the equation.

$$d = \sqrt{(x_2 - x_1)^2 + (y_2 - y_1)^2}$$

The result is the distance formula.

◄

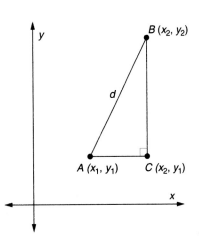

y

B (x_2, y_2)

d

A (x_1, y_1) C (x_2, y_1)

x

GEOMETRY: WHO, WHERE, WHEN

The scarecrow in the classic movie *The Wizard of Oz* tried to recite the Pythagorean Theorem as proof of intelligence when he received his "brains." But he stated it incorrectly.

An **altitude of a prism** is a segment perpendicular to the bases with one endpoint on each base. The length of an altitude is the *height of the prism*.

A *diagonal* of a rectangular prism is a segment whose endpoints are opposite vertices. If you know the height of a rectangular prism and the length and width of its bases, you can find the length of a diagonal by using the Pythagorean Theorem twice.

EXAMPLE 3

Find the length of diagonal $\overline{BH}$ in the rectangular prism shown.

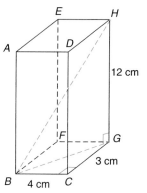

Solution

Diagonal $\overline{BH}$ is the hypotenuse of right triangle BGH. You know that the length of leg $\overline{HG}$ is 12 cm. You can find the length of leg $\overline{BG}$ by applying the Pythagorean theorem to right triangle BGC.

$$CG^2 + BC^2 = BG^2$$
$$3^2 + 4^2 = BG^2$$
$$25 = BG^2$$
$$5 = BG$$

Use the lengths HG and BG to find the length of hypotenuse $\overline{BH}$.

$$HG^2 + BG^2 = BH^2$$
$$12^2 + 5^2 = BH^2$$
$$169 = BH^2$$
$$13 = BH$$

So the length of diagonal $\overline{BH}$ is 13 cm. ◀

The **altitude of a pyramid** is the segment from the vertex perpendicular to the base. The length of the altitude is the *height of the pyramid*. For a regular pyramid, there also is a **slant height**, which is the height of a lateral face.

EXAMPLE 4

In the regular hexagonal pyramid shown, the height VP is 12 cm and $RP = 6$ cm. Find the slant height VR.

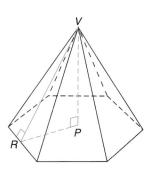

Solution

$$RP^2 + VP^2 = VR^2$$
$$6^2 + 12^2 = VR^2$$
$$180 = VR^2$$

The slant height is $\sqrt{180}$, or about 13.4 cm. ◀

Find the length of the unknown side of each right triangle. Then find the perimeter and area. Round to the nearest tenth.

1.
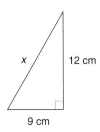
x   12 cm

9 cm

2.   5 in.
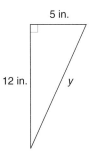
12 in.   y

3.
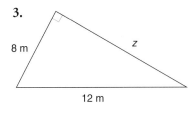
8 m   z

12 m

4. **WRITING MATHEMATICS** When using the distance formula, is it important which of the two points you call (x_1, y_1)? Explain.

Find the length of the segment connecting each pair of points. Round to the nearest tenth.

5. $(1, 1), (7, 9)$

6. $(1, -3), (9, 15)$

7. $(-2, 4), (5, -8)$

8. **HEALTH AND FITNESS** A rectangular pool for swimming laps has length 48 ft and width 14 ft. A life preserver must be located nearby and have a rope at least as long as the diagonal of the pool. Find the length of the diagonal.

Find the length of the unknown measure(s) in each solid figure.

9.
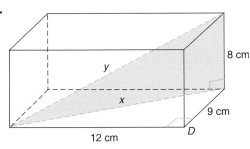
8 cm
y
x
9 cm
12 cm   D

10.
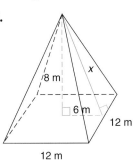
8 m   x
6 m   12 m
12 m

Find the length of the unknown side of each right triangle. Then find the perimeter and area. Round to the nearest tenth.

11.
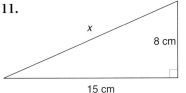
x   8 cm

15 cm

12.
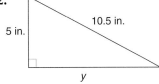
10.5 in.
5 in.
y

13.
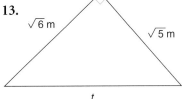
$\sqrt{6}$ m $\sqrt{5}$ m
t

14. **WRITING MATHEMATICS** Explain how you could find the perimeter of a rhombus whose diagonals measure 12 m and 16 m. Find the perimeter.

Find the length of the segment connecting each pair of points. Round to the nearest tenth.

15. $(4, 1), (2, 9)$ **16.** $(-2, -3), (-8, 15)$ **17.** $(4, -4), (4, 12)$

18. SPORTS A football field has length 100 yd, with an extra 10 yd at each end for the end zones. The width of the field is 45 yd. Find the length of the diagonal of the field. Round to the nearest yard.

Find the length of the unknown measure(s) in each solid figure.

19.

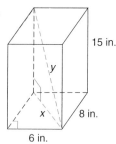

15 in.

y

x

8 in.

6 in.

20.

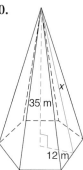

35 m

x

12 m

21. PYRAMIDS The Great Pyramid of King Khufu at Giza in Egypt has a square base with sides of length 230 m. The height of the pyramid is 147 m. Find the slant height *s* of the pyramid, to the nearest meter.

EXTEND

A **pythagorean triple** is a set of three positive integers *a*, *b*, and *c* that satisfy the equation $a^2 + b^2 = c^2$. If *p* and *q* are positive integers with $p > q$, you can find the legs *a* and *b* and the hypotenuse *c* of a right triangle by substituting into the formulas below.

$$a = p^2 - q^2 \qquad b = 2pq \qquad c = p^2 + q^2$$

22. What triple is formed when $p = 5$ and $q = 3$? Show that the numbers you found satisfy the equation $a^2 + b^2 = c^2$.

23. Pythagorean triples that have no common factor are called *primitive* triples.
Do $p = 5$ and $q = 2$ form a primitive Pythagorean triple? Explain.

24. Find two pairs of primitive Pythagorean triples.

You can use the Fibonacci sequence 1, 1, 2, 3, 5, 8, 13, 21, . . . to generate Pythagorean triples.

25. Pick four consecutive Fibonacci numbers. Show that the numbers you find by applying the following form a Pythagorean triple.

Let leg *a* be the product of the first and fourth numbers.
Let leg *b* be twice the product of the middle two numbers.
Let the hypotenuse *c* be the sum of the squares of the middle two numbers.

26. Show that the area of the right triangle formed by the numbers you found in Exercise 25 is equal to the product of the original four Fibonacci numbers you used.

THINK CRITICALLY

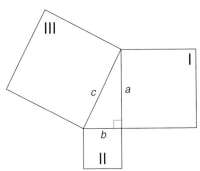

27. WRITING MATHEMATICS Explain the relationship among the areas of squares I, II and III in the diagram at the right.

Use the Pythagorean Theorem for Exercises 28–29.

28. Find the area of the isosceles trapezoid shown below.

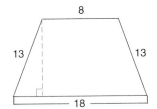

29. Derive a formula for the area of an equilateral triangle in terms of its side s.

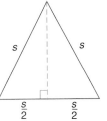

30. GEOMETRIC CONSTRUCTION You can use the Pythagorean Theorem to construct an irrational spiral, as shown.

Begin with a line segment of length 1 unit. Use a compass and straightedge to construct an irrational spiral up to a segment with length $\sqrt{5}$.

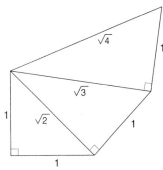

ALGEBRA AND GEOMETRY REVIEW

31. The measures of the angles of $\triangle ABC$ are in the ratio 2:3:5. Find the number of degrees in the smallest angle.

32. STANDARDIZED TESTS Which equation represents the graph of a horizontal line?

   A. $x = y$ **B.** $y = 7$ **C.** $x = -2$ **D.** $y = x + 1$

33. Find the slope of a line that is perpendicular to the graph of $2x + 5y = 17$.

34. List the capital letters of the English alphabet that have vertical symmetry.

35. STANDARDIZED TESTS Which of the following irrational numbers lie between 9 and 11?

   A. $\sqrt{65}$ **B.** $\sqrt{80}$ **C.** $\sqrt{89}$ **D.** $\sqrt{125}$

36. Find the perimeter of the right triangle in which the length of the hypotenuse is 34 cm and the length of one leg is 16 cm.

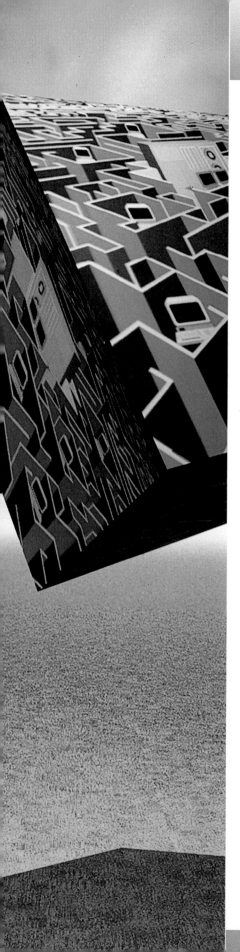

The Fourth Dimension

One day, Katrina and Zachery noticed pictures of points, lines, polygons, and solids on the chalkboard. Although the pictures were flat, they saw features that differentiate one shape from another. Some objects have length and breadth. Other have length, breadth, and depth.

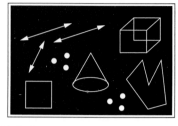

From their observations, they made the table below.

	Length	Breadth	Depth
Point	✗	✗	✗
Line	✓	✗	✗
Polygon	✓	✓	✗
Solid	✓	✓	✓

1. Sketch a plane and two points on it. Join them to make a segment. Place a third point on the plane not collinear with the first two. Join them to form a triangle. Place a fourth point not coplanar with the three points. Join the four points to illustrate a pyramid.

2. **WRITING MATHEMATICS** Using your sketches from Question 1, describe how to build up three-dimensional space from a point and moving through one- and two-dimensional space.

3. Copy and complete the table below.

Dimensional Analysis of a Cube		
Number of Faces	Shape of Face	Dimension of Face
Number of Edges	Shape of Edge	Dimension of Edge
Number of Vertices	Shape of Vertex	Dimension of Vertex

Four-dimensional space is space one dimension beyond length, width, and depth. The *hypercube* is the four-dimensional analog of the three-dimensional cube. You can model a hypercube by starting with a three-dimensional cube and then placing six cubes on its faces. These cubes share a square as an edge.

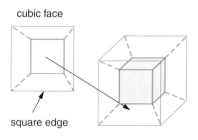

cubic face

square edge

4. MODELING Make your own model of a hypercube.

5. THINKING CRITICALLY How many vertices does one face of a hypercube have? How many edges does that face have?

Suppose that you have an ant farm, a thin rectangular solid filled with sand or dirt. In the farm, the ants can move left, right, up, and down. They may not, however move from side to side as the ant farm is not wide enough.

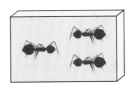

6. WRITING MATHEMATICS Contrast the freedom the ants in the farm have with the freedom that tiny dots living in a one-dimensional world do not have. Contrast the freedom you have in a three-dimensional world with the freedom the ants do not have.

Just as you can use coordinates to study geometry in one-, two-, and three-dimensional space, you can represent a point in four-dimensional space as an ordered quadruple of real numbers (x, y, z, t).

When you organize the formulas for the distance d between two points in one-, two-, and three-dimensional space, you get this list.

$P(x_1), Q(x_2)$ $\qquad\qquad d = \sqrt{(x_2 - x_1)^2}$

$P(x_1, y_1), Q(x_2, y_2)$ $\qquad d = \sqrt{(x_2 - x_1)^2 + (y_2 - y_1)^2}$

$P(x_1, y_1, z_1), Q(x_2, y_2, z_2)$ $\quad d = \sqrt{(x_2 - x_1)^2 + (y_2 - y_1)^2 + (z_2 - z_1)^2}$

7. THINKING CRITICALLY What formula do you think will give the distance d between $P(x_1, y_1, z_1, t_1)$ and $Q(x_2, y_2, z_2, t_2)$?

8. A *hypersphere* is the set of all points in four-dimensional space a fixed distance from a fixed point. Show that each point in the list below is on the hypersphere with center $O(0, 0, 0, 0)$ and radius 1.

$A(1, 0, 0, 0)$ $\qquad B(0, 1, 0, 0)$ $\qquad C(0, 0, 1, 0)$
$D(-1, 0, 0, 0)$ $\qquad E(0, -1, 0, 0)$ $\qquad F(0, 0, -1, 0)$

9. Neglecting the fourth coordinate, points A through F lie on the three-dimensional sphere with equation $x^2 + y^2 + z^2 = 1$. Do you think a diameter of a hypersphere is a three-dimensional sphere? Explain.

A *hyperplane* is the four-dimensional analog of a three-dimensional plane. An equation of the form $ax + by + cz + dt = e$, with a, b, c, d, and e real numbers not all 0, is an equation of a hyperplane.

$$\begin{cases} x + 3y - 2z + 2t = 0 \\ y + z - t = 0 \\ -3z + 7t = 1 \\ t = 1 \end{cases}$$

10. Use $t = 1$ and the third equation in the list to find z.

11. Use similar reasoning to find y then x. Give the coordinates of the point in four-dimensional space where the four hyperplanes meet.

8.8 The Converse of the Pythagorean Theorem

Explore/Working Together

● Work with a partner. Examine the 12 triangles below. In each triangle, the two shorter sides are *a* and *b* and the longest side is *c*. Each triangle is drawn so that side *a* is opposite $\angle A$, *b* is opposite $\angle B$, and *c* is opposite $\angle C$, which is the largest angle in each triangle.

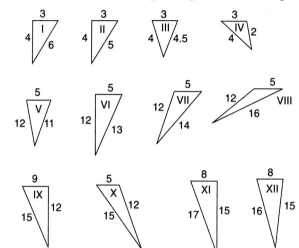

1. Copy and complete the following table for the 12 triangles.

△	a	b	$a^2 + b^2$	c	c^2	Is △ABC right?
I	3	4	25	6	36	no
II						

2. Make a conjecture about the relationship between $a^2 + b^2$ and c^2 and whether or not a right triangle is formed.

Build Understanding

● In Explore you used the converse of the Pythagorean Theorem to determine if each △ABC is a right triangle.

THEOREM 8.7   CONVERSE OF PYTHAGOREAN THEOREM

If *a*, *b*, and *c* are the lengths of the sides of a triangle such that $a^2 + b^2 = c^2$, then the triangle is a right triangle.

You can prove the converse by applying the SSS Congruence Postulate.

Given $\triangle ABC$ with $a^2 + b^2 = c^2$

Prove $\triangle ABC$ is a right triangle.

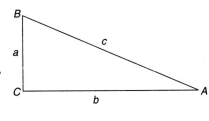

Proof Construct another triangle, $\triangle DEF$, so that $\triangle DEF$ is a right triangle with legs a and b and hypotenuse h.

Since $\triangle DEF$ is a right triangle, you know that $a^2 + b^2 = h^2$.

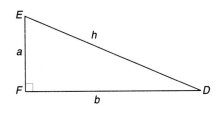

Using the information given and substitution, you know $c^2 = h^2$. Since c and h are both positive, $c = h$.

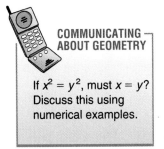

By the SSS Congruence Postulate, $\triangle ABC$ is congruent to $\triangle DEF$. Since corresponding parts of congruent triangles are congruent, $\angle C \cong \angle F$.

Since $\angle F$ is a right angle, $\angle C$ is a right angle. So $\triangle ABC$ is a right triangle. ◄

EXAMPLE 1

Determine if $\triangle ABC$ with sides $a = 5$ cm, $b = 12$ cm, and $c = 15$ cm is a right triangle.

Solution

First check that the lengths of the three sides satisfy the triangle inequality. No triangle is formed if they do not.

$$5 + 12 \overset{?}{>} 15 \qquad 5 + 15 \overset{?}{>} 12 \qquad 12 + 15 \overset{?}{>} 5$$

$$17 > 15 \text{ True} \qquad 20 > 12 \text{ True} \qquad 27 > 5 \text{ True}$$

The given lengths do satisfy the triangle inequality.

To use the converse of the Pythagorean Theorem, you need to compare the values of a, b, and c.

$$a^2 + b^2 = c^2$$

$$5^2 + 12^2 \overset{?}{=} 15^2 \quad \text{Substitute for } a, b, \text{ and } c.$$

$$25 + 144 \overset{?}{=} 225$$

$$169 \neq 225$$

So $\triangle ABC$ is not a right triangle. ◄

In many real world situations, such as construction, it is important to know if an angle is a right angle.

EXAMPLE 2

CONSTRUCTION Often builders assemble new stud walls on the floor. They then lift up the finished stud wall and nail it in place. Determine if the stud wall in the diagram has right-angled corners.

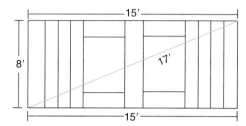

Solution

Since $8^2 + 15^2 = 17^2$, the stud wall has right-angled corners. ◀

In Explore you may have conjectured that the relationship between $a^2 + b^2$ and c^2 can tell you whether a triangle is acute or obtuse. The following theorems can be used to classify triangles as acute or obtuse.

THEOREM 8.8

If the square of the length of the longest side of a triangle is greater than the sum of the squares of the lengths of the other two sides, then the triangle is an obtuse triangle.

THEOREM 8.9

If the square of the length of the longest side of a triangle is less than the sum of the squares of the lengths of the other two sides, then the triangle is an acute triangle.

THINK BACK

In an *acute* triangle, all of the angles are acute. Each angle measures less than 90°.

In an *obtuse* triangle, the largest angle is obtuse. It measures more than 90° but less than 180°.

EXAMPLE 3

The measures of the sides of a triangle are 6 in., 8 in., and 9 in. Determine if the triangle is right, obtuse, or acute.

Solution

$$6^2 + 8^2 \stackrel{?}{=} 9^2 \qquad \text{Substitute values into } a^2 + b^2 = c^2.$$

$$100 > 81$$

Since $a^2 + b^2 > c^2$, the triangle is an acute triangle. ◀

Determine if the measures given form a triangle. If a triangle is formed, determine whether or not it is a right triangle.

1. 5, 7, 9   **2.** 5, 12, 13   **3.** 9, 3, 4   **4.** 4, 4, 4   **5.** 12, 9, 15

6. WRITING MATHEMATICS Explain how the relative sizes of the angles of a triangle are related to the relative sizes of its sides.

The measures given form a triangle. Classify each triangle as right, acute, or obtuse.

7. 8, 8, 8   **8.** 4.5, 6, 9   **9.** 1, 2, $\sqrt{3}$ **10.** 5, 12, 12   **11.** 5, 12, 14

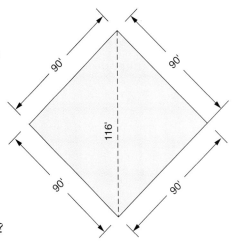

SPORTS Some members of the Van Buren High School girls softball team are using stakes and string to lay out a practice infield. They measured out four 90-ft baselines at what appear to be right angles, as shown at the right.

12. The girls found the measure of a diagonal to be 116 ft. Are the baselines at right angles? Explain.

13. What should the diagonal measure for the baselines to be at a right angle? Round to the nearest tenth.

14. Should the other diagonal measure more or less than your answer to Question 13?

15. What type of angle is formed at first base on the field shown?

PRACTICE

Determine if the measures given form a triangle. If a triangle is formed, determine whether or not it is a right triangle.

16. 1, 1, 2   **17.** 1, $\sqrt{3}$, 2   **18.** 9, 41, 40   **19.** 11, $\sqrt{3}$, 13   **20.** 8, 15, 18

The measures given form a triangle. Classify each triangle as right, acute, or obtuse.

21. 7, 25, 24   **22.** 2, 3.1, 4   **23.** 6, 8, 10.1   **24.** 6, 8, 9.9   **25.** 3, 4, 4.99

26. WRITING MATHEMATICS Discuss which triangles, acute, right, or obtuse, can be isosceles or equilateral.

27. ART Athleen is framing a picture and needs to cut out a rectangular piece of glass. The sides must be 11 in. by 14 in. She places an 11-in. long piece of masking tape at roughly a right angle to another piece of tape that is 14 in. long. The diagonal measures just over 18 in. Should she move the pieces of tape closer together or farther apart to achieve the right angle?

28. LANDSCAPING To strengthen a sagging gate, Max runs a metal wire down one diagonal between two horizontal sections of wood, as shown. How long should the wire be to make sure the gate's corners are right angles? Round to the nearest tenth.

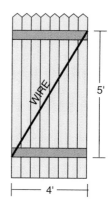

EXTEND

Use the distance formula and any theorem from this lesson for Exercises 29–32.

29. The vertices of $\triangle ABC$ are $A(2, 14)$, $B(12, 12)$, and $C(7, 8)$. Determine if $\triangle ABC$ is a right triangle.

30. The vertices of $\triangle EFG$ are $E(0, 9)$, $F(4, 8)$, and $G(1, 6)$. Determine if $\triangle EFG$ is acute, obtuse, or right.

31. Determine if $\angle ACD$ in the solid is a right angle.

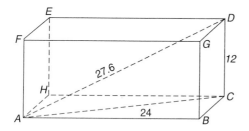

32. Determine if the quadrilateral shown at the right is a rhombus.

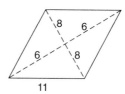

33. GEOMETRIC CONSTRUCTION Construct a triangle whose sides measure 8 cm, 15 cm, and 17 cm. Use a protractor to find the measure of the largest angle in the triangle.

If the sum of the lengths of two sides of a triangle is equal to the length of the third side, the triangle is called a *degenerate* triangle. Such a triangle appears as a line segment.

34. GEOMETRIC CONSTRUCTION Use a compass and straightedge to show that segments that have length 3 in., 4 in., and 7 in. can only form a degenerate triangle.

THINK CRITICALLY

35. Look for a pattern and guess and check to determine whether the following sentence should be completed with the word *always*, *sometimes*, or *never*.

A triangle with side lengths $\sqrt{a}$, $\sqrt{b}$, and $\sqrt{a + b}$ is __?__ a right triangle.

36. How many different right triangles can have a hypotenuse of length 20 cm?

37. Writing Mathematics Bruce framed out a stud wall. He measured both pairs of opposite sides and found that opposite sides were congruent. He then measured both diagonals and found that they were congruent. Explain why it is valid for Bruce to conclude that the four corners of this stud wall are right angles.

Manipulatives Cut out 7 strips of cardboard, each measuring 2 cm by 25 cm. Use a hole puncher to punch a hole near both ends of each strip. With 3 of the strips, form a triangle using a paper fastener in the hole at each vertex. With the other 4 strips, form a quadrilateral using a paper fastener at each vertex.

38. Show that you can transform the quadrilateral by changing its angles.

39. Can you also transform the triangle by changing its angles?

40. Discuss how your findings in Exercises 38–39 relate to your explanation in Exercise 37.

ALGEBRA AND GEOMETRY REVIEW

41. Find the mean, median and mode of the scores 87, 98, 77, 80, and 87.

42. $\triangle ABC$ is isosceles with $AB = BC$ and $m\angle B = 50°$. Find the measure of an exterior angle at A.

43. Find the measure of each angle of a regular nonagon.

44. Arrange the following numbers in ascending order.

$$\sqrt{28}, \quad 1.3, \quad 14\%, \quad \frac{1}{4}, \quad 0.2, \quad 0.099999, \quad 1^8$$

45. Solve for x: $|x - 2| = 7$

46. Find the area of the trapezoid with bases 10 cm and 18 cm, and height 15 cm.

47. Standardized Tests Select the type of triangle formed by the lengths 3 m, 4 m, and 6 m.

    A. acute   **B.** obtuse   **C.** right   **D.** equilateral

PROJECT *Connection*   Work as a class to choose a design for the proposed school garden. Continue to work on project as a class.

1. Use the scale drawing and the fact sheet to prepare a detailed estimate of the cost of building the garden. Check prices at local nurseries and home improvement stores.

2. With the appropriate school personnel, discuss having students, parents, and community members donate their time and tools to build all or part of the garden.

3. Discuss possible methods for raising the money needed for the project. List different strategies, with pros and cons for each type of fund-raising activity.

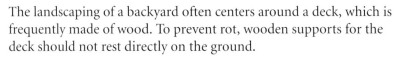

The landscaping of a backyard often centers around a deck, which is frequently made of wood. To prevent rot, wooden supports for the deck should not rest directly on the ground.

When constructing a deck, a builder may lay out a series of holes that are filled with concrete to form a *footing*. A *pier* is then placed in the concrete and the deck supports rest on the pier, above ground.

Despite all of today's technology, builders often use a centuries-old method to make sure that rectangular decks have true, right-angled corners.

To lay out the footings and piers, builders use tape measures and a right triangle relationship. To find a right angle, a builder lays out a multiple of a 3-4-5 triangle, adjusting the placement of the sides until a triangle is formed.

Decision Making

1. Explain whether the builder's method for making right angles is an example of the Pythagorean Theorem or of the converse of the Pythagorean Theorem.

2. Suzanne lays out two sides of a triangle and finds that $a^2 + b^2 \neq c^2$. Use the contrapositive of the Pythagorean Theorem to explain why she knew she didn't have a right angle.

3. Julie and Carlos are laying out string to mark their deck footings. They mark off a quadrilateral that is 20 ft by 34 ft. Carlos says that since both pairs of opposite sides are congruent, the deck must be a rectangle and have four true right angles. Julie disagrees. She says that the pairs of equal opposite sides do not guarantee a rectangle. Who is correct? Explain.

4. When a deck builder uses the Pythagorean technique to mark off a quadrilateral and all measurements are made carefully, the deck will be a rectangle. Explain how a builder determines the location of the four corner piers.

5. Explain how to verify the piers in Question 4 support a rectangular deck.

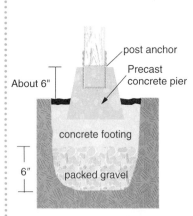

post anchor

Precast
concrete pier

About 6"

concrete footing

6"

packed gravel

Focus on Reasoning
Using Laws of Logic

8.9

Discussion

● Isaac, Louisa, and Franz were discussing conditional statements. Isaac developed the following conditional statement.

> If a figure is a square with a perimeter of 20 cm, then the length of each side of the figure is 5 cm.

1. Write a proof of Isaac's conditional statement.

2. If Louisa creates a square with a perimeter of 20 cm, what can be concluded about each side?

3. If Franz creates a trapezoid with a perimeter of 20 cm, what can be concluded about each side of the trapezoid?

Isaac also develops this conditional statement.

> If each side of a square is 5 cm, then the area of the square is 25 cm².

4. Write a proof of Isaac's new conditional statement.

5. What can be concluded about the area of Louisa's square?

SPOTLIGHT ON LEARNING

WHAT? In this lesson you will learn
• to apply the logical Laws of Detachment and Syllogism.

WHY? The law of detachment can help you solve problems in careers where logical thinking is critical.

Build a Case

● Isaac, Louisa, and Franz have just explored two important laws of logic. One of them is the *Law of Detachment*.

> **LAW OF DETACHMENT**
>
> If $p \Rightarrow q$ is a true conditional and p is true, then q is true.

This law means if you have a true conditional, and for a situation the hypothesis is true, then the conclusion is also true.

6. For Isaac's first conditional statement, identify the hypothesis p, and the conclusion q.

7. Explain how you used the Law of Detachment in Questions 2 and 3.

8. WRITING MATHEMATICS Create a true conditional statement. Then create a situation which makes the hypothesis true. Use the Law of Detachment to explain the conclusion you can make about the situation.

Write a proof using the Law of Detachment.

9. Given: $\overleftrightarrow{AB} \parallel \overleftrightarrow{CD}$

    Conclusion: $\angle AFG \cong \angle FGD$

10. Given: $\triangle KLM$ and $\triangle PQR$

    Conclusion: $\triangle KLM \cong \triangle PQ$

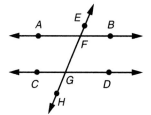

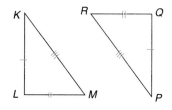

The other law of logic explored in Discussion is the *Law of Syllogism.*

> ── LAW OF SYLLOGISM ──
>
> If $p \Rightarrow q$ and $q \Rightarrow r$ are true , then $p \Rightarrow r$ is a true conditional.

 CHECK UNDERSTANDING

What mathematical property is analogous to the Law of Syllogism?

This law means if you have two true conditional statements, and the conclusion of the first conditional statement is the hypothesis of the second, then the hypothesis of the first and the conclusion of the second can be combined to form one true conditional statement.

11. For Isaac's two conditional statements, identify p, q, and r. Write the conditional statement $p \Rightarrow r$.

12. Explain how to use the Law of Syllogism in Question 11.

13. WRITING MATHEMATICS Create two true conditional statements where the conclusion of the first is the hypothesis of the second. Explain how to use the Law of Syllogism to create a third true conditional statement from the first two.

14. Identify the steps where the Law of Detachment or the Law of Syllogism are used in this argument.

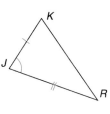

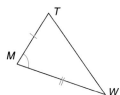

1. If two sides and the included angle of one triangle are congruent to two sides and the included angle of another triangle, the triangles are congruent.

2. $\overline{JK} \cong \overline{MT}$, $\overline{JR} \cong \overline{MW}$, and $\angle J \cong \angle M$

3. Thus, $\triangle JKR \cong \triangle MTW$.

4. CPCTC

5. Thus, $\overline{KR} \cong \overline{TW}$.

EXTEND AND DEFEND

15. The diagram at the right shows one set of objects contained in a larger set of objects.
    a. Write a conditional statement that relates the set of squares to the set of parallelograms.
    b. Suppose you know that polygon *P* is a square. What can you conclude by the law of detachment about *P*?

16. To be free to go to the movies, Michael must first help with food shopping, complete his school work, and clean his room.
    a. Write a conditional statement that relates task completion to his freedom to go to the movies.
    b. What can Michael conclude about his freedom if he completes all three of his three tasks? if he completes the first two tasks but not the third one?

17. Michael does not go to the movies. Can you logically reason that Michael did not complete at least one required task? Explain your response.

For Exercises 18–19, refer to the flow proof below.

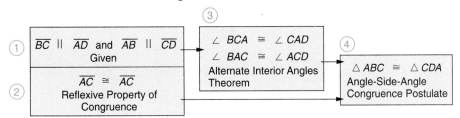

18. Explain how to use the Law of Detachment to conclude step 3 from step 1.

19. Explain how to use the Law of Detachment to conclude step 4 with the help of step 2 and step 3.

Now You See It
Now You Don't

Areas In Motion
Without blinking, look directly at the black rectangle and count to 25. After finishing the count, look at a dark wall or another dark object. What do you notice?

This effect is called an afterimage, and it is the basis for most optical illusions involving motion.

• • • CHAPTER REVIEW • • •

VOCABULARY

Choose the word from the list that correctly completes each statement.

1. The longest side of a right triangle is called its __?__ .

2. If a triangle has sides a, b, and c and $a^2 + b^2 > c^2$, then the triangle is __?__ .

3. If the __?__ of a rhombus are a and b, then its area is $\frac{1}{2}ab$.

4. If a triangle has sides a, b, and c and $a^2 + b^2 < c^2$, then the triangle is __?__ .

5. If the __?__ of a rectangle are a and b, then its area is ab.

a. acute

b. obtuse

c. hypotenuse

d. sides

e. diagonals

Lesson 8.1 PERIMETER AND AREA   pages 435–440

- The perimeter P of a rectangle with length l and width w can be found using $P = 2l + 2w$.
- The area A can be found using $A = lw$.

Find the area and perimeter of the following polygons.

6.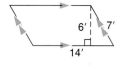
17 m
8.5 m

7.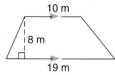
14′

8.
5"
6"
7"
4"

9.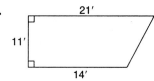
12 cm
2 cm
9 cm
5 cm

Lessons 8.2 and 8.3 AREAS OF PARALLELOGRAMS, TRIANGLES, AND TRAPEZOIDS pages 441–453

- The area A of a parallelogram with base b and height h can be found using $A = bh$.

- The area of a triangle with base b and height h can be found using $A = \frac{1}{2}bh$.

- The area A of a trapezoid with height h and bases b_1 and b_2 can be found using $A = \frac{1}{2}h(b_1 + b_2)$.

Find the area of the following polygons.

10.
6′ 7′
14′

11.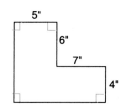
40″
100″

12.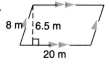
8 m 6.5 m
20 m

13.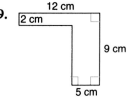
15 cm
10 cm

14.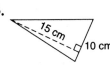
10 m
8 m
19 m

15.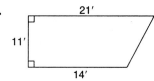
21′
11′
14′

16. One base of a trapezoid is five more than the other base. The height is 2 cm and the area is 90 cm². Find the length of each base.

484 CHAPTER 8 **Perimeters and Areas of Polygons**

- You can approximate the area A under the graph of a function $y = f(x)$.

Approximate the areas of each of the following regions using the trapezoidal rule.

17. The area between the curve $y = x^2 + 5$, the x-axis, and $x = 1$ and $x = 3$. Use four trapezoids.

18. The area bounded by the graph of $y = x^3 + 1$, the x-axis, and $x = 1$ and $x = 4$. Use $h = 1$.

- You can find the area of an irregular polygon using Pick's Theorem where $A = \frac{1}{2}B + I - 1$.

19. Use subtraction of areas to find the area of polygon *GREAT* where $G(11, 1)$, $R(10, 7)$, $E(4, 9)$, $A(2, 7)$, and $T(4, 2)$.

20. Find the area of polygon *GREAT* using Pick's Theorem.

21. A computer generates 900 points inside a 20 by 25 rectangle. One hundred eighty of these points fall within region R inside of the rectangle. Approximate the area of region R.

- In a right triangle with sides a and b and hypotenuse c, $a^2 + b^2 = c^2$.

- Conversely, suppose c is the length of the longest side of a triangle with side lengths a, b, and c. If $a^2 + b^2 < c^2$, then the triangle is obtuse. If $a^2 + b^2 > c^2$, then the triangle is acute.

Find the missing side in each of the following right triangles. Leave irrational answers in radical form.

22.

23.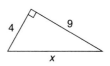

24.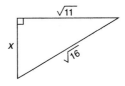

Determine if the following sets of three numbers could represent the sides of a right triangle, an obtuse triangle, an acute triangle, or no triangle at all.

25. $3, 3, \sqrt{18}$ 26. $4, 7, 12$ 27. $5, 12, 12$ 28. $5, 12, 16$ 29. $5, 12, 19$

- Law of Detachment: If $p \Rightarrow q$ is a true conditional and p is true, then q is true.

- Law of Syllogism: If $p \Rightarrow q$ and $q \Rightarrow r$ are true, then $p \Rightarrow r$ is a true conditional.

30. What conclusion follows from Statements 1 and 2 below? What law did you use?

   1. If an animal is a dog, then it barks.   **2.** Fido is a dog.   **3.** If an animal barks, then it has fur.

31. What conclusion follows from Statements 1 and 3 above? What law did you use?

CHAPTER ASSESSMENT

CHAPTER TEST

Find the area and perimeter of each region.

1.
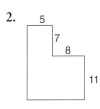

2.

3. **WRITING MATHEMATICS** Explain the steps required to find the area of a square whose perimeter is given.

4. **MODELING** Use a geometric model to find the product of $(x + 7)$ and $(x + 2)$.

Find the area of each of the following polygons.

5.

6.

7.

8.
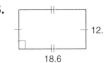

9.

10.
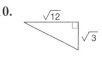

11. Find the area of *GLAD* with coordinates $G(3, 4)$, $L(12, 4)$, $A(15, 9)$, and $D(6, 9)$.

12. One base of a trapezoid is 7 in longer than the other. The height of the trapezoid is 10 in., and its area is 95 in.^{2} Find the length of each base.

13. Polygon *PRIMES* has vertices $P(1, 3)$, $R(2, 8)$, $I(7, 8)$, $M(10, 5)$, $E(10, 0)$, and $S(6, 2)$.

    a. Use Pick's Theorem to find the area.
    b. Use subtraction of areas to find the area.

14. Graph the parabola $y = x^2 + 3$. Shade the region bounded by the x-axis, the lines $x = 1$ and $x = 7$, and the parabola. Use the trapezoidal rule with three trapezoids to approximate the shaded area.

15. **WRITING MATHEMATICS** Explain how you could improve your approximation in Question 14.

Find the missing side of each of the following right triangles. Leave irrational answers in radical form.

16.   17.   18.
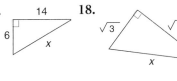

Determine whether the following sets of lengths could form an acute triangle, a right triangle, an obtuse triangle, or no triangle at all.

19. 8, 15, 25   20. 8, 15, 17

21. 8, 15, 16   22. 8, 15, 19

23. A right triangle has leg 7 and hypotenuse 25. Find the area of the triangle.

24. A triangle's height is eight less than its base. Its area is 16.5. Find the base of this triangle.

25. A trapezoid has bases with lengths 17 and 29, and height equal to the mean of the lengths of the bases. Find the area of the trapezoid.

26. **STANDARDIZED TESTS** A square has area 100 m^2. What is the length of its diagonal?

    A. 10 m   B. 20 m

    C. $10\sqrt{2}$ m   D. $25\sqrt{2}$ m

27. Prove the diagonals of a rhombus form four congruent right triangles.

PERFORMANCE ASSESSMENT

TO WRITE TWO RIGHT Find two right triangles such that both of the triangles have sides whose lengths are integers and both triangles have one leg with length 24. The triangles cannot be congruent.

SHEARING On graph paper, draw the line segment that connects $A(1, 4)$ and $B(11, 4)$. Sketch the line $y = 7$. If a point C is on the line $y = 7$. draw acute, right and obtuse triangles ABC. Identify the point C you used in each triangle, and find the area of each triangle. If C is moved along $y = 7$, (this is called shearing), what do you know about the area of the resulting triangles? Use a geoboard and rubber bands to show the movement of point C and its affect on the area of triangle ABC.

TRAPPED BY THE TRAPEZOIDAL RULE Use the trapezoidal rule to approximate the area between the curve $y = x^2 - 9$ and the x-axis between $x = 1$ and $x = 5$. Use $h = 1$. Then, graph the function, and explain what adjustment must be made in the trapezoidal rule formula to improve the approximation, still using $h = 1$.

AREA ART Create an abstract, nonconvex polygonal design on graph paper. The design must have at least twelve sides. Find the area of the polygon and draw five rectangles that have the same area as the polygon. Discuss which rectangles look like they have more area than the polygon and which rectangles have less area than the polygon.

PROJECT ASSESSMENT

PROJECT *Connection* Your class is going to make a presentation about the entire concept of the garden. Include all stages of research, planning and preparation. Not everybody will speak at the formal presentation, but everyone should take part in preparing it. Posters need to be made, handouts need to be created and photocopied, and all the people who contributed time and knowledge need to be recognized. It can be a lasting testament to ingenuity, creativity, mathematics, and teamwork!

1. As a class, make an outline of your presentation.

2. Assign different jobs, prepare presentation materials and rehearse the presentation.

3. Arrange to give your presentation to a school authority. Give the presentation, and after the presentation, discuss the possibility of actually building the garden at your school.

4. Set up a time line with your school authorities for implementing your ideas for a commemorative garden.

CUMULATIVE REVIEW

Fill in the blank.

1. Any parallelogram whose area can be calculated by finding half the product of the lengths of its diagonals is a(n) ___?___.

2. An angle whose measure is 180° is called a(n) ___?___ angle.

3. A figure that can be turned about a single point and be the same as the original figure at least twice during one complete turn is said to have ___?___.

4. In the equation $y = mx + b$, m represents the ___?___ and b represents the ___?___.

Find the area of each figure.

5.
6 cm
8 cm
10 cm

6.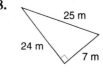
5 m   7 m
14 m

7.
7 ft
1.5 ft

8.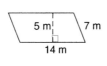
25 m
24 m
7 m

9. WRITING MATHEMATICS Describe how the formulas for the area of a trapezoid and the area of a triangle are related.

10. In the diagram, two parallel lines are cut by a transversal. Find the measure of $\angle 1$ and $\angle 2$.

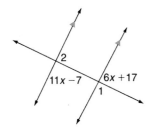
2
$11x - 7$ $6x + 17$
1

11. STANDARDIZED TESTS The point $(-1, 3)$ lies on a line that has a slope of -2. Which of the following points also lies on the line?

   A. $(-3, 3)$ **B.** $(-1, 1)$

   C. $(-3, 4)$ **D.** $(2, -3)$

   E. none of these

Consider the statement:

   All squares are rhombuses.

12. Write the statement in if-then form. Tell whether the statement is true or false.

13. Write the converse of the statement. Tell whether the converse is true or false.

14. Write the inverse of the statement. Tell whether the inverse is true or false.

15. Write the contrapositive of the statement. Tell whether the contrapositive is true or false.

Determine if each set of three lengths forms a triangle. If a triangle is formed, classify it as acute, right, or obtuse.

16. 4, 7, 10   17. 6, 12, 21

18. $\sqrt{3}, \sqrt{19}, 4$ 19. 7, 6, 8

20. WRITING MATHEMATICS In the figure below, explain how it could be proven that $\angle A \cong \angle X$.

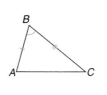
B
A   C

X
Y
Z

21. Find the perimeter and area of the rectangle.

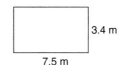
3.4 m
7.5 m

• • • STANDARDIZED TEST • • •

QUANTITATIVE COMPARISON In each question, compare the quantity in Column 1 with the quantity in Column 2. Select the letter of the correct answer from these choices.

A. The quantity in Column 1 is greater.
B. The quantity in Column 2 is greater.
C. The two quantities are equal.
D. The relationship cannot be determined by the information given.

Notes: In some questions, information which refers to one or both columns is centered over both columns. A symbol used in both columns has the same meaning in each column. All variables represent real numbers. Most figures are not drawn to scale.

Column 1	Column 2
1.	$x < y < 90°$
complement of x	supplement of y
2.	$x > y$

perimeter of	perimeter of

3. an obtuse triangle with sides of length a, b, and c, with c being the longest side

$a^2 + b^2$	c^2

4. an isosceles triangle

the sum of the measures of the base angles	the measure of the vertex angle

5.

$p^2 - m^2$	n^2

	Column 1	Column 2
6.	area of	area of

7.

the supplement of an interior angle of a regular octagon	the supplement of an interior angle of a regular decagon

8. the point $(0, y)$

the distance from the point to the point $(-4, 0)$	the distance from the point to the point $(4, 0)$

9. points $P(3, -5)$ and $Q(-1, -2)$

the magnitude of $\vec{PQ}$	the magnitude of $\vec{QP}$

10. A segment in Quadrant II is rotated 180° about the origin.

the product of the coordinates of any point on the original segment	the product of the coordinates of the corresponding point on the image segment

11.	area of	area of

  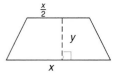

12.

the measure of the smallest angle in an acute triangle	the measure of the smallest angle in an obtuse triangle

13.

the x-coordinate of any point on the x-axis	the y-coordinate of any point on the y-axis

9 Similarity

Take a Look
AHEAD

Make notes about things that look new.

- How is the symbol for similarity related to the congruence symbol?
- What criteria for triangle similarity will you study in this chapter?

Make notes about things that look familiar.

- What does it mean to say that two polygons are similar?
- What does it mean to say that one geometric figure is the image of another?

DATA Activity

SKILL FOCUS

▶ Use estimation.

▶ Compare numbers.

▶ Write numbers as fractions, decimals, and percents.

▶ Determine ratios.

▶ Use ratios to compare.

▶ Write and solve an equation.

Population Density

The *population* of a country or other area is the total number of people who live in it. The number of people per square mile is known as the *population density*.

As Earth's population rapidly increases, *overpopulation*—the point at which there is more life on our planet than the planet is able to support—is a concern.

Many regions with the greatest population density are not able to supply adequate food for their populations.

GEOMETRYWORKS

HELP WANTED

Maps

In this chapter, you will see how:

- **CARTOGRAPHERS** use angle measure and projections to create maps that preserve the integrity of different properties of Earth's actual surface.
  (Lesson 9.4, page 514)

- **AERIAL PHOTOGRAPHERS** use triangulation techniques to produce photographs that focus on different views of a region.
  (Lesson 9.8, page 543)

WORLD POPULATION		
Region	**Population (millions)**	**Area (millions of mi²)**
Asia	3,257	10.644
Africa	677	11.707
North America	443	9.360
South America	305	6.883
Antarctica	0	6.000
Europe	513	1.905
Oceana	28	3.284
Former USSR	285	8.647

Use the table to answer the following questions.

1. Find the population density of each region of the world.

2. Which region has the greatest population density? Which has the lowest?

3. The total land mass of Earth is about 57.9 million mi². Find the percent of the total land mass of Earth represented by each region. Round to the nearest tenth.

4. Which region has the greatest percent of land mass? Which has the least?

5. Which region is most likely to reach overpopulation first? Explain.

6. **WORKING TOGETHER** Investigate three countries that are overpopulated and not able to supply adequate food for their people. Examine the average daily temperature and location in relation to the equator for each country. Make a bar graph for temperature and another for location to show how these countries compare to the U.S. Can you draw any conclusions about a country's ability to meet its food demands?

Create a Map

A map can be as simple as a sketch of how to get to a friend's house or as complicated as one of the world showing continents, countries, cities, and topographical features such as mountains.

Maps are important when you take a road trip. They are vital for air and sea navigators.

In this project, you will create an enlargement of a map while maintaining the size and shape of the actual land mass.

PROJECT GOAL

To create an enlargement of a map.

Getting Started

Work with a group.

1. Decide which continent or country to study. The location you choose should include several different regions such as countries, states, or provinces.

2. Discuss how to obtain a map of the chosen region.

3. Discuss with your group the type of surface you will use to create your map. Consider creating it on a large piece of butcher's paper, a painter's canvas, or on a wall or floor in or around your school.

4. Discuss the materials you will use to create your map. Consider supplies such as colored pencils, markers, or paint.

PROJECT Connections

Lesson 9.3, page 509:
  Learn a grid technique for enlarging or reducing a drawing.
Lesson 9.5, page 521:
  Apply the grid technique to prepare a tracing for enlargement.
Lesson 9.6, page 531:
  Plan a color scheme for your map.
Lesson 9.8, page 542:
  Plan the construction of your enlarged map.
Chapter Assessment, page 547:
  Work together as a group to create an enlarged map.

Internet Connection

www.learninggeometry.com

Focus on Reasoning
Proportional Reasoning

9.1

Discussion

● "I'm in charge of the set design for the school play," Jim told his study group. "One set needs a large U.S. flag as a backdrop, but each time I make a sketch, it doesn't look right." Jim put his sketch of the outline of the flag on the table.

4 cm
15 cm

1. Describe what is wrong with Jim's sketch. Sketch your own U.S. flag.

Jim continued, "I've been working from a picture of the flag that I got out of an atlas. I want to make a small sketch look good before I start on the backdrop."

"The flag in the atlas picture is 19 cm by 10 cm. Just add or subtract the same amount from the base and height to draw the flag," suggested Cathy.

2. Try Cathy's suggestion. Subtract 8 cm from each dimension given in Jim's atlas picture and draw an outline of the flag. Does the new drawing appear to be correct? Explain.

"Think about this another way. If you wanted to increase the number of cookies a recipe makes, you have to increase each of the ingredients," explained Cathy.

3. Would you increase all ingredients by the same amount? Explain.

"What if we cut each of the measurements of the flag in half, or double or triple them instead?" suggested Joe.

4. Use Joe's suggestion. Cut each dimension in half and draw the flag again. Does the outline of the flag appear to be correct now?

Build A Case

● A **ratio** is a comparison of two quantities by division.

For example, a ratio that compares the dimensions of the flag in the picture that Jim had is

$$\frac{\text{height}}{\text{base}} = \frac{10}{19} = 10 : 19$$

You can write the standard American flag ratio as $\frac{1}{1.9}$ or $1 : 1.9$.

5. What is the ratio of the height to the base of your flag drawing in Question 1? in Question 2? in Question 4?

A statement that two ratios are equal is called a **proportion**.

6. Determine which of your ratios in Question 5 is equal to the standard American flag ratio. Write a proportion.

Proportions are usually written in one of two ways:

$$a : b = c : d \quad \text{or} \quad \frac{a}{b} = \frac{c}{d}$$

where a, b, c, and d are real numbers and $b \neq 0$ and $d \neq 0$.

In either of the forms above, b and c are called the **means** of the proportion and a and d are called the **extremes**. You can use the multiplication property of equality to show that in any proportion the product of the means is equal to the product of the extremes: $ad = bc$. The products ad and bc are called **cross products**.

7. Find the cross products in the proportion of Question 6. What do you notice?

> **CROSS-PRODUCT PROPERTY OF PROPORTIONS**
>
> **The product of the means of a proportion is equal to the product of the extremes.**

8. Use the Cross-Product Property of Proportions to test the dimensions in Jim's sketch.

$$\frac{15}{4} \overset{?}{=} \frac{1}{1.9}$$

9. Explain why Jim's sketch didn't look proportional.

You can also use the Cross-Product Property of Proportions to solve a proportion for an unknown term. For example, if Jim wanted the base of the flag on the set to be 4 m, he could compute the height of the flag using the following proportion.

$$\frac{\text{height}}{\text{base}} = \frac{1}{1.9} = \frac{x}{4}$$

10. Use the Cross-Product Property of Proportions to write an equation from the proportion above.

11. Solve the equation for x. Round to the nearest tenth.

12. If you substitute your value for x, are the cross products equal?

The blue area on a U.S. flag is called the *union*. The ratio of the height of the union to the height of the flag is 7 : 13.

13. Using your value of x from Question 11 for the height of the flag, what should be the height of the union?

14. Using the correct proportions, sketch the outline of a flag if the height of the flag is 5 cm. Include the union in your sketch.

Using what you know about proportions, you can explore some other properties of proportions.

Use the proportion $\dfrac{5}{9.5} = \dfrac{1}{1.9}$ for Questions 15–17.

15. Rewrite the proportion by taking the reciprocal of each side. Is the new relation a proportion? Explain.

16. Rewrite the proportion by interchanging the means. Is the new relation a proportion? Explain.

THINK BACK

Two numbers are reciprocals if their product is 1.

17. Addition was used to rewrite the proportion as shown below. Are the new relations proportions? Explain.

$$\dfrac{5 + 9.5}{9.5} \overset{?}{=} \dfrac{1 + 1.9}{1.9} \qquad \dfrac{5 + 1}{1} \overset{?}{=} \dfrac{9.5 + 1.9}{1.9}$$

PROPERTIES OF PROPORTIONS

For any nonzero real numbers a, b, c, and d:

If two ratios are equal, then their reciprocals are also equal.

If $\dfrac{a}{b} = \dfrac{c}{d}$, then $\dfrac{b}{a} = \dfrac{d}{c}$. **Reciprocal Property**

If $\dfrac{a}{b} = \dfrac{c}{d}$, then $\dfrac{a + b}{b} = \dfrac{c + d}{d}$ and $\dfrac{a + c}{c} = \dfrac{b + d}{d}$.

EXTEND AND DEFEND

Solve each proportion. Check your result.

18. $\dfrac{8}{12} = \dfrac{x}{9}$

19. $\dfrac{x + 2}{6} = \dfrac{x + 1}{4}$

20. $\dfrac{x + 1}{x - 1} = \dfrac{x + 4}{x}$

Set up and solve a proportion.

21. Recall the slope of a line is a ratio of the rise to the run. If a line has a slope of $\dfrac{3}{5}$, and you have chosen two points on the line that have a rise of 15, what is the run?

22. EDUCATION For field trips, the Brighton School requires that the teacher to student ratio be at least 2 : 15. If there are 40 students on a field trip, what is the least number of teachers needed to supervise?

23. HOME ECONOMICS The recipe below makes 60 cookies. Convert the amount of each ingredient so that the recipe will make 90 cookies.

$1\frac{1}{2}$ cups flour $\frac{1}{2}$ teaspoon salt

1 cup sugar   1 teaspoon baking soda

1 cup butter   1 teaspoon vanilla

2 eggs

24. MEDICINE For many medications, the amount that should be administered to a child is based on the ratio of the child's weight to that of an average adult. The average adult weight often used is 150 lb. If an adult dose of a medication is 60 mg, how much should a 40-lb child receive?

25. ANGLE MEASURE In a certain regular polygon, the measures of an interior angle and an exterior angle are in the ratio 3 : 2. Name the type of polygon. Describe your reasoning.

CHEMISTRY A certain concentrated cleaning agent is supposed to be mixed with water in the ratio 2 : 5.

26. If you use 1 cup of cleaner, how much water should you add?

27. If you want 3 L of mixed cleaner, how much water and cleaner should you mix?

Now You See It
Now You Don't

IS SEEING REALLY BELIEVING?

In which figure at the right is the horizontal segment longer, figure I or figure II?

Your eyes might fool you into believing that the horizontal segment in figure I is longer. However, if you measure each segment with a ruler, you will find that the horizontal segment in figure I actually is congruent to the horizontal segment in figure II.

Which dimension of the black figure at the right is larger—its height h or its width w?

Optical illusions like these show why you cannot simply say that two segments are congruent "because they look congruent." As you continue to study geometry, you will learn several methods for *proving* that segments are congruent.

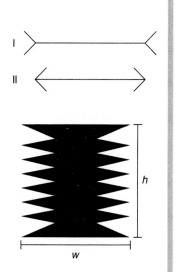

9.2 Geometry Workshop
The Golden Ratio

Think Back/Working Together

● Work with a group. Each person in the group should draw an isosceles triangle that has base angles of 72°.

   1. Find the measure of the vertex angle. Explain your reasoning.

   2. Measure the lengths of the base and the legs of your triangle. Find the ratio of the measure of a leg to the measure of the base. Write the ratio as a decimal rounded to the nearest thousandth.

   3. Compare answers with your group members and then share your group's findings with the class. What observation can you make about the ratios?

Explore

● You have computed the ratio known as the **golden ratio**, about 1.6.

This numerical value was discovered in Ancient Greece and applied particularly in architectural structures that were considered most visually appealing. Expressing its prized status in its name, the golden ratio has endured over the centuries as a symbol of perfection.

Look at the rectangles pictured below.

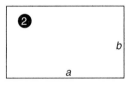

   4. Choose the rectangle that you find "most pleasing to look at." Compare your choice with your group members and then share your group's findings with the class. Record the class results.

TECHNOLOGY TIP

You can use geometry software to construct rectangles that are approximately golden.

5. For each of the four rectangles shown on the previous page, measure sides a and b and calculate the ratio $a : b$. Record your results in a table.

6. According to your calculations, which of the four rectangles has the golden ratio?

In 1876 a study conducted by the German psychologist Gustav Feshner found that 75% of his subjects chose as the most visually pleasing rectangular shape a rectangle that was in the golden ratio, known as a **golden rectangle**.

7. Compare the results from your class to Feshner's results.

Make Connections

To find the exact value of the golden ratio, you must construct a golden rectangle.

8. Construct a 2-in. square. Label it $ABCD$. Find the midpoint of $\overline{AB}$. Label it E. Find EB.

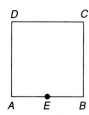

THINK BACK

For a right triangle whose legs measure a and b and whose hypotenuse measures c, the Pythagorean Theorem states that $a^2 + b^2 = c^2$.

9. Draw $\overrightarrow{AB}$. Locate point F on $\overrightarrow{AB}$ so $EF = EC$.

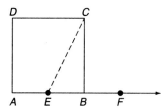

10. Find EC. 11. Find AF. 12. Draw $\overrightarrow{DC}$.

13. At F construct a line perpendicular to $\overrightarrow{AB}$ and label the point at which it intersects $\overrightarrow{DC}$ as point G.

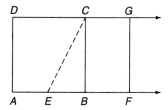

In the bottom figure, $ADGF$ is a golden rectangle. The golden ratio is $\dfrac{AF}{AD}$.

14. What is the exact value of the golden ratio? Give a decimal approximation rounded to the nearest millionth.

The Parthenon is an ancient Greek temple that is generally considered to be the finest structure built by that civilization.

Today an incomplete ruin remains as shown in the photograph.

15. Describe parts of the Parthenon that you think are examples of the golden rectangle.

The Greeks also used the golden ratio in sculpture, believing that the golden ratio exists in certain parts of the human body.

16. Copy and complete the table below. Record the indicated measurements for each member of your group. Write the ratios in decimal form.

17. Calculate average ratios for your group.

   A: distance from elbow to fingertip

   B: distance from shoulder to fingertip

   C: height from floor to mid abdomen

   D: total height

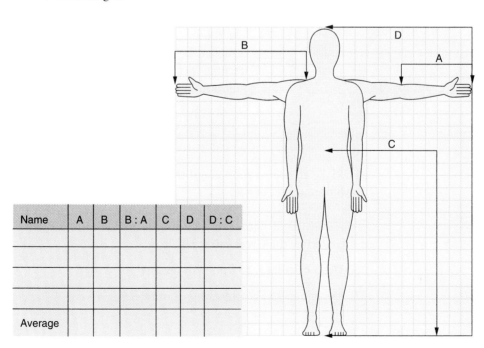

Name	A	B	B : A	C	D	D : C
Average						

Geometry Workshop

GEOMETRY: WHO, WHERE, WHEN

The repeating process of adding on a new square to a golden rectangle to get a new, larger golden rectangle is analogous to the growth process demonstrated by the *nautilus*, a sea animal whose soft body is partly covered with a coiled shell.

The nautilus shell grows longer and wider to make room for the growing animal within, but it grows at one end only. Each new section is increased in size so that the overall shape remains similar.

18. Make a conjecture about measurements on your index finger (the finger next to your thumb) that you think might yield the golden ratio.

19. Take these measurements to verify your conjecture.

20. Share your group's results with the class. Are the ratios similar for most of the groups? Does your class possess ideal Greek proportions?

Summarize

21. **WRITING MATHEMATICS** Reconsider your construction of the golden rectangle. Describe how $\overline{BC}$ divides golden rectangle *ADGF*. How is *CGFB* related to *ADGF*? Describe what happens if you repeat this division on *CGFB*, and then continue the process.

22. **THINK CRITICALLY** How can the process you wrote about in Question 21 be done in reverse? Draw a diagram to illustrate.

GOING FURTHER Recall the *Fibonacci sequence*.

1, 1, 2, 3, 5, 8, 13, 21, 34, 55, 89, 144, . . .

23. How are the terms of the Fibonacci sequence determined?

24. The following ratios are formed from successive terms of the Fibonacci sequence. Find the decimal form of each, rounded to the nearest millionth.

a. 34 : 21   b. 144 : 89
c. 377 : 233   d. 610 : 377
e. 987 : 610   f. 1597 : 987

25. Make an observation about the ratios of successive terms of the Fibonacci sequence.

GOING FURTHER The exact value of the golden ratio is often designated by the Greek letter phi, ϕ.

26. Use a calculator to find $\frac{1}{\phi}$ in decimal form.

27. Describe what is interesting about ϕ and $\frac{1}{\phi}$.

Explore

● Study the four rectangles below.

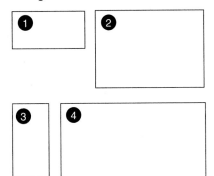

1. Which of the rectangles appear to be congruent?

2. Which of the rectangles appear to be similar?

3. In what way are similar figures like congruent figures?

4. In what way are similar figures different from congruent figures?

Build Understanding

● **Similar polygons** are polygons whose vertices can be paired in such a way that corresponding angles are congruent and corresponding sides are proportional.

Below are two similar quadrilaterals. The symbol ~ means "is similar to." So you write $ABCD \sim WXYZ$. As with congruence, the order of the letters indicates which sides and angles are corresponding.

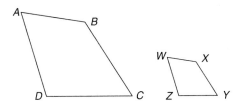

Since you are given that the quadrilaterals are similar, you know

$$\angle A \cong \angle W \qquad \angle B \cong \angle X \qquad \angle C \cong \angle Y \qquad \angle D \cong \angle Z$$

$$\frac{AB}{WX} = \frac{BC}{XY} = \frac{CD}{YZ} = \frac{DA}{ZW}$$

EXAMPLE 1

Verify that the two triangles shown are similar. Write a statement of similarity.

Solution

Check that the corresponding angles are congruent.

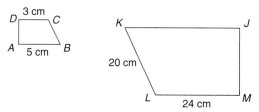

From the given measures,
$\angle A \cong \angle D$ and $\angle B \cong \angle E$.

Since the sum of the angles in each triangle must be 180°, $\angle C$ and $\angle F$ each measure 27° and $\angle C \cong \angle F$.

Check that the corresponding sides are proportional.

$$\frac{AB}{DE} = \frac{6 \text{ in.}}{9 \text{ in.}} = \frac{2}{3} \qquad \frac{BC}{EF} = \frac{12 \text{ in.}}{18 \text{ in.}} = \frac{2}{3} \qquad \frac{AC}{DF} = \frac{8 \text{ in.}}{12 \text{ in.}} = \frac{2}{3}$$

The two criteria of similarity are met, so $\triangle ABC \sim \triangle DEF$. ◄

The ratio of the lengths of corresponding sides of similar polygons is also called the **scale factor** of the similar polygons.

EXAMPLE 2

In the figure, $ABCD \sim JKLM$, find the scale factor of the quadrilaterals.

Solution

Use the similarity statement to identify the ratios of the corresponding sides.

$$\frac{AB}{JK} \qquad \frac{BC}{KL} \qquad \frac{CD}{LM} \qquad \frac{DA}{MJ}$$

Since the measures of corresponding sides $\overline{CD}$ and $\overline{LM}$ are given, you can use that ratio to determine the scale factor.

$$\frac{CD}{LM} = \frac{3 \text{ cm}}{24 \text{ cm}} = \frac{1}{8}$$

The scale factor of the similar quadrilaterals is 1 : 8. ◄

In Example 2, since both figures have the same unit of measure, the scale factor is independent of the unit of measure.

A scale drawing is an important application of similar polygons. Road maps and blueprints are examples of scale drawings.

The unit of measure in a scale drawing can be different from the unit of measure of the object it represents. For example, a map may be measured in inches while the region it represents may be in miles.

To interpret a scale drawing, you must know the scale ratio, including its unit of measure.

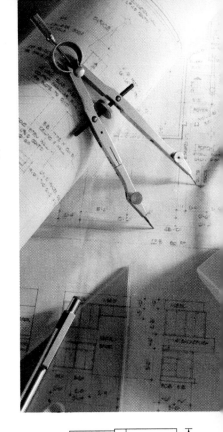

EXAMPLE 3

CONSTRUCTION An architect has drawn a blueprint to show the floor plan for a 3-bedroom home in the Caribbean. If the scale ratio is 1 cm : 4 m, find the dimensions of Bedroom 2.

Solution

Measure the dimensions of Bedroom 2 in the blueprint. In the blueprint, the dimensions of Bedroom 2 are:

length = 1.8 cm   width = 1.2 cm

Identify the ratio that the scale represents.

$$\text{scale ratio} = \frac{\text{blueprint measure}}{\text{actual measure}} = \frac{1 \text{ cm}}{4 \text{ m}}$$

Use the scale ratio to write a proportion for each unknown dimension.

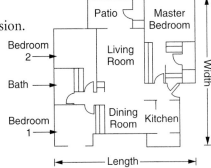

Let ℓ = the actual length
$$\frac{1 \text{ cm}}{4 \text{ m}} = \frac{1.8 \text{ cm}}{\ell \text{ m}}$$
$$1(\ell) = 4(1.8)$$
$$\ell = 7.2 \text{ m}$$

Let w = the actual width
$$\frac{1 \text{ cm}}{4 \text{ m}} = \frac{1.2 \text{ cm}}{w \text{ m}}$$
$$1(w) = 4(1.2)$$
$$w = 4.8 \text{ m}$$

So the actual dimensions of Bedroom 2 are 7.2 m by 4.8 m. ◀

TRY THESE

Are the polygons similar? Explain. If the polygons are similar, write a statement of similarity.

1.
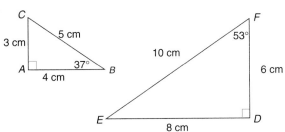

2.

3. WRITING MATHEMATHICS Consider the statement: All squares are similar. Is this statement true or false?. Write a paragraph to support your answer.

Find the scale factor of each set of similar figures. Justify your answers.

4. $ABCD \sim WXYZ$

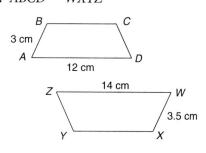

5. $\triangle ABC \sim \triangle KLM$

6. $\triangle CDE \sim \triangle XYZ$
$CD = 8$ cm, $EC = 3$ cm, $YZ = 10$ cm, and $YX = 12$ cm

7. CARTOGRAPHY The scale ratio of the map is 1 in. : 140 mi. The city of Pueblo is south of Denver. The distance between them on this map is 0.75 in. Write a proportion and find the actual distance between Denver and Pueblo.

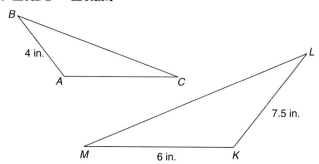

PRACTICE

Are the polygons similar? Explain. If the polygons are similar, write a statement of similarity.

8.

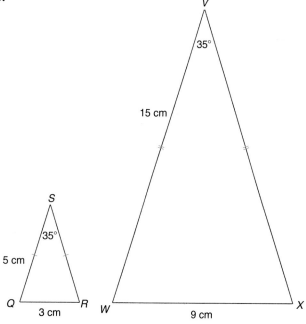

9.

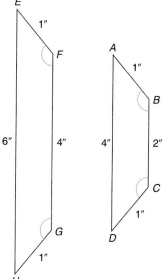

10. WRITING MATHEMATICS Consider the conditional statement: If two polygons are congruent, then they are similar. Discuss whether or not this conditional statement and its converse are true. Justify your conclusions.

Find the scale factor of each set of similar figures. Justify your answers.

11. $ABCD \sim MLKJ$

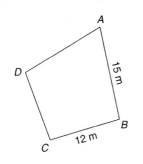

12. $\triangle XYZ \sim \triangle TSQ$

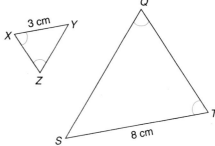

13. $MNOP \sim SMQR$

14. $\triangle ABC \sim \triangle ADE$

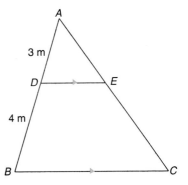

15. TOY DESIGN It is illegal to use the actual dimensions when producing a copy of any U.S. paper currency. An actual U.S. bill measures $6\frac{1}{8}$ in. by $2\frac{9}{16}$ in.

A toy company wants to manufacture play money with the same proportions as U.S. bills. If the company produces rectangular bills that are 5 cm by 2.1 cm, will this play money meet the company's specifications without violating the law? Explain.

Tell whether two figures of the given types are always, sometimes, or never similar.

16. rectangles

17. right triangles

18. equilateral triangles

19. equiangular triangles

20. isosceles trapezoids

21. regular hexagons

22. a right triangle and an equilateral triangle

23. a right triangle and an isosceles triangle

Find all the angle measures and side lengths for each figure of the given similar pairs. Round decimal answers to the nearest hundredth.

24. *ABCDE ~ STUVW*

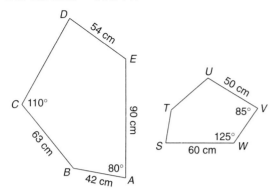

25. △*ABE ~* △*DBC*

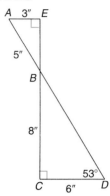

26. *QRST ~ UVWX*

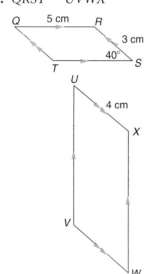

27. △*WXY ~* △*WVZ*

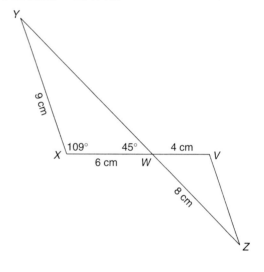

REAL ESTATE Realtors often use a scale drawing to answer questions about room sizes of the homes they are trying to sell.

The scale drawing shown is of a 2-bedroom condominium. The scale factor is 1 in. : 16 ft.

28. Prospective buyers want to know if an entertainment center they own will fit in the living room on the wall opposite the door. If the piece is 8 ft long, will it fit? Explain.

29. The buyers are planning to carpet Bedroom 2. How many square feet of carpet will they need? Answer to the nearest square foot.

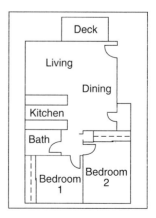

Find the lengths *x* and *y* in each pair of similar figures.

30. △ABC ~ △DBE

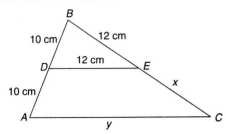

31. △ABC ~ △BDC

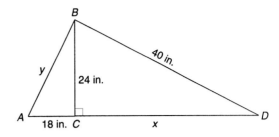

32. △ABC ~ △AED

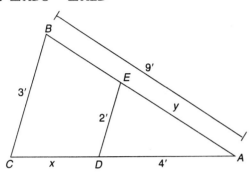

33. △JKL ~ △MNL

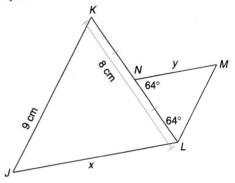

EXTEND

34. Are the polygons similar? Justify your answer.

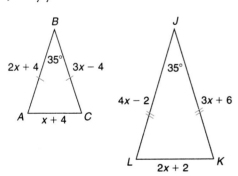

35. Given △JKL ~ △MNO, find the value of *x*.

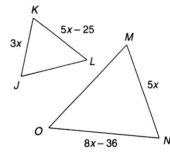

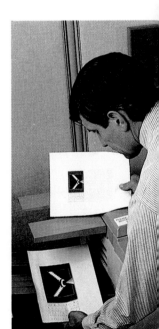

BUSINESS Many photocopiers can reduce or enlarge an item being copied. The scale factor of the copy to the original is expressed as a percent.

36. If you want the original to be one-half the size of the copy, at what percent do you set the copier?

37. Can you set the copier to make a reduction that reduces each side of the rectangular original by 1 in.? Explain.

SURVEYING A surveyor can use similarity to determine inaccessible distances.

A surveyor has set up the diagram shown to find the distances BC and CD across a river. In the diagram, $\triangle ACE \sim \triangle BCD$.

38. Are $\overline{AB}$ and $\overline{DE}$ the sides of any triangle in the diagram? Explain.

39. Find x and y.

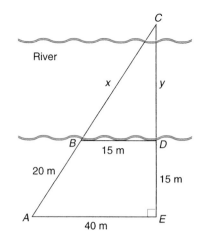

THINK CRITICALLY

PUBLISHING Standard paper sizes, called the A-series, are used for printing in the field of publishing. The A-series, sizes A0 through A10, is pictured at the right. The actual size of the entire page size A0 is 841 mm wide and 1189 mm long.

The next size in the series is obtained by folding the previous size in half.

40. Using the short side as the width and the long side as the length, make a table showing the lengths and widths of the sizes A0 through A10.

41. Use the data in the table to determine if the A-series rectangles are similar. Explain your conclusion.

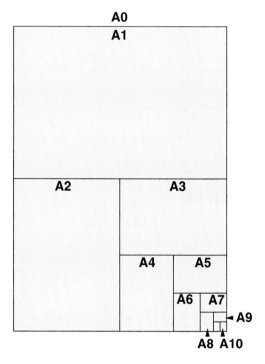

DRAFTING A pantograph is a tool used to create similar objects in a drawing. When working on mechanical drawings, drafters often use pantographs.

In a pantograph, four bars are connected, as in the drawing shown, so that $ABCD$ is a parallelogram and points E, A, and F are collinear.

To enlarge a map or picture, place the tracing point of the pantograph at A and a pencil at F. As you trace the picture with the tracing point at A, the pencil draws an enlarged copy at F. So, $\triangle EAD \sim \triangle EFC$.

42. Given $ED = 3$ cm and $DC = 12$ cm, what is the scale factor of the similar triangles?

43. Given $EA = 4$ cm, what is the length of $\overline{AF}$?

44. If the original picture is 13 cm wide, how wide should the enlarged picture be? Justify your answer.

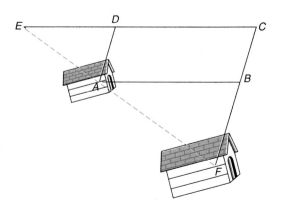

ALGEBRA AND GEOMETRY REVIEW

Graph each line.

45. $y = 2x + 3$

46. $3x - y = 4$

47. $x = 4$

Solve for x.

48. $\dfrac{x + 6}{3} = \dfrac{5x}{9}$

49. $\dfrac{2x + 1}{9} = \dfrac{12 - 7x}{6}$

50. $\dfrac{2(3x - 4)}{7} = \dfrac{-4(10 - x)}{-4}$

51. STANDARDIZED TESTS The bases of a trapezoid measure 6 m and 10 m, and the height is 8 m. The area is

   A. $32\ m^2$ **B.** $52\ m^2$ **C.** $64\ m^2$ **D.** $128\ m^2$

Solve each equation.

52. $x^2 + 6 = 86$

53. $3x^2 - 4 = 290$

54. $(x - 4)^2 = 225$

Find all the angle measures and side lengths for each figure of the given similar pairs.

55. $ABCDE \sim JKLMN$

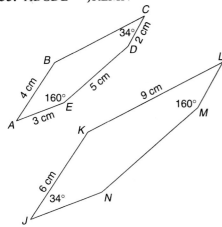

56. trapezoid $EFGH \sim$ trapezoid $QRST$

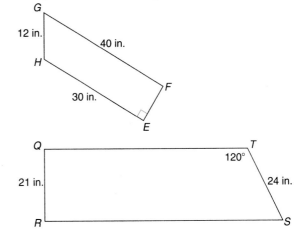

PROJECT *Connection*   People in a wide variety of careers are often asked to make a drawing that is larger than an original but maintains the same proportions. Using proportional grids is an effective method for accomplishing this.

1. Choose a simple drawing.

2. On the original, draw a $\frac{1}{2}$ cm or 1 cm grid.

3. On a clean sheet of paper, draw a new grid that has the same number of grid squares as the original, but the grid size is larger. For example, you may want to use a 2 cm grid.

4. Transfer the drawing box-by-box to the new grid. That is, sketch the contents of a single box from the original drawing to the corresponding box of the larger grid.

5. For a cleaner look, erase the grid on the new drawing.

9.4

Dilations

Think Back

Recall that a *transformation* moves each point of a plane figure according to some rule, mapping the original figure onto an *image* in a one-to-one correspondence of points.

In each of the three diagrams $\triangle ABC$ has been transformed to result in the image $\triangle A'B'C'$.

1. In the reflection transformation shown, name the line of reflection.

2. Describe the translation shown.

3. Describe the rotation shown.

A transformation that preserves distance is called an *isometry*.

4. Which of the transformations shown are isometries?

You may name the vertices of a polygon in clockwise or in counterclockwise order to establish the *orientation* of the polygon.

5. Which of the transformations shown preserves orientation?

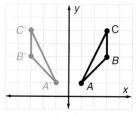

reflection

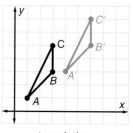

translation

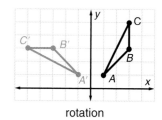

rotation

Explore

Work with a partner. Each partner needs to trace the golden rectangle *ABCD* shown. Partner A chooses a point *X* that is outside of *ABCD*. Partner B chooses a point *X* that is inside *ABCD*. Each partner then draws a ray, $\overrightarrow{XA}$. Use a compass to find *XA*. Copy the segment using a dashed line, creating $\overline{AA'}$ so that *A* is the midpoint of $\overline{XA'}$.

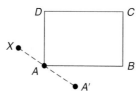

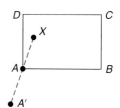

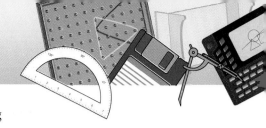

6. Draw $\overline{XB}$, $\overline{XC}$, $\overline{XD}$. Copy segments $\overline{XB}$, $\overline{XC}$, $\overline{XD}$, creating $\overline{BB'}$, $\overline{CC'}$, $\overline{DD'}$.

7. How is your image $A'B'C'D'$ different from your partner's? How is it the same?

8. Is your image $A'B'C'D'$ also a golden rectangle? Is your partner's? Justify your answers.

Make Connections

You and your partner have each performed a type of transformation called a *dilation*. $A'B'C'D'$ is the *dilation image* of $ABCD$.

9. Do your dilations preserve orientation? Explain.

10. Are your dilations isometries? Explain.

The ratio of the length of a side of the dilation image to the length of the corresponding side in the original figure is called the *scale factor*.

11. Find the scale factor of your dilation. Compare it to your partner's.

12. Create another dilation by following these steps.

    Draw any $\triangle ABC$.
    Choose a point inside or outside $\triangle ABC$ and label it O.
    With a dashed line, draw $\overrightarrow{OA}$, $\overrightarrow{OB}$, $\overrightarrow{OC}$.
    Bisect $\overline{OA}$, $\overline{OB}$. $\overline{OC}$. Label the midpoints A', B', C'.
    Draw $\triangle A'B'C'$.

13. Determine the scale factor of this dilation. Is the scale factor less than or greater than 1?

14. Make an observation about the relationship between how the scale factor compares to 1 and whether the image is larger or smaller than the original figure.

The *center of dilation* is the point from which the dilation is created.

15. What was the center of your dilation for the rectangle? What was your partner's?

16. What was the center of your dilation for the triangle?

17. Does the center of dilation affect the size of the dilation image? Explain.

18. Does the center of dilation affect the location of the dilation image? Explain.

A dilation with center *C* and positive scale factor *k* is a transformation in a plane that maps every point *P* of the plane to a point *P'* so that

- If the point *P* is not point *C*, then $\overrightarrow{CP'}$ and $\overrightarrow{CP}$ are collinear and $CP' = k(CP)$.

- If the point *P* is point *C*, then $P = P'$.

An enlargement is a dilation with scale factor *k* such that $k > 1$. A reduction is a dilation with scale factor *k* such that $0 < k < 1$.

TECHNOLOGY TIP

When using geometry software for a dilation, you need to choose a center point for the dilation.

As with other transformations, you can perform a dilation transformation in the coordinate plane.

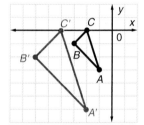

In the diagram, $\triangle A'B'C'$ is the image of $\triangle ABC$ after a dilation.

19. What is the center of the dilation?

20. Compare each image to its original vertex. What is the scale factor?

21. Consider another dilation.

Draw any $\triangle DEF$ on the coordinate plane, labeling the points and the coordinates of the points. With $(0, 0)$ as the center of dilation, create a dilation that has a scale factor of 3. Label the dilation image $\triangle D'E'F'$. Write the coordinates of D', E', and F'.

22. In general, if point P' is the image of $P(x, y)$ under a dilation with its center at zero and the scale factor *k*, where $k > 0$, write the coordinates of P'.

Summarize

23. WRITING MATHEMATICS Consider the four types of transformations: reflection, translation, rotation, and dilation.

Discuss the ways in which these transformations are alike and the ways in which they are different. Include in your discussion a comparison of how these transformations behave with respect to orientation. Mention also which of these transformations are isometries and which are not.

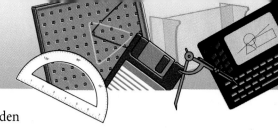

THINKING CRITICALLY Reconsider your dilation of the golden rectangle *ABCD* from Explore.

24. What relationship between the opposite sides of the rectangle is not preserved under the dilation?

25. What relationship between the opposite sides of the rectangle is preserved under the dilation?

26. What relationship between the consecutive sides of the rectangle is preserved under the dilation?

GOING FURTHER So far you have worked with scale factors that are positive, $k > 0$.

27. For what values of k are the dilations reductions?

28. For what values of k are the dilations enlargements?

29. What do you think would be true about the image if the scale factor of the dilation were negative, $k < 0$?

On graph paper, copy $\triangle PQR$.

30. Using $(0, 0)$ as the center of dilation and a scale factor of 2, draw $\triangle P'Q'R'$, the dilation image of $\triangle PQR$. Label the coordinates of the vertices.

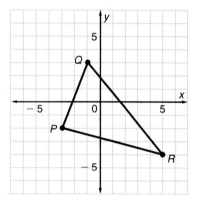

31. On the same graph, draw $\triangle P''Q''R''$ the image of $\triangle PQR$ under a dilation with center at $(0, 0)$ and a scale factor of -2. Label the coordinates of the vertices.

32. Compare $\triangle P''Q''R''$ to $\triangle P'Q'R'$.

33. What do you think would be true about the image if the scale factor of the dilation were 1 or -1?

34. Draw a new figure on the coordinate plane. Test your hypothesis and describe your findings.

35. WRITING MATHEMATICS Explain how to determine if an image is a dilation of another figure.

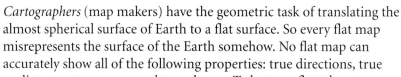

Career
Cartographer

Cartographers (map makers) have the geometric task of translating the almost spherical surface of Earth to a flat surface. So every flat map misrepresents the surface of the Earth somehow. No flat map can accurately show all of the following properties: true directions, true distances, true areas, and true shapes. To better reflect these properties, cartographers use various projections.

Decision Making

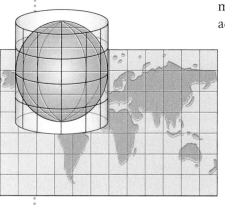

MERCATOR PROJECTION Imagine a transparent globe with a light in the center shining the image of the Earth out, and a sheet of paper wrapped around the globe so that it touches the Earth at the equator. After the image is traced, the paper is unwrapped and laid flat.

1. Will the lines of latitude and longitude remain parallel? Will they be evenly spaced?

2. What will happen to the continents as you move closer to the North or South poles? Why?

Mercator maps, or *cylindrical projections*, are used for navigation because any straight line drawn on the map represents a true direction.

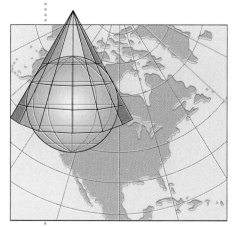

CONIC PROJECTION Imagine a cone placed over the globe. The image of Earth is again projected out and traced. Then the cone is cut open and laid flat.

3. What will the lines of latitude and longitude look like?

Conic projection maps are used when the areas of the continents need to be accurately represented.

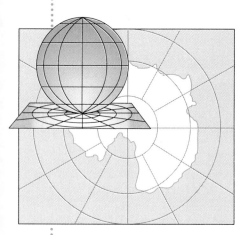

AZIMUTHAL PROJECTION Imagine a globe sitting on a piece of paper. The light inside the globe shines the image out onto the paper.

4. What will the lines of latitude and longitude look like?

5. Where will the most distortion occur on this map? Why?

6. How much of the Earth can be represented at a time on this map? Why?

Azimuthal projection maps serve well for radio work and aviation.

9.5 Perimeter and Area of Similar Polygons

Explore / Working Together

1. Copy rectangle *ABCD* onto graph paper. Label the length of each side and the measure of the perimeter.

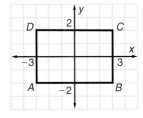

2. On the same graph draw three new rectangles, each similar to *ABCD*, and for which the scale factor of the new rectangle to *ABCD* is 2 : 1, 3 : 1, and 4 : 1, respectively.

   For each of the new rectangles, label the length of each side and the measure of the perimeter.

3. Copy and complete the following table using the data you have collected about the three pairs of similar rectangles. Enter the ratios in simplest form.

Scale Factor	2 : 1	3 : 1	4 : 1
Ratio of Perimeters			

4. Make an observation about the relationship between the scale factor and the ratio of the perimeters.

5. Do you think the same relationship is true for the scale factor and the ratio of the areas? Why?

6. Verify your conjecture by finding the area of each of the four rectangles. Add a new row to your table called "Ratio of Areas" and fill in the table.

7. What is the relationship between the scale factor of two similar polygons and the ratio of their areas?

Build Understanding

Since the lengths of the sides of similar polygons have a constant ratio and those sides are used to determine perimeter and area, it follows that the ratios of the perimeters and areas should be related to the scale factor. These relationships are presented in the following two theorems.

> **THEOREM 9.1**
>
> If two polygons are similar with a scale factor of $a : b$, then the ratio of their perimeters is $a : b$.

Given Similar rectangles with a scale factor of $a : b$.

Prove The ratio of their perimeters is $a : b$.

Proof Consider these rectangles.

$$P_1 = 2(x + w) \quad \text{and} \quad P_2 = 2(y + z)$$

$$\frac{P_1}{P_2} = \frac{2(x + w)}{2(y + z)} = \frac{x + w}{y + z}$$

Since the scale factor is $\frac{a}{b}$,

$$\frac{x}{y} = \frac{a}{b} \quad \text{and} \quad \frac{w}{z} = \frac{a}{b}$$

$$x = \frac{ay}{b} \quad \text{and} \quad w = \frac{az}{b}$$

Substitute for x and w into the perimeter ratio. Simplify.

$$\frac{P_1}{P_2} = \frac{\frac{ay}{b} + \frac{az}{b}}{y + z}$$

$$= \frac{\frac{a(y + z)}{b}}{y + z} \qquad \text{Factor out } a \text{ in numerator.}$$

$$= \frac{a(y + z)}{b} \cdot \frac{1}{y + z} \qquad \text{Simplify complex fraction.}$$

$$= \frac{a}{b} \qquad \text{Reduce.}$$

So the ratio of the perimeters is the same as the scale factor. ◄

> **THEOREM 9.2**
>
> If two polygons are similar with a scale factor of $a : b$, then the ratio of their areas is $a^2 : b^2$.

CHECK UNDERSTANDING

In Example 1, could the scale factor be written as 8 : 12 or 2 : 3? Explain.

How would rewriting the scale factor affect the ratio of the perimeters and the ratio of the areas?

EXAMPLE 1

Given $\triangle ABC \sim \triangle DEF$. What is the scale factor? What is the ratio of the perimeters? What is the ratio of the areas?

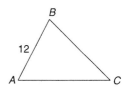

Solution
The scale factor is 12 : 8 or 3 : 2. The ratio of the perimeters is also 3 : 2. The ratio of the areas is $3^2 : 2^2$ or 9 : 4.

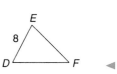

The manner in which the perimeters and areas of similar polygons is related to the scale factor applies to different real world situations.

EXAMPLE 2

INTERIOR DECORATING Mrs. Telesco is buying a rectangular area rug for her living room. She has chosen a particular design and fabric. In that design and fabric, a 3 ft by 5 ft rug costs $149. What would be a reasonable cost for the same design and fabric in a 9 ft by 15 ft size?

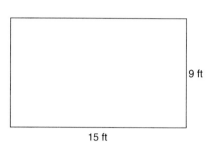

Solution

Examine the ratios of the corresponding sides of the rectangles.

$$\frac{\text{large}}{\text{small}} = \frac{9 \text{ ft}}{3 \text{ ft}} = \frac{15 \text{ ft}}{5 \text{ ft}} = \frac{3}{1}$$

The scale factor is constant and the rectangles are similar. So the ratio of the areas is $3^2 : 1^2$.

3 ft

5 ft

9 ft

15 ft

The larger rug is 9 times the size of the smaller rug. A reasonable cost for the larger rug would be 9(149) or about $1341. ◄

If you know the ratio of the areas of similar polygons, you can work backwards to determine the scale factor.

EXAMPLE 3

The ratio of the areas of the similar figures shown is 4 : 225. Find the scale factor. Find JK.

A   5 cm   B

2.5 cm

D   C

J   K

M   L

Solution

$$\frac{a^2}{b^2} = \frac{4}{225}$$

$$\frac{a}{b} = \frac{2}{15}$$ Take square root of both sides.

The scale factor is $\frac{2}{15}$.

$\overline{JK}$ corresponds to $\overline{AB}$.

$$\frac{AB}{JK} = \frac{5}{JK}$$ Substitute for AB.

$$\frac{2}{15} = \frac{5}{JK}$$ Substitue scale factor for ratio of sides

$$2(JK) = 5(15)$$ Cross product property

$$JK = 37.5$$ Solve for JK.

JK is 37.5 cm. ◄

Find the scale factor. Find the ratio of the perimeters and of the areas.

1. △ABC ~ △DEF

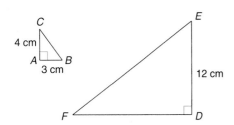

2. JKLMN ~ XWVZY

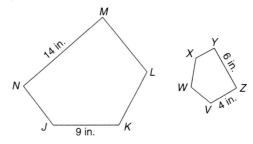

3. WRITING MATHEMATICS Discuss the relationship between the ratio of the areas of two similar figures and the ratio of the perimeters. Justify your answer.

SPORTS A baseball diamond is really a square with each side measuring 90 ft. A regulation softball diamond is a square with each side measuring 60 ft.

4. Are baseball diamonds and softball diamonds similar? Justify your answer.

5. What is the ratio of the distance a player must run around a baseball field to the distance a player must run around a softball field?

6. The grassy infields of baseball and softball fields are well cared for and require fertilizer. If a groundskeeper uses 10 lb of fertilizer on the infield of a baseball diamond, how much should be used on the infield of a softball diamond?

Consider a figure similar to each figure below with the given area. Find the dimensions of this figure that correspond to the dimensions in the drawing.

7. $A = 135 \text{ cm}^2$

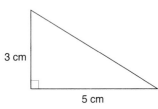

8. $A = 75 \text{ in.}^2$

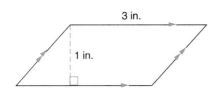

9. $A = 5.5 \text{ cm}^2$

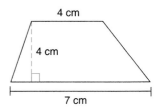

10. In the diagram at the right, △JKL ~ △PQR. The ratio of the areas of the similar triangles is 4 : 9. Find the scale factor and determine PR.

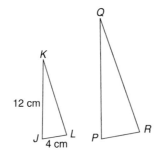

PRACTICE

11. Copy and complete the table below.

Scale Factor	3 : 4				1 : 5		3 : 8	
Ratio of Perimeters		2 : 7		1 : 6				
Ratio of Areas			9 : 100			9 : 16		1 : 3

Find the scale factor. Find the ratios of the perimeters and of the areas.

12. △XYZ ~ △ABC

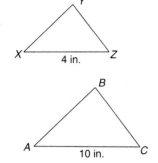

13. ABCDEF ~ JKLMNO

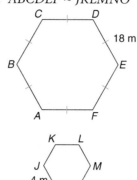

14. △ABC ~ △BDC

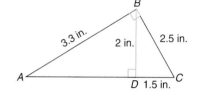

15. WRITING MATHEMATICS If two similar polygons have the same area, must they be congruent? Justify your answer.

INTERIOR DESIGN To coordinate the decor of a master bedroom and an adjoining dressing alcove, a decorator has suggested using the same wallpaper on one wall of each room. The wall in the bedroom is 12 ft high and 8 ft wide. The wall in the dressing alcove is 8 ft high and $5\frac{1}{3}$ ft wide.

16. Are the two walls similar? If so, what is the scale factor?

17. The decorator estimated the cost for papering the alcove wall, including the paper and the labor to hang it, at about $150. What would be a reasonable estimated cost for papering the bedroom wall?

GARDENING For her flower garden this year, Mrs. Lowry wants to lay out a rectangular region that has twice as much area as the rectangular region she had last year. Last year, Mrs. Lowry's garden measured 12 ft by 20 ft.

18. Mr. Lowry suggests that they double the length of each side of last year's region. Do you agree or disagree? Explain.

19. If Mrs. Lowry wants to adjust both of the old dimensions by the same value, explain how she could get the area she wants for her new garden.

For each pair of similar polygons, find the measure of the indicated side.

20. The areas of two similar polygons are in the ratio 16 : 25. If the length of a side of the smaller polygon is 28 mm, find the length of the corresponding side of the larger polygon.

21. The areas of two similar polygons are 108 cm^2 and 192 cm^2. If the length of a side of the larger polygon is 8 cm, find the length of the corresponding side of the smaller polygon.

EXTEND

In the diagram, *ABCD* ~ *FGHI*.

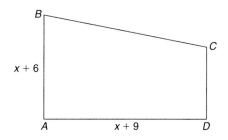

22. Find *x*. What is the scale factor of the quadrilaterals?

23. If the perimeter of the smaller quadrilateral is $3x - 2$, find the perimeter of the larger.

24. If the area of the larger quadrilateral is $x^2 + 7x - 18$, find the area of the smaller.

25. Prove the following theorem for the special case of a rectangle. If two polygons are similar with a scale factor of $a : b$, the ratio of their areas is $a^2 : b^2$.

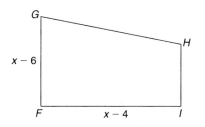

CARTOGRAPHY   On Phil's map of Kansas, 1 cm represents 45 km.

26. The map shape of Kansas is similar to the actual shape. How can you determine the scale factor between the similar shapes? What is the scale factor?

27. What information would you need to know in order to use the scale factor to calculate the actual area of Kansas? If you knew this information, how would you calculate the area of Kansas? In what unit would your answer be?

THINK CRITICALLY

28. In the diagram $\triangle ABC \sim \triangle ADE$. What is the ratio of the area of trapezoid *BCED* to the area of $\triangle ABC$?

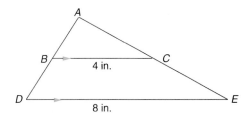

29. The scale factor for two similar polygons is 2 : 3. The sum of the areas of the polygons is 143 cm². Find the area of each polygon.

30. The shortest sides of two similar polygons have lengths of 5 in. and 12 in. Find the length of the shortest side of a similar polygon whose area equals the sum of the areas of the first two polygons.

31. One of two similar polygons has an area 25 percent more than that of the other. What is the ratio of the perimeters of the polygons?

ALGEBRA AND GEOMETRY REVIEW

Find the slope of the line that passes through each pair of points.

32. $(2, 3)$ and $(-6, 1)$ **33.** $(-4, 0)$ and $(-2, 1)$ **34.** $(-3, 5)$ and $(2, 5)$

Find the perimeter of each polygon.

35. a regular hexagon with side lengths of 6 cm

36. an equilateral triangle with side lengths of 3 in.

Find the sum of the exterior angles of each polygon.

37. triangle   **38.** quadrilateral   **39.** pentagon

40. STANDARDIZED TESTS If the ratio of the perimeters of two similar polygons is 4 : 9, then the ratio of the areas of the polygons is

  A. 4 : 9   **B.** 2 : 3   **C.** 16 : 81   **D.** 8 : 18

PROJECT *Connection*   For this activity, you will be applying the grid technique you used in the last project connection to get the map your group has chosen ready for enlargement.

1. Obtain a copy of the map your group has chosen for enlargement. Each group member needs a copy.

2. Each group member should trace the map onto a sheet of paper. Include country, state, or province boundaries.

3. Choose a grid size for the map. As you may have noticed from the last project connection, a grid size that is too large will leave lots of the drawing to be free-handed, while a grid size too small will be tedious to transfer.

4. Using a straightedge, draw the grid on your tracing of the map.

Geometry Excursion

The Four Color Map Problem

Maps have always fascinated topologists because of certain properties they possess with respect to color. The intriguing question concerns the *least* number of colors required to color a map so that no two regions sharing a common border have the same color.

Consider the flat map at the right.

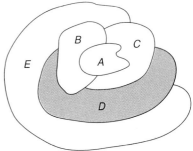

1. Would the map-coloring condition be satisfied if region *B* and region *C* were colored with the same color? Explain.

2. What other regions, if any, can also be colored red? Explain.

Trace the flat maps shown below. Color each with as few colors as possible with no two adjacent regions the same color.

3.

4.

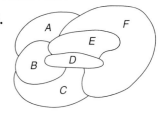

5. A flat map of eight neighboring states is shown below. As colored so far, three colors are sufficient to distinguish among seven of the eight states. How would you color the state of Virginia? Explain.

Is it possible to draw a flat map that requires more than four colors? This problem became well known after it was first presented to the London Mathematical Society in 1848 by the English mathematician Arthur Cayley.

Although mathematicians were convinced that only four colors were necessary, it was not until 1976 that two American mathematicians Kenneth Appel and Wolfgang Haken of the University of Illinois, with the aid of a powerful computer, were able to offer a definitive proof for the Four Color Map Theorem.

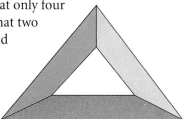

6. **WRITING MATHEMATICS** In your own words, write a statement of the Four Color Map Theorem. Explain how the diagram above illustrates this theorem.

Draw a flat map with seven regions that requires the number of colors indicated.

7. only 2 colors   8. only 3 colors   9. 4 colors

10. **WRITING MATHEMATICS** Describe the strategy you used in order to force your map to need more colors.

THE MOEBIUS STRIP At the right, is an ordinary loop, which has two sides and two edges. It is impossible to travel on this loop from one side to the other without crossing over one of the edges. At the far right is a strip that contains a *half twist*. This twisted strip, known as the *Moebius strip*, has only one side and one edge.

The rectangle divided into 8 regions as shown is from a paper presented in 1910 by the German mathematician Heinrich Tietze. Below is a list of the numbered regions in pairs.

5		6	
3	1		2
	4		

1-2, 1-3, 1-4, 1-5, 1-6, 2-3, 2-4, 2-5, 2-6, 3-4, 3-5, 3-6, 4-5, 4-6, 5-6

11. Are there any pairs of numbered regions that do not share part of their borders? If so, which are they?

On one side of a strip of paper, sketch a rectangle like the one above to fill the strip. Turn the strip over and trace your divided rectangle with its numbered regions. Give the strip a half twist to make it into a Moebius strip. Now the unnumbered region at the lower right becomes part of region 5 and the unnumbered region at the lower left becomes part of region 6.

12. Are there any pairs of regions on the Moebius strip map that do not share parts of their borders? If so, which are they?

13. How many colors are needed to color this map on the Moebius strip?

TOPOLOGY started as a branch of geometry, but during the second quarter of the 20th century it underwent such generalization and became involved with so many other branches of mathematics that is now more properly considered a fundamental division of mathematics.

Today topology may be roughly defined as the mathematical study of continuity.

9.6 Proving Triangles Similar

Explore / Working Together

- Each group member needs to draw a triangle.

  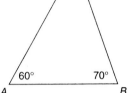

  1. Use a ruler to draw a segment $\overline{AB}$.

  2. Use a protractor to draw a 60° angle at A and a 70° angle at B.

  3. Extend the rays of the angles until they intersect. Label the point of intersection C.

  4. What must be true about $\angle C$ in each of the triangles drawn?

  5. Are the triangles that your group created congruent? Explain.

  6. Are the triangles similar? Justify your answer.

Build Understanding

- Recall, two triangles are similar if all pairs of corresponding angles are congruent and all pairs of corresponding sides are proportional.

  Just as there are theorems and postulates about congruent triangles, there are theorems and postulates to determine if two triangles are similar.

  > **ANGLE-ANGLE (AA) SIMILARITY POSTULATE**
  >
  > **POSTULATE 23** **If two angles of one triangle are congruent to two angles of another triangle, then the triangles are similar.**

  > **THEOREM 9.3** **SIDE-SIDE-SIDE (SSS) SIMILARITY THEOREM**
  >
  > If corresponding sides of two triangles are proportional, then the triangles are similar.

  > **THEOREM 9.4** **SIDE-ANGLE-SIDE (SAS) SIMILARITY THEOREM**
  >
  > If an angle of one triangle is congruent to an angle of a second triangle, and the lengths of the sides including these angles are proportional, then the triangles are similar.

EXAMPLE 1

Are the triangles similar? Justify your answer. If the triangles are similar, write a similarity statement.

a.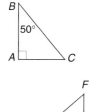

b.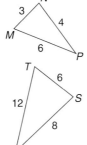

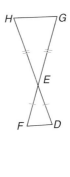

c.

Solution

a. In $\triangle ABC$, since $m\angle A = 90°$ and $m\angle B = 50°$, then $m\angle C = 40°$.

In $\triangle EFD$, you know $m\angle E = 90°$ and $m\angle D = 40°$.

So two angles of one triangle are congruent to two angles of the other triangle, $\angle C \cong \angle D$ and $\angle A \cong \angle E$.

By the AA Similarity Postulate, $\triangle ABC \sim \triangle EFD$.

b. Pair the lengths of the sides of the triangles so that you see a common ratio.

$$\frac{MN}{TS} = \frac{3}{6} = \frac{1}{2} \qquad \frac{NP}{SU} = \frac{4}{8} = \frac{1}{2} \qquad \frac{PM}{UT} = \frac{6}{12} = \frac{1}{2}$$

So the corresponding sides of the triangles are in proportion.

$$\frac{MN}{TS} = \frac{NP}{SU} = \frac{PM}{UT}$$

By the SSS Similarity Postulate, $\triangle MNP \sim \triangle TSU$.

c. The vertical angles formed by the intersecting line segments are congruent, $\angle DEF \cong \angle HEG$.

In $\triangle DEF$, the including sides of $\angle DEF$ are congruent, $\frac{DE}{EF} = 1$.

In $\triangle HEG$, the including sides of $\angle HEG$ are congruent, $\frac{HE}{GF} = 1$.

So an angle of one triangle is congruent to an angle of the other triangle, and the including sides are proportional.

By the SAS Similarity Postulate, $\triangle DEF \sim \triangle HEG$. ◄

> **CHECK UNDERSTANDING**
>
> Write the similarity statement of Example 1b in another way. Justify your statement.

The methods for proving that two triangles are similar are also useful in proving other properties of similar triangles.

THEOREM 9.5

Corresponding medians of similar triangles are proportional to the corresponding sides.

THEOREM 9.6

Corresponding altitudes of similar triangles are proportional to the corresponding sides.

EXAMPLE 2

Find the length of $\overline{BD}$.

Solution

By the AA Similarity Postulate, $\triangle ABC \sim \triangle YZV$.

$\overline{AC}$ and $\overline{YV}$ are corresponding sides of similar triangles. Determine the ratio of their lengths.

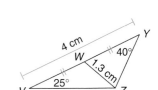

$$\frac{AC}{YV} = \frac{3+3}{4} = \frac{6}{4} = \frac{3}{2}$$

So the scale factor is 3 : 2.

You know that $\overline{BD}$ and $\overline{WZ}$ are medians. So $BD : WZ = 3 : 2$ by Theorem 9.5.

$\dfrac{3}{2} = \dfrac{BD}{WZ}$	Write a proportion.
$\dfrac{3}{2} = \dfrac{BD}{1.3}$	Substitute for WZ.
$2(BD) = 3(1.3)$	Cross Product Property of Proportions
$BD = 1.95$	

The length of $\overline{BD}$ is 1.95 cm. ◀

Similar triangles are useful in modeling real-world situations.

EXAMPLE 3

ENVIRONMENTALISM The highest mountain on the Atlantic coast of the United States may one day be on Staten Island, New York where a mountain is being constructed from garbage. The plan for the completed mountain is shown on the next page in a plane figure. The scale of the drawing is 1 in. : 637 ft. Find the approximate height of the proposed mountain.

$\frac{11}{16}$ in.

18°

Solution

The drawing of the garbage mountain is similar to a two-dimensional slice of the figure of the actual garbage mountain. Let h represent actual height in feet.

$$\frac{\text{height in drawing}}{\text{actual height}} = \frac{1 \text{ in.}}{637 \text{ ft}}$$

$$\frac{\frac{11}{16}}{h} = \frac{1}{637}$$

$$637\left(\frac{11}{16}\right) = h$$

$$437.9 = h$$

So the height of the proposed mountain of garbage is about 438 ft. ◄

TRY THESE

Are the triangles similar? Justify your answer. If the triangles are similar, write a similarity statement.

1.

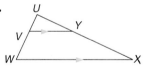

2.

3.

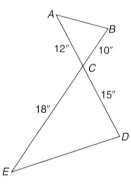

Find the length of x in each figure.

4.

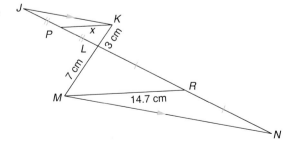

5.

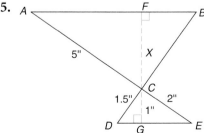

6. **WRITING MATHEMATICS** Name the triangle similarity methods that have corresponding triangle congruence methods. Discuss why the AA similarity method does not have a corresponding congruence method.

CINEMATICS The drawing below shows how film is projected onto a screen.

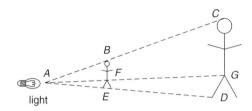

7. What must be true about $\overline{BE}$ and $\overline{CD}$?

8. What must be true about $\triangle ABE$ and $\triangle ACD$?

9. In the film, the person is 2 cm tall. If the projection lamp is 5 cm away from the film, and the projector is 8 m away from the wall, how tall should the image of the person be?

10. Use the drawing to explain why film images are distorted on the screen if the projector is not perpendicular to the wall.

PRACTICE

Are the triangles similar? Justify your answer. If the triangles are similar, write a similarity statement.

11.
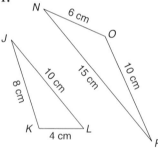

12.
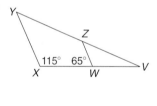

13.
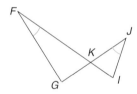

14.
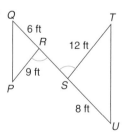

15.
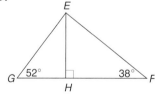

16.
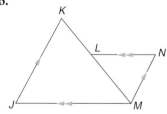

17. **WRITING MATHEMATICS** Explain why the AA Similarity Postulate requires only two pairs of congruent angles.

Find the length of x in each figure.

18.

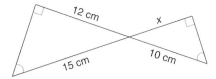
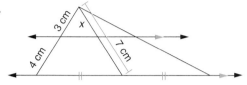

19.

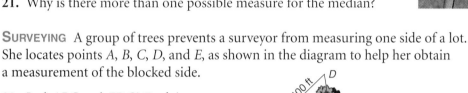

Draw two triangles so that $\triangle ABC \sim \triangle XYZ$, $AB = 18$ cm, and $XY = 14$ cm. The measure of a median of one of the triangles should be 6 cm.

20. What are the possible measures of the other median?

21. Why is there more than one possible measure for the median?

<small>**SURVEYING**</small> A group of trees prevents a surveyor from measuring one side of a lot. She locates points A, B, C, D, and E, as shown in the diagram to help her obtain a measurement of the blocked side.

22. Is $\triangle ABC \sim \triangle EDC$? Explain.

23. Find DE

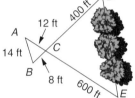

<small>**PHYSICS**</small> When an object is placed on a ramp, part of its weight w (which is a downward force) is directed along the ramp as a sliding force f. These forces are represented by vectors with lengths proportional to w and f.

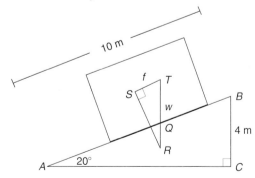

24. Is $\triangle ABC \sim \triangle RTS$ Explain.

25. Find $m\angle RTS$, the angle between the vectors.

26. If $w = 50$ lb, find the value of f.

EXTEND

27. Prove the following.

Given $\overline{AB} \perp \overline{AE}, \overline{AE} \perp \overline{ED}$

Prove $\triangle ABC \sim \triangle DEC$

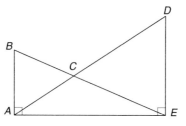

SIGHTING-BY-EYE METHOD To estimate the distance to an object, you need two objects that are about the same distance from you. With your left eye closed, line up your outstretched arm with the first object. With your right eye closed, sight the other object with your other arm.

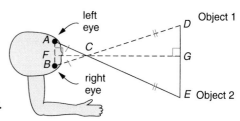

28. What is true about $\triangle ABC$ and $\triangle EDC$? Justify your answer.

You can measure the distance between your eyes, AB, and the distance from your eye to your hand, FC. Then you can estimate the distance between the two objects.

29. If $AB = 7$ cm, $FC = 70$ cm, and $DE = 7$ m, approximately how far is it from F to G? Justify your answer.

THINK CRITICALLY

To prove both the SSS and SAS Similarity Theorems, refer to the diagrams shown.

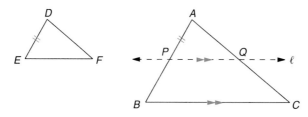

30. Prove the SSS Similarity Theorem.

Given $\dfrac{AB}{DE} = \dfrac{AC}{DF} = \dfrac{BC}{EF}$

Prove $\triangle ABC \sim \triangle DEF$

Plan for Proof Introduce line ℓ parallel to $\overline{BC}$ and intersecting $\overline{AB}$ at P so that $\overline{AP} \cong \overline{DE}$. Then show $\triangle APQ \sim \triangle ABC$. Use the resulting proportion $\dfrac{AP}{AB} = \dfrac{AQ}{AC}$ with a given proportion to show $\overline{AQ} \cong \overline{DF}$. Similarly, $\overline{PQ} \cong \overline{EF}$. Then $\triangle APQ \cong \triangle DEF$. Use corresponding congruent angles to show $\triangle ABC \sim \triangle DEF$.

31. Prove the SAS Similarity Theorem.

Given $\angle A \cong \angle D, \dfrac{AB}{DE} = \dfrac{AC}{DF}$

Prove $\triangle ABC \sim \triangle DEF$

Plan for Proof Introduce line ℓ and show $\triangle APQ \sim \triangle ABC$. Show $\overline{AQ} \cong \overline{DF}$. Then $\triangle APQ \cong \triangle DEF$. Use corresponding congruent parts to show $\triangle ABC \sim \triangle DEF$.

32. Prove that corresponding altitudes of similar triangles are proportional to the corresponding sides.

CONSTRUCTION Two supporting poles are to be connected with wires. The state building code requires the point at which the wires cross to be at least 14 ft off the ground.

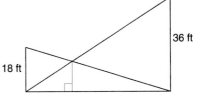

36 ft

18 ft

33. Explain whether or not the plan shown in the diagram meets the specification.

34. The architect wants the wires to meet exactly 14 ft off the ground. Can you reposition the poles in the diagram so that the wires will meet 14 ft off the ground? Explain.

ALGEBRA AND GEOMETRY REVIEW

The diagram shows two lines cut by a transversal.

35. Name all pairs of alternate interior angles.

36. Name all pairs of same side interior angles.

37. Name all pairs of corresponding angles.

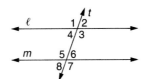

Solve each equation.

38. $3(x - 4) + 6x = 2x - 8$

39. $6x - 8(x - 3) = 18$

40. STANDARDIZED TESTS The measures of two supplementary angles are in the ratio 3 : 2. The measure of the smaller of the two angles is

A. 36°   **B.** 18°   **C.** 108°   **D.** 72°

Simplify the radicals.

41. $\sqrt{80}$ **42.** $\sqrt{120}$ **43.** $\sqrt{75}$

PROJECT *Connection*   You will be choosing a color scheme for your map. Your scheme must use the least number of colors possible and satisfy certain restrictions.

1. Read about the four color map problem on page 522. State the restrictions required for coloring a map.

2. Determine the least number of colors needed for your map.

3. Decide on the colors you will use. Lightly pencil in the names of the colors on the areas of your map.

4. Check to be sure no two regions that share a boundary have the same color. Ask another group member to verify your scheme.

5. Color in your map.

Indirect Measurement of Height

Some objects are too tall to be measured directly with a tape measure or other tool. **Indirect measurement** uses mathematical relationships with known measures to determine inaccessible heights rather than using a direct measurement tool.

Problem

The Smiths have hired a tree removal service to cut down a tree. To be certain that the tree won't hit the fence or the house, the workers need to know approximately how tall the tree is. The workers measure the length of the shadow cast by the tree and find that it is 38 ft long. At the same time, a worker who is 6 ft tall casts a shadow whose length is 7 ft. Find the approximate height of the tree.

Explore the Problem

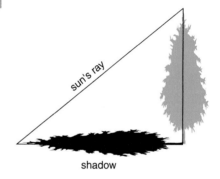

Copy the sketches above. Label the lengths that you know.

1. How are the segments representing the tree and the man positioned with respect to the ground? Mark the sketch.

2. How are the sun's rays positioned? What angles are congruent as a result? Mark the sketch.

3. What can you conclude about the two triangles? Why?

4. Write and solve a proportion to find the height of the tree. Round answer to the nearest foot.

5. WRITING MATHEMATICS The tree is in the center of the Smith's yard, which is a 40 ft by 30 ft rectangle. Several of the workers insist that the tree cannot be cut down without trimming the top first because it is likely that the tree will hit the house or the fence. Other workers claim that it is safe to cut the tree down as it is. Can the tree be cut down without trimming first? Explain.

6. When would this *shadow method* of measurement not be effective?

Investigate Further

The rope on a school flagpole needs to be replaced. Margo Finney, the custodian, will tie the new rope to the old and pull it through the upper pulley. To estimate how much rope she needs, Margo must approximate the height of the pole. It is not a sunny day, so she cannot use the shadow method. Margo will use the *mirror method* to find the height. She will place a mirror on the ground, between herself and the flagpole, so when she looks into the center of the mirror, she will see the top of the flagpole.

PROBLEM
SOLVING PLAN

• Understand
• Plan
• Solve
• Examine

7. Draw line segment $\overline{AB}$ to represent the ground. To represent Margo, draw $\overline{AD}$ perpendicular to $\overline{AB}$. To represent the flagpole, draw $\overline{BE}$ perpendicular to $\overline{AB}$. Place point C on the ground about where you think the mirror should go. Draw $\overline{DC}$ and $\overline{CE}$.

$\angle ACD$ is the *angle of incidence* and $\angle BCE$ is the *angle of reflection*. A law of physics states that the angle of incidence and the angle of reflection are congruent. Adjust your diagram, if necessary.

8. What must be true about $\triangle DCA$ and $\triangle ECB$? Why?

9. Margo is 5.5 ft tall. She has placed the mirror on the ground 6 ft from herself and 32 ft from the foot of the flagpole. Find the approximate height of the flagpole.

10. How long of a rope should Margo use? Justify your answer.

11. WRITING MATHEMATICS Discuss the advantages and disadvantages of both the shadow method and mirror method. Which method do you think would be easier to use? Explain.

12. Explain some factors that make the results of using each method only an approximation.

Apply the Strategy

13. EGYPTOLOGY About 3000 years ago, the Greek geometer Thales measured the height of the Great Pyramid in Egypt. Taking simultaneous measurements, he measured the length of the shadow of the Great Pyramid and the length of the shadow of a nearby pole. Use the drawing to write a proportion Thales might have used to find the height of the Great Pyramid.

height of pyramid $\frac{1}{2}$ base of pyramid   shadow of pyramid   shadow of pole

14. **HOME IMPROVEMENT** A painter needs to know the height of a building to estimate the amount of paint needed for the front side. When the building casts an 18 ft shadow, the 6 ft tall painter casts a 3 ft long shadow. How tall is the building?

15. **TOURISM** Each time Old Faithful in Yellowstone National Park erupts, rangers could estimate the height of the geyser by comparing it to the height of a tree. First the rangers locate a tree of the same height. The shadow of the tree is 93 ft at the same time that the shadow of a 6 ft ranger is 4 ft. Find the height of the tree to the nearest foot.

16. **CIVIL ENGINEERING** A city ordinance prohibits a tree over 20 m from being cut down unless it poses a danger. When planning a new driveway, the city engineer, Roy Kahn, determines which trees can be removed and which cannot. To judge the height of a tree, Roy uses the mirror method. Roy, who is 2 m tall, places the mirror on the ground 1.5 m from himself and 17 m from the tree. Should Roy allow this tree to be cut down to clear a path for the driveway? Explain.

17. **AERONAUTICS** A bush pilot flies naturalists into remote areas where there are no airports. To determine if a safe take-off can be executed, the pilot needs to know approximately how tall the trees are at the edge of a meadow. To judge the height of the tallest tree, pilot Jan Bruk, who is 1.75 m tall, places a mirror on the ground 2.5 m from herself and 25 m from the tree. If Jan needs a 20 m clearance, is a safe take-off possible? Explain.

18. **LUMBERING** A lumber company wants to harvest trees about 100 ft tall. To estimate the height of trees in a new section, supervisor Tony Gomez, who is 6 ft tall, places a mirror 4 ft from himself and 50 ft from a tree on the edge of the section. Should Tony declare that this section is ready for harvesting? Explain.

INSTRUMENTATION A *hypsometer* is used for determining heights by indirect measurement. You can make a hypsometer from a piece of cardboard, a drinking straw, a length of string, and a washer. Use equal spacing for the unit markings on $\overline{AB}$ and $\overline{BC}$. The diagram shows how you can use a hypsometer to measure the height of a tree.

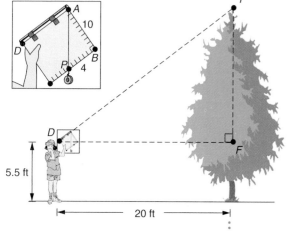

19. Explain why $\triangle ABP \sim \triangle DFT$.

20. As shown in the diagram, $AB = 10$ units, $BP = 4$ units, and $DF = 20$ ft. Find FT. Find the height of the tree. Explain your calculation.

REVIEW PROBLEM SOLVING STRATEGIES

Starlight, Starbright

1. Two congruent equilateral triangles overlap and form a six pointed star. The intersection of the two triangles determine the vertices of a regular hexagon. If the area of the hexagon is 60 units^2, what is the area of each original triangle?

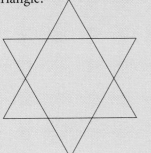

SUNSHINE the Magician

2. Mr. Sunshine, the magician, does a mindreading trick in his magic act. A spectator uses the diagram below to "randomly" arrive at a geometric term. First, she places her finger on any one of the five words. She next moves in the direction of the arrows around the diagram spelling her selected term, one letter for each new word on which she lands. When the spelling is complete she concentrates on the final word under her finger. The magician then reads her mind and tells her the thought of word. No matter which word the spectator chooses first, Mr. Sunshine knows the final term. How does he do it? Can you tell what the word is?

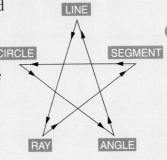

Off to the Races

3. ABCD is a square field, 400 yd on each side. Point E is on side DC, 100 yd from C. Joe and Leah are going to have a race. Joe starts at D and Leah starts at A. Joe runs from D directly to C, but Leah is going to run from A to E and then to C. Both runners maintain a constant rate of speed. When Leah gets to E, Joe is only 70 yd from C. Who wins the race and by how many yards?

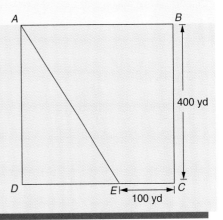

Proportional Segments

Explore/Working Together

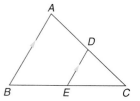

- Work with a partner. Each partner should draw a △ABC. Draw $\overline{DE} \parallel \overline{AB}$.

  1. Measure AD, DC, BE, and EC. Find the values of $\frac{AD}{DC}$ and $\frac{BE}{EC}$. Write both ratios as decimals rounded to the nearest tenth.

  2. Make an observation about the values of the ratios. Compare your results with those of your partner and classmates.

Build Understanding

- Explore illustrates a proportionality property of triangles.

> **THEOREM 9.7** **TRIANGLE PROPORTIONALITY THEOREM**
>
> If a line parallel to one side of a triangle intersects the other two sides, then it divides those sides proportionally.

Given △ABC, $\overline{DE} \parallel \overline{AB}$

Prove $\frac{DA}{CD} = \frac{EB}{CE}$

Plan for Proof Use similar triangles.

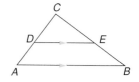

Proof

Statements	Reasons
1. $\overline{DE} \parallel \overline{AB}$	1. Given
2. $\angle CDE \cong \angle CAB$, $\angle CED \cong \angle CBA$	2. Corresponding Angles Postulate
3. $\triangle CDE \sim \triangle CAB$	3. AA Similarity Postulate
4. $\frac{CA}{CD} = \frac{CB}{CE}$	4. Definition of similar triangles
5. $CD + DA = CA$; $CE + EB = CB$	3. Segment Addition Postulate
6. $\frac{CD + DA}{CD} = \frac{CE + EB}{CE}$	6. Substitution Property of Equality
7. $1 + \frac{DA}{CD} = 1 + \frac{EB}{CE}$	7 Algebra
8. $\frac{DA}{CD} = \frac{EB}{CE}$	8. Subtraction Property of Equality

EXAMPLE 1

In the diagram of $\triangle ACE$, $\overline{BD} \parallel \overline{AE}$,
$CD = 3$ cm, $CB = 4$ cm, and $BA = 2$cm.
Find CE.

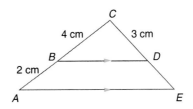

Solution

Apply the Triangle Proportionality Theorem.

Method 1 Use a proportion involving CE.

$$\frac{CE}{CD} = \frac{CA}{CB}$$

$$\frac{CE}{3} = \frac{4+2}{4}$$

$$\frac{CE}{3} = \frac{6}{4}$$

$$4(CE) = 18$$

$$CE = \frac{18}{4}$$

$$CE = 4.5 \text{ cm}$$

Method 2 Find DE. Then add to get CE.

$$\frac{CD}{DE} = \frac{CB}{BA}$$

$$\frac{3}{DE} = \frac{4}{2}$$

$$4(DE) = 6$$

$$DE = \frac{6}{4}$$

$$DE = 1.5$$

$$CE = 1.5 + 3 \quad \text{or} \quad 4.5 \text{ cm}$$

By either method, CE = 4.5 cm   ◄

When three parallel lines have two transversals, the dotted auxiliary
line shown in the diagram produces two triangles. Applying the
Triangle Proportionality Theorem to the triangles results in the
following statement.

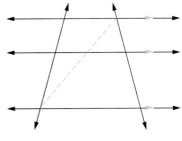

> **COROLLARY 9.8**   If three parallel lines are cut by two
> transversals, then they divide the transversals
> proportionally.

Refer to the above diagram when you write the proof of this corollary as
an exercise.

EXAMPLE 2

Use the diagram to find RS.

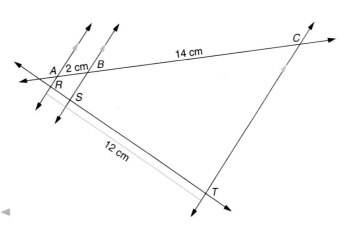

Solution

$$\frac{RS}{RT} = \frac{AB}{AC}$$

$$\frac{RS}{12} = \frac{2}{2+14}$$

$$16(RS) = 24$$

$$RS = \frac{24}{16} \text{ or } 1.5 \text{ cm} \quad ◄$$

Applying the converse of Theorem 9.7 is useful in understanding proportional segments of triangles.

THEOREM 9.9   **CONVERSE OF TRIANGLE PROPORTIONALITY THEOREM**

If a line divides two sides of a triangle proportionally, then it is parallel to the third side.

EXAMPLE 3

Use the given diagram to answer each question.

   a. Is $\overline{KM} \parallel \overline{JN}$?

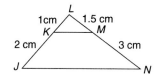

   b. Is $\overline{QS} \parallel \overline{PT}$?

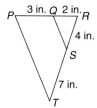

Solution

Test the given lengths to see if they form a proportion.

   a. $\dfrac{1}{2} \overset{?}{=} \dfrac{1.5}{3}$ The cross products are equal, $1(3) = 2(1.5)$.

   By the Converse of the Triangle Proportionality Theorem, $\overline{KM} \parallel \overline{JN}$.

   b. $\dfrac{2}{3} \overset{?}{=} \dfrac{4}{7}$ The cross products are not equal, $2(7) \neq 3(4)$.

   The segments are not parallel.   ◀

Another way to create proportional segments is by bisecting an angle of a triangle.

THEOREM 9.10   **TRIANGLE ANGLE-BISECTOR THEOREM**

If a ray bisects an angle of a triangle, then it divides the opposite side into segments proportional to the other two sides of the triangle.

Given $\triangle ABC$, with $\overline{AP}$ bisecting $\angle A$.

Prove $\dfrac{BP}{PC} = \dfrac{AB}{AC}$

Plan for Proof Draw a line through B parallel to $\overrightarrow{AP}$. Extend $\overrightarrow{CA}$ to intersect that line, at Q.

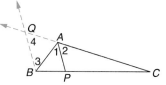

Since $\overline{BQ} \parallel \overline{AP}$, then $\dfrac{BP}{PC} = \dfrac{AQ}{AC}$, $\angle 2 \cong \angle 4$, and $\angle 1 \cong \angle 3$. It follows that $\angle 3 \cong \angle 4$ and $AQ = AB$. Use substitution to reach the conclusion.

Refer to the Plan for Proof of Theorem 9.10 when you do Exercise 20.

Proportional segments are applied in different real world situations.

CARPENTRY To support a book shelf, a carpenter is constructing a wooden brace as shown in the diagram.

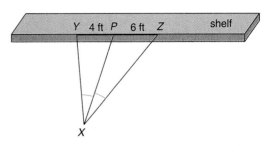

The perimeter of $\triangle XYZ$ is 30 ft. What are the lengths of its sides?

Solution

Since $\overline{XP}$ bisects $\angle X$ of $\triangle XYZ$, apply the Triangle Angle-Bisector Theorem.

Step 1:
Use the value $YZ = 4 + 6 = 10$ and the given value of the perimeter to express one of the remaining two sides in terms of the other.

$$\text{perimeter} = 30$$

$$YX + ZX + YZ = 30$$

$$YX + ZX + 10 = 30$$

$$YX + ZX = 20$$

$$ZX = 20 - YX$$

Step 2:
Write a proportion.
$$\frac{YP}{PZ} = \frac{YX}{ZX}$$

$$\frac{4}{6} = \frac{YX}{20 - YX}$$

$$6(YX) = 4(20 - YX)$$

$$6(YX) = 80 - 4(YX)$$

$$10(YX) = 80$$

$$YX = 8$$

$$ZX = 20 - 8$$

$$ZX = 12$$

The measures of the sides are 8 cm, 10 cm, and 12 cm.

Find the unknown lengths in each diagram.

1.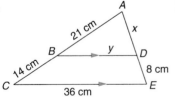

2.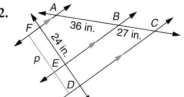

3.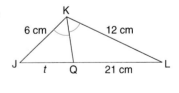

4. **WRITING MATHEMATICS** The Triangle Proportionality Theorem allows you to write many different proportions. How can you tell that the proportion you are setting up is correct? Include a labeled diagram with specific proportions.

5. In $\triangle ABC$, draw $\overline{DE}$ to intersect $\overline{AB}$ at D and $\overline{AC}$ at E.
   If $AD = 30$ mm, $DB = 12$ mm, $AE = (x + 3)$ mm, and $EC = (x - 3)$ mm, find the value of x that will make $\overline{DE} \parallel \overline{BC}$.

Housing The studs that form the frame for a house are parallel and equally spaced along the floor support so that $\overline{AB} \cong \overline{BC} \cong \overline{CD}$.

6. What must be true about $\overline{EF}$, $\overline{FG}$, and $\overline{GH}$? Justify your answer.

7. If the relation you stated about $\overline{EF}$, $\overline{FG}$, and $\overline{GH}$ were not true, how would the house framing be affected?

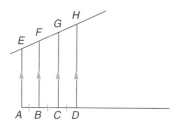

PRACTICE

Find the unknown lengths in each diagram.

8.

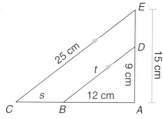

9.

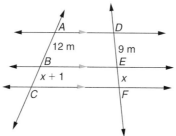

10.

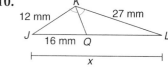

11. **Writing Mathematics** Discuss how the Triangle Angle-Bisector Theorem when applied to the bisector of the vertex angle of an isosceles triangle results in a special ratio.

12. In $\triangle RST$ draw $\overline{VW}$ to intersect $\overline{RS}$ at V and $\overline{RT}$ at W. If $TW = 15$ in., $WR = 18$ in., $SV = (x + 3)$ in., and $VR = (x + 5)$ in., find the value of x that will make $\overline{VW} \parallel \overline{ST}$.

Astronomy Two comets are on a collision course if they continue on the paths shown in the diagram.

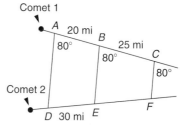

13. What is true about $\overline{AD}$, $\overline{BE}$, and $\overline{CF}$? Explain.

14. How far did Comet 2 travel from E to F? Explain.

15. Are the comets traveling at the same speed? Explain.

16. Can the collision point be found using the Corollary of the Triangle Proportionality Theorem? Explain.

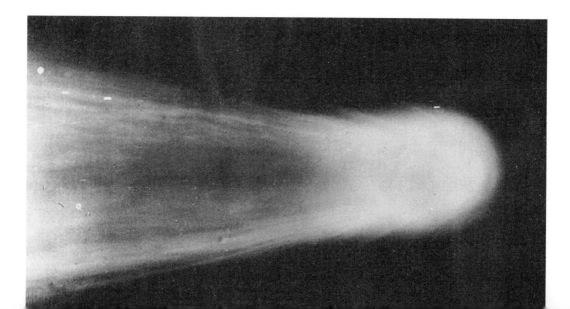

17. One side of a triangle is 3 cm longer than another. The ray bisecting the angle between these sides divides the opposite sides into 5 cm and 3 cm segments. Find the perimeter of the triangle.

EXTEND

Refer to the diagrams and plans for proof in the Build Understanding section.

18. Prove the Corollary 9.8.

19. Prove the Converse of Triangle Proportionality Theorem.

20. Prove the Triangle Angle-Bisector Theorem.

21. Prove the Triangle Midsegment Theorem. The segment joining the midpoint of two sides of a triangle is parallel to the third side and half its length. Refer to the diagram at the right where $\overline{EF}$ is extended to G so that $\overline{FE} \cong \overline{FG}$. Show $\triangle BFE \cong \triangle CFG$. Show $AEGC$ is a parallelogram.

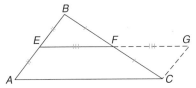

THINK CRITICALLY

GEOMETRIC CONSTRUCTION Follow these steps to divide $\overline{AB}$ into 3 segments of equal length.

Using point P not on $\overline{AB}$, draw $\overrightarrow{AP}$.

With the compass tip at A, make an arc to intersect $\overrightarrow{AP}$ at D.

With the same compass setting, make additional arcs on $\overrightarrow{AP}$ by placing the compass tip at D to get E and at E to get F. Draw $\overline{FB}$.

At E, construct a line parallel to $\overline{FB}$. Use G to label the point at which this line intersects $\overline{AB}$.

At D, construct a line parallel to $\overline{FB}$. Use H to label the point at which this line intersects $\overline{AB}$.

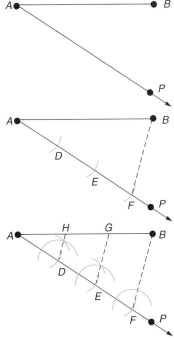

22. Explain why $AH = HG = GB$.

23. Use the procedure to divide a line segment into 4 segments of equal length.

24. Suppose you have a piece of lined notebook paper and a pencil. Using just these items, describe how you could divide a strand of dry spaghetti into nine congruent parts. Justify your answer.

25. Use paper folding to divide a segment into three equal segments.

26. Use paper folding to divide a segment into four equal segments.

27. WRITING MATHEMATICS Describe a different paper-folding construction for Exercise 26.

Multiply the polynomials.

28. $(x + 3)(x + 5)$

29. $(2y + 1)(2y - 1)$

30. $(x + 4)(x^2 + 5x - 1)$

Find the distance between the two points.

31. $(3, 5)$ and $(2, -10)$

32. $(0, -5)$ and $(4, 2)$

33. $(-1, 3)$ and $(-4, -2)$

Find the measure of the third side of right triangle ABC **where** C **is the right angle.**

34. $a = 3$ m, $c = 5$ m

35. $a = 5$ in., $b = 7$ in.

36. $b = 6$ cm, $c = 9$ cm

Find the value of x **that makes each statement a proportion.**

37. $\dfrac{3}{x} = \dfrac{x}{12}$

38. $\dfrac{x}{5} = \dfrac{20}{x}$

39. $\dfrac{7}{x + 1} = \dfrac{x + 1}{7}$

40. STANDARDIZED TESTS In $\triangle ABC$, point D is on $\overline{CA}$ and E is on $\overline{CB}$ so that $\overline{DE} \parallel \overline{AB}$. If $CD = 2$ ft, $DA = 7$ ft, and $CE = 5$ ft, then EB is

A. 17.5 ft

B. 19.5 ft

C. 21.5 ft

D. 22.5 ft

PROJECT *Connection*

Cartographers use a variety of methods when adapting the size of their drawings. The grid method, combined with a light table, can provide a surface which allows for an accurate reproduction. Your group may want to consider using a light table or some other light source to complete the enlargement of your map. You need to finalize a plan for constructing the enlargement of your map.

1. Finalize with your group the surface that you will use.

2. Determine the size of the grid you will use for the enlargement. Be sure the chosen surface will accommodate the entire map.

3. Write a plan that includes all of the following.

   • The nature and size of the surface to be used.

   • The size of grid that you will use and what size the final map will be.

   • How your group will create perpendicular lines for the grid on the surface.

   • A list of the materials you will need.

   • An estimate of the amount of time it will take to create your map and a schedule for completing tasks.

Surveyors, architects, civil engineers, tax assessors, and naturalists are some of the many professionals who use aerial photographs in their jobs. Aerial photographs provide a current pictorial view of a region and can show features that do not appear on maps.

Decision Making

There are three different ways in which aerial photographs are taken.

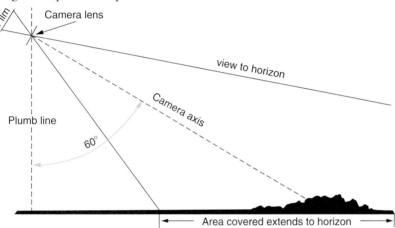

High Oblique Technique

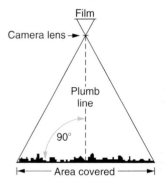

Vertical Technique

1. For each of the different techniques illustrated, determine if the two triangles in the drawing are similar. Explain your conclusion.

The aerial photographs that result from the three techniques have different advantages and disadvantages related to the amount of area covered and the kind of distortion that is in the photograph.

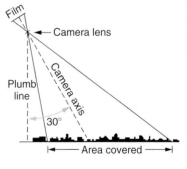

Low Oblique Technique

2. Identify which of the aerial photographs below is vertical, which is low oblique, and which is high oblique. Explain your conclusions.

a.

b.

c.

CHAPTER REVIEW

VOCABULARY

Choose the word from the list that correctly completes each statement.

1. A ___?___ is a comparison of two quantities by division.

    a. extremes

2. In proportion $a : b = c : d$, a and d, the first and fourth terms, are called ___?___.

    b. similar

3. In proportion $a : b = c : d$, b and c the second and third terms, are called ___?___.

    c. means

4. Two polygons are ___?___ if their vertices can be paired in such a way that corresponding angles are congruent and corresponding sides are in proportion.

    d. ratio

Lessons 9.1 and 9.2 PROPORTIONAL REASONING AND THE GOLDEN RATIO pages 493–500

- In a proportion the product of the means is equal to the product of the extremes.

Solve each proportion.

5. $\dfrac{x}{24} = \dfrac{5}{15}$

6. $\dfrac{x + 1}{5} = \dfrac{2x + 3}{4}$

7. $\dfrac{9}{x} = \dfrac{x}{4}$

8. On a field trip the ratio of students to teacher must be less than $7 : 1$. Twenty-five students are going to the art museum on a field trip, how many teachers are needed?

Lessons 9.3 and 9.4 SIMILAR POLYGONS AND DILATIONS pages 501–513

- Corresponding sides of similar polygons are proportional.

- When the scale factor is less than 1, the dilation is a reduction. When the scale factor is greater than 1, the dilation is an enlargement.

Determine if the figures are similar. If they are similar, find the scale factor of the figures.

9.
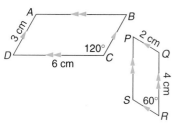

10.
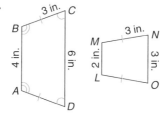

11.
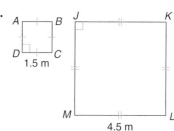

12. Draw $\triangle ABC$ with $A(-1, 3)$, $B(2, 1)$, $C(3, -2)$. Using $(0, 0)$ as the center of dilation, draw a dilation with scale factor 3. Give the coordinates of the vertices of the image of $\triangle A'B'C'$.

- If the scale factor of two similar polygons is $a : b$, then the ratio of the perimeters of the polygons is $a : b$, and the ratio of the areas is $a^2 : b^2$.

13. Copy and complete the table.

Scale Factor	1 : 4		
Ratio of Perimeters		2 : 3	
Ratio of Areas			4 : 100

Lessons 9.6 and 9.7   PROVING TRIANGLES SIMILAR AND INDIRECT MEASUREMENT pages 525–535

- Triangles may be proven similar by using AA Similarity Postulate, SSS Similarity Theorem, or SAS Similarity Theorem.

- Similar triangles have corresponding medians and corresponding altitudes are proportional to the corresponding sides.

Determine if the triangles are similar. Justify your answer. If they are similar, write similarity statement.

14.
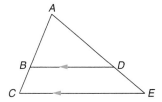

15.
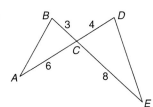

16.
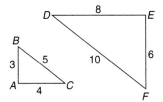

17. The scale factor of two similar triangles is 4 : 5. The length of a median in the smaller triangle is 14 cm, what is the length of the corresponding median in the larger triangle?

18. The length of the shadow of a 6 ft man is 4.8 ft. Find the height of a pole that casts a 22 ft shadow.

Lesson 9.8   PROPORTIONAL SEGMENTS   pages 536–542

- If a line parallel to one side of a triangle intersects the other two sides, then it divides those sides proportionally.

Find the unknown value.   **Determine if $\overline{AB} \parallel \overline{CD}$.**

19.
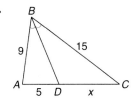

20.

21.
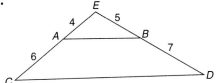

CHAPTER ASSESSMENT

CHAPTER TEST

Solve each proportion.

1. $\dfrac{x}{81} = \dfrac{4}{36}$

2. $\dfrac{5}{4} = \dfrac{5x}{x+3}$

3. A chemical cleaner is to be mixed with water at a ratio of 3 : 20. If you use 5 c of cleaner, how many cups of water should you use?

4. Draw and label two similar figures. Explain why they are similar.

5. **WRITING MATHEMATICS** Explain why two squares are always similar.

Are the polygons similar? Justify your answer. If they are similar, write a similarity statement.

6.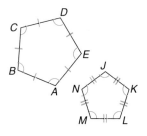

7.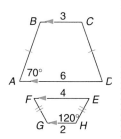

8. Given $ABCD \sim JKLM$. Find the scale factor.

 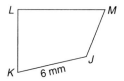

9. **STANDARDIZED TESTS** The coordinates of $\triangle ABC$ are $A(-5, 2)$, $B(-3, -1)$, $C(2, 4)$. If $(0, 0)$ is the center of dilation and the scale factor is 3, name the coordinates of the dilation image of A'?

   A. $A\left(-\dfrac{5}{3}, \dfrac{2}{3}\right)$ B. $A'(-2, 5)$

   C. $A'(-15, 6)$ D. can't be determined

10. The scale factor of two similar figures is 3 : 5. What is the ratio of the perimeters? of the areas?

11. A carpet that is 3 ft by 5 ft costs $65. The company produces a larger carpet with the same materials that is 9 ft by 15 ft. What should be the price of the larger carpet?

Are the triangles similar? Justify your answer. If they are similar write a similarity statement.

12.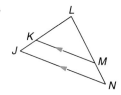

13.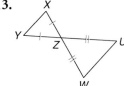

14. To find the height of a tree, Wyatt places a mirror so that he can see the top of the tree in the mirror. The mirror is 3 ft from him and 7.5 ft from the tree. If Wyatt is 5.5 ft tall, how tall is the tree?

Find the value of x.

15.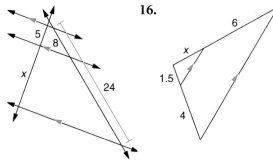

16.

17. Given: $\dfrac{AC}{AE} = \dfrac{BD}{BE}$

   Prove: $\triangle CED \sim \triangle AEB$

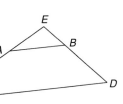

PERFORMANCE ASSESSMENT

WRITE ABOUT IT Write an essay which describes what you have learned about similar triangles and congruent triangles. Compare and contrast the two concepts. Include ways you know you can prove triangles congruent or similar.

USE GRAPH PAPER Draw a triangle on graph paper and perform a series of dilations to the triangle using different scale factors and the origin as the center of dilation. Indicate the order in which the dilations were done and the scale factors used. Determine what scale factor would need to be used in order to do a dilation of the first triangle to create the last triangle directly.

LOOK AT THINGS IN A DIFFERENT WAY Draw two similar triangles on graph paper. Use the distance formula to show that the sides have the same ratio.

RESEARCH Research applications of the golden ratio and appearances of the golden rectangle in nature. Locate photographs or illustrations to include with your research.

PROJECT ASSESSMENT

 Work together as a group to construct your enlarged map. Here are some suggestions for how to proceed.

1. Choose one tracing from among those done by the individual group members. Agree as to the choice of grid size and overall appearance.

2. Consolidate suggestions from the plans of the individual group members to finalize one plan for the entire group. Agree on the surface for the enlarged map, the color scheme, and the materials to be used for coloring.

3. Finalize the method the group will use to achieve perpendicular grid lines on the surface chosen for the enlarged map.

4. List the tasks that need to be performed, such as ruling the grid lines, transferring the images in the grid squares, and coloring the final enlargement.

   Select individual group members, or perhaps partners, to perform these tasks. Use the maps produced by the individuals to identify their special talents.

5. Invite parents to view the group works produced by the class. Plan a presentation to explain the mathematics behind the project.

CUMULATIVE REVIEW

Fill in the blank.

1. Polygons whose corresponding angles are congruent and whose corresponding sides are all proportional are ____?____ polygons.

2. A line segment whose endpoints are the midpoints of two sides of a triangle is called a(n) ____?____.

3. The relationship in a right triangle in which the sum of the squares of the lengths of the two legs is equal to the square of the length of the hypotenuse is called the ____?____.

4. A quadrilateral with opposite sides parallel and congruent can be classified as a(n) ____?____.

In each problem, $\triangle ABC \sim \triangle DEF$. Find x and y.

5.

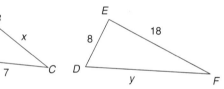

6.

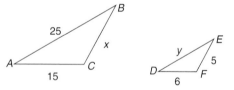

State the property that justifies each conclusion.

7. If $x = 20$, then $\frac{x}{4} = 5$.

8. If $x - y = 7$, then $x - y + 3 = 10$.

9. If $x = y$ and $y = z$, then $x = z$.

Find the area and perimeter of each rectangle.

10.

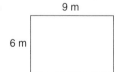

9 m

6 m

11.

8.3 cm

1.6 cm

12. **WRITING MATHEMATICS** Explain why any triangle must have at least two acute angles.

Find the area of each figure.

13.

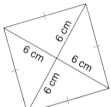

6 cm 6 cm 6 cm 6 cm

14.

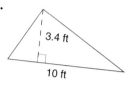

3.4 ft

10 ft

15. A person 5 feet tall casts a shadow that is 8 feet 9 inches long. The person is standing next to a flagpole, and the flagpole casts a shadow that is 21 feet long. How tall is the flagpole?

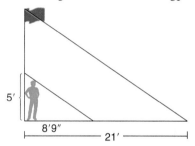

5′

8′9″

21′

16. **STANDARDIZED TESTS** Which of the following statements is not true about an isosceles trapezoid?

    A. The diagonals are congruent.

    B. The length of the midsegment is half the sum of the lengths of the two bases.

    C. The base angles are congruent.

    D. The diagonals bisect each other.

17. Determine the measure of $\angle 1$.

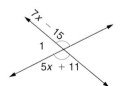

$7x - 15$

1

$5x + 11$

18. Find the coordinates of the midpoint of the segment whose endpoints are $(-3, 4)$ and $(5, -3)$.

STANDARDIZED TEST

STANDARD FIVE-CHOICE Select the best choice for each question.

1. In the drawing below △*ABC* ~ △*LMN*. Which of the following statements is true?

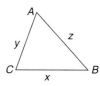 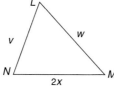

I. $y = 2v$ **II.** $\dfrac{y}{z} = \dfrac{v}{w}$

II. The perimeter of the larger triangle is twice the perimeter of the smaller triangle.

- **A.** I only
- **B.** II only
- **C.** III only
- **D.** II and III only
- **E.** I, II, and III

2. The point $(-1, 2)$ lies on a line that has a slope of $\dfrac{2}{3}$. Which of the following points also lies on the line?

- **A.** $(1, 5)$
- **B.** $(-3, -1)$
- **C.** $(2, 4)$
- **D.** $(-4, 4)$
- **E.** Two of the above

3. The point (x, y) lies in the first quadrant. Which of the following is the image of the point under a reflection in the *x*-axis?

- **A.** $(x, -y)$
- **B.** $(-x, y)$
- **C.** $(-x, -y)$
- **D.** $(y, -x)$
- **E.** None of these

4. Which of the following pairs represent complementary angles?

- **A.** 87° and 93°
- **B.** 13° and 70°
- **C.** 50° and 40°
- **D.** 10° and 90°
- **E.** None of these

5. Which of the following figures does not have an area of 10 square units?

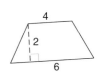

- **E.** Two of the above

6. Which of the following statements is not true concerning similar figures?

- **A.** The corresponding angles are congruent.
- **B.** The corresponding sides are proportional.
- **C.** The ratio of the perimeters equals the ratio of the lengths of corresponding sides.
- **D.** The ratio of the areas equals the ratio of the lengths of corresponding sides.
- **E.** All are true

7. Which of the following pairs of points could be the endpoints of the base of an isosceles triangle that has a midsegment parallel to the base with length 5?

- **A.** $(-1, 6)$ and $(7, 0)$
- **B.** $(-4, 5)$ and $(0, 2)$
- **C.** $(0, -3)$ and $(5, 2)$
- **D.** $(-2, -3)$ and $(8, 7)$
- **E.** None of these

8. Given a line and a point not on the line, how many lines can be drawn through the point parallel to the given line?

- **A.** 0
- **B.** 1
- **C.** 2
- **D.** Infinite
- **E.** Cannot be determined

10 Applications of Similarity and Trigonometry

Take a Look AHEAD

Make notes about things that look familiar.

- What angle measures describe special right triangles?
- How many examples can you find that have triangles that are similar to each other?
- How are ratios used in this chapter?

Make notes about things that look new.

- Find three names of special types of angles used in real world applications.
- What does it mean to "solve a triangle"? What methods can you use?

DATA Activity

Open House Today

The construction of new homes is considered to be an *economic indicator*. An economic indicator is a factor that can help describe the strength of the economy. In years or months when new housing starts are few, the economy is described to be in a state of slow growth or sluggish. When the number of housing starts are high, the economy is considered to be growing or healthy.

SKILL FOCUS

- ▶ Read and interpret data.
- ▶ Determine percent change.
- ▶ Solve multi-step problems.
- ▶ Graph data.

GEOMETRYWORKS

HELP WANTED

Construction & Carpentry

In this chapter, you will see how:

- **DRAFTERS** use special right triangles to create parallel lines and angles of a given measure. (Lesson 10.3, page 569)

- **CARPENTERS** use trigonometry to form the necessary angles for a proper fit. (Lesson 10.5, page 579)

- **SURVEYORS** use geometry to measure land and determine property boundaries. (Lesson 10.6, page 586)

New Housing Starts, thousands			
Year	Number of units started	Year	Number of units started
1900	189	1981	1084
1910	387	1982	1062
1920	247	1983	1703
1925	937	1984	1749
1930	330	1985	1742
1935	221	1986	1806
1940	603	1987	1621
1945	326	1988[2]	1488
1950	1952	1989	1376
1955	1646	1990	1193
1960[1]	1296	1991	1014
1965	1510	1992	1200
1970	1469	1993	1288
1975	1171	1994	1457
1980	1292	1995	1354

[1]Figures prior to 1960 do not include farm housing.
[2]Figures for 1988 and after do not include public housing.

Use the table to answer the following questions.

1. What was the percent increase in housing starts from 1985 to 1986?

2. What was the percent decrease in housing starts from 1986 to 1987?

3. What would these numbers indicate about the economy of the nation from 1985 to 1987?

4. Compare housing starts in 1987 to start-ups in 1988. Does this comparison give an accurate indication of the economy at that time? Why?

5. Look at the low statistics in 1930, 1935 and 1945. What historical events could explain this data?

6. **WORKING TOGETHER** Gather housing start statistics for your community. Select a specific span of years to study. Make a graph of the data.

PROJECT BUILDING TO SCALE

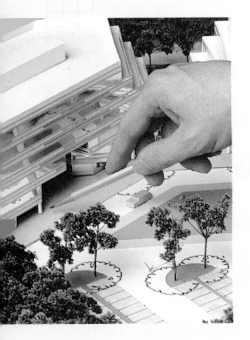

In the process of designing a new building, complex, or structure, an architect may build a scale model of a design. Scale models are miniature structures usually made of foamboard or other lightweight sturdy material that provide a three-dimensional view of the finished structure.

In this project, you will work in groups to collect data and create a scale model.

PROJECT GOAL

To build a scale model of a building or playing field.

Getting Started

Work in groups of three to five students.

1. Imagine flying in a plane or helicopter above your school. On paper, sketch the buildings on your school grounds as you would see them from above.

2. Add any parking lots and playing fields which surround the school.

3. Compare sketches among members of your group. Add anything missing from your sketch.

4. Estimate the dimensions of your school buildings and write them on your drawing. Be sure to include the height of the buildings.

5. Estimate and record the dimensions of any playing fields and parking lots in your drawing.

PROJECT Connections

Lesson 10.3, page 568:
Use a clinometer and a special right triangle to measure height.

Lesson 10.6, page 585:
Use a transit to measure lengths.

Lesson 10.9, page 599:
Make a scale drawing of your design.

Chapter Assessment, page 603:
Make a scale model of your design.

Internet Connection

www.learninggeometry.com

Geometry Workshop

Fractals

Think Back

- For each pattern of points, draw the next figure of the pattern. Then write a function rule to generalize the pattern.

   1.

   2.

Explore

- **3.** Construct a large equilateral triangle as shown in the figure labeled *stage 0* below. Shade the interior of the triangle lightly.

- **4.** Locate the midpoint of each side of the triangle. Draw a new triangle whose vertices are these midpoints.

- **5.** Erase the shading from the interior of the new triangle, as shown in *stage 1* below.

- **6.** Three shaded triangular regions remain. Repeat Questions 4 and 5 on each. The result should look like *stage 2* below.

stage 0

stage 1

stage 2

stage 3

stage 4

Geometry Workshop

GEOMETRY: WHO, WHERE, WHEN

Fractals were used to create computer images of the entire Genesis planet in the motion picture *Star Trek II: The Wrath of Khan*.

7. Copy and complete this table for the pattern of triangles shown on the previous page.

stage	0	1	2	3	4
number of shaded triangles	▢	▢	▢	▢	▢

8. What numerical pattern(s) do you see in the table?

9. How many shaded triangles will there be at stage 6? stage 10?

10. Write a rule for the function *t* that relates the number of shaded triangles to the number *n* of the stage.

Make Connections

The triangle pattern in Explore is called the *Sierpinski triangle*. It is an example of a complex geometric structure called a *fractal*.

In 1980, Benoit B. Mandelbrot, a mathematician at the IBM Watson Research Center, presented his observations of the structure called the *Mandelbrot set*, which is pictured at the left. Mandelbrot coined the term *fractal* to describe structures like these.

You can describe a fractal informally as a structure that displays increasing complexity as it is viewed more and more closely. A fractal is created by performing infinitely many repetitions, or *iterations*, of a geometric process. So the Sierpinski triangle actually is the result of infinitely many iterations of the steps in Questions 4 and 5.

Below are stages 0 and 1 of a fractal called the *Sierpinski carpet*.

11. Write a description of the steps you follow to proceed from stage 0 to stage 1.

12. Perform two iterations of the steps outlined in Question 11.

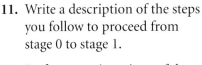

stage 0   stage 1

13. Write a rule for the function *s* that relates the number of shaded squares to the number *n* of the stage.

Draw stages 2 and 3 of each fractal. Describe a function related to the fractal and give a rule for the function.

14.

stage 0   stage 1

15.

stage 0   stage 1

GEOMETRY: WHO, WHERE, WHEN

The Sierpinski triangle and Sierpinski carpet are named after the Polish mathematician Waclaw Sierpinski (1882–1969). Sierpinski first introduced his observations of the triangle in 1916.

An underlying theme of all fractals is *self-similarity*. A figure is **self-similar** if some part of it contains a replica of the whole.

For example, suppose a book cover shows a picture of itself. This means the picture of the book cover shows a picture of itself, and so on to infinity. So, in the figure at the right, the interior of each blue circle contains a replica of the whole book cover. These circular regions are approaching a single point, so the book cover is *self-similar at one point*.

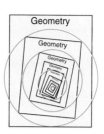

Now consider the Sierpinski triangle as shown at the right. Each circled region shown contains at least one replica of the whole triangle. In fact, if you draw a circle around a region containing *any* shaded part, it will contain at least one replica of the whole triangle. So the Sierpinski triangle is *self-similar at every point*, or **strictly self-similar**.

Tell whether each fractal is strictly self-similar. Explain.

16.
stage 0 ———————
stage 1 ——   ——
stage 2 – –   – –

17.
stage 0   stage 1   stage 2

COMMUNICATING ABOUT GEOMETRY

The fractal pattern in Question 14 is called the *Koch curve*. How is it related to the geometric structure below, which is called the *Koch snowflake*?

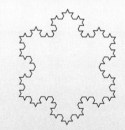

Is the Koch curve self-similar? Is it strictly self-similar?

Is the Koch snowflake self-similar? Is it strictly self-similar?

Summarize

18. **WRITING MATHEMATICS** Explain the difference between *similarity*, *self-similarity*, and *strict self-similarity*.

19. **GOING FURTHER** Suppose the length of each side of the stage 0 Sierpinski triangle is 1 unit. What is the total perimeter of the shaded triangles in stage n?

20. **GOING FURTHER** Suppose the area of the stage 0 Sierpinski triangle is 1 square unit. What is the total area of the shaded triangles in stage n?

21. **MODELING** A *regular tetrahedron* has exactly four faces, each bounded by an equilateral triangle. The *Sierpinski tetrahedron* in the photograph at the right is a fractal pattern of regular tetrahedrons. Make a model of the Sierpinski tetrahedron. What stage did you construct? How many tetrahedrons are in your model? How many would be in stage n?

Explore

- You will need graph paper, a straightedge, and scissors.

1. Use graph paper to draw a large triangle ABC so that $\angle C$ is a right angle and $\overline{BC}$ is the longer leg. Label $\angle A$, $\angle B$, and $\angle C$.

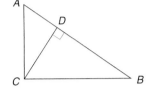

2. Draw altitude $\overline{CD}$. Label D.

3. Make an exact copy of $\triangle ABC$ with altitude $\overline{CD}$. Label the two triangles formed. Place all labels inside the triangle.

4. Cut $\triangle ABC$ into two triangles along $\overline{CD}$.

5. Arrange your triangles so that the right angle is in the same position in each triangle. What angle in each triangle corresponds with $\angle A$ of $\triangle ABC$? $\angle B$ of $\triangle ABC$? $\angle C$ of $\triangle ABC$?

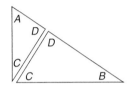

6. Use your answer to Question 5 to help you draw conclusions about $\triangle ABC$, $\triangle ADC$, and $\triangle BDC$.

7. Explain how you can justify your conclusions. Write a similarity statement for each conclusion.

8. Draw an altitude from $\angle D$ to $\overline{BC}$.

9. Test your conclusion from Question 7 with the triangles you formed in Question 8.

Build Understanding

- In Explore, you reviewed the Angle-Angle Similarity Postulate by using the altitude to the hypotenuse of a right triangle to create similar triangles. As a result, you discovered a characteristic of a right triangle.

> **THEOREM 10.1**
>
> If the altitude is drawn to the hypotenuse of a right triangle, then the two triangles formed are similar to the original triangle and to each other.

You can use corresponding parts of similar triangles and proportions to find unknown measures.

EXAMPLE 1

Solve for *a* when $XW = 3$ and $WZ = 12$.

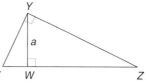

Solution

From Theorem 10.1, $\triangle XWY \sim \triangle YWZ$.
Use corresponding sides to write a proportion containing *a*.

$$\frac{XW}{WY} = \frac{WY}{WZ} \qquad \frac{\text{Side of } \triangle XWY}{\text{Corresponding side of } \triangle YWZ}$$

$$\frac{3}{a} = \frac{a}{12} \qquad \text{Substitute.}$$

$$36 = a^2 \qquad \text{Solve for } a.$$

$$a = 6 \quad \text{or} \quad a = -6$$

Since you are finding the length of a segment, a solution of -6 is not reasonable, so $WY = 6$. ◄

CHECK UNDERSTANDING

Name the three similarity statements associated with the figure in Example 1. In each case, list the corresponding parts.

The **geometric mean** of two positive numbers *a* and *b* is the number *x* such that $\frac{a}{x} = \frac{x}{b}$ where *x* is positive. When you solve this proportion for *x*, you find $x = \sqrt{ab}$. In Example 1, side *a* equals the principal square root of the product of 12 and 3. So, *WY* is the geometric mean of segments *XW* and *WZ*.

THINK BACK

Every positive real number has a positive square root and a negative square root. The positive square root is called the *principal square* root of the number.

EXAMPLE 2

Find the geometric mean between 3 and 8.

Solution

$$x = \sqrt{3 \cdot 8} = \sqrt{24} = 2\sqrt{6} \qquad x = \sqrt{ab}$$

If you estimate the geometric mean, $x \approx 4.9$ ◄

The definition of geometric mean leads to the following corollaries.

COROLLARY 10.2   If the altitude is drawn to the hypotenuse of a right triangle, then its length is the geometric mean of the lengths of the two segments of the hypotenuse.

COMMUNICATING ABOUT GEOMETRY

Explain to a friend how $\sqrt{24}$ is simplified and then estimated in Example 2.

COROLLARY 10.3   If the altitude is drawn to the hypotenuse of a right triangle, then the length of each leg is the geometric mean of the length of the hypotenuse and the length of the segment of the hypotenuse that is adjacent to the leg.

These corollaries allow you to write proportions in similar triangles without identifying the corresponding sides.

EXAMPLE 3

Find each unknown length.

a.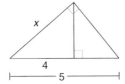

b.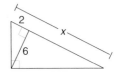

Solution

a. By Corollary 10.3, you can determine that x is the geometric mean of the hypotenuse and the segment of the hypotenuse with a length of 4.

$$x = \sqrt{4 \cdot 5} = \sqrt{20} = 2\sqrt{5}$$

b. By Corollary 10.2, you know 6 is the geometric mean of the two segments of the hypotenuse.

$$\frac{2}{6} = \frac{6}{x - 2} \qquad \text{Write a proportion.}$$

$$2x - 4 = 36$$

$$2x = 40$$

$$x = 20$$

The similarity relationships can be used to measure indirectly.

EXAMPLE 4

MEASUREMENT To approximate the height of a flagpole, Chirag holds a book near his eye so that one edge lines up with the top of the flagpole and the adjacent edge lines up with the bottom of the flagpole. Chirag's eye is 5.5 ft from the ground and he is standing 15 ft from the flagpole. Find the height of the flagpole.

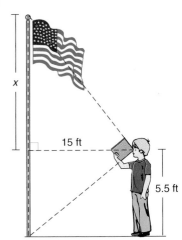

Solution

$$\frac{5.5}{15} = \frac{15}{x}$$

$$5.5x = 225$$

$$x \approx 41$$

The estimated height of the flagpole is 46.5 ft, the sum of x and 5.5.

Copy each statement and fill in the blank.

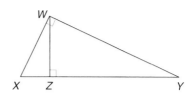

1. $m\angle X = m\angle$ __?__ and $m\angle Y = m\angle$ __?__

2. $\triangle XYW \sim$ __?__ $\sim$ __?__

3. __?__ is the geometric mean of XZ and ZY.
   WX is the geometric mean of __?__ and __?__.

Find the geometric mean.

4. 2, 3   5. 10, 7   6. 1, 16   7. 6, 14   8. $5n, 20n \ (n > 0)$

Find the unknown lengths.

9. 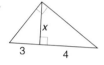   10.   11. 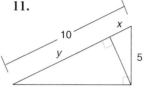   12.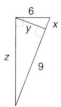

DESIGN Omar is in charge of tent design for a manufacturer. He has designed the tent in the figure at the right. Unlike the tents in the photograph, the front of the tent is not symmetric with respect to the front support pole.

13. What is the length of the front pole to the nearest tenth of a foot? The peak at the top of the tent is a right angle.

14. **WRITING MATHEMATICS** Alicia in marketing feels the pole should be at least 6 ft long so they can advertise that a person can walk into the tent without stooping. Does Omar's design meet this criteria? What are some of the drawbacks to his design?

15. Write a two-column proof of Theorem 10.1: If the altitude is drawn to the hypotenuse of a right triangle, then the two triangles formed are similar to the original triangle and to each other.

Copy each statement and fill in the blank.

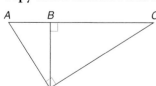

16. $m\angle A = m\angle$ __?__; $m\angle C = m\angle$ __?__

17. $\triangle ABD \sim$ __?__ $\sim$ __?__

18. __?__ is the geometric mean of AB and AC.

Find the geometric mean.

19. 7, 28   20. 14, 15   21. 3, 15   22. $7t, 14t \ (t > 0)$

Find the unknown lengths.

23.

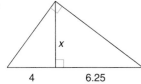

24.

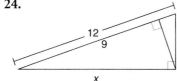

25.

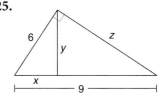

26. MEASUREMENT Stanley wants to find the distance across a pond. He climbs a 40-ft observation tower. When he reaches the top he sights along the two sides of a carpenter's square as shown, locating points B and C. Stanley determines C is 5 ft from A. What is the distance across the pond?

EXTEND

27. Recall the Pythagorean Theorem: In a right triangle, the square of the length of the hypotenuse c is equal to the sum of the squares of the lengths a and b of the legs, or $a^2 + b^2 = c^2$. Complete the proof of the Pythagorean theorem.

Given Right $\triangle ABC$ with altitude $\overline{CD}$

Prove $a^2 + b^2 = c^2$

Proof

Statements	Reasons
1. Right $\triangle ABC$ with altitude $\overline{CD}$	**1.**
2. $\dfrac{x}{a} = \dfrac{a}{c}, \dfrac{y}{b} = \dfrac{b}{c}$	**2.**
3.	**3.** Cross Product Property of Proportions
4. $a^2 + b^2 = cx + cy$	**4.**
5.	**5.** Distributive Property
6. $a^2 + b^2 = c^2$	**6.**

The **geometric mean of three numbers** is the cube root of the product of the three numbers. Find the geometric mean.

28. 2, 4, 8   **29.** 2, 5, 6   **30.** 4, 6, 5

BIOLOGY A starfish is in the shape of a pentagram. Geometric means are found in the shape of the starfish and the pentagram.

31. In a pentagram, AB is the geometric mean between BC and AC. Write a proportion to illustrate this relationship.

32. Begin with the relationship given in Exercise 31 and use the fact that $AB = CD$ to prove that AC is the geometric mean between AB and AD.

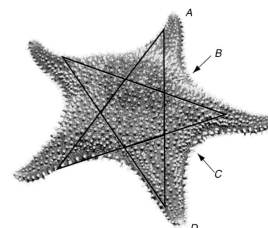

33. Prove Corollary 10.2: If the altitude is drawn to the hypotenuse of a right triangle, then its length is the geometric mean of the lengths of the two segments of the hypotenuse.

34. Prove Corollary 10.3: If the altitude is drawn to the hypotenuse of a right triangle, then the length of each leg is the geometric mean of the length of the hypotenuse and the length of the segment of the hypotenuse that is adjacent to the leg.

THINK CRITICALLY

35. **USING ALGEBRA** The arithmetic mean between two numbers a and b is defined to be. $\frac{a + b}{2}$. Prove that the arithmetic mean must always be greater than or equal to the geometric mean. (Hint: work backward, then write the proof).

36. **USING ALGEBRA** When is the arithmetic mean equal to the geometric mean?

ALGEBRA AND GEOMETRY REVIEW

37. The measure of one angle of a right triangle is four times the measure of a second angle. Find all the possible measures of the angles of the triangle.

38. The sum of the measures of five angles of a hexagon is 678°. Find the measure of the sixth angle.

39. The perimeter of a right triangle is 60 and the sum of the legs is 34. Find the length of each side.

40. In what kind of regular polygon is the ratio of the measure of an interior angle to the measure of an exterior angle 3 to 1?

Solve. Round your answer to the nearest hundredth.

41. $0.5698 = \frac{4.5}{x}$

42. $\frac{8}{x} = 2.4209$

43. $0.87 = \frac{3}{x}$

Find the decimal equivalent of each fraction. Round to the nearest ten-thousandth.

44. $\frac{8}{15}$

45. $\frac{3}{7}$

46. $\frac{5}{9}$

Use parallelogram ABCD pictured at the right to find the following.

47. AD

48. $m\angle C$

49. $m\angle B$

50. DE

51. AB

52. Area of $ABCD$

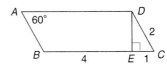

53. **STANDARDIZED TESTS** In $\triangle CDE$, DG is the geometric mean of CG and GE. The length of CE is

   A. 2.25   B. 3.25   C. 5   D. 6.25

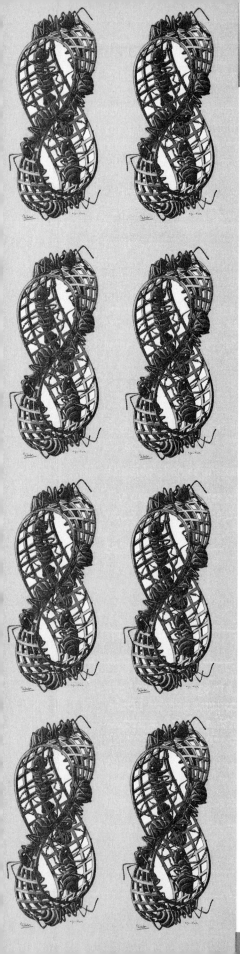

The Möbius Strip

1. Cut a strip of paper 11 in. long and 1 in. wide. Tape the ends together forming a loop in the shape of a cylinder, similar in appearance to a belt that you might wear around your waist. Examine the loop. How many surfaces does it have? Explain your answer.

2. Cut another strip with the same dimensions as the first. This time before taping the ends together, turn one end over 180°, giving the strip a half-twist. Tape the ends together forming a loop similar to the one at the right.

3. Put your pencil on your strip, halfway between the edges and draw a line down the middle parallel to the edges. Continue drawing around the strip until you return to your starting point. Examine the strip and the line. What do you notice?

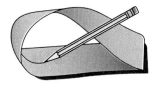

The loop in Question 2 is called a *Möbius strip*, after its inventor Augustus Ferdinand Möbius. The "one-sidedness" property of a Möbius strip has a practical application. Consider a belt used to drive two wheels. Notice it is twisted and then joined to make a Möbius strip.

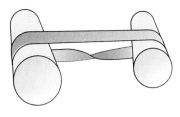

4. THINKING CRITICALLY What are advantages to using a belt that models the Möbius strip?

5. Use scissors to cut the Möbius strip along the line you drew in Question 3. Describe the result.

6. Cut a third strip in the same dimensions as the first. This time put a whole-twist in the strip by turning one end 360° before you tape it to the other end.

7. Do you think this loop is one-sided like the Möbius strip or does it have two sides like the cylindrical loop? Verify your conjecture by drawing a line around the loop halfway between the edges.

8. Use scissors to cut the loop along the line you drew in Question 7. Describe the result.

9. Construct another Möbius strip, a loop with a half-twist. Cut this strip parallel to one edge about one-third of the way from the edge. Describe what happens when you get around the strip one time. Continue to cut until you get back to where you started. Describe the result.

10. Construct a loop with a full twist. Cut this loop parallel to one edge about one-third of the way from the edge. How is the result similar to the result in Question 9? How is it different?

It is possible to construct loops with more than one or two half-twists. Use newspaper to make longer strips. This will make it easier to construct the loops and do the necessary cutting.

11. Construct a loop with three half-twists. With scissors cut it halfway between the edges parallel to the edges. Describe the result.

12. Construct a loop with four half-twists (two full-twists). Cut it halfway between the edges parallel to the edges. Describe the result and compare it to the result in Question 11.

13. Construct a loop with five half-twists. Cut it halfway between the edges parallel to the edges. Describe the result and compare it to the result in Question 11.

14. Make a conjecture about the relationship between the number of half-twists in your loop and the result obtained when the loop is cut through a line running parallel to the edges.

NINETEENTH CENTURY mathematics enjoyed a tremendous growth spurt. People made discoveries in virtually all areas of mathematics and created new areas for mathematical study. Augustus Ferdinand Möbius (1790–1868) was one of the German mathematicians who made mathematics an exciting subject. Artists have used the Möbius strip as the subject of many drawings, printings, and sculptures over the years. Each tile of the photograph on page 562 is M. C. Esher's work using the Möbius strip.

10.3 Special Right Triangles

Explore

Use the figure at the right for Questions 1–4.

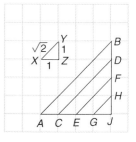

1. In $\triangle XYZ$, what is m$\angle Z$? m$\angle X$? m$\angle Y$?

2. Use the Pythagorean Theorem to verify the measure of XY.

3. Explain why $\triangle ABJ$, $\triangle CDJ$, $\triangle EFJ$, and $\triangle GHJ$ are similar to $\triangle XYZ$.

4. Without applying the Pythagorean Theorem, find GH, EF, CD, and AB.

Use the figure at the right for Questions 5–6.

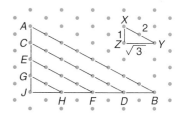

5. In $\triangle XYZ$, what is m$\angle Z$? m$\angle X$? m$\angle Y$?

6. Repeat Questions 2–4.

Build Understanding

In Explore Questions 1–4, you worked with isosceles right triangles that are also named 45°-45°-90° triangles. This name is given because the measures of the angles are 45°, 45°, and 90°.

You can use a 45°-45°-90° triangle and the Pythagorean theorem to find the special relationship of the length of the hypotenuse and the length of the sides.

$$a^2 + b^2 = c^2 \qquad \text{Pythagorean Theorem}$$
$$a^2 + a^2 = c^2 \qquad \text{Isosceles triangle, so } a = b. \text{ Substitute.}$$
$$2a^2 = c^2$$
$$\sqrt{2a^2} = c \qquad \text{Take square root of both sides.}$$
$$a\sqrt{2} = c \qquad \text{Simplify radical.}$$

From this fact, you can derive Theorem 10.4.

THEOREM 10.4	**45°-45°-90° THEOREM**

In a 45°-45°-90° triangle, the length of the hypotenuse is $\sqrt{2}$ times the length of a leg.

If you know the length of one side of a 45°-45°-90° triangle, you can find the length of all sides.

EXAMPLE 1

Find the unknown lengths.

a.
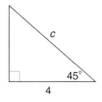

b.
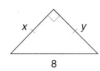

COMMUNICATING ABOUT GEOMETRY

Explain how you can use Theorem 10.4 to justify the following statement.

If the length of one side of a square is n units, then the length of a diagonal of the square is $n\sqrt{2}$ units. Include a figure to illustrate your explanation.

Solution

a. The theorem states hypotenuse = side · $\sqrt{2}$. So by substitution $c = 4\sqrt{2}$.

b.

$$c = x\sqrt{2}$$ Hypotenuse = side · $\sqrt{2}$

$$8 = x\sqrt{2}$$ Substitute for c.

$$\frac{8}{\sqrt{2}} = x$$ Divide each side by $\sqrt{2}$.

$$\frac{\sqrt{2}}{\sqrt{2}} \cdot \frac{8}{\sqrt{2}} = x$$ Rationalize the denominator.

$$4\sqrt{2} = x$$

Since the triangle is isosceles and $x = 4\sqrt{2}$, $y = 4\sqrt{2}$. ◄

The 45°-45°-90° triangle relationship is used in real world applications.

EXAMPLE 2

HIKING Keenan and Joey use a directional compass to find the distance across a river. They select an object directly across the river

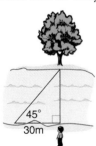

and take a compass reading to that object. Keenan then walks along the river in a direction perpendicular to his original line of sight until the compass reading has changed by 45°. Joey measures the distance he walked as 30 m. What is the width of the river? Explain.

Solution

The distance walked and the width of the river are equal since they are the legs of a 45°-45°-90° triangle. The width of the river is 30 m. ◄

CHECK UNDERSTANDING

Draw an equilateral triangle. Use the length of each side as *n* units. Draw an altitude. What type of triangles are formed? Looking at one of the smaller triangles, write the expression for the length of each side in terms of *n*.

In Explore, you also investigated the relationship of the 30°-60°-90° triangle. You can use the following theorem to find the length of the sides of a 30°-60°-90° triangle, when you know the length of one side.

> **THEOREM 10.5   30°-60°-90° THEOREM**
>
> In a 30°-60°-90° triangle, the length of the hypotenuse is twice the length of the shorter leg, and the length of the longer leg is $\sqrt{3}$ times the length of the shorter leg.

EXAMPLE 3

Find *x* and *y*.

a.

b.

Solution

a. shorter leg
$$2 \cdot x = 6$$ 2 · shorter leg = hypotenuse
$$x = 3$$

longer leg
$$y = \sqrt{3} \cdot x$$ longer leg = $\sqrt{3}$ · shorter leg
$$y = 3\sqrt{3}$$ Substitute for *x*.

b. shorter leg
$$\sqrt{3} \cdot x = 7\sqrt{3}$$ $\sqrt{3}$ · shorter leg = longer leg
$$x = 7$$

hypotenuse
$$y = 2 \cdot x$$ Hypotenuse = 2 · shorter leg
$$y = 14$$ Substitute for *x*.

TRY THESE

Find the unknown lengths.

1.

2.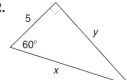

3.

4.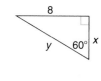

5. The longest side of a 30°-60°-90° triangle is 15. Find the lengths of the other sides.

6. **WRITING MATHEMATICS** Describe the ways that Exercises 1 and 3 are similar and the ways they are different.

7. **PLUMBING** When connecting two pipes of different elevations, a 90° joint is often not desirable. Instead, the pipes are laid so that a different angled joint can be used. The difference in height between the two pipes is 3 ft. How long is the connector pipe?

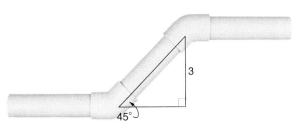

PRACTICE

Find the unknown lengths.

8.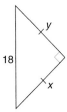

9.

10.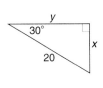

11.

12.

13.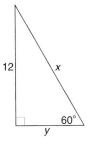

14.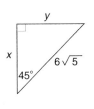

15.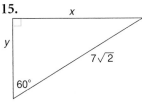

16. Find the length of a diagonal of a square if each side is 4 in.

17. Find the length of an altitude of an equilateral triangle where each side is 10 cm.

18. **WRITING MATHEMATICS** In the figure $\angle P$ is labeled correctly. Explain how you can determine that at least two of the other measures are labeled incorrectly. What do you think are the correct lengths of the sides?

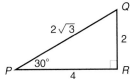

19. **BASEBALL** A professional baseball diamond is a square in which each side is 90 ft. Find the distance from first base to third base to the nearest foot.

20. **HOUSE PAINTING** A 24-ft ladder leans against a house and the angle it makes with the ground is 60°. How high up the side of the house will the ladder reach? Round the answer to the nearest tenth of a foot.

21. Prove Theorem 10-5: In a 30°-60°-90° triangle, the length of the hypotenuse is twice the length of the shorter leg and the length of the longer leg is $\sqrt{3}$ times the length of the shorter leg.

EXTEND

22. **HISTORY** The *Wheel of Theodorus* may be used to find segments whose lengths are irrational numbers. Find the unknown lengths in this drawing. Copy the drawing and identify a 45°-45°-90° triangle and a 30°-60°-90° triangle.

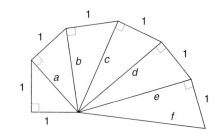

23. In regular hexagon *ABCDEF* the perimeter is 12. Find the lengths of diagonals $\overline{AD}$ and $\overline{BF}$.

MANUFACTURING Metal punches are used to stamp holes in sheet metal. To manufacture a triangular shaped punch, a cylindrical bar is ground down to the shape of the triangle. The cross-section drawing shows the machinist how much of the cylinder to remove.

24. The machinist needs to know the radius of the bar. Find the radius if the center is *E* and each side of the triangle is 1 in. Explain your solution.

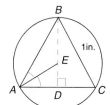

25. A square punch is going to be ground from a circular bar with a radius of 1 in. What is the largest size square punch that can be produced? Explain.

THINK CRITICALLY

26. Each edge of a cube is *r*. Explain how you would find the length of segment *AB* in terms of *r*. (Hint: Draw $\overline{BC}$.)

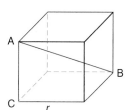

27. **WRITING MATHEMATICS** The figure is a regular octagon divided into several nonoverlapping regions. Show how to use these regions to find the area of the octagon.

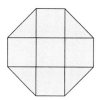

PROJECT *Connection*   A clinometer is used to measure the angle at which you must look up to see the top of a structure. When you know this angle and the horizontal distance to the structure, you can estimate the height of the structure.

1. To make a sighting tool, tape a straw onto the straight edge of a protractor. Tie one end of a 1-ft piece of thread through the hole in the protractor and tie a paper clip or some type of weight to the other end. Tie the knot so the thread can swing back and forth.

2. When the protractor is level and the straw is parallel to the ground, where is the thread?

3. Stand in front of a building. Look through the straw to the top of the structure.

4. Note the angle of the thread. Walk forward or backward until the thread changes 45° from its initial position. Measure the distance to the object.

5. Find the height of the structure. Include the height from the ground to your eye.

A drafter works closely with architects and engineers to prepare technical drawings that are used in construction, manufacturing and engineering. During the design process a drafter takes sketches from the architect or engineer and prepares precise drawings.

Drafting is a skill that requires geometry. Drafters work at a computer using CAD software (Computer Aided Drafting) or at drafter's board using special paper and drafting tools such as a T-square, drafting triangles, and a compass.

Decision Making

Drafting triangles come in two sizes and are models of the special right triangles in this lesson.

1. Name the angle measures of each drafting triangle.

2. Using heavy paper, create each size drafter's triangle.

Most of the measurement, scale, and computational work is automated if the drafter uses a computer. However, a drafter who still draws by hand uses drafting triangles and a T-square to do many of the constructions you have done with a compass and straightedge.

3. In Lesson 3.5, you used special right triangles to draw parallel lines. Describe how to use a ruler and drafting triangles to draw parallel lines.

4. Use a drafting triangle and a ruler to draw $\overline{AB}$ as shown. Describe how you can use the drafting triangle to draw a line perpendicular to $\overline{AB}$.

A drafter also draws objects on paper so they appear three-dimensional. This is achieved by positioning the object at a 30° angle from the x-axis.

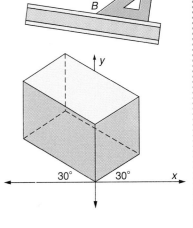

5. Draw a set of axes lightly so they can be erased when you are finished. Use your drafting triangle and a ruler to draw a right square prism. Recall that the opposite edges of a face are parallel.

Exploring Trigonometric Ratios

Think Back

For Questions 1–4, refer to the diagram of △RST.

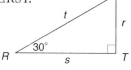

1. Which side is the hypotenuse?
   Which side is the leg opposite ∠R?
   Which side is the leg adjacent ∠R?

2. What is the difference between the labels R and r?

3. Using the given values for r, copy and complete the table.

4. Describe any patterns you see in the table.

	r	s	t	$\frac{r}{s}$	$\frac{r}{t}$	$\frac{s}{t}$
1	$\sqrt{3}$	2	$\frac{1}{\sqrt{3}} = \frac{\sqrt{3}}{3}$	$\frac{1}{2}$	$\frac{\sqrt{3}}{2}$	
2						
3						
4						

Explore

You need a centimeter ruler, a protractor, and a calculator.

5. Draw and label a large △XYZ in which m∠X = 65°, m∠Y = 25°, and m∠Z = 90°. Label each angle with its measure.

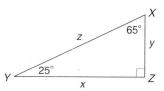

6. Write the label x on side YZ, the label y on side XZ, and the label z on side XY.

7. Label your triangle with the Roman numeral I.

8. Repeat Questions 6 and 7 three times. Each time, make the triangle a different size but maintain the same angle measures. Label the new triangles II, III, and IV.

9. What is the relationship among your triangles?

10. Copy and complete the table. Measure the sides of each triangle to the nearest millimeter.

11. Give each ratio in the last three columns in decimal form, rounded to the nearest tenth.

	x	y	z	$\frac{x}{y}$	$\frac{x}{z}$	$\frac{y}{z}$
△I						
△II						
△III						
△IV						

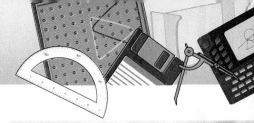

12. Describe any patterns you see in the table.

13. Complete the following conditional statement in as many ways as you can.

If an acute angle of one right triangle is congruent to an acute angle of another right triangle, then the ratio ___?___ .

Make Connections

- In Explore, the patterns you discovered and the conditional statements you completed describe three trigonometric ratios. **Trigonometric ratios** are ratios of the lengths of two sides of a right triangle.

14. Prove the following conjecture.

If an acute angle of one right triangle is congruent to an acute angle of another right triangle, then the ratio of the length of the leg opposite the angle to the length of the leg adjacent to the angle is the same in each triangle.

Given $\triangle RST$ and $\triangle XYZ$ are right triangles.
$\angle R \cong \angle X$

Prove $\dfrac{r}{s} = \dfrac{x}{y}$

Proof

Statements	Reasons
1. ___?___	**1.** Given
2. $\triangle RST \sim \triangle XYZ$	**2.** ___?___
3. $\dfrac{r}{x} = \dfrac{s}{y}$	**3.** ___?___
4. ___?___	**4.** Property of Proportions

The trigonometric ratio in Question 14 is the *tangent ratio*.

$$\text{tangent of } \angle A = \tan A = \frac{\text{length of side opposite } \angle A}{\text{length of side adjacent to } \angle A}$$

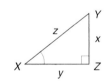

15. For $\triangle ABC$, write tan A as a fraction using the lowercase labels for the sides of the triangle.

For an acute angle of a given size, the tangent ratio is always the same number, no matter how large or how small the triangle.

Two other important trigonometric ratios associated with right triangles are the *sine ratio* and the *cosine ratio*.

$$\text{sine of } \angle A = \sin A = \frac{\text{length of side opposite } \angle A}{\text{length of hypotenuse}}$$

$$\text{cosine of } \angle A = \cos A = \frac{\text{length of side adjacent to } \angle A}{\text{length of hypotenuse}}$$

16. For $\triangle ABC$ in Question 15, write $\sin A$ and $\cos A$ as fractions.

Write a proof of each statement.

17. If an acute angle of one right triangle is congruent to an acute angle of another right triangle, then the sines of the two angles are equal.

18. If an acute angle of one right triangle is congruent to an acute angle of another right triangle, then the cosines of the two angles are equal.

Measure the sides of the triangle to the nearest millimeter. Calculate $\tan Q$, $\sin Q$, and $\cos Q$ to the nearest hundredth.

19.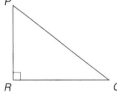

20.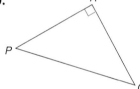

Summarize

21. WRITING MATHEMATICS Restate in your own words the definition of the tangent ratio, the sine ratio, and the cosine ratio.

22. WRITING MATHEMATICS In $\triangle JKL$, $\angle L$ is a right angle, $\tan J = \frac{9}{40}$, and $\sin J = \frac{9}{41}$. Your friend says this means that $KL = 9$, $JL = 40$, and $JK = 41$. Do you agree or disagree? Explain your reasoning.

23. THINKING CRITICALLY Is it possible for the sine of an acute angle in a right triangle to be equal to its cosine? Is it possible for the sine of an angle to be equal to its tangent? Justify your answers.

24. GOING FURTHER Prove this statement: If the tangent of an acute angle of one right triangle is equal to the tangent of the acute angle of another right triangle, then the angles are congruent.

10.5 The Tangent Ratio

Explore

1. Explain why $\triangle AZB \sim \triangle CZD \sim \triangle EZF$.

2. Find each ratio in lowest terms. What is the relationship among the ratios?

   a. $\dfrac{AB}{ZB}$ **b.** $\dfrac{CD}{ZD}$ **c.** $\dfrac{EF}{ZF}$

3. Suppose you extend line ℓ to form a similar $\triangle GZH$ in which $GH = 36$ units. What would be the length of $\overline{ZH}$?

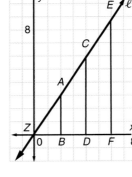

Repeat Questions 1–3 for each figure below.

4.
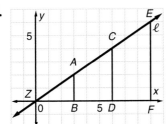

5.
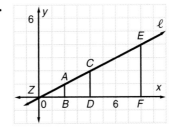

6. Make a conjecture about what happens to the ratios as $m\angle EZF$ becomes smaller. Test your conjecture by drawing a new set of figures on grid paper and finding the ratios.

7. As the triangles in the drawing get larger, does the ratio of the rise and the run, or the ratio of the measure of the legs change?

Build Understanding

The word trigonometry comes from a Greek term meaning "measure of triangles." A **trigonometric ratio** is a ratio of the lengths of two sides of a right triangle. One common trigonometric ratio is tangent. For the right triangle shown, the tangent of $\angle A$ is written **tan A**.

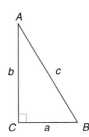

TANGENT RATIO

For any acute angle A in a right triangle, the tangent ratio is

$$\tan A = \frac{\text{length of the side opposite } \angle A}{\text{length of the side adjacent to } \angle A}$$

EXAMPLE 1

Find tan A and tan B. Write each ratio as a fraction and as a decimal. Round your answer to the nearest ten-thousandth and use the $\approx$ symbol if necessary.

Solution

$$\tan A = \frac{21}{20} = 1.05 \qquad \tan B = \frac{20}{21} \approx 0.9524 \qquad \blacktriangleleft$$

In Explore, you saw that the tangent of an angle depends only on the measure of the angle. It does not depend on the size of the triangle. So an acute angle of a given measure has a unique tangent ratio. You can use a calculator to find tangent values given an angle value.

EXAMPLE 2

Find the tangent value of each angle. Round to the nearest ten-thousandth.

a. tan 25°   **b.** tan 80°

Solution

a. tan 25° $\approx$ 0.4663   **b.** tan 80° $\approx$ 5.6713 $\blacktriangleleft$

PROBLEM SOLVING TIP

Before entering data for trigonometric ratios be sure that the angle mode on your calculator is set for degrees (DEG).

If you know the measure of the angle and the length of one side, use the tangent ratio to find the length of the other side.

EXAMPLE 3

Find the value of y to the nearest hundredth.

Solution

$$\tan 64° = \frac{y}{12}$$

$$y = 12 \, (\tan 64°) \approx 24.60$$

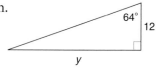

$\blacktriangleleft$

CHECK UNDERSTANDING

List the keystroke sequence your calculator requires for you to solve Example 4.

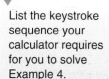

Every tangent ratio has a unique acute angle associated with it. If you know the tangent ratio, you can work backwards to find the measure of the angle by using the inverse tangent function $\tan^{-1}$ on your calculator.

EXAMPLE 4

Find the measure of each angle. Round to the nearest degree.

a. $\tan A = \dfrac{5}{12}$ **b.** $\tan B = 2.1445$

Solution

a. $m\angle A = \tan^{-1} \dfrac{5}{12} \approx 23°.$ **b.** $m\angle B = \tan^{-1} 2.1445 \approx 65°$ $\blacktriangleleft$

An **angle of elevation** is an angle whose vertex is at the eye of an observer and whose sides are a horizontal line and the observer's line of sight upward to an object. An **angle of depression** is an angle whose vertex is at the eye of an observer and whose sides are a horizontal line and the observer's line of sight downward to an object.

For example, a person in a hot-air balloon is the vertex of the angle of depression and a person on the ground looking up to the balloon is the vertex of the angle of elevation. Trigonometry is useful for finding an angle measure or a distance when you are unable to measure it directly.

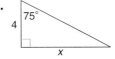

EXAMPLE 5

FORESTRY A forest ranger stands in a lookout tower 250 ft tall. She sights a forest fire at an angle of depression of 4°. How far is the fire from the base of the tower?

Solution

The distance to the fire is AB. Since $\overline{DC} \parallel \overline{AB}$, $\angle CDB$ and $\angle ABD$ are congruent alternate interior angles. So, $m\angle CDB = m\angle ABD = 4°$.

$$\tan 4° = \frac{250}{AB}$$

$$AB = \frac{250}{\tan 4°}$$

$$AB \approx 3575$$

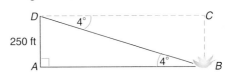

The fire is about 3575 ft away.   ◄

> **COMMUNICATING ABOUT GEOMETRY**
>
> In Example 5, you can also find AB by solving the equation
>
> $$\tan 86° = \frac{AB}{250}.$$
>
> Explain how this equation is obtained. Why might you prefer to use this equation?

TRY THESE

Find tan A and tan B. Write each ratio as a fraction and as a decimal. Round your answer to the nearest ten-thousandth if necessary.

1.

2.

3.
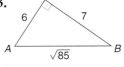

4.

Use a calculator to find the values. Round to the nearest ten-thousandth.

5. tan 25°

6. tan 47°

7. tan 65°

8. tan 60°

Find x. Round to the nearest hundredth.

9.

10.

11.
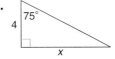

12.
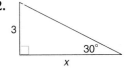

Find m∠A. Round your answer to the nearest degree.

13. $\tan A = 0.4245$ **14.** $\tan A = 8.1443$ **15.** $\tan A = \dfrac{14}{3}$ **16.** $\tan A = 5.\overline{3}$

17. 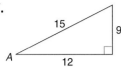   **18.** 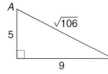   **19.** 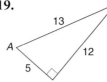   **20.**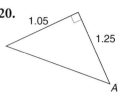

21. WRITING MATHEMATICS On grid paper, draw three different right triangles for which the tangent of one acute angle is 0.8. What is the tangent of the other acute angle of each triangle? Explain how you arrived at the three right triangles you drew.

SIGHTSEEING Jaime is standing at street level and looking at the top of a building. Use the figure at the right to explain what each measure represents.

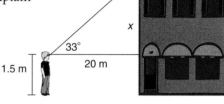

22. 1.5 m   **23.** 20 m   **24.** 33°

25. x **26.** $x + 1.5$

27. Show how to use the tangent ratio to find the value of x.

28. What is the height of the building to the nearest tenth of a meter?

29. NAVIGATION The angle of elevation from a ship to the top of a 150 ft lighthouse is 9°. How far away from the lighthouse is the ship?

PRACTICE

Find tan *C* and tan *D*. Write each ratio as a fraction and as a decimal. Round your answer to the nearest ten-thousandth if necessary.

30. 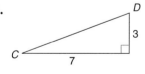   **31.** 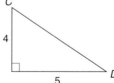   **32.**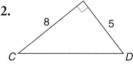

Find *x* in each figure. If necessary, round angles to the nearest degree and lengths to the nearest tenth.

33.   **34.**    **35.** 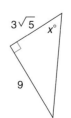   **36.**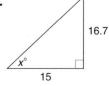

Find _b_ in each figure. If necessary, round angles to the nearest degree and lengths to the nearest tenth.

37.

38.

39.

40.

41. NATURE The sun shines on a tree 25 m tall, so that a shadow 18 m long is cast. To the nearest degree, find the angle the sun's rays make with the ground.

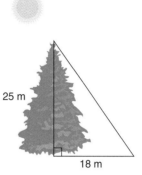

25 m

18 m

42. WRITING MATHEMATICS Suppose you are at the top of an observation tower and you see an object at ground level in the distance. You want to use the tangent ratio to find the distance to the object. What measurement do you need to know? Explain.

43. CONSTRUCTION The *grade* of a driveway is a measure of its steepness. It is defined by the ratio $\frac{\text{rise}}{\text{run}}$, given as a percent. The driveway shown in the figure below rises 6 ft for every 100 ft of run. This means it has a grade of $\frac{6}{100}$ or 6%.

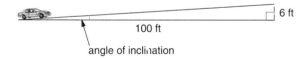

100 ft

6 ft

angle of inclination

The angle that the driveway forms with the horizontal is called the *angle of inclination*. What is the angle of inclination of a driveway whose grade is 6% rounded to the nearest tenth?

44. TRANSPORTATION The tailgate of a truck is 1.12 m from the ground. How long should a ramp be so the incline from the ground up to the tailgate is 9.5°? Round your answer to the nearest tenth.

EXTEND

Find the area of each figure. Round to the nearest tenth.

45.
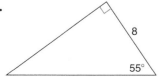

8

55°

46.
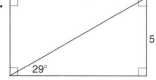

5

29°

47.
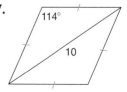

114°

10

VECTORS You can identify the *direction* of a vector on a coordinate plane by finding the angle it forms with the positive *x*-axis, measured in a counterclockwise direction. To determine the measure of this angle, you can use the tangent ratio.

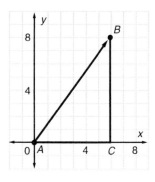

48. Refer to the figure on the coordinate axes to find each.

    a. length of $\overline{AC}$?   **b.** length of $\overline{BC}$?

    c. tangent of $\angle BAC$?   **d.** measure of $\angle BAC$?

    e. direction of $\overrightarrow{AB}$?   **f.** magnitude of $\overrightarrow{AB}$?

A vector has initial point P and terminal point Q as given. Find the direction of $\overrightarrow{PQ}$ to the nearest degree and the magnitude of $\overrightarrow{PQ}$.

49. $P(0, 0)$, $Q(6, 1)$ **50.** $P(0, 0)$, $Q(1, 6)$ **51.** $P(3, 2)$, $Q(5, 9)$ **52.** $P(-3, -5)$, $Q(-1, 4)$

53. Let $\overrightarrow{u} = (-2, 8)$ and $\overrightarrow{v} = (4, -3)$. Find the magnitude and direction of $\overrightarrow{u} + \overrightarrow{v}$.

THINK CRITICALLY

54. One person facing north and looking up spots an airplane at an angle of elevation of 26° from the ground. At the same time another person is standing 10 mi away also facing north and looking up spots the same airplane at an angle of elevation of 62° from the ground. What is the altitude of the airplane in feet?

55. The length, width, and height of a rectangular room are 12 ft, 9 ft, and 10 ft, respectively. Find the measure of $\angle ABC$ to the nearest degree.

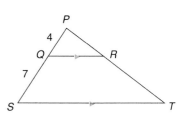

56. WRITING MATHEMATICS Explain what happens to the tangent ratio of an acute angle of a right triangle as the measure of that angle increases.

57. WRITING MATHEMATICS What do you think is the least possible value of the tangent ratio? What do you think is the greatest possible value? Explain your reasoning.

58. Do you think $\tan A + \tan B = \tan (A + B)$? Justify your answer.

ALGEBRA AND GEOMETRY REVIEW

59. Solve $\frac{3}{4}x \leq 12$. Check your answer.

60. Are lines $3x + 2y = 4$ and $y = \frac{3}{2}x + 5$ parallel?

61. Find the perimeter and area of a 30°-60°-90° triangle if the longer leg is 12 cm.

62. Use the figure at the right to find the ratio of the area of $\triangle PQR$ to the area of $\triangle PST$.

63. Find the length of a diagonal of a square if its area is 100 m^2.

A phrase often heard from a carpenter is "Measure twice, cut once." Carpenters make certain a measurement is exact before cutting a piece of wood because the cut can not be erased and tried again. Therefore, *laying out* angles for a certain cut is an important task.

A carpenter uses a framing square and fence, pictured at the right, to lay out angles. The *tongue* and the *blade* of the framing square model two sides of a right triangle. The fence is placed on the framing square so that the fence and the tongue form an acute angle, pictured as $\angle A$. The placement of the fence determines lengths for the tongue and blade which can be used to find m$\angle A$.

tongue

fence

$\angle A$

blade

DECISION MAKING

1. Think of a 45°-45°-90° triangle, how can you set the fence on the tongue and blade in order to make an angle of 45°? Explain.

The reference table helps the carpenter set the fence in the correct position on the tongue and the blade in order to form the correct angle.

Angle	Tongue, inches	Blade, inches
30	12	$20\frac{7}{8}$
54	12	$8\frac{25}{32}$
60	12	$6\frac{15}{16}$
70	12	$4\frac{3}{8}$
72	12	$3\frac{7}{8}$

2. Use column one and three to determine how a change in the angle measure affects the blade length.

3. Estimate the angle measure when the fence is set so that the tongue length is 12 in. and the blade is set on 5 in. Explain your estimation.

4. Use a trigonometric ratio to find the angle measure formed by the tongue and blade lengths in Question 3. Round to the nearest degree.

5. Explain how to verify the angle measures given in the reference table.

6. If a carpenter wants to make a 75° angle and the fence is set at 12 in. on the tongue, where should the fence be set on the blade? Describe how to find the length.

10.6 The Sine and Cosine Ratios

Explore

1. Explain how you can find the length of the hypotenuse of each triangle.

2. Copy and complete the table below.

3. As the measure of the angle with its vertex at point Z increases, what happens to each ratio in the table?

4. What do you think is the greatest possible value of each ratio? the least possible value? Justify your answers.

	length of vertical leg ÷ length of hypotenuse	length of horizontal leg ÷ length of hypotenuse
△AZB		
△AZC		
△AZD		
△AZE		
△AZF		
△AZG		

Build Understanding

In Explore, you worked with two more important trigonometric ratios. Unlike the tangent ratio, the sine and cosine ratios include the length of the hypotenuse as the denominator. For the right triangle shown, the sine of ∠A is written **sin A** and the cosine of ∠A is written **cos A**.

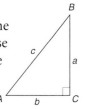

SINE RATIO AND COSINE RATIO

For any acute angle A in a right triangle, the ratios are

$$\sin A = \frac{\text{length of side opposite } \angle A}{\text{length of hypotenuse}}$$

$$\cos A = \frac{\text{length of side adjacent to } \angle A}{\text{length of hypotenuse}}$$

EXAMPLE 1

Write each ratio as a fraction and as a decimal. Round to the nearest ten-thousandth and use the $\approx$ symbol if necessary.

a. Find sin A, cos A, and tan A.

b. Find sin B, cos B, and tan B.

Solution

a. $\sin A = \dfrac{16}{20} = 0.8$ $\dfrac{\text{opposite}}{\text{hypotenuse}}$

$\cos A = \dfrac{12}{20} = 0.6$ $\dfrac{\text{adjacent}}{\text{hypotenuse}}$

$\tan A = \dfrac{16}{12} \approx 1.3333$ $\dfrac{\text{opposite}}{\text{adjacent}}$

b. $\sin B = \dfrac{12}{20} = 0.6$ $\cos B = \dfrac{16}{20} = 0.8$ $\tan B = \dfrac{12}{16} = 0.75$ ◄

PROBLEM SOLVING TIP

You can use a **mnemonic** to help you remember the trigonometric ratios. The first letter of each word in the rhyme below represents the first letter of the ratios.

Some Old Hound
Came A Hopping
Through Our Alley

Just as with the tangent ratio, an acute angle of a given measure has a unique sine ratio and a unique cosine ratio associated with it. When given the sine or cosine value of an acute angle, you can use the appropriate function on your calculator to find the measure of the angle.

EXAMPLE 2

Find the value of w to the nearest hundredth.

Solution

$\cos 52° = \dfrac{14}{w}$ $\dfrac{\text{adjacent}}{\text{hypotenuse}}$

$w = \dfrac{14}{\cos 52°}$

$w \approx 22.74$ ◄

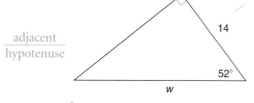

TECHNOLOGY TIP

Be sure your calculator is in degree mode.

Every sine or cosine ratio has a unique angle associated with it. You can work backwards to find the measure of the angle by using the inverse sine function $\sin^{-1}$ or the inverse cosine function $\cos^{-1}$ on your calculator.

EXAMPLE 3

Find m$\angle R$. Round to the nearest degree.

Solution

$\sin R = \dfrac{6}{10}$ $\dfrac{\text{opposite}}{\text{hypotenuse}}$

$m\angle R = \sin^{-1} \dfrac{6}{10} \approx 37°$ ◄

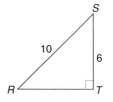

COMMUNICATING ABOUT GEOMETRY

How would the solution of Example 3 be different if you were asked to find the measure of $\angle S$?

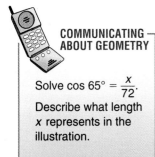

The sine and cosine ratios are used to solve real world problems.

EXAMPLE 4

COMMUNICATIONS An engineer plans the installation of a 125 ft microwave relay tower. The tower must be supported by guy wires which are anchored to the ground on each side of the tower. The shortest guy wire has a length of 72 ft and forms a 65° angle with the ground at the point of the anchor. How far up the tower will the first guy wire be secured to the tower? Round the length to nearest foot.

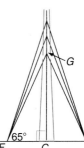

Solution

The guy wire is the hypotenuse of right triangle FCG. Since you are finding the side opposite the given angle, use the sine ratio.

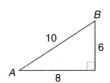

$$\sin 65° = \frac{CG}{72}$$

$$72(\sin 65°) = CG$$

$$65.25 \approx CG$$

The shortest guy wire will be secured to the tower about 65 ft up from the base of the tower.

TRY THESE

Find the trigonometric ratio in the figure. Write each ratio as a fraction and as a decimal. Round decimal answers to the nearest ten-thousandth if necessary.

1. $\sin A$
2. $\cos A$
3. $\tan A$

4. $\sin B$
5. $\cos B$
6. $\tan B$

Use a calculator to find each value. Round to the nearest ten-thousandth.

7. $\sin 40°$
8. $\sin 78°$
9. $\cos 40°$
10. $\cos 78°$

Find x to the nearest hundredth.

11.

12.

13.

14.
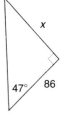

Find the m∠A to the nearest degree.

15. $\sin A = 0.5463$ **16.** $\cos A = 0.2318$ **17.** $\cos A = \dfrac{5}{9}$ **18.** $\sin A = 0.1\overline{6}$

19.   **20.** 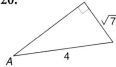   **21.**   **22.**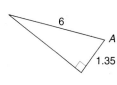

23. SPORTS A paraskier is being towed behind a boat on a 250 ft rope. If the angle of elevation is 32°, approximately how high is the paraskier from the surface of the water?

24. WRITING MATHEMATICS Describe the type of geometric problems you can solve by using trigonometric ratios. Explain how you decide which trigonometric ratio to use.

PRACTICE

Refer to the triangle at the right. Match each fraction with one or more of the trigonometric ratios a–f. There will be more than one match for some fractions and no matches for others.

25. $\dfrac{8}{17}$ **26.** $\dfrac{15}{17}$ **a.** $\sin J$
b. $\cos J$

27. $\dfrac{8}{15}$ **28.** $\dfrac{17}{8}$ **c.** $\tan J$
d. $\sin K$

29. $\dfrac{17}{15}$ **30.** $\dfrac{15}{8}$ **e.** $\cos K$
f. $\tan K$

Find x in each figure. If necessary, round angles to the nearest degree and lengths to the nearest tenth.

31.   **32.** 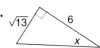   **33.** 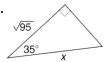   **34.**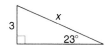

35. 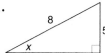   **36.** 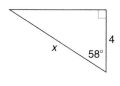   **37.**   **38.**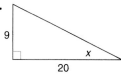

39. WRITING MATHEMATICS State as many facts as you can about the figure at the right. Justify each fact.

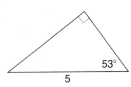

40. AGRICULTURE Aisha needs to buy a conveyor to haul bales of hay to the loft of his barn. The loft is 15 ft above the ground. One conveyor he is considering has legs which sets it 3 ft above the ground and can be raised to a maximum angle of 35°. Can this conveyor reach his loft? Explain.

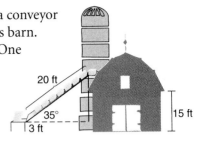

41. AVIATION An airplane takes off and climbs at a steady 18° angle. After flying along a path of 2 mi, how much altitude has the plane gained in feet? Round to the nearest foot.

42. HISTORY It is said the physicist Galileo dropped objects off a tower in Pisa to disprove Aristotle's claim that objects fall at speeds proportional to their weight. Use the figure to find at what angle from the vertical the tower leans? Round your answer to the nearest degree.

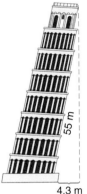

Copy and complete the chart of trigonometric ratios for the angles associated with the special right triangles.

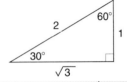

Angle Measure	Tangent		Sine		Cosine	
	exact	approximate	exact	approximate	exact	approximate
43. 30°	$\dfrac{1}{\sqrt{3}} = \dfrac{\sqrt{3}}{3}$	0.5774				
44. 45°						
45. 60°						

EXTEND

Find the area and perimeter of each figure. Round to the nearest tenth.

46.

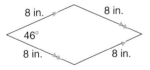

47.

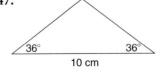

48.

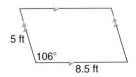

49. Prove that $\tan x = \dfrac{\sin x}{\cos x}$.

50. VECTORS A vector on a coordinate plane has magnitude 8 and forms an angle of 25° with the positive *x*-axis. The initial point of the vector is the origin. Find the coordinates of the terminal point of the vector, rounded to the nearest tenth.

THINK CRITICALLY

51. Each lateral face of the regular pyramid at the right is an isosceles triangle with base angles that each measure 77° and legs that each measure 12 in. Find the slant height and altitude of the pyramid. Round answers to nearest tenth.

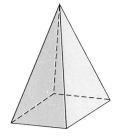

Let *x* be the measure of an acute angle of a right triangle.

52. Write inequalities to describe the range of possible values for sin *x* and cos *x*.

53. For what values of *x* is sin *x* greater than cos *x*?

54. Is it ever true that sin *x* is equal to cos *x*?

55. Use a variable expression to make a true statement: sin *x* = cos _?_

ALGEBRA AND GEOMETRY REVIEW

Find sin *A*, cos *A*, and tan *A*. Round your answer to the nearest ten-thousandth.

56.

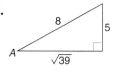

57.

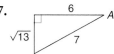

58.

Solve and check.

59. $9x + 8 = 10x$

60. $3(5.5 - x) = 8x$

61. $2(-3x + 5) = -2x - 17$

Work with a partner and use a simple surveyor's transit to estimate a distance.

1. Attach a straw to the small hole of a protractor using a paper clip. The straw is your telescope so it must be able to pivot.

2. To use your device to find the distance an object is from you, have one person stand at a point *A* perpendicular to the object at point *C*. Then have the other person stand about two arm lengths from the first person, point *B* in the figure. Measure $\overline{AB}$.

3. The person at point *A* holds the protractor parallel to $\overline{AB}$ and sets the straw at 90° so it points towards the object.

4. Pass the device to the person at point B, who pivots the straw to sight the object at point *C*. Read ∠*ABC*.

5. Use trigonometry to find the distance *AC*.

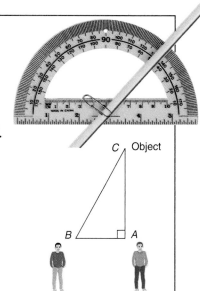

Land measurements are necessary to establish property boundaries, to construct a building or road, or to make a map. A surveyor measures the land using a transit. *Triangulation* is a method surveyors use to measure distance. A base line of known length is established and the angles from each end of the baseline to the object are measured. The drawing illustrates a triangulation.

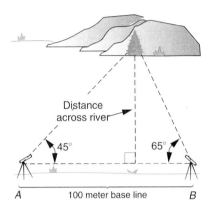

Distance across river

45°   65°

A   100 meter base line   *B*

DECISION MAKING

1. If *x* is the distance from the vertex of the 45° angle to the altitude, what expression can be used for the other segment of the base line?

2. If the altitude, or the distance from the tree to the base line is *d*, write two trigonometric equations using *d* and *x*.

3. How wide is the river to the nearest hundredth of a meter?

The Chiaramonte Office Complex is being built near the intersection of Frye Road and Burns Avenue. The two streets will be connected .

4. Find the length of the altitude to the new road.

5. Find the length of the new road.

Frye Rd.

Burns Ave.   772 yd

45°   new road   75°

Focus on Reasoning
Checking Reasonableness of Solutions

Discussion

Work with a partner.

Three students individually used trigonometric ratios to find the distance between R and T. Shawn wrote the equations below.

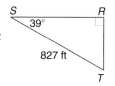

$$\tan 39° = \frac{RT}{827} \qquad RT = 827 \tan 39° \approx 670$$

1. Can Shawn expect to get the correct answer for RT? Explain your response.

Katrina wrote the following two equations.

$$\sin 39° = \frac{RT}{827} \qquad RT = 827 \sin 39° \approx 520$$

2. Can Katrina expect to get the correct answer for RT? Explain your response.

Liu wrote the following as his equations.

$$\cos 51° = \frac{RT}{827} \qquad RT = 827 \cos 51° \approx 520$$

3. Can Liu expect to get the correct answer for RT? Explain your response.

4. How can you tell from the answers Shawn, Katrina, and Liu got that someone's work is incorrect? Whose work is it?

5. As a check, Katrina found $m\angle T = 51°$. She then wrote and solved $\sin 51° = \frac{SR}{827}$. She got $SR \approx 642$. What inequality theorem about sides and angles of triangles helps you know that her work may be correct?

6. Suppose Katrina solves her equation correctly and writes the complete calculator display. Do you think her answer can be considered reasonable? Explain your response.

7. Shawn needed some confirming evidence that the sides of $\triangle RST$ are approximately equal to 520 ft, 642 ft, and 827 ft. Explain how to use the Pythagorean Theorem as a check on the reasonableness of these estimated lengths.

Build a Case

After working through a trigonometry problem, Shawn, Katrina, and Liu decided to identify ways to assure their results are reasonable.

8. Shawn argued that if one acute angle of a right triangle is smaller than the other, then the length of the side opposite the first must be smaller than the length of the side opposite the second. What theorem is Shawn considering as a check on reasonableness?

9. Katrina solved a problem involving a right triangle. She found that the length of one leg was greater than the length of the hypotenuse. What can she conclude about her work?

Refer to the diagram at the right for Exercises 10–13.

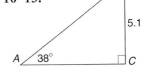

10. Approximate AC to the nearest tenth.

11. As a check on reasonableness, approximate AC by using a different trigonometric ratio or angle.

12. As a reasonableness check, approximate AB to the nearest tenth. Use the Pythagorean Theorem as a reasonableness check.

13. With the angle measures in $\triangle ABC$ and your calculations in Questions 10 and 12, use a triangle inequality theorem as a check on reasonableness.

Solve each problem. Give lengths to the nearest tenth. Give degree measures to the nearest whole number. Use a triangle inequality theorem, a different trigonometric ratio, or the Pythagorean Theorem as a reasonableness check on your work.

14. Given $m\angle A = 40°$ and $AC = 12.4$, find BC.

15. Given $m\angle B = 42°$ and $AC = 9.3$, find AB.

16. Given $m\angle A = 33°$ and $AB = 4.7$, find BC and $m\angle B$.

17. Given $m\angle B = 12°$ and $AC = 8.5$, find AB and $m\angle A$.

18. Given $m\angle B = 61°$ and $BC = 12.4$, find AC and $m\angle A$.

19. Given $m\angle A = 42°$ and $AC = 9.3$, find BC and $m\angle B$.

20. Katrina knows that $m\angle A = 54°$ and $AC = 13$ in right triangle $\triangle ABC$. She reasoned that $m\angle B = 36°$ and, using trigonometric ratios, that $AB \approx 17.9$ and $BC \approx 10.5$. Explain how you know that her answers cannot be reasonable.

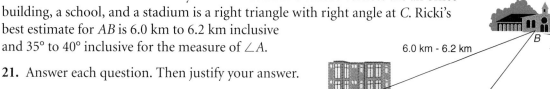

EXTEND AND DEFEND

MARGIN OF ERROR Ricki is fairly certain that $\triangle ABC$ whose vertices are an office building, a school, and a stadium is a right triangle with right angle at C. Ricki's best estimate for AB is 6.0 km to 6.2 km inclusive and 35° to 40° inclusive for the measure of $\angle A$.

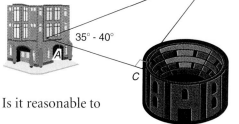

6.0 km - 6.2 km

35° - 40°

21. Answer each question. Then justify your answer.

  a. Suppose Ricki uses 35° for m$\angle A$ and 6.0 for AB. Is it reasonable to expect she will get the smallest estimate of BC?

  b. Suppose Ricki uses 40° for m$\angle A$ and 6.2 for AB. Is it reasonable to expect she will get the largest estimate of BC?

22. Estimate BC using each set of measurements.
  a. m$\angle A = 35°$ and $AB = 6.0$ **b.** m$\angle A = 40°$ and $AB = 6.2$

23. If Ricki estimates BC using m$\angle A = 37.5°$ and $AB = 6.1$, is it reasonable to assume her answer will be between the answers for Questions 22a and 22b? Explain.

24. Describe and apply two different methods to approximate AC to the nearest whole number. Apply a third method to check the reasonableness of your solution.

25. **WRITING MATHEMATICS** Suppose you are given right triangle $\triangle ABC$, the measure of one of its acute angles and the length of one of its sides. Summarize various methods for finding the measures of other parts of the triangle and for checking the reasonableness of your work.

Now You See It
Now You Don't

TRI-THESE

Is the dot midway between the tip of the triangle and the base of the triangle?

Is the crossbar midway between the tip of the triangle and its base?

Explain what you see.

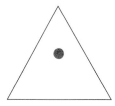

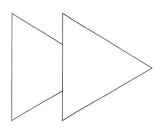

Problem Solving File

Choosing an Appropriate Method

In many real world problems, there is more than one method you can use to find a solution.

Problem

ENGINEERING Sheila is a civil engineer who designs elevations that use an open ditch for water drainage. Her client needs a ditch that is 4 m wide. To prevent erosion, she determines the best design has the angle measure shown. What will be the depth of the ditch?

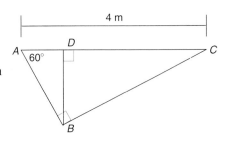

Explore the Problem

1. Identify the geometric figures you see in the illustration of the ditch. Name the segment that represents the depth of the ditch.

2. The illustration is drawn to scale. Determine the scale and show how you can use the scale to find the depth of the ditch.

3. The construction workers who will dig the ditch need the depth accurate to the nearest hundredth of a meter. Do you think your answer from Question 2 provides the required measure. Explain.

In Questions 4–7, you may work only with the measures given in the illustration. Analyze each method individually. Determine whether you can solve the problem using *only that method*. Explain.

4. the Pythagorean Theorem

5. the geometric mean

6. special right triangles

7. trigonometric ratios

8. Describe a combination of two or more methods that could be used to find the depth. Then use this combination to write a detailed solution of the problem. What will be the depth of the ditch, rounded to the nearest hundredth of a meter?

9. How could you solve the problem by using a combination of methods different from those you used in Question 8. Use this combination to write a detailed solution. What will be the depth of the ditch, rounded to the nearest hundredth of a meter?

10. Are the answers to Questions 8 and 9 the same?, If not, explain what might account for any differences.

11. WRITING MATHEMATICS Which method or combination of methods do you prefer for solving the problem? Explain.

PROBLEM
SOLVING PLAN

- Understand
- Plan
- Solve
- Examine

Investigate Further

In some problems involving right triangles, you must solve the right triangle to obtain a solution. To **solve a right triangle**, you must find the measures of all the angles and sides. To do this, you often use a combination of methods.

Triangle *XYZ* represents a wooden panel that encloses one side of a basement staircase. A carpenter records the two measurements shown. To be sure that the panel is cut correctly, the carpenter needs to know the measures of all the angles and sides of the panel.

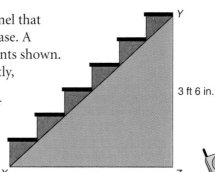

3 ft 6 in.

One way to solve the problem is to begin by finding *XY*.

12. Use the Pythagorean Theorem to find *XY*. Round the measure to the nearest inch.

13. Find m∠*X* and m∠*Y*. Round the measures to the nearest degree.

14. Summarize your results. Copy and complete this list.

$XZ = $ _?_ $\qquad$ $YZ = $ _?_ $\qquad$ $XY \approx $ _?_

$m\angle X \approx $ _?_ $\qquad$ $m\angle Y \approx $ _?_ $\qquad$ $m\angle Z = $ _?_

Another way to solve the problem is to first find m∠*X*.

15. Use the tangent ratio to find m∠*X*. Round the measure to the nearest degree.

16. Find m∠*Y* to the nearest degree. Use the sine ratio to find *XY* to the nearest inch.

17. Summarize your results as described in Question 14.

18. Compare your results in Questions 14 and 17. Are they the same? If not, what might account for any differences? Which method did you prefer?

COMMUNICATING
ABOUT GEOMETRY

Work with a partner. List several methods to consider when solving problems that involve right triangles. Can you describe problems that you know are best solved using any particular method?

GEOMETRY: WHO, WHERE, WHEN

The German physicist Wilhelm Roentgen discovered x-rays in 1895 by accident.

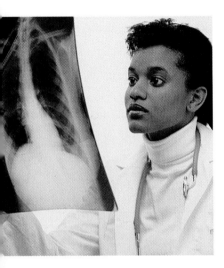

Apply the Strategy

Solve each problem using an appropriate method or combination of methods.

19. **MEDICINE** Gabriel requires an x-ray of his spine. He will need to be positioned so that his spine is at a 45° angle with the horizontal. The length from his neck to the base of his spine is 2 ft. How high must the back of his neck be raised? Round to the nearest inch.

20. **CONSTRUCTION** A 15 ft pole is supported with two guy wires. One wire is secured at the ground to form an angle of 42°. The other wire is secured at the ground to form an angle of 48°. Find the distance from where each guy wire is secured in the ground to the base of the pole. Find the length of each wire. Round all measures to the nearest tenth.

In Questions 21–22, solve each right triangle . Round angle measures to the nearest degree and lengths of sides to the nearest tenth.

21.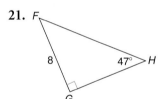

22.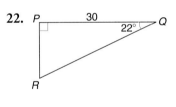

23. **HOME IMPROVEMENT** Craig purchased a ladder to use when he paints his house. The safety book which accompanied the ladder stated that the measure of the angle that an unsecured ladder makes with the ground should not be less than 75°. He decides he must place the ladder and then check the measurements for safety. The ladder is 16 ft and he places the base of the ladder 3 ft from the house. To the nearest tenth, what is the height at which the ladder rests against the house? Is this a safe position?

24. **WRITING MATHEMATICS** Suppose your friend calls you on the telephone requesting help with his geometry homework. He explains that he is to solve the triangle, but the handout was dropped on the bus and now only a few labels are clear. He knows that $\triangle ABC$ has $\angle C$ as a right angle, and the measures of two other parts. Your friend thinks he can solve the triangle with that information. Do you agree or disagree? Explain your reasoning.

REVIEW PROBLEM SOLVING STRATEGIES

1. A circle with circumference 1 in. is rolled once, without slipping, around the outside of an octagon with perimeter 10 in. How many revolutions does the circle make in its trip around the octagon? Use these questions to help you solve the problem.

   a. What would the answer be if the octagon were flattened out to a straight line?

   b. When the circle rolls around the octagon, it must also "turn" through an exterior angle at each vertex of the octagon. What is the sum of the measures of these eight exterior angles? How many additional revolutions must the circle make to account for the additional turning required?

Ring Around the Octagon

Doesn't That Take the Cake?

3. Janie has baked a 2′ by 3′ rectangular shaped cake to serve at her investment club meeting. Chuckie sees the cake and before Janie can stop him he cuts a 1′ by 1′ square piece out of the corner and eats it. Janie finds the remainder of her cake and decides to cut it apart, reassemble the pieces in a square and re-ice it. She makes only two straight slices and is able to form a square. A diagram of the partially eaten cake is shown below. Copy it. Draw in where Janie makes the slices. Cut out the pieces and make the square.

   Hint: What is the area of the square? What would then be the length of a side? What triangle has the

We Need a "Cosiner"

2. Draw an acute triangle ABC. Draw altitudes $\overline{BP}$ and $\overline{CQ}$. If $AP = x$, $PC = 2x - 6$, $AQ = 8$ and $QB = x - 5$, find the value of the cosine of $\angle A$.

Explore

SPOTLIGHT ON LEARNING

WHAT? In this lesson you will learn
• to find the area of regular polygons.

WHY? Finding the area and perimeter of regular polygons can help you solve problems in transportation and building trades.

THINK BACK

A *regular polygon* is a convex polygon that is both equilateral and equiangular.

● You need paper, a ruler, scissors, and tape.

1. Trace the regular octagon at the right onto a sheet of paper and cut it out.

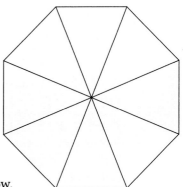

2. Draw the diagonals on your octagon. Cut your octagon into pieces along the diagonals. What is the shape of each piece? How are the pieces related to each other?

3. Rearrange the pieces as shown below. Tape them together.

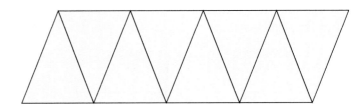

4. What type of quadrilateral did you form in Question 3?

5. Sketch an altitude on one of your triangles.

6. Let the variable a represent the length of the altitude. Let p represent the perimeter of the octagon. Write the area of the quadrilateral in terms of a and p. Justify your answer.

GEOMETRY: WHO, WHERE, WHEN

The Greek mathematician Zenodorus (ca 200 B.C.–140 B.C.) showed that among polygons with equal perimeter and number of sides, a regular polygon has the greatest area.

Build Understanding

● The **center** of a regular polygon is the point that is equidistant from its vertices. A segment whose endpoints are the center and a vertex is called a **radius** of the polygon. A segment whose endpoints are the center and the midpoint of a side is called an **apothem**. An apothem is perpendicular to the side it intersects.

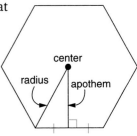

All the radii of a regular polygon are congruent to each other, as are all the apothems. The common length of the radii is referred to as *the* radius. The common length of the apothems is *the* apothem.

A **central angle** of a regular polygon is an angle whose vertex is the center of the polygon and whose sides contain two consecutive vertices.

EXAMPLE 1

Find the measure of each central angle, the radius, and the apothem of a regular pentagon in which the measure of each side is 12 in.

Solution

A regular pentagon has five sides. With all its radii drawn, there are five congruent isosceles triangles. The vertex angle of each is a central angle. So the measure of each central angle is $\frac{360°}{5}$, or 72°.

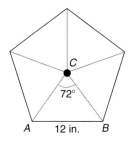
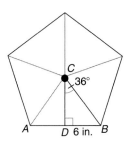

COMMUNICATING ABOUT GEOMETRY

In the regular pentagon in Example 1, how do you know the five triangles are isosceles? How do you know they are congruent?

In the figures above, $\angle ACB$ is a central angle. When apothem $\overline{CD}$ is drawn, it forms right triangle CBD with the labeled measures. Apothem $\overline{CD}$ is a leg of this triangle, and radius $\overline{CB}$ is the hypotenuse. So you can use trigonometry to find their lengths.

$$\tan 36° = \frac{DB}{CD} \qquad \sin 36° = \frac{DB}{CB}$$

$$\tan 36° = \frac{6}{CD} \qquad \sin 36° = \frac{6}{CB}.$$

$$\tan 36°(CD) = 6 \qquad \sin 36°(CB) = 6$$

$$CD = \frac{6}{\tan 36°} \qquad CB = \frac{6}{\sin 36°}$$

$$CD \approx 8.3 \qquad CB \approx 10.2$$

So the apothem is about 8.3 in., and the radius is about 10.2 in. ◄

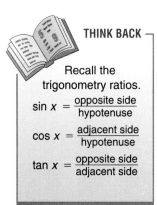

THINK BACK

Recall the trigonometry ratios.

$\sin x = \dfrac{\text{opposite side}}{\text{hypotenuse}}$

$\cos x = \dfrac{\text{adjacent side}}{\text{hypotenuse}}$

$\tan x = \dfrac{\text{opposite side}}{\text{adjacent side}}$

As you have seen, the radii of a regular polygon divide it into congruent isosceles triangles. The apothem of the polygon is the altitude of each triangle. By writing an expression for the sum of the areas of the triangles, you can arrive at this theorem.

THEOREM 10.6	**AREA OF A REGULAR POLYGON**

The area A of a regular polygon equals half the product of its apothem a and perimeter p, or $A = \frac{1}{2}ap$.

EXAMPLE 2

Find the area of a regular hexagon whose perimeter is 108 cm.

Solution

The perimeter is 108 cm. Since each of the six
sides is congruent, the length of one side is
$(108 \div 6)$ cm, or 18 cm.

Since $\triangle CTS$ is a 30°-60°-90° triangle, the
apothem $CT = 9\sqrt{3}$ cm.

Use the area formula from Theorem 10.6.

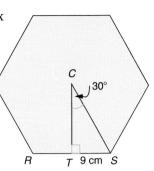

$$A = \frac{1}{2}ap$$

$$A = \frac{1}{2}(9\sqrt{3})(108) = 486\sqrt{3}$$

The area of this hexagon is about 842 cm². ◄

You can use areas of regular polygons to solve many real world
problems.

EXAMPLE 3

LANDSCAPE ARCHITECTURE The floor of a gazebo is shaped like a
regular octagon. The plans indicate that the distance from the door to
the back of the gazebo is 14 ft. What is the area of the floor?

Solution

The distance from front to back is
twice the apothem. So the apothem
is 7 ft.

Find ZY.

$$\tan 22.5° = \frac{ZY}{CZ}$$

$$\tan 22.5° = \frac{ZY}{7}$$

$$ZY = 7(\tan 22.5°) \approx 2.9$$

Since ZY is about 2.9 ft, XY is about 2(2.9) ft, or about 5.8 ft.

The perimeter of the floor is about 8(5.8) ft, or about 46.4 ft.

$$A = \frac{1}{2}ap \qquad \text{Area formula for a regular polygon.}$$

$$A \approx \frac{1}{2}(7)(46.4) \approx 162.4$$

So the area of the floor of the gazebo is about 162 ft². ◄

Each regular polygon has one side with the given measure. Find the measure of a central angle of the polygon, its radius, and its apothem. When necessary, round angle measures to the nearest degree and lengths to the nearest tenth.

1.

8 in.

2.

10 cm

3.

2 ft

4.

6 m

Find the area of a polygon with the given dimension. Round your answer to the nearest whole number.

5. a regular hexagon in which the length of one side is 8 in.

6. a regular nonagon with perimeter 54 cm

7. a regular pentagon with apothem 3 m

8. a regular octagon with radius 4 yd

9. TRANSPORTATION A standard stop sign is made in three sizes. The length of the side of the smallest sign is 12.5 in. Find the area of the stop sign.

10. WRITING MATHEMATICS How are a radius and apothem of a regular polygon alike? How are they different?

11. WRITING MATHEMATICS Explain how the area formula for a regular pentagon is related to the area formula for a triangle.

PRACTICE

Each figure is a regular polygon with center at point C. Find the perimeter and area of each. Round perimeters to the nearest tenth and areas to the nearest whole number.

12.

C
8 cm

13.

C
6 ft

14.
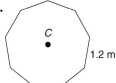
C
1.2 m

15. WRITING MATHEMATICS Look back at your work and solutions for Exercises 12–14. How was your work alike? How was it different?

16. WRITING MATHEMATICS The figure at the right is a regular polygon with center at point C. List as many facts about the polygon as you can. Justify each fact in your list.

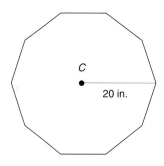
C
20 in.

Find the indicated dimension of each polygon. Round lengths to the nearest tenth and areas to the nearest whole number.

17. the apothem of a regular decagon with perimeter 60 ft

18. the radius of a regular octagon with one side of length 3 in.

19. the area of a regular pentagon with radius 8 m

20. the perimeter of a regular hexagon with apothem 12 cm

21. the apothem of a square with one side that measures 11 cm

22. the area of a square with radius 15 yd

23. the area of a regular polygon with one side of length 4 ft and a central angle of measure 30°

24. the perimeter of a regular polygon with radius 24 mm and a central angle of measure 90°

25. HOME IMPROVEMENT Luis needs to cut a regular octagon out of the siding of a house to insert a new octagonal window. Each side of the window is 40 cm long. Find the area of the window.

26. WRITING MATHEMATICS Explain which window would let in more light, the octagonal window in Exercise 25 or a rectangular window that measures 80 cm by 100 cm?

EXTEND

GOVERNMENT The United States Department of Defense is housed in Arlington, VA, just across the Potomac River from Washington, D.C. Their office building is called the Pentagon and is composed of five buildings. From an aerial view, you can see that the outer boundary of each building is a regular pentagon. The buildings are *nested* so that each building has the same center and the innermost building has the smallest area. The five buildings are connected by ten corridors.

27. At the center of the innermost building, there is a courtyard that has an apothem measuring 245 ft. Find the perimeter and area of the courtyard.

28. The outer boundary of the largest building is a regular pentagon whose perimeter is approximately 1 mi and whose apothem is about 730 ft. Find the area of a regular pentagon with the same dimensions as the outermost building of the Pentagon.

29. The Pentagon has five stories each having approximately 741,159 ft² office space. Suppose the Pentagon was one building having an area equal to your answer to Exercise 28. Would the Department of Defense have as much office space if the Pentagon had been designed as a one story building the size of the outermost building, without outdoor corridors and a courtyard?

30. WRITING MATHEMATICS Draw a diagram of the Pentagon and include all the measurements you can determine. Describe the Pentagon in your own words.

31. GARDENING A gardening catalog lists a kit that consists of eight wooden boards. You assemble the boards to form the border of a flower bed shaped like a regular octagon. Two kits are available. In one kit, each board is 12 in. long. In the other kit, each board is 24 in. long. How many times larger is the area of the flower bed enclosed with the 24-in. boards than the bed enclosed with the 12-in. boards?

THINK CRITICALLY

32. In Explore, a regular octagon was *dissected* into triangles by cutting it along its radii. When the pieces were rearranged, a parallelogram was formed. What figure would be formed if you performed the same dissection and rearrangement on a regular 35-gon? (*Hint:* Try some simpler cases and look for a pattern.)

LOOKING AHEAD The figure at the right is a right pentagonal prism. Point C is the center of the base. Round answers to nearest square inch.

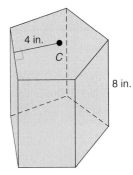

33. What is the area of each base of the prism?

34. What is the area of each lateral face of the prism?

35. The table below appears in a book of mathematics tables. This instruction is printed next to the table.

To approximate the area of a regular polygon, multiply the square of the length of one side by the appropriate factor from the table.

Explain how the "polygon factors" in the table were derived.

Polygon Factors									
No. of Sides	3	4	5	6	7	8	9	10	11	12
Factor	0.433	1.000	1.720	2.598	3.634	4.828	6.182	7.694	9.366	11.196

PROJECT *Connection*   Make a scale drawing of a playing field or building at your school.

1. Choose the building or field for which your group will make a drawing.

2. Make a sketch of the building or field and identify each of the measures needed to make your drawing.

3. Collect data by measuring directly or by using learned techniques that allow for indirect measurement.

4. Decide on a scale for your drawing. Make sure the drawing will fit on your paper.

5. Draw your building to scale.

• • • CHAPTER REVIEW • • •

Choose the word from the list that correctly completes each statement.

1. A __?__ is the ratio of the lengths of two sides of a right triangle.

   a. radius

2. For any acute angle A in a right triangle __?__ $= \dfrac{\text{length of side opposite } \angle A}{\text{length of hypotenuse}}$.

   b. tan A

3. Performing many repetitions of a geometric process is called __?__.

   c. trigonometric ratio

4. A(n) __?__ is a segment whose endpoints are the center and the midpoint of a side of a regular polygon.

   d. sin A

5. A(n) __?__ is a segment whose endpoints are the center and a vertex of a regular polygon.

   e. apothem

   f. iteration

Lesson 10.1   FRACTALS   pages 553–555

- A fractal can be described informally as a structure that displays increasing complexity as it is viewed more and more closely.

6. **WRITING MATHEMATICS** Explain self-similarity. Compare a fractal that is self-similar at one point and a fractal that is strictly self-similar.

Lesson 10.2   PROPERTIES OF RIGHT TRIANGLES   pages 556–561

- When the altitude is drawn to the hypotenuse of a right triangle, three similar triangles are formed.
- The geometric mean can be used to find missing sides of a right triangle with an altitude drawn to the hypotenuse of a right triangle.

Use the figure at the right for Questions 7 and 8.

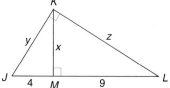

7. $\triangle JKL \sim \triangle$ ____ $\sim \triangle$ ____

8. Find the value of x, y, z.

Lesson 10.3   SPECIAL RIGHT TRIANGLES   pages 564–568

- When the measures of the angles of a triangle are 30°, 60°, and 90° or 45°, 45°, and 90°, special side relationships exist.

Find the unknown lengths.

9.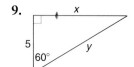

10.

11.

12.

- For an acute angle A in a right triangle, $\tan A = \dfrac{\text{length of side opposite } \angle A}{\text{length of side adjacent to } \angle A}.$

- For an acute angle A in a right triangle, $\sin A = \dfrac{\text{length of the side opposite } \angle A}{\text{length of the hypotenuse}}.$

- For an acute angle A in a right triangle, $\cos A = \dfrac{\text{length of side adjacent to } \angle A}{\text{length of hypotenuse}}.$

Find the value of x.

13.

14.

15.
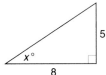

16.

17.

18.
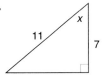

- Solving a right triangle means to find all angle measures and side lengths.
- To check for reasonableness use a different strategy or verify that known relationships hold.

Solve each right triangle. Explain how you checked the reasonableness of your solution.

19.

20.
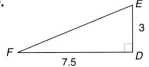

21.
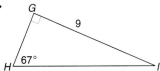

- For a regular polygon, the area is $A = \dfrac{1}{2}ap$ where a is the length of the apothem and p is the perimeter.

Find the area of each regular polygon. Round to the nearest whole number.

22.

23.

24.
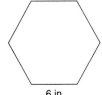

CHAPTER ASSESSMENT

CHAPTER TEST

Use the figure at the right for Questions 1–2.

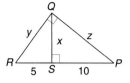

1. **STANDARDIZED TESTS** $\triangle PQR \sim \triangle$ _?_

   I. *QSR*   II. *QSP*   III. *PSQ*   IV. *RSQ*

   A. I only   **B.** II and III

   C. I and III   **D.** II and IV

   E. none of the above

2. Find *x*, *y*, *z*.

3. **WRITING MATHEMATICS** Discuss the relationships of the sides of a right triangle with acute angle measures of 30° and 60°.

Find the value of *x* and *y*. Round angle measures to the nearest degree and side lengths to the nearest tenth.

4.

5.

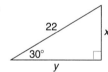

6.

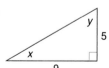

7.

8.

9.

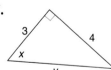

10. **WRITING MATHEMATICS** Describe two different ways to find the value of *x* in the figure at the right.

11. Find *x* and *y* in the drawing using two different methods. Method 1: First find *x* using the Angle-Sum Theorem for Triangles. Next find *y* using the cosine ratio. Method 2: First find *y* using the sine ratio. Next find *x* using the cosine ratio. Compare your results. Which method do you prefer?

Below are stages 0, 1 and 2 of a fractal.

12. Write a description of the steps you follow to proceed from stage 0 to 1.

13. Use the steps outlined in Question 12 to draw stage 3 of the fractal.

14. How many squares would be shaded in Stage 4 of this fractal?

15. **STANDARDIZED TESTS** What is the measure of the central angle of a regular pentagon?
    A. 36°   **B.** 45°   **C.** 18°   **D.** 72°

Find the area of each regular polygon. Round to the nearest whole number.

16.

8 in.

17.

4 cm

18.

18 ft

19.

17 m

20.

7 cm

21.

2 in.

PERFORMANCE ASSESSMENT

FRACTAL FEVER Design and name your own fractal. Show each stage of the fractal up to the 6th stage. Include color. Write a paragraph describing how the fractal is created.

MEASUREMENT ANALYSIS Choose three indirect measurement methods. Measure three different objects using all three methods. Write a report which compares and contrasts the methods. Discuss any discrepancies in answers and detail which method you believe is most accurate.

GRAPH IT! Create a graph of the sine, cosine and tangent ratios. Discuss whether each ratio is a function.

PROJECT ASSESSMENT

PROJECT *Connection*   With your group, you will be making a scale model of a building, parking lot or playing field at your school. Consider using the same building, parking lot or playing field that you used for your scale drawing.

1. Brainstorm a list of materials that you will need in order to create an accurate, creative and interesting scale model.

2. Discuss the measurements that you are still missing in order to complete the project.

3. Obtain those measurements.

4. Create your scale model.

• • • CUMULATIVE REVIEW • • •

Fill in the blank.

1. In a right triangle, the ratio of the length of the side opposite an acute angle to the length of the side adjacent to the same acute angle is the __?__ ratio.

2. In a right triangle, the ratio of the length of the side opposite an acute angle to the length of the hypotenuse is the __?__ ratio.

3. A __?__ is formed when two ratios are set equal to each other.

4. A three-dimensional figure created by connecting each point on a polygon to a point P not in the plane of the polygon is called a(n) __?__ .

Find the unknown lengths.

5.

6.

Classify each triangle by its sides and its angles.

7.

8.

9. **WRITING MATHEMATICS** Explain how to determine more points on a line when one point and the slope are known.

Find x in each triangle. Round to the nearest tenth if necessary.

10.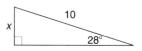

11.

Find the image of the segment that has endpoints at $(-1, 5)$ and $(4, 2)$ under each transformation.

12. reflection in the y-axis

13. rotation of 270° about the origin

14. translation of 10 units right and 6 units down

Find the geometric mean of each pair of numbers.

15. 9 and 16

16. 2 and 98

17. **STANDARDIZED TESTS** Which of the following statements is true about a 30°–60°–90° right triangle?

   A. The hypotenuse is three times the length of the side opposite the 30° angle.

   B. The hypotenuse is twice as long as the side opposite the 60° angle.

   C. The length of the side opposite the 60° angle is twice the length of the side opposite the 30° angle.

   D. The length of the side opposite the 30° angle is half the length of the hypotenuse.

   E. None are true.

18. In the figure, two parallel lines are cut by a transversal. Find x.

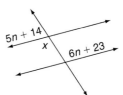

19. In the figure, find x.

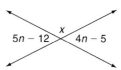

Solve each proportion.

20. $\dfrac{x}{28} = \dfrac{13}{52}$

21. $\dfrac{3}{x} = \dfrac{x}{27}$

STANDARDIZED TEST

STUDENT-PRODUCED ANSWERS Solve each question and on the answer grid write your answer at the top and fill in the ovals.

Notes: Mixed numbers such as $1\frac{1}{2}$ must be gridded as 1.5 or 3/2. Grid only one answer per question. If your answer is a decimal, enter the most accurate value the grid will accommodate.

1. Determine the coordinates of point *A*. Grid the product of the coordinates.

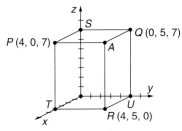

2. Refer to the figure in Question 1. To the nearest tenth, find the distance from *R* to *S*.

3. For the rectangles shown below, grid the ratio of the perimeter of rectangle *ABCD* to the perimeter of rectangle *EFGH* as a fraction.

4. For the rectangles shown above in Question 3, grid the ratio of area of rectangle *ABCD* to the area of rectangle *EFGH* as a fraction.

5. In the figure below, $\angle X$ is bisected by ray *XY*. Determine *WY* to the nearest tenth.

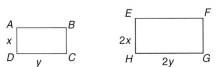

6. The two shorter sides of an acute triangle have lengths 8 and 11. To the nearest tenth, grid the greatest possible length for the third side.

7. The parallelogram and the trapezoid below have the same area. Find the height of the parallelogram.

8. In the proportion below, grid the value of *mn*.
$$\frac{m}{7} = \frac{8}{n}$$

9. Find the length of the hypotenuse in the right triangle shown.

10. The ratio of the perimeters of two rectangles is 2 to 3. Determine the ratio of their areas. Grid the answer as a fraction.

11. If the two lines $4x + 6y = 11$ and $nx - 4y = 7$ are to be parallel, what is the value of $|n|$? Grid the answer as a fraction if necessary.

12. An architect drew a blueprint of a house, using a scale of 2 cm : 9 ft. What would be the square footage of a rectangular room that on the blueprint is 3 cm by 4 cm?

13. The figure below is a regular pentagon. Ray *AB* bisects $\angle A$. What is the measure of $\angle 1$?

14. The vertex angle of an isosceles triangle measures 30°. What is the measure of one of the base angles?

11 Geometry of the Circle

Take a Look AHEAD

Make notes about things that look familiar.
- Name the parts of a circle that you know.
- What do you call the perimeter of a circle?
- In what way have you already connected area and probability?

Make notes about things that look new.
- What ratio does the number π express?
- In what ways will you apply proportional reasoning?
- Name two regions of a circle that have an arc as part of their boundary.

DATA Activity

What Gives You the Rights?

The U.S. government makes it possible for people who create original pieces of work to register their creations, either in their own name or in the name of an organization. Inventors apply for a *patent* to register their inventions. Authors, artists, and songwriters can *copyright* their material. Logos and slogans can be protected by applying for a *trademark*. Registering an original piece of work gives the owners certain rights as to how their creations are used by others.

SKILL FOCUS

- ▶ Add, subtract, multiply, and divide real numbers.
- ▶ Compare numbers.
- ▶ Round numbers.
- ▶ Find the percent one number is of another.

Inventions

In this chapter, you will see how:

- **TOOL DESIGNERS** use arc length to increase efficiency. (Lesson 11.2, page 619)
- **RECORDING TECHNICIANS** use linear speed to improve sound quality. (Lesson 11.6, page 639)
- **SEISMOLOGISTS** use circles to locate the epicenter of earthquakes. (Lesson 11.7, page 645)

TECHNOLOGY PATENTS			
Year	Patents to U.S. Residents	Patents owned by U.S. Corporations	Patents owned by U.S. Individuals
1984	38,367	29,999	8,911
1985	39,555	31,181	9,265
1986	38,126	29,490	9,477
1987	43,520	33,726	10,887
1988	40,496	31,437	10,122
1989	50,185	38,664	13,028
1990	47,393	36,091	12,544
1991	51,183	39,134	13,207
1992	52,254	40,308	12,751
1993	53,235	41,827	12,281
1994	56,067	44,043	12,805

1. Which years show a decline from the previous year in the number of patents issued to U.S. residents?

2. In what years did U.S. corporations own less than three times the number of patents as did U.S. individuals?

3. As of June 1995, U.S. corporations own 1,124,286 patents. The two corporations with the most patents are General Electric with 22,650 and IBM with 16,651. What percent of the total number of patents are held by GE and IBM combined?

4. **WORKING TOGETHER** Obtain the statistics for the number of trademarks and copyrights issued from 1984–1994. Display your information in a chart that makes the differences visible for the number of patents, trademarks, and copyrights.

The Gadget Games

The circle is one of the most familiar of all geometric figures. Many of the appliances and gadgets we use in our lives have the circle as an essential component in their structure.

The properties of the circle are used to design gears, cranks, pulleys, and a long list of mechanical devices. Think about some of the mechanical things you use and whether they could work without the circle.

In this project you will work cooperatively to create two gadgets, both of which use the circle in their design. Your inventions will then compete with those of the other groups at a special type of olympics called *The Gadget Games*.

PROJECT GOAL

To design and test models that use circles to function efficiently.

Getting Started

Work in groups of six students. Each group will design and build a self-powered car and a support structure.

1. A group will work in two teams of three students each. One team will work on the car and the other team on the support structure.

2. Each team should make a list of people and things that can be used as information resources. Consider teachers in your school that teach other subjects such as science and industrial arts.

3. Make a schedule for team meetings.

4. Read all of the Project Connections before you begin working.

5. As a class, choose a date and location for *The Gadget Games*. Make a list of materials that will be needed.

> ## PROJECT *Connections*
>
> **Lesson 11.2**, page 618:
>   Design and build a self-powered car.
> **Lesson 11.3**, page 625:
>   Use tubular design to build a support structure.
> **Lesson 11.6**, page 638:
>   Test your invention and revise it based on the trials.
> **Chapter Assessment**, page 657:
>   Stage "The Gadget Games".

Internet Connection

www.learninggeometry.com

Geometry Workshop

Exploring Circles

Think Back

● Draw a point at the center of a sheet of paper. Label it *O*.
Draw a horizontal line through point *O*.

Sketch and describe the set of all points whose distance from point *O* is 3 in. and that

1. are on the line   **2.** are in the plane   **3.** are in space

Explore/Working Together

● The set of all points in a plane that are a given distance from a given point in the plane is a **circle**. The given point is the **center** of the circle, and the given distance is the **radius**.

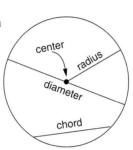

A segment whose endpoints are on the circle is called a **chord**. A chord that contains the center of the circle is a **diameter** of the circle. The length of a diameter is often referred to as *the* diameter.

The perimeter of a circle is called its **circumference**.

Work in a group. You need string, a metric ruler, and five circular objects of different sizes. Use objects such as lids or a wheel.

4. For each of the circular objects, cut a length of string equal to the circumference. This is length *C*. Cut another length of string equal to the diameter. This is length *d*. Keep lengths *C* and *d* paired for each object.

5. Before using a ruler to measure, compare the paired lengths. Which of the following statements best describes the relationship between *C* and *d*?

    A. $C \approx 1d$ **B.** $C \approx 2d$ **C.** $C \approx 3d$ **D.** $C \approx 4d$

6. Use a metric ruler to measure each pair of strings to the nearest tenth of a centimeter. Record your measures in a table like the one shown at the right. Determine the ratio of *C* to *d*.

7. Find the mean value of your five measures for $\frac{C}{d}$. Compare your mean value for this ratio with those of other groups.

SPOTLIGHT ON LEARNING

WHAT? In this lesson you will learn
• to relate the circumference of a circle to its diameter.
• to relate the area of a circle to its radius.

Why? Knowing how a circle's measures are related can help you solve problems in auto maintenance, sports, and carpentry.

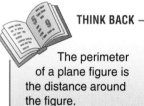

THINK BACK

The perimeter of a plane figure is the distance around the figure.

Object	C	d	$\frac{C}{d}$
1.			
2.			
3.			
4.			
5.			

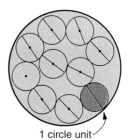

8. Notice that the circumference of a circle is measured in the same linear unit as the diameter. Recall that the areas of squares and rectangles are measured in *square units*. Use the diagram below to explain why the area of a circle is not measured in *circle units*.

1 circle unit

9. The radius of each circle below is *r*. The area of each square is r^2. The area of the circle is represented by *A*. Which of the following statements appears to be true?

A. $A \approx r^2$ **B.** $A \approx 2r^2$ **C.** $A \approx 3r^2$ **D.** $A \approx 4r^2$

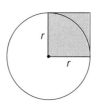
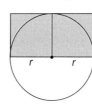
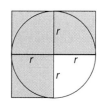
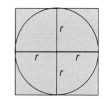

TECHNOLOGY TIP

To use geometry software to construct a circle, you may need to select a center and a radius. Or, you can select two points. One point will be the center of the circle; the other point will be a point on the circle.

Continue to work together. You need a compass and a sheet of grid paper.

10. Each group member should draw a circle on the grid paper. These circles should be of different sizes. Use radii of 4, 5, 6, 8, and 9 units.

11. Count the number of squares that lie entirely in the interior of your circle. The diagram is an example of a circle of radius 7 units.

Note from the diagram that some *pieces* of squares remain in the interior of the circle. Look at the shaded pieces. If the same colored pieces are joined, they approximate one square unit.

12. Pair partial squares to approximate one square unit. Shade paired pieces and record the number of pairs for the entire circle.

13. Add your answers from Questions 11 and 12. Express the area of your circle in square units.

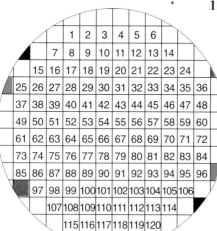

14. Record your group's results in a table like the one shown at the right. Notice that the data for the circle of radius 7 units has been entered. Determine the ratio of A to r^2.

15. Find the mean value of your six measures for $\frac{A}{r^2}$. Compare your mean value for this ratio with those of other groups.

r	r^2	A	$\frac{A}{r^2}$
4			
5			
6			
7	49	154	3.14
8			
9			

Make Connections

● In Questions 4–7, you observed that for circles of varying size the ratio of a circle's circumference to its diameter is always the same. This ratio is denoted by π, the Greek letter *pi*.

16. Use the relation between the circumference and diameter to find (in terms of π) the circumference of the circle whose diameter is 12 m.

The number π has an exact value. Its approximate value is 3.14 or $\frac{22}{7}$.

17. Using 3.14 as an approximation for π, find to the nearest tenth of a meter the circumference of the circle whose diameter is 12 m.

In Questions 8–15, you observed that for circles of varying size the ratio of a circle's area to the square of its radius is always the same. This ratio has the same value as the ratio of the circumference to the diameter, namely π.

18. Use the relation between the area and the radius to find in terms of π the area of the circle whose radius is 6 m.

19. Using 3.14 as an approximation for π, find the area to the nearest tenth of a square meter of the circle whose radius is 6 m.

Summarize

● **20.** WRITING MATHEMATICS If you know the length of the radius of a circle, explain how you can find the measures of the circumference and area. Discuss how you can state an exact value for each of these measures and how you can state an approximate value.

21. THINKING CRITICALLY Suppose a bicylist peddles so that each wheel of the bicycle makes 100 revolutions in a minute. The wheels have a diameter of 22 in. How far will the bicycle travel in 1 min?

22. GOING FURTHER Express the circumference of a circle in terms of the radius and π. Use this equation to express the radius in terms of the circumference and π. Then express the area of a circle in terms of its circumference and π.

GEOMETRY: WHO, WHERE, WHEN

The Chinese Tsu Ch'ung-chih (about A.D. 470), gave the rational value of $\frac{355}{113}$ for π, which is correct to the 6th decimal place.

In 1674 the German Gottfried Wilhelm von Leibniz expressed π as the limit of an infinite series. This discovery continues as the classically simple theoretical description.

In 1761 the exact nature of the number was established as irrational. A later proof showed that π is not an algebraic number. It is called *transcendental*.

Modern computers are able to calculate π with extreme accuracy.

Explore/Working Together

● Work in a group. You need a compass, scissors, a piece of cardboard, and masking tape.

Draw and cut out a large cardboard circle. Label the center of the circle *O*. Draw diameter $\overline{AB}$.

Place a long strip of masking tape on the floor in a straight line. Use *C* to label one endpoint of the strip.

Place the cardboard circle on the tape so that point *A* is on point *C*. Think of the circle as a wheel and roll it along the tape strip. Use *D* to mark the place on the tape where point *A* again meets the tape.

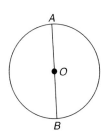

1. What does the length *CD* represent with respect to the wheel?

How would you find the distance the wheel traveled

2. in a half-turn?   **3.** in a quarter-turn?   **4.** in a 60°-turn?

5. in a 40°-turn?   **6.** in a 120°-turn?   **7.** in a 135°-turn?

Build Understanding

● A **radius of a circle** is a segment, or the length of the segment, whose endpoints are the center of the circle and a point on the circle. A **chord of the circle** is a line segment whose endpoints are on a circle. A **diameter of a circle** is a chord, or the length of the chord, that contains the center. The measure of a diameter is twice the measure of a radius. You name a circle by its center. Circle *O*, written as ⊙ *O*, shows radius $\overline{OA}$, diameter $\overline{AB}$, and chord $\overline{CD}$.

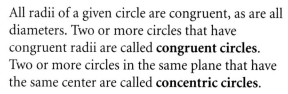

All radii of a given circle are congruent, as are all diameters. Two or more circles that have congruent radii are called **congruent circles**. Two or more circles in the same plane that have the same center are called **concentric circles**.

The set of all points in the plane of a circle whose distance from the center is less than the radius is the *interior of the circle* and whose distance from the center is greater than the radius is the *exterior of the circle*. Since $\overline{PI} < r$, point I is in the interior of ⊙ *P*. Since $\overline{OE} > r$, point *E* is in the exterior of ⊙ *P*.

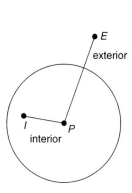

The perimeter of a circle is called its **circumference**. In all circles, the ratio of the circumference to the diameter is equal to the same number, which is denoted by π. That is, for a circle with circumference C and diameter d, $\dfrac{C}{d} = \pi$.

THEOREM 11.1	**CIRCUMFERENCE OF A CIRCLE**

> The circumference C of a circle equals the product of π and its diameter d, or $C = \pi d$.

The number π is an irrational number. Recall that an irrational number has a nonterminating, nonrepeating decimal. When you use a calculator the value of π is calculated to many decimal places.

The rational approximations commonly used for π are 3.14 and $\dfrac{22}{7}$.

EXAMPLE 1

The diameter of a circle is 24 m. Find the circumference.

Solution

$C = \pi d$ Circumference formula

$C = \pi \cdot 24$ Substitute 24 for d.

$C \approx 75.4$ Round to the nearest tenth.

The circumference of the circle is exactly 24π m. The circumference is *approximately* 75.4 m.

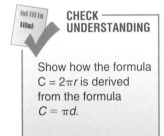
Since the diameter d of a circle is twice its radius r, an alternative formula for the circumference is $C = 2\pi r$.

EXAMPLE 2

FORESTRY The tallest living trees in the world are the redwoods of California and Oregon. What is the radius to the nearest foot of a redwood whose circumference is 21 ft?

Solution

$C = 2\pi r$ Circumference formula

$21 = 2\pi r$ Substitute 21 for C.

$\dfrac{21}{2\pi} = r$ Division Property of Equality

$3 \approx r$ Round to the nearest foot.

The radius of the tree is exactly $\dfrac{21}{2\pi}$ ft. To the nearest foot, the radius is about 3 ft.

CHECK
UNDERSTANDING

Explain why you must use three points to name a semicircle.

Two distinct points on a circle divide the circle into two parts called **arcs**. The points are the **endpoints** of each arc.

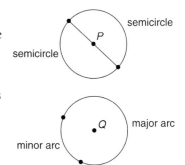

When the endpoints of the arcs also are endpoints of a diameter, each arc is a **semicircle**. Otherwise, a *minor arc* and a *major arc* are formed. A **minor arc** is smaller than a semicircle. A **major arc** is larger than a semicircle.

A **central angle** of a circle is an angle in the plane of the circle whose vertex is the center of the circle. In the figure, $\angle AOB$ is a central angle of $\odot O$.

In relation to $\angle AOB$, minor arc AB, written $\overset{\frown}{AB}$, is an **intercepted arc**. This means that its endpoints lie on the sides of the angle and all its other points lie in the interior of the angle. The major arc consists of points A and B and all points on the circle that are in the exterior of $\angle AOB$. To distinguish it from the minor arc, name the major arc by its endpoints and one other of its points. Major arc ACB is written $\overset{\frown}{ACB}$.

To find the length of an arc, you use the measure of the central angle that intercepts it and apply proportional reasoning.

--- ARC LENGTH ---

COROLLARY 11.2   The ratio of the length of a minor arc of a circle to the circumference equals the ratio of the degree measure of the central angle that intercepts the arc to 360°. The length of a semicircle is half the circumference.

EXAMPLE 3

Find the length of $\overset{\frown}{FG}$. Answer to the nearest tenth.

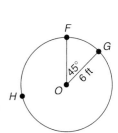

Solution

$C = 2\pi r = 2\pi(6) = 12\pi$ ft $r = 6$

Use the Arc Length Corollary to write a proportion. Let a represent the length of $\overset{\frown}{FG}$.

arc length → $\dfrac{a}{12\pi} = \dfrac{45}{360}$ ← central angle
entire circle →   ← entire circle

$$a = \frac{45}{360} \cdot 12\pi = \frac{3\pi}{2}$$

So the length of $\overset{\frown}{FG}$ is 4.7 ft to the nearest tenth.   ◄

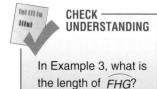

CHECK
UNDERSTANDING

In Example 3, what is the length of $\overset{\frown}{FHG}$?

Name the indicated parts of ⊙ *Z*, which is shown at the right.

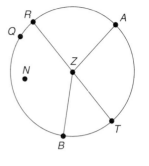

1. the center
2. four radii
3. a diameter

4. two points in the interior
5. a point in the exterior

6. five points on the circle
7. four minor arcs

8. four major arcs
9. two semicircles

10. four central angles and their intercepted arcs

Find the circumference of each circle. Give each answer in exact form and as a decimal rounded to the nearest tenth.

11.
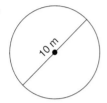
10 m

12.
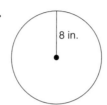
8 in.

13.
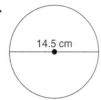
14.5 cm

14.
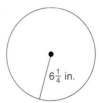
$6\frac{1}{4}$ in.

Find the radius and diameter of a circle with the given circumference. Give each answer in exact form and as a decimal rounded to the nearest tenth.

15. $C = 200\pi$ in.
16. $C = 600$ cm
17. $C = 41.8$ m
18. $C = 17\frac{3}{5}$ ft

Find the length of the indicated arc in each circle. Give each answer in exact form and as a decimal rounded to the nearest tenth.

19. minor arc *AB*
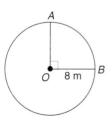
8 m

20. semicircle *ATB*
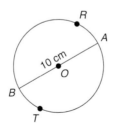
10 cm

21. minor arc *AB*
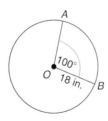
100°
18 in.

22. major arc *ADB*
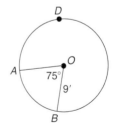
75°
9'

23. **WRITING MATHEMATICS** Discuss how an arc is like a segment and how it differs.

AUTO MAINTENANCE When new tires are installed on an automobile, the wheels must be balanced. To help evenly distribute the weight of the wheel, a mechanic clamps balance weights shaped liked arcs onto the wheel. A special balancing apparatus aids the mechanic in determining where on the wheel to place the weights.

24. For a wheel of diameter 14 in., a mechanic is using balance weights with a 10°-central angle that intercepts the arc of the weight. Find the arc length of the weight. Round to the nearest tenth of an inch.

PRACTICE

25. **WRITING MATHEMATICS** $\overline{ZX}$ is a diameter of $\odot O$ at the right. Is there a diameter $\overline{ZQ}$ in this circle? Explain. Use the vocabulary of this lesson to name as many other parts of the circle as you can.

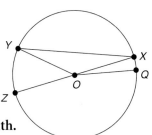

Find the circumference of a circle with the given radius _r_ or diameter _d_. Give each answer in exact form and as a decimal rounded to the nearest tenth.

26. $d = 20$ m

27. $r = 110$ ft

28. $d = 8$ cm

29. $r = 9$ in.

30. $d = 53.8$ mm

31. $d = 7\frac{1}{8}$ cm

32. $r = 62.9$ in.

33. $r = 10\frac{4}{5}$ yd

Find the radius and diameter of a circle with the given circumference. Give each answer in exact form and as a decimal rounded to the nearest tenth.

34. $C = 400\pi$ in.

35. $C = 900$ cm

36. $C = 98.6$ m

37. $C = 21\frac{2}{3}$ ft

Find the length of minor arc _RS_ in each circle. Give each answer in exact form and as a decimal rounded to the nearest tenth.

38.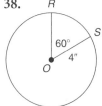

39.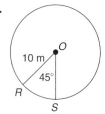

40.

41.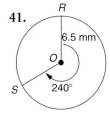

SPORTS Part of a basketball court is shown.

42. The free-throw line is the diameter of a circle. The length of the free-throw line is 12 ft. Find the circumference of this circle, to the nearest foot.

43. The basketball rim is a circle whose circumference is about 56.55 in. Find the radius of the rim, to the nearest tenth of an inch.

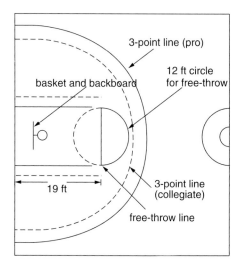

MUSIC On stage, drummers are usually positioned behind other performers. To give them visibility, drummers often stand the drum on a riser.

44. The figure at the right gives a top view of the riser used by the drummer of the Just Sixties classic-rock band. The drummer plans to buy a string of lights to hang around the semicircular edge of the riser. The lights cost $8 per foot and must be purchased in whole foot lengths only. Find the cost of these lights for this drum riser.

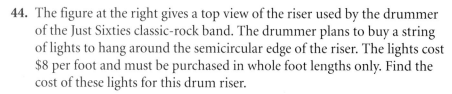

616 CHAPTER 11 **Geometry of the Circle**

EXTEND

Find the perimeter of each figure. Round to the nearest tenth of a unit.

45.

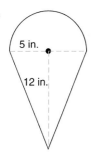

5 in.

12 in.

46.

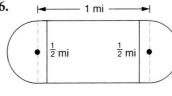

1 mi

$\frac{1}{2}$ mi $\frac{1}{2}$ mi

47.

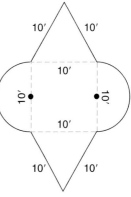

10'   10'

10'

10'   10'

10'

10'   10'

48. CONSTRUCTION A company that is installing vinyl siding needs to put a vinyl edge around the arched window shown at the right. Find the distance around the window. Round to the nearest tenth of a foot.

49. In the figure, *ABCD* is a square. Find the circumference of the circle. Round to the nearest tenth of a centimeter.

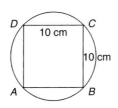

D   10 cm   C

10 cm

A   B

THINK CRITICALLY

50. The length of a minor arc of a circle is equal to the radius of the circle. Find, to the nearest degree, the measure of the central angle that intercepts the arc.

51. Determine whether each statement is true. Justify your answer by giving specific examples.

a. If two circles are equal in circumference, then they are congruent.

b. If two squares are equal in perimeter, then they are congruent.

c. If two rectangles are equal in perimeter, then they are congruent.

Replace each ___?___ with *always*, *sometimes*, or *never* to make a true statement.

52. In any given circle, chord *AB* is ___?___ longer than arc *AB*.

53. In any given circle, a radius ___?___ has the same length as a chord.

SPORTS A trainer is running laps with an athlete on the track shown. The trainer is running on the inside lane and the athlete is running on the outside lane.

54. In one lap around the track, which runner goes farther? Explain.

55. In one lap around the track, how much farther does one runner go than the other if each runner travels along the inside edge of his lane?

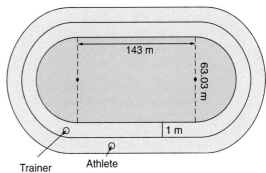

143 m

63.03 m

1 m

Trainer   Athlete

ALGEBRA AND GEOMETRY REVIEW

QUANTITATIVE COMPARISON In each question compare the quantity in Column 1 with the quantity in Column 2. Select the letter of the correct answer from these choices:

A. The quantity in Column 1 is greater.
B. The quantity in Column 2 is greater.
C. The two quantities are equal.
D. The relationship cannot be determined by the information given.

Column 1	Column 2
56. The circumference of a circle with radius 4 cm	The perimeter of a square with side 4 cm
57. $\sqrt{52}$	7
58. The slope of the line $y = 2x + 9$	The multiplicative inverse of $\frac{1}{2}$
59. The number of degrees in a base angle of an isosceles triangle whose vertex angle measures 54°	The number of square inches in the area of a rhombus with diagonals 10 in. and 15 in.
60. The number of feet in the perimeter of a right triangle with legs 9 ft and 12 ft	$\sqrt{36^2}$
61. The number of square feet in the area of a parallelogram with base 10 ft and height 23 ft	The number of square feet in the area of a right triangle with legs 10 ft and 23 ft
62. The height of an equilateral triangle of side 20 m	The length of the leg opposite the 60°-angle in a 30°-60°-90° triangle with hypotenuse 20 m

---

PROJECT *Connection*   Your team is going to build a four-wheeled, self-powered car. You may not use an electric motor or a battery-powered motor. There are no restrictions on size or weight. Your mission is to design a car that can win a distance competition on a smooth floor.

1. Obtain information on building model cars and related projects. Members should share research with the team.

2. Set up meetings with resource people to discuss possible methods of harnessing power. You may try to harness power from a spring, inflated balloon, rubber band, and fan blade.

3. Make a list of materials needed. It's best to have a large inventory of materials. Share the responsibilities for obtaining supplies.

4. Plan a construction schedule. Decide which parts can be built separately by different team members. Allow drying time for items that are glued.

5. Create a circular logo to be displayed on your car. Choose a name.

Career
Tool Designer

Tool Designers develop tools for many specialized uses.

Nails, *screws*, and *bolts* are used to hold things together. A *nut* is used on a bolt or screw for tightening. A *wrench* is a tool used to tighten and loosen nuts.

So that it can be turned in tight spaces, the handle of a wrench may be designed in an angular fashion, called *offset*. Tool designers have found that a 15°-offset minimizes the arc needed to use the wrench, making it more effective in tight spaces.

Decision Making

The faces of a nut come in different geometric shapes, commonly square or hexagonal. Hexagonal-shaped nuts are called *hex-nuts*.

1. The hexagonal face of a hex-nut has rotation symmetry. State the smallest number of degrees needed to rotate the hexagon about its center in order to show this symmetry.

Suppose a mechanic is using a wrench without an offset to turn a hex-nut in the counterclockwise direction.

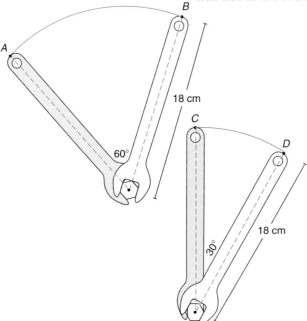

2. Refer to the diagram in which the central angle that intercepts $\overset{\frown}{AB}$ measures 60°. Find the length of $\overset{\frown}{AB}$. Round your answer to the nearest tenth of a centimeter.

3. When working with a wrench that has a 15°-offset, the mechanic does not need to use the full 60°-arc. Instead, the mechanic can flip the wrench after the hex-nut is turned only 30°. Refer to the second diagram. Find the length of $\overset{\frown}{CD}$. Round your answer to the nearest tenth of a centimeter.

4. Use the results of Questions 2 and 3 to compare the amount of clearance the offset wrench needs to turn the hex-nut to the amount of clearance the wrench without the offset needs.

11.2 **Circumference and Arc Length** **619**

11.3 Areas of Circles and Parts of Circles

Explore/Working Together

- Work with a partner.

    1. Use a compass to draw a circle that takes up an entire sheet of paper. Cut the circle out and fold it into 16 congruent parts. Unfold the circle. Use a straightedge to outline the crease lines. Cut out the 16 congruent wedges. Arrange them to form a figure that resembles a parallelogram.

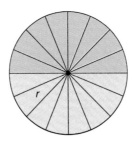 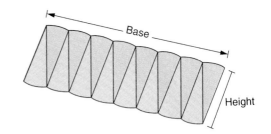

Base

Height

r

    2. How is the area of the "parallelogram" related to the area of the circle?

Let the radius of the circle equal *r*.

    3. Express the height of the "parallelogram" in terms of *r*.

    4. Express the circumference of the circle in terms of *r*.

    5. How is the base of the "parallelogram" related to the circumference of the circle? Express the base of the "parallelogram" in terms of *r*.

    6. Express the area of the "parallelogram" in terms of *r*.

    7. What can you conclude about the area of the circle?

Build Understanding

- In Explore, since the "parallelogram" was formed from the wedges of the circle, the two figures have the same area. The result leads to the following theorem.

> **THEOREM 11.3   AREA OF A CIRCLE**
>
> The area *A* of a circle equals the product of π and the square of its radius *r*, or $A = \pi r^2$.

EXAMPLE 1

LANDSCAPING A rotating lawn sprinkler waters a circular area that is 60 ft in diameter. Find the area of the lawn watered by the sprinkler.

Solution

Find the length of the radius. Since *d* is 60 ft, *r* is 30 ft.

$$A = \pi r^2 \qquad \text{Area formula}$$
$$= \pi(30^2) \qquad \text{Substitute 30 for } r.$$
$$= 900\pi \qquad \text{Simplify.}$$
$$\approx 2827.43 \qquad \text{Substitute for } \pi \text{ and multiply.}$$

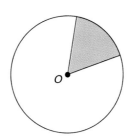

The area covered by the sprinkler is exactly 900π ft². Rounded to the nearest hundred, the area is approximately 2800 ft². ◄

A **sector** of a circle is the region bounded by two radii of the circle and their intercepted arc. Just as the length of an arc of a circle is a fractional part of the circumference of the circle, the area of a sector is a fractional part of the area of the circle.

> **THEOREM 11.4   AREA OF A SECTOR**
>
> The ratio of the area of a sector of a circle to the area of the circle equals the ratio of the degree measure of the central angle of the sector to 360°.

EXAMPLE 2

Find the area of the shaded sector. Answer to the nearest tenth.

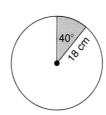

Solution

Find the area of the circle using the radius 18 cm.
$$A = \pi r^2 = \pi(18^2) = 324\pi \text{ cm}^2$$

Use Theorem 11.4 to write a proportion. Let *a* represent the area of the shaded sector.

area of sector → $\dfrac{a}{324\pi} = \dfrac{40}{360}$ ← central angle
entire circle → $\phantom{\dfrac{a}{324\pi}}$ ← entire circle

$$a = \frac{40}{360} \cdot 324\pi \qquad \text{Multiply both sides by } 324\pi.$$
$$a = 36\pi$$

So the area of the sector is 113.1 cm² to the nearest tenth. ◄

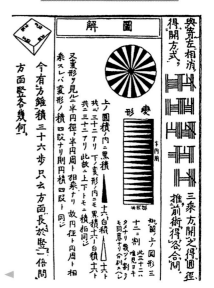

A **segment** of a circle is a region bounded by an arc and the chord having the same endpoints as the arc. You can find the area of the shaded segment in the figure at the right by subtracting the area of the triangle from the area of the sector.

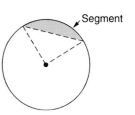
Segment

EXAMPLE 3

Find the area of the shaded segment. Round to the nearest tenth.

Solution

area of right $\triangle AOB = \dfrac{1}{2}(10)(10) = 50 \text{ cm}^2$

$$\text{area of sector } AOB = \dfrac{90}{360}(\pi r^2) \qquad \text{Theorem 11.4}$$

$$= \dfrac{1}{4}(\pi \cdot 10^2) \qquad \text{Simplify. Substitute 10 for } r.$$

$$= \dfrac{1}{4}(100\pi)$$

$$= 25\pi$$

area of segment = area of sector − area of triangle
$$= 25\pi - 50$$
$$\approx 28.5$$

Rounded to the nearest tenth, the area is 28.5 cm².

◄

TRY THESE

Find the area of a circle with the given radius or diameter. Give each answer in exact form and as a decimal rounded to the nearest tenth.

1. $r = 12$ m
2. $d = 10$ ft
3. $r = 12.5$ cm
4. $d = 20.4$ in.

Find the radius of a circle with the given area. Give each answer in exact form and as a decimal rounded to the nearest tenth.

5. $A = 121\pi$ in.²
6. $A = \dfrac{25}{4}\pi$ cm²
7. $A = 89$ m²
8. $A = 63\dfrac{3}{5}$ ft²

Find the area of each shaded sector. Give each answer in exact form.

9.
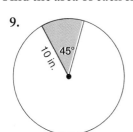
45°
10 in.

10.
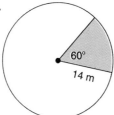
60°
14 m

11.
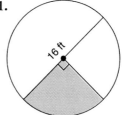
16 ft

12.
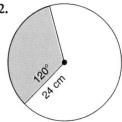
120°
24 cm

13. WRITING MATHEMATICS Compare a sector of a circle to a segment of a circle. How are they alike? How are they different?

In ⊙O at the right, m∠AOB = 60° and OA = 8 in.

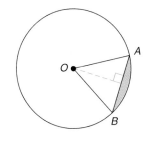

14. Explain why △AOB is equilateral.

15. The dashed line is an altitude of △AOB. What is its length?

Calculate the area of each figure in ⊙O.

16. △AOB   **17.** sector AOB   **18.** the shaded segment

Use subtraction of areas to find the area of the shaded region of each figure.

19. 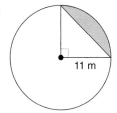   **20.** 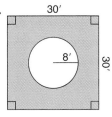   **21.** 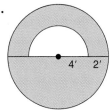   **22.**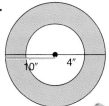

23. BUSINESS At Mackin's Pizzeria, the diameter of a small pizza is 12 in. and the diameter of a large pizza is 14 in. Which has the greater area: a slice that is one-sixth of a small pizza or a slice that is one-eighth of a large pizza? Justify your answer.

PRACTICE

Find the area of a circle with the given radius or diameter. Give each answer in exact form and as a decimal rounded to the nearest tenth.

24. $r = 31$ m   **25.** $d = 17$ ft   **26.** $r = 16.3$ cm   **27.** $d = 45.4$ in.

Find the radius of a circle with the given area. Give each answer in exact form and as a decimal rounded to the nearest tenth.

28. $A = 225\pi$ ft^2 **29.** $A = \dfrac{100}{49}\pi$ m^2 **30.** $A = 78$ m^2 **31.** $A = 45\dfrac{2}{3}$ in.^{2}

Find the area of each shaded sector. Give each answer in exact form.

32. 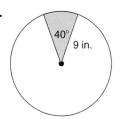   **33.** 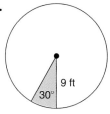   **34.** 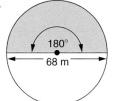   **35.**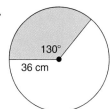

36. WRITING MATHEMATICS Draw ⊙P with radius 2 in. Label $\overline{AC}$ as a diameter. Draw $\overline{PB}$ perpendicular to $\overline{AC}$ with B on the circle. Draw $\overline{AB}$. State as many facts as you can about the figure.

Use subtraction of areas to find the area of the shaded region of each figure.

37.
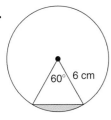

38.
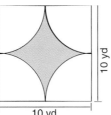

39.
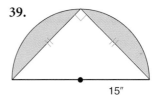

40.
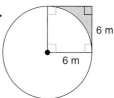

41. **AUTOMOBILE DESIGN** A rear-windshield wiper has a rubber blade that is 18 in. long. The wiper clears an area that is a part of a sector, as shown below. Use subtraction of areas to find the area cleared by the wiper. Round to the nearest square inch.

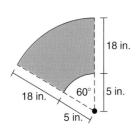

42. **HOME IMPROVEMENT** The arched window frame shown below is composed of a rectangle and a semicircle. The frame is to be filled in with a stained glass panel. Find the area of the glass panel. Round to the nearest square inch.

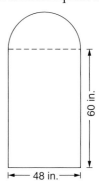

EXTEND

Find the area of each shaded segment. Give each answer as a decimal to the nearest tenth.
Hint: **You will need to use trigonometry.**

43.
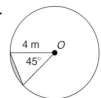

44.

45.
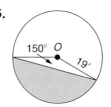

46.
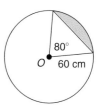

47. Find the circumference of a circle whose area is 49π ft^2.

48. Find the area of a circle whose circumference is 16π m. Round to the nearest tenth.

49. A sector of a circle has area 20π cm^2. The area of the circle is 300π cm^2. Find the number of degrees in the central angle of the sector.

THINK CRITICALLY

50. Write the inverse of the given statement. Determine whether the inverse is true or false.

    If the areas of two circles are equal, then the circumferences are equal.

51. Determine whether the given statement is true or false. Justify your answer.
If two chords of a given circle are congruent, then the segments bounded by the chords are equal in area.

The ratio of the radii of two circles is $a : b$.

52. What is the ratio of their circumferences?

53. What is the ratio of their areas?

54. In the figure below, $\triangle ABC$ is a right triangle. Three semicircular regions, I, II, and III, have been constructed so that the diameter of each is one of the sides of the triangle. Show that the area of region III is the sum of the areas of regions I and II.

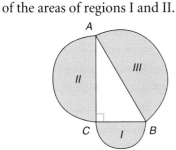

55. Below, the shaded region bounded by the two concentric circles is an *annulus*. Its width w is the difference between the radii of the concentric circles. The diameter of the inner circle is d. Show that a formula for the area A of the annulus is $A = \pi dw + \pi w^2$.

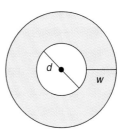

ALGEBRA AND GEOMETRY REVIEW

56. Find the axis of symmetry of the graph of the equation $y = 2x^2 - 12x + 11$.

57. Find the slope of the line represented by the equation $2x + 5y = 10$.

58. Find the perimeter of a regular pentagon whose sides measure 14.2 dm.

59. The angles of a triangle are in the ratio $1 : 2 : 7$. Find the measure of the largest angle.

60. STANDARDIZED TESTS Which set cannot represent the lengths of the sides of a triangle?

    A. 3, 4, 2   **B.** 3, 4, 5   **C.** 3, 4, 6   **D.** 3, 4, 8

PROJECT *Connection* Using only 100 straws and two 1-foot circles of corrugated cardboard, your team is to build a support structure that will hold as much weight as possible.

1. Set up meetings with resource people to discuss possible designs for your support structure. Find out why bicycle frames are tubular, and not made of flat steel.

2. Gather enough materials to build several structures for different trials. To test your structure, gather a series of stackable weights such as books or bricks. Use a scale to find the weight of each object and record the weights.

3. Write a description of your design including reasons for each feature.

4. Create a circular logo to be displayed on your completed structure. Choose a name.

11.4 Focus on Reasoning
Limits

Discussion

In the figures below, the vertices of each polygon lie on a circle. You say that each polygon is **inscribed** in the circle. Each circle is **circumscribed** about the polygon.

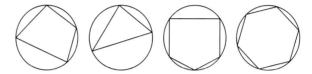

Ramiro, Katrina, Adam, and Felice are investigating the perimeters of regular polygons inscribed in circles. They drew several circles of the same size and inscribed a series of regular polygons in them.

Ramiro said there is no limit to the number of sides the polygon can have. So he reasoned that there is no limit to the perimeter of the polygon.

1. Do you agree or disagree with Ramiro's reasoning? Explain.

Katrina made this observation: As the number of sides of a regular polygon increases, the length of each side decreases. So she said the perimeter will be the same, no matter how many sides there are.

2. Do you agree or disagree with Katrina's reasoning? Explain.

Adam noticed that each side of the polygon is a chord of the circle. He said that any chord is shorter than the minor arc with the same endpoints. So he said that the perimeter of a regular polygon must be shorter than the circumference of its circumscribed circle.

3. Justify Adam's conclusion about the length of each chord.

4. How does Adam's conclusion about each side of the polygon lead to his conclusion about the perimeter of the polygon?

Felice made this statement: As the number of sides of the regular polygon increases, its perimeter gets closer and closer to the circumference of the circumscribed circle.

5. Do you agree or disagree with Felice's statement? Explain.

COMMUNICATING ABOUT GEOMETRY

Explain why it is possible to inscribe any regular polygon in a circle.

Hint: What is the definition of *center of a regular polygon*?

Build a Case

In the Discussion, Adam and Felice began to explore an important mathematical concept called a *limit*. That is, as the number of sides of the regular polygon increased, they found that the perimeter *approached* the circumference of the circumscribed circle. However, they know the perimeter will never equal the circumference, nor will it exceed the circumference.

In the questions that follow, you will see how limits help to verify the circumference and area formulas for a circle.

First consider the regular octagon inscribed in circle P, shown below.

6. Find m$\angle APB$ and m$\angle APC$.

7. Copy and complete.

$$\sin \angle APC = \frac{AC}{PA}$$

$$\sin \underline{\ ?\ } = \frac{AC}{r}$$

$$\underline{\ ?\ } \approx \frac{AC}{r}$$

$$\underline{\ ?\ } \approx AC$$

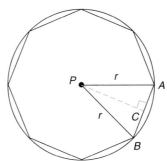

8. Write an expression that represents AB.

9. Write an expression that represents the perimeter of the regular octagon.

Work with a partner.

10. Use the method from Questions 6–9. Copy and compete the table.

11. As the number of sides of the regular polygon increases, what is the limit of the length of one side?

12. As the number of sides of the regular polygon increases, what is the limit of the perimeter?

THINK BACK

The sine of an acute angle of a right triangle is

length of opposite side
―――――――――――
length of hypotenuse

The cosine is

length of adjacent side
――――――――――――
length of hypotenuse

GEOMETRY: WHO, WHERE, WHEN

In 1722, the Japanese mathematician Takebe used a regular polygon of 1024 sides and the concept of a limit to calculate π correct to 41 decimal places.

Regular Polygons Inscribed in a Circle of Radius r		
Number of Sides	**Length of One side**	**Perimeter**
3	▪	▪
4	▪	▪
5	▪	▪
6	▪	▪
8	▪	▪
10	▪	▪
12	▪	▪
20	▪	▪
60	▪	▪
120	▪	▪

13. WRITING MATHEMATICS Explain how the limits from Questions 11 and 12 verify the circumference formula for a circle.

Recall an apothem of a regular polygon is a segment whose endpoints are the center of the polygon and the midpoint of a side. The area of the polygon equals half the product of its apothem and perimeter.

14. Add two columns to your table from Question 10. Label them *Apothem* and *Area*. Complete these two columns. *Hint*: You will need to use the cosine ratio to find the apothems.

15. As the number of sides of the regular polygon increases, what happens to the relationship between the radius of the circle and the apothem of the polygon? What is the limit of the area?

16. WRITING MATHEMATICS Explain how your answers to Question 15 verify the area formula for a circle.

EXTEND AND DEFEND

A **sequence** is a set of numbers arranged in a specific order. The numbers are called the **terms** of the sequence. A sequence that has no last term is an **infinite sequence**. For example, each perimeter you investigated in Question 10 is one term of an infinite sequence of perimeters. The same is true of the areas in Question 14.

Although an infinite sequence continues without end, you sometimes will observe that the terms get closer and closer to a particular number. That number is called the *limit* of the sequence.

Give the next four terms of each sequence. Then give the limit of the sequence, if it exists. If you think there is no limit, explain.

17. $\dfrac{1}{4}, \dfrac{1}{9}, \dfrac{1}{16}, \dfrac{1}{25}, \ldots$

18. $0.2, 0.02, 0.002, 0.0002, \ldots$

19. $2, 4, 6, 8, \ldots$

20. $6\dfrac{1}{3}, 6\dfrac{1}{9}, 6\dfrac{1}{27}, 6\dfrac{1}{81}, \ldots$

21. $100, 50, 25, 12.5, \ldots$

22. $1\dfrac{1}{2}, 1\dfrac{3}{4}, 1\dfrac{7}{8}, 1\dfrac{15}{16}, \ldots$

The first figure below is an equilateral triangle with sides of length one unit. Each figure after that is formed by joining the midpoints of sides as shown. Suppose this pattern is continued.

23. Write the first six terms of a sequence that represents the total of the perimeters of all the triangles in each figure.

24. Give the limit of this sequence, if it exists.

FRACTALS Each set of figures represents stages 0 through 3 of a fractal. Identify one or more sequences that can represent each fractal. Write the first six terms of each sequence you identify and give its limit, if the limit exists.

25.

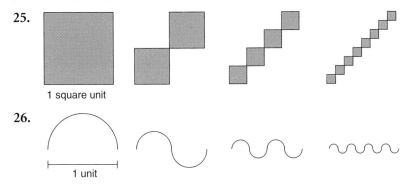

1 square unit

26.

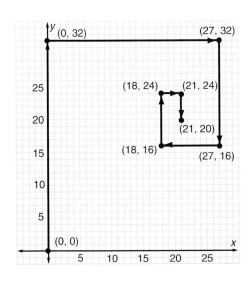

1 unit

COORDINATE GEOMETRY On the set of coordinate axes below, a set of vectors is positioned head-to-tail. Together they form a "spiral" that begins at (0, 0) and approaches a specific point.

27. Describe how the terminal point of each vector is determined.

28. Draw two more vectors using your answer to Question 27. Name the terminal point of the last vector drawn.

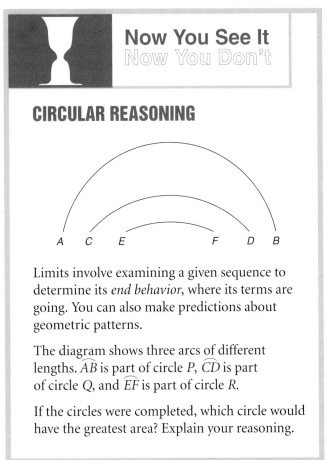

Now You See It
~~Now You Don't~~

CIRCULAR REASONING

Limits involve examining a given sequence to determine its *end behavior*, where its terms are going. You can also make predictions about geometric patterns.

The diagram shows three arcs of different lengths. $\overset{\frown}{AB}$ is part of circle P, $\overset{\frown}{CD}$ is part of circle Q, and $\overset{\frown}{EF}$ is part of circle R.

If the circles were completed, which circle would have the greatest area? Explain your reasoning.

Optimizing Perimeter and Area

You have already calculated the perimeters and areas of plane figures. Now you will find figures that have the greatest area or least perimeter under certain conditions. Moreover, you will consider the best possible solution for the situation, called the *optimal solution*.

Problem

Hanover High is sponsoring H^4: *Hands Helping Hanover's Homeless*. The students plan to join hands to enclose the largest area possible. A local company will donate 1¢ for every square foot of ground the students enclose. Suppose all 612 students participate. What shape should they form? What will be the donation?

Explore the Problem

The distance the students can span when holding hands is the perimeter of the plane figure. To arrive at the optimal solution, you need to find the maximum area associated with this perimeter.

1. Ten students are selected at random. When they stand with arms outstretched holding hands, they span a distance of 50 ft. What is the mean distance that one person's outstretched arms span?

2. Why is it better to use ten students to determine the mean distance rather than just measure one student?

3. Why is it better to randomly select students rather than pick specific students?

4. What is the approximate length of the 612 people with arms outstretched holding hands?

To examine a rectangle with perimeter of 3060 ft. Create a spreadsheet like the one shown to record and analyze data.

5. Explain the formulas used for length and area.

6. Continue to complete the chart until you can find the dimensions of the rectangle that will give the greatest area. Give the dimensions of that rectangle.

Width	Length	Area
w	$1530 - w$	$w(1530 - w)$
750	780	585000
755	775	585125

As a result of their findings, the students' decided to explore another regular polygon with perimeter 3060 ft.

7. Draw an equilateral $\triangle ABC$. The perimeter of $\triangle ABC$ is 3060 ft. Find its area.

8. Which is greater, the area of $\triangle ABC$ or that of the square from Question 6?

9. Copy the table shown. Determine the apothem for each regular polygon listed. Use the apothem and Area of Regular Polygon Theorem to complete the table.

Regular Polygons with Perimeter 3060 ft	
Number of sides	Area
5	
6	
10	
12	
15	
20	

10. As the number of sides increases, what happens to the area?

11. Make a conjecture about the type of figure that gives the maximum area for a perimeter of 3060 ft.

12. Find the area of a circle with circumference 3060 ft.

13. What shape should the students use to enclose the maximum area? What amount will the company donate?

Investigate Further

- In the preceding problem, you discovered two important facts.

  Of all plane figures with a given perimeter, the circle has the greatest area.

  Of all plane figures with a given area, the circle has the least perimeter.

  Demonstrating the truth of this statement requires methods that are beyond the scope of this course. However, the following problem illustrates one application of it.

TECHNOLOGY TIP

You can use geometry software to construct a circle with circumference equal to the perimeter of a polygon. Construct any polygon and measure its perimeter. Then construct a circle and measure its circumference. Change the size of the circle until the circumference equals the perimeter of the polygon.

The residents of an apartment complex want to build an enclosure in which their pets can run and exercise safely. They want its area to be 1500 ft². What is the least amount of fencing they will need?

14. Justify each step in the set of equations shown at the right.

$$A = \pi r^2$$
$$1500 = \pi r^2$$
$$\frac{1500}{\pi} = r^2$$
$$\sqrt{\frac{1500}{\pi}} = r$$

15. Find the value of r to the nearest whole number.

16. What is the least amount of fencing that will be needed for the enclosure?

17. WRITING MATHEMATICS A circle would use the least amount of fencing, but it might not be the *optimal* shape for the pet enclosure. Why? What shape would you recommend? Explain.

Apply the Strategy

18. ENERGY CONSERVATION A solar collector needs to be framed in steel to protect it against the weather. Suppose the area of a collector is to be 75 ft². Find the shape and dimensions of the plane figure that will require the least amount of framing for the perimeter. Round to the nearest foot.

19. FARMING Two farmers need to enclose 720 ft² of land for their plant research. One farmer suggests using a rectangle that is 18 ft by 40 ft. The other farmer suggests a circle. Which do you think is a better choice? Explain. State the dimensions for your choice.

ANTIQUITIES A restorer of antique furnishings is planning to install an octagonal window in a 19th-century building. She plans to frame the window with 400 cm of carved wood molding.

20. Make a conjecture about the type of octagonal shape she should use if she wants the window to let in the most light.

21. What are the dimensions and area of the optimal octagon?

22. WRITING MATHEMATICS Suppose you want to find the maximum rectangular area enclosed by 100 m of fencing. At the right is the graph of the equation $y = x(50 - x)$. Explain how this graph can help you solve the problem.

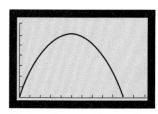

Xmin = 0   Ymin = 0
Xmax = 60   Ymax = 800
Xsci = 5   Ysci = 100

REVIEW PROBLEM SOLVING STRATEGIES

A Square Peg in a ◯ Hole.

Round

1. Which fits better, a round peg in a square hole or a square peg in a round hole? To answer this question consider the cross sections of the two situations. On the left is a square with a side of length d and a circle incribed in it. On the right is a circle with diameter d and a square inscribed in it. Find whether a larger percent of the circle is filled by the square or a larger percent of the square is filled by the circle.

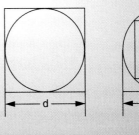

TWO'S COMPANY, THREE'S A CIRCLE

2. In the coordinate plane, any circle which passes through $(-2, -2)$ and $(1, 4)$ cannot also pass through $(x, 1000)$. What is the value of x?

Pythagoras Could Have Worked This One!

3. Regular hexagon *ABCDEF* is inscribed in Circle *M*. Point *P* is on the circle. Find the sum of the squares of the lengths of the segments drawn from *P* to each of the vertices of the hexagon.

If you want some help getting started, answer the following questions.

a. Draw $\overline{CF}$. What is the measure of $\angle CPF$?

b. Draw $\overline{AD}$. What is the measure of $\angle APD$?

c. Draw $\overline{BE}$. What is the measure of $\angle BPE$?

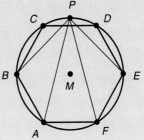

Explore/ Working Together

• In the section of the coordinate plane bounded by the large dots, consider only those points whose coordinates are integers.

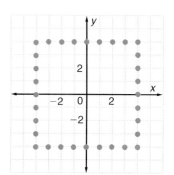

1. How many such points are in this region?

Suppose one of these points, (x, y), is chosen at random. Work with a partner to decide on the probability of each event.

2. It is in the third quadrant.

3. $x \leq 0$ 4. $y \leq 3$

5. $x \geq 4$ 6. $y \leq x$

Build Understanding

• In Lesson 2.2, you saw that you could solve some probability problems using *linear models* based on relationships among segments. You can solve other probability problems using *area models* that involve relationships among regions. Rather than counting outcomes, you calculate area of regions. Suppose that a region A contains a smaller region B. The probability P that a randomly chosen point in A is in B is given by

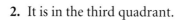

$$P = \frac{\text{area of } B}{\text{area of } A}$$

EXAMPLE 1

A dart lands on a random point on the circular target shown. What is the probability that it lands on the 3 in. square bull's-eye?

Solution

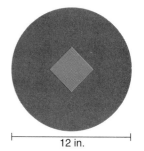

12 in.

$$P(\text{dart hits bull's-eye}) = \frac{\text{area of square}}{\text{area of circle}}$$

$$= \frac{3^2}{\pi(6^2)} \approx 0.08$$

The probability that the dart will hit the bull's-eye is about 0.08 or 8%.

Area models are generally useful for situations in which you are trying to land an object in a given region.

EXAMPLE 2

AMUSEMENT In a game at a fair, you toss a penny onto a table that has been divided into 1-in. squares. If the penny lands without touching any line, you win. Otherwise, you lose your penny. What are your chances of winning on one toss of a penny? The radius of a penny is about $\frac{3}{8}$ in.

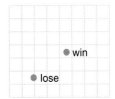

Solution

You need only consider one square in which the penny could land.

To avoid touching a line, the center of the penny must be more than $\frac{3}{8}$ in. from each side of the square.

To win, the center of the penny must land in the shaded region. The shaded region is a square whose side measures $\frac{2}{8}$ in., or $\frac{1}{4}$ in.

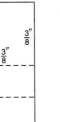

$$P(\text{win}) = \frac{\text{area of shaded region}}{\text{area of the square}} = \frac{\frac{1}{4} \cdot \frac{1}{4}}{1 \cdot 1} = \frac{\frac{1}{16}}{1} = \frac{1}{16}$$

◄

COMMUNICATING ABOUT GEOMETRY

In Example 2, suppose you tossed a coin that had a radius a little larger than that of the penny. Would the probability of winning increase or decrease?

What if the radius of the coin were greater than $\frac{1}{2}$ in.?

TRY THESE

A dart lands at a random point on the target shown. What is the probability that it lands in the shaded area?

1.
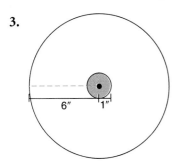

2.
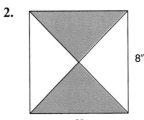

3.
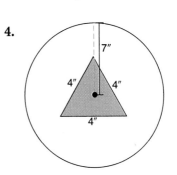

4.

A circular disc with radius 10 mm is thrown onto a checkerboard. To win, the disc must land inside a square and not touch a line. The side of the square has the given measure. What is the probability of winning?

5. 15 mm   **6.** 26 mm   **7.** 30 mm   **8.** 40 mm

9. WRITING MATHEMATICS In Exercises 5–8, you need consider only one square in calculating a probability for the entire checkerboard. Explain why this is so.

10. GAMES The dart board shown has five concentric circular regions. The diameter of the innermost circle is 1 in. The radii of the other circles increase by 1 in. as you move outward. When a dart lands in a numbered region, the player receives the score shown for the region. Suppose a dart lands at a random point on the board. What is the probability that its score is greater than 5 points?

Suppose a dart lands at a random point on each target. A dart is a winner if it lands in the shaded region. Give dimensions for the shaded region of each target so that the probability of winning with the one dart is 25%.

11.

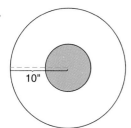

10"

12.

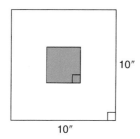

10"

10"

PRACTICE

A graphing calculator is programmed to generate a random point (x, y) where $0 \le x \le 10$ and $0 \le y \le 8$. Find the probability that the point will be in each shaded region.

13.

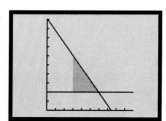

14.

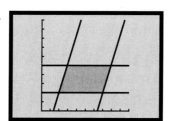

15.

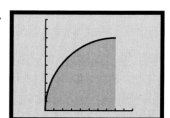

16. WRITING MATHEMATICS Write a probability problem that you can solve using a linear model and another problem that you can solve using an area model. Show how to solve each problem.

17. Barbara bought a new racquetball. The surface area of the racquetball is approximately 24 cm². A rectangular price sticker on the ball has area of 2 cm². If the ball is randomly hit against a wall, what is the probability that the price sticker will come in contact with the wall on any specific shot?

A circular disc with radius 6 mm is thrown onto a checkerboard. To win, the disc must land inside a square and not touch a line. The side of the square has the given measure. What is the probability of winning?

18. 24 mm   **19.** 10 mm   **20.** 18 mm   **21.** 48 mm

SKYDIVING A skydiver jumps from a plane and lands in a square field that is 1.5 mi on each side. In each corner of the field, there is a large tree. The ropes of the chute will get tangled in the tree if the parachutist lands within 0.1 mi of its trunk.

22. Draw a diagram of the field. What is the probability that the parachutist will land in the field without getting caught in a tree? Round to the nearest percent.

23. What would you say is the theoretical probability that the parachutist can land on the point that is at the exact center of the square? Explain your reasoning.

EXTEND

A circular spinner is divided into eight congruent sectors, as shown. The radius of the circle is 6 in. Find each probability.

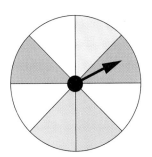

24. P(red)   **25.** P(white)   **26.** P(green or yellow)

27. P(not blue)   **28.** P(orange)   **29.** P(not purple)

30. Suppose the radius of the circle were 12 in. Explain how this would affect your answers to Exercises 24–29.

Use the shaded area in the first quadrant of the coordinate plane for Exercises 31–33.

31. How many points are in this region?

32. Write an equation to describe the location of all points whose coordinates have a sum of 6.

33. Suppose a point is chosen at random from this region. Find the probability that the sum of its coordinates is

    a. greater than or equal to 6   **b.** less than 6

    c. less than or equal to 3   **d.** greater than 4

    e. less than 0   **f.** less than or equal to 12

Two six-sided number cubes, numbered 1 to 6, are rolled and the numbers are added.

34. How many possible outcomes are there?

35. How many outcomes are there in which the sum is 6?

36. Use two number cubes to find the probability of each condition in Exercise 33 parts a–f.

37. WRITING MATHEMATICS Compare the coordinate-plane questions in Exercise 33 to the number cube questions in Exercises 34–36. How are the exercises alike? How do they differ?

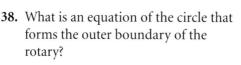

HIGHWAY ENGINEERING The figure at the right shows an engineer's plan for a traffic rotary that will help ease the flow of vehicles through a busy intersection.

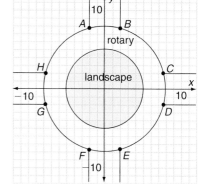

38. What is an equation of the circle that forms the outer boundary of the rotary?

39. The labeled points identify the places where the roads intersect the rotary. What are the coordinates of these points?

40. Write a mathematical sentence to represent the green landscaped area at the center of the rotary.

THINK CRITICALLY

41. The three spinners shown are congruent. In each spinner, region 1 has a different area. However, for each spinner, $P(1) = \frac{1}{4}$. Explain how this is possible.

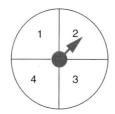 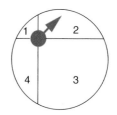 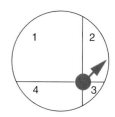

SCIENCE A scientist is conducting an experiment on color preference of a fly. She lets 1000 flies loose in a room with tiled walls, floor, and ceiling. Each tile is a 1-ft square. There are 3600 tiles in all—1800 black, 900 red, 200 yellow, and 700 pink. The colored tiles are evenly distributed throughout the room.

42. At any given moment, there are 500 flies flying around, and 300 are on yellow tiles. Use probability and area to explain whether or not you think the flies have a color preference, or if they land randomly.

PROJECT *Connection*   Your models need to be tested and revised to optimize their performance.

1. The team that built the model car should test its performance at the location agreed upon for the competition. Make several trial runs. Record the distances traveled.

2. The team that built the support structure should test its performance using the weights agreed upon for the competition. Since the groups in the class must use the same weights, these trials might have to take place at different times. Remember that several support structures with different designs should be built since they will be destroyed at the trials. Make sure you know how heavy the weights are. Record the weight at which each design fails.

3. Revise the car and the support structure based on the results at the trial runs.

4. Make a detailed diagram of the final version of the model that will compete. Keep your group's designs a secret from the other groups until the actual competition.

In 1857, Leon Scott of France invented the first device to record sound. Twenty years later, Thomas Edison's phonograph became the first device to record and play back sound.

In the early 1900s, circular records became popular. They evolved over the next few decades and became classified by the number of revolutions per minute (rpm) at which they operated. Seven-inch, 45-rpm records and twelve-inch, $33\frac{1}{3}$-rpm records were the rage of the 1950s and 1960s. In the 1970s, 8-track cartridges and cassette tapes became popular. In the 1980s, the compact disc (CD) revolutionized the recording industry. With each advance, sound quality improved. Recording technicians must keep pace with the changing technology of the industry.

Decision Making

1. Twelve-inch, long-playing (LP) records spin at $33\frac{1}{3}$ rpm. Through how many degrees does an LP spin each second?

Suppose a phonograph's needle is on an LP record. To the nearest inch, how many inches of the record travel past the needle per second when the distance of the needle from the center of the record is

2. 5 in.?   3. 4 in.?   4. 3 in.?   5. 2 in.?   6. 1 in.?

The number of inches of the record that travel past the needle per second is called the *linear speed* of the record.

7. From the results of Questions 2–6, make an observation about the linear speed of an LP with respect to the location of the needle.

To play a CD, a laser beam is used instead of a needle. The location of the beam determines the number of rpm's at which the CD plays.

8. When the laser beam is 4.5 inches from the center of a CD, the CD spins at approximately 200 rpm. To the nearest inch, what is the linear speed of the CD at this point?

9. When the laser beam is 1.8 inches from the center of a CD, the CD spins at approximately 500 rpm. To the nearest inch, what is the linear speed of the CD at this point?

10. What do you notice about the linear speed of a CD?

11.6 **Using Area to Find Probability** **639**

11.7 Equations for Circles

Explore

● Draw coordinate axes on grid paper. Use a compass to construct a circle with center at $(0, 0)$ and with radius r equal to 10 units.

1. Refer to your graph. Copy and complete the table at the right.

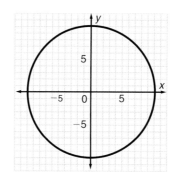

(x, y)	$x^2 + y^2$	Is (x, y) on the circle?
$(-6, 8)$	▨	▨
$(10, 3)$	▨	▨
$(8, 6)$	▨	▨
$(-4, 7)$	▨	▨
$(-10, 0)$	▨	▨
$(6, 8)$	▨	▨
$(11, -2)$	▨	▨
$(-6, 8)$	▨	▨
$(0, -10)$	▨	▨
$(-7, -8)$	▨	▨

2. Use your results from Question 1. Complete this statement: If a point (x, y) is on a circle with center at $(0, 0)$ and with radius ___?___, then $x^2 + y^2 = $ ___?___.

3. Complete the statement in Question 2 for a circle with radius 4.

Build Understanding

● Recall you can write an equation for any line on a coordinate plane. The same is true of circles. The figure at the right shows a circle with its center at the origin. By the definition of a circle, the distance between the center and any point $P(x, y)$ on the circle is equal to the radius r. So you can apply the distance formula as follows.

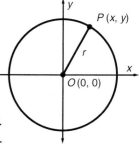

$d = \sqrt{(x_2 - x_1)^2 + (y_2 - y_1)^2}$ Use the distance formula.

$r = \sqrt{(x - 0)^2 + (y - 0)^2}$ Substitute.

$r = \sqrt{x^2 + y^2}$ Simplify.

$r^2 = x^2 + y^2$ Square each side of the equation.

A circle of radius r with its center at the origin can be represented by an equation of the form $x^2 + y^2 = r^2$.

To write an equation of a circle with center at the origin, you need to know the length of the radius.

Write an equation for the circle with center at (0, 0) and diameter 32.

Solution
Since the diameter is 32, the radius is 16.

$$x^2 + y^2 = r^2 \qquad \text{Use the equation of a circle with center} \\ (0, 0) \text{ and radius } r.$$

$$x^2 + y^2 = 16^2 \qquad \text{Substitute.}$$

$$x^2 + y^2 = 256 \qquad \text{Simplify.} \qquad \blacktriangleleft$$

When you know an equation for a circle, you can determine the radius.

EXAMPLE 2

Find the radius of the circle with equation $x^2 + y^2 = 75$.

Solution

$$r^2 = 75 \qquad \text{Use equation of a circle.}$$

$$r = \pm\sqrt{75} \qquad \text{Take the square root of each side.}$$

$$r = \pm 5\sqrt{3} \qquad \text{Simplify.}$$

Since r is a length, it cannot be negative. So $r = 5\sqrt{3}$, which is about 8.7. $\qquad \blacktriangleleft$

TECHNOLOGY TIP

To obtain a circular graph that does not appear distorted, you must use a square viewing window. That is, the ratio of (Ymax − Ymin) to (Xmax − Xmin) must be equal to the ratio of screen height to screen width.

You can check that you have correctly identified the radius of the circle whose equation is given in Example 2 by graphing it with a graphing utility. You first have to solve the equation for y.

$$x^2 + y^2 = 75$$

$$y^2 = 75 - x^2$$

$$y = \pm\sqrt{75 - x^2}$$

Then graph these two equations.

$$y_1 = \sqrt{75 - x^2}$$

$$y_2 = -\sqrt{75 - x^2}$$

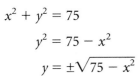

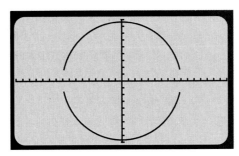

Xmin = −14.1   Ymin = −9.3
Xmax = 14.1   Ymax = 9.3
Xscl = 1   Yscl = 1

Not every circle on a coordinate plane is centered at the origin. In the figure, the center of the circle is $C(h, k)$. However, it remains true that the distance between the center and any point $P(x, y)$ on the circle is equal to the radius r. So you can once again apply the distance formula.

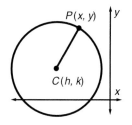

$d = \sqrt{(x_2 - x_1)^2 + (y_2 - y_1)^2}$ Use the distance formula.

$r = \sqrt{(x - h)^2 + (y - k)^2}$ Substitute.

$r^2 = (x - h)^2 + (y - k)^2$ Square each side of the equation.

> The equation $(x - h)^2 + (y - k)^2 = r^2$ is called the *standard form* of the equation of a circle of radius r with its center at (h, k).

EXAMPLE 3

ELECTRONICS Find the equation of the *woofer* shown on the sketch made by an audio designer for a stereo speaker cabinet.

Solution

The circle has its center at $(6, 7)$ and a radius of 5.

$(x - h)^2 + (y - k)^2 = r^2$

$(x - 6)^2 + (y - 7)^2 = 25$

The equation of the circular cutout for the woofer is

$(x - 6)^2 + (y - 7)^2 = 25.$ ◄

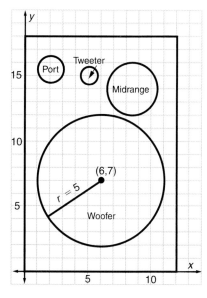

When you know an equation for a circle, you can determine the center and the radius.

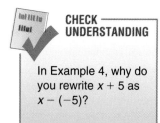

CHECK UNDERSTANDING

In Example 4, why do you rewrite $x + 5$ as $x - (-5)$?

EXAMPLE 4

A circle has equation $(x + 5)^2 + (y - 14)^2 = 121$. Find its center and radius.

Solution

$(x + 5)^2 + (y - 14)^2 = 121$ Use the equation of a circle.

$(x - (-5))^2 + (y - 14)^2 = 11^2$ Rewrite the given equation.

The center of the circle is at $(-5, 14)$ and the radius is 11. ◄

Write an equation in standard form for each circle.

1.
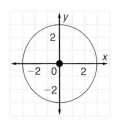

2.
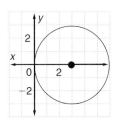

3.
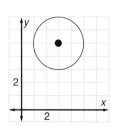

4.
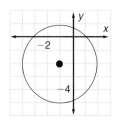

Find the center and radius of the circle with the given equation. Express irrational answers in simplest radical form.

5. $x^2 + y^2 = 144$

6. $x^2 + y^2 = 60$

7. $2x^2 + 2y^2 = 10$

8. $(x - 3)^2 + (y - 15)^2 = 1$

9. $(x + 11)^2 + (y - 5)^2 = 4$

10. $(x + 2)^2 + (y + 3)^2 = 72$

11. **WRITING MATHEMATICS** The equation of $\odot A$ is $(x - 4)^2 + (y + 6)^2 = 225$. Give equations for two different circles that are congruent to $\odot A$ and two different circles that are concentric with $\odot A$. Justify your answers.

12. **PLUMBING** The builders of an apartment complex plan to have some of the building materials pre-cut by machine. At the right, you see their template for the sub-floor of a bathroom shower stall. Write an equation for the circle that forms the drain opening.

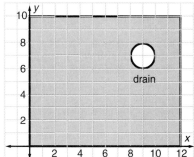

PRACTICE

Write an equation for each circle described.

13. center, $(0, 0)$; radius, 18

14. center, $(0, 0)$; radius, 1

15. center, $(0, 0)$; diameter, 40

16. center, $(1, 2)$; radius, 4

17. center, $(-9, -5)$; radius, 9

18. center, $(-8, 4)$; diameter, 16

Find the center and radius of the circle with the given equation. Express irrational answers in simplest radical form.

19. $x^2 + y^2 = 162$

20. $3x^2 + 3y^2 = 54$

21. $(x - 1)^2 + (y - 10)^2 = 81$

22. $(x - 8)^2 + (y + 2)^2 = 50$

23. $x^2 + (y - 5)^2 = 100$

24. $(x - 3)^2 + y^2 = 169$

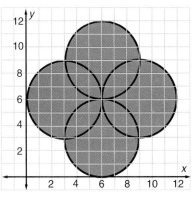

25. **DESIGN** At the right a design for a stained glass sun-catcher is drawn on a set of coordinate axes. An industrial engineer must write instructions for a machine that will cut the glass. Write an equation for each of the four circles in the design.

EXTEND

Find the circumference and area of a circle with the given equation. Give answers in exact form and as a decimal rounded to the nearest tenth.

26. $x^2 + y^2 = 36$ **27.** $x^2 + y^2 = 18$ **28.** $(x - 3)^2 + (y - 5)^2 = 81$

Write an equation for each circle described.

29. The circle with area 64π cm^2 and center at $(1, 4)$.

30. The circle with circumference 100π m and center at $(-2, 7)$.

31. A circle that has a diameter with endpoints at $(1, 5)$ and $(7, 11)$.

32. The equation of a circle is $x^2 - 10x + 25 + y^2 + 2y + 1 = 49$. Find the center and radius.

33. On a set of coordinate axes, graph the inequality $x^2 + y^2 < 25$.

THINK CRITICALLY

34. Square $ABCD$ has vertices $A(0, 8)$, $B(8, 8)$, $C(8, 0)$ and $D(0, 0)$. Write an equation of its circumscribed circle.

35. Solve $(x - 2)^2 + (y + 9)^2 = 4$ for y.

Consider the equation $(x - 3)^2 + (y + 4)^2 = k$.

36. Suppose $k = 64$. How many solutions does the equation have?

37. For what value(s) of k does the equation have exactly one solution? no solution?

TRANSFORMATIONS **Consider a circle with an equation of the form $x^2 + y^2 = r^2$. Give an equation for its image under each transformation.**

38. a reflection across the x-axis   **39.** a rotation through 90° about the center

40. translation right a units and up b units   **41.** a reflection across the line $y = x$

42. Repeat Exercises 38–41, this time considering a circle with an equation of the form $(x - h)^2 + (y - k)^2 = r^2$.

ALGEBRA AND GEOMETRY REVIEW

43. Find the x-intercepts of the parabola $y = x^2 - 8x - 20$.

44. STANDARDIZED TESTS Which ordered pair is a solution of the inequality $y < 4x - 2$?

     A. $(0, -2)$ **B.** $(5, 24)$ **C.** $(3, 10)$ **D.** $(\pi, 10)$

45. Find the perimeter of a square whose area is 121 ft^2.

46. Find the diameter of the circle whose equation is $x^2 + (y - 3)^2 = 121$.

47. Draw a rectangular prism. How many planes of symmetry does this solid have?

Career
Seismologist

When an earthquake occurs, the sudden and violent movement beneath the surface causes shock waves, called *seismic waves*, to radiate through the earth. The strength and direction of these waves can be measured by a sensitive instrument called a *seismograph*.

Seismologists, scientists who study earthquakes, use the seismograph to determine the *epicenter* of an earthquake, the point on the earth's surface likely to suffer the most damage. The epicenter is directly above the *focus* of the earthquake, the point in the earth where the quake starts.

Seismologists locate earthquakes by studying the time intervals at which the different seismic *waves* reach a number of seismographic stations. They use circles as a model for recording the data.

Decision Making

Primary waves, those energy waves released first, travel at an approximate speed of 4.8 mi/sec. *Secondary waves* travel slower, at about 2.75 mi/sec.

1. To the nearest ten-thousandth of a second, how much longer than a primary wave does it take a secondary wave to travel 1 mi?

Suppose secondary waves reach Seismic Station A 50 sec after the primary waves.

2. To the nearest mile, how far away is the epicenter of the earthquake from Station A?

3. Using your result from Question 2 describe, in geometric terms, the set of all points that are this distance from Station A.

Suppose the secondary waves reach Station B 2 min after the primary waves and reach Station C 3.5 minutes after the primary waves.

4. To the nearest mile, how far is the epicenter from Station B? from Station C?

5. The findings of Stations A, B, and C are recorded in the diagram shown at the right. Where is the epicenter of the earthquake?

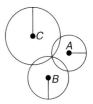

6. Explain why it is not appropriate to use only two seismic stations to find an epicenter.

11.7   **Equations for Circles**   **645**

Explore/Working Together

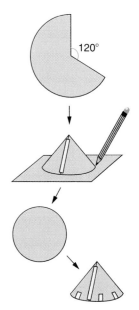

You will need heavy paper, a compass, a protractor, a ruler, tape, and scissors.

1. Draw a circle with radius 10 cm. In this circle, draw a central angle with measure 120°. Cut along the sides of the central angle and discard the smaller sector.

2. Bring the straight edges of the larger sector together and tape them to form the lateral surface of a cone.

3. Trace the circular base of this figure onto a sheet of paper. Cut out the circle. Tape the circle to the lateral surface to complete the cone.

4. Measure the height of the cone and the diameter of the circular base.

5. Work with a partner. Construct a cylinder with the same height as the cone whose bases are congruent to the base of the cone.

THINK BACK

The *height* of any geometric figure is the length of an altitude.

Build Understanding

A **cylinder** is a three-dimensional figure that consists of two parallel congruent circular regions, called **bases**, and a **lateral surface** that connects the boundaries of the bases. The **axis** of a cylinder is the segment whose endpoints are the centers of the bases. An **altitude** is a segment perpendicular to the bases with one endpoint in the plane of each base. In a **right cylinder**, the axis also is an altitude. A cylinder that is not a right cylinder is called an **oblique cylinder**.

right cylinder   oblique cylinder

EXAMPLE 1

Refer to the cylinder at the right. Draw a net for this cylinder and label its dimensions.

Solution

The cylinder has two circular bases each with a radius of 4 cm.

The lateral surface of the cylinder is rectangular. The width of the rectangle is the height of the cylinder, or 12 cm. Its length is the circumference of the bases.

$C = 2\pi r$ Circumference formula

$C = 2\pi(4) = 8\pi$

$C \approx 25.1$

The net is shown at the right.   ◀

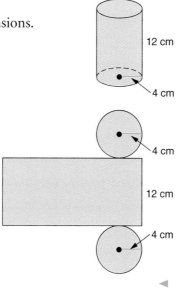

THINK BACK

A *net* is a two-dimensional figure that, when folded, forms the surface of a three-dimensional figure.

A **cone** is a three-dimensional figure that consists of a circular face, called the **base,** a point called the **vertex** that is not in the plane of the base, and a **lateral surface** that connects the vertex to each point on the boundary of the base. The **axis** of a cone is the segment whose endpoints are the vertex and the center of the base. The **altitude** is the segment from the vertex perpendicular to the plane of the base. In a **right cone,** the axis and the altitude are the same segment. A cone that is not a right cone is called an **oblique cone.**

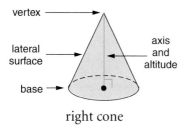

right cone

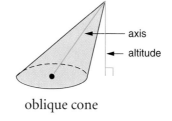

oblique cone

A right cone has one additional characteristic to consider. In a right cone, the distance from the vertex to any point on the boundary of the base is constant. This is called the **slant height** of a right cone.

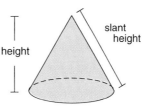

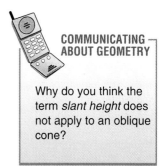

COMMUNICATING ABOUT GEOMETRY

Why do you think the term *slant height* does not apply to an oblique cone?

The radius of a base of a cylinder or a cone also is considered the radius of the cylinder or the radius of the cone. Similarly, the diameter and circumference of a base are considered the diameter and circumference of the cylinder or cone.

EXAMPLE 2

The figure shown is a right cone.
Find its slant height *s*.

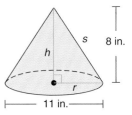

Solution

The segment labeled *s* is the hypotenuse of
a right triangle whose legs are the height *h*
and the radius *r* of the cone.

$$r^2 + h^2 = s^2 \qquad \text{Pythagorean Theorem.}$$
$$(5.5)^2 + 8^2 = s^2 \qquad \text{Substitute.}$$
$$94.25 = s^2 \qquad \text{Simplify.}$$
$$9.7 \approx s \qquad \text{Take the square root of each side.}$$

The slant height *s* is about 9.7 in. ◄

THINK BACK

The common
length of all the radii
is called *the* radius.

The common length
of all the diameters is
called *the* diameter.

The diameter is twice
the radius.

A **sphere** is the set of all points in space that are a given distance
from a given point. The given point is the **center** of the sphere. A
segment whose endpoints are the center of the sphere and a point on
the sphere is a **radius** of the sphere. A **chord** of a sphere is a segment
whose endpoints are points on the sphere. A **diameter** of a
sphere is a chord that contains the center.

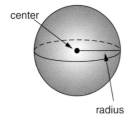
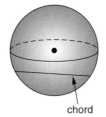
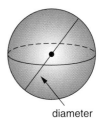

center   radius   chord   diameter

EXAMPLE 3

EARTH SCIENCE The shape of the Earth is not a
perfect sphere, but it is very close to one. Its
diameter at the equator is about 7926 mi.
What are the shape and area of the
cross section of Earth at the equator?

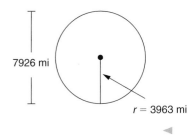

Solution

The cross section is a circle.

$$A = \pi r^2 \qquad \text{circle area formula}$$
$$A = \pi (3963)^2$$
$$A = 15{,}705{,}369\pi$$
$$A \approx 49{,}339{,}872$$

7926 mi

r = 3963 mi

The area of the cross section is about
49,339,872 mi², or nearly fifty million
square miles. ◄

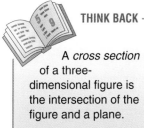

THINK BACK

A *cross section*
of a three-
dimensional figure is
the intersection of the
figure and a plane.

The equator is a model of a *great circle*. A **great circle** is a cross section of a sphere whose center is the center of the sphere. Any great circle divides the sphere into two congruent parts. Each part is called a **hemisphere**.

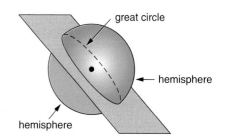

great circle

hemisphere

hemisphere

Draw a net for each right cylinder and label its dimensions.

1.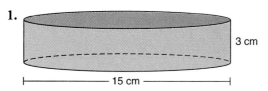

3 cm

15 cm

2.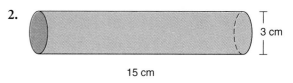

3 cm

15 cm

Identify the radius, diameter, circumference, height, and slant height of each right cone. If necessary, round to the nearest tenth.

3.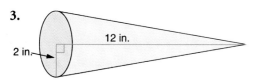

12 in.

2 in.

4.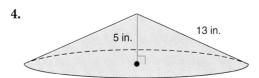

5 in.

13 in.

5. PACKAGING A soup can is shaped like a right cylinder with radius 4 cm and height 12 cm. You must design a label to cover the side of the can completely without overlap. Identify the shape of the label and give its dimensions.

6. ARCHITECTURE The building that houses an indoor ice skating rink is shaped like a hemisphere with diameter 20 yd. Ice covers the floor of the building to within 10 ft of its edges. What is the area covered by the ice?

7. WRITING MATHEMATICS How are a right cylinder and a right cone alike? How are they different? Name as many likeneses and differences as you can.

8. In the figure below, plane $\mathcal{N}$ intersects a right cylinder and is parallel to the bases. Sketch the cross section formed and find its area.

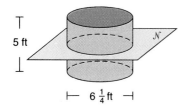

5 ft

$\mathcal{N}$

$6\frac{1}{4}$ ft

9. The figure below is a net for a right cone. Sketch the cone. On your sketch, label the radius, height, and slant height.

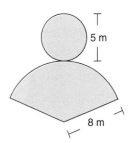

5 m

8 m

PRACTICE

In the figures below, a plane intersects a right cylinder and right cone through their axes and intersects a sphere through a great circle. In each case, sketch the cross section formed and find its area.

10.
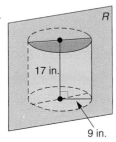
R
17 in.
9 in.

11.
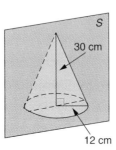
S
30 cm
12 cm

12.
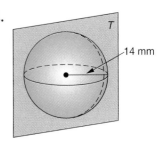
T
14 mm

Draw a net for a right cylinder of the given description and label its dimensions. When necessary, round to the nearest tenth.

13. diameter = 10 cm, height = 8 cm

14. radius = 4 yd, height = 15 yd

15. radius = 1 ft, height = 7 in.

16. circumference = 16π m, height = 5 m

Find the value of *x* in each right cone. Round to the nearest tenth.

17.
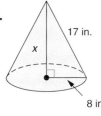
17 in.
x
8 in.

18.
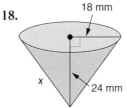
18 mm
x
24 mm

19.

11 ft
10 ft
x

20.
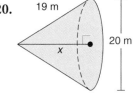
19 m
x
20 m

21. SOCIAL STUDIES In 1964 and 1965, a World's Fair was held in New York City. The symbol of this fair was a model of Earth called the Unisphere. The diameter of the Unisphere is 120 ft. Find the circumference of a great circle of the Unisphere.

22. EARTH SCIENCE The radius of Earth at the equator is about 3963 mi. Suppose you could travel at an average speed of 40 mi/h. How many days would it take to travel around the equator?

23. WRITING MATHEMATICS In Chapter 3, you learned that *oblique lines* are two lines that intersect and are not perpendicular. What is the connection between this term and the terms *oblique cylinder* and *oblique cone*?

EXTEND

TRANSFORMATIONS **For each type of figure, give the number of planes of symmetry and the number of axes of symmetry.**

24. right cylinder   **25.** right cone   **26.** oblique cylinder   **27.** oblique cone   **28.** sphere

COORDINATE GEOMETRY **In a three-dimensional coordinate system, the standard form of the equation of a sphere with radius r and center (h, j, k) is $(x - h)^2 + (y - j)^2 + (z - k)^2 = r^2$.**

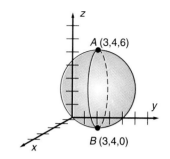

29. The equation of a sphere is $(x - 5)^2 + (y + 2)^2 + z^2 = 64$. Identify its center, radius, diameter, and the circumference of a great circle.

30. At the right is a sphere with two points of a great circle labeled. Write the equation of the sphere in standard form.

THINK CRITICALLY

LOOKING AHEAD **A *conic section* is a cross section formed when a plane intersects a double right cone. For example, the figure at the right shows that a circle is the conic section formed when the plane is parallel to the bases of the cones. Make a sketch showing how to position the plane to form each of these conic sections.**

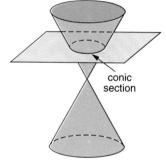

conic section

31. ellipse   **32.** hyperbola   **33.** parabola

34. Is it possible for the intersection of the plane and the double cone to be anything other than a circle, ellipse, hyperbola, or parabola? Explain.

35. Sketch all possible types of cross sections formed when a plane intersects a right cylinder.

ALGEBRA AND GEOMETRY REVIEW

36. Find the slope of a line perpendicular to the line with equation $y = 2x + 15$.

37. Find the distance between the points $A(-1, 9)$ and $B(7, 13)$.

38. Draw and label a net for a right cylinder with radius 6 in. and height 10 in.

39. The measures of two angles of a triangle are 53° and 17°. Classify this triangle by its sides and by its angles.

40. The lengths of the legs of a right triangle are 8 ft and 15 ft. Find its area and perimeter.

Replace each __?__ with *always*, *sometimes*, or *never* to make a true statement.

41. If two angles are congruent, they are __?__ supplementary.

42. If two triangles are congruent, they are __?__ similar.

The Arbelos of Archimedes

The diagram at the right shows three circles, two of which nest snugly inside the third one. The picture may represent a cross section of pipe with two smaller-diameter pipes inside it or it may represent part of a corporate logo. Notice that the sum of the diameters of the smaller circles equals the diameter of the largest one.

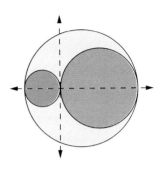

Refer to the diagram at the right. From the diagram, $r_1 + r_2 = R$.

1. Write expressions involving r_1, r_2, R, and π for the semicircumferences of the three semicircles. Then write and simplify a formula for the total length L around the shaded region. What variable(s) does your expression contain?

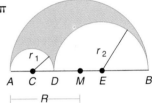

The shaded region outside the two smaller semicircles and inside the largest semicircle is called the *arbelos of Archimedes*. Some people also call the arbelos the *cobbler's knife* or *shoemaker's knife*.

Consider moving C along $\overline{AB}$. Examine changes in the semicircle with center C and radius r_1 and the semicircle with center E, and radius r_2. Describe changes to the distance around the arbelos.

2. Jamie argued that the total distance around the arbelos was unaffected by the movement of C along $\overline{AB}$. Explain how you know from your answer to Question 1 that she is right.

Just as you can explore the distance around an arbelos, you can explore its area. Remember that the area of the shaded region is the area of the largest semicircle less the sum of the areas of the two smaller ones.

3. Write expressions involving r_1, r_2, R, and π for the areas of the three semicircles. Then write and simplify an expression for the area of the arbelos. What variable(s) does your answer contain?

4. What can you conclude about the area of the arbelos determined by r_1, r_2, and R?

5. URBAN PLANNING To the nearest whole number, approximate the area of the grassy region in green that surrounds the two wading pools shown in blue.

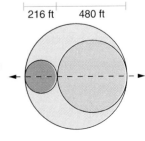

In the diagram, $\overline{DF} \perp \overline{AB}$ at D and $\overline{DF}$ is a diameter of $\odot X$. Several relationships between the area of $\odot X$ and the area of the arbelos follow.

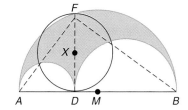

- To find the area of $\odot X$, you need FD.

- If you can show that $\triangle AFB$ is a right triangle with altitude $\overline{FD}$, you can write $(FD)^2 = (AD)(DB)$. This will give FD.

- You can then apply an area formula to $\odot X$ with radius $\dfrac{FD}{2}$.

To show $\angle AFB$ is a right angle. Make a diagram showing isosceles triangles $\triangle FMB$ and $\triangle FMA$.

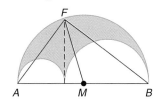

6. How do you know $m\angle FMA = 2m\angle FBM$?

7. How do you know $m\angle FAM = m\angle AFM$?

8. Use the results from Questions 6 and 7 to find $m\angle AFM + m\angle MFB$. What does this sum tell you about $\triangle AFB$?

9. Find FD and the area of $\odot X$. Relate its area to that of the arbelos.

LAWN MOWER Sunee is going to mow the lawn. She notices that the spout to the gasoline can consists of an outer cylinder that contains two inner cylinders as shown in the drawing. The larger inner cylinder is for gasoline flow and the smaller one is an air vent. The size of the inner cylinders affects the rate of flow of gasoline.

10. What geometric figure is formed by a cross section of the spout determined by a plane perpendicular to the axes of the cylinders?

11. As the diameters of the inner cylinders change, the volume of each of the cylinders change. What affect does changing the diameters have on the total volume of the two inner cylinders?

ARCHIMEDES (287–212 B.C.) was perhaps one of the greatest mathematicians of all time. He was one of the first people to investigate the arbelos.

Archimedes spent countless hours drawing diagrams in the sand during his investigations.

• • • CHAPTER REVIEW • • •

Choose the word from the list that correctly completes each statement.

1. A line segment that connects two points on a circle is a(n) __?__ .

2. A plane figure formed by two radii and an arc of a circle is called a(n) __?__ .

3. A portion of a circle that connects two points on the circle is called a(n) __?__ .

4. The largest chord in any circle is called its __?__ .

a. sector

b. chord

c. diameter

d. arc

Lesson 11.1 and 11.2   CIRCLE AND CIRCUMFERENCE AND ARC LENGTH   pages 609–619

- The circumference C of a circle with diameter d and radius r is $C = \pi d$, and the area A is $A = \pi r^2$.
- The ratio of the length of an arc of a circle to its circumference equals the ratio of the degree measure of the arc to 360°.

Give the radius, diameter, circumference, and area of each circle. Round to nearest tenth, if necessary.

Find the length of $\overarc{AB}$ and $\overarc{ACB}$. Round to the nearest tenth, if necessary.

5.

6 m

6.

8 ft

7.

A 40° B O 18 in. C

9. Find the circumference of a circle whose area is $81\pi^2$. Round to the nearest foot.

Lesson 11.3   AREAS OF CIRCLES AND PARTS OF CIRCLES   pages 620–625

- The ratio of the area of a sector of a circle to its area equals the ratio of the degree measure of the central angle to 360°. The area of a segment of a circle can be computed by finding the area of the associated sector and subtracting the area of the associated triangle.

Find the area of the shaded region. Round to the nearest tenth, if necessary.

8.

120° O 3 ft

9.

10° 9 m

10.

7 cm

11.

165° 3.2 m

Lesson 11.4   LIMITS   pages 626–629

- The limit of a sequence is a number that the terms of the sequence keep getting closer to.

12. The nth term of a sequence is $\dfrac{4n}{n+2}$. Find the limit of this sequence as n increases.

- For a given perimeter, a circle is the plane figure that maximizes the area. For a given perimeter and a given polygon, the regular polygon maximizes the area.
- For a given area, a circle is the plane figure that minimizes the perimeter. For a given area and a given polygon, the regular polygon minimizes the perimeter.

13. A rectangular garden is 6400 ft². Find the perimeter of the rectangle that requires the least amount of fencing.

14. How much more area would a circle with circumference 36π enclose than a square with perimeter 36π? Round to the nearest integer.

- Region A represents event A and region B represents event B, calculate area to detemine probability.

A randomly thrown dart is confined to land within region S. What is the probability that it lands in the shaded region A? Round to the nearest hundredth.

15.

16.

17.

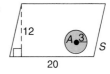

- The equation of a circle with center (h, k) and radius r is $(x - h)^2 + (y - k)^2 = r^2$.

Find the center and radius of each circle.   **Write the equation of each circle.**

18. $(x - 3)^2 + (y - 6)^2 = 36$ **20.** center $(1, 6)$; radius 14

19. $(x + 5)^2 + (y - 1)^2 = 81$ **21.** center $(-9, 2)$; diameter 10

- The altitude, slant height, and base radius of a right circular cone form a right triangle.
- A sphere is the locus of points in three dimensions that are equidistant from a fixed point called its center.
- The net for a right circular cylinder is two circles and a rectangle.

22. Find the measure of the altitude of a cone with radius 10 m and slant height 26 m.

23. Find the length of the rectangular net used to form the side of a cylinder with radius 12 ft and altitude 5 ft. Round to the nearest tenth.

24. Find the circumference of a great circle of a sphere with diameter 15 cm. Answer to nearest tenth.

CHAPTER TEST

Find the radius, diameter, area, and circumference. Round to the nearest tenth, if necessary.

1.

2.

3. WRITING MATHEMATICS Use a diagram and explain why, in any circle with radius r and diameter d, $d^2 > 2r^2$.

4. A circle has circumference 16π meters. Find the area of the circle in terms of π.

Find the length of $\overarc{AB}$. Round to the nearest tenth, if necessary.

5.

6.

Find the area of the shaded sector. Round to the nearest tenth, if necessary.

7.

8.

9. WRITING MATHEMATICS Use the variables in the figure at the right to explain the steps necessary to find the area of the shaded segments.

10. Given $\overline{AD}$ and $\overline{AB}$ are tangent to circle C at D and B.
Prove $\overline{AC}$ bisects $\angle DAB$

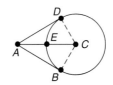

11. Find the number of degrees in the central angle of the yellow sector.

A. 4.5°   **B.** 9°
C. 18°   **D.** 22.5°

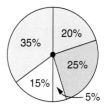

12. How much less area is enclosed by the largest-area rectangle with perimeter 60 ft than by a circle with circumference 60 ft? Round to the nearest square foot.

13. FITNESS A health club is planning to lay out a work-out area with 1 ft by 1 ft square padded tiles. What is the minimum perimeter of the work out area if it will be covered with 400 tiles that cannot be cut?

14. A computer will randomly generate a point within a 5-in. diameter circle. Inside the circle is a square with side 2 in. What is the probability that the point will land in the square? Round to the nearest hundredth.

Write an equation of the circle.

15. Center (0, 0); radius 7

16. Center (4, 6); radius 9

17. Center (1, −3); diameter 12

18. Find the slant height of a right circular cone whose base has a diameter 16 m and whose altitude is 9 m. Round to the nearest hundredth.

PERFORMANCE ASSESSMENT

ENERGY CONSERVATION A local hamburger stand cooks 3-in. diameter hamburgers on a 12-in. by 24-in. grill. The burgers sell for $0.79 each. How many hamburgers could be cooked in the grill space wasted by the fact that the burgers are circular and not square? What percent larger is a square burger with sides 3 in. as compared to the round burger with diameter 3 in.? If the store decides to increase the burger size, and increase the price proportionately, how much will the burger cost?

CIRCULAR REASONING Find the center and radius of the circle with equation

$$x^2 + 6x + y^2 + 8y + 34 = 25$$

Then, describe the graph of $x^2 + 6x + y^2 + 8y + 34 = -25$ and explain your reasoning.

DISCOVERING RADIANS Throughout this chapter, you used degrees to measure angles. A **radian** is also used to measure angles. Radians will be used extensively when you study higher mathematics. A central angle of 1 radian intercepts an arc whose length is equal to the length of the radius.

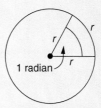

Find the number of degrees in one radian (*Hint:* Use the circumference formula.)

TRYING TRIANGLES Triangle *RST* has vertices *R*(1, 2) and *S*(9, 2). Vertex *T* is on a semicircle with diameter *RS*. Find the coordinates of *T* that maximize the area of triangle *RST*.

PROJECT ASSESSMENT

PROJECT *Connection*   Hold a competition called *The Gadget Games* to judge the merits of the inventions. Here are some suggestions for how to proceed.

1. Decide on the qualities for which each invention is to be judged. For example, the distance the self-powered car can travel is important. Is the appearance of the car to be important? Plan events. Should the model cars compete in a race?

2. Decide on judges for each event. Judges must measure and record data for each team's performance. Decide on prizes.

3. Stage the games and determine a winner for each event.

4. Have the class vote on a logo and name for each winning model.

5. As a class, share design strategies. Each team should have a written explanation of the role the circle played in the model. Discuss an optimal design for each model by taking the best features of each group's models.

CUMULATIVE REVIEW

Fill in the blank.

1. A line segment whose endpoints are on a circle is called a(n) __?__.

2. The perimeter of a circle is called the __?__ of the circle.

3. A region of a circle bounded by two radii and the intercepted arc is called a __?__ of a circle.

4. In a regular polygon, a segment whose endpoints are the center and the midpoint of a side is called a(n) __?__.

Find the circumference of each circle. Give each answer in exact form and as a decimal rounded to the nearest tenth.

5.
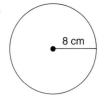
8 cm

6.
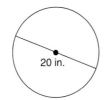
20 in.

Find the area of each circle. Give each answer in exact form and as a decimal rounded to the nearest tenth.

7.
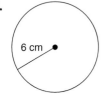
6 cm

8.
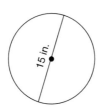
15 in.

9. **STANDARDIZED TESTS** Which of the following statements is not always true about a kite?
   A. Two pairs of adjacent sides are congruent.
   B. The diagonals are perpendicular.
   C. It is a quadrilateral.
   D. It is a rhombus.
   E. The sum of the measures of the angles is 360°.

Write the equation of each circle described.

10. center at $(4, -1)$, radius of 8

11. center at $(-3, 0)$, diameter of 14

12. the circle that has a diameter with endpoints at $(-2, 5)$ and $(6, -1)$

Find x in each triangle below. Round angles to the nearest degree and lengths to the nearest tenth if necessary.

13.

8
6
$x°$

14.
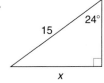
24°
15
x

15.
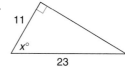
11
$x°$
23

16.

45°
25
x

17. **WRITING MATHEMATICS** Describe two possible methods for determining x in Question 16.

Consider the statement:
   Congruent figures are similar.

18. Write the statement in if-then form. Tell whether the statement is true or false.

19. Write the converse of the statement. Tell whether the converse is true or false.

20. Write the inverse of the statement. Tell whether the inverse is true or false.

21. Write the contrapositive of the statement. Tell whether the contrapositive is true or false.

22. Find the perimeter and area of the rectangle shown.

15 cm
2 cm

STANDARD FIVE-CHOICE Select the best choice for each question.

1. Which of the following is the equation of a circle that has its center at $(7, -5)$ and diameter of 10?

   A. $(x + 7)^2 + (y - 5)^2 = 10$
   B. $(x + 7)^2 + (y - 5)^2 = 100$
   C. $(x - 7)^2 + (y + 5)^2 = 5$
   D. $(x - 7)^2 + (y + 5)^2 = 10$
   E. $(x - 7)^2 + (y + 5)^2 = 25$

2. Which of the following is not true about the right triangle shown?

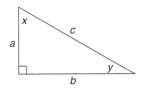

   A. $\sin y = \dfrac{a}{c}$
   B. $\cos x = \dfrac{a}{c}$
   C. $\tan x = \dfrac{a}{b}$
   D. $\sin x = \dfrac{b}{c}$
   E. All are true.

3. Which of the following pairs of vectors is perpendicular?

   A. vector $p = (12, -4)$ and vector $q = (-2, 6)$
   B. vector $p = (2, -2)$ and vector $q = (-2, 2)$
   C. vector $p = (3, 6)$ and vector $q = (-8, -4)$
   D. vector $p = (6, -3)$ and vector $q = (-2, -4)$
   E. none of these

4. If the image of a point rotated about the origin $180°$ is $(-2, 1)$, which of the following are the coordinates of the original point?

   A. $(-2, -1)$
   B. $(2, 1)$
   C. $(2, -1)$
   D. $(-2, -1)$
   E. $(-2, 1)$

5. In the diagram, two parallel lines are cut by a transversal. Which pair of angles is congruent because they are alternate interior angles?

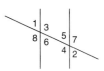

   A. 5 and 6   B. 7 and 8
   C. 3 and 6   D. 2 and 5
   E. 4 and 8

6. Which of the following regular polygons does not have an area of 60 to the nearest square unit?

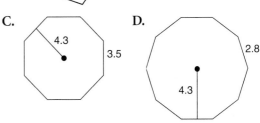

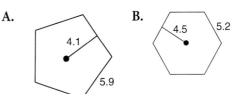

   E. All have an area of 60 square units.

7. Which of the following could be a net for the figure shown?

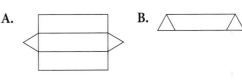

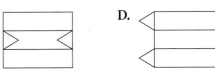

   E. Two of these.

12 Circles: Lines, Segments, and Angles

Take a Look AHEAD

Make notes about things that look familiar.

- What do you recall about circles and central angles?
- Look through the chapter to see where addition and subtraction are used to find measures of angles.

Make notes about things that look new.

- Look at the vocabulary used in the theorems of this chapter. Is the term inscribed angle new to you?
- Describe what you think adjacent areas are.

DATA Activity

SKILL FOCUS

▶ Compare numbers.
▶ Determine ratios.
▶ Solve ratio problems.
▶ Convert measurements.

Basic "Train" ing

Toy trains have been capturing the imagination of children over the centuries. The fascination of the rails doesn't necessarily end with childhood. The popularity of model railroading clubs throughout the country is evidence that model railroads are not just for kids. They engage enthusiasts of all ages.

TOYS AND GAMES

In this chapter, you will see how:

- **PLAYGROUND DESIGNERS** use tangent circles to design safe circular play areas. (Lesson 12.2, page 674)
- **BICYCLE DESIGNERS** use circles and tangents to design chainrings and sprockets. (Lesson 12.4, page 687)
- **TOY ENGINEERS** use chords and similarity to design toys and games. (Lesson 12.7, page 706)

Model Trains					
Scale Name	Scale Ratio model:actual	Track Gauge	Scale Name	Scale Ratio model:actual	Track Gauge
Z	1 : 220	6.5 mm	S	1 : 64	0.875 in.
N	1 : 160	9 mm	O	1 : 48	1.25 in.
HO	1 : 87		Gn3	1 : 22.5	1.75 in.

Model trains are built to scale. Each feature of a model train is in proportion to the train it is modeling. The most popular is the HO scale.

1. Which model trains are the largest? How can you tell?

On an actual train track the distance between the railheads is called the gauge. In most of North America, the gauge is 4 ft 8.5 in.

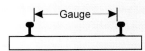

2. If 1 in. = 25.4 mm, how many millimeters to the nearest tenth are equivalent to 4 ft 8.5 in? Round.

3. Use your answer from Question 2 to find the track gauge for HO scale tracks. Round to the nearest tenth.

4. Find the length in mm of a 55-ft freight car in S scale.

5. Find the length in mm of a 55-ft box car in Z scale.

6. How tall would a scale model of yourself be in Gn3 scale?

7. **WORKING TOGETHER** A *layout* is a scale-modeled "train town" that includes such things as houses, roads, cars, and factories, all in correct scale. Find the actual heights of five famous buildings in the United States. Find their height in HO scale. Which buildings could fit on a layout that sits on the floor of a room with an 8-ft ceiling? Which scale best accommodates tall buildings?

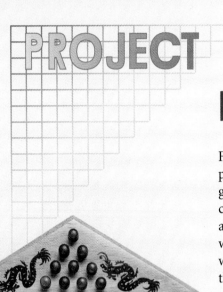

Fun for All Ages

For all of recorded history, children and adults have played games, in particular board games. Many games have stood the test of time. Chinese checkers, for example, originated in 1880, and is still played in households around the world. Game boards are geometric designs where the circle is often the focal point of the design. In some games, the winning strategy means being the first to reach the center of the circle.

PROJECT GOAL

To research the use of circles in games played by different cultures.

Getting Started

Work in groups of three to five students.

1. Have a brainstorming session to make a list of games, such as board games, video games, card games, etc. Do not include outdoor games; they will be considered separately.

2. Organize your list into categories. Decide which two categories your group will investigate. Remember to consider the use of circles when making your choice.

3. Work just with your group's categories. Extend your lists to include games that may be played by a specific culture or country.

4. Use family and friends, as well as the library and the internet to find games to include in your lists.

PROJECT Connections

Lesson 12.3, page 680:
Research and select from one of your categories a game that uses circles as part of the design.

Lesson 12.4, page 686:
Research and select from your other category a game that uses circles as part of the winning strategy.

Lesson 12.7, page 705:
Research and select one outdoor game that uses circles as part of the playing field or as part of the winning strategy.

Chapter Assessment, page 717:
Present your selected games and explain how circles are an integral part of the game.

Internet Connection

www.learninggeometry.com

12.1 Tangents to Circles

Explore

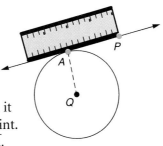

- You need a compass, ruler, and protractor.

  1. Draw a circle with center Q and a point P outside of $\odot$ Q. With your pencil at P, place the ruler against the pencil point and position it so that it touches $\odot$ Q in just one point. Label that point A. Draw $\overrightarrow{PA}$ and $\overline{QA}$.

  2. Find PA and m$\angle QAP$.

  3. Draw a second line containing P and touching $\odot$ Q in just one point. Mark that point B.

  4. Find PB and m$\angle QBP$.

  5. Make a conjecture about $\angle QAP$ and $\angle QBP$.

  6. Make a conjecture about PA and PB.

Build Understanding

- A line can intersect a circle in two points, one point, or not at all. The lines you drew in Explore are lines that intersect a circle in just one point each.

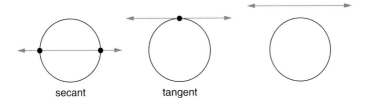

  secant   tangent

  A **secant** to a circle is a line that intersects a circle in two points. A **tangent** to a circle is a line in the plane of the circle that intersects the circle in just one point. This point is called the **point of tangency**. The following theorem states a relationship of tangents and radii.

  > **THEOREM 12.1**
  >
  > **If a line is tangent to a circle, then it is perpendicular to the radius drawn to the point of tangency.**

SPOTLIGHT ON LEARNING

WHAT? In this lesson you will learn
- to use properties of tangents.

WHY? Tangents to circles can help you solve problems about safety, design, and distance.

TECHNOLOGY TIP

You can use geometry software to make the figure for the Explore activity. Use the circle tool to draw circle O, and use the point tool to draw point P. Then look for the *Point on Object* command to draw points A and B on circle O.

Using the Pythagorean Theorem and Theorem 12.1, you can find the length of a tangent segment or the radius of a circle.

EXAMPLE 1

In the figure at the right, $\overrightarrow{RP}$ is tangent to $\odot Q$ at R. Find the radius of $\odot Q$.

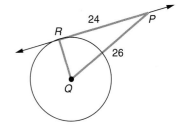

Solution

Since $\overrightarrow{RP}$ is tangent to $\odot Q$ at R, $\angle QRP$ is a right angle.

$$26^2 = (QR)^2 + 24^2 \qquad (QP)^2 = (QR)^2 + (RP)^2$$
$$676 = (QR)^2 + 576 \qquad \text{Substitute}$$
$$100 = (QR)^2 \qquad \text{Take square root of both sides.}$$
$$10 = QR$$

The radius of $\odot Q$ is 10. ◄

CHECK UNDERSTANDING

In Example 2 suppose that you are given $QR = 10$ and $QP = 26$. Rewrite the solution so as to find RP.

You can use the converse of the Pythagorean Theorem and contrapositive of Theorem 12.1 to decide whether a given line is tangent to a given circle.

EXAMPLE 2

Determine whether $\overline{SP}$ is tangent to $\odot Q$ at S.

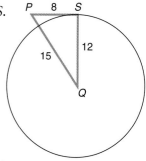

Solution

If $\triangle QPS$ is a not a right triangle, then $\overline{SP}$ is not tangent to $\odot Q$ at S. Apply the converse of the Pythagorean Theorem to test whether $\angle QSP$ is a right angle.

$$(QP)^2 \stackrel{?}{=} (QS)^2 + (SP)^2$$
$$15^2 \stackrel{?}{=} 12^2 + 8^2$$
$$225 \stackrel{?}{=} 208$$

Since $(QP)^2 \neq (QS)^2 + (SP)^2$, $\triangle QPS$ is not a right triangle. So, $\angle QSP$ is not a right angle, and $\overline{SP}$ is not tangent to $\odot Q$ at S. ◄

PROBLEM SOLVING TIP

Recall that if a conditional is true, then its contrapositive is true also. If a line is not perpendicular to a radius at its endpoint, the line is not tangent to the circle at that point.

Corollary 12.2 follows from Theorem 12.1.

> **COROLLARY 12.2** If two segments from the same exterior point of a circle are tangent to the circle, then they are congruent.

Congruent segments can be used to solve problems involving real world models of tangents to circles.

Example 3

LINES OF SIGHT The figure shows distances measured from the base of a lighthouse to a circular pond. What is the radius of the pond?

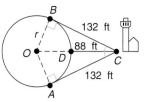

Solution

$$(OB)^2 + (BC)^2 = (OC)^2$$ Pythagorean Theorem
$$r^2 + 132^2 = (r + 88)^2$$ $OC = r + 88$
$$r^2 + 17{,}424 = r^2 + 176r + 7744$$ Simplify.
$$9680 = 176r$$ Solve for r.
$$55 = r$$

The radius of the pond is 55 ft. ◄

The converse of Theorem 12.1 is also true and its proof follows.

THEOREM 12.3

In a plane, if a line is perpendicular to a radius of a circle at its endpoint on the circle, then the line is tangent to the circle at that point.

Given $\overline{PT}$ is a radius of ⊙ P, $\overleftrightarrow{AB} \perp \overline{PT}$

Prove $\overleftrightarrow{AB}$ is tangent to ⊙ P at T

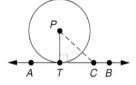

Proof Let point C be any point on $\overleftrightarrow{AB}$ other than point T. Draw $\overline{PC}$. By definition of perpendicular, $\angle PTC$ is a right angle. So, PC is a hypotenuse of a right triangle. By the Unequal Angles Theorem $PC > PT$. Therefore, point C is in the exterior of the circle and point T is the only point of intersection for ⊙ P and $\overleftrightarrow{AB}$. ◄

A tangent to two coplanar circles is a **common tangent** of the circles. A *common internal tangent* intersects the segment whose endpoints are the centers of the circles. A *common external tangent* does not.

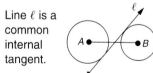

Line ℓ is a common internal tangent.

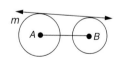

Line m is a common external tangent.

Tangent circles are circles in the same plane that are tangent to the same line at the same point. The circles are *externally tangent* if all points of one circle, except for the point of tangency, lie in the exterior of the other. Otherwise, they are *internally tangent*.

internally tangent circles

externally tangent circles

Refer to the figure at the right. Tell whether $\overleftrightarrow{AB}$ is tangent to $\odot$ P at T.

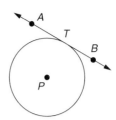

1. $PT = 5$, $AT = 12$, and $PA = 13$

2. $PT = 5.0$, $AT = 11.0$, and $PA = 13.0$

3. $PT = 3.2$, $BT = 4.1$, and $PB = 5$

4. $PT = 2.5$, $BT = 2.5$, and $PB = 2.5\sqrt{2}$

5. Refer to the figure at the right above. Suppose that $\overleftrightarrow{AB}$ is tangent to $\odot$ P at T. Suppose also that $PT = 15$ and $PA = 18$. Find AT.

6. **ACCIDENT INVESTIGATION** A police officer is investigating an accident where a driver loses control as he rounds a circular bend in the road and skids off the roadway along a tangent to the road. How far does the driver skid from the point where he leaves the roadway?

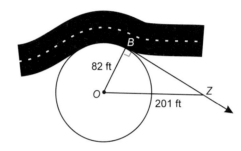

7. **WRITING MATHEMATICS** In your own words, summarize the relationship between a tangent to a circle and the radius of the circle to the point of tangency.

PRACTICE

Refer to the figure at the right. Tell whether $\overleftrightarrow{LM}$ is tangent to $\odot$ C at N.

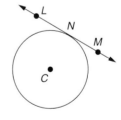

8. $CN = 12$, $LN = 12$, and $CL = 12\sqrt{2}$

9. $CN = 3.8$, $LN = 4.2$, and $CL = 4.5$

10. $CN = 1$, $NM = 1$, and $CM = 2$

11. $CN = 3$, $NM = 3\sqrt{3}$, and $CM = 6$

12. Draw $\odot$ C. Draw $\overline{NA}$ tangent to $\odot$ C at point A. Draw $\overline{NB}$ tangent to $\odot$ C at point B. Let $CA = 12$, and $CN = 20$. Find NA and NB.

13. **ROAD CONDITIONS** When a motorcycle drives through water and mud, the water and mud is thrown from a tire along a line tangent to the tire. Suppose $RQ = 13$ in. and $PR = 20$ in. How far does water and mud move between Q and P?

14. **Writing Mathematics** Describe how to construct a tangent to a circle given a point on it.

15. In the figure at the right, $\overline{AD}$ and $\overline{BC}$ are common tangents to $\odot X$ and Y. Also, $PD = 18$, $AP = 5x$, and $PC = 4x + 10$. Find PB.

16. In $\odot E$, $\overline{JH}$ is tangent to $\odot E$ at J. Given that $JH = 20$ and $EH = 25$, approximate the circumference and the area of $\odot E$ to the nearest whole numbers.

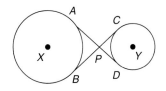

Coordinate Geometry A circle whose center is C has equation $(x - 1)^2 + (y - 8)^2 = 25$.

17. Find the radius of the circle and write the coordinates of its center.

18. Show that $P(4, 12)$ is on the circle.

19. Find the slope of $\overline{CP}$ and the slope of the tangent to $\odot C$ at P.

20. Write an equation for the tangent to the circle at P.

EXTEND

21. **Machinery** Two wheels are linked by a belt as shown. In the diagram, $\odot X$ and $\odot Y$ have $\overline{AC}$ and $\overline{BD}$ as common tangents. Let $AX = 6$ in., $CY = 4$ in., and X and Y are 12 in. apart. Find AC and BD.

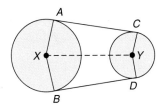

22. **Coordinate Geometry** Tell whether the circle whose equation is $x^2 + y^2 = 64$ and the circle whose equation is $(x - 5)^2 + y^2 = 9$ are internally tangent, externally tangent, or neither. Justify your response.

23. At Y, $\odot A$ and $\odot B$ are internally tangent with $\overleftrightarrow{XY}$ as the common tangent. Let $AX = 4.67$, $BX = 4.17$, and $XY = 3.60$. How far apart are the centers of $\odot A$ and $\odot B$?

24. Rework Question 23 given that $\odot A$ and $\odot B$ are externally tangent with $\overleftrightarrow{XY}$ as common tangent.

A tangent line to a sphere is a line that touches the sphere in just one point. The tangent plane T to a sphere at a point is the set of all tangent lines at that point.

25. In the diagram at the right, $\overline{CP}$ is a radius and $\mathcal{T}$ is the tangent plane to the sphere at P. Suppose $PX = 2(CP)$. Write a formula for CX in terms of CP.

26. Rework Question 25 given that $PX = k(CP)$, where $k > 0$.

In each figure below, the circles all have radius 1 and are tangent to one another as shown. Find the area of each shaded region.

27.

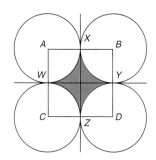

28.

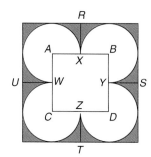

THINK CRITICALLY

In the figure at the right, $\odot A$, $\odot B$, and $\odot C$ are tangent to one another in pairs and $\overrightarrow{PZ}$ and $\overrightarrow{PT}$ are common tangents to the three circles at X, Y, and Z and at R, S, and T, respectively.

29. Show that $\overline{XA} \parallel \overline{YB} \parallel \overline{ZC}$ and that $\overline{RA} \parallel \overline{SB} \parallel \overline{TC}$.

30. Show that $\triangle XAP$, $\triangle YBP$, and $\triangle ZCP$ are similar and that $\triangle RAP$, $\triangle SBP$, and $\triangle TCP$ are also similar.

31. Write a statement that tells how PX, PY, and PZ and PR, PS, and PT are related to one another.

32. Write a theorem that relates the line through the centers of two congruent circles and any common tangents the circles may have. Illustrate your statement and prove it.

33. Write an indirect proof of Theorem 12.1.

   If a line is tangent to a circle, it is perpendicular to the radius drawn to the point of tangency.

34. Prove Corollary 12.2: If two segments from the same exterior point of a circle are tangent to it, then they are congruent.

ALGEBRA AND GEOMETRY REVIEW

Find the image of each point under the reflection across the specified line.

35. $P(-2, 5)$; x-axis
36. $P(-2, 5)$; y-axis
37. $P(2, -5)$; $y = x$
38. $P(-2, 5)$; $y = x$

Use the substitution method to solve each system of equations.

39. $\begin{cases} y = 3x \\ 2x - y = 10 \end{cases}$
40. $\begin{cases} x = 0.5y \\ x + 4y = 6 \end{cases}$
41. $\begin{cases} x = y \\ 3x - 3y = 0 \end{cases}$

42. $\begin{cases} 2x - y = 1 \\ x = 4y \end{cases}$
43. $\begin{cases} -5x + y = -2 \\ y = 5x \end{cases}$
44. $\begin{cases} x = 7y \\ y = 4x \end{cases}$

45. In $\odot Q$, $\overline{SP}$ is a tangent at point P. Given that $SP = 12.0$ and $QS = 20.0$, approximate QP to the nearest 1 tenth of a unit.

12.2 Central Angles and Arcs

Explore

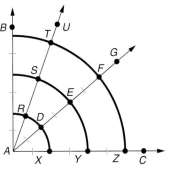

● Draw a figure like the one at the right. Label it as shown. In the figure, $\angle BAC$ is a right angle and the three arcs are concentric and equally spaced apart. The angles with vertex A may have any measure you choose.

1. How are the lengths of $\overset{\frown}{DX}$, $\overset{\frown}{EY}$, and $\overset{\frown}{FZ}$ related to the circumference of their respective circles?

2. Find m$\angle GAC$. Does m$\angle GAC$ depend on the lengths of $\overset{\frown}{DX}$, $\overset{\frown}{EY}$, and $\overset{\frown}{FZ}$?

3. Repeat Questions 1 and 2 for $\angle UAC$ and the lengths of $\overset{\frown}{RX}$, $\overset{\frown}{SY}$, and $\overset{\frown}{TZ}$.

SPOTLIGHT ON LEARNING

WHAT? In this lesson you will learn
● to find the degree measure of central angles and arcs.

WHY? Central angles and arcs can help you solve problems about statistics, anthropology, and recreation.

Build Understanding

● In Explore, you probably discovered that although $\overset{\frown}{DX}$, $\overset{\frown}{EY}$, and $\overset{\frown}{FZ}$ increase in length, they are the same fraction of a complete circle as m$\angle GAC$ is a fraction of 360°. Because of this, it is reasonable to assign to $\overset{\frown}{DX}$, $\overset{\frown}{EY}$, and $\overset{\frown}{FZ}$ the degree measure of $\angle GAC$. Just as m$\angle ABC$ denotes the degree measure of $\angle ABC$, m $\overset{\frown}{DX}$ denotes the degree measure of $\overset{\frown}{DX}$. In the diagram above, m $\angle GAC = 40°$. So,

$$\text{m } \overset{\frown}{DX} = \text{m } \overset{\frown}{EY} = \text{m } \overset{\frown}{FZ} = 40°$$

In general, the **degree measure of an arc** is defined as follows: The degree measure of a minor arc is the degree measure of the central angle that intercepts it. The degree measure of a major arc is 360° minus the degree measure of the minor arc that makes up the rest of the circle. The degree measure of a semicircle is 180°.

Two arcs are **adjacent** if they are part of the same circle and intersect at exactly one point or intersect at two points and together make up the entire circle.

> ── ARC ADDITION POSTULATE ──
>
> **POSTULATE 24**   The measure of an arc formed by two adjacent arcs is the sum of the measures of the two arcs.

EXAMPLE 1

In ⊙ O, $\overline{DB}$ is a diameter. Find m $\overset{\frown}{AC}$ and m $\overset{\frown}{EA}$.

Solution

To find m $\overset{\frown}{AC}$, apply the Arc Addition Postulate.

$$m \overset{\frown}{AC} = m \overset{\frown}{AB} + m \overset{\frown}{BC}$$
$$= 49° + 90° = 139°$$

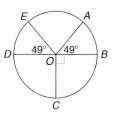

To find m $\overset{\frown}{EA}$, write and solve an equation.

$$m \overset{\frown}{EA} + m \overset{\frown}{ECA} = 360°$$
$$m \overset{\frown}{EA} + 2(49°) + 2(90°) = 360°$$
$$m \overset{\frown}{EA} = 82°$$

$\begin{cases} m\angle DOC = m\angle BOC = 90° \\ m\angle EOD = m\angle AOB = 49° \end{cases}$

So, m $\overset{\frown}{AC}$ = 139° and m $\overset{\frown}{EA}$ = 82°. ◄

Recall a circle graph is one method of displaying data. Circle graphs are often called pie charts. Each "slice of the pie" is bound by a central angle and its arc.

EXAMPLE 2

COMMUTING This table indicates how Americans traveled to work in 1990. Show these data in a circle graph.

Method	Percent
Drive alone	73.2%
Carpool	13.4%
Public Trans.	5.3%
Walking	3.9%
Other	4.2%

Solution

To create each slice of the pie, change each percent to a degree measure. To find each degree measure, change each percent to a decimal and then multiply by 360.

Method	Degrees
Drive alone	0.732(360) ≈ 264
Carpool	0.134(360) ≈ 48
Public Trans.	0.053(360) ≈ 19
Walking	0.039(360) ≈ 14
Other	0.042(360) ≈ 15

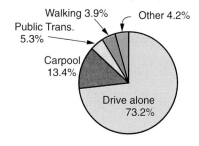

Methods for getting to work (1990)

Draw a circle and mark its center. Draw a central angle whose measure is 264°. Label the slice formed "Drive alone." Proceed in this way for each other category. ◄

Arcs of the same circle or of congruent circles that have the same measure are called **congruent arcs**. To denote the congruence, you can use ≅ . Use central angles to determine if two arcs are congruent.

> **THEOREM 12.4**
>
> In the same circle, or in congruent circles, two minor arcs are congruent if and only if the central angles that intercept them are congruent.

TRY THESE

Find the measure of each arc in ⊙ *C*. In the figure,
$\overline{RU}$ is a diameter of ⊙ *C* and $\overrightarrow{CZ}$ bisects ∠ *RCV*.

1. $\widehat{VR}$ 2. $\widehat{VZ}$

3. $\widehat{RT}$ 4. $\widehat{ZS}$

5. $\widehat{UT}$ 6. $\widehat{ZSU}$

7. **AUTOMOBILE MANUFACTURING** The table at the right indicates the distribution of automobile manufacturing in 1950 by country. Draw a circle graph to represent the data.

8. **WRITING MATHEMATICS** Briefly describe how the Arc Addition Postulate is similar to but different from the Segment Addition Postulate and the Angle Addition Postulate.

Automobile Manufacture (1950)	
Origin	**Percent**
United States	76%
Canada	4%
Europe	19%
Other	1%

PRACTICE

Find the measure of each arc in ⊙ *C*.
In the figure, $\overrightarrow{CZ}$ bisects ∠ *RCM*.

9. $\widehat{PN}$ 10. $\widehat{QN}$

11. $\widehat{RP}$ 12. $\widehat{RM}$

13. $\widehat{RZ}$ 14. $\widehat{RPN}$

15. **WRITING MATHEMATICS** Explain why Theorem 12.4 would not be true if "In the same circle," or in congruent circles" is deleted from the statement of the theorem.

16. **AUTOMOBILE MANUFACTURING** The table at the right indicates the distribution of automobile manufacturing in 1992 by country. Draw a circle graph to represent the data.

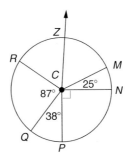

Automobile Manufacture (1992)	
Origin	**Percent**
United States	21%
Canada	4%
Europe	36%
Japan	26%
Other	13%

17. SCHOOL SPIRIT In designing a circular school logo, Debbie decided to subdivide each quarter of the circular design into three arcs as shown. In her plan, she wants to be sure that

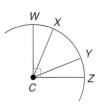

$$m \, \widehat{WX} = m \, \widehat{YZ} = \frac{1}{2} \, (m \, \widehat{XY}).$$

What should she make the measure of ∠*WCX*?

18. ANTHROPOLOGY One place where the iguana can be found is in South Africa. The map shows a location in South Africa whose latitude is 30° S and 25° E. One degree of arc along a great circle such as the prime meridian, corresponds to about 60 nautical miles along that circle. At 30° S one degree of arc along a parallel of latitude corresponds to about 52 nautical miles. How many miles south of the equator and east of the prime meridian is the location shown?

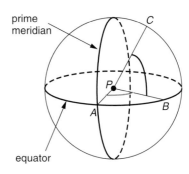

In Questions 19–21, circle ⊙ *P* has equation $x^2 + y^2 = 169$.

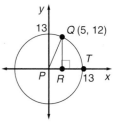

19. Show that *Q*(5, 12) is on ⊙ *P*.

20. In △*PQR* use the tangent and inverse tangent to find m∠*QPR*. Then find m $\widehat{QT}$.

21. Use Question 20 to find the length of $\widehat{QT}$.

22. Congruent circles with centers at the vertices of equilateral triangle △*XYZ* are tangent to one another. Find the measure of each arc in red.

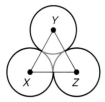

23. Congruent circles with centers at the vertices of square *ABCD* are tangent to one another in pairs. Show that the arcs in red are congruent.

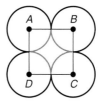

24. Based on Questions 22 and 23, make a generalization about the measures of the arcs determined by *n* congruent circles with centers at the *n* vertices of a regular *n*-gon and tangent to one another in pairs. Justify your conjecture.

25. CONSTRUCTIONS Salli wants to copy a given minor arc $\overset{\frown}{AB}$ in $\odot P$. Devise a construction using only straightedge and compass that will enable her to do that.

26. An equation for $\odot P$ is $x^2 + y^2 = 100$. Points $P(-6, 8)$ and $Q(6, 8)$ are on $\odot P$. Is $\overset{\frown}{PQ}$ a minor arc, a semicircle, or a major arc? Justify your response.

TRANSFORMATIONS The figure at the right shows a rotation of the plane through $x°$ about point P.

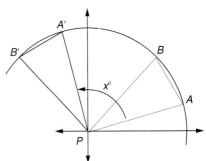

27. The figure shows the rotation of $\overline{AB}$ through $x°$ about point P. Use the diagram to show that a rotation through $x°$ about point P preserves arc length and arc measure.

28. Use the figure to show that if two arcs in the same circle are congruent, the chords they determine are congruent.

THINK CRITICALLY

Exercises 29 and 30 together form the two parts of Theorem 12.4. Prove each statement.

29. In a circle, or in congruent circles, if two minor arcs are congruent, then the central angles that intercept them are congruent.

30. In a circle, or in congruent circles, if the central angles that intercept two minor arcs are congruent, then the arcs are congruent.

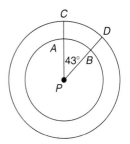

31. The Arc Length Corollary states that the ratio of the length L of a minor arc of a circle with radius r to the circumference of the circle equals the degree measure $d°$ of the central angle that intercepts the arc to 360°. Write this corollary as a proportion involving L, r, and d. Then solve for L. Use the equation to rewrite the theorem.

32. Two concentric circles with center P have radii m and n, where $m > n$. m $\angle APB$ = m$\angle CPD$ = 43°. The length of $\overset{\frown}{AB}$ is 12.4 in. and the length of $\overset{\frown}{CD}$ is 124.5 in. How are the radii of the circles related?

ALGEBRA AND GEOMETRY REVIEW

Factor each quadratic expression.

33. $x^2 - 14x - 15$ **34.** $x^2 + 16x + 64$ **35.** $x^2 - x - 30$

36. $x^2 - 121$ **37.** $2x^2 + 11x - 21$ **38.** $x^2 - 16x + 64$

39. In $\triangle KLM$, m$\angle KLM$ = 44° and m$\angle LMK$ = 65°, find m$\angle MKL$.

40. In $\triangle ABC$, m$\angle ABC$ = 32°, m$\angle BCA$ = 55°, and m$\angle CAB$ = 93°. List the lengths of the sides in order from least to greatest.

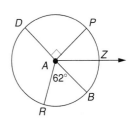

41. In $\odot A$, $\overline{DB}$ is a diameter and $\overrightarrow{AZ}$ bisects $\angle PAB$. Find m $\overset{\frown}{PZ}$, m $\overset{\frown}{ZR}$, and m $\overset{\frown}{DR}$.

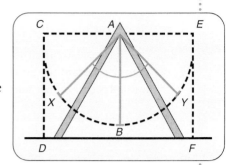

Designing and installing children's play equipment is not child's play. Much thought is given to determining the rectangular area that indicates the danger area for those not playing on the swings. There should be enough space surrounding each piece of equipment that children can move safely from one play area to another.

Decision Making

The diagrams below show a swing set and a seesaw.

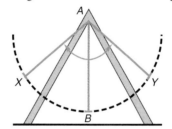

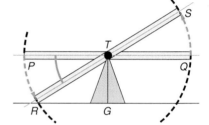

1. An installer sets up a swing set with $\overline{AB}$ measuring 7 ft. Given that a child swings 11.5 ft from Y to X on the swing set, approximate the degree measure of the arc of the child's swing to the nearest whole number of degrees.

2. A child at P moves through $\overset{\frown}{PR}$ while on a seesaw whose length is $PQ = RS = 10.0$ ft. Given that m $\overset{\frown}{PR} = 31°$, approximate the distance the pivot point T is above the ground to the nearest tenth of a foot.

The diagram at the left shows a design for four safe play areas. The circles centered at A and C have radius 12 ft and the circles centered at B and D have radius 15 ft.

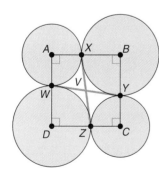

3. Show that $AB = BC = CD = DA$.

4. Identify what kind of quadrilateral $WXYZ$ is. Justify your answer.

12.3 Chords and Arcs

Explore

- Draw a large circle and label its center P. Draw two congruent nonintersecting chords $\overline{AB}$ and $\overline{CD}$.

  1. Find m$\angle APB$ and m$\angle CPD$. What appears to be true about $\angle APB$ and $\angle CPD$?

  2. Locate the midpoint of $\overline{AB}$. Label it as point X. Locate the midpoint of $\overline{CD}$. Label it as point Y. What appears to be true about $\overline{OX}$ and $\overline{OY}$?

  3. Draw another circle and label its center as point P. Draw a chord $\overline{LM}$ that is not a diameter. Locate and label the midpoint N of $\overline{LM}$. Draw the chord through point N perpendicular to $\overline{LM}$. What can you say about this chord?

Build Understanding

- In Explore, you began to see a relationship between congruent chords and the measures of the central angles they determine.

THEOREM 12.5

In the same circle, or in congruent circles, two minor arcs are congruent if and only if their corresponding chords are congruent.

The following is a proof of one part of Theorem 12.5: In the same circle, or in congruent circles, if two minor arcs are congruent, then their corresponding chords are congruent.

Given $\odot C \cong \odot Z$; $\overset{\frown}{AB} \cong \overset{\frown}{XY}$

Prove $\overline{AB} \cong \overline{XY}$

Plan for Proof Use corresponding parts of congruent triangles.

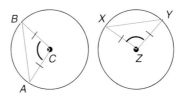

Proof All radii of a given circle and of congruent circles are congruent, so $\overline{CA} \cong \overline{CB} \cong \overline{ZX} \cong \overline{ZY}$. If two minor arcs of congruent circles are congruent, then the central angles that intercept them are congruent. So $\angle ACB \cong \angle XZY$. Then, by the SAS Congruence Postulate, $\triangle ABC \cong \triangle XYZ$. Thus, by CPCTC, $\overline{AB} \cong \overline{XY}$. Using similar reasoning, if $\overset{\frown}{AB}$ and $\overset{\frown}{XY}$ are minor arcs of the same circle, it can be shown that $\overline{AB} \cong \overline{XY}$. ◄

There are many useful theorems about chords. One of them is below.

The following is a proof of Theorem 12.6.

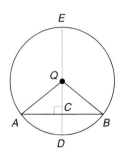

Given $\overline{ED}$ is a diameter of $\odot Q$.
$\overline{ED}$ is perpendicular to chord $\overline{AB}$ at point C.

Prove $\overline{AC} \cong \overline{BC}$; $\overset{\frown}{AD} \cong \overset{\frown}{BD}$

Plan for Proof Show that $\triangle AQC \cong \triangle BQC$. Use CPCTC and Theorem 12.4.

Proof Write a paragraph proof.

Since $\overline{ED} \perp \overline{AB}$, and perpendicular lines form four right angles, $\angle QCA$ and $\angle QCB$ are right angles. From the definition of right triangle, it follows that $\triangle QCA$ and $\triangle QCB$ are right triangles.

All radii of a given circle are congruent, so $\overline{QA} \cong \overline{QB}$. By the Reflexive Property, $\overline{QC} \cong \overline{QC}$. So, by the HL Congruence Theorem, $\triangle AQC \cong \triangle BQC$. Therefore, by CPCTC, $\overline{AC} \cong \overline{BC}$.

By CPCTC, it also is true that $\angle AQD \cong \angle BQD$. If the central angles that intercept two minor arcs are congruent, then the arcs are congruent. Therefore, $\overset{\frown}{AD} \cong \overset{\frown}{BD}$. ◄

EXAMPLE 1

MANUFACTURING The diameter of a circular piece of sheet metal is 26 cm. A machinist wants to make a straight 10-cm cut across it. What will be the distance between the center of the sheet of metal and the cut edge?

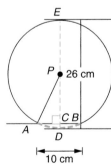

Solution

Draw a diagram. The given information implies that $\triangle APC$ is a right triangle. By Theorem 12.6, $AC = CB = 5$ cm. Since you are given that $ED = 26$ cm, you know $PA = PD = 13$ cm. So, you can apply the Pythagorean Theorem to find PC.

$$(PC)^2 + (AC)^2 = (PA)^2$$
$$(PC)^2 + 5^2 = 13^2$$
$$PC = 12$$

The distance between the center and the cut edge is 12 cm. ◄

Another useful theorem about chords is stated below.

THEOREM 12.7

In the same circle, or in congruent circles, two chords are congruent if and only if they are equidistant from the center.

CHECK UNDERSTANDING

Suppose in Example 2 that $\overline{LM}$ and $\overline{PQ}$ are 8 in. long. Verify that they are closer to R than if they were 6 in. long.

EXAMPLE 2

The diameter of $\odot R$ is 10 in. Chords $\overline{LM}$ and $\overline{PQ}$ are 6 in. long. How far are they from R?

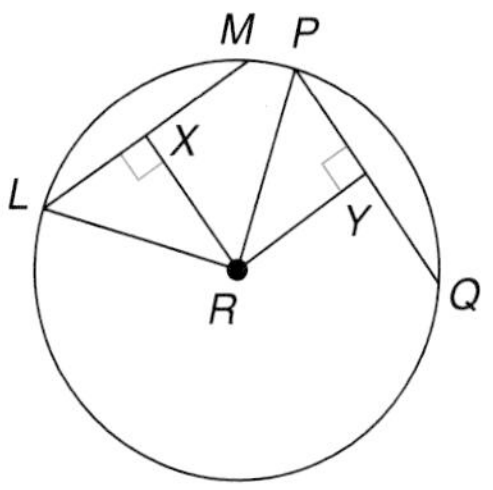

Solution

Since $\overline{LM} \cong \overline{PQ}$, then by Theorem 12.7, they are the same distance from R.

$(RX)^2 + (LX)^2 = (RL)^2$ Apply the Pythagorean Theorem.

$(RX)^2 + 3^2 = 5^2$ $LX = \frac{1}{2}LM = 3$

$RX = 4$

Thus, $\overline{LM}$ and $\overline{PQ}$ are each 4 in. from the center of $\odot R$.

A third useful theorem about chords relates one chord to another chord that bisects it.

THEOREM 12.8

If one chord of a circle is the perpendicular bisector of another, then the first chord is a diameter of the circle.

CHECK UNDERSTANDING

What justification is there for stating that $ED = EB + BD$?

EXAMPLE 3

LANDSCAPING Approximate the radius of the circular plot shown, $BD = 184$ ft.

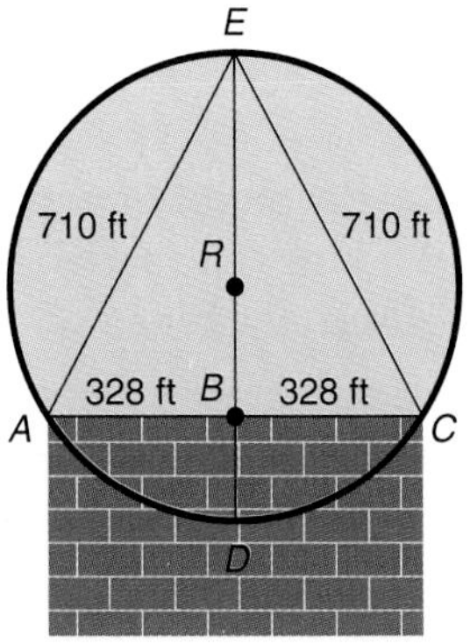

Solution

Since $\overline{ED}$ bisects $\overline{AC}$, $\overline{ED}$ is a diameter of $\odot R$, Thus, $ER = \frac{1}{2}ED = \frac{1}{2}(EB + BD)$. So, you can find ER if you can find EB.

$(EB)^2 + (AB)^2 = (AE)^2$

$(EB)^2 + 328^2 = 710^2$

$EB = \sqrt{710^2 - 328^2} \approx 630$

Therefore, $ER \approx \frac{1}{2}(630 + 184)$, or $ER \approx 407$.

The circular grass plot has a radius of about 407 ft.

In ⊙ *P*, $\overline{CD}$ is a diameter and $\overline{CD} \perp \overline{AB}$ at point *E*. Find each length. **Round answers to the nearest tenth.**

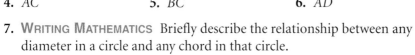

1. *EB*
2. *QE*
3. *EC*
4. *AC*
5. *BC*
6. *AD*

7. **WRITING MATHEMATICS** Briefly describe the relationship between any diameter in a circle and any chord in that circle.

8. **LANDSCAPING** In the circular pond modeled by ⊙ *P*, an architect wants to build a small wooden bridge spanning points *K* and *L*. Approximately how far from the center of the pond will the bridge be built? Give your answer to the nearest whole foot.

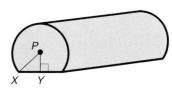

9. **AGRICULTURAL CONSTRUCTION** An agricultural manager is planning to build a storage facility that is shaped like a cylinder but with a flat bottom. The diagram at the right shows a picture of the proposed building. The base of the building is to be 18 ft along its front. The center of the front is to be 12 ft above ground level. Approximate the radius of the circular front and the height of the building to the nearest foot.

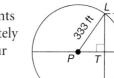

PRACTICE

In ⊙ *P*, $\overline{XY}$ is a diameter, $\overline{XY} \perp \overline{AB}$ at point *N*, and *AB* = 5.3. Find each length. **Round answers to the nearest tenth.**

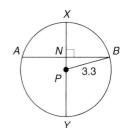

10. *AN*
11. *NP*
12. *NX*
13. *XB*
14. *XA*
15. *PY*

16. **WRITING MATHEMATICS** Summarize two different ways to tell if two minor arcs in a given circle are congruent.

17. **METAL WORKING** A metal worker cuts a steel drum whose base has a diameter of 30.0 in. The cut across the front is 28.0 in. long. How deep is the basin the worker has made? Give you answer to the nearest tenth of an inch.

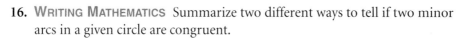

18. **CONCRETE WORK** Find the radius of a circular concrete patio that is adjacent to a house 32 ft long. The back of the house is a chord of the circle. The distance between the center of the circle and the back of the house is 15 ft.

19. Denzl is making a sign for an ecology project. The concentric circles have diameters 36.8 in. and 30.8 in. The width of the strip is the difference in the radii of the circles. The two edges of the diagonal are the same distance from the center of the circles. Approximate the length of each edge of the diagonal to the nearest tenth of an inch.

No Unauthorized Dumping

EXTEND

Use the figure below for Exercises 20–22. Round answers to the nearest tenth.

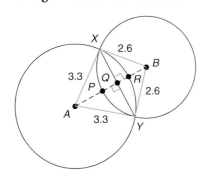

20. Two circles with center at points *A* and *B* intersect as shown. The length of $\overline{XY}$ joining the points of intersection is 3.9 units. How far apart are points *A* and *B*?

21. Approximate *AP*, the distance between points *A* and the circle with center *B*.

22. Approximate *BR*, the distance between *B* and the circle with center points *A*.

A plot of land in the shape of quadrilateral *ABCD* has vertices on a circle whose radius is 404 m. The diagram shows the distance between the center of the circle and each side of quadrilateral *ABCD*.

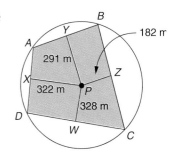

23. Use the fact that $\overline{AD}$ is a chord of $\odot$ *P* and $\overline{XP} \perp \overline{AD}$ to find *AX*. Give your answer to the nearest meter.

24. Find the approximate value of the perimeter of quadrilateral *ABCD*. Give your answer to the nearest meter.

In the figures, $\overline{PD} \perp \overline{AB}$. Use trigonometric ratios in Exercises 25–28.

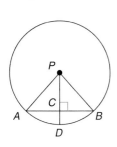

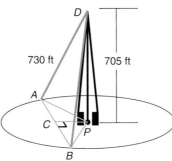

25. Given that *PA* = 10 and m∠*APB* = 50°, find *PC* and *AB*.

26. Given that *PA* = 10 and m∠*APB* = 130°, find *PC* and *AB*.

27. Support cables join the top of the tower to anchor locations on the ground. The measure of ∠*APB* is 120°. Find the distance between points *A* and *B*.

28. Use the result of Exercise 27 to find the shortest distance between the base of the tower and the segment joining points *A* and *B*.

An equation for a circle with center C(5, 4) and radius 8 is $(x - 5)^2 + (y - 4)^2 = 64$. **Approximate the length of each chord to the nearest tenth.**

29. The chord determined by the circle and the *x*-axis

30. The chord determined by the circle and the *y*-axis

31. The chord determined by the circle and the line $x = 3$

32. Approximate the distance between each side of a square inscribed in a circle of radius 10 and the center of the circle.

THINK CRITICALLY

33. A central angle in a circle of radius *r* has measure *x*. Generalize your work in Exercise 25 and 26 to write formulas for *PC* and *AB* in terms of *r* and *x*.

34. Prove this part of Theorem 12.5: In a circle, or in congruent circles, if the chords that correspond to two minor arcs are congruent, then the arcs are congruent.

35. Prove this part of Theorem 12.7: In a circle, or in congruent circles, if two chords are congruent, then they are equidistant from the center of the circle.

36. Prove this part of Theorem 12.7: In a circle, or in congruent circles, if two chords are equidistant from the center of the circle, then they are congruent.

37. Prove Theorem 12.8: If one chord of a circle is the perpendicular bisector of another, then the first chord is a diameter of the circle.

ALGEBRA AND GEOMETRY REVIEW

Find the measure of an interior angle in each regular polygon.

38. pentagon

39. nonagon

40. dodecagon

Solve each proportion for *x*.

41. $\dfrac{18}{15} = \dfrac{42}{x}$

42. $\dfrac{52}{x} = \dfrac{x}{13}$

43. $\dfrac{36}{x} = \dfrac{-6}{5}$

44. Find the length of a chord 5 cm from the center of a circle whose radius is 8 cm.

PROJECT *Connection* | Select one of your categories, and then select one game for your group to research. Focus your research on the design.

1. Investigate the origin of your game. Is it played only by a specific culture or country? If so, what features of the game limit its widespread use.

2. Sketch the design of the game. Highlight the use of circles and parts of circles.

3. Can you locate any tangents to the circles used? Can you locate any arcs or chords in the design?

12.4 Inscribed Angles

Explore

● Draw a large circle and label its center *S*. Draw $\overset{\frown}{AB}$ whose measure is less than 180° and ∠*ASB*. Locate and mark *P*, *Q*, and *R* on the circle but not on $\overset{\frown}{AB}$. Draw ∠*APB* ∠*AQB*, and ∠*ARB*.

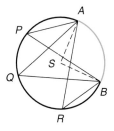

    1. Find m $\overset{\frown}{AB}$ by finding m∠*ASB*.

    2. Find m∠*APB*, m∠*AQB*, and m∠*ARB*.

    3. Write a statement that tells how m∠*APB*, m∠*AQB*, and m∠*ARB* are related to m $\overset{\frown}{AB}$.

Build Understanding

● An **inscribed angle** in a circle is an angle whose vertex is on the circle and whose sides contain chords of the circle.

inscribed angle

not an inscribed angle

not an inscribed angle

The Explore activity leads to the following theorem.

> **THEOREM 12.9**
>
> If an angle is inscribed in a circle, then its measure is half the measure of its intercepted arc.

To prove Theorem 12.9, you must consider three cases.

Case I

The center of the circle is on one side of the angle.

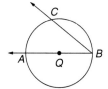

Case II

The center of the circle is in the interior of the angle.

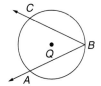

Case III

The center of the circle is in the exterior of the angle.

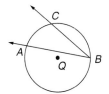

A proof of Case I follows.

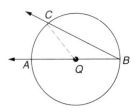

Given $\angle ABC$ is inscribed in circle Q.

Prove $m\angle ABC = \frac{1}{2}m \ \overset{\frown}{AC}$

Plan for Proof Draw $\overline{QC}$ and apply the Exterior Angle Theorem for Triangles.

Proof
Draw $\overline{QC}$. All radii of a circle are congruent, so $QB = QC$. Then, by definition, $\triangle CQB$ is isoscles. By the Base Angles Theorem, $m\angle ABC = m\angle QCB$. Let $m\angle ABC = m\angle QCB = x$. Then, by the Exterior Angle Theorem for Triangles, $m\angle AQC = m\angle ABC + m\angle QCB = x + x = 2x$. The measure of an arc is equal to the measure of the central angle that intercepts it, so $m \ \overset{\frown}{AC} = 2x$. By the Substitution Property, $m \ \overset{\frown}{AC} = 2(m\angle ABC)$. Therefore, using properties of equality, $m\angle ABC = \frac{1}{2}m \ \overset{\frown}{AC}$. ◄

You will prove Case II and Case III in Exercises 21 and 22.

EXAMPLE 1

Find $m\angle CDB$ and $m\angle CDA$.

Solution

Both $\angle CDB$ and $\angle CDA$ are inscribed angles. Apply Theorem 12.9.

$$m\angle CDB = \frac{1}{2}m \ \overset{\frown}{BC}$$

$$= \frac{1}{2}(55°) = 27.5°$$

$$m\angle CDA = \frac{1}{2}m \ \overset{\frown}{ABC}$$

$$= \frac{1}{2}(m \ \overset{\frown}{AB} + m \ \overset{\frown}{BC}) \qquad \text{Use the Arc Addition Postulate.}$$

$$= \frac{1}{2}(107° + 55°) = \frac{1}{2}(162°) = 81°$$

Thus, $m\angle CDB = 27.5°$ and $m\angle CDA = 81°$. ◄

Two corollaries follow from Theorem 12.9.

> **COROLLARY 12.10** If two inscribed angles in a circle intercept the same arc or congruent arcs, then the angles are congruent.
>
> **COROLLARY 12.11** If an inscribed angle in a circle intercepts a semicircle, then the angle is a right angle.

EXAMPLE 2

The Mathesons are having a wrought iron
gate installed on their property. The
diagram shows a decorative piece for the
top of the gate. In the diagram, line ℓ is a
line of symmetry. m $\overset{\frown}{AF}$ = 40°,
m $\overset{\frown}{AB}$ = 100°, and m $\overset{\frown}{BC}$ = m $\overset{\frown}{DC}$.
Find m∠1 and m∠2.

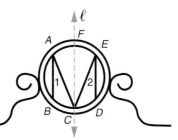

Solution

Since line ℓ is a line of symmetry for the circle, $\overset{\frown}{FAC}$ is a semicircle,
and m $\overset{\frown}{FAC}$ = 180°. It is given that m $\overset{\frown}{AF}$ = 40° and m $\overset{\frown}{AB}$ = 100°.
Write and solve an equation to find m $\overset{\frown}{BC}$.

$$\text{m } \overset{\frown}{AF} + \text{m } \overset{\frown}{AB} + \text{m } \overset{\frown}{BC} = 180°$$
$$40° + 100° + \text{m } \overset{\frown}{BC} = 180°$$
$$\text{m } \overset{\frown}{BC} = 40°$$

CHECK UNDERSTANDING

In Example 2, find
m∠ACF and m∠ECF.

Since ∠1 is an inscribed angle that intercepts $\overset{\frown}{BC}$,
m∠1 = $\frac{1}{2}$(m $\overset{\frown}{BC}$) = $\frac{1}{2}$(40°) = 20°.

It is given that m $\overset{\frown}{BC}$ = m $\overset{\frown}{DC}$, since ∠2 is an inscribed angle that
intercepts $\overset{\frown}{DC}$, by Corollary 12.10, m∠2 = m∠1 = 20°. ◄

Another corollary to Theorem 12.9 relates to quadrilaterals.

> **COROLLARY 12.12** If a quadrilateral is inscribed in a
> circle, then its opposite angles are supplementary.

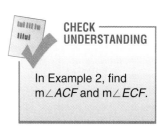

The figure at the right illustrates this corollary.

m∠A + m∠C = 180°
m∠B + m∠D = 180°

EXAMPLE 3

Quadrilateral *LMNP* is inscribed in ⊙ *Q*.
Points *M* and *P* are endpoints of a diameter, and
m∠M = 113°. Find m∠L, m∠N, and m∠P.

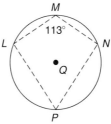

Solution

Since ∠MNP is an inscribed angle that intercepts
a semi-circle, m∠N = 90°. The same is true of
∠PLM and m∠L. So, m∠N + m∠L = 180°.

By Corollary 12.12, ∠M and ∠P also are supplementary.

$$\text{m}\angle P + \text{m}\angle M = 180°$$
$$\text{m}\angle P + 113° = 180°$$
$$\text{m}\angle P = 67°$$

Therefore, m∠L = m∠N = 90°, and m∠P = 67°. ◄

CHECK UNDERSTANDING

State the theorem
about kites that allows
you to conclude that
$\overline{MP}$ is a line of
symmetry.

In ⊙ *P*, $\overline{BD}$ is a diameter. Find each measure.

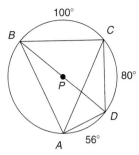

1. m∠*BAC*
2. m∠*ABD*
3. m∠*CAD*
4. m∠*BCD*
5. m∠*CBA*
6. m∠*BCA*

7. **WRITING MATHEMATICS** Explain how Corollary 12.12 provides another way to prove that the sum of the measures of the angles in a quadrilateral inscribed in a circle is 360°.

8. **HOME IMPROVEMENTS** Janet is having a stained glass window made. Points *P*, *Q*, *R*, *S*, and *T* are collinear and the arcs are parts of concentric circles with center *R*. Segment UR is a line of symmetry. Find m∠*PUT*, m∠*QVS*, m∠*QVR*, and m∠*RVS*. Explain why △*PUR* ≅ △*TUR* and △*QVR* ≅ △*SVR*.

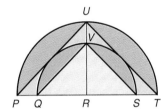

9. Quadrilateral *EFGH* is inscribed in a circle. Given that opposite angles are congruent, find the measures of the four angles and classify the quadrilateral.

In ⊙ *C*, $\overline{AD}$ and $\overline{BE}$ are diameters. Find each measure.

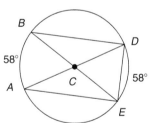

10. m∠*BDE*
11. m∠*AED*
12. m∠*DAE*
13. m∠*BED*
14. m∠*ADE*
15. m∠*DCE*

16. **WRITING MATHEMATICS** Draw a diagram to illustrate this statement. *In a circle, two inscribed angles intercept congruent arcs.* Write a statement that tells how the two inscribed angles are related to each other.

17. **SCHOOL SPIRIT** Students in the North High School sophomore class are designing a school flag. In ⊙ *P*, $\overline{AD}$ and $\overline{BC}$ are parallel congruent chords. Explain why m∠1 ≅ m∠2. Then find m∠1 and m∠2.

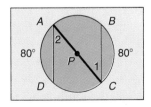

18. Quadrilateral *WXYZ* is inscribed in a circle. For each pair of opposite angles, one angle has a measure that is twice the measure of the other. Find the measures of the four angles and classify the quadrilateral.

19. In ⊙ *C*, $\overline{XZ}$ is a diameter. Use trigonometric ratios to find *XY*, *YZ*, and the perimeter of △*XYZ*.

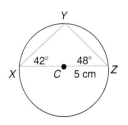

20. Use one of your answers from Question 19 to find the length of the altitude of △*XYZ* from *Y*. Then find the area of △*XYZ*.

Refer to Theorem 12.9. Use the given figures to prove Case II and Case III. (Hint: Use the results from the proof of Case I on page 682.)

21. Given ∠*ABC* is inscribed in circle *Q*.

Prove $m\angle ABC = \frac{1}{2}m\,\widehat{AC}$

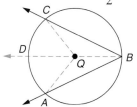

22. Given ∠*ABC* is inscribed in circle *Q*.

Prove $m\angle ABC = \frac{1}{2}m\,\widehat{AC}$

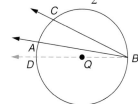

Find *x*. Then find m $\widehat{AB}$, m $\widehat{BC}$, and m $\widehat{ADC}$.

23.

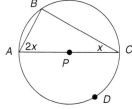

24.

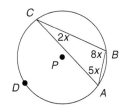

EXTEND

25. COMMERCIAL ART Members of the Amateur Archery Association are creating a logo for the club. In ⊙ *P*, $\overrightarrow{BZ}$ passes through *P* and is a line of symmetry for *ABCP*. Given m∠*APC* = 110°, find m∠*BAC*, m∠*ABP*, m∠*ABC*, and m∠*PAB*.

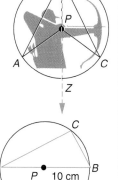

26. In ⊙ *P*, $\overline{AB}$ is a diameter and *AC* = 2(*BC*). Explain how you know that △*ABC* is a right triangle. Use the Pythagorean Theorem to find *AC* and *BC*. Use trigonometric ratios to find m∠*CAB* and m∠*CBA*.

27. An observer at point *A* watches cars *B* and *C* start at point *A* and travel around the circular track whose radius is 90 ft as shown. Car *B* travels at 4 ft/s and car *C* travels at 6 ft/s. Find the observer's angle of view, ∠*BAC*, after the cars travel for 35 s.

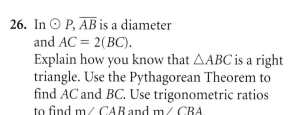

For Exercise 28–30, use a circle whose equation is $(x-8)^2 + (y-10)^2 = 100$.

28. Verify that *A*(14, 18), *B*(−2, 10), and *C*(18, 10) are on the circle.

29. Use slopes to show that ∠*BAC* is a right angle.

30. From Question 29, what can you say about $\overline{BC}$? Verify your response.

The diagram shows a sphere whose radius, $\overline{QR}$ is 10 cm. Plane *J* intersects the sphere determining a circular cross section containing points *R* and *S*. Points *P*, *Q*, *R*, *S*, and *C* lie on a plane perpendicular to plane *J*.

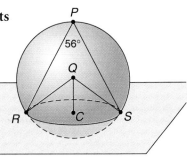

31. Find m∠*RQS*. Use a trigonometric ratio to find *RC*, the radius of ⊙ *C*. Then find the area of ⊙ *C*.

32. Find *CQ*, the distance between the circular cross section containing *C* and a plane through the center of the sphere.

THINK CRITICALLY

33. Prove Corollary 12.10: If two inscribed angles in a circle intercept the same arc or congruent arcs, then the angles are congruent.

34. Prove Corollary 12.11: If an inscribed angle in a circle intercepts a semicircle, then it is a right angle.

35. Prove Corollary 12.12: If a quadrilateral is inscribed in a circle, then its opposite angles are supplementary.

36. Suppose you mark four points *A*, *B*, *C*, and *D* on a circle in such a way that m $\overset{\frown}{AB}$ = m $\overset{\frown}{BC}$ = 50° and m $\overset{\frown}{CD}$ = m $\overset{\frown}{DA}$ = 130°. Show that quadrilateral *ABCD* is a kite.

37. Prove that a kite inscribed in a circle must have two right angles.

38. Points P_1, P_2, P_3, . . . , P_{n-1}, and P_n are placed on a circle in such a way that they are equally spaced apart. Describe the polygon that is inscribed in the circle when each point is connected to the next one and P_n is connected to P_1.

ALGEBRA AND GEOMETRY REVIEW

Find the slope of the line with the given equation.

39. $y = 4x - 1$ **40.** $3y = 27x + 11$ **41.** $2y = 14 - 6x$ **42.** $2x + 3y = 30$

Write an equation for the axis of symmetry of the parabola whose equation is given.

43. $y = x^2 + 4x - 6$ **44.** $y = 2x^2 - 20x + 3$ **45.** $y = 3x^2 - 12x + 11$

46. In ⊙ *Q*, point *P* is on ⊙ *Q* not on $\overset{\frown}{AB}$ or $\overset{\frown}{BC}$, m $\overset{\frown}{AB}$ = 73°, and m $\overset{\frown}{BC}$ = 52°. Find m∠*APC*.

PROJECT *Connection*   Select one game from your other category to research. Focus on a winning strategy and the use of circles.

1. Investigate the origin of your game. Is it played only by a specific culture or country? If so, what features of the game limit its widespread use.

2. Describe the objective of the game. Explain the use of circles and parts of circles to the winning strategy.

3. Does any move in the game model a tangent to a circle used? Is any move in the game along an arc or a chord?

Career
Bicycle Designer

A bicycle is a symphony of circles designed for people of all sizes and all ages. Approximately 800 million bicycles are in use around the world today. Bicycles and tricycles are not only used for recreation, but also for transportation. In some parts of the world, such as Beijing, China, bicycling is a common mode of transportation for people of all ages.

A bicycle is designed with its rider and function in mind. Bicycle designers use circles, circumference, tangents, central angles, arcs, chords, and inscribed angles when creating a new model.

Decision Making

1. Name as many parts of a bicycle or tricycle as possible that involve circles. Give the bicycle terminology and then give the mathematical terminology.

2. On most tricycles, the pedals are connected to the center of the front wheel. For each complete revolution of the pedals, how many revolutions does the front wheel make?

Bicycles have a circular gear called the chainring, attached to the pedals. A chain connects it to a circular gear, called a sprocket, on the rear wheel. These gears have teeth which are formed from arcs of circles. These teeth, in conjunction with the diameter of the chainring and sprocket determine the gear ratio.

3. What is the ratio of the number of teeth in the chainring to the number of teeth in the sprocket for a bicycle that has a 48-tooth chainring and a 12-tooth sprocket?

4. For the bicycle in Question 3, how many times will the rear wheel turn for each complete turn of the pedals? How far will this bicycle travel for each complete turn of the pedals if it has a 27-in. diameter wheel? Round to the nearest inch.

5. What is the ratio of the number of teeth in the chainring to the number of teeth in the sprocket for a bicycle that has a 28-tooth chainring and a 28-tooth sprocket? Explain how this bicycle is similar to the tricycle in Question 2.

6. Parts of the bicycle's chain wrap around the chainring and sprocket, and part of the chain remains linear. What is the geometric name for segments *AB* and *CD*?

Locating the Center of a Circle

Think Back/Working Together

1. Make a list of circular objects in everyday use on which the center of the circle is marked or evident.

2. Make a list of circular objects in everyday use on which the center of the circle is not marked or evident.

Answer each question about the geometry of a circle.

3. What must be true of all diameters in a given circle?

4. What would happen if you drew two chords in a circle and then drew the perpendicular bisector of each?

Explore

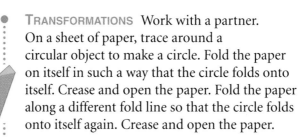

TRANSFORMATIONS Work with a partner. On a sheet of paper, trace around a circular object to make a circle. Fold the paper on itself in such a way that the circle folds onto itself. Crease and open the paper. Fold the paper along a different fold line so that the circle folds onto itself again. Crease and open the paper.

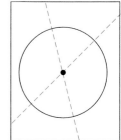

5. What do you think is true about the point where the two fold lines intersect?

6. Place the point of your compass on the point where the fold lines intersect. Open the compass so that the pencil point is on the circle. Draw a circle. Does it coincide with the original circle?

On a second sheet of paper, trace around another circular object. Repeat the instructions above.

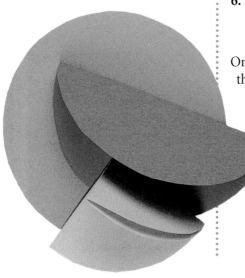

7. What do you think is true about the point where the two fold lines intersect?

8. Repeat Question 6 with the new folded circle.

9. Do you think that folding a circle on paper is a reliable way to locate its center? Explain your response.

10. Describe situations in which a circle with no center marked is given to you and folding is not a practical way to find its center.

Make Connections

- In addition to relying on the symmetry of a circle to locate its unmarked center, you can apply any one of a variety of theorems about circles.

Theorem 12.8 states: If one chord of a circle is the perpendicular bisector of another, then the first chord is a diameter of the circle.

11. Suppose that, in a given circle, chord $\overline{RS}$ is part of the perpendicular bisector of chord $\overline{AB}$. Explain how you know the center of the circle must lie on $\overline{RS}$.

12. Suppose that, in the same circle, chord $\overline{XY}$ is part of the perpendicular bisector of chord $\overline{CD}$. Explain how you know the center must lie on $\overline{XY}$.

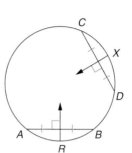

13. What conclusion about the center of a circle can you draw from Questions 11 and 12?

14. Trace a circular object on paper. Draw two nonparallel chords. Use what you learned in Questions 11–13 along with the construction of perpendicular bisectors to locate its center.

15. **ARCHAEOLOGY** An archaeologist found a broken plate at a dig. The plate appears to be circular. Explain how the archaeologist can use three points to find the radius of the plate.

16. **ARCHAEOLOGY** Suppose the archaeologist puts the plate on a sheet of paper and traces around the edge. How can folding be used to find the center and radius of the plate?

Corollary 12.11 states:

If an inscribed angle in a circle intercepts a semicircle, then it is a right angle.

17. **CARPENTRY** To locate the center of a circle, a carpenter can use carpenter's square. Its use is illustrated at the right. Explain how you know the blue line contains a diameter of the circle.

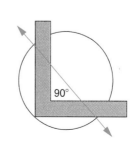

18. **CARPENTRY** Explain how the carpenter can use the square twice on a given circle to locate the center of the circle.

A woodworker who uses a lathe can locate the center of a circle inscribed in a square face of a long block of wood by using the square that circumscribes the circle.

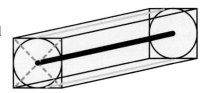

Refer to the diagram below for Questions 19 through 22.

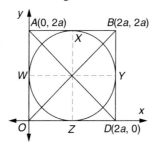

19. **COORDINATE GEOMETRY** Two diameters of the circle are $\overline{WY}$ and $\overline{XZ}$. Find the coordinates of the intersection of $\overline{WY}$ and $\overline{XZ}$ by finding the midpoints of $\overline{WY}$ and $\overline{XZ}$.

20. **COORDINATE GEOMETRY** The diagonals of square $ABDO$ are $\overline{AD}$ and $\overline{OB}$. Find the coordinates of the point at which $\overline{AD}$ and $\overline{OB}$ intersect.

21. What can you conclude from your answers to Questions 19 and 20?

22. **WRITING MATHEMATICS** Write a statement that relates the diagonals of a square to the center of the circle inscribed in it. Then write a brief justification of your statement.

You can use a ruler and the fact that a diameter is the longest chord in a circle to locate a circle's center.

23. Trace around a circular object.
    a. Place the 0 mark of a ruler on the circle. Pivot the ruler with the 0 mark fixed until you get the greatest measure you can for a chord of the circle. Draw the chord resulting from the greatest measurement you can read.
    b. Repeat the instructions a second time starting with a different point on the circle.
    c. Explain how you know that the intersection of the two chords is the center of the circle.
    d. **WRITING MATHEMATICS** Write a statement that relates the chords of maximum length to the center of the given circle. Then write a brief justification of your statement.

When you use geometry software to create a circle, the program will give you the circle and its center. Since the center is marked, you need not find it. However, you can use the software to test the validity of the methods you learned for finding the center.

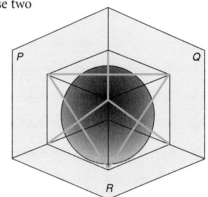

24. Using your geometry software, sketch a circle and place three points A, B, and C on it. Draw $\overline{AB}$ and $\overline{BC}$. Locate and mark their midpoints. Then construct the perpendicular bisectors of the chords. Do they meet at the center of the circle? Drag one of the points on the circle around it. Do the new perpendicular bisectors still intersect at the center?

25. Draw a new circle in your software program. Locate and mark four points A, B, C, and D on it. Explain how to use two inscribed angles to locate the center.

26. The figure at the right shows a solid sphere snugged up against two walls and the floor of a room. The figure also shows a cube circumscribed about the solid sphere and diagonals of three faces of the cube. Devise a method for finding the radius of the ball.

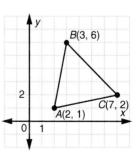

27. Write a statement about the dimensions of the cube and the radius of the ball in Question 26.

Summarize

28. **WRITING MATHEMATICS** In this workshop you saw many ways to find the center of a given circle whose center is unmarked. Summarize what you have learned.

29. **THINKING CRITICALLY** In the figure at the right, consider $\overline{AC}$ and $\overline{BC}$ as two chords of the same circles. Find the coordinates of the midpoints of $\overline{AC}$ and $\overline{BC}$. Then find equations for the perpendicular bisectors of $\overline{AC}$ and $\overline{BC}$. Solve the resulting system of equations to find the coordinates of the center of the circle.

30. **GOING FURTHER** Devise a method of your own that you can use to find the center of an unmarked circle of moderate size, that is, a circle that fits on a sheet of paper.

Geometry Excursion

Inversion in a Circle

In this book, you have learned much about transformations. What you read here describes a new transformation of all points in a plane except one. This new transformation, called an *inversion in a circle,* relies on a circle and its center to assign to each point an image. Suppose you have $\odot Q$ with radius r in a plane, and point P distinct from point Q in the plane. The image of P under an inversion in the circle is the point P' on $\overrightarrow{QP}$ such that $(QP)(QP') = r^2$.

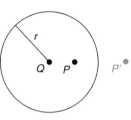

Given $\odot Q$ and point P inside its interior and different from Q, you can locate point P' as follows. Draw $\overrightarrow{QP}$ and a line through P perpendicular to $\overrightarrow{QP}$. Locate point T where the perpendicular to $\overrightarrow{QP}$ intersects $\odot Q$. Then draw the tangent to $\odot Q$ at point T. The image P' of P is the point where $\overrightarrow{QP}$ intersects the tangent to $\odot Q$ at T.

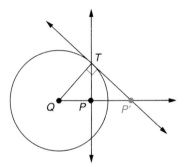

In Corollary 10.3, you learned that the length of the altitude drawn to the hypotenuse of a right triangle is the geometric mean of the lengths of the two segments of the hypotenuse. Refer to the figure above.

1. Let r represent the radius of $\odot Q$. Apply Corollary 10.3 to $\triangle QTP'$ and altitude $\overline{PT}$ to prove that $(QP)(QP') = r^2$.

2. Suppose that point P is on $\odot Q$. When you follow the instructions for locating point P', what do you discover?

3. Write a generalization, based on your answer to Question 2, about the image of any point on a given circle under an inversion in the circle.

Given $\odot Q$ and point P in its exterior, you can locate P' as follows. Draw the circle that has $\overline{QP}$ as a diameter and the line containing the points where they meet. Point P' is where that line meets $\overline{QP}$.

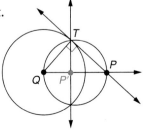

4. Explain how you know that $\triangle QTP$ is a right triangle. Explain how you know that $\overline{TP'}$ is perpendicular to $\overline{QP}$.

5. Let r represent the radius of $\odot Q$. Prove that $(QP)(QP') = r^2$.

Write $(QP)(QP') = r^2$ as $QP' = \dfrac{r^2}{QP}$ and as $QP' = \left(\dfrac{r}{QP}\right)(r)$.

What can you say about the location of point P' relative to $\odot Q$ given each condition?

6. $QP < r$ **7.** $QP > r$ **8.** $QP = r$

Suppose that line m intersects $\odot Q$ at points R and S. Assume that line m does not pass through the center of $\odot Q$.

9. What do your answers to Questions 1–8 tell you about R' and S', the images of R and S under an inversion in $\odot Q$?

10. What does your answer to Question 9 tell you about the images of all points on line m between R and S under an inversion in $\odot Q$?

11. Explain how you know that the image of a line m intersecting $\odot Q$ in two points R and S cannot be a line. Justify your conjecture.

12. Make a conjecture about the inversion image of line k that lies in the plane of $\odot Q$ and does not intersect the circle. Write an argument that shows that your conjecture is reasonable.

13. Make a conjecture about the inversion image of line n that lies on the plane of $\odot Q$ and intersects $\odot Q$ in just one point. What geometric figure do you think that image is?

14. WRITING MATHEMATICS Summarize what you have learned about inversion. Include a diagram involving $\odot Q$ with radius 2 in., points 0.5 in., 1 in. and 1.5 in. from Q, and the images of those points under an inversion in $\odot Q$.

15. THINKING CRITICALLY Consider an inversion in $\odot Q$ with radius 1. All points on the circle with center Q and radius $d = 0.25$ will have images on a circle centered at Q with radius $z = 4$. Images of points on the circle centered at Q and radius $d = 0.5$ will have images on the circle centered at Q and radius $z = 2$. How does the graph at the right picture the function that assigns points on circles centered at Q the distance of the images from Q?

Preimage Distance d from Q	Image Distance z From Q
0.25	4
0.5	2
1	1
2	0.5
4	0.25

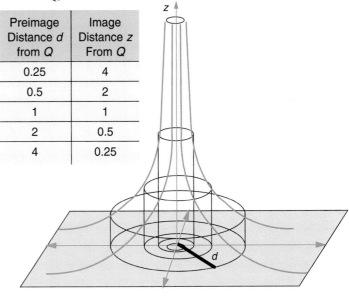

12.6 Angles Formed by Chords, Secants, and Tangents

Explore

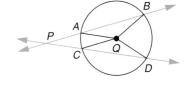

Draw circle Q and $\overleftrightarrow{AB}$ and $\overleftrightarrow{CD}$ intersecting at point P outside the circle. Then draw radii $\overline{QA}$, $\overline{QB}$, $\overline{QC}$, and $\overline{QD}$.

1. Find $m\angle BPD$, $m\angle AQC$, and $m\angle BQD$.

2. Calculate $m\angle BQD - m\angle AQC$.

3. How is $m\angle BPD$ related to your answer in Question 2?

4. Repeat Questions 1 and 2 for a new circle diagram. Can you draw the same conclusion that you drew in Question 3? Explain.

Build Understanding

The following theorem expands on what you learned in Explore.

THEOREM 12.13

If two secants, one secant and one tangent, or two tangents intersect in the exterior of a circle, then the measure of the angle formed is equal to half the difference of the measures of the intercepted arcs.

Illustration of Theorem 12.13		
Two secants		$m\angle P = \dfrac{1}{2}(m\,\widehat{BD} - m\,\widehat{AC})$
One secant and one tangent		$m\angle P = \dfrac{1}{2}(m\,\widehat{AC} - m\,\widehat{AB})$
Two tangents		$m\angle P = \dfrac{1}{2}(m\,\widehat{ACB} - m\,\widehat{AB})$

The following is the proof of Theorem 12.13 for the case in which a circle is intersected by two secants.

> If two secants intersect in the exterior of a circle, then the measure of the angle formed is equal to half the difference of the measures of the intercepted arcs.

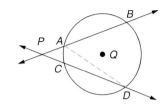

Given ⊙ Q and secants $\overrightarrow{AB}$ and $\overrightarrow{CD}$ intersecting at P outside ⊙ Q

Prove $m\angle APC = \frac{1}{2}(m\, \widehat{BD} - m\, \widehat{AC})$

Plan for Proof Draw $\overline{AD}$. Then use what you know about inscribed angles and exterior angles of triangles.

Proof Write a paragraph proof.

Since $\angle BAD$ and $\angle ADC$ are inscribed angles, by Theorem 12.9 $m\angle BAD = \frac{1}{2} m\, \widehat{BD}$ and $m\angle ADC = \frac{1}{2} m\, \widehat{AC}$.

Since $\angle BAD$ is an exterior angle of $\triangle PAD$, you can conclude $m\angle BAD = m\angle APC + m\angle ADC$ by the Exterior-Angle Theorem for Triangles.

Using properties of equality, $m\angle APC = m\angle BAD - m\angle ADC$. So by the Substitution Property $m\angle APC = \frac{1}{2} m\, \widehat{BD} - \frac{1}{2} m\, \widehat{AC}$. ◄

When you apply Theorem 12.13, be sure that you subtract the smaller arc measure from the larger one.

COMMUNICATING ABOUT GEOMETRY

Discuss whether you could write a legitimate proof of Theorem 12.13 if, instead of drawing $\overline{AD}$, you draw $\overline{CB}$.

CHECK UNDERSTANDING

In Example 1, would you get the correct answer if you calculated $\frac{1}{2}(m\, \widehat{SU} - m\, \widehat{TV})$? Explain.

EXAMPLE 1

Given the information in the diagram at the right, find $m\angle SRU$.

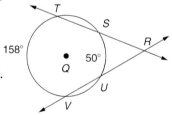

Solution
Secants $\overleftrightarrow{TR}$ and $\overleftrightarrow{VR}$ intersect in the exterior of ⊙ B so apply Theorem 12.13.

$$m\angle SRU = \frac{1}{2}(m\, \widehat{TV} - m\, \widehat{SU})$$

$$= \frac{1}{2}(158° - 50°)$$

$$= 54°$$

Thus, $m\angle SRU = 54°$. ◄

When two secants intersect in the interior of a circle, you can find the measures of the angles they form by using Theorem 12.14.

12.6 **Angles Formed by Chords, Secants, and Tangents** **695**

THEOREM 12.14

If two secants intersect in the interior of a circle, then the measure of each angle formed is half the sum of the measures of the arcs intercepted by the angle and its vertical angle.

EXAMPLE 2

MODEL RAILROADING Bruce wants to build a turntable for the locomotives in his train set. His plan is shown in the diagram at the right. Find m$\angle LMQ$.

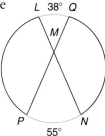

Solution
From the diagram, $\overline{LN}$ and $\overline{PQ}$ are two chords that intersect at M, in the interior of the circular turntable. Apply Theorem 12.14.

$$m\angle LMQ = \frac{1}{2}(m\,\overset{\frown}{LQ} + m\,\overset{\frown}{PN})$$

$$= \frac{1}{2}(38° + 55°) = 46.5°$$

Therefore, m$\angle LMQ = 46.5°$. ◄

The following theorem applies to the situation in which two lines intersect a circle and one of them is a tangent.

THEOREM 12.15

If a secant and a tangent to a circle intersect at the point of tangency, then the measure of each angle formed is half the measure of the intercepted arc.

EXAMPLE 3

In the figure at the right, $\overleftrightarrow{TK}$ is a tangent to $\odot P$ at point I and $\overleftrightarrow{IM}$ is a secant that intersects $\overleftrightarrow{TK}$ at point I. Find m$\angle KIM$ and m$\angle TIM$.

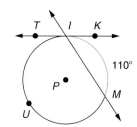

Solution
Apply Theorem 12.15 using $\overleftrightarrow{TK}$ and $\overleftrightarrow{IM}$.
$$m\angle KIM = \frac{1}{2}(m\,\overset{\frown}{IM}) = \frac{1}{2}(110°) = 55°$$

The intercepted arc for $\angle TIM$ is $\overset{\frown}{IUM}$.

$$m\,\overset{\frown}{IUM} = 360° - 110° = 250°$$

Thus, m$\angle TIM = \frac{1}{2}(m\,\overset{\frown}{IUM}) = \frac{1}{2}(250°) = 125°$. ◄

Find the measure of each angle.

1. ∠KAD   2. ∠ADG   3. ∠DGK   4. ∠GKA

5. **WRITING MATHEMATICS** Explain the similarities and differences between Theorem 12.9, which gives the measure of an inscribed angle in terms of the measure of its intercepted arc, and Theorem 12.13.

6. **AUTOMOTIVE MECHANICS** A chock is used as a wedge to hold an automobile in place. In the diagram below, $\overrightarrow{PY}$ is tangent to the tire at point B and $\overrightarrow{ZA}$ is tangent to the tire at point A. Given m $\overarc{AB}$ = m∠BDA = 46°, use Theorem 12.15 to find m∠APB.

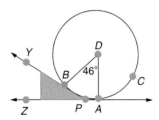

7. **PHOTOGRAPHY** In the diagram below ∠BEA is called the *angle of view*. Given the information in the diagram, find the angle of view. What would the angle of view be under the conditions that m $\overarc{AB}$ = 84° and m $\overarc{CD}$ = 22°?

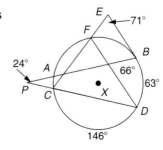

PRACTICE

Find the measure of each angle.

8. ∠GAH   9. ∠EDF

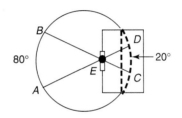

10. **WRITING MATHEMATICS** Draw ⊙ Q, points X and Y on ⊙ Q, and tangent segments $\overline{XP}$ and $\overline{YP}$. How does letting m∠XQY get closer to 180° effect m∠XPY? Draw several diagrams to illustrate your response.

Refer to the figure at the right, $\overline{BE}$ is tangent to ⊙ X at B.

11. Find m $\overarc{AC}$.

12. Find m $\overarc{FB}$.

13. Find m $\overarc{AF}$.

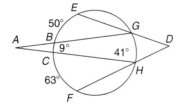

In the figure at the right, $\overset{\frown}{PX} = 81°$, $\overset{\frown}{XZ} = 30°$, $\overset{\frown}{ZY} = 51°$, and $\overset{\frown}{YQ} = 41°$.

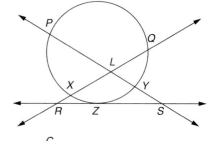

14. Find m∠*PLX*.   **15.** Find m∠*QRS*.

16. Find m∠*PLQ*.   **17.** Find m∠*PSR*.

Use the figure of ⊙ *O* and pentagon *ABCDE* for Exercises 18 and 19.

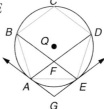

18. Suppose that you inscribe regular pentagon *ABCDE* in ⊙ *Q* and draw $\overrightarrow{GA}$ and $\overrightarrow{GE}$ tangent to ⊙ *Q* at points *A* and *E*, respectively. Explain how you know that m∠*CQD* = m∠*DQE* = m∠*EQA* = m∠*AQB* = m∠*BQC*. Use the equality you explained to find m∠*AGE*.

19. Find m∠*BFA*.

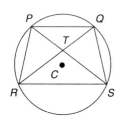

20. The figure at the left shows circle ⊙ *C* with parallel chords $\overline{PQ}$ and $\overline{RS}$. Given that m $\overset{\frown}{PR}$ = 72°, justify the statement that m $\overset{\frown}{QS}$ = 72°. Then find m∠*PTR*.

21. Use the information in Exercise 20 to find m $\overset{\frown}{PQ}$ and m $\overset{\frown}{RS}$ in the figure. Then find m∠*RTS*. Explain how to find m∠*RTS* a different way.

EXTEND

22. In the figure at the right, $\overrightarrow{PX}$ and $\overrightarrow{PZ}$ are tangent to ⊙ *C* at points *X* and *Z*, respectively. Given that m $\overset{\frown}{XYZ}$ = *x*°, show that m∠*XPZ* = *x*° − 180°.

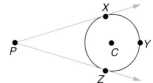

23. LINES OF SIGHT Jonathon and Marsha used a pair of protractors joined to make a *clinometer* to sight a spherical tank some distance away. In the diagram, $\overleftrightarrow{PX}$ and $\overleftrightarrow{PZ}$ are tangent to the tank at points *X* and *Z*, respectively. Given that m $\overset{\frown}{AC}$ = 24° and m $\overset{\frown}{BD}$ = 45°, find m∠*XPZ*. Then find m $\overset{\frown}{XZ}$ and m $\overset{\frown}{XYZ}$.

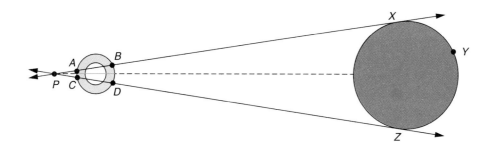

24. Find the measure of each angle in $\triangle PXY$.

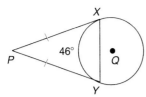

25. Find $m\angle YPX$.

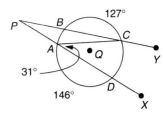

THINK CRITICALLY

26. In the diagram at the right, $\overrightarrow{PF}$ contains Q and bisects $\angle BPD$. Show that $m\angle BPD = m\,\widehat{BF} - m\,\widehat{AE}$ and $m\angle BPD = m\,\widehat{ABD} - 180°$.

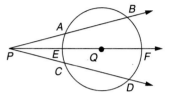

27. Prove part of Theorem 12.13: If one secant and one tangent intersect in the exterior of a circle, then the measure of the angle formed is equal to half the difference of the measures of the intercepted arcs.

28. Prove part of Theorem 12.13: If two tangents intersect in the exterior of a circle, then the measure of the angle formed is equal to half the difference of the measures of the intercepted arcs.

29. Prove Theorem 12.14: If two secants intersect in the interior of a circle, then the measure of each angle formed is half the sum of the measure of the arcs intercepted by the angle and its vertical angle.

30. Prove Theorem 12.15: If a secant and a tangent to a circle intersect at the point of tangency, then the measure of each angle formed is half the measure of the intercepted arc. (*Hint:* You must prove the theorem for three cases—a right angle, an acute angle, and an obtuse angle.)

The diagram at the right shows a regular hexagon inscribed in a circle. It also shows a pair of diagonals joining alternate vertices P_1 and P_3 and P_2 and P_4.

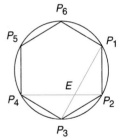

31. Find $m\angle P_1 E P_2$.

32. Write an argument to show that the measure of the acute angle formed by any pair of diagonals joining alternate vertices is $m\angle P_1 E P_2$.

ALGEBRA AND GEOMETRY REVIEW

Find the radius and the coordinates of the center of each circle with the given equation.

33. $(x + 4)^2 + (y - 7)^2 = 81$

34. $(x - 3)^2 + y^2 = 121$

Find the value of each trigonometric ratio.

35. $\sin A$

36. $\cos B$

37. $\cos A$

38. $\sin B$

39. $\tan A$

40. $\tan B$

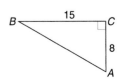

41. Points A, B, C, and D lie on $\odot Q$ and $\overline{AC}$ and $\overline{BD}$ intersect in the interior of $\odot Q$ at E. Find $m\angle AEB$ given that $m\,\widehat{AB} = 46°$ and $m\,\widehat{CD} = 132°$.

Segments Formed by Chords, Secants, and Tangents

Explore/Working Together

● Work in a group of three or four students. Draw a circle with a radius of at least 8 cm. Draw two chords $\overline{AB}$ and $\overline{CD}$ that intersect in the interior of the circle. Label the point of intersection E.

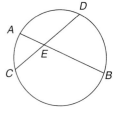

1. Carefully measure AE, EB, CE, and ED.

2. Calculate (AE)(EB) and (CE)(ED). Compare results with others in your group. Make a conjecture about the products?

3. Based on your conjecture in Question 2, write a proportion involving AE, EB, CE, and ED.

4. The proportion in Question 3 should suggest that two triangles formed by the four points on the circle and the point of intersection are similar. Write the similarity.

Build Understanding

● In Explore, you probably discovered the relationship stated in Theorem 12.16.

> **THEOREM 12.16**
>
> **If two chords of a circle intersect in the interior of the circle, then the product of the lengths of the segments of one chord equals the product of the lengths of the segments of the other chord.**

The following is a proof of Theorem 12.16.

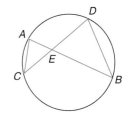

Given Chords $\overline{AB}$ and $\overline{CD}$ intersect at point E.

Prove $(AE)(EB) = (CE)(ED)$

Plan for Proof Draw $\overline{AC}$ and $\overline{DB}$ and show that $\triangle AEC \sim \triangle DEB$.

Proof

By the Vertical Angles Theorem, $m\angle AEC = m\angle DEB$. Since $\angle A$ and $\angle D$ are both inscribed angles that intercept $\overset{\frown}{BC}$, $m\angle A = m\angle D$. Then $\triangle AEC \sim \triangle DEB$ by the AA Similarity Postulate. By the definition of similar polygons, $\dfrac{AE}{ED} = \dfrac{CE}{EB}$. Therefore, by the Cross Product Property, $(AE)(EB) = (CE)(ED)$. ◄

Your knowledge of solving linear equations will help you solve problems involving intersecting chords.

EXAMPLE 1

In $\odot S$, $MR = 10$, $RN = 2$, and $PR = 4$. Find RQ.

Solution

By Theorem 12.16, the products of the lengths of the two parts of $\overline{PQ}$ equals the product of the lengths of the two parts of $\overline{MN}$.

$$(PR)(RQ) = (MR)(RN)$$
$$4(RQ) = (10)(2)$$
$$4(RQ) = 20 \qquad \text{Divide each side by 4.}$$
$$RQ = 5$$

So, $RQ = 5$.

When secants and tangents intersect at a point in the exterior of a circle, some special segments are formed. In the figure at the right, $\overline{PB}$ and $\overline{PD}$ are called *secant segments*. The parts of these segments that lie in the exterior of the circle, $\overline{PA}$ and $\overline{PC}$, are *external secant segments*. Similarly, $\overline{PE}$ is called a *tangent segment*.

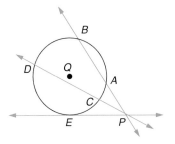

THEOREM 12.17

If two secants intersect in the exterior of a circle, then the product of the lengths of one secant segment and its external segment is equal to the product of the lengths of the other secant segment and its external segment.

COMMUNICATING ABOUT GEOMETRY

In the figure for Theorem 12.17, discuss what happens to $\overline{AB}$, $\overline{CD}$, $\overline{AP}$, and $\overline{CP}$ when you move points B and C closer together.

THEOREM 12.18

If a secant and a tangent intersect in the exterior of a circle, then the product of the lengths of the secant segment and its external segment is equal to the square of the length of the tangent segment.

Theorem 12.17

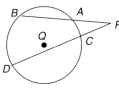

$(PA)(PB) = (PC)(PD)$

Theorem 12.18

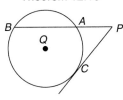

$(PA)(PB) = (PC)^2$

In this lesson, you have now seen three theorems that you can use to solve problems involving circles.

EXAMPLE 2

In $\odot Q$, $RG = 13$, $RK = 3$, and $JL = 12$. Find x.

Solution

By Theorem 12.17, $(RG)(RK) = (RL)(RJ)$.

$$(13)(3) = (x + 12)(x)$$

$$x^2 + 12x = 39$$

$$x^2 + 12x - 39 = 0$$

$$x = \frac{-12 \pm \sqrt{12^2 - 4(1)(-39)}}{2(1)} \quad \text{Quadratic Formula}$$

$$x \approx 2.66 \text{ or } -14.66$$

Since x cannot be a negative number, reject -14.66 as a value for x. Thus, $x \approx 2.66$. ◄

Example 3 illustrates the application of Theorem 12.18.

EXAMPLE 3

WELDING Students in a welding class at Kings Park High School are participating in a local government project to build bicycle racks at a local recreational area. The diagram at the right shows a side view of their plan. They want to make a weld at point D along $\overline{BP}$. How high up from the ground should they make the weld?

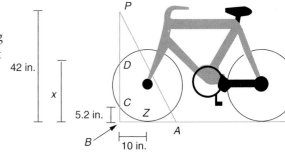

Solution

Since $\overline{BZ}$ is a tangent segment and $\overline{BD}$ is a secant segment, choose and apply Theorem 12.18.

$$(BD)(CB) = (BZ)^2$$

$$5.2x = 10^2$$

$$x = \frac{100}{5.2}$$

$$x \approx 19.2$$

The students should make the weld about 19.2 in. above the ground and along $\overline{BP}$. ◄

Refer to the figure at the right.

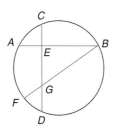

1. Given $AE = 1.5$, $EB = 3.5$, and $CE = 1.2$, find ED to the nearest tenth of a unit.

2. Given $FG = 1.3$, $DG = 1.5$, and $CG = 3.7$, find GB to the nearest tenth of a unit.

3. WRITING MATHEMATICS Write Theorem 12.16 in such a way that the similarity of the two triangles formed by a pair of intersecting chords is the heart of the theorem and the equation about products is an immediate result of the theorem.

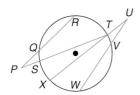

Refer to the figure at the left. Give your answer to the nearest tenth of a unit.

4. Given $PR = 5.5$, $PS = 1.5$, and $PT = 7.0$, find PQ.

5. Given $UW = 6.5$, $UT = 2.0$, and $UX = 7.6$, find UV.

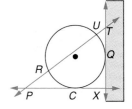

6. CAN STORAGE A clerk needs to construct a horizontal shelf $\overline{PX}$ for the storage of cans as shown at the right. The shelf will be supported by a brace from the shelf to the wall. Given $PR = 1.9$ in. and $PC = 4.0$ in., find PU to the nearest tenth of an inch.

PRACTICE

Refer to the figure at the right.

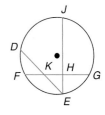

7. Given $FH = 2.4$, $HG = 1.5$, and $HE = 1.6$, find HJ to the nearest tenth of a unit.

8. Given $FH = 2.4$, $HG = 1.5$, and $JH = 3.4$, find HE to the nearest tenth of a unit.

9. WRITING MATHEMATICS The diagram that accompanies Theorem 12.17 shows two overlapping triangles when you add $\overline{AC}$ and $\overline{DB}$. Write Theorem 12.17 so that it states a triangle similarity and proportion relating corresponding sides.

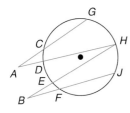

Refer to the figure at the left. Give your answer to the nearest tenth of a unit.

10. Given $CG = 3.8$, $AD = 1.9$, and $AH = 7.4$, find AC.

11. Given $BE = 2.2$, $EH = 5.5$, and $BJ = 6.7$, find BF.

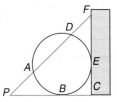

12. BARREL STORAGE A factory worker wants to build a shelf to hold empty cylindrical barrels as shown at the right. Given that $PB = 33$ cm and $AP = 20.0$ cm, find PD to the nearest centimeter.

13. **Logo Design** Students at the Dandberg High School plan to submit a proposed logo to the Circle Kite Company. Quadrilateral $ABCD$ is a kite inscribed in $\odot P$, $\overline{AC}$ is a diameter, and $AE = 3$ cm and $EC = 12$ cm. Find BC. Use the fact about kites, $BE = ED$, to find BD.

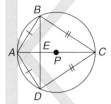

14. **Instrument Storage** In the Staff Music Store, drums are stored on a rack joined to the wall and suspended from the ceiling by a length of chain. You can see these in the diagram at the right. Find AB, the height above the rack at which the drum extends over the edge of the rack.

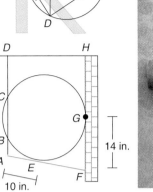

Find x. Round to the nearest tenth if necessary.

15.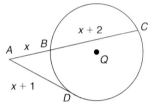

16.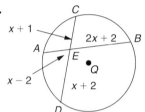

17.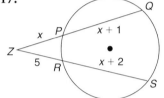

EXTEND

In Exercises 18–20, refer to the figure at the right. In it, $\overline{AD}$ is a diameter of $\odot Q$, $\overline{QE} \perp \overline{AD}$, and $QA = 10$. Round answers to the nearest tenth.

18. Given $QE = 2$ and $CE = BE$, find CE and BE.

19. Given $AB = CD$ and $CE = BE = 6$, find QE.

20. Given $AE = ED = 14$ and $CE = BE$, find CE and BE.

In Exercises 21–23, refer to the figure at the right. In it, $\overline{AB} \parallel \overline{CD}$, $\overline{EF} \parallel \overline{GH}$, and $m\angle RPQ = 90°$. Round answers to the nearest hundredth.

21. Given $PQ = 3$, $PR = 15$, $AP = QB$, and $AP = 2(EP)$, find AP.

22. Given $PQ = 3$, $PR = 17$, $CR = SD$, and $CR = 3(RF)$, find CR.

23. Given $PR = 14$, $CR = RS = SD$, $EP = RF = \frac{1}{7}(PR)$, and $CR = 2(EP)$, find CR and EP.

24. **Coordinate Geometry** Given the circle with equation $x^2 + y^2 = 100$, show $W(6, 8)$, $X(6, -8)$, $Y(8, 6)$, and $Z(-8, 6)$ are on it. To the nearest tenth, find the coordinates of the point V where $\overline{WX}$ and $\overline{YZ}$ meet.

THINK CRITICALLY

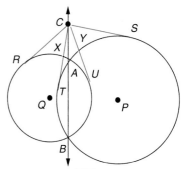

25. Prove Theorem 12.17: If two secants intersect in the exterior of a circle, then the product of the lengths of one secant segment and its external segment is equal to the product of the lengths of the other secant segment and its external segment.

26. Prove Theorem 12.18: If a secant and a tangent intersect in the exterior of a circle, then the product of the lengths of the secant segment and its external segment is equal to the square of the length of the tangent segment.

27. In the figure at the right, $\odot Q$ and $\odot P$ intersect at points A and B. The segments $\overline{CR}$ and $\overline{CU}$ are tangent segments to $\odot Q$, and the segments $\overline{CT}$ and $\overline{CS}$ are tangent segments to $\odot P$. Show that $CT = CU$. Then explain why $CR = CT = CU = CS$.

28. Secant $\overleftrightarrow{DE}$ intersects $\odot Q$ at points E and Y, with point Y between points E and D. Secant $\overleftrightarrow{DF}$ intersects $\odot P$ at points F and Z, with point Z between points F and D. Must it be true that $EY = FZ$? Justify your response.

ALGEBRA AND GEOMETRY REVIEW

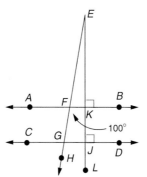

Find the measure of each angle.

29. $\angle AFE$ 30. $\angle EGJ$

31. $\angle FEJ$ 32. $\angle CGF$

33. $\angle CGH$ 34. $\angle LJC$

Find the specified percent of each number.

35. 15% of 60   36. 120% of 60   37. 18.5% of 250

38. 0.5% of 1200   39. 53% of 136   40. 200% of 12

41. In $\odot O$, chords $\overline{AB}$ and $\overline{CD}$ intersect at E. Given $AE = 4$, $EB = 6$, and $CE = 10$, find ED.

PROJECT *Connection*   Make a list of games that are played outdoors. Select one game for your group to research.

1. Investigate the origin of your game. Is it played only by a specific culture or country? If so, what features of the game limit its widespread use?

2. Sketch the playing field for the game. Highlight the use of circles and parts of circles.

3. Can you locate any tangents to the circles used? Can you locate any arcs or chords in the design?

4. Describe the objective of the game. Explain the use of circles and parts of circles to the winning strategy.

5. Does any move in the game model a tangent to a circle used? Is any move in the game along an arc or a chord?

Career
Toy Engineer

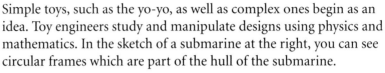

Simple toys, such as the yo-yo, as well as complex ones begin as an idea. Toy engineers study and manipulate designs using physics and mathematics. In the sketch of a submarine at the right, you can see circular frames which are part of the hull of the submarine.

Given measurements in circle I and a relationship between corresponding arc measures in circles, an engineer can find the corresponding cross-member lengths in the other circles.

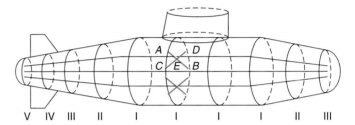

Decision Making

Refer to $\odot Q$ with radius r and $\odot Q'$ with radius 80% of r and the specification table at the right.

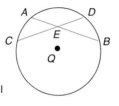

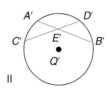

1. Show that $\triangle AEC \sim \triangle A'E'C'$ and $\triangle DEB \sim \triangle D'E'B'$.

2. Using the result of Question 1, find $A'E'$, $E'D'$, $C'E'$, and $E'B'$.

Circle III has radius 80% of that of circle II. Its cross members intercept arcs with the same measures as those in circle II.

3. Do you think the lengths of the cross members in circle III will be 80% of the lengths of the cross members in circle II? Justify your response.

4. Find the lengths of the four cross members in circle III.

Circle IV has radius 80% of that of circle III. Its cross members intercept arcs with the same measures as those in circle III. Circle V has radius 80% of that of circle IV. Its cross members intercept arcs with the same measures as those in circle IV.

5. Find the lengths of the four cross members in circles IV and V.

6. The work you did in the questions above suggest a theorem about similar circles, pairs of chords in each that intercept arcs of equal measure, and the lengths of the parts of those corresponding chords. Write such a theorem. Illustrate it with a diagram and equation. Then provide a proof of your theorem.

Specifications

$AE = ED = $ **1.7 in.**
$CE = EB = $ **3.0 in.**

$m\,\overset{\frown}{CA} = m\,\overset{\frown}{C'A'} = 38°$

$m\,\overset{\frown}{AD} = m\,\overset{\frown}{A'D'} = 76°$

$m\,\overset{\frown}{DB} = m\,\overset{\frown}{D'B'} = 38°$

12.8 Focus on Reasoning

Truth Tables

Discussion

- Work in a group of three or four students.

Yolanda, Jeke, and Andy were walking home from school one day. Yolanda wondered how she would keep straight all the theorems they have learned. "To make matters worse," said Jeke "most of the theorems are in *if-then* form." "What do you mean?" asked Andy. Jeke replied "Not only do we have to remember the parts of the theorems, we must remember their order."

Yolanda asked their mathematics teacher to help them find out when a conditional statement is true and when it is not. Ms. Gomez introduced Yolanda to *truth tables*. The truth table at the right indicates when the conditional *If p, then q* is true and when it is not.

Truth Table for If *p*, then *q*.		
p	*q*	If *p*, then *q*.
T	T	T
T	F	F
F	T	T
F	F	T

1. Jeke looked at the first row of the truth table. Does it make sense to say a conditional is true when both hypothesis and conclusion are true? Explain your response.

2. Does it make sense to say that a conditional is false when the hypothesis is true but the conclusion is false? In your response, consider *If 2 is an even number, then 3(2) is an odd number.*

3. Jeke had a hard time accepting the third row of the truth table. Why do you think a conditional is true when the hypothesis is false but the conclusion is true? Explain your response.

4. When Andy examined the truth table, he saw four rows. Explain how you know that four rows are needed to determine all the possibilities for the truth or falseness of *If p, then q.*

Determine whether each statement is true or false. Explain your answer.

5. If $\angle ABC$ is a right angle inscribed in $\odot O$, then $\overline{AC}$ is a diameter.

6. If $\overline{AC}$ is a diameter in $\odot O$, then $\angle ABC$ is a right angle inscribed in $\odot O$.

Build a Case

To make a truth table to represent other types of conditional statements, you need to keep in mind that, if a statement is true, then its negation is false. In other words, when p is true, $\sim p$ is false. When p is false, $\sim p$ is true.

Andy recalled that Ms. Gomez said that proving the contrapositive of an *if-then* statement is as good as proving the statement directly. To help them study the claim, they reproduced the truth table for *If p, then q* as shown.

Truth Table for If p, then q.

p	q	If p, then q.
T	T	T
T	F	F
F	T	T
F	F	T

Truth Table for If $\sim q$, then $\sim p$.

p	q	$\sim q$	$\sim p$	If $\sim q$ then $\sim p$.
T	T			
T	F			
F	T			
F	F			

7. Copy and complete the third and fourth columns of truth table at the left.

8. Use Columns 3 and 4 to fill in the fifth column.

9. How do the first, second, and fifth columns of your table from Question 8 compare with the first, second, and third columns of the truth table for *If p, then q*?

10. Andy decided that Ms. Gomez was right to say that proving the contrapositive of an *if-then* statement is as good as proving the *if-then* statement itself. How do the truth tables for *If p, then q* and *If ~ q, then ~ p* help you and Andy believe Ms. Gomez?

Make a truth table for each variation on *If p, then q*. The table at the right can help you get started.

p	q	$\sim p$	$\sim q$
T	T		
T	F		
F	T		
F	F		

11. If q, then p.

12. If $\sim p$, then $\sim q$.

13. If $\sim q$, then p.

14. If q, then $\sim p$.

15. Jeke worked Exercise 11 and made a conjecture about the truth of an *if-then* statement and the truth of its converse. Is proving the converse of an *if-then* statement as good as proving the *if-then* statement itself? Explain your response.

Refer to the figure at the right. In it, $\overrightarrow{AB} \parallel \overrightarrow{CD}$.

16. Write an *if-then* statement about the lines and angles shown.

17. Would proving the contrapositive, inverse, or converse of your statement be as good as proving the statement itself?

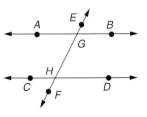

EXTEND AND DEFEND

The Side-Angle-Angle Congruence Theorem states the following.

If two angles of one triangle are congruent to two angles of another and the nonincluded side of the first triangle is congruent to the nonincluded side of the second, then the two triangles are congruent.

To determine the truth value of an *if-then* statement with a *compound hypothesis*, you need to know when two statements joined by *and* are true simultaneously and when they are not. In some *if-then* statements, the hypothesis contains statements joined by *or*. The table at the right shows truth values for statements like these.

	Truth Table for *p and q* and *p or q*.		
p	*q*	*p* and *q*	*p* or *q*
T	T	T	T
T	F	F	T
F	T	F	T
F	F	F	F

Use the table at the right for Questions 18 and 19.

18. Copy the table. Fill the third column.

19. Fill in the fifth column.

20. When is the Side-Angle-Angle Congruence Theorem false?

21. HIGHWAY SAFETY When is *If you drive defensively and obey the traffic laws, then you will not have an accident* true?

22. Make a truth table for *If p or q, then r.*

		Truth Table for *If p and q, then r.*		
p	*q*	*p* and *q*	*r*	If *p* and *q* then *r.*
T	T		T	
T	T		F	
T	F		T	
T	F		F	
F	T		T	
F	T		F	
F	F		T	
F	F		F	

23. PROSPERITY PROSPECTS When is *If you win a lottery or get an excellent education, you will become prosperous* false?

Now You See It
Now You Don't

ROUND AND ROUND

Can you find the center of this circle? How did you find it?

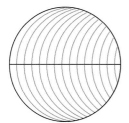

Are you looking inside a tube or at the top of a beach ball?

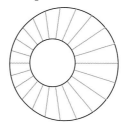

Using Secants and Tangents

How Earth appears to the eye depends in part on where you are. If you are in a space shuttle, Earth appears to be spherical and the horizon appears curved and circular. If however, you are close to the surface of Earth, perhaps on a ship at sea or atop a city building or lighthouse, Earth appears to be planar and the horizon appears to resemble a straight line.

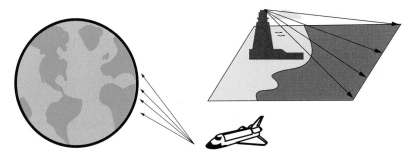

Problem

This summer Samantha and her brother Michael plan to visit New York City. One of the places they want to visit is the World Trade Center. Its height is 1368 ft. Standing in the observation deck above the 110th floor, how far can they see to the horizon?

Explore the Problem

Samantha drew the diagram at the left below to model Earth and the World Trade Center. Knowing the radius of Earth is about 3813 mi, she then drew the diagram at the right below. In it, $\overrightarrow{PC}$ is tangent to $\odot Q$ at C, and $\overline{AB}$ is a diameter.

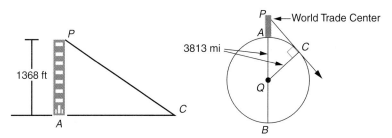

1. Explain how she knows that $\overline{AB}$ is a diameter.

2. Using your knowledge of secants and tangents, classify $\overline{PB}$, $\overline{PA}$, and $\overline{PC}$.

3. Use one of the theorems you learned in this chapter to write an equation involving PB, PA, and PC that you can use to find PC.

4. Michael realized that before they could use the equation from Question 3, all the measurements had to be in the same units. Convert 1368 ft to miles.

5. Now solve the equation in Question 3 to determine how far they can see to the horizon from atop the World Trade Center in New York City. Round to the nearest tenth of a mile.

Investigate Further

● Generalize the method used in Explore the Problem for finding the distance to the horizon from atop any tall landmark.

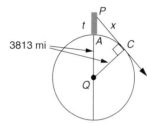

6. Let t represent the height of a tall structure in feet. Let x represent the distance to the horizon in miles. Write and solve an equation in terms of t and the radius of the Earth to find x.

7. Use your formula for x in terms of t from Question 6 to find how far you can see to the horizon from atop the Empire State Building. Its height is 1250 ft.

You can use the Pythagorean Theorem to generalize the method.

$$\left(3813 + \frac{t}{5280}\right)^2 = x^2 + 3813^2$$

8. Use algebra to show that $x^2 = 2(3813)\left(\frac{t}{5280}\right) + \left(\frac{t}{5280}\right)^2$.

9. Explain why $x = \sqrt{2(3813)\left(\frac{t}{5280}\right) + \left(\frac{t}{5280}\right)^2}$.

10. Use the formula from Question 9 to find the distance to the horizon from atop the World Trade Center and from atop the Empire State Building.

11. If t is small, you can write $x = \sqrt{2(3813)\left(\frac{t}{5280}\right)}$. Simplify this expression for x.

12. Use the formula from Question 11 to determine the distance to the horizon from atop the Statue of Liberty. Its height is about 305 ft.

13. Writing Mathematics Describe why someone would want to generalize a method used to solve a specific problem.

Apply the Strategy

Sketch and use a diagram as necessary to solve each problem.

14. SPACE TRAVEL On May 5, 1961, Alan Shepard became the first American in space. His spacecraft achieved an altitude of 116.5 mi above Earth. How far could he see to the horizon of Earth?

15. FITNESS The diagram at the right shows a bicycle used as an exercise device. The diameter of the wheel is 26 in. Supports $\overline{QD}$ and $\overline{QP}$ keep the rear wheel off the floor. You are given that $PA = 9$ in. Metal bar $\overline{BP}$ is tangent to the rear wheel at B and serves as a source of resistance to pedaling. Find the length of $\overline{BP}$.

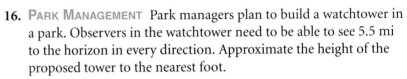

16. PARK MANAGEMENT Park managers plan to build a watchtower in a park. Observers in the watchtower need to be able to see 5.5 mi to the horizon in every direction. Approximate the height of the proposed tower to the nearest foot.

17. AIRLINE TRAVEL An airliner is flying at an altitude of 35,000 ft. To the nearest mile, approximate the distance the pilot can see to the horizon.

18. CAR BUILDING Gregory and June are building a toy car. They want the brake $\overline{BD}$ to extend 2.8 in. beyond the wheel. That is, $ED = 2.8$. What should be the length of the brake?

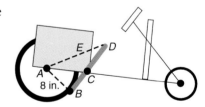

The *curvature of a circle* is a measure of how "flat" the circle is and is defined as the reciprocal of the radius of the circle.

19. Find the curvature of a circle whose radius is 1 mi, 2 mi, 3 mi, and 4 mi. Then find the curvature of Earth, whose radius is about 3813 mi.

20. WRITING MATHEMATICS Draw a small circle, a second circle with twice the radius, and a third circle with four times the radius. Do your answers from Question 19 and your diagrams suggest that the reciprocal of the radius is a good definition of curvature?

Review Problem Solving Strategies

Turning the Tables

1. Rachel and Jessica are decorating tables for a math club meeting at their school. The table tops are circles with 4-ft diameters. In the center of each table they have placed four strips of crepe paper, overlapping at right angles as shown. Each paper strip is 12 in. long and 3 in. wide. What is the area of the table they do not cover?

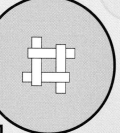

Hexagon in a Cube

2. A regular hexagon is inscribed in a cube, as shown, so that each vertex of the hexagon is the midpoint of an edge of the cube. Each edge of the cube is 12 cm long. What is the area of the hexagon?

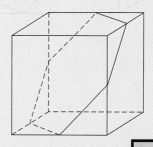

Can She Do It?

3. Ollie drew an isosceles triangle with base length 10 and each leg of length 13. Ingrid said that she could draw another isosceles triangle with legs of length 13, a base with length different from 10, but with the same area as Ollie's triangle. Ollie did not think this was possible. What do you think?

CHAPTER REVIEW

. . . . . .

VOCABULARY

Choose the word from the list that correctly completes each statement.

1. The measure of a(n) __?__ angle is one half the measure of its intercepted arc.

2. A line that intersects a circle in one point is called a(n) __?__ .

3. Arcs of the same circle that have the same measure are __?__ arcs.

4. Two arcs of the same circle that have exactly one point in common are __?__ arcs .

a. adjacent

b. congruent

c. inscribed

d. tangent

Lesson 12.1 TANGENTS TO CIRCLES   pages 663–668

- If a line is tangent to a circle, then it is perpendicular to the radius drawn to the point of tangency.
- If two segments from the same exterior point are tangent to a circle, then they are congruent.

Draw a figure to determine if $\overline{PA}$ is tangent to $\odot O$ at P on $\odot O$.

5. $OP = 6, PA = 8, AO = 12$ 6. $OP = 9, PA = 40, AO = 41$ 7. $OP = 3, PA = 3\sqrt{3}, AO = 6$

Lesson 12.2 and 12.3 CENTRAL ANGLES, CHORDS AND ARCS   pages 669–680

- In the same circle, or in congruent circles, two minor arcs are congruent if and only if the central angles that intercept them are congruent
- If a diameter of a circle is perpendicular to a chord, then the diameter bisects the chord and its arc.

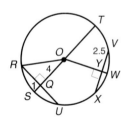

Find the length of the following segments in the figure.

8. $\overline{OS}$ 9. $\overline{OT}$

10. $\overline{RQ}$ 11. $\overline{VX}$

Lesson 12.4 INSCRIBED ANGLES   pages 681–687

- If an angle is inscribed in a circle, then its measure is half the measure of its intercepted arc.

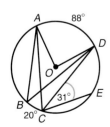

Find the measure of each of the following.

12. $\angle ABD$ 13. $\angle AOD$

14. $\angle ACD$ 15. $\angle BDC$

Lesson 12.5   LOCATING THE CENTER OF A CIRCLE

pages 688-691

- You can find the center of a circle by finding the intersection of two diameters, finding the intersection of the perpendicular bisectors of two nonparallel chords, and constructing inscribed right angles to create intersecting diameters.

Trace three congruent circles onto different sheets of paper.

16. Find the center of one circle by finding the intersection of two diameters, the center of another by finding the intersection of the perpendicular bisectors of two nonparallel chords, and the center of the third by constructing inscribed right angles to create intersecting diameters.

Lesson 12.6 and 12.7   ANGLES AND SEGMENTS FORMED BY CHORDS, SECANTS, AND TANGENTS

pages 692-706

- Angles that are formed by two tangents, two secants, or a tangent and a secant that intersect in the exterior of a circle are measured by half the difference of the degree measures of their intercepted arcs. Angles formed by two chords intersecting in a circle are measured by half the sum of the degree measures of their intercepted arcs. An angle formed by a tangent and a chord is measured by half the degree measure of its intercepted arc.
- If two chords intersect in a circle, the product of the lengths of the segments of one chord equal the product of the lengths of the segments of the other chord. If two secants are drawn from an exterior point of a circle, then the product of the lengths of one secant segment and its external segment is equal to the product of the lengths of the other secant segment and its external segment. If a secant and a tangent are drawn from an exterior point of a circle, the product of the lengths of the secant segment and its external segment is equal to the square of the length of the tangent segment.

Find the measure of each of the following.

17. $\angle BPE$ 18. $\angle BCE$ 19. $\angle BCA$

20. $\overline{AB}$ 21. $\overline{FP}$ 22. $\overline{PB}$

Lesson 12.8   TRUTH TABLES

pages 707-709

- Truth tables can help you determine when a conditional statement is true and when it is not.

23. Copy and complete the truth table.

p	q	$\sim p$	$\sim q$	If $\sim p$, then $\sim q$.
T	T			
T	F			
F	T			
F	F			

Lesson 12.9   USING SECANTS AND TANGENTS

pages 710-713

- Secants, tangents, chords, and arcs can be used to model and solve real life situatons.

24. A helicopter is hovering at an altitude of 100 feet. The radius of Earth is about 3813 mi. To the nearest mile, how far can the helicopter pilot see to the horizon?

CHAPTER ASSESSMENT

CHAPTER TEST

Find x to the nearest tenth.

1.

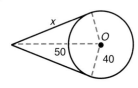

2.

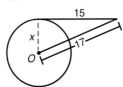

In the figure, $\overline{AB}$ is a diameter of $\odot Q$. Find the degree measure of each arc.

3. $\overset{\frown}{AC}$

4. $\overset{\frown}{BC}$

5. $\overset{\frown}{DB}$

6. $\overset{\frown}{AD}$

7. STANDARDIZED TESTS In a circle graph about students, one sector represents 55%. What is the degree measure of the central angle of this sector?

A. 27.5° **B.** 55° **C.** 162° **D.** 198°

8. WRITING MATHEMATICS Explain why a chord that bisects another chord in the same circle is not necessarily a diameter.

In the figure, m∠AOB = m∠EOF. Find the length of each segment.

9. $\overline{AC}$

10. $\overline{AB}$

11. $\overline{CD}$

12. $\overline{EF}$

13. The diameter of $\odot O$ is 30 in., and the length of chord AB in $\odot O$ is 14 in. What is the distance from $\overline{AB}$ to the center of $\odot O$ to the nearest tenth?

14. Point P is on $\odot R$, point M is on $\overrightarrow{PM}$, $RP = 7$, $PM = 24$, and $MR = 26$. Is $\overline{PM}$ tangent to $\odot R$? Explain.

Find the measure of each arc or angle in $\odot O$.

15. $\overset{\frown}{RU}$

16. $\angle RTS$

17. $\overset{\frown}{TRU}$

18. $\angle TSU$

19. Use a compass and straightedge to find the center of a circle. Explain the procedure you used.

Find the measure of each arc or angle in $\odot O$.

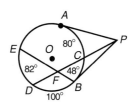

20. $\overset{\frown}{AE}$

21. $\overset{\frown}{AD}$

22. $\angle APD$

23. $\angle PBE$

24. Prove that, given regular octagon $ABCDEFGH$ inscribed in $\odot O$, m∠ADB is 22.5°.

Find x in each.

25.

26.

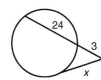

27. Make a truth table for the statement *If not p, then q.*

28. FORESTRY The radius of Earth is about 3813 mi. How many feet above Earth's surface must a watchtower be so you can see 10 mi to the horizon from its top?

PERFORMANCE ASSESSMENT

TRAPPED FAIR AND SQUARE You are going to inscribe a square in a circle and any isosceles trapezoid in a congruent circle. Shade in the areas outside of these quadrilaterals but inside the circle. Using diagrams show which shape can create the larger shaded area.

CUSTOMIZE YOUR CIRCLE Write an equation in standard form of a circle whose center is not $(0, 0)$. Graph your circle. Select four points on the circle, one point in the interior of the circle, and one point in the exterior of the circle. Label the center G. Label the points on the circle A–D. Label the exterior point E and the interior point F.

Use your equation and your selected points to write equations in slope-intercept form of lines that satisfy the following criteria.

a) a tangent at any point A–D

b) a secant through any two points A–D

c) a tangent and a secant that intersect at point E

d) two secants that intersect at point F

e) two secants that do not intersect

USING TANGENT SEGMENTS Circle O is inscribed in a quadrilateral whose consecutive sides have lengths a, b, c, and d. Explain why $a + c = b + d$.

ALL AROUND THE GEOBOARD Use a circle geoboard like the one shown below to illustrate the vocabulary terms in this chapter. Make a list of the terms that you can model using the circle geoboard. Make a list of the terms that you are unable to model using the circle geoboard. Explain the limitations of the geoboard that prevent you from doing the model.

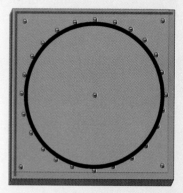

Review the theorems and corollaries in the chapter. Try to illustrate all the theorems that use the vocabulary terms you successfully modeled.

PROJECT ASSESSMENT

PROJECT Connection   Organize a presentation day, followed by time to play some of the games.

1. Each group presents a summary of each of the three games researched in Project Connections. Include in the presentation the games' origin and the cultures or countries that primarily play the games.

2. Show your sketch of the game whose design is based on circles.

3. Demonstrate how to play the game whose winning strategy involves circles.

4. Explain the rules and how to play the outdoor game you researched.

5. Discuss as a class the origin of the game and those who play the game. Determine if this game is played in other cultures by a different name. Talk about what makes a game fun, popular among many ages, and able to stand the test of time.

6. After the last group has presented their games, display all games. Take some time to play the other groups' games.

Fill in the blank.

1. A line that intersects a circle in two points is called a(n) __?__.

2. An angle whose vertex is on a circle and whose sides pass through two other points of the circle is called a(n) __?__ angle.

3. The __?__ of an acute angle in a right triangle is the ratio of the length of the side adjacent to the angle to the length of the hypotenuse.

4. A triangle in which all the sides are congruent is called a(n) __?__ triangle.

Find x in each figure below.

5. $m \overset{\frown}{AB} = 112°$

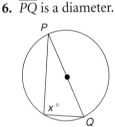

6. $\overline{PQ}$ is a diameter.

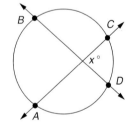

7. $m \overset{\frown}{AC} = 84°$ and $m \overset{\frown}{BC} = 191°$

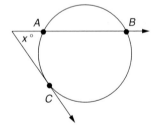

8. $m \overset{\frown}{AB} = 103°$ and $m \overset{\frown}{CD} = 25°$

Write the equation of each line described in standard form.

9. passes through $(4, 3)$ and $(-1, -3)$

10. passes through $(-2, 7)$ and parallel to the y-axis

11. passes through $(1, -2)$ and perpendicular to the line $2x - 3y = 9$

12. **STANDARDIZED TEST** Which of the following pairs of triangles are not necessarily congruent?

A.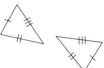

B.

C.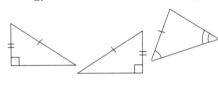

D.

E. All are congruent.

In the target shown, the bullseye has a radius of 3 inches, and each concentric circle has a radius that is 3 inches more than the next smaller circle. An archer shoots an arrow that hits the target. Determine the probability of each event.

13. hitting the bullseye

14. hitting the shaded area

15. hitting one of the outer two circles

16. **WRITING MATHEMATICS** Describe the relationship between a secant, a chord, and a diameter.

17. Find the image of the segment shown under a glide reflection in the line $y = 2$ and along the translation vector $v(1, -3)$.

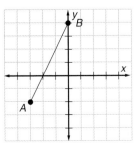

QUANTITATIVE COMPARISON In each question, compare the quantity in Column 1 with the quantity in Column 2. Select the letter of the correct answer from these choices:

A. The quantity in Column 1 is greater.
B. The quantity in Column 2 is greater.
C. The two quantities are equal.
D. The relationship cannot be determined by the information given.

Notes: In some questions, information which refers to one or both columns is centered over both columns. A symbol used in both columns has the same meaning in each column. All variables represent real numbers. Most figures are not drawn to scale.

Column 1	Column 2
1. the radius of a circle with a circumference of 31.4 units	the radius of a circle with an area of 314 square units

2. Point Q is between points P and R

PQ	QR

3.

$a < b$

$\sin x$	$\cos x$

4. the average of the measures of the three angles in an equilateral triangle | the average of the measures of the three angles in an obtuse triangle

5.

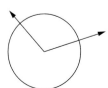

the measure of the central angle | the measure of the intercepted arc

Column 1	Column 2
6. area of	area of

 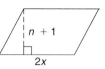

7. the angle formed by the endpoint of a radius and a line tangent to the circle at that endpoint | an angle that has its vertex on the circle and whose rays pass through the endpoints of a diameter

8. $a > b$

x	y

9. the perimeter of a rectangle with length x and width y | the perimeter of a non rectangular parallelogram with base x and height y

10.

the length of the hypotenuse if $y = 30°$ | the length of the hypotenuse if $y = 45°$

11. $a < x, b < y$

the distance from the origin to the point (a, b) | the distance from the origin to the point (x, y)

12. the measure of the angle formed by two vectors that have a dot product of 0 | the measure of the angles formed by two lines whose slopes have a product of -1

Take a Look
AHEAD

Make notes about things that look new.
- Which three-dimensional figures are grouped with prisms? Which are grouped with pyramids?
- List the three-dimensional figures that look new to you.

Make notes about things that look familiar.
- What three-dimensional figures do you recognize? What are their names?
- Area is a two-dimensional measure. How do you think area relates to a three-dimensional figure?
- Explain how volume is different from weight.

DATA Activity

SKILL FOCUS

Paper or Plastic?

Most containers used to ship, shelve, or sell items are made of either paper or plastic. The demand for paper and plastic packaging has soared over the last twenty years. There are three types of packaging that account for the majority of paper and plastic use: rigid packaging, flexible packaging, and food-container packaging. The table shows the demand of packaging since 1980.

- ▶ Construct a bar graph.
- ▶ Analyze trends and make predictions based on a set of data.
- ▶ Find the percent of a number.
- ▶ Determine percent of increase.

Packaging

In this chapter, you will see how:

- **PACKAGING ENGINEERS** use surface area to minimize packaging costs.
  (Lesson 13.4, page 743)

- **PACKAGING DESIGNERS** use volume to create protective packaging.
  (Lesson 13.6, page 755)

- **PACKING STORE OWNERS** use girth to calculate maximum shipping sizes.
  (Lesson 13.7, page 762)

PACKAGING DEMAND FOR PAPER/PAPERBOARD AND PLASTICS		
Year	**Paper and Paperboard** (billions of pounds)	**Plastics** (billions of pounds)
1980	58.13	4.2
1991	72.99	7.61
1996 (est.)	81.91	9.06

1. Construct a bar graph to illustrate the data.

2. What trends do the table and the bar graph show about the data?

A 1993 report entitled "Plastics Versus Paper to 1996" predicted that the demand in the U.S. for plastic packaging would increase by 3.5% per year while demand for paper and paperboard packaging would increase 2.3% per year.

3. Use these rates to verify the estimated demand for 1996.

4. At these rates, what should be the demand for these products in the year 2000?

5. **WORKING TOGETHER** The *biodegradability* of a particular material is its ability to decompose under natural conditions.

Research this topic as it relates to both paper and plastic. Is one of these substances more biodegradable than the other? Can chemical changes be made to either substance to affect its biodegradability?

Find out how, and at what cost, the industry is making packaging more ecologically safe. How does the manufacture of plastic differ from the manufacture of paper? Is the use of one of these substances more ecologically preferable?

Ship Shape

A *package* is a way of getting a product from one place to another in a condition that is acceptable for its intended use. Most all purchased goods are in some form of a package. Often, it is the packaging that promotes the purchase of the product.

The packaging industry seeks to fulfill the needs of producers, manufacturers, retailers, and consumers. Packaging designers and engineers look for creative and innovative package treatments. Packages come in all shapes and sizes and in a variety of media, such as paper, wood, glass, plastic, and aluminum.

In this project, you will examine the role that packaging plays in our daily lives.

PROJECT GOAL

To examine the geometric properties of the shapes of packages and containers.

Getting Started

Work in a group.

1. Contact a company that produces packaged products or is directly involved in moving, shipping, or cargo transport. Ask for information about the containers that are used. Obtain specific data about measurements, shapes, and packaging materials used.

2. Obtain additional information. Browse the Internet or the library for material related to the packaging industry.

3. Prepare two lists of packaged products, one of products that are sold by weight and the other of those that are sold by volume. These lists will be used in the first Project Connection.

PROJECT *Connections*

Lesson 13.4, page 742:
Compare packaging of items sold by weight to those of items sold by volume.

Lesson 13.6, page 754:
Investigate the factors that determine moving costs. Plan the efficient packing of a truck used for moving.

Lesson 13.9, page 775:
Investigate shipping costs.

Chapter Assessment, page 779:
Prepare a presentation summarizing your findings about variations in packaging materials and shapes, and shipping costs.

Internet Connection

www.learninggeometry.com

13.1 Geometry Workshop
Platonic Solids and Euler's Formula

Think Back

The figures below show three patterns of regular polygons.

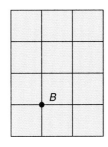
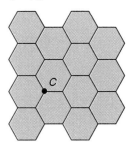

SPOTLIGHT ON LEARNING

WHAT? In this lesson you will learn
- to identify polyhedrons and regular polyhedrons.
- to apply Euler's Formula.

Why? Examining the characteristics of polyhedrons can help you understand and utilize geometric models for a variety of everyday objects.

1. Explain why each pattern is a *regular tessellation*.

2. What is the sum of the measures of the angles with vertices at point *A*? point *B*? point *C*? Justify your answers.

3. Explain why it is not possible to make a regular tessellation of regular pentagons.

Explore

Work in a group with two or three other students. You will need a supply of flexible straws and tape.

4. To join any two straws, you pinch the end closest to the bendable section of one straw and slide it inside the opposite end of the other straw, as shown below. Using this method, join twenty-four straws to model six squares.

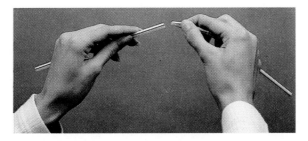

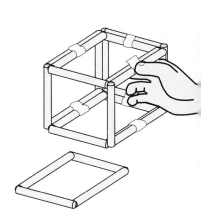

5. To join any two straw squares, you align a side of one with a side of the other and tape the sides together at their midpoints. Using this method, tape the six squares together to model a cube.

Geometry Workshop

GEOMETRY: WHO, WHERE, WHEN

The Platonic solids are named after the Greek philosopher Plato, who lived in the fourth century, B.C. He and his followers wrote extensively about the solids, making the following associations with the fundamental "elements" of the universe.

tetrahedron: fire
cube: earth
octahedron: air
icosahedron: water

They believed the dodecahedron represented the entire universe.

The five figures shown below are called the *Platonic solids*. All the faces of each solid are bounded by congruent regular polygons.

regular hexahedron (cube)   regular tetrahedron   regular octahedron

regular dodecahedron   regular icosahedron

6. In Questions 4 and 5, you made a model of a cube. Using the same method, make models of the other four Platonic solids.

7. How are the Platonic solids like regular tessellations? How are they different?

8. The figure at the right shows a solid of a different type. It is called a *snub cube*. How does this figure appear to be like a Platonic solid? How does it appear to be different?

Make Connections

A **polyhedron** is a three-dimensional figure formed by flat surfaces that are bounded by polygons and joined in pairs along their sides. These surfaces together completely enclose a region of space, which is called the interior of the polyhedron.

COMMUNICATING ABOUT GEOMETRY

Explain why cylinders, cones, and spheres are *not* polyhedrons.

Each of the flat surfaces of a polyhedron is called a **face**. A segment at which two faces intersect is an **edge**. A point at which three or more edges intersect is a **vertex**.

The prisms and pyramids you have studied in previous chapters are all examples of polyhedrons.

A polyhedron is **convex** if no plane that contains a face also contains a point in its interior. Otherwise, the polyhedron is **concave**.

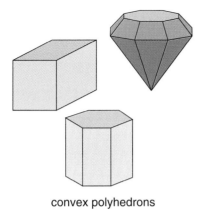

convex polyhedrons   concave polyhedrons

9. WRITING MATHEMATICS Describe the difference in physical appearance between a convex and a concave polyhedron.

A **regular polyhedron** is a convex polyhedron in which the faces are bounded by congruent regular polygons and the same number of edges intersect at each vertex. Although there are infinitely many types of regular *polygons*, there are exactly five regular polyhedrons. These are the **Platonic solids**, which you modeled in Explore.

10. Examine your models of the regular tetrahedron, regular octahedron, and regular icosahedron. How many equilateral triangles intersect at each vertex of each polyhedron?

11. Is it possible to form a polyhedron in which fewer equilateral triangles intersect at each vertex? more? Explain.

12. Examine your model of a cube. How many squares intersect at each vertex?

13. Is it possible to form a polyhedron in which fewer squares intersect at each vertex? more? Explain.

14. Examine your model of the regular dodecahedron. How many regular pentagons intersect at each vertex?

15. Is it possible to form a polyhedron in which fewer regular pentagons intersect at each vertex? more? Explain.

16. Is it possible to form a polyhedron in which three regular hexagons intersect at each vertex? Explain.

17. Explain why it is impossible to form a polyhedron whose faces are bounded by regular polygons of six or more sides.

GEOMETRY: WHO, WHERE, WHEN

Just as you can circumscribe a circle about a regular polygon, you can circumscribe a sphere about a regular polyhedron. In the tenth century A.D., the Arab mathematician Abû'l-Wefâ produced a geometric study in which he showed how to locate the vertices of the regular polyhedrons on their circumscribed spheres.

THINK BACK

The measure of each angle of a regular n-gon is $\dfrac{(n-2)180°}{n}$.

Examine your models further. Copy and complete this table.

Polyhedron	Type of Faces	Number of Faces (F)	Number of Edges (E)	Number of Vertices (V)
18. regular tetrahedron	equilateral triangles	▦	▦	▦
19. cube	▦	▦	▦	▦
20 regular octahedron	▦	▦	▦	▦
21. regular dodecahedron	▦	▦	▦	▦
22. regular icosahedron	▦	▦	▦	▦

23. Make a conjecture about the relationship between F, E, and V. Describe the relationship in your own words.

The following relationship called Euler's Formula can be proved true.

> **EULER'S FORMULA**
>
> If F is the number of faces, E is the number of edges, and V is the number of vertices of a convex polyhedron, then $F - E + V = 2$.

GEOMETRY: WHO, WHERE, WHEN

There is some evidence that the relationship of F, E, and V was known to the great Greek mathematician Archimedes as early as 225 B.C. However, it was Swiss mathematician Leonhard Euler who, in 1751, became the first to prove it true.

Summarize

24. **WRITING MATHEMATICS** Explain how you can extend the concept of a tessellation to demonstrate that there are exactly five regular polyhedrons.

25. **THINKING CRITICALLY** Write expressions that represent the number of faces, number of edges, and number of vertices of a prism with bases bounded by n-gons. Show that Euler's Formula applies to the prism.

26. **THINKING CRITICALLY** Repeat Question 25 for a pyramid whose base is bounded by an n-gon.

27. **GOING FURTHER** Using string as shown at the left, find the center of each face on your model of a cube. Connect the centers with string. Which Platonic solid do you see?

28. **GOING FURTHER** Two Platonic solids that can be related by the method shown in Question 27 are considered *dual*. Identify all other duals among the Platonic solids.

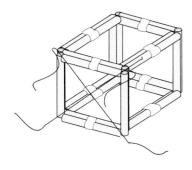

13.2 Focus on Reasoning
Using Venn Diagrams

Discussion

- Work with a partner.

After studying many different three-dimensional figures, Denise, Debbie, and Charles decided to find a classification scheme for the five Platonic solids. Debbie's first effort was to show all five solids as different from one another. She used the *Venn diagram* above. In a Venn diagram, you picture the *universal set* of objects as a rectangle and a subset of objects as a closed region.

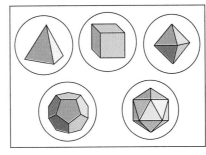

SPOTLIGHT ON LEARNING

WHAT? In this lesson you will learn
- to construct a Venn diagram to show how groups of objects relate to one another.

WHY? Venn diagrams can help you solve problems in research and sampling.

1. According to Debbie's diagram, is it possible for a tetrahedron to be in the same set as the octahedron?

2. Charles argued that Debbie's diagram would not work well as a memory device to remember the characteristics of each of the solids. Do you agree? Explain your response.

3. Denise noted that the tetrahedron and the octahedron are similar and yet different. How are the tetrahedron and the octahedron similar and how are they different?

Refer to the Venn diagram at the right.

4. The diagram shows three distinct and separate groupings. On what basis was the diagram constructed?

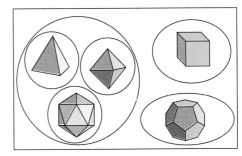

5. Explain how you know that the three Platonic solids grouped together in the largest circular region belong together and how you know the three smaller circular regions inside the largest one do not intersect.

6. Looking at the diagram, Charles concluded that he could remember an octahedron as a Platonic solid made up of eight triangular regions. How could you use the diagram to remember the characteristics of a cube? a tetrahedron?

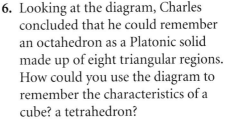

Build a Case

● The diagram shows how five sets of numbers in the real number system relate to one another.

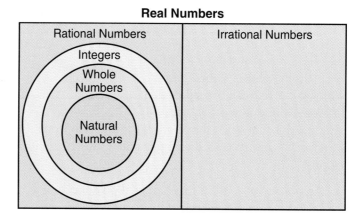

Real Numbers

7. How do you know the region representing irrational numbers does not intersect the region representing the rational numbers?

8. How do you know the region representing the natural numbers is contained in the regions representing the whole numbers, integers, and rational numbers?

9. The number -3 is an integer. Given the Venn diagram, what other descriptions of -3 can you state?

Denise, Debbie, and Charles decided to represent the eight solids shown at the left by using a Venn diagram. They began by sketching the Venn diagram at the right.

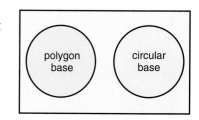

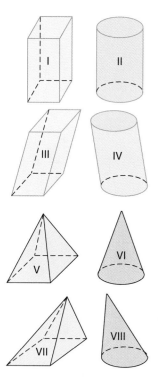

10. What four solids will the left circle be used to represent?

11. What four solids will the right circle represent?

12. Choose how to classify the four solids you listed in Question 10. Inside the left circle, represent them by two circles that do not intersect.

13. Choose how to classify the four solids you listed in Question 11. Inside the right circle, represent them by two circles that do not intersect.

In Questions 10–13, you classified the solids by distinguishing one type of base from another. You can classify the solids in other ways.

14. Make a new Venn diagram to classify the solids by starting with a separation of right solids from oblique ones.

15. Make a third Venn diagram to classify the solids by first separating prism from cylinder, cylinder from pyramid, and pyramid from cone.

16. Which Venn diagram helps you best remember how the solids are similar and how they are different?

EXTEND AND DEFEND

Debbie and Charles realized they could represent membership in two sets simultaneously, **intersection**, and membership in one set or another set, **union**, using Venn diagrams.

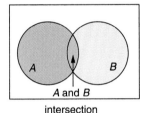

intersection

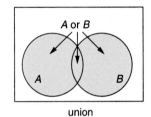

union

Make a Venn diagram to show how the sets are related.

17. Set *A*: multiples of 2   Set *B*: multiples of 5

18. Set *A*: isosceles triangles   Set *B*: right triangles

19. Set *A*: kites   Set *B*: rhombuses   Set *C*: squares

20. Set *A*: regular octagons   Set *B*: concave octagons   Set *C*: convex octagons

21. Suppose *A* is the set of quadrilaterals whose sides are congruent and *B* is the set of quadrilaterals whose angles are congruent. What can you say about the quadrilaterals that are in both set *A* and set *B*? set *A* but not set *B*? set *B* but not set *A*?

The *complement* of a set *A* is the set of all members of the universal set that are not in set *A*. If you have two sets *A* and *B*, the members of *A* not in *B* is the *relative complement* of *B* in *A*. The members of *B* not in *A* is the relative complement of *A* in *B*.

Make a Venn diagram to show how sets *A* and *B* are related. Describe the relative complement of *B* in *A* and of *A* in *B*.

22. Set *A*: rectangles   Set *B*: rhombuses

23. Set *A*: isosceles triangles   Set *B*: equilateral triangles

24. Match each number with the letter that represents the corresponding part of the Venn diagram.

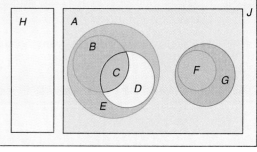

1. isosceles trapezoid
2. triangle
3. parallelogram
4. rectangle
5. quadrilateral
6. rhombus
7. square
8. trapezoid
9. polygon

CLUB MEMBERSHIP The table shows the results of a recent survey of 90 students on club membership. The Venn diagram is the beginning of one representation of club membership.

Club	Membership
Science	45
Music	44
Boosters	43
Science and Music	19
Science and Boosters	17
Music and Boosters	15
All three clubs	9

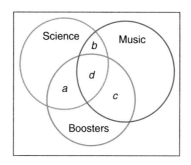

$$\begin{cases} b + d = 19 \\ a + d = 17 \\ c + d = 15 \\ d = 9 \end{cases}$$

25. Explain how you know from the data in the table and the Venn diagram that the four equations at the right above represent multiple club membership.

26. Use the four equations to find a, b, c, and d. What can you conclude about club membership from a? from b? from c?

27. How many students in the survey belong to the Science Club but no other club?

28. Make a Venn diagram that shows all the membership possibilities for the 90 students.

Now You See It
Now You Don't

PLATONIC PREDICAMENT

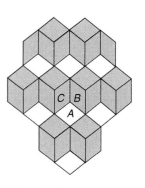

1. How many cubes are there? Are there seven cubes? or eight cubes?

2. Is A the top of a cube, or is it the bottom of the cube having C and B as faces? Can you see it both ways?

Explore/Working Together

● Pictured below are plans for a set of gift boxes. Boxes I and II are shaped like regular prisms. The shape of box III is a right cylinder. In all three boxes, the 4-in. segment is a radius of the base.

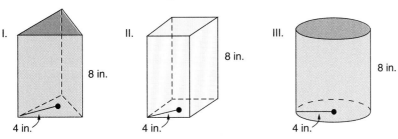

I. 8 in. 4 in.

II. 8 in. 4 in.

III. 8 in. 4 in.

Work with a partner. Suppose you must design a decorative label to cover all of the top of each box.

1. How would these labels be alike? How would they be different?

2. Calculate the area of each label.

Suppose you must design a decorative label to wrap around the sides of each box.

3. How would these labels be alike? How would they be different?

4. Calculate the area of each label.

Build Understanding

● Throughout this book, you have worked with three-dimensional figures that completely enclose, or *bound*, a region of space. The boundary of this type of figure is called its **surface**. The area of the surface is called the **surface area** *SA* of the figure.

When finding a surface area, a useful strategy is drawing a *net*. Calculate the area of each surface. Then add the areas to find the surface area of the figure.

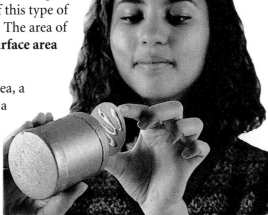

SPOTLIGHT ON LEARNING

WHAT? In this lesson you will learn
● to find surface areas of right prisms and right cylinders.

WHY? Knowing about surface areas of right prisms and right cylinders will help you solve problems related to industrial arts, graphic design, and manufacturing.

THINK BACK

A *net* is a two-dimensional figure that, when folded, forms the surface of a three-dimensional figure.

EXAMPLE 1

Find the surface area of the right triangular prism.

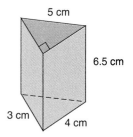

Solution

The faces of the prism consist of two triangular bases and three rectangular lateral faces. Draw a net like the one below. Use the net to find the areas of all the faces.

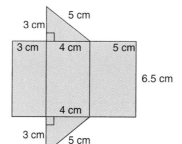

area of the two bases

$$= 2\left[\frac{1}{2}(4)(3)\right]$$
$$= 12$$

area of the three lateral faces

$$= 6.5(3) + 6.5(4) + 6.5(5)$$
$$= 78$$

surface area = area of bases + area of lateral faces

$$SA = 12 + 78$$
$$SA = 90$$

The surface area of the prism is 90 cm^2. ◄

The sum of the areas of the lateral faces of a prism is called the **lateral area** of the prism. When the prism is a right prism, as in Example 1, you can think of the lateral faces joined together into one large rectangular region. The width of this region is the height of the prism. Its length is the perimeter of a base. This leads to a theorem about lateral area.

THEOREM 13.1	**LATERAL AREA OF A RIGHT PRISM**

The lateral area LA of a right prism equals the product of the perimeter p of a base and the height h of the prism, or $LA = ph$.

Since the two bases of a right prism are congruent, you can find the area of both bases by doubling the area of one. This leads to the following theorem about surface area.

THEOREM 13.2	**SURFACE AREA OF A RIGHT PRISM**

The surface area SA of a right prism equals the sum of its lateral area LA and twice the area of one of its bases B, or $SA = LA + 2B = ph + 2B$.

EXAMPLE 2

PACKAGING The cracker box shown is a regular prism. Find the amount of cardboard on its surface.

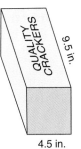

Solution

You need to find the surface area of the box. There are two square bases and four rectangular lateral faces.

The perimeter p of the base is 4(4.5) in. or 18 in.
The height h of the prism is 9.5 in.
The area of a base B is (4.5)(4.5) in.^{2}, or 20.25 in.^{2}

$$SA = ph + 2B$$
$$SA = (18)(9.5) + 2(20.25) = 211.5$$

The amount of cardboard on the surface is about 212 in.^{2} ◄

THINK BACK

A regular prism is a right prism whose bases are regular polygons.

The formulas for the lateral area and surface area of a right cylinder are closely related to the formulas for a right prism.

THEOREM 13.3 LATERAL AREA OF A RIGHT CYLINDER

The lateral area LA of a right cylinder with radius r equals the product of the circumference of a base and the height h of the cylinder, or $LA = 2\pi rh$.

THINK BACK

A cylinder has two circular bases that are congruent and parallel to each other.

The curved surface that connects the boundaries of the bases is the lateral surface of the cylinder.

In a right cylinder, any segment on the lateral surface is perpendicular to the bases.

THEOREM 13.4 SURFACE AREA OF A RIGHT CYLINDER

The surface area SA of a right cylinder with radius r equals the sum of its lateral area LA and the area of its bases, or $SA = LA + 2B = 2\pi rh + 2\pi r^2$.

EXAMPLE 3

Find the lateral area and surface area of the right cylinder shown.

12 m

6.5 m

Solution

The radius r of the cylinder is 6.5 m.
The height h of the cylinder is 12 m.
The area of the base B is $\pi(6.5)(6.5)$ m^2, or about 132.73 m^2.

$$LA = 2\pi rh \qquad\qquad SA = LA + 2B$$
$$LA = 2\pi(6.5)(12) \qquad SA \approx 490.09 + 2(132.73)$$
$$LA \approx 490.09 \qquad\qquad SA \approx 755.55$$

The lateral area of the cylinder is about 490 m^2.
The surface area is about 756 m^2. ◄

CHECK UNDERSTANDING

A net for the cylinder in Example 3 would look like this.

What are the dimensions of the circles? of the rectangle?

Refer to the regular prism shown at the right for Exercises 1–6.

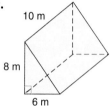

1. Sketch a net for the prism.

2. Describe the types of geometric figures that form the net.

3. Label the dimensions of each figure in the net.

4. Show how to use the net to find the lateral area and the surface area of the prism.

5. Show how to use the formulas you learned in this lesson to find the lateral area and surface area of the prism.

6. WRITING MATHEMATICS Compare your work in Exercises 4 and 5. Which method for finding the areas did you like better? Explain.

Find the lateral area and the surface area of each right prism or right cylinder. Assume that all quadrilaterals are rectangles. Round your answers to the nearest whole number.

7.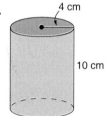
4 cm
10 cm

8.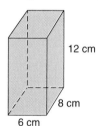
12 cm
8 cm
6 cm

9.
10 m
8 m
6 m

10.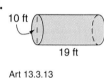
10 ft
19 ft

Art 13.3.13

11. FOOD PACKAGING Find the area of the label on a can of cranberry sauce that has a height of $4\frac{1}{8}$ in. and a radius of $1\frac{1}{2}$ in. Assume that the label completely surrounds the can, and that there is no overlap. Round your answer to the nearest whole number.

PRACTICE

The figure below is a net for a three-dimensional figure. You may assume that all angles that appear to be right angles are right angles.

12. What is the name of the three-dimensional figure that this net will form?

13. Describe the surface of the three-dimensional figure.

14. Show how to use the net to find the lateral area and the surface area of the figure.

15. Use the formulas lesson to find the lateral area and surface area of the figure.

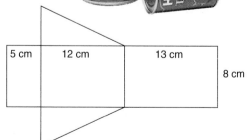

5 cm   12 cm   13 cm   8 cm

Find the lateral area and the surface area of each right prism or right cylinder. Assume that all quadrilaterals are rectangles. Round your answers to the nearest whole number.

16.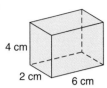

4 cm
2 cm
6 cm

17.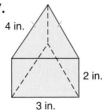

4 in.
2 in.
3 in.

18.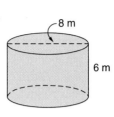

8 m
6 m

19.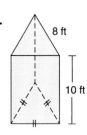

8 ft
10 ft

Find the lateral area and surface area of each right prism described.

20. The bases of the prism are equilateral triangles with sides of length 16 cm. The height of the prism is 15 cm.

21. The bases of the prism are rectangles with length 5 in. and width 8 in. The height of the prism is 22.5 in.

22. The bases of the prism are right triangles with hypotenuse of length 13 m and one leg of length 12 m. The height of the prism is 20 m.

23. WRITING MATHEMATICS Explain the difference between the lateral area and the surface area of a right prism or cylinder.

INDUSTRIAL ARTS In class, Zenobia and Maseo each made a wooden paperweight, with dimensions as shown.

Zenobia's Paperweight

Maseo's Paperweight

24. Zenobia is going to use 1-in. square tiles to decorate her paperweight by completely covering it, with no overlap. How many tiles will she need?

25. Maseo intends to use stamps that are each 1-in. square to completely cover his paperweight, with no overlap. He has collected 250 such stamps. Does he have enough stamps? Explain.

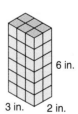

6 in.
3 in.   2 in.

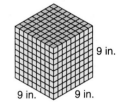

9 in.
9 in.   9 in.

EXTEND

GRAPHIC DESIGN A designer plans to make two different cardboard mobiles. For each mobile, the designer first makes a template on graph paper to show the construction of its faces. Use the description of each mobile to determine its surface area.

26. The first mobile will be in the shape of a cube. The template for one of the faces has coordinates $(0, 0)$, $(0, 90)$, $(90, 90)$, and $(90, 0)$.

27. The second mobile will be in the shape of a rectangular prism (all the faces of the prism are rectangles). The template for a base has coordinates $(0, 0)$, $(0, 60)$, $(100, 60)$, and $(100, 0)$. The template for one of the lateral surfaces has coordinates $(0, 0)$, $(0, 200)$, $(100, 0)$, and $(100, 200)$. The template for the other lateral surface has coordinates $(0, 0)$, $(60, 0)$, $(0, 200)$, and $(60, 200)$.

28. The height of a regular pentagonal prism is 10 in. The length of one side of a base is 4 in. Find the lateral area and surface area of the prism.

MANUFACTURING A company manufactures drums for the disposal of hazardous waste materials. Each drum is shaped like a right cylinder with a height of 90 cm and a diameter of 40 cm. Determine if it is possible to cut the three faces of the drum from a single rectangular sheet of metal with the given measurements. Explain your answers.

29. 30 cm by 500 cm   **30.** 95 cm by 150 cm

31. 95 cm by 200 cm

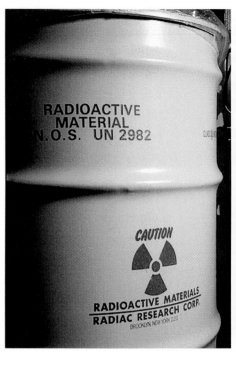

THINK CRITICALLY

32. A right cylinder and a cube both have the same height h. Write an expression for the radius r of the cylinder in terms of the height at which both lateral areas will be equal.

33. Is there a cube that has the same number of cubic inches in its volume as square inches in its surface area? Explain.

34. Sixty-four small cubes whose edges are 1 in. are stacked to form a large cube. A tunnel is then formed in the large cube by removing four small cubes in a vertical stack from the interior. Find the surface area of the resulting solid.

ALGEBRA AND GEOMETRY REVIEW

35. Given $\triangle ABC$
           D is the midpoint of $\overline{AC}$.
           $\angle ADB \cong \angle CDB$

    Prove $\triangle ADB \cong \triangle CDB$

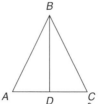

36. STANDARDIZED TESTS Any two medians of a(an) __?__ triangle are congruent.

  A. isosceles   **B.** equilateral   **C.** scalene   **D.** right

37. Is the following statement *true* or *false*? Explain.

     If $p \Rightarrow q$, then $q \Rightarrow p$.

38. In right triangle ABC, $\angle C$ is a right angle, $AC = 30$ in., and $AB = 35$ in. Find BC to the nearest hundredth of an inch.

39. Determine the slope of the line that passes through the points $(-3, 5)$ and $(8, 6)$.

40. Solve $-3x + 5 > 2x - 15$ and graph the solution.

41. Write $(h - 3)(h + 1)^2$ as a polynomial.

42. Simplify $\dfrac{1 + \sqrt{3}}{1 - \sqrt{3}}$.

13.4 Volumes of Prisms and Cylinders

Explore / Working Together

- Work with a partner. Decide how many cubes are in each stack. Write the steps that you followed to arrive at each answer.

1.
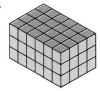

2.
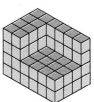

3.

4.
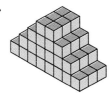

Build Understanding

- The number associated with each stack in Explore could be called the *volume* of the stack. The **volume** of any three-dimensional figure is the number of nonoverlapping cubic units contained in its interior.

  When calculating volume, some of the formulas you use are based on the following assumptions.

  VOLUME OF A CUBE POSTULATE

  POSTULATE 25 The volume V of a cube is the cube of the length e of one edge, or $V = e^3$.

  VOLUME CONGRUENCE POSTULATE

  POSTULATE 26 If two three-dimensional figures are congruent, then they have the same volume.

  VOLUME ADDITION POSTULATE

  POSTULATE 27 The volume of a three-dimensional figure is the sum of the volumes of all its nonoverlapping parts.

COMMUNICATING ABOUT GEOMETRY

The volume control of a radio, television, or other electronic device allows you to regulate the loudness of the sound you hear. How do you think this meaning of *volume* is related to the geometric meaning of the word?

PACKAGING A small, open-top box is shaped like a rectangular prism. The width of the box is 4 in., its length is 6 in., and its height is 5 in. What is its volume?

Solution

The dimensions of the box are given in inches. Imagine that you are going to pack the box with cubes whose edges are 1 in. long.

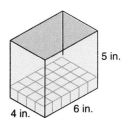

To cover the bottom of the box completely, you need a layer of $(4 \cdot 6)$ cubes, or 24 cubes.

To fill the box completely, you need five layers. This would be a total of $(5 \cdot 24)$ cubes, or 120 cubes.

So the volume of the box is 120 in.^{3}

In Example 1, you might have recognized a familiar formula for the volume V of a rectangular prism with length ℓ, width w, and height h.

$$V = \ell w h$$

The same reasoning that leads to this formula for rectangular prisms can be used to justify the following theorem for all prisms.

THEOREM 13.5   VOLUME OF A PRISM

The volume V of a prism equals the product of the area of a base B and the height h, or $V = Bh$.

EXAMPLE 2

Find the volume of the right prism shown at the right.

Solution

Each base of the prism is a triangle with area $\frac{1}{2}(18)(9)$ ft, or 81 ft^2.

So the base B of the prism is 81 ft^2.

The height h of the prism is $13\frac{1}{2}$ ft.

Use the volume formula for a prism.

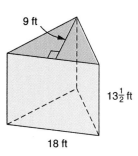

$$V = Bh = (81)\left(13\frac{1}{2}\right) = 1093\frac{1}{2}$$

The volume of the prism is $1093\frac{1}{2}$ ft^3.

As the number of sides of a regular polygon increases, it looks more and more like a circle. As the number of sides of the bases increases, a regular prism looks more and more like a right cylinder.

| THEOREM 13.6 | VOLUME OF A RIGHT CYLINDER |

The volume *V* of a right cylinder with radius *r* equals the product of the area of a base *B* and the height *h*, or
$$V = Bh = \pi r^2 h.$$

EXAMPLE 3

HEATING EQUIPMENT The figure shows a tank used to store home heating oil. Its shape is a right cylinder. What is its volume?

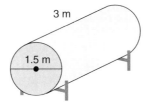

Solution
Each base of the cylinder is a circle with diameter 1.5 m.

The radius *r* of the cylinder is 0.75m.
The height *h* of the cylinder is 3 cm.

$$V = \pi r^2 h = \pi (0.75)^2 (3) \approx 5.3$$

The volume of the tank is about 5.3 m³. ◄

An important concept called *Cavalieri's Principle* guarantees that the volume formulas apply to all prisms and cylinders, not just right prisms and right cylinders. You can examine Cavalieri's Principle in depth in the Geometry Excursion on page 768.

TRY THESE

Find the volume of each right prism or right cylinder. Assume that all quadrilaterals are rectangles. Round your answers to the nearest whole number.

1.
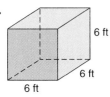

2.
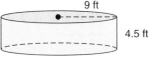

3.
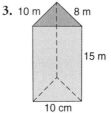

4.
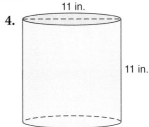

5. **GLASSWARE DESIGN** A water glass has the shape of a right cylinder. Its inside diameter is 6 cm, and its height is 13 cm. Find its volume to the nearest cubic centimeter.

The figures below are a net for a right cylinder and a rectangular prism. Find the volume of the figure formed when each net is folded.

6.

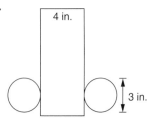
4 in.

3 in.

7.

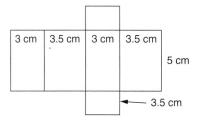

| 3 cm | 3.5 cm | 3 cm | 3.5 cm |

5 cm

3.5 cm

Find the volume and the surface area of each stack of cubes. The length of an edge of each cube is 1 unit.

8.

9.

10.

11.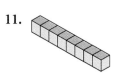

12. **WRITING MATHEMATICS** Explain the difference between the surface area and the volume of a three-dimensional figure.

PRACTICE

Find the volume of each right prism or right cylinder. You may assume that all quadrilaterals which appear to be rectangles are rectangles. Round your answers to the nearest whole number.

13.
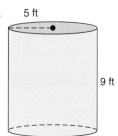
5 ft

9 ft

14.
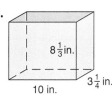
$8\frac{1}{3}$ in.

10 in.

$3\frac{1}{4}$ in.

15.
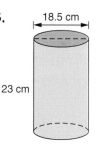
18.5 cm

23 cm

16.
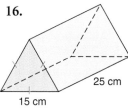
25 cm

15 cm

17.
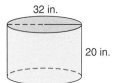
32 in.

20 in.

18.
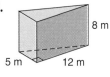
8 m

5 m   12 m

19.
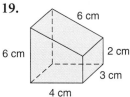
6 cm

6 cm

2 cm

3 cm

4 cm

20.
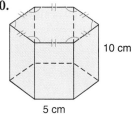
10 cm

5 cm

21. The area of the base of a right prism is 25.5 ft^2. The height of the prism is 8 ft. Find the volume.

22. ARCHITECTURAL DESIGN A 19th-century mansion once had two marble pillars in front of it, each in the shape of a regular octagonal prism. The measure of a side of each octagon was 1.5 ft and the height of each pillar was 25 ft. A restorations expert wishes to replace the pillars. To the nearest cubic foot, how much marble is needed for the two new pillars?

23. WRITING MATHEMATICS Write a problem that you can solve by finding the volume of a right prism. Write a different problem that you can solve by finding the surface area of a right prism. Show how to solve the problems you wrote.

EXTEND

24. CONSTRUCTION The diagram at the right shows a section of a concrete drainage pipe, which is in the shape of a hollowed-out right cylinder. To the nearest cubic centimeter, how much concrete was used to make this section of the pipe?

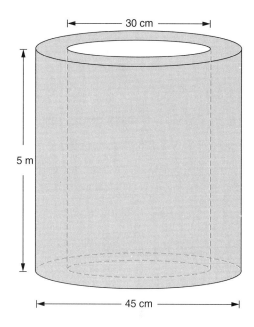

HOME IMPROVEMENT A plumber is installing a new hot water heater in the Kiernan's basement. The heater is in the shape of a right cylinder with height of 4 ft. The vertical interior walls of the heater are coated with a layer of low-density polyurethane insulation that is 1.5 in. thick. Including the insulation, the outside diameter of the cylinder is 25 in.

25. What is the inside diameter of the water heater?

26. When filled to capacity, what is the volume of the water heater to the nearest cubic inch?

27. What is the volume of the polyurethane insulation to the nearest cubic inch?

28. A cylinder and a cube each have a height of 25 cm. For what radius will the volume of the cylinder equal the volume of the cube? Answer to the nearest tenth of a centimeter.

29. Find the volume of a cube whose surface area is 37.5 m^2.

30. STEEL MANUFACTURING Steel weighs about 0.283 lb/in.^{3} Find the weight of the steel I-beam shown below. Assume that all angles are right angles. Round to the nearest pound.

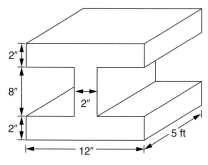

THINK CRITICALLY

31. The height of a right cylinder is a, and its diameter is b. Is this cylinder equal in volume to another right cylinder whose height is b and whose diameter is a? Explain.

32. MODELING Explain how you can use Algeblocks to model the product $(x + 1)^3$. Sketch your model.

33. LOGIC Determine whether the following statement and its converse are true or false. Justify your answer.
    If two three-dimensional figures are equal in volume, then they are equal in surface area.

34. LOOKING AHEAD What is the effect on the volume and lateral area of a right cylinder if its height is doubled? if its radius is doubled?

ALGEBRA AND GEOMETRY REVIEW

35. The measure of an angle is 6° more than five times its supplement. Find the angle.

36. What is the sum of the interior angles of a 13-sided figure?

37. Evaluate the determinant. $\begin{vmatrix} 5 & -2 \\ -3 & 6 \end{vmatrix}$

38. Write 0.00000000467 in scientific notation.

39. STANDARDIZED TESTS Which of the equations represents the line that is perpendicular to $y = -3x + 6$ at the point where $y = -3x + 6$ intersects the y-axis?

    A. $y = -\frac{1}{3}x + 6$ **B.** $y = -\frac{1}{3}x - 6$ **C.** $y = \frac{1}{3}x + 6$ **D.** $y = \frac{1}{3}x - 6$

PROJECT Connection

Work with the two lists that your group has prepared. The "volume list" should have items such as milk and bleach. The "weight list" should have items such as cereal and bread crumbs.

1. Gather 10 empty containers from products on each list.

2. Are there any common packaging shapes or materials used for the "volume products"? for the "weight products? Explain your findings. Write a summary.

3. Select 3 packages from each group of products. How does the packaging help to deliver the product to the consumer in the way that it was originally intended for use? How does the packaging preserve and protect the product during shipping, on the store shelf, and in your home? Write a summary.

4. Visually estimate the volume of all 20 packages. Arrange in order from smallest volume to largest volume.

5. From a science lab, obtain a device that can measure volume. Use sand or pebbles to get an accurate volume measurement of each container. Compare your results with your visual estimates.

A *packaging engineer* uses plastics, metals, glass, paper, and other materials to create containers for a variety of products. Presented with packaging problems from manufacturers, producers, designers, and consumers, the packaging engineer must understand how to create a package that is good for the environment as well as for the product.

Recently, many products have been introduced to the market in smaller, more compact forms. Concentrates of products such as juice and liquid detergent result in cost-saving packaging.

Decision Making

The size of a package is not always an accurate measure of the amount inside. Typically, cereal is sold by weight and not volume.

Consider a cereal box with the following dimensions.

length: 8.5 in.　　width: 2.75 in.　　height: 12 in.

1. What is the volume of the box? What is the surface area?

2. Copy and complete the chart. Read the instructions below.

length (in.)	width (in.)	height (in.)	surface area (in.²)	volume (in.³)
8.5	2.75	12		
9.5	2.75	12		
8.5	2.75	11		

The first line of the chart uses the given dimensions of the cereal box. Complete the first line with your results from Question 1.

For the second line of the chart, increase the original length by 1 in. Keep the original width and height.

For the third line of the chart, decrease the original height by 1 in. Keep the original length and width.

Like a packaging engineer, you are interested in minimizing the amount of cardboard used but maintaining the same volume of the cereal box.

3. Continue your chart. Keep the volume constant. Try different values for the length, width, and height. Calculate the surface areas.

4. Using the results from your chart, describe the *ideal* cereal box.

Problem Solving File

Maximizing Volume

In many problems, you need to find the greatest possible volume of a container under given conditions. Often those conditions involve the dimensions of the material from which the container is made.

Problem

A group of students at Rose High School want to build planters shaped like open-top rectangular prisms. For each planter, they will work with a sheet of metal 15 in. long and 8 in. wide, and they will use the pattern below. To form the box, they will cut along the solid segments and fold along the dashed segments. What value of x will make the planter with the maximum volume?

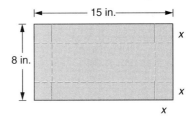
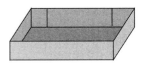

Explore the Problem

Work with a partner.

1. Write variable expressions in x that represent the length, width, and height of the planter.

2. Use your answers to Question 1 to write an expression that represents the volume V of the planter.

3. Explain why it must be true that $0 < x < 4$.

4. The spreadsheet below lists possible values of x in *increments* of 0.5. Copy and complete the spreadsheet.

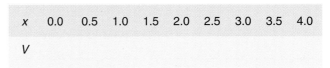

x	0.0	0.5	1.0	1.5	2.0	2.5	3.0	3.5	4.0
V									

5. For which value of x is the value of V the greatest?

6. Make a new spreadsheet listing values of x less than and greater than the number you identified in Question 5. Use increments of 0.05. For which value of x is the value of V the greatest?

7. How does changing the length of the cut affect the size of the planter?

8. Make a third spreadsheet that will give a better approximation for the value of x that maximizes the volume. For which value of x in this spreadsheet is the value of V the greatest?

9. Find the length, width, height, and volume of the planter with the maximum volume. Round all answers to the nearest tenth.

10. One student solved the problem by using a graphing utility to graph

$$y = x(8 - 2x)(15 - 2x)$$

The student then used the tracing feature to find the value of x that maximizes this function. Is this solution different from the solution obtained in Questions 1–8? Explain.

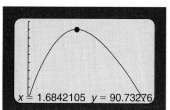

$x = 1.6842105 \quad y = 90.73276$

Xmin = 0	Ymin = 0
Xmax = 4	Ymax = 100
Xscl = 1	Yscl = 10

> **PROBLEM SOLVING PLAN**
>
> - Understand
> - Plan
> - Solve
> - Examine

> **PROBLEM SOLVING TIP**
>
> You may want to use a different problem solving method as a check on your solution. For example, you can use the graph from the graphing utility to verify the spreadsheet solution.

Investigate Further

A second group of students is designing a closed-top box for a trail mix they plan to sell. They have rectangular sheets of cardboard that are 15 in. long and 12 in. wide. They plan to use the pattern shown below to create a box shaped like a rectangular prism.

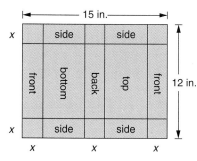

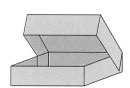

11. What is the range of possible values for x? Explain.

12. Write variable expressions in x that represent the length, width, height, and volume of the closed box.

13. Find the maximum volume of the box. Round to the nearest tenth.

14. **WRITING MATHEMATICS** Compare the problems about the planter and the trail mix box. How are they alike? How are they different?

Apply the Strategy

● Round answers to the nearest tenth. Unless noted otherwise, you may assume all three-dimensional shapes are rectangular prisms.

15. METAL WORKING A baking pan will be made from a square sheet of metal with sides of length 12 in. The pan will be shaped by cutting four congruent squares from the sheet, one at each corner, then folding and soldering the metal that remains. What dimensions create the pan with maximum volume?

16. ENGINEERING A section of water duct will be made from a rectangular sheet of metal that is 128 cm long and 40 cm wide, folded as shown at the right. What dimensions of the duct maximize the amount of water it can carry?

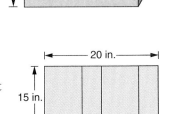

17. PACKAGING Jamie plans to make a rectangular "tube" by folding a rectangular sheet of cardboard that is 20 in. long and 15 in. wide along the blue segments shown at the right. What is the maximum volume of the tube?

18. PACKAGING Refer to Question 17. Suppose Jamie decides to simply bend the cardboard to make a right cylindrical tube with height 15 in. Will its volume be greater than or less than the maximum volume of the rectangular tube? Justify your answer.

19. OFFICE SUPPLIES Peter is making a letter tray from a rectangular sheet of copper that is 18 in. long and 12 in. wide. He will cut and fold the copper along the red and blue lines shown at the right. What dimensions to the nearest tenth will maximize the volume of the tray?

20. WRITING MATHEMATICS In Question 19, explain why the dimensions that maximize the volume are *not* practical.

PROBLEM SOLVING TIP

In Question 16, you must write an expression for y in terms of x.

COMMUNICATING ABOUT GEOMETRY

Do you prefer to solve problems about maximizing volume by using a spreadsheet, a graphing utility, or a combination of both? Explain.

REVIEW PROBLEM SOLVING STRATEGIES

A Prime Solid

1. In a right rectangular prism, the length of each edge is an integer. The length of the longest edge is less than twice the length of the shortest edge. If the volume of the solid is 1001, what is its suface area?

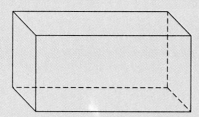

Splish, Splash

2. Artie Meedies tosses a rock into a cylindrical tank of water. The diameter of the tank is 3 ft. He observes that the level of the water rises $1\frac{1}{2}$ in. What is the volume of the rock?

3 ft

I scream, You scream, We all scream for Ice Cream

3. The ice cream cone below consists of ice cream in the shape of a sphere of diameter 3 in. and a cone with base diameter 3 in. and height $5\frac{3}{4}$ in. If the ice cream in the cone is allowed to completely melt into the cone, will it spill out of the cone?

13.6 Surface Areas of Pyramids and Cones

Explore

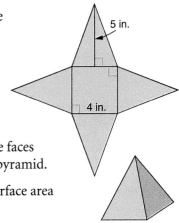

- You need heavy paper, scissors, tape, and a ruler.

1. The figure at the right is a net for a regular pyramid. Carefully draw the net on heavy paper.

2. What is the area of each triangle? What is the total area of all four triangles?

3. What is the area of the square?

4. Cut out the net, fold it, and tape the faces together to make the model of the pyramid.

5. What are the lateral area and the surface area of the pyramid?

6. What is the height of the pyramid? Did you need to know the height in order to find the lateral area and surface area?

Build Understanding

- Recall that the height of any pyramid is the length of an altitude. When a pyramid is regular, however, it also has a *slant height*. This is the height of one of its lateral faces.

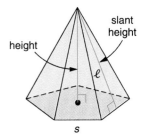

The **lateral area** of a pyramid is the sum of the areas of its lateral faces. The slant height, together with the perimeter of the base, plays an important role in finding the lateral area of a regular pyramid. For example, consider the lateral faces of the regular pentagonal pyramid above when they are arranged like this.

perimeter of base

Use the sum of the areas of the triangles and algebra to find the lateral area.

$$\angle A = \frac{1}{2}\ell s + \frac{1}{2}\ell s + \frac{1}{2}\ell s + \frac{1}{2}\ell s + \frac{1}{2}\ell s \qquad \text{Area of triangle: } A = \frac{1}{2}bh$$

$$\angle A = \frac{1}{2}\ell(s + s + s + s + s) \qquad \text{Use distributive property.}$$

$$\angle A = \frac{1}{2}\ell p \qquad \text{Let } p = \text{the perimeter of the base.}$$

This algebraic process leads to the following theorem.

THEOREM 13.7   **LATERAL AREA OF A REGULAR PYRAMID**

The lateral area LA of a regular pyramid equals half the product of the slant height ℓ of the pyramid and the perimeter p of its base, or $LA = \frac{1}{2}\ell p.$

TECHNOLOGY TIP

You can use geometry software to create patterns for pyramids. If the pyramid is right, the triangles used for the lateral surface will all be congruent isosceles triangles.

EXAMPLE 1

Find the lateral area of the regular pyramid shown.

Solution

The base of the pyramid is a square. The length of one side is 8 ft. So the perimeter p of the base is 4(8) ft, or 32 ft.

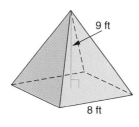

The height of the pyramid is 9 ft. To find its slant height ℓ, find the hypotenuse of the right triangle shown in blue. Use the Pythagorean Theorem.

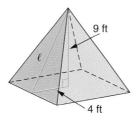

$$\ell^2 = 4^2 + 9^2$$
$$\ell = \sqrt{4^2 + 9^2}$$
$$\ell = \sqrt{97}$$

Now apply the formula for the lateral area of a regular pyramid.

$$LA = \frac{1}{2}\ell p$$
$$LA = \frac{1}{2}(\sqrt{97})(32)$$
$$LA \approx 157.58$$

The lateral area of the pyramid is about 158 ft^2. ◄

To find the surface area of a regular pyramid, you add the area of its base to the lateral area.

EXAMPLE 2

The figure at the right is a regular hexagonal pyramid. Find its surface area.

Solution

The slant height ℓ is 7 in. The perimeter p of the base is 6(4) in., or 24 in.

The base is shown below the pyramid. Its area B is $\frac{1}{2}(2\sqrt{3})(24)$ in., or $24\sqrt{3}$ in.

Now apply the formula for surface area.

$$SA = \frac{1}{2}\ell p + B$$

$$SA = \frac{1}{2}(7)(24) + 24\sqrt{3} \approx 125.57$$

The surface area is about 126 in.^{2} ◄

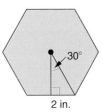

7 in.

4 in.

30°

2 in.

As the number of sides of the base increases, a regular pyramid approaches a cone in its appearance.

    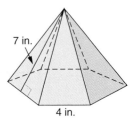

Therefore, the formulas for the lateral area and surface area of a right cone are closely related to the formulas for a regular pyramid.

> THEOREM 13.9   LATERAL AREA OF A RIGHT CONE
>
> The lateral area LA of a right cone of radius r equals half the product of the circumference of its base and the slant height ℓ of the cone, or $LA = \frac{1}{2} \cdot 2\pi r\ell = \pi r\ell$.

> THEOREM 13.10   SURFACE AREA OF A RIGHT CONE
>
> The surface area SA of a right cone of radius r and slant height ℓ equals the sum of its lateral area LA and the area of its base B, or $SA = LA + B = \pi r\ell + \pi r^2$.

EXAMPLE 3

FOOD PACKAGING The frozen yogurt cone shown is enclosed in a paper wrapper shaped like a right cone. Find the lateral area and surface area of the wrapper.

Solution

The diameter of the cone is 7.6 cm, so the radius r is 3.8 cm. The slant height ℓ of the cone is 13.3 cm. The area of the base B is $\pi(3.8)(3.8)$ cm^2, or about 45.36 cm^2.

$$LA = \pi r\ell \qquad\qquad SA = LA + B$$
$$LA \approx \pi(3.8)(13.3) \qquad SA \approx 158.78 + 45.36$$
$$LA \approx 158.78 \qquad\qquad SA \approx 204.14$$

The lateral area of the wrapper is about 318 cm^2. The surface area is about 363 cm^2. ◄

7.6 cm

13.3 cm

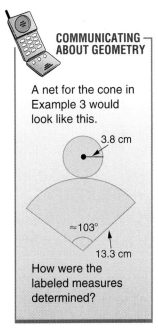

COMMUNICATING ABOUT GEOMETRY

A net for the cone in Example 3 would look like this.

3.8 cm

≈103°

13.3 cm

How were the labeled measures determined?

TRY THESE

1. **WRITING MATHEMATICS** The figures below are regular pyramids. Compare the procedures you would use to find the lateral area of each. Which steps would be the same? Which would be different?

a.

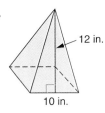

5 in.

4 in.

b.

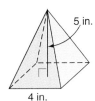

5 in.

4 in.

Find the lateral area and surface area of each regular pyramid or right cone. Round your answers to the nearest whole number.

2.

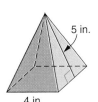

12 in.

10 in.

3.

15 cm

10 cm

4.

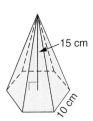

10 ft

15 ft

5.

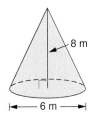

8 m

6 m

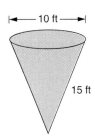

6. The height of a right cone is 13 in. Its diameter is 8 in. Find the lateral area.

7. The height of a regular pyramid is 30 cm. The base is a 20-cm square. Find the lateral area.

8. **MANUFACTURING** Tulip Inc. manufactures paper cups for water cooler dispensers. Each cup is in the shape of a right cone with diameter 7.5 cm and height 9 cm. How much paper is used to make each cup? (Assume no overlap.)

9. The figure shown at the right is a *composite* of a cube and a regular pyramid. How many faces does it have? What type of figure bounds each face?

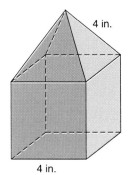

4 in.

4 in.

10. Carefully draw a net for the figure. Cut out your net, fold it, and tape the faces together to form the composite figure.

11. Find the surface area of the figure.

PRACTICE

12. WRITING MATHEMATICS One figure in part a is a regular pyramid. Part b is a right cone. Compare the procedures you would use to find the lateral area of each. Which steps would be the same? Which would be different?

a.

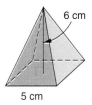

6 cm

5 cm

b.

6 cm

5 cm

Find the lateral area of each regular pyramid or right cone. Round your answers to the nearest whole number.

13.

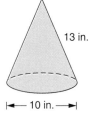

13 in.

|◄— 10 in. —►|

14.

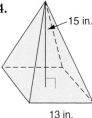

15 in.

13 in.

15.

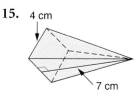

4 cm

7 cm

16.

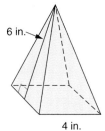

6 in.

4 in.

Find the surface area of each regular pyramid or right cone. Round your answers to the nearest whole number.

17.

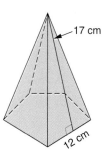

17 cm

12 cm

18.

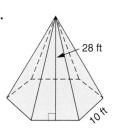

28 ft

10 ft

19.

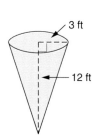

3 ft

12 ft

20.

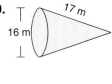

17 m

16 m

21. The height of a right cone is 35 cm. Its diameter is 16 cm. Find the lateral area.

22. The slant height of a right cone is 18.75 in. Its radius is 9 in. Find the lateral area.

23. The slant height of a regular pyramid is 1 ft. The base is an equilateral triangle whose side measures 9.75 in. Find the surface area.

24. ARCHITECTURE The Great Pyramid of Khufu is in Giza, Egypt. When it was built, its height was 147 m and the base was a 230-m square. What was the lateral area?

25. COSTUME DESIGN For a production of *Ivanhoe*, the hats for the period costumes of the women characters are in the shape of a right cone. The height of the cone is 15 in. and the diameter is 7 in. How much fabric would be needed to cover each hat?

EXTEND

Find the surface area of each composite figure. You may assume that the parts of the figures are regular prisms, regular pyramids, right cylinders, and right cones. Round your answers to the nearest whole number.

26.
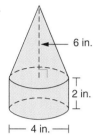

27.
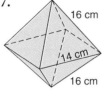

28.
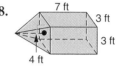

29.
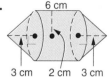

30. KITES The kite shown below has four parts, which are joined at their corners. Each part is shaped like a tetrahedron. The length of each edge of the tetrahedrons is 12 in. Find the surface area of fabric needed to make this kite.

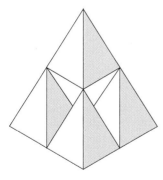

31. The slant height of a right cone and the slant height of a square pyramid are both 12 cm. The side of the base of the pyramid measures 10 cm. The lateral area of the cone is equal to the lateral area of the pyramid. What is the radius of the cone?

32. The slant height of a square pyramid and the slant height of a cone are equal in measure. The side of the base of the pyramid measures 20 in. The radius of the cone is 12 in. The surface area of the pyramid is equal to the surface area of the cone. What is the measure of the slant height of each figure?

THINK CRITICALLY

33. What is the surface area of the right cone whose net is shown?

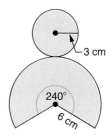

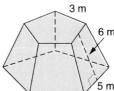

A *frustum* of a right pyramid or cone is the figure formed when a cross section that is parallel to the base is joined with all points of the pyramid or cone that lie between the cross section and the base. Find the surface area of each frustum.

34.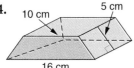

35.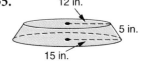

36.

37. Suppose the radius of a cone is doubled. What effect does this have on the lateral area of the cone?

38. Suppose the slant height of a pyramid is halved. What effect does this have on the lateral area of the pyramid?

ALGEBRA AND GEOMETRY REVIEW

39. STANDARDIZED TESTS If the diagonals of a quadrilateral bisect each other, then the quadrilateral must be a
   A. square   **B.** rectangle   **C.** rhombus   **D.** parallelogram

40. The side of an equilateral triangle measures 18 cm. Find the measure of the altitude.

41. The degree measures of the interior angles of a quadrilateral are x, $x + 15$, $x + 20$, and $x + 45$. Find the measure of the largest angle.

42. Solve for x: $\dfrac{4}{5} = \dfrac{14}{x}$ **43.** Factor: $xy - 5x + 3y - 15$ **44.** Solve for x: $\sqrt{6 - x} = x$

PROJECT *Connection*   A moving company may be hired to pack up and move the entire contents of a house.

1. Find the web page of a nationwide moving company on the Internet, or contact a local moving company. Obtain information on how this company prices moving and shipping. Ask for a listing of the sizes and types of cartons that are used for shipping. Is the price to move the carton based on volume or is it based on weight?

2. Obtain the dimensions of a truck used for moving. Find the volume of the storage area. Write dimensions of 12 different sized cartons. Draw up a plan for packing the truck so that little or no space is wasted.

3. What are the most common shapes of packages that are used in moving? Why do you think that this is so?

A *packaging designer* combines graphic arts with a knowledge of packaging materials and manufacturing processes. These designers must also have a direct link to marketing departments so that the packaging design meets the perceived needs of the consumer. One of the most important aspects of packaging design is to assure the safe transportation of goods from the production line to the consumer.

Decision Making

A bottler of mineral and spring water packages the product in glass bottles. It has been determined that a suitable container has the shape of a rectangular prism with the dimensions shown. There should be no more than 12 bottles packaged per container.

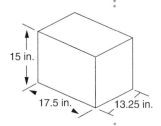

The packaging designer must create a polystyrene interior case to protect the glass bottles during shipping.

1. On a large sheet of paper, sketch a rectangle to represent the base of the carton. The bottles are no more than 3.75 in. in diameter. Using a compass, create a plan for the placement of the bottles so that they are equally spaced in the carton. How much space will there be between bottles?

A sketch of the polystyrene insert is shown below.

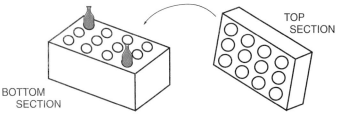

BOTTOM SECTION

TOP SECTION

2. Twelve cylinders are cut from the inside of this polystyrene insert. The height of each cylinder is 12 in. What is the maximum volume of each cylindrical hole?

3. Calculate the amount of polystyrene material that would be needed to make one such protective package. Explain your method.

4. Polystyrene is an expensive packaging material. The designer must therefore build into the plan ways in which the materials can be minimized without sacrificing the safety of the product. How might the design of the polystyrene container be altered to cut down on the amount of material used?

Explore/Working Together

● Work in a group with two or three other students. You will need heavy paper, scissors, tape, a ruler, and some dry cereal.

1. Carefully draw the two nets shown below.

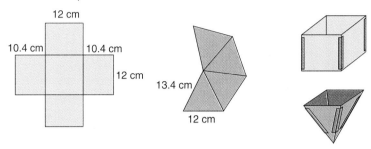

2. Cut out the nets. Fold and tape them to model an "open" prism and pyramid, as shown to the right of the nets.

3. What is the relationship between the heights of the prism and the pyramid? What is the relationship between their bases?

4. Fill the pyramid with some dry cereal. Pour the cereal into the prism. Repeat this process until the prism is full. How many times did you have to fill the pyramid?

5. What do you think is the relationship between the volume of a pyramid and the volume of a prism? Write the relationship in your own words and as an algebraic formula.

Build Understanding

In the figure below, the prism and pyramid have congruent bases and are equal in height. Clearly the volume of the prism is greater than the volume of the pyramid. In fact, the volume of the prism is exactly three times the volume of the pyramid.

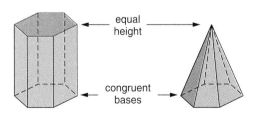

This relationship between the volumes is stated formally in the following theorem.

THEOREM 13.11   **VOLUME OF A PYRAMID**

The volume V of a pyramid equals one third the product of the area of the base B and the height h, or $V = \frac{1}{3}Bh$.

EXAMPLE 1

The figure at the right is a pyramid with a rectangular base. Find its volume.

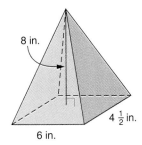

8 in.

$4\frac{1}{2}$ in.

6 in.

Solution

The base of the pyramid is a rectangle with area $(6)\left(4\frac{1}{2}\right)$ in.^{2}, or 27 in.^{2}

Use the volume formula for a pyramid.

$$V = \frac{1}{3}\,Bh = \frac{1}{3}(27)(8) = 72$$

The volume of the pyramid is 72 in.^{3} ◄

In Explore, you verified Theorem 13.11 experimentally. If you performed a similar experiment with a cylinder and a cone, you would arrive at a similar result. That is, when a cylinder and a cone have congruent bases and are equal in height, the volume of the cylinder is three times the volume of the cone.

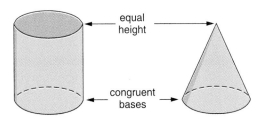

equal height

congruent bases

THEOREM 13.12   **VOLUME OF A CONE**

The volume V of a cone of radius r equals one third the product of the area of the base B and the height h, or $V = \frac{1}{3}\,Bh = \frac{1}{3}\pi r^2 h.$

Notice that the volume formulas for pyramids and cones are not restricted to regular pyramids and right cones. Because of *Cavalieri's Principle*, they apply to all pyramids and cones. You can examine why this is true in the Geometry Excursion on page 768.

EXAMPLE 2

The figure at the right is a right cone. Find its volume.

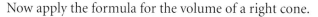

Solution

The radius r is 4.55 cm.

The slant height of the cone is 6.6 cm. To find its height h, find the length of the vertical leg of the right triangle. Use the Pythagorean Theorem.

$$(6.6)^2 = h^2 + (4.55)^2$$

$$h^2 = (6.6)^2 - (4.55)^2$$

$$h = \sqrt{(6.6)^2 - (4.55)^2} = \sqrt{22.8575}$$

Now apply the formula for the volume of a right cone.

$$V = \frac{1}{3}\pi r^2 h$$

$$V = \frac{1}{3}(\pi)(4.55)^2(\sqrt{22.8575}) \approx 103.65$$

The volume of the cone is about 104 cm³.   ◄

You can use the volume formulas for pyramids and cones to solve many real world problems.

EXAMPLE 3

INDUSTRIAL DESIGN A water storage tank is to be shaped like an inverted cone with a depth of 5 m. The tank must be able to hold 50 m³ of water. What must be the diameter of the tank?

Solution

$$\frac{1}{3}\pi r^2 h = V \qquad \text{Use the cone volume formula.}$$

$$\frac{1}{3}\pi r^2(5) = 50 \qquad \text{Substitute 50 for } V \text{ and 5 for } h.$$

$$\frac{5}{3}\pi r^2 = 50 \qquad \text{Simplify.}$$

$$r^2 = \frac{30}{\pi} \qquad \text{Solve for } r^2.$$

$$r = \sqrt{\frac{30}{\pi}} \qquad \text{Take the square root of each side.}$$

$$d = 2\sqrt{\frac{30}{\pi}} \qquad \text{The diameter of a cone is twice the radius.}$$

$$d \approx 6.18$$

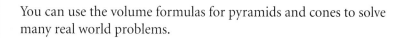

The diameter of the tank must be at least 6.18 m.   ◄

CHECK UNDERSTANDING

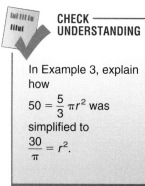

In Example 3, explain how $50 = \frac{5}{3}\pi r^2$ was simplified to $\frac{30}{\pi} = r^2$.

Find the volume of each right pyramid or right cone. Assume that any quadrilaterals are rectangles. Round your answers to the nearest whole number.

1.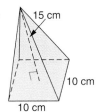

2.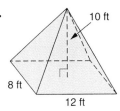

3.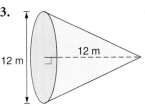

4.

5. The diameter of a right cone is 13.5 ft. The height is 22 ft. Find the volume.

6. The base of a right pyramid is a rectangle with length 5.5 in. and width 8.75 in. The height of the pyramid is 12 in. Find the volume.

7. The volume of a square pyramid is 15 m³. The height is 5 m. Find the dimensions of the base.

8. The volume of a right cone with a 7 mm radius is 820 mm³. Find the height. Find the slant height. Round your answers to the nearest tenth.

Each figure below is a net for a right cone and a square pyramid. Find the volume of the figure formed when each net is folded.

9.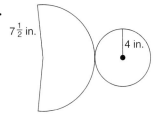

10.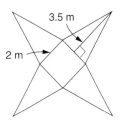

11. **WRITING MATHEMATICS** How is finding the volume of a cone similar to finding the volume of a pyramid? How is it different? How is finding the volume of a cone similar to finding the volume of a cylinder? How is it different?

12. **MARKETING** Ye Olde Ice Cream Shoppe sells novelty cones that are completely filled with ice cream, with no ice cream jutting over the top. The small cone, which costs $1.25, is 5 in. high and its diameter is 2 in. The large cone, which costs $2.50, is 7 in. high and its diameter is 3 in. Which is the better buy? Justify your answer.

The figure at the right is a composite of two right cones that share a common base.

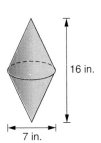

13. What would be the volume of a right cylinder whose diameter and height are equal to those of the composite figure?

14. What is the volume of the composite figure?

PRACTICE

Find the volume of each regular pyramid or right cone. Answer to the nearest whole number.

15.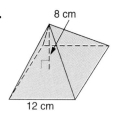
8 cm
12 cm

16.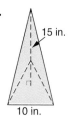
15 in.
10 in.

17.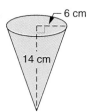
6 cm
14 cm

18.
13 in.
10 in.

19.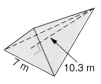
7 m
10.3 m

20.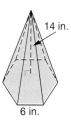
14 in.
6 in.

21.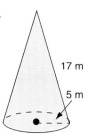
17 m
5 m

22.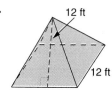
12 ft
12 ft

23. The volume of an equilateral triangular pyramid is $\dfrac{32\sqrt{3}}{3}$ ft³. The height is 8 ft. Find the length of the side of the equilateral triangle.

24. The volume of a right cone with a 16 cm diameter is 256π cm³. Find the height. Find the slant height. Round your answers to the nearest tenth.

25. **WRITING MATHEMATICS** The figure at the right is a regular pyramid. List as many facts about it as you can. Justify each fact in your list.

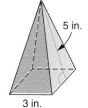
5 in.
3 in.

26. **GARDENING SUPPLIES** Mr. Borghese has a container that he wishes to fill with a special fertilizer for his roses. The container is in the shape of a right cone. The diameter of the container is 12 ft. The height is 5.5 ft. The fertilizer costs $2.25 per ft³. How much will it cost Mr. Borghese to fill the container?

27. **PACKAGING DESIGN** To package a new makeup kit, the designer for a cosmetics company is considering the shape of a regular pyramid. The perimeter of the base is to be 36 in. The height of the pyramid is to be 10 in. For which shaped base will the package hold more, a square or an equilateral triangle? How much more?

EXTEND

Find the volume of each composite figure. You may assume that the parts of the figures are regular prisms, regular pyramids, right cylinders, and right cones.

28.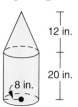
12 in.
20 in.
8 in.

29.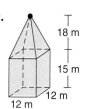
18 m
15 m
12 m
12 m

30.
All edges are 12 in.

31. The figure below shows a cone inscribed in a cube. The length of each edge of the cube is 8 in. Find the volume of the space between the cone and the cube.

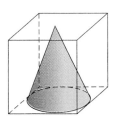

32. The diagram below shows a pyramid in a three-dimensional coordinate system. Find the volume of the pyramid.

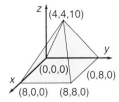

THINK CRITICALLY

Each figure is a frustum of a right pyramid or a right cone. Find its volume.

33.

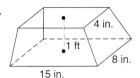
4 in.
1 ft
8 in.
15 in.

34.

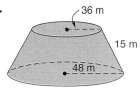
36 m
15 m
48 m

35.

4 ft
8 ft
10 ft

36. In a two-chamber sand timer like the one shown at the right, sand passes from one chamber to the other at a rate of 5 in.^{3}/min. The sand forms a pile shaped like a right cone whose diameter is twice its height. Suppose all the sand is in one chamber. You turn the timer and the sand begins to fall into the empty chamber at the bottom. What is the height of the pile in the bottom chamber after three minutes?

37. LOOKING AHEAD What is the effect on the volume of a square pyramid if its height is doubled? What is the effect if the length of all four sides of its base are doubled?

38. The figure at the right shows a pyramid that has been cut from a rectangular prism. What is the relationship between the volume of the prism and the volume of the pyramid? Justify your answer.

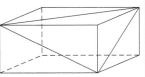

ALGEBRA AND GEOMETRY REVIEW

Find the distance between the pairs of points.

39. $(-8, 5)$ and $(3, -7)$

40. $(1, 2, 3)$ and $(-5, 0, 6)$

41. Solve the system. $\begin{aligned} x - y &= 1 \\ 2x - y &= 8 \end{aligned}$

42. Simplify. $\dfrac{\dfrac{6}{x} - 5}{-3 - \dfrac{3}{x}}$

43. STANDARDIZED TESTS After a reflection across the line $y = x$, the image of $(-3, 2)$ is in

   A. Quadrant I   **B.** Quadrant II   **C.** Quadrant III   **D.** Quadrant IV

Career
Packing Store Owner

In response to the ever-growing need for packaging and shipping, a new type of business has emerged throughout the country. *Packing and mailing stores*, often owned and operated by local individuals, have found a place in the shipping market.

Consumers bring items to these stores where they can be packed and sent directly out from that location. These stores usually offer the consumer a choice of shipper, with a variety of shipping options.

The owner of a packing store must be knowledgeable about shipping rules, costs, package options, and materials.

Decision Making

The *girth* of a package is the perimeter of an end of the box, the face with the smaller dimensions. Shippers use the sum of the measures of the length and the girth of a package when calculating maximum shipping sizes.

A packing store lists the following options for inside package measurements.

Small Box: $11\frac{1}{4}'' \times 10\frac{5}{8}'' \times 1\frac{1}{8}''$ Medium Box: $11\frac{1}{2}'' \times 13'' \times 2\frac{3}{8}''$

Large Box: $17\frac{1}{2}'' \times 12\frac{3}{8}'' \times 3''$ Economy Box: $17'' \times 15'' \times 13''$

Triangular Tube: $38'' \times 6'' \times 6'' \times 6''$

For each of the box sizes listed at this store, determine the following measures. Round your answers to the nearest hundredth of a unit.

1. the sum of the length and the girth

2. the volume

3. the surface area

The *International Air Transport Association* uses a measurement known as the *dimensional weight* or *volumetric weight* of a package. The dimensional weight is calculated by dividing the volume of the package in cubic inches by 166. If the magnitude of the dimensional weight is greater than the actual weight of the package in pounds, you are charged by the dimensional weight of the package rather than the actual weight.

4. Determine the dimensional weight for each of the boxes listed.

Explore/Working Together

● Work in a group with two or three other students. You will need a small ball or other spherical object, a tape measure, a compass, paper, scissors, and tape.

Recall that a great circle of a sphere is any cross section of the sphere whose center is the center of the sphere.

1. Use the tape measure to find the circumference of a great circle of the sphere. Calculate the radius of the great circle.

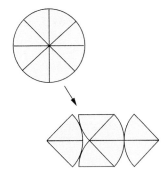

2. On paper, draw a circle whose radius is equal to the radius of the great circle. Calculate its area.

3. Cut out the circle. Fold it to form eight congruent sectors as shown at the right. Cut the sectors apart.

4. Rearrange the sectors into the pattern shown at the right. Tape the pattern together.

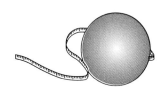

5. Tape the pattern to the surface of the ball, as shown at the right. What fraction of the surface of the sphere does the pattern appear to cover?

6. What do you think is the surface area of the sphere? Justify your response.

Build Understanding

● The radius of a great circle is equal to the radius of the sphere. So, for a sphere of radius r, the area of any great circle is πr^2.

great circle

In Explore, you probably discovered a special relationship between the area of a great circle and the surface area of a sphere. That is, the area of any great circle of a sphere is one fourth its surface area. This is stated formally as the following theorem.

THINK BACK

A *sphere* is the set of all points in space that are a given distance from a given point. The given point is the *center* of the sphere, and the given distance is the *radius*.

THEOREM 13.13   **SURFACE AREA OF A SPHERE**

The surface area SA of a sphere equals the product of 4π and the square of the radius r, or $SA = 4\pi r^2$.

EXAMPLE 1

NATURE An orange is spherical in shape, with a radius of 4 cm. What is the area of peel on this orange?

Solution

One way to measure the area of peel is to determine the surface area of the orange.

The radius of the orange is 4 cm. Substitute 4 for r in the formula for the surface area of a sphere.

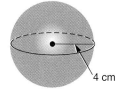

4 cm

$$SA = 4\pi r^2$$
$$SA = 4\pi(4)^2$$
$$SA = 64\pi$$
$$SA \approx 201$$

The area of peel is about 201 cm^2.   ◄

Imagine a sphere and its interior divided into many "pyramids" like those shown at the right. Notice that the height of each pyramid equals the radius r of the sphere. So the volume of each pyramid is $\frac{1}{3}Br$, where B is the area of its base.

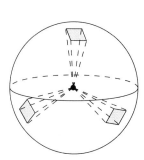

Suppose there are n of these pyramids. This means you can identify their volumes as $\frac{1}{3}B_1 r, \frac{1}{3}B_2 r, \frac{1}{3}B_3 r, \ldots, \frac{1}{3}B_n r$. The volume of the sphere is equal to the sum of the volumes of all these pyramids.

r

Volume V of sphere

$$V = \frac{1}{3}B_1 r + \frac{1}{3}B_2 r + \frac{1}{3}B_3 r + \ldots + \frac{1}{3}B_n r$$

$$V = \frac{1}{3}r(B_1 + B_2 + B_3 + \ldots + B_n) \quad \text{Use the Distributive Property.}$$

$$V = \frac{1}{3}r(4\pi r^2) \qquad B_1 + B_2 + B_3 + \ldots + B_n = 4\pi r^2$$

$$V = \frac{4}{3}\pi r^3 \qquad \text{Simplify.}$$

This result is stated formally in the following theorem.

THEOREM 13.14	VOLUME OF A SPHERE

The volume V of a sphere equals the product of $\frac{4}{3}\pi$ and the cube of the radius r, or $V = \frac{4}{3}\pi r^3$.

THINK BACK

A *chord* of a sphere is a segment whose endpoints are points on the sphere. A *diameter* of a sphere is a chord that contains the center. The common length of all the diameters is called *the* diameter of the sphere. The diameter is twice the radius.

EXAMPLE 2

HORTICULTURE The figure at the right is a sketch of a proposed greenhouse that is to be shaped like a hemisphere. What is the amount of space inside this greenhouse?

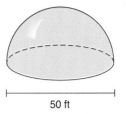

50 ft

Solution

To determine the amount of space, you must find the volume of the hemisphere.

Since, the diameter is 50 ft, the radius is 25 ft.
Substitute 25 for r in the formula for the volume of a sphere.

$$V = \frac{4}{3}\pi r^3 = \frac{4}{3}\pi(25)^3 = \frac{62,500}{3}\pi$$

The volume of a sphere with radius 25 ft is about 65,449.85 ft³. So the volume of the *hemisphere* is half of $\frac{62,500}{3}\pi$ ft³.

The amount of space inside is about 32,725 ft³.

◄

THINK BACK

Any great circle of a sphere divides the sphere into two congruent parts. Each part is called a *hemisphere*.

TRY THESE

Find the surface area and the volume of a sphere with the given radius r or diameter d. Give the answers in exact form and rounded to the nearest whole number.

1. $r = 10$ cm

2. $d = 16$ in.

3. $d = 11\frac{1}{2}$ in.

4. $r = 3\frac{3}{4}$ ft

5. $r = 3$ yd

6. $d = 1$ m

7. RECREATION The radius of a fully inflated beach ball is 10 in. What is the amount of air inside? Round the answer to the nearest whole number.

8. SPORTS Tennis balls are covered by a fabric made of wool and artificial fibers. The diameter of a tennis ball is 6.35 cm. Calculate the amount of fabric needed for three tennis balls. Round the answer to the nearest whole number.

9. **POTTERY** The figure at the right shows a clay dish. It is shaped like a right cylinder and has an indentation shaped like a hemisphere. Find the amount of clay needed to make the dish.

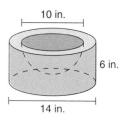
10 in.
6 in.
14 in.

10. **WRITING MATHEMATICS** Many real world problems involve objects shaped like spheres. Explain how you know whether to find the surface area or the volume of the sphere to solve such a problem.

PRACTICE

Find the surface area and the volume of each sphere. Give the answers in exact form and rounded to the nearest whole number.

11.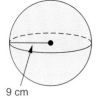
9 cm

12.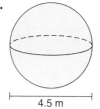
4.5 m

13.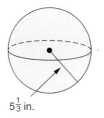
$5\frac{1}{3}$ in.

14.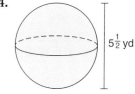
$5\frac{1}{2}$ yd

15. **WRITING MATHEMATICS** Explain how to find the surface area and the volume of a hemisphere when you know its radius.

16. **AGRICULTURE** The figure at the right shows a grain silo shaped like a composite of a right cylinder and a hemisphere. Find the amount of grain that can be stored in this silo when it is filled to capacity.

15 m
9 m

ASTRONOMY The planets of our solar system are nearly spherical in shape. Use the data in the chart below to find the surface area and the volume of each planet. Give the answers in scientific notation.

	Planet	Average Distance from Sun, mi	Equatorial Diameter, mi
17.	Mercury	35,900,000	3,031
18.	Venus	67,200,000	7,520
19.	Earth	92,960,000	7,926
20.	Mars	141,600,000	4,222
21.	Jupiter	483,600,000	88,700
22.	Saturn	886,700,000	74,800
23.	Uranus	1,783,000,000	32,200
24.	Neptune	2,794,000,000	30,800
25.	Pluto	3,666,100,000	1,423

EXTEND

26. SPORTS Three spherical tennis balls are packed so they fit snugly in a can shaped like a right cylinder. The diameter of each ball is 6.35 cm. Find the amount of empty space between the balls and the can.

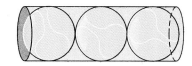

FOOD In the figure at the right, a spherical scoop of ice cream rests on an ice cream cone that is shaped like a right cone.

27. The part of the ice cream above the top of the cone is shaped like a hemisphere. Find the total surface area of the ice cream and the cone.

28. Suppose the ice cream melts. Will it fit inside the cone? Justify your answer. (Assume that melted and frozen ice cream have equal volume.)

$3\frac{1}{4}$ in.

$5\frac{1}{2}$ in.

Give each answer in exact form and rounded to the nearest whole number.

29. The circumference of a great circle of a sphere is 100π mm. Find the surface area and the volume of the sphere.

30. Find the volume of a sphere whose surface area is 100π mm^2.

31. Find the surface area of a sphere whose volume is 100π mm^3.

THINK CRITICALLY

32. BUBBLES A spherical soap bubble with radius r hits a flat surface. Its shape becomes a hemisphere. The surface area of the hemisphere is equal to the surface area of the original sphere, What is the relationship between the volume of the hemisphere and the volume of the original sphere?

33. A *small circle* of a sphere is a cross section of the sphere whose center is *not* the center of the sphere. Suppose the radius of a sphere is 10 in. Find the area of a small circle of this sphere whose center lies 6 in. from the center of the sphere.

34. LOOKING AHEAD What is the effect on the surface area and volume of a sphere when its radius is doubled?

ALGEBRA AND GEOMETRY REVIEW

35. Find the slope of the line that passes through the points $A(-3, 8)$ and $B(4, 8)$.

36. What is the slope of the line that is parallel to the line $y = -10$?

37. What is the slope of the line that is parallel to the line $x = 3$?

A sphere is inscribed in a cube. The length of each edge of the cube is 12 cm.

38. Find the surface area and volume of the cube and the sphere.

39. Find the amount of space between the cube and the sphere.

40. Solve $|3y - 2| \leq 5$. Graph the solution.

Cavalieri's Principle

At the right is an oblique prism and a right prism. Their bases are congruent rectangles with area B. The height of each is h. By Theorem 13.5, the volume of each is Bh.

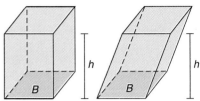

Therefore, although these two prisms are quite different in shape, they are equal in volume. How can this be possible?

1. Suppose you have a stack of twelve books. The stack is shaped like the right prism above. Each book is 8 in. wide, 11 in. long, and 1 in. deep. What is the volume of the stack?

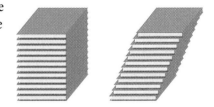

2. Imagine that you push the stack so it takes on a shape like the oblique prism. Has the volume of the stack changed? Explain.

3. What is the area of the cover of the first book in each stack? the fifth book? the twelfth book? The nth book?

The stacks of books provide an illustration of an important geometric concept.

CAVALIERI'S PRINCIPLE

POSTULATE 28   If two solids lying between parallel planes have the same height and all cross sections at equal distances from their bases have equal areas, then the solids have equal volume.

---

4. Explain how parallel planes $\mathcal{R}$, $\mathcal{S}$, and $\mathcal{T}$ illustrate Cavalieri's Principle for the prisms shown at the top of the page.

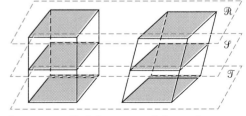

Cavalieri's Principle provides an important link between right and oblique prisms. That is, whether a prism is right or oblique, you can use the same formula, $V = Bh$, to find its volume. Cavalieri's Principle provides a similar connection between right and oblique cylinders, right and oblique pyramids, and right and oblique cones.

Find each volume. If necessary, round to the nearest tenth.

5. prism

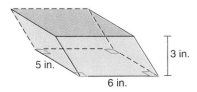

5 in.

6 in.

3 in.

6. cylinder

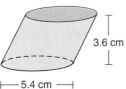

3.6 cm

—5.4 cm—

7. pyramid

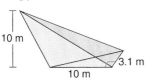

10 m

10 m

3.1 m

8. cone

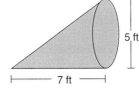

5 ft

—7 ft—

9. WRITING MATHEMATICS Each solid shown at the right is a prism with height 5. Do you think Cavalieri's Principle applies to these prisms? Explain your reasoning.

5

4

4

4

8

Figure I below is a sphere with radius *r*. Figure II is a solid formed when two conical sections are removed from the interior of a cylinder of radius *r* and height 2*r*.

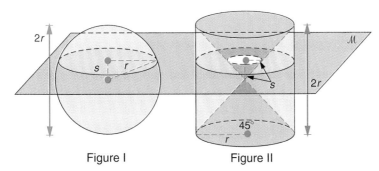

2*r*

s   *r*

s

2*r*

45°

r

𝓜

Figure I   Figure II

10. Plane 𝓜 is parallel to the bases of the cylinder and intersects each solid at a distance *s* from its center. What is the shape of the cross section formed when the plane intersects each solid?

11. Express the area of each cross section in terms of *r* and *s*.

12. Show how you can use Cavalieri's Principle and your answer to Question 11 to derive the formula for the volume of a sphere.

13.9 Similarity in Three Dimensions

Explore/Working Together

You need heavy paper, scissors, tape, a ruler, and some dry cereal.

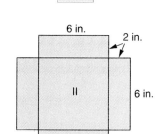

3 in. 1 in.
I 3 in.

6 in. 2 in.
II 6 in.

1. At the right are nets for two open-top boxes. Draw the nets on heavy paper and label them I and II. Cut out each net, fold it, and tape it to make the box.

2. Measure the perimeter of the base of each box. Write the ratio $\dfrac{\text{perimeter I}}{\text{perimeter II}}$ in simplest form.

3. Calculate the area of each net. Write $\dfrac{\text{area I}}{\text{area II}}$ in simplest form.

4. Fill box I with dry cereal. Pour the cereal into box II. Repeat this process until box II is full. How many times did you fill box I? Write the ratio $\dfrac{\text{volume I}}{\text{volume II}}$ in simplest form.

5. Describe a pattern among the ratios in Questions 2–4.

6. Suppose you constructed a box III so that it is 9 in. wide, 9 in. long, and 3 in. high. Make a conjecture about the ratios $\dfrac{\text{perimeter I}}{\text{perimeter III}}$, $\dfrac{\text{area I}}{\text{area III}}$, and $\dfrac{\text{volume I}}{\text{volume III}}$.

Build Understanding

Two solids are **similar solids** if they have the same shape and all corresponding linear measures are in proportion. The ratio of the linear measures is called the **scale factor** of the solids.

The boxes in Explore are models of similar solids. While investigating them, you probably discovered the relationships between similar solids that are stated in the following theorem.

THEOREM 13.15

If two solids are similar with scale factor $a : b$, then the ratio of the perimeters of corresponding faces is $a : b$, the ratio of the areas of corresponding faces is $a^2 : b^2$, and the ratio of their volumes is $a^3 : b^3$.

EXAMPLE 1

The two right cylinders shown at the right are similar. Find each ratio.

a. $\dfrac{\text{circumference of cylinder I}}{\text{circumference of cylinder II}}$

b. $\dfrac{\text{lateral area of cylinder I}}{\text{lateral area of cylinder II}}$

c. $\dfrac{\text{volume of cylinder I}}{\text{volume of cylinder II}}$

2 ft   1 yd

Solution

Use the labeled diameters to find the scale factor of the solids.

$$\text{scale factor} = \frac{\text{diameter of cylinder I}}{\text{diameter of cylinder II}} = \frac{2 \text{ ft}}{1 \text{ yd}} = \frac{2 \text{ ft}}{3 \text{ ft}} = \frac{2}{3}$$

So the scale factor $a : b$ is $2 : 3$. Apply Theorem 13.15.

a. $\dfrac{\text{circumference of cylinder I}}{\text{circumference of cylinder II}} = \dfrac{a}{b} = \dfrac{2}{3}.$

b. $\dfrac{\text{lateral area of cylinder I}}{\text{lateral area of cylinder II}} = \dfrac{a^2}{b^2} = \dfrac{2^2}{3^2} = \dfrac{4}{9}.$

c. $\dfrac{\text{volume of cylinder I}}{\text{volume of cylinder II}} = \dfrac{a^3}{b^3} = \dfrac{2^3}{3^3} = \dfrac{8}{27}.$ ◄

CHECK UNDERSTANDING

In Example 1, why is it necessary to rewrite 1 yd as 3 ft before determining the scale factor?

Knowing how lengths, areas, and volumes of similar solids are related can simplify solutions of some real world problems.

EXAMPLE 2

MANUFACTURING A manufacturer makes small paper cups shaped like right cones with a radius of 1.5 in. The surface area of each cup is 15.8 in.^{2} The manufacturer plans to make a similar large paper cup with a radius of 2 in. What will be its surface area?

COMMUNICATING ABOUT GEOMETRY

Work together to write a solution of Example 2 that does not involve the use of Theorem 13.15. Do you agree that using the theorem simplifies the solution of the problem?

Solution

Use the given radii to find the scale factor of the solids.

$$\text{scale factor} = \frac{\text{radius of small cup}}{\text{radius of large cup}} = \frac{1.5}{2} = \frac{3}{4}$$

By Theorem 13.15, the ratio of the surface areas will be $3^2 : 4^2$, or $9 : 16$. Use this ratio to write and solve an equation.

$\begin{aligned}\text{small cup} \rightarrow \\ \text{large cup} \rightarrow\end{aligned} \quad \dfrac{9}{16} = \dfrac{15.8}{x} \quad \begin{aligned}\leftarrow \text{small cup} \\ \leftarrow \text{large cup}\end{aligned}$

$$9x = 16(15.8)$$

$$x \approx 28.1$$

1.5 in.   2 in.

The surface area of the large paper cup will be about 28.1 in.^{2} ◄

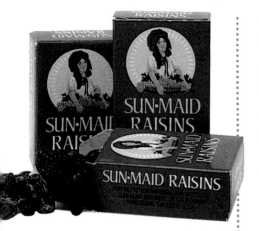

You also can use Theorem 13.15 to solve problems involving other measures related to length, area, and volume.

EXAMPLE 3

PACKAGING A company currently sells only snack-size boxes of raisins. They are investigating the possibility of selling a new family-size box. The dimensions of both boxes are shown below. The weight of the raisins in each snack-size box is 2 oz. What might the company expect to be the weight of the raisins in a family-size box?

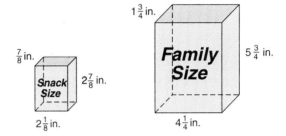

Solution

The weight of a substance is directly proportional to its volume.

Each dimension of the family-size box is twice the corresponding dimension of the snack-size box.

$$2\left(\frac{7}{8}\right) = 1\frac{3}{4} \qquad 2\left(2\frac{1}{8}\right) = 4\frac{1}{4} \qquad 2\left(2\frac{7}{8}\right) = 5\frac{3}{4}$$

This means that the scale factor of snack-size to family-size is 1 : 2. So, by Theorem 13.15, the ratio of the volumes is $1^3 : 2^3$, or 1 : 8.

The company might expect the weight of the raisins in the family-size box to be 8(2) oz, or 16 oz. This is 1 lb of raisins. ◄

TRY THESE

Determine whether the solids in each pair are similar. Justify your answer.

1. regular pyramids

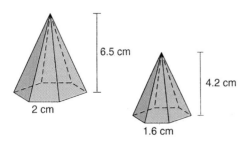

2. rectangular prisms

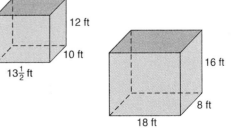

The two figures in each pair are similar. For each pair, find these ratios.

a. $\dfrac{\text{perimeter of base of figure I}}{\text{perimeter of base of figure II}}$
b. $\dfrac{\text{surface area of figure I}}{\text{surface area of figure II}}$
c. $\dfrac{\text{volume of figure I}}{\text{volume of figure II}}$

3. regular prisms

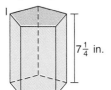

$7\frac{1}{4}$ in. $3\frac{5}{8}$ in.

4. right cones

5.1 m   8.5 m

5. Toys A toy manufacturer makes spherical rubber balls in two sizes. The ratio of the radius of the small ball to the radius of the large ball is 3 : 4. The amount of rubber on the surface of the small ball is 400 cm². What amount of rubber is on the surface of the large ball?

6. Nutrition To make small, medium, and jumbo muffins, a bakery uses baking cups with diameters of $1\frac{3}{4}$ in., $2\frac{3}{4}$ in., and $3\frac{1}{4}$ in., respectively. A medium bran muffin has 150 calories. Assume that the small, medium, and jumbo muffins are similar in shape. How many calories would you expect the small and jumbo bran muffins to have?

7. Writing Mathematics Draw and label two right cones that are similar with a scale factor of 1 : 4. Explain how Theorem 13.14 can help you find the surface areas and volumes of the cones.

PRACTICE

In each exercise, you are given one ratio associated with two similar square pyramids. Copy and complete the table.

	scale factor	ratio of heights	ratio of base perimeters	ratio of lateral areas	ratio of surface areas	ratio of volumes
8.	1 : 5	▩	▩	▩	▩	▩
9.	▩	▩	2 : 7	▩	▩	▩
10.	▩	▩	▩	▩	64 : 1	▩
11.	▩	▩	▩	▩	▩	8 : 27

12. **TRAVEL** A luggage company makes a "Traveler" suitcase that is similar in shape to its "Overnighter." The depth of the Traveler is 15 in., and its volume is 6750 in.³ The depth of the Overnighter is 9 in. What is the volume of the Overnighter?

STATUES Two solid brass statues are cast from similar molds. The scale factor of the statues is 2 : 5.

13. The cost of the brass for the small statue is $150. What would you expect to be the cost of the brass for the large statue?

14. The cost of a protective finish for the surface of the small statue is $12. What would you expect to be the cost of the protective finish for the large statue?

15. **WRITING MATHEMATICS** Sketch a solid that is similar to the rectangular prism shown at the right. Explain how you know that the solids are similar. Then identify as many ratios as you can that describe the relationship between the two solids.

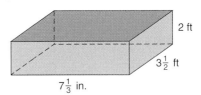

7⅓ in.   3½ ft   2 ft

EXTEND

SCALE MODELS The employees of a science museum are building a scale model of the space shuttle. The scale of the model will be 1 in. : 2 ft.

16. The distance from the tip of one wing of the actual shuttle orbiter to the tip of the other wing is about 26 yd. What will be the corresponding distance on the model?

17. One of the protective tiles on the actual shuttle orbiter covers 36 in.² What area will be covered by the corresponding tile on the model?

18. The large orange external fuel tank of the actual shuttle holds about 1,564,999 lb of liquefied fuel. What amount of this same fuel would fill the external tank of the model?

19. **WRITING MATHEMATICS** The weight of the actual shuttle orbiter when empty is about 165,000 lb. Can you use this information to calculate the weight of the model orbiter? Explain your reasoning.

20. Three right cones are similar. The scale factor relating the first cone to the second is 1 : 2. The scale factor relating the second cone to the third is 1 : 3. What is the ratio between the diameters, surface areas, and volumes of the first cone and the third cone? Justify your answers.

21. The scale factor of two spheres is 1 : 6. The surface area of the larger sphere is 600 in.² What is the radius of the smaller sphere?

THINK CRITICALLY

Replace each __?__ with *always*, *sometimes*, or *never* to make a true statement.

22. Two right cylinders are __?__ similar.

23. Two spheres are __?__ similar.

24. Two rectangular prisms are __?__ similar.

25. Two cubes are __?__ similar.

26. A cone is __?__ similar to a cylinder.

27. A solid is __?__ similar to itself.

A rectangular prism has length ℓ, width w, and height h. Describe the effect of each action on the volume of the prism. Is the resulting prism similar to the original prism?

28. doubling ℓ

29. doubling both ℓ and w

30. doubling ℓ, w, and h

31. doubling ℓ and tripling h

32. multiplying ℓ by n

33. multiplying ℓ by p, w by q, and h by r

ALGEBRA AND GEOMETRY REVIEW

34. The measure of each central angle of a regular n-gon is $5°$. Find the value of n.

35. STANDARDIZED TESTS The scale factor of two similar containers is $1 : 2$. Express the liquid capacity of the larger container as a percent of the liquid capacity of the smaller container.

    A. 12.5%   **B.** 25%   **C.** 50%   **D.** 800%

36. Evaluate the expression $a^2b^3 - 2ab^3 + 3a^2b$ for $a = 2$ and $b = -\frac{1}{4}$.

37. When $x = -20$, $y = 8$. If y varies directly as x, find x when $y = 28$.

38. A building casts a shadow that is 480 ft long. At the same time, a person who is 5 ft 6 in. tall casts a shadow that is 4 ft long. What is the height of the building?

PROJECT *Connection* The United States Postal Service, as well as private delivery services, ship letters and packages around the world. These businesses compete for the millions of consumer's dollars spent each year to ship items.

1. Gather information about regulations and costs for shipping packages through the United States Postal Service and through two other delivery services.

2. For each delivery service, are the costs to ship packages determined by distance, volume, weight, or a combination of characteristics? Explain.

3. Select a distant city. Suppose you want to ship a box so that it arrives in this city in one day. The box has length 2 ft, width 1.5 ft, and height 10 in. Its weight is 5 lb. Compare the costs to send the package through each service. Suppose the delivery day is Saturday. Is this possible? Does it affect the cost?

···· CHAPTER REVIEW ····

VOCABULARY

Choose the word from the list that correctly completes each statement.

1. The __?__ of any three-dimensional figure is the number of nonoverlapping cubic units contained in its interior.

2. The __?__ height is the height of one of the lateral faces of a regular pyramid.

3. The ratio of the linear measures of two similar solids is the __?__ of the solids.

4. The segment in which two faces of a three-dimensional figure intersect is known as its __?__ .

a. volume

b. scale factor

c. edge

d. slant

Lesson 13.1   PLATONIC SOLIDS AND EULER'S FORMULA   pages 723–726

- Three-dimensional figures formed by flat surfaces that are bounded by polygons and joined in pairs along their sides as polyhedra. A polyhedron has an interior, edges, faces, and vertices.

5. A solid has 8 vertices and 15 edges. How many faces does it have.

6. A solid has 15 edges and 10 faces. How many vertices does it have?

7. A solid has 9 vertices and 14 faces. How many edges does it have?

Lesson 13.2   USING VENN DIAGRAMS   pages 727–730

- A Venn diagram is a illustration of a universal set of objects and its subsets.

8. What do all three figures have in common?

9. What characteristic separates the two sets?

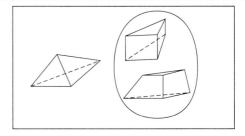

Lessons 13.3 and 13.4   SURFACE AREAS AND VOLUMES OF PRISMS AND CYLINDERS   pages 731–743

- The area of the surface of a three-dimensional object is known as the surface area.
- The sum of the areas of the lateral faces of a three-dimensional figure is its lateral area.
- The volume of any three-dimensional figure is the number of nonoverlapping cubic units contained in its interior.

10. A triangular prism has a 6-8-10 right triangle base and a height of 15 units. Determine its surface area, lateral area, and volume.

11. The radius of the base of a cylinder is 5.2 in. Its height is 8.5 in. Determine its surface area, lateral area, and volume.

● In many applications, it is necessary to determine the greatest possible volume of a container under certain conditions. Often those conditions involve the dimensions of the material from which the container is made.

12. An open top box is to be manufactured from a 16 in. by 30 in. piece of tin by cutting congruent squares from each of the four corners and folding up the sides. What size should the squares be in order to obtain the maximum volume?

● The slant height of a regular pyramid is the height of one of its lateral faces.

13. A regular pyramid has a square base with a 6 in. edge. The height of the pyramid is 4 in. Determine the surface area, lateral area, and volume of this pyramid.

14. The diameter of the base of a right cone is 14 cm. The height of the cone is 14 cm. Determine the surface area, lateral area, and volume of this cone.

● The surface area of a sphere equals the product of 4π and the square of radius r, or $SA = 4\pi r^2$.

● The volume of a sphere equals the product of $\frac{4}{3}\pi$ and the cube of radius r, or $V = \frac{4}{3}\pi r^3$.

15. Find the volume of the sphere shown.

16. A sphere has a diameter of 22 cm. Determine its surface area and volume.

● Two solids are similar solids if they have the same shape and all corresponding linear measures are in proportion. The ratio of the linear measures is called the scale factor of the solids.

17. A manufacturer makes open top boxes in the shape of rectangular prisms with a height of 12 in. The volume is 1200 in³. The manufacturer plans to make a similar open top box with a height of 18 in. What will be the volume?

CHAPTER ASSESSMENT

CHAPTER TEST

Copy and complete the chart using Euler's formula.

	Polyhedron	Faces	Edges	Vertices
1.	Octogonal Prism	10	24	
2.	Hexagonal Pyramid	7		7
3.	Pentagonal Prism		15	10
4.	Dodecagonal Pyramid	13	24	

5. **STANDARDIZED TESTS** In the Venn diagram, circle *X* contains all possible scalene triangles and circle *Y* contains all possible equiangular triangles. Which region(s) contains no triangles?

    A. 1 only
    B. 2 only
    C. 3 only
    D. 1 and 3
    E. none

    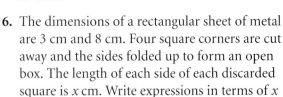

6. The dimensions of a rectangular sheet of metal are 3 cm and 8 cm. Four square corners are cut away and the sides folded up to form an open box. The length of each side of each discarded square is *x* cm. Write expressions in terms of *x* for the height, width, and length of the box.

FOOD PACKAGING Determine the lateral area of the label of the following products packaged in right cylindrical containers. Assume there is no overlap on the label that completely surrounds the container from top to bottom.

7. Sliced Peaches   height = $4\frac{3}{8}$ in., radius = 2 in.

8. Cranberry Sauce height = *x* in., radius = $1\frac{1}{2}$ in.

9. Tuna   height = *x*, diameter = $(3x + 1)$

10. A regular hexagonal prism has a base whose sides measure 6 in. and whose height measures 20 in. Determine the surface area, lateral area, and volume of the prism.

HISTORY The Great Pyramid of Choeps is near Cairo, Egypt. When it was built, it had a square base that was 230 m on each side and a height of 137 m.

11. What was the surface area of the base of the pyramid?

12. What was the lateral area of the pyramid?

13. **WRITING MATHEMATICS** A cone and a cylinder are the same height. The area of the bases are equal. Explain why Cavalieri's Principle would guarantee that the volumes of these two figures are not the same measure. What is the relationship of the volume of the cone to the volume of the cylinder?

14. Find the volume of a cone whose radius is $3\frac{1}{2}$ ft and height is 5 ft.

15. Find the approximate volume of a regular tetrahedron with a side of 6 cm and an approximate height of 4.9 cm.

16. The New England Soup Company packages their chicken broth in two sizes. The smaller cylindrical can has a diameter of 3 in. and a height of 4 in. The smaller and larger cans are similar with a scale ratio of 5:9. Determine the area of the paper label surrounding the larger size can.

17. **ASTRONOMY** Consider the moon a sphere. The radius of the moon is approximately 1080 mi. Determine the surface area and the volume of the moon. Express each answer in scientific notation.

PERFORMANCE ASSESSMENT

BUILDING BLOCKS Draw one figure that has a volume of 20 cubic unit blocks and a surface area of 48 square unit blocks.

MODELING WITH ALGEBLOCKS I Examine 4 of the basic Algeblocks pieces. Write an algebraic representation of the surface area and the volume of each piece.

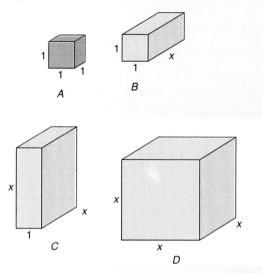

PLATONIC SOLIDS IN NATURE Research the forms of the crystals of salt, chrome alum, and sodium sulphantimoniate. Which of the Platonic solids can be found in these crystal structures? Two of the Platonic solids are not found in crystals but can be found in the skeletons of *radiolaria*, which are microscopic sea animals. Research these creatures to determine which of the Platonic solids can be observed in this sea life.

MODELING WITH ALGEBLOCKS II A cube is built with a total of 27 blocks. The area of one of its faces is $x^2 + 4x + 4$. Build the cube using only four different sizes of Algeblocks. Make a sketch of your model and label all of the pieces.

PROJECT ASSESSMENT

PROJECT *Connection* Prepare a presentation on the aspects of the packing industry that you have researched in this project. Relate the materials used for packaging to the nature of the product. Comment on the different shapes of the packages. Summarize procedures and costs related to moving and shipping.

Discuss possible careers in the packaging industry. Research colleges and universities that offer degree programs leading to a career in packaging. Consider how technological advancements are changing the industry.

··· CUMULATIVE REVIEW ···

Fill in the blank.

1. The region of a circle bounded by an arc and the chord having the same endpoints as the arc is called a(n) ___?___ of the circle.

2. An angle in a regular polygon whose vertex is at the center of the polygon and whose sides contain two consecutive vertices is called a(n) ___?___ angle.

3. The ___?___ of an acute angle in a right triangle is the ratio of the length of the side opposite the angle to the length of the hypotenuse.

4. A parallelogram that has all right angles is called a(n) ___?___.

Find the surface area of each figure. Round to the nearest whole number.

5.

6.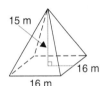

Find the volume of each figure. Round to the nearest whole number.

7.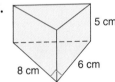

8.

Find the complement of each angle with the given degree measure.

9. $27°$ 10. $83.3°$ 11. $n°$

Find the supplement of each angle with the given degree measure.

12. $118°$ 13. $10.4°$ 14. $(n + 30)°$

Give the image of the point $(4, -2)$ under each transformation.

15. reflection in the x-axis

16. rotation of $180°$ about the origin

17. translation of 2 units up and 6 units left

18. reflection in the line $y = -1$

19. **WRITING MATHEMATICS** Explain why showing that two corresponding angles of two triangles are congruent is sufficient to prove that the triangles are similar.

20. Find the surface area and volume of the sphere.

Find x in each figure.

21.

22.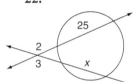

Find the center and radius of each circle.

23. $x^2 + y^2 = 49$

24. $(x - 4)^2 + (y + 2)^2 = 15$

25. $x^2 + 6x + 9 + y^2 - 10y + 25 = 100$

26. $(-8, 2)$ and $(8, -10)$ are endpoints of a diameter

Find the area of each figure.

27.

28.

STANDARD FIVE-CHOICE Select the best choice for each question.

1. A segment has endpoints $(8, -3)$ and $(2, -1)$. Which of the following is the equation of its perpendicular bisector?

   A. $3x - y = 17$
   B. $x + 3y = -1$
   C. $3x - y = 27$
   D. $3x - y = 7$
   E. Cannot be determined

2. Which of the following values could be used for x and y in the figure below?

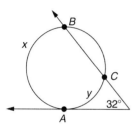

   A. $x = 170°, y = 138°$
   B. $x = 160°, y = 96°$
   C. $x = 164°, y = 36°$
   D. $x = 150°, y = 118°$
   E. None of these

3. In each figure, $m\angle 1 = m\angle 2$, and $\angle 1$ and $\angle 3$ are supplementary. Which of the following does not necessarily show two parallel lines cut by a transversal?

   A.   B.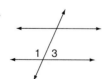

   C.   D.

   E. All necessarily show parallel lines

4. Which of the following pairs of points could be the endpoints of the base of an isosceles triangle that has a midsegment of length 10?

   A. $(-2, 12)$ and $(14, 0)$
   B. $(-8, 10)$ and $(0, 4)$
   C. $(0, -6)$ and $(10, 4)$
   D. $(-4, -6)$ and $(16, 14)$
   E. None of these

5. Which of the following sets of numbers cannot be used as the lengths of the three sides of an obtuse triangle?

   A. 5, 6, and 10
   B. 3, 5, and 7
   C. 11, 15, and 8
   D. 9, 2, and 7
   E. All can be used

6. Two solids are similar with a scale factor of $5 : 7$. What is the ratio of their volumes?

   A. $5 : 7$
   B. $25 : 49$
   C. $125 : 343$
   D. $15 : 21$
   E. None of these

7. Which of the following solids has a volume of at least 500 cubic units?

   I.   II.

         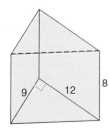

   A. I only   III.
   B. III only
   C. I and II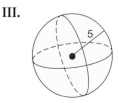
   D. I and III
   E. None of them

Algebra
Quick Notes

Order of Operations

Use the order of operations to evaluate expressions with more than one operation.

> **ORDER OF OPERATIONS**
>
> 1. Perform operations within grouping symbols first.
> 2. Perform all calculations involving exponents.
> 3. Multiply or divide in order from left to right.
> 4. Add or subtract in order from left to right.

Linear Equations

When one side of an equation has a variable with a coefficient of 1, the value of the expression on the other side of the equal symbol is the solution of the equation. Use the properties that follow to solve linear equations. When linear equations contain parentheses, you may need to use the distributive property. Use more than one property when the equation contains more than one operation or variables on both sides of the equal symbol.

> **PROPERTIES OF EQUALITY**
>
> For all real numbers a, b, and c:
>
> If $a = b$, then $a + c = b + c$. Addition   If $a = b$, then $ca = cb$.   Multiplication
>
> If $a = b$, then $a - c = b - c$. Subtraction   If $a = b$ and $c \neq 0$, then $\dfrac{a}{c} = \dfrac{b}{c}$. Division

> **DISTRIBUTIVE PROPERTY OF MULTIPLICATION OVER ADDITION**
>
> For all real numbers a, b, and c:
>
> $$a(b + c) = ab + ac \text{ and } (b + c)a = ba + ca$$

Linear Functions and Their Graphs

A linear function is a function that can be represented by a non-vertical line. The graphs show transformations of the parent linear function $y = x$.

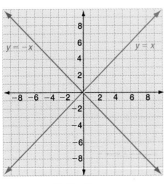

Reflection

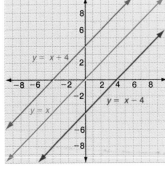

Vertical Translation

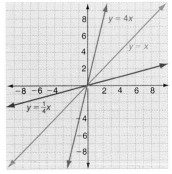

Change in Slope

Linear Inequalities

You can solve an inequality by getting the variable alone on one side of the inequality symbol, just as you do with equations. Use the following properties.

PROPERTIES OF INEQUALITY

For all real numbers a, b, and c:

If $a > b$, then $a + c > b + c$ and $a - c > b - c$.

If $a < b$, then $a + c < b + c$ and $a - c < b - c$.

If $a > b$ and $c > 0$, then $ac > bc$ and $\dfrac{a}{c} > \dfrac{b}{c}$.

If $a > b$ and $c < 0$, then $ac < bc$ and $\dfrac{a}{c} < \dfrac{b}{c}$.

Systems of Linear Equations

An ordered pair that is a solution of all the equations in a system of equations in two variables is a solution of that linear system. The system can solved by graphing both lines and locating their intersection, by using the substitution method, or by using the elimination method.

Substitution Method
1. Solve one equation for one variable in terms of the other variable.
2. Substitute for that variable in the other equation.

Elimination Method
1. Write the equations in standard form, $Ax + By = C$.
2. If no coefficients are identical or opposite coefficients, multiply one or both equations so that you have identical or opposite coefficients for one of the variables.
3. Add or subtract equations so that you have one equation in one variable.
4. Solve that equation.
5. Use the solution for that variable to find the solution for the other variable,

Quadratic Functions and Their Graphs

The standard form of a quadratic equation is $ax^2 + bx + c = 0$ and may have

- **one solution** if the graph of its corresponding function touches the x-axis at only one point.
- **two solutions** if the graph of its corresponding function crosses the x-axis at two points.
- **no solutions** if the graph of its corresponding function neither touches or crosses the x-axis.

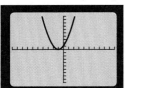

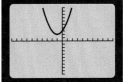

You can find the solution to a quadratic equation by using the *quadratic formula*.

QUADRATIC FORMULA

For a quadratic equation of the form $ax^2 + bx + c = 0$ where a, b, and c are real numbers and $a \neq 0$,

$$x = \frac{-b \pm \sqrt{b^2 - 4ac}}{2a}$$

The graphs show reflection, dilation, and translation of the parent quadratic function $y = x^2$.

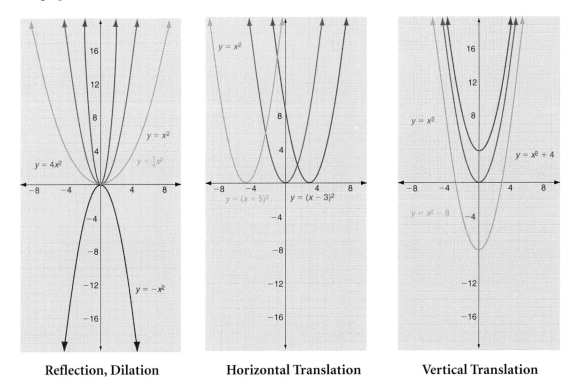

| Reflection, Dilation | Horizontal Translation | Vertical Translation |

Radical Expressions

A radical expression is in simplest form when the radicand (expression under the radical symbol) contains no perfect square factors other than 1 and no fractions. A simplified expression cannot have a denominator that contains a radical. Use these properties to simplify radical expressions.

PROPERTIES OF SQUARE ROOTS

For all real numbers a, b, where $a \geq 0$:

If $b \geq 0$, then $\quad \sqrt{ab} = \sqrt{a} \cdot \sqrt{b}.$ **Product Property**

If $b > 0$, then $\quad \sqrt{\dfrac{a}{b}} = \dfrac{\sqrt{a}}{\sqrt{b}}.$ **Quotient Property**

To eliminate a radical expression from a denominator, you must *rationalize the denominator* by multiplying both the numerator and denominator by the same expression.

Like radicals are radical expressions with the same radicand and are combined in a way similar to combining like terms. Use the distributive property to add or subtract radicals.

Solving Radical Equations

- A **radical equation** has a radical with a variable in the radicand. To solve radical equations use the principle of squaring.

  > PRINCIPLES OF SQUARING
  >
  > If the equation $a = b$ is true, then the equation
  >
  > $a^2 = b^2$ is also true.

Rational Expressions

- To add or subtract expressions with the same denominator, add or subtract the numerators. If the denominators are different, find the *least common denominator (LCD)* before adding or subtracting. Use the LCD to rewrite each expression as an equivalent rational expression having the LCD as its denominator.

  A complex rational expression contains one or more rational expressions in its numerator or denominator. To simplify, first determine the LCD of all the rational expressions in both the numerator and denominator. Then multiply the numerator and denominator by the LCD.

Rational Equations

- A **rational equation** contains one or more rational expressions. Use these steps to solve.
  1. Multiply both sides of the rational equation by the LCD to eliminate the denominators.
  2. Solve the resulting equation.
  3. Check each solution in the original equation. A value that makes a denominator of an expression in the original equation equal to zero is extraneous, and therefore, not a solution.

Measures of Central Tendency

- The *mean* is appropriate to use when all the data are approximately equal.

  > The mean or *average* is the sum of the data divided by the number of items of data.

  The **range** is the difference between the greatest and least values of data. When the range is large compared to the values themselves, the *median* may better represent the data.

  > The median is the middle value when the data are arranged in numerical order. When there are two middle values, the median is the average of the two.

  The *mode*, like the median, may be appropriate when the mean is not.

  > The mode is the element that occurs most often in the set. A set may have no mode, one mode, or several modes. If a set of data has two modes, the set is bimodal.

Geometry
Quick Notes

Geometry Basics

All geometric figures are made up of at least one point.

point   line   ray   line segment   angle
(or segment)

About Lines

Lines in a plane can be either parallel to each other or they can intersect each other.

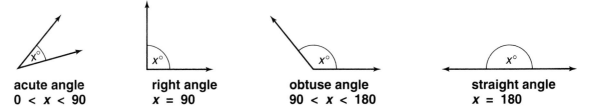

parallel lines   **intersecting lines**   **perpendicular lines**

About Angles

Angles are measured in degrees.

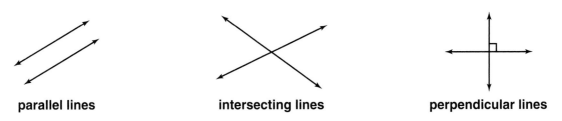

acute angle   **right angle**   **obtuse angle**   **straight angle**
$0 < x < 90$ $x = 90$ $90 < x < 180$ $x = 180$

Complementary and Supplementary Angles

Two angles are complementary if the sum of their measures is exactly 90°.

Two angles are supplementary if the sum of their measures is exactly 180°.

About Triangles

Triangles are three-sided plane figures. They can be classified according to the measures of their sides or their angles.

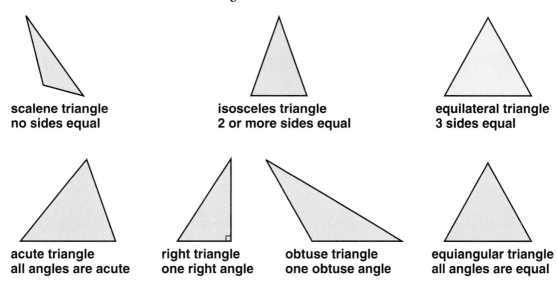

scalene triangle
no sides equal

isosceles triangle
2 or more sides equal

equilateral triangle
3 sides equal

acute triangle
all angles are acute

right triangle
one right angle

obtuse triangle
one obtuse angle

equiangular triangle
all angles are equal

About Quadrilaterals

Quadrilaterals are four-sided plane figures. Each figure in the diagram has all the properties of the figures preceding it, including the properties listed with that figure.

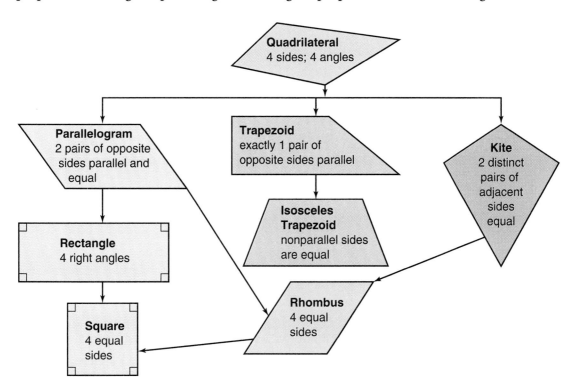

Quadrilateral
4 sides; 4 angles

Parallelogram
2 pairs of opposite sides parallel and equal

Trapezoid
exactly 1 pair of opposite sides parallel

Kite
2 distinct pairs of adjacent sides equal

Rectangle
4 right angles

Isosceles Trapezoid
nonparallel sides are equal

Square
4 equal sides

Rhombus
4 equal sides

About Other Polygons

Polygons are plane figures made up of segments and angles. Triangles and four-sided figures are also polygons.

pentagon

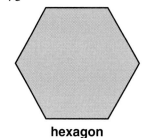
hexagon

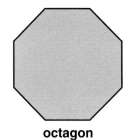
octagon

Perimeter Formulas

In the following formulas, l = length, w = width, s = side, and P = perimeter.

Perimeter of a rectangle $P = 2l + 2w$
Perimeter of a square $P = 4s$

Area Formulas

In the following formulas, b = base, B = long base, h = height, l = length, w = width, s = side, and A = area.

Area of a parallelogram $A = bh$
Area of a rectangle $A = lw$
Area of a square $A = s^2$
Area of a trapezoid $A = \frac{1}{2}(B + b)h$
Area of a triangle $A = \frac{1}{2}bh$

About Circles and Spheres

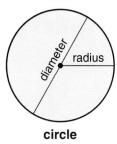
circle

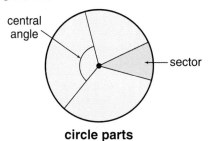
circle parts

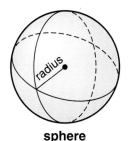
sphere

Circle Formulas

Circumference of a circle $C = 2\pi r$ or $C = \pi d$
Area of a circle $A = \pi r^2$

Sphere Formulas

Area of a sphere $A = 4\pi r^2$
Volume of a sphere $V = \frac{4}{3}\pi r^3$

About Geometric Solid Figures

Geometric solid figures are made up of plane polygons. Below are some geometric solid right figures.

cone

pyramid

cylinder

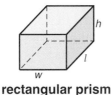

rectangular prism

cube

Base

The cone and the pyramid have one base. The cylinder and the prism have two bases and they are parallel. The cone and cylinder have circular bases. The base of a pyramid or prism can be any polygonal shape.

Lateral Surface

The lateral surface is the side or sides of the solid figure other than a base. The cone and cylinder have one lateral surface. The lateral surface of a pyramid is made up of triangles. The lateral surface of a right prism is made up of rectangles.

Slant Height

The slant height of a cone is measured from the vertex of the cone to the edge of its base. The slant height of a pyramid is measured from the vertex to the center of one side of the base.

Formulas

Total surface area of a right circular cone	$T = \pi r(l + r)$
Volume of a cone	$V = \frac{1}{3}\pi r^2 h$
Total surface area of a right cylinder	$T = 2\pi r(r + h)$
Volume of a cylinder	$V = \pi r^2 h$
Total surface area of a rectangular prism	$T = 2(lw + lh + wh)$
Volume of a rectangular prism	$V = lwh$
Total surface area of a cube	$T = 6s^2$
Volume of a cube	$V = s^3$

Technology

Quick Notes

- **For additional features and instructions, consult your user's manual.**

THE KEYBOARD The feature accessed when a key is pressed is shown in white on the key. To access the features in blue above each key, first press the 2ND key. To access what is in white above the keys, first press ALPHA.

CALCULATIONS Calculations are performed in the Home Screen. This screen may be returned to at any time by pressing 2ND QUIT. The calculator evaluates according to the order of operations. Press ENTER to calculate. For 3 + 4 × 5 ENTER, the result is 23. You can replay the previous line by pressing 2ND ENTER. Use the arrow keys to edit.

Displaying Graphs

- GRAPH FEATURE To enter an equation, press Y=. Enter an equation such as $y = 2x + 3$ using the X, T, θ key for x. Then press GRAPH to display the graph in the viewing window.

VIEWING WINDOW To set range values for the viewing window, press WINDOW. Press ▶ to access FORMAT where you can choose features such as Grid Off or Grid On.

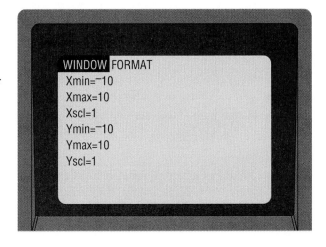

ZOOM FEATURE Press ZOOM and then 6 (Standard) to set a standard viewing window. Press ZOOM 1 (Box) to highlight a particular area and zoom in on that part of the graph. Press ZOOM 8 (Integer) to set values for a friendly window.

TRACE FEATURE Pressing TRACE places the cursor directly on the graph and shows the x- and y-coordinates of the point where the cursor is located. You can move the cursor along the graph using the right and left arrow keys.

TABLE FEATURE Press 2ND TBLSET to set up the table. Then press 2ND TABLE to see a table of values for each equation.

INTERSECTION FEATURE To determine the coordinates of the point of intersection of two graphs, press 2ND CALC, then 5 (Intersect). The calculator will then prompt you to identify the first graph. Use the right and left arrow keys to move the cursor to the first graph, close to the point of intersection. Repeat to identify the second graph and get the coordinates of the point of intersection.

Statistics

- ENTERING DATA Enter data into lists by pressing STAT 1 (Edit).

  CALCULATING STATISTICS Return to the Home Screen by pressing 2ND QUIT. To calculate the mean of List 1, press 2ND LIST ▶ (MATH) 3 (Mean) 2ND L1 ENTER. To calculate the median, choose 4 instead of 3. To calculate statistics, press STAT ▶ (CALC) 1 (1 - Var Stats) ENTER. You can also choose 2 to calculate two variable statistics and 5 to calculate linear regression. To see the lower quartile of a boxplot, press VARS 5 (Statistics) ▶ ▶ ▶ (BOX) 1 (Q1). Use 2 for median and 3 for the upper quartile.

  GRAPHING DATA To graph your data, press 2ND STATPLOT and choose a scatter plot, a line graph, a boxplot, or a histogram. Then choose the data list. Then press GRAPH to draw the graph.

Using the Casio CFX-9800G Graphing Calculator

- **For additional features and instructions, consult your user's manual.**

  THE KEYBOARD The feature accessed when a key is pressed is shown in white on the key. To access the features in gold above each key, first press the SHIFT key. To access what is in red above the keys, first press ALPHA.

  THE MAIN MENU This is the screen you see when you first turn the calculator on. Highlight and press EXE or press the number to choose the menu item. You can access the Main Menu at any time by pressing the MENU key.

  PERFORMING CALCULATIONS Calculations are performed by pressing 1 (for COMPutations) in the Main Menu. The calculator evaluates according to the order of operations. Press EXE (for EXEcute) to calculate. For $3 + 4 \times 5$ EXE, the result is 23. You can replay the previous line by pressing ◀. Use the arrow keys to edit.

Displaying Graphs

- COMP MODE Press 1 in the Main Menu. To enter an equation, press GRAPH. Enter an equation such as $y = 2x + 3$ using the X, θ, T key for x. Then press EXE to display the graph in the viewing window.

  GRAPH MODE Press 6 (GRAPH) in the Main Menu. Then press AC. Use the up or down arrows to choose a location to store the equation. Enter the equation as above. Then press F6 (DRW) to display the graph in the viewing window.

VIEWING WINDOW To set range values for the viewing window, press RANGE. You can use F1 (INIT) to set standard values for a viewing window.

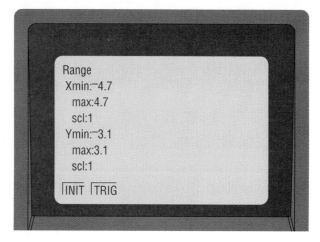

Range
Xmin:⁻4.7
   max:4.7
   scl:1
Ymin:⁻3.1
   max:3.1
   scl:1
|INIT |TRIG

ZOOM FEATURE Press SHIFT F2 (ZOOM) to access this feature. Press F1 (BOX) to highlight a particular area and zoom in on that part of the graph. Press F5 (AUT) to set range values for a friendly window.

TRACE FEATURE Pressing SHIFT F1 (TRACE) places the cursor directly on the graph and shows the x- and y-coordinates of the point where the cursor is located. You can move the cursor along the graph using the right arrow keys.

TABLE FEATURE Press 8 (TABLE) in the Main Menu. Press AC to clear the screen. Then press F1 (RANGE FUNC). Select the function. Then press F5 (RNG) to set up the table. Press F6 (TBL) to see the table of values. You can press SHIFT QUIT to return to a previous screen.

INTERSECTION FEATURE To determine the coordinates of the point of intersection of two graphs, press SHIFT, then 9 (G-SOLV). Then press F5 (ISCT). The calculator will then prompt you to identify the graphs.

Statistics

- **ENTERING DATA** From the Main Menu, press 3 (SD). Press SHIFT SET UP and select STOre for S-data. Then press EXIT AC to clear the screen. Enter data, pressing F1 after each entry.

  CALCULATING STATISTICS To calculate the mean ($\bar{x}$), press F4 (DEV) F1 ($\bar{x}$) EXE. To calculate the median, press F4 (DEV) F4 ▼ and F2 (Med) EXE.

  GRAPHING DATA To graph your data, press SHIFT SET UP and select DRAW for S-graph. Press EXIT and then GRAPH EXE.

Using the Hewlett-Packard 38G Graphing Calculator

- **For additional features and instructions, consult your user's manual.**

  THE KEYBOARD The feature accessed when a key is pressed is shown in yellow on the key. To access the features in green above each key, first press the green key. To access what is in red below the keys, first press A...Z. The blank keys at the top of the keyboard are used for the menu items at the bottom of the screen. Pressing the Menu key at the far right will return menu items to the screen.

CALCULATIONS Calculations are performed in the Home Screen. This screen may be returned to at any time by pressing HOME. The calculator evaluates according to the order of operations. Press ENTER to calculate. For $3 + 4 \times 5$ ENTER, the result is 23. You can edit a previous line by using the arrow keys to choose a line you want to edit. Press the Copy key. Use the arrow keys to edit.

Displaying Graphs

- GRAPH FEATURE To enter an equation, press LIBrary. Select Function. Press ENTER. Press the Edit key to enter an equation such as $y = 2x + 3$, using the X, T, θ key for x. Then press PLOT to display the graph in the viewing window.

  VIEWING WINDOW Press the green key and VIEWS. Select Auto Scale to get a friendly window, or you can press the green key and PLOT to change the range in the viewing window. Next, press PLOT to show the graph.

  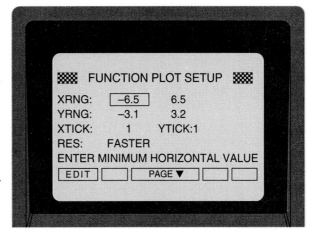

  ZOOM FEATURE Press the Menu key. Press the Zoom key. Then select Box... to highlight a particular area and zoom in on that part of the graph.

  TRACE FEATURE Pressing the Trace key puts the cursor directly on the graph and shows the x- and y-coordinates of the point of its location. You can move the cursor along the graph using the right and left arrow keys.

  TABLE FEATURE To display a table of values for the graphed equation, press NUMber.

  INTERSECTION FEATURE To determine the coordinates of the point of intersection of two graphs, move the cursor close to the point of intersection. Press the Function key and select Intersection. This verifies one of your equations. Press ENTER. Then press the Function key and select Intersection to verify the other equation. Press ENTER.

Statistical Features

- ENTERING DATA Data may be entered into a list by pressing LIB and selecting Statistics.

  CALCULATING STATISTICS To calculate mean, median, upper, and lower quartiles of the data entered in C1, press the Stats key and use the arrow keys to access all the information.

  GRAPHING DATA To graph your data, press the green key and then PLOT. Then press the Choos key. Select BoxWhisker or Histogram and ENTER. Then press the green key and VIEWS. Select Auto Scale.

Technology

Quick Notes

Using the Texas Instruments TI-92 Geometry

● **For additional features and instructions, consult your user's manual.**

THE KEYBOARD The feature accessed when a key is pressed is shown in white on the key. To access the features in yellow above a key, first press the 2ND key, then the key desired. To access what is in green above the keys, first press ◆, then the desired key. To access capital letters on the keyboard, use the ↑ key. The CAPS key allows all letters to be entered uppercase.

CALCULATIONS Calculations are performed in the Home Screen. This screen may be returned to at any time by pressing APPS, then 1 or ◆ HOME. The calculator evaluates according to the order of operations. Press ENTER to calculate. For $3 + 4 \times 5$ ENTER, the result is 23. You can replay a previous calculation by highlighting it with the arrow keypad and pressing ENTER. Use the arrow keypad or ← key to edit.

CONSTRUCTIONS Constructions are performed in the Geometry Screen. A new geometry screen may be accessed at any time by pressing APPS, 8, 3. In the NEW dialog box press the down arrow, name your construction. Use up to 8 characters, letters only and press ENTER twice.

POINTS To construct a new point, press F2 then 1. Position the cursor with the arrow keypad and press ENTER. To construct a point on an object already on the screen, press F2 then 2. Select the place on the object where you want your point with the arrow keypad and press ENTER.

To find the point of intersection of two objects already on the screen, press F2, then 3. Select each object with the arrow keypad and press ENTER. To construct the midpoint of a segment, press F4 then 3. Select the segment with the arrow keypad and press ENTER.

LINES, RAYS AND SEGMENTS To construct a new line, press F2 then 4. Designate the two points through which the line will pass with the arrow keypad and press ENTER after each. To construct a new ray, press F2, then 6. Designate the endpoint and another point through which the ray will pass. Move the cursor with the arrow keypad and press ENTER after each. To construct a new line segment, press F2 then 5. Designate the end points of the segment and press ENTER after each.

PARALLEL LINES To construct a line parallel to another line, ray or segment, press F4 then 2. Select the line, ray, or segment with the arrow keypad and press ENTER. Then select the point through which the parallel line will pass with the arrow keypad and press ENTER.

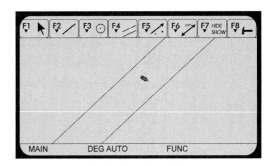

PERPENDICULAR LINES To construct a line perpendicular to another line, ray, or segment, press F4 then 1. Select the point on the line ray or segment through which you want the perpendicular to pass with the arrow keypad and press ENTER. To construct the perpendicular bisector of a line segment, press F4 then 4. Select the segment through which the perpendicular bisector passes, and press ENTER.

POLYGONS To construct a new triangle, press F3 then 3. Select the three vertices with the arrow keypad and press ENTER after each. To construct a new polygon, press F3 then 4. Select the vertices with the arrow keypad and press ENTER after each. When you have selected the final vertex, press ENTER a second time. To construct a new regular polygon, press F4 then 5. Select the center point of the polygon with the arrow keypad and press ENTER. Use the arrow keypad to determine the radius of the circumscribed circle and press ENTER. Then use the arrow keypad to determine the number of sides of the polygon and press ENTER. The number in braces near the center point will change as you move the cursor.

CIRCLES AND ARCS To construct a new circle, press F3 then 1. Use the arrow keypad to select the center and press ENTER. Then determine the radius with the arrow keypad and press ENTER. To construct a new arc, press F3 then 2. Select the initial endpoint, the curvature point and the final endpoint with the arrow keypad and press ENTER after each.

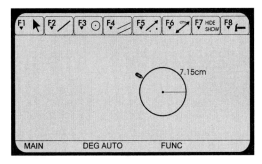

ANGLES To construct and angle, construct a line, ray or segment, and a second line ray or segment that intersects the first.

MEASUREMENTS To measure the length of a line segment, press F6 then 1. Select the line segment to be measured with the arrow keypad and press ENTER.

To measure an angle, press F6 then 3. Use the arrow keypad to select a point on one terminal side, the vertex point and a point on the other terminal side and press ENTER after each.

To find the area of an enclosed region, press F6 then 2. Select the region with the arrow keypad and press ENTER.

TRANSFORMATIONS To transform any object, move a marked point on that object. Select the point with the arrow keypad. Press and hold the HAND key. While holding the HAND key, move the point with the arrow keypad.

LABELS To label a point, press F7 then 4. Select the point with the arrow keypad and press ENTER. Label the point with one letter and press ENTER.

POSTULATES

Postulates About Points, Lines, and Planes

Postulate 1 Unique Line Postulate Through two distinct points, there is exactly one line. (p. 66)

Postulate 2 A line contains at least two distinct points. (p. 66)

Postulate 3 Unique Plane Postulate Through three noncollinear points, there is exactly one plane. (p. 66)

Postulate 4 A plane contains at least three noncollinear points. (p. 66)

Postulate 5 If two distinct points lie in a plane, then the line joining them lies in that plane. (p. 66)

Postulate 6 If two distinct planes intersect, then their intersection is a line. (p. 66)

Postulates About Segments and Angles

Postulate 7 Ruler Postulate The points on a line can be paired, one-to-one, with the real numbers so that any point is paired with 0 and any other point is paired with 1. The real number that corresponds to a point is the *coordinate* of that point. The *distance* between two points on the line is equal to the absolute value of the difference of their coordinates. (p. 72)

Postulate 8 Segment Addition Postulate If point C is between points A and B, then $AC + CB = AB$. (p. 73)

Postulate 9 Protractor Postulate Let O be a point on $\overleftrightarrow{AB}$ such that O is between A and B. Consider $\overrightarrow{OA}$, $\overrightarrow{OB}$, and all the rays that can be drawn from O on one side of $\overleftrightarrow{AB}$. These rays can be paired with the real numbers from 0 to 180 in such a way that:
1. $\overrightarrow{OA}$ is paired with 0 and $\overrightarrow{OB}$ is paired with 180.
2. If $\overrightarrow{OP}$ is paired with x and $\overrightarrow{OQ}$ is paired with y, then the number paired with $\angle POQ$ is $|x - y|$. This number is called the measure, or the degree measure, of $\angle POQ$. (p. 92)

Postulate 10 Angle Addition Postulate If point B is in the interior of $\angle AOC$, then m$\angle AOB +$ m$\angle BOC =$ m$\angle AOC$. (p. 94)

Postulate 11 Linear Pair Postulate If two angles form a linear pair, then they are supplementary. (p. 94)

Postulates About Parallel and Perpendicular Lines

Postulate 12 Corresponding Angles Postulate If two parallel lines are cut by a transversal, then the pairs of corresponding angles are congruent. (p. 139)

Postulate 13 Parallel Postulate Through a point not on a line, there is exactly one line parallel to the given line. (p. 148)

Postulate 14 Converse of Corresponding Angles Postulate If two lines are cut by a transversal so that corresponding angles are congruent, then the lines are parallel. (p. 148)

Postulate 15 Two nonvertical lines are parallel if and only if their slopes are equal. (p. 163)

Postulate 16 Two nonvertical lines are perpendicular if and only if the product of their slopes is -1. (p. 163)

Postulates About Triangle Congruence

Postulate 17 Side-Side-Side (SSS) Congruence Postulate If three sides of one triangle are congruent to three sides of another triangle, then the triangles are congruent. (p. 275)

Postulate 18 Side-Angle-Side (SAS) Congruence Postulate If two sides and the included angle of one triangle are congruent to two sides and the included angle of another triangle, then the triangles are congruent. (p. 275)

Postulate 19 Angle-Side-Angle (ASA) Congruence Postulate If two angles and the included side of one triangle are congruent to two angles and the included side of another triangle, then the triangles are congruent. (p. 275)

Postulates About Area

Postulate 20 Area of a Square Postulate The area A of a square is the square of the length s of one side, or $A = s^2$. (p. 435)

Postulate 21 Area Congruence Postulate If two polygons are congruent, then they have the same area. (p. 435)

POSTULATES/THEOREMS

Postulate 22 Area Addition Postulate   The area of a region is the sum of the areas of all its nonoverlapping parts. (p. 435)

Postulates About Similarity

Postulate 23 Angle-Angle (AA) Similarity Postulate
If two angles of one triangle are congruent to two angles of another triangle, then the triangles are similar. (p. 524)

Postulates About Arcs

Postulate 24 Arc Addition Postulate   The measure of an arc formed by two adjacent arcs is the sum of the measures of the two arcs. (p. 669)

Postulates About Volume

Postulate 25 Volume of a Cube Postulate   The volume V of a cube is the cube of the length e of one edge, or $V = e^3$. (p. 737)

Postulate 26 Volume Congruence Postulate   If two three-dimensional figures are congruent, then they have the same volume. (p. 737)

Postulate 27 Volume Addition Postulate   The volume of a three-dimensional figure is the sum of the volumes of all its nonoverlapping parts. (p. 737)

Postulate 28 Cavalieri's Principle   If two solids lying between parallel planes have the same height and all cross sections at equal distances from their bases have equal areas, then the solids have equal volume. (p. 768)

THEOREMS

Theorems About Segments and Angles

Theorem 2.1 Midpoint Theorem   If point M is the midpoint of $\overline{AB}$, then $AM = \frac{1}{2}AB$ and $MB = \frac{1}{2}AB$. (p. 117)

Theorem 2.2 Angle Bisector Theorem   If $\overrightarrow{BX}$ is the bisector of $\angle ABC$, then $m\angle ABX = \frac{1}{2} m\angle ABC$ and $m\angle XBC = \frac{1}{2} m\angle ABC$. (p. 117)

Theorem 2.3 Congruent Supplements Theorem   Supplements of congruent angles are congruent. (p. 118)

Theorem 2.4 Congruent Complements Theorem   Complements of congruent angles are congruent. (p. 118)

Theorem 2.5   All right angles are congruent. (p. 118)

Theorem 3.1 Vertical Angles Theorem   If two angles are vertical angles, then they are congruent. (p. 129)

Theorem 3.2   If two lines are perpendicular, then they intersect to form four right angles. (p. 130)

Theorems About Parallel Lines and Transversals

Theorem 3.3 Alternate Interior Angles Theorem   If two parallel lines are cut by a transversal, then the pairs of alternate interior angles are congruent. (p. 140)

Theorem 3.4 Alternate Exterior Angles Theorem   If two parallel lines are cut by a transversal, then the pairs of alternate exterior angles are congruent. (p. 140)

Theorem 3.5 Same-Side Interior Angles Theorem   If two parallel lines are cut by a transversal, then the pairs of same-side interior angles are supplementary. (p. 141)

Theorem 3.6 Perpendicular Transversal Theorem   If a transversal is perpendicular to one of two parallel lines, then it is perpendicular to the other. (p. 142)

Theorem 3.7 Converse of Alternate Interior Angles Theorem   If two lines are cut by a transversal so that alternate interior angles are congruent, then the lines are parallel. (p. 149)

Theorem 3.8 Converse of Alternate Exterior Angles Theorem   If two lines are cut by a transversal so that alternate exterior angles are congruent, then the lines are parallel. (p. 149)

Theorem 3.9 Converse of Same-Side Interior Angles Theorem   If two lines are cut by a transversal so that same-side interior angles are supplementary, then the lines are parallel. (p. 149)

Theorem 3.10   If two lines are parallel to a third line, then they are parallel to each other. (p. 150)

Theorem 3.11   If two coplanar lines are perpendicular to a third line, then they are parallel to each other. (p. 150)

Theorems About Transformations

Theorem 4.1   A reflection is an isometry. (p. 211)

Theorem 4.2   A translation is an isometry. (p. 219)

Theorem 4.3   If $\ell \parallel m$, then a reflection over line ℓ followed by a reflection over line m is a translation. If P'' is the image of P after the two reflections, then PP'' is perpendicular to ℓ and $PP'' = 2d$, where d is the distance between ℓ and m. (p. 220)

Theorem 4.4   A rotation is an isometry. (p. 225)

Theorem 4.5   If two lines, ℓ and m, intersect at point O, then a reflection over line ℓ followed by a reflection over line m is a rotation about point O. The measure of the angle of rotation is $(2x)°$, where $x°$ is the measure of the acute or right angle between ℓ and m. (p. 226)

Theorem 4.6   A composition of two or more isometries is an isometry. (p. 231)

Theorems About Angles of Polygons

Theorem 5.1 Angle-Sum Theorem for Triangles   The sum of the measures of the angles of a triangle is 180°. (p. 256)

Theorem 5.2 Exterior-Angle Theorem for Triangles   The measure of an exterior angle of a triangle is equal to the sum of the measures of the two remote interior angles. (p. 257)

Theorem 5.3 Angle-Sum Theorem for Polygons   The sum of the measures of the angles of a convex polygon with n sides is $(n - 2)180°$. (p. 260)

Corollary 5.4   The measure of each angle of a regular n-gon is $\dfrac{(n - 2)180°}{n}$. (p. 261)

Theorem 5.5 Exterior-Angle Theorem for Polygons   The sum of the measures of the exterior angles of a convex polygon, one angle at each vertex, is 360°. (p. 261)

Corollary 5.6   The measure of each exterior angle of a regular n-gon is $\dfrac{360°}{n}$. (p. 261)

Theorems About Triangle Congruence

Theorem 5.7 Angle-Angle-Sided (AAS) Congruence Theorem   If two angles and a nonincluded side in one triangle are congruent to two angles and the corresponding nonincluded side in another triangle, then the triangles are congruent. (p. 288)

Theorem 5.8 Hypotenuse-Leg (HL) Congruence Theorem   If the hypotenuse and one leg of a right triangle are congruent to the hypotenuse and one leg of another right triangle, then the triangles are congruent. (p. 290)

Theorems About Properties of Triangles

Theorem 6.1 Perpendicular Bisector Theorem   If a point lies on the perpendicular bisector of a segment, then the point is equidistant from the endpoints of the segment. (p. 316)

Theorem 6.2 Converse of Perpendicular Bisector Theorem   If a point is equidistant from the endpoints of a segment, then the point lies on the perpendicular bisector of the segment. (p. 317)

Theorem 6.3 Angle Bisector Theorem   If a point lies on the bisector of an angle, then the point is equidistant from the sides of the angle. (p. 319)

Theorem 6.4 Converse of Angle Bisector Theorem   If a point is equidistant from the sides of an angle, then the point lies on the bisector of the angle. (p. 319)

Theorem 6.5 Base Angles Theorem   If two sides of a triangle are congruent, then the angles opposite those sides are congruent. (p. 325)

Corollary 6.6 If a triangle is equilateral, then it is equiangular. (p. 326)

Corollary 6.7 The measure of each angle of an equilateral triangle is 60°. (p. 326)

Theorem 6.8 Converse of Base Angles Theorem If two angles of a triangle are congruent, then the sides opposite those angles are congruent. (p. 327)

Corollary 6.9 If a triangle is equiangular, then it is equilateral. (p. 327)

Theorem 6.10 Triangle Midsegment Theorem The segment joining the midpoints of two sides of a triangle is parallel to the third side, and its length is half the length of the third side. (p. 336)

Theorems About Triangle Inequalities

Theorem 6.11 Unequal Sides Theorem If two sides of a triangle are unequal in length, then the measure of the angle opposite the longer side is greater than the measure of the angle opposite the shorter side. (p. 352)

Theorem 6.12 Unequal Angles Theorem If two angles of a triangle are unequal in measure, then the side opposite the larger angle is longer than the side opposite the smaller angle. (p. 353)

Theorem 6.13 Triangle Inequality Theorem The sum of the lengths of any two sides of a triangle is greater than the length of the third side. (p. 353)

Theorem 6.14 Hinge Theorem If two sides of one triangle are congruent to two sides of another triangle, and the included angle of the first triangle is greater than the included angle of the second, then the third side of the first triangle is longer than the third side of the second. (p. 358)

Theorem 6.15 Converse of Hinge Theorem If two sides of one triangle are congruent to two sides of another triangle, and the third side of the first triangle is longer than the third side of the second, then the included angle of the first triangle is larger than the included angle of the second. (p. 359)

Theorems About Properties of Quadrilaterals

Theorem 7.1 If a quadrilateral is a parallelogram, then its opposite sides are congruent. (p. 378)

Theorem 7.2 If a quadrilateral is a parallelogram, then its opposite angles are congruent. (p. 378)

Theorem 7.3 If a quadrilateral is a parallelogram, then its consecutive angles are supplementary. (p. 378)

Theorem 7.4 If a quadrilateral is a parallelogram, then its diagonals bisect each other. (p. 378)

Theorem 7.5 If both pairs of opposite sides of a quadrilateral are congruent, then the quadrilateral is a parallelogram. (p. 384)

Theorem 7.6 If both pairs of opposite angles of a quadrilateral are congruent, then the quadrilateral is a parallelogram. (p. 384)

Theorem 7.7 If an angle of a quadrilateral is supplementary to both consecutive angles, then the quadrilateral is a parallelogram. (p. 384)

Theorem 7.8 If the diagonals of a quadrilateral bisect each other, then the quadrilateral is a parallelogram. (p. 386)

Theorem 7.9 If one pair of opposite sides of a quadrilateral are congruent and parallel, then the quadrilateral is a parallelogram. (p. 386)

Theorem 7.10 A parallelogram is a rhombus if and only if its diagonals are perpendicular. (p. 392)

Theorem 7.11 A quadrilateral is a rhombus if and only if each diagonal bisects a pair of opposite angles. (p. 392)

Theorem 7.12 A parallelogram is a rectangle if and only if its diagonals are congruent. (p. 392)

Theorem 7.13 A parallelogram is a square if and only if its diagonals are both perpendicular and congruent. (p. 393)

Theorem 7.14 Midsegment Theorem for Trapezoids The midsegment of a trapezoid is parallel to each base, and its length is equal to half the sum of the lengths of the bases. (p. 397)

Theorem 7.15 If a quadrilateral is an isosceles trapezoid, then its base angles are congruent. (p. 398)

Theorem 7.16 If a quadrilateral is an isosceles trapezoid, then its diagonals are congruent. (p. 398)

Theorem 7.17 If the diagonals of a trapezoid are congruent, then the trapezoid is isosceles. (p. 399)

Theorem 7.18 If one pair of base angles of a trapezoid are congruent, then the trapezoid is isosceles. (p. 399)

Theorem 7.19 If a quadrilateral is a kite, then its diagonals are perpendicular. (p. 404)

Theorem 7.20 Every kite has a diagonal that bisects the angles at its endpoints. (p. 405)

Theorems About Area

Theorem 8.1 Area of a Parallelogram The area A of a parallelogram equals the product of a base b and the height h to that base, or $A = bh$. (p. 441)

Theorem 8.2 Area of a Triangle The area A of a triangle equals half the product of a base b and the height h to that base, or $A = \frac{1}{2}bh$. (p. 442)

Theorem 8.3 Area of a Rhombus The area A of a rhombus equals half the product of the lengths of its diagonals d_1 and d_2, or $A = \frac{1}{2}d_1 d_2$. (p. 443)

Theorem 8.4 Area of a Trapezoid The area A of a trapezoid equals half the product of its height h and the sum of its bases b_1 and b_2, or $A = \frac{1}{2}h(b_1 + b_2)$. (p. 448)

Theorem 8.5 Pick's Theorem If a polygon on a lattice has B boundary points and I interior points, then its area A is $A = \frac{1}{2}B + I - 1$. (p. 450)

Theorem 8.6 Pythagorean Theorem If a triangle is a right triangle with legs of length a and b and hypotenuse of length c, then $a^2 + b^2 = c^2$. (p. 466)

Theorem 8.7 Converse of Pythagorean Theorem If a, b, and c are the lengths of the sides of a triangle such that $a^2 + b^2 = c^2$, then the triangle is a right triangle. (p. 474)

Theorem 8.8 If the square of the lengths of the longest side of a triangle is greater than the sum of the squares of the lengths of the other two sides, then the triangle is an obtuse triangle. (p. 476)

Theorem 8.9 If the square of the length of the longest side of a triangle is less than the sum of the squares of the lengths of the other two sides, then the triangle is an acute triangle. (p. 476)

Theorems About Similarity

Theorem 9.1 If two polygons are similar with a scale factor of $a{:}b$, then the ratio of their perimeters is $a{:}b$. (p. 516)

Theorem 9.2 If two polygons are similar with a scale factor of $a{:}b$, then the ratio of their areas is $a^2{:}b^2$. (p. 516)

Theorem 9.3 Side-Side-Side (SSS) Similarity Theorem If corresponding sides of two triangles are proportional, then the triangles are similar. (p. 524)

Theorem 9.4 Side-Angle-Side (SAS) Similarity Theorem If an angle of one triangle is congruent to an angle of a second triangle, and the lengths of the sides including these angles are proportional, then the triangles are similar. (p. 524)

Theorem 9.5 Corresponding medians of similar triangles are proportional to the corresponding sides. (p. 526)

Theorem 9.6 Corresponding altitudes of similar triangles are proportional to the corresponding sides. (p. 526)

Theorem 9.7 Triangle Proportionality Theorem If a line parallel to one side of a triangle intersects the other two sides, then it divides those sides proportionally. (p. 536)

Corollary 9.8 If three parallel lines have two transversals, then they divide the transversals proportionally. (p. 537)

Theorem 9.9 Converse of Triangle Proportionality Theorem If a line divides two sides of a triangle proportionally, then it is parallel to the third side. (p. 538)

Theorem 9.10 Triangle Angle-Bisector Theorem If a ray bisects an angle of a triangle, then it divides the opposite side into segments proportional to the other two sides of the triangle. (p. 538)

Theorems About Right Triangles

Theorem 10.1 If the altitude is drawn to the hypotenuse of a right triangle, then the two triangles formed are similar to the original triangle and to each other. (p. 556)

Corollary 10.2 If the altitude is drawn to the hypotenuse of a right triangle, then its length is the geometric mean of the lengths of the two segments of the hypotenuse. (p. 557)

Corollary 10.3 If the altitude is drawn to the hypotenuse of a right triangle, then the length of each leg is the geometric mean of the length of the hypotenuse and the length of the segment of the hypotenuse that is adjacent to the leg. (p. 557)

Theorem 10.4 45°−45°−90° Theorem In a 45°−45°−90° triangle, the length of the hypotenuse is $\sqrt{2}$ times the length of a leg. (p. 564)

Theorem 10.5 30°−60°−90° Theorem In a 30°−60°−90° triangle, the length of the hypotenuse is twice the length of the shorter leg, and the length of the longer leg is $\sqrt{3}$ times the length of the shorter leg. (p. 566)

Theorem 10.6 Area of a Regular Polygon The area A of a regular polygon equals half the product of its apothem a and perimeter p, or $A = \frac{1}{2}ap$. (p. 595)

Theorems About Circles

Theorem 11.1 Circumference of a Circle The circumference C of a circle equals the product of π and its diameter d, or $C = \pi d$. (p. 613)

Corollary 11.2 Arc Length The ratio of the length of a minor arc of a circle to the circumference equals the ratio of the degree measure of the central angle that intercepts the arc to 360°. The length of a semicircle is half the circumference. (p. 614)

Theorem 11.3 Area of a Circle The area A of a circle equals the product of π and the square of its radius r, or $A = \pi r^2$. (p. 620)

Theorem 11.4 Area of a Sector The ratio of the area of a sector of a circle to the area of the circle equals the ratio of the degree measure of the central angle of the sector to 360°. (p. 621)

Theorem 12.1 If a line is tangent to a circle, then it is perpendicular to the radius drawn to the point of tangency. (p. 663)

Corollary 12.2 If two segments from the same exterior point of a circle are tangent to a circle, then they are congruent. (p. 664)

Theorem 12.3 In a plane, if a line is perpendicular to a radius of a circle at its endpoint on the circle, then the line is tangent to the circle at that point. (p. 665)

Theorem 12.4 In the same circle, or in congruent circles, two minor arcs are congruent if and only if the central angles that intercept them are congruent. (p. 671)

Theorem 12.5 In the same circle, or in congruent circles, two minor arcs are congruent if and only if their corresponding chords are congruent. (p. 675)

Theorem 12.6 If a diameter of a circle is perpendicular to a chord, then the diameter bisects the chord and its arc. (p. 676)

Theorem 12.7 In the same circle, or in congruent circles, two chords are congruent if and only if they are equidistant from the center. (p. 677)

Theorem 12.8 If one chord of a circle is the perpendicular bisector of another, then the first chord is a diameter of the circle. (p. 677)

Theorem 12.9 If an angle is inscribed in a circle, then its measure is half the measure of its intercepted arc. (p. 681)

Corollary 12.10 If two inscribed angles in a circle intercept the same arc or congruent arcs, then the angles are congruent. (p. 682)

Corollary 12.11 If an inscribed angle in a circle intercepts a semicircle, then the angle is a right angle. (p. 682)

Corollary 12.12 If a quadrilateral is inscribed in a circle, then its opposite angles are supplementary. (p. 683)

Theorem 12.13 If two secants, one secant and one tangent, or two tangents intersect in the exterior of a circle, then the measure of the angle formed is equal to half the difference of the measures of the intercepted arcs. (p. 694)

Theorem 12.14   If two secants intersect in the interior of a circle, then the measure of each angle is half the sum of the measures of the arcs intercepted by the angle and its vertical angle. (p. 696)

Theorem 12.15   If a secant and a tangent to the circle intersect at the point of tangency, then the measure of each angle formed is half the measure of the intercepted arc. (p. 696)

Theorem 12.16   If two chords intersect in the interior of a circle, then the product of the lengths of the segments of one chord equals the product of the lengths of the segments of the other chord. (p. 700)

Theorem 12.17   If two secants intersect in the exterior of a circle, then the product of the lengths of one secant segment and its external segment is equal to the product of the lengths of the other secant segment and its external segment. (p. 701)

Theorem 12.18   If a secant and a tangent intersect in the exterior of a circle, then the product of the lengths of the secant segment and its external segment is equal to the square of the length of the tangent segment. (p. 701)

Theorems About Surface Area and Volume

Theorem 13.1 Lateral Area of a Right Prism   The lateral area LA of a right prism equals the product of the perimeter p of a base and the height h of the prism, or $LA = ph$. (p. 732)

Theorem 13.2 Surface Area of a Right Prism   The surface area SA of a right prism equals the sum of its lateral area LA and twice the area of one of its bases B, or $SA = LA + 2B = ph + 2B$ (p. 732)

Theorem 13.3 Lateral Area of a Right Cylinder   The lateral area LA of a right cylinder with radius r equals the product of the circumference of a base and the height h of the cylinder, or $LA = 2\pi rh$. (p. 733)

Theorem 13.4 Surface Area of a Right Cylinder   The surface area SA of a right cylinder with radius r equals the sum of its lateral area LA and the area of its bases, or $SA = LA + 2B = 2\pi rh + 2\pi r^2$. (p. 733)

Theorem 13.5 Volume of a Prism   The volume V of a prism equals the product of the area of a base B and the height h, or $V = Bh$. (p. 738)

Theorem 13.6 Volume of a Right Cylinder   The volume V of a right cylinder with radius r equals the product of the area of a base B and the height h, or $V = Bh = \pi r^2 h$. (p. 739)

Theorem 13.7 Lateral Area of a Regular Pyramid   The lateral area LA of a regular pyramid equals half the product of the slant height ℓ of the pyramid and the perimeter p of its base, or $LA = \frac{1}{2}\ell p$. (p. 749)

Theorem 13.8 Surface Area of a Regular Pyramid   The surface area SA of a regular pyramid of slant height ℓ and base perimeter p equals the sum of its lateral area LA and the area of its base B, or $SA = LA + B = \frac{1}{2}\ell p + B$. (p. 750)

Theorem 13.9 Lateral Area of a Right Cone   The lateral area LA of a right cone of radius r equals half the product of the circumference of its base and the slant height ℓ of the cone, or $LA = \frac{1}{2} \cdot 2\pi r\ell = \pi r\ell$. (p. 750)

Theorem 13.10 Surface Area of a Right Cone   The surface area SA of a right cone of radius r and slant height ℓ equals the sum of its lateral area LA and the area of its base B, or $SA = LA + B = \pi r\ell + \pi r^2$. (p. 750)

Theorem 13.11 Volume of a Pyramid   The volume V of a pyramid equals one third the product of the area of the base B and the height h, or $V = \frac{1}{3}Bh$. (p. 757)

Theorem 13.12 Volume of a Cone   The volume V of a cone of radius r equals one third the product of the area of the base B and the height h, or $V = \frac{1}{3}Bh = \frac{1}{3}\pi r^2 h$. (p. 757)

Theorem 13.13 Surface Area of a Sphere   The surface area SA of a sphere equals the product of 4π and the square of the radius r, or $SA = 4\pi r^2$. (p. 764)

Theorem 13.14 Volume of a Sphere   The volume V of a sphere equals the product of $\frac{4}{3}\pi$ and the cube of the radius r, or $V = \frac{4}{3}\pi r^3$. (p. 765)

Theorem 13.15   If two solids are similar with scale factor $a : b$, then the ratio of the perimeters of corresponding faces is $a : b$, the ratio of the areas of corresponding faces is $a^2 : b^2$, and the ratio of their volumes is $a^3 : b^3$. (p. 770)

GLOSSARY

• • **A** • •

acute angle (p. 93) An angle whose measure is greater than 0° and less than 90°.

acute triangle (p. 256) A triangle with three acute angles.

adjacent angles (p. 94) Two angles in the same plane that share a common side and a common vertex, but have no interior points in common.

adjacent arcs (p. 669) Arcs that are part of the same circle and intersect at exactly one point.

alternate exterior angles (p. 140) When two lines are intersected by a transversal, alternate exterior angles are two nonadjacent exterior angles on opposite sides of the transversal.

alternate interior angles (p. 140) When two lines are intersected by a transversal, alternate interior angles are two nonadjacent interior angles on opposite sides of the transversal.

altitude of a cone (p. 647) The segment from the vertex perpendicular to the plane of the base.

altitude of a cylinder (p. 646) A segment perpendicular to the bases with one endpoint in the plane of each base.

altitude of a parallelogram (p. 441) Any segment perpendicular to the line containing a base from any point on the opposite side.

altitude of a prism (p. 468) A segment perpendicular to the bases with one endpoint on each base.

altitude of a pyramid (p. 468) The segment from the vertex perpendicular to the base.

altitude of a trapezoid (p. 448) Any segment perpendicular to the line containing one base from any point on the opposite base.

altitude of a triangle (pp. 333, 443) A perpendicular segment from a vertex to the line containing the opposite side.

angle (pp. 31, 92) The figure formed by two rays that have a common endpoint.

angle bisector (p. 95) The ray that divides the angle into two congruent angles.

angle of depression (p. 575) An angle whose vertex is at the eye of an observer and whose sides are a horizontal line and the observer's line of sight downward to an object.

angle of elevation (p. 575) An angle whose vertex is at the eye of an observer and whose sides are a horizontal line and the observer's line of sight upward to an object.

angle of rotation (p. 224) See *rotation*.

apothem of a regular polygon (p. 594) A segment, or the length of the segment, whose endpoints are the center of the polygon and the midpoint of a side.

arc (pp. 48, 614) An unbroken part of a circle. Any two distinct points on a circle divide the circle into two arcs. These points are the *endpoints* of each arc.

area of a plane figure (p. 435) The number of nonoverlapping square units contained in the interior of the figure.

auxiliary figure (pp. 212, 257) A point, line, ray, or segment added to a figure to aid in a proof.

axis of a cone (p. 647) The segment whose endpoints are the vertex and the center of the base.

axis of a cylinder (p. 646) The segment whose endpoints are the centers of the bases.

axis of symmetry (p. 18) If a three-dimensional figure can be turned around a line so it coincides with its original position two or more times during a complete turn, the line is an axis of symmetry.

• • **B** • •

base angles of an isosceles triangle (p. 324) The angles opposite the legs.

base angles of a trapezoid (p. 397) Two consecutive angles whose vertices are endpoints of a single base form a pair of base angles.

base of a cone or cylinder (pp. 646, 647) See *cone* and *cylinder*.

base of an isosceles triangle (p. 324) See *isosceles triangle*.

base of a parallelogram (p. 441) Any side of a parallelogram may be considered a base. The word *base* also refers to the length of that side.

base of a prism or pyramid (pp. 419, 420) See *prism* and *pyramid*.

base of a triangle (p. 442) The side of a triangle to which an altitude is drawn. The word *base* also refers to the length of that side.

bases of a trapezoid (p. 397) The parallel sides, or the lengths of the parallel sides.

bearing of an object (p. 160) The measure, taken in a clockwise direction, of the angle whose vertex is an observer's location and whose sides are a ray along a *reference meridian* and a ray along the observer's line of sight to the object. If the reference meridian is the longitudinal line through the geographic North Pole, the bearing is a *true bearing*. If the reference meridian is the longitudinal line through the magnetic pole, the bearing is a *magnetic bearing*.

between (p. 73) On a number line, a point *C* is between points *A* and *B* if the coordinate of *C* is between the coordinates of *A* and *B*.

biconditional statement (p. 105) The statement formed by combining a conditional statement and its converse, when both are true, and using the connector *if and only if*. Also called a *biconditional*.

bisector of an angle (p. 95) See *angle bisector*.

bisector of a segment (pp. 49, 74, 88) See *segment bisector*.

• • C • •

center of a circle (pp. 48, 609) The point in the plane of the circle that is equidistant from all the points of the circle.

center of gravity (p. 332) See *centroid of a triangle*.

center of a regular polygon (p. 594) The point that is equidistant from all the vertices of the polygon.

center of rotation (p. 224) See *rotation*.

center of a sphere (pp. 154, 648) The point that is equidistant from all the points of the sphere.

center of symmetry (p. 10) If a plane figure can be turned around a point so it coincides with its original position two or more times during a complete turn, the point is its center of symmetry.

central angle of a circle (p. 614) An angle in the plane of the circle whose vertex is the center of the circle.

central angle of a regular polygon (p. 595) An angle whose vertex is the center of the polygon and whose sides contain two consecutive vertices.

centroid of a triangle (p. 332) The point of concurrency of the medians. Also called the *center of gravity*.

chord of a circle (p. 612) A segment whose endpoints are points on the circle.

chord of a sphere (p. 648) A segment whose endpoints are points on the sphere.

circle (pp. 48, 609, 612) The set of all points in a plane that are a given distance from a given point in the plane.

circumcenter of a triangle (p. 332) The point of concurrency of the perpendicular bisectors of the sides.

circumference of a circle (pp. 609, 613) The perimeter of the circle.

circumscribed circle (pp. 332, 626) A circle is circumscribed about a polygon if each vertex of the polygon lies on the circle.

circumscribed polygon (p. 332) A polygon is circumscribed about a circle if each side of the polygon intersects the circle at exactly one point.

collinear points (p. 66) Points that lie on the same line. Points that do not lie on the same line are called *noncollinear points*.

common tangent (p. 665) A line that is tangent to each of two circles in the same plane. A *common internal tangent* intersects the segment whose endpoints are the centers of the circles; a *common external tangent* does not.

compass (p. 48) A tool used to draw circles and arcs.

compass-and-straightedge construction (p. 48) A precise drawing of a geometric figure made with the aid of only two tools: a compass and an unmarked straightedge.

complementary angles (p. 93) Two angles whose measures have a sum of 90°. Each angle is called the *complement* of the other.

composition of transformations (p. 231) The process of performing a transformation on a figure followed by a second transformation on the figure's image.

concave polygon (p. 196) A polygon in which a line that contains a side of the polygon also contains a point in its interior. Also called a *nonconvex polygon*.

concave polyhedron (p. 725) A polyhedron in which a plane that contains a face of the polyhedron also contains a point in its interior.

concentric circles (p. 612) Circles in the same plane that have the same center.

conclusion (p. 53) In a conditional statement, the part that follows *then*.

concurrent lines (p. 332) Two or more lines that intersect at a single point. The point of intersection is called the *point of concurrency*.

conditional statement (p. 53) A statement that can be written in *if-then* form. Also called a *conditional*.

cone (pp. 16, 647) A three-dimensional figure that consists of a circular face, called the *base*, a point called the *vertex* that is not in the plane of the base, and a *lateral surface* that connects the vertex to each point on the boundary of the base.

congruent angles (p. 95) Angles that are equal in measure.

congruent arcs (p. 671) Arcs of the same circle or of congruent circles that are equal in measure.

congruent circles (p. 612) Circles that have congruent radii.

congruent figures (p. 33) Figures that have exactly the same shape and size.

congruent polygons (p. 211) Polygons whose sides and angles can be placed in a correspondence such that corresponding sides are congruent and corresponding angles are congruent.

congruent segments (p. 74) Segments that are equal in length.

conic section (p. 651) The cross section formed when a plane intersects a double right cone.

conjecture (p. 85) A generalization that results from inductive reasoning. Also called an *educated guess*.

contrapositive (p. 106) The statement that results when the hypothesis and conclusion of a conditional statement are both negated, then interchanged.

converse (p. 104) The statement that results when the hypothesis and conclusion of a conditional statement are interchanged.

convex polygon (p. 196) A polygon in which no line that contains a side of the polygon contains a point in its interior.

convex polyhedron (p. 725) A polyhedron in which no plane that contains a face of the polyhedron contains a point in its interior.

coordinate plane (p. 41) A number plane formed by two perpendicular number lines that intersect at their origins.

coordinate(s) of a point (pp. 40, 72) The real number or numbers that correspond to the point. On a number line, the coordinate of each point is a single real number. On a coordinate plane, each point has an ordered pair (x, y) of real-number coordinates. In a three-dimensional coordinate system, the coordinates of each point are an ordered triple of real numbers, (x, y, z).

coplanar points (p. 66) Points that lie in the same plane. Points that do not lie in the same plane are called *noncoplanar points*.

corollary (p. 261) A theorem that follows easily from a previously proved theorem.

corresponding angles (p. 139) When two lines are intersected by a transversal, corresponding angles are two angles that are in similar positions relative to the lines and the transversal.

cosine (pp. 572, 580) The cosine of an acute angle of a right triangle is the ratio of the length of the side adjacent to the angle to the length of the hypotenuse.

counterexample (p. 53) A counterexample to a conditional statement is an instance for which the hypothesis is true and the conclusion is false. In general, a counterexample is an example that shows a conjecture to be false.

cross products of a proportion (p. 494) The product of the means and the product of the extremes. In the proportion $a : b = c : d$, the cross products are ad and bc.

cross section (p. 17) The intersection of a three-dimensional figure and a plane.

cube (pp. 16, 724) A regular polyhedron with six faces, each bounded by a square. Also called a *regular hexahedron*.

cylinder (pp. 16, 646) A three-dimensional figure that consists of two parallel congruent circular regions, called *bases*, and a *lateral surface* that connects the boundaries of the bases.

• • **D** • •

decagon (p. 196) A polygon with ten sides.

deductive reasoning (p. 116) The process of reasoning logically from an accepted hypothesis to show that a desired conclusion must be true.

definition (p. 89) A statement of the meaning of a word, term, or phrase.

degree measure of an angle (pp. 31, 92) A unique positive number less than or equal to 180 that is paired with the angle.

degree measure of an arc (p. 669) The degree measure of a minor arc is the degree measure of the central angle that intercepts it. The degree measure of a major arc is 360° minus the degree measure of the minor arc that makes up the rest of the circle. The degree measure of a semicircle is 180°.

degree of a vertex (p. 78) In a network, the number of edges that have that vertex as an endpoint. An *even vertex* is one whose degree is even. An *odd vertex* is one whose degree is odd.

diagonal of a polygon (p. 197) A segment whose endpoints are nonconsecutive vertices.

diameter of a circle (pp. 609, 612) A chord, or the length of the chord, that contains the center of the circle.

diameter of a sphere (p. 648) A chord, or the length of the chord, that contains the center of the sphere.

dihedral angle (pp. 138, 321) The figure formed by two half-planes that have a common edge.

dilation (p. 512) A dilation with center C and positive scale factor k is a transformation in a plane that maps every point P of the plane to a point P' such that: (1) If the point P is not point C, then $\overline{CP'}$ and $\overline{CP}$ are collinear $\overrightarrow{AB}$ and $CP' = k(CP)$; (2) If the point P is point C, then P' is P.

direction of a vector (p. 578) The measure, taken in a counterclockwise direction, of the angle the vector forms with the positive x-axis.

dissection (p. 334) The process of separating a geometric shape into "pieces" and rearranging the pieces to form a different shape.

distance between a point and a line (or plane) (p. 159) The length of the perpendicular segment from the point to the line (or plane).

distance between two parallel lines (or planes) (p. 159) The distance between one line (or plane) and any point on the other line (or plane).

distance between two points (number line) (pp. 40, 72) The absolute value of the difference of the coordinates of the two points.

distance formula on a coordinate plane (p. 41) The distance d between points with coordinates (x_1, y_1) and (x_2, y_2) is given by
$$d = \sqrt{(x_2 - x_1)^2 + (y_2 - y_1)^2}$$

distance formula in three dimensions (p. 42) The distance d between points with coordinates (x_1, y_1, z_1) and (x_2, y_2, z_2) is given by
$$d = \sqrt{(x_2 - x_1)^2 + (y_2 - y_1)^2 + (z_2 - z_1)^2}$$

dodecahedron (p. 724) A polyhedron with twelve faces.

dot product of vectors (p. 178) For vectors $\vec{u} = (a_1, b_1)$ and $\vec{v} = (a_2, b_2)$, the dot product $\vec{u} \cdot \vec{v}$ is the real number $a_1 a_2 + b_1 b_2$.

• • E • •

edge of a dihedral angle (p. 138) The common edge of the faces of the angle.

edge of a half-plane (p. 138) See *half-plane*.

edge of a network (p. 78) See *network*.

edge of a polyhedron (p. 724) A segment that is the intersection of two faces.

endpoint (pp. 30, 67, 92, 614) See *ray*, *segment*, and *arc*.

endpoint of a kite (p. 405) The points where the line of symmetry intersects the kite.

enlargement (p. 512) A dilation with scale factor k such that $k > 1$.

equal vectors (p. 176) Vectors that have the same magnitude and the same direction.

equiangular polygon (p. 197) A polygon whose angles are all congruent.

equilateral polygon (p. 197) A polygon whose sides are all congruent.

equilateral triangle (p. 256) A triangle with three congruent sides.

exterior angle of a polygon (p. 196) An angle that forms a linear pair with an angle of the polygon.

exterior angles (p. 139) In the figure at the right, lines ℓ and m are intersected by transversal t. The exterior angles are $\angle 1$, $\angle 2$, $\angle 7$, and $\angle 8$.

externally tangent circles (p. 665) See *tangent circles*.

extremes of a proportion (p. 494) In the proportion $a : b = c : d$, a and d are called the extremes.

• • F • •

face of a dihedral angle (p. 138) One of the two half-planes that form the angle.

face of a polyhedron (p. 724) One of the flat surfaces.

figurate number (p. 81) A number that can be modeled by a specific pattern of points.

figure (pp. 9, 65) Any set of points.

frieze pattern (p. 242) A pattern that extends infinitely in two opposite directions in such a way that the pattern can be mapped onto itself by a translation in either direction. Also called a *strip pattern*.

frustum of a pyramid or cone (p. 754) The figure formed when a cross section parallel to the base is joined with all points of the pyramid or cone that lie between the cross section and the base.

function (p. 80) When one quantity depends upon another, the first quantity is said to be a function of the second.

• • G • •

geometric mean (p. 557) The geometric mean of two positive numbers a and b is the number x such that $\frac{a}{x} = \frac{x}{b}$, where x is positive.

geometric model (p. 7) A geometric figure that represents a real life object.

glide reflection (p. 232) A composition of transformations that consists of the reflection of a figure across a line followed by a translation of the image in the direction of the line.

golden ratio (p. 497) The ratio $\frac{1 + \sqrt{5}}{2}$.

golden rectangle (p. 498) A rectangle for which the ratio $\frac{\text{length}}{\text{width}}$ is equal to the golden ratio.

great circle (pp. 154, 649) A cross section of a sphere whose center is the center of the sphere.

• • H • •

half-plane (p. 138) A line that lies in a plane separates the points of the plane not on the line into two distinct regions called half-planes. The line is the *edge* of each half-plane.

height of a parallelogram (p. 441) The length of an altitude.

hemisphere (p. 649) A great circle of a sphere divides the sphere into two congruent parts, each of which is called a hemisphere.

heptagon (p. 196) A polygon with seven sides.

hexagon (pp. 82, 196) A polygon with six sides.

hexahedron (p. 724) A polyhedron with six faces.

horizon line (p. 23) A horizontal line that contains the vanishing point(s) of a drawing.

hypotenuse (p. 290) The side of a right triangle that is opposite the right angle.

hypothesis (p. 53) In a conditional statement, the part that follows *if*.

• • I • •

icosahedron (p. 724) A polyhedron with twenty faces.

image (p. 210) See *transformation*.

incenter of a triangle (p. 332) The point of concurrency of the angle bisectors.

indirect measurement (p. 532) Determining an unknown measure by using mathematical relationships among known measures rather than using a *direct* measurement tool such as a ruler or protractor.

indirect reasoning (p. 347) A type of reasoning in which all possible conclusions but one are eliminated, with the result that the one remaining conclusion must be true.

inductive reasoning (p. 85) The process of looking for patterns among a set of data and using the patterns to make a generalization.

infinite sequence (p. 628) A sequence that has no last term.

initial point of a vector (p. 176) The vector denoted $\overrightarrow{PQ}$ begins at point P and ends at point Q. Point P is the initial point.

inscribed angle (p. 681) An angle whose vertex is on a circle and whose sides contain chords of the circle.

inscribed circle (p. 332) A circle is inscribed in a polygon if each side of the polygon intersects the circle at exactly one point.

inscribed polygon (pp. 332, 626) A polygon is inscribed in a circle if each vertex of the polygon lies on the circle.

intercepted arc (p. 614) An arc of a circle is intercepted by an angle if its endpoints lie on the sides of the angle and all its other points lie in the interior of the angle.

interior angles (p. 139) In the figure at the right, lines ℓ and m are intersected by transversal t. The interior angles are $\angle 3$, $\angle 4$, $\angle 5$, and $\angle 6$.

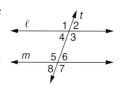

internally tangent circles (p. 665) See *tangent circles.*

intersection of geometric figures (pp. 17, 65) The set of points common to all the figures.

inverse (p. 106) The statement that results when the hypothesis and conclusion of a conditional statement are both negated.

isometric drawing (p. 25) A drawing of a three-dimensional object in which parallel edges of the object are shown as parallel line segments in the drawing.

isometry (p. 211) A transformation in which a figure and its image are congruent. Also called a *rigid transformation.*

isosceles trapezoid (pp. 374, 398) A trapezoid in which the legs are congruent.

isosceles triangle (pp. 256, 324) A triangle with at least two congruent sides. The two congruent sides are called *legs.* The third side is the *base.*

• • K • •

kite (p. 404) A convex quadrilateral in which two distinct nonoverlapping pairs of consecutive sides are congruent.

• • L • •

lateral area of a prism (p. 732) The sum of the areas of its lateral faces.

lateral edge of a prism or pyramid (pp. 419, 420) The intersection of two lateral faces.

lateral faces of a prism or pyramid (pp. 419, 420) See *prism* and *pyramid.*

lateral surface of a cone or cylinder (pp. 646, 647) See *cone* and *cylinder.*

legs of an isosceles triangle (p. 324) See *isosceles triangle.*

legs of a right triangle (p. 290) The sides opposite the acute angles.

legs of a trapezoid (p. 397) The nonparallel sides.

length of a segment (pp. 40, 72) The distance between the endpoints of the segment.

line (pp. 9, 65) A set of points that extends without end in two opposite directions. This is one of the basic undefined terms of geometry.

linear pair (p. 94) Two adjacent angles whose noncommon sides are opposite rays.

line of reflection (p. 210) See *reflection.*

line of symmetry (pp. 7, 10) A line that divides a plane figure into two parts that are mirror images of each other.

line segment (pp. 30, 67) See *segment.*

line symmetry (pp. 7, 10) See *reflection symmetry of a plane figure.*

locus (p. 314) The set of all points that satisfy a given condition or a set of given conditions. The plural of locus is *loci.*

• • M • •

magnitude of a vector (p. 176) The size of the vector. The magnitude of a vector $\overrightarrow{AB}$ is the length of $\overline{AB}$.

major arc (p. 614) An arc that is larger than a semicircle.

mapping (p. 210) See *transformation.*

means of a proportion (p. 494) In the proportion $a : b = c : d$, b and c are called the means.

measure of an angle (pp. 31, 92) See *degree measure of an angle.*

measure of an arc (p. 669) See *degree measure of an arc.*

median of a triangle (p. 332) A segment whose endpoints are a vertex of the triangle and the midpoint of the opposite side.

midpoint of a segment (pp. 49, 74) The point that divides the segment into two congruent segments.

midsegment of a trapezoid (p. 397) The segment whose endpoints are the midpoints of the legs.

midsegment of a triangle (p. 336) A segment whose endpoints are the midpoints of two sides.

minor arc (p. 614) An arc that is smaller than a semicircle.

• • N • •

net (p. 18) A two-dimensional figure that, when folded, forms the surface of a three-dimensional figure.

network (p. 78) A finite set of points, called the *vertices* or *nodes* of the network, joined by segments or arcs, called the *edges* of the network. Also called a *graph.*

***n*-gon** (p. 196) A polygon with *n* sides.

nonagon (p. 196) A polygon with nine sides.

non-Euclidean geometry (pp. 47, 155) A geometry in which one or more postulates are not the same as those of Euclidean geometry. The *classical non-Euclidean geometries* assume a parallel postulate that is fundamentally different from Euclid's.

number line (p. 40) A line whose points are placed in correspondence with the set of real numbers.

• • **O** • •

oblique cone (p. 647) A cone that is not a right cone.

oblique cylinder (p. 646) A cylinder that is not a right cylinder.

oblique line and plane (p. 137) A line and a plane that are neither parallel nor perpendicular.

oblique lines (p. 131) Two lines that intersect and are not perpendicular.

oblique planes (p. 137) Two planes that are neither parallel nor perpendicular.

oblique prism (p. 419) A prism that is not a right prism.

oblique pyramid (p. 420) A pyramid that is not a right pyramid.

oblong numbers (p. 81) The terms of the sequence 2, 6, 12, 20, 30, ... The function f that relates the value of an oblong number to its order n in the sequence is $f(n) = n(n + 1)$.

obtuse angle (p. 93) An angle whose measure is greater than 90° and less than 180°.

obtuse triangle (p. 256) A triangle with one obtuse angle.

octagon (p. 196) A polygon with eight sides.

octahedron (p. 724) A polyhedron with eight faces.

octant (p. 42) One of the eight regions into which a three-dimensional coordinate system is divided by the x-, y-, and z-axes.

one-point perspective (p. 24) A drawing with one vanishing point has one-point perspective.

opposite of a vector (p. 180) A vector that has the same magnitude as the given vector, but opposite direction.

opposite rays (p. 92) $\overrightarrow{BA}$ and $\overrightarrow{BC}$ are opposite rays if point B is between points A and C.

ordered pair (p. 41) On a coordinate plane, an ordered pair (x, y) is the pair of real numbers that corresponds to a point.

ordered triple (p. 42) In a three-dimensional coordinate system, an ordered triple (x, y, z) is the triple of real numbers that corresponds to a point.

order of rotation symmetry (p. 10) The number of times a plane figure coincides with its original position during a complete turn around its center of symmetry.

orientation (p. 211) The order in which the vertices of a polygon are named, either *clockwise* or *counterclockwise*.

origin of a coordinate plane (p. 41) The point where the axes intersect.

origin of a number line (p. 40) The point that corresponds to zero.

orthocenter of a triangle (p. 333) The point of concurrency of the lines that contain the altitudes.

orthogonal drawing (p. 24) A drawing that depicts a three-dimensional object by detailing three views of the object. Also called an *orthographic drawing*.

• • **P** • •

paper-folding construction (p. 49) A precise representation of a geometric figure made by folding creases on a sheet of paper.

parallel line and plane (p. 136) A line and a plane that do not intersect.

parallel lines (pp. 23, 131) Coplanar lines that do not intersect.

parallelogram (pp. 178, 374, 377) A quadrilateral with two pairs of parallel sides.

parallel planes (p. 136) Planes that do not intersect.

parallel vectors (p. 176) Vectors that have either the same direction or opposite directions.

pentagon (pp. 82, 196) A polygon with five sides.

perimeter of a plane figure (p. 435) The distance around the figure. The perimeter of a polygon is the sum of the lengths of its sides.

perpendicular bisector of a segment (p. 157) A line, ray, or segment that is perpendicular to the given segment at its midpoint.

perpendicular line and plane (p. 137) A line and a plane are perpendicular if and only if the line is perpendicular to two distinct lines in the plane at their point of intersection.

perpendicular lines (pp. 41, 130) Two lines that intersect to form a right angle.

perpendicular planes (p. 137) Two planes are perpendicular if and only if one plane contains a line that is perpendicular to the other plane.

perpendicular vectors (p. 178) Vectors whose directions are perpendicular.

perspective drawing (p. 24) A drawing made on a two-dimensional surface in such a way that three-dimensional objects appear true-to-life.

plane figure (p. 9) A set of points that lie in the same plane. Also called a *two-dimensional figure*.

plane of symmetry (p. 16) A plane that divides a three-dimensional figure into two parts that are mirror images of each other.

plane symmetry (p. 16) See *reflection symmetry of a three-dimensional figure*.

Platonic solid (pp. 724, 725) See *regular polyhedron*.

point (pp. 9, 65) A location. This is one of the basic undefined terms of geometry.

point of concurrency (p. 332) See *concurrent lines*.

point of tangency (p. 663) See *tangent to a circle*.

polygon (p. 195) A plane figure formed by three or more segments such that each segment intersects exactly two others, one at each endpoint, and no two segments with a common endpoint are collinear. Each segment is a *side* of the polygon. The common endpoint of two sides is a *vertex*. The polygon completely encloses a region of the plane, called its *interior*.

polyhedron (p. 724) A three-dimensional figure formed by flat surfaces that are bounded by polygons and joined in pairs along their sides. The polyhedron completely encloses a region of space, called its *interior*.

polyomino (pp. 216, 217) A plane figure formed by a set of unit squares arranged so each square shares a common side with at least one other square.

postulate (p. 66) A statement that is accepted as true without proof.

preimage (p. 210) See *transformation*.

prism (p. 419) A three-dimensional figure that consists of two parallel *bases* that are congruent polygonal regions and *lateral faces* that are bounded by parallelograms connecting corresponding sides of the bases.

probability of an event (p. 74) The likelihood that the event will occur.

proof (p. 116) A set of justified statements organized in logical order so that together they form a convincing argument that a given statement is true. The justification for a statement may be a definition, a property from algebra, a postulate, or a previously proved theorem.

proportion (pp. 34, 494) A statement that two ratios are equal.

protractor (pp. 31, 92) A tool used to find the degree measure of an angle.

pyramid (p. 420) A three-dimensional figure that consists of a polygonal face, called a *base*, a point called the *vertex* that is not in the plane of the base, and triangular *lateral faces* that connect the vertex to each side of the base.

Pythagorean triple (p. 470) A set of three positive integers a, b, and c that satisfy the equation $a^2 + b^2 = c^2$.

· · **Q** · ·

quadrant (p. 41) One of the four regions into which a coordinate plane is divided by the x- and y-axes.

quadrilateral (p. 196) A polygon with four sides.

· · **R** · ·

radius of a circle (pp. 48, 609, 612) A segment, or the length of the segment, whose endpoints are the center of the circle and a point of the circle.

radius of a regular polygon (p. 594) A segment, or the length of the segment, whose endpoints are the center and a vertex of the polygon.

radius of a sphere (p. 648) A segment, or the length of the segment, whose endpoints are the center of the sphere and a point on the sphere.

ratio (pp. 34, 493) A comparison of two quantities by division.

ray (pp. 30, 92) Part of a line that begins at one point and extends without end in one direction. The point is called the *endpoint* of the ray.

rectangle (pp. 6, 374, 391) A quadrilateral with four right angles.

rectangular prism (pp. 16, 468) A prism with rectangular bases.

reduction (p. 512) A dilation with scale factor k such that $0 < k < 1$.

reflection (p. 210) A reflection in a line ℓ is a transformation that maps every point P to a point P' such that: (1) if P is not on ℓ, then ℓ is the perpendicular bisector of $\overline{PP'}$; (2) if P is on ℓ, then P' is P. Line ℓ is the *line of reflection*, and P' is the *reflection image* of P.

reflection symmetry of a plane figure (pp. 7, 10, 212) A plane figure has reflection symmetry if it can be divided along a line into two parts that are mirror images of each other. Also called *line symmetry*.

reflection symmetry of a three-dimensional figure (p. 16) A three-dimensional figure has reflection symmetry if it can be divided along a plane into two parts that are mirror images of each other. Also called *plane symmetry*.

regular polygon (p. 197) A polygon that is both equilateral and equiangular.

regular polyhedron (p. 725) A convex polyhedron in which the faces are bounded by congruent regular polygons and the same number of edges intersect at each vertex. Also called a *Platonic solid*.

regular prism (p. 419) A right prism whose bases are regular polygonal regions.

regular pyramid (p. 420) A right pyramid whose base is a regular polygonal region.

regular tessellation of the plane (p. 268) A tessellation in which each shape is a regular polygon and all the shapes are congruent.

rhombus (pp. 374, 391) A quadrilateral with four congruent sides.

right angle (p. 93) An angle whose measure is equal to 90°.

right cone (p. 647) A cone in which the axis and the altitude are the same segment.

right cylinder (p. 646) A cylinder in which the axis also is an altitude.

right prism (p. 419) A prism whose lateral edges are perpendicular to the bases.

right pyramid (p. 420) A pyramid in which the base has rotation symmetry and the segment from the vertex to the center of symmetry is perpendicular to the base.

right triangle (p. 256) A triangle with one right angle.

rigid transformation (p. 211) See *isometry*.

rotation (p. 224) A rotation through $a°$ about a point O is a transformation that maps every point P to a point P' such that: (1) if point P is different from point O, then $OP' = OP$ and $m\angle POP' = a°$; (2) if point P is point O, then P' is P. Point O is the *center of rotation*, $a°$ is the *angle of rotation*, and P' is the *rotation image* of P.

rotation symmetry of a plane figure (p. 10) A plane figure has rotation symmetry if it can be turned around a point so it coincides with its original position two or more times during a complete turn.

rotation symmetry of a three-dimensional figure (p. 18) A three-dimensional figure has rotation symmetry if it can be turned around a line so it coincides with its original position two or more times during a complete turn.

ruler (pp. 29, 72) A tool used to find the length of a segment.

• • • **S** • • •

same-side exterior angles (p. 145) When two lines are intersected by a transversal, same-side exterior angles are two exterior angles on the same side of the transversal.

same-side interior angles (p. 141) When two lines are intersected by a transversal, same-side interior angles are two interior angles on the same side of the transversal.

scale drawing (p. 34) A two-dimensional drawing that is similar to the object it represents. The ratio of the size of the drawing to the actual size of the object is the *scale* of the drawing.

scale factor of similar polygons (p. 502) The ratio of the lengths of corresponding sides.

scale factor of similar solids (p. 770) The ratio of corresponding linear measures.

scale model (p. 35) A three-dimensional figure whose surfaces are similar to the corresponding surfaces of the object it represents. The ratio of the size of the model to the actual size of the object is the *scale* of the model.

scalene triangle (p. 256) A triangle with no congruent sides.

secant of a circle (p. 663) A line that intersects a circle at two points.

sector (p. 621) A region bounded by two radii of a circle and their intercepted arc.

segment (pp. 30, 67) Part of a line that begins at one point and ends at another. The points are called the *endpoints* of the segment.

segment bisector (pp. 49, 74) Any segment, ray, line, or plane that intersects the segment at its midpoint.

segment of a circle (p. 622) A region bounded by an arc and the chord having the same endpoints as the arc.

semicircle (p. 614) An arc whose endpoints are the endpoints of a diameter.

sequence (p. 628) A set of numbers, called *terms*, arranged in a specific order.

side of an angle (pp. 31, 92) One of the two rays that form the angle.

side of a polygon (p. 195) See *polygon*.

similar figures (p. 34) Figures that have the same shape, but not necessarily the same size.

sequence (p. 628) A set of numbers, called *terms*, arranged in a specific order.

side of an angle (pp. 31, 92) One of the two rays that form the angle.

side of a polygon (p. 195) See *polygon*.

similar figures (p. 34) Figures that have the same shape, but not necessarily the same size.

similar polygons (p. 501) Polygons whose vertices can be paired in such a way that corresponding angles are congruent and corresponding sides are in proportion.

similar solids (p. 770) Two solids are similar if they have the same shape and all corresponding linear measures, such as heights and radii, are in proportion.

sine (pp. 572, 580) The sine of an acute angle of a right triangle is the ratio of the length of the side opposite the angle to the length of the hypotenuse.

skew lines (p. 131) Lines that are noncoplanar.

slant height of a regular pyramid (p. 468) The height of a lateral face.

slant height of a right cone (p. 647) The distance from the vertex of the cone to any point on the boundary of the base.

slope-intercept form of a linear equation (p. 167) The form $y = mx + b$, where m is the slope of the line and b is its y-intercept.

slope of a line (p. 161) On a coordinate plane, the ratio of the change in value of the y-coordinates (the *rise*) to the corresponding change in the x-coordinates (the *run*), as measured between any two points on the line. For a nonvertical line containing points (x_1, y_1) and (x_2, y_2), the slope m is given by the formula
$$m = \frac{\text{rise}}{\text{run}} = \frac{\text{change in } y}{\text{change in } x} = \frac{y_2 - y_1}{x_2 - x_1}.$$

small circle of a sphere (p. 648) A cross section of the sphere whose center is not the center of the sphere.

solid figure (p. 16) A three-dimensional figure that consists of all its surface points and all the points the surface encloses. Also called a *solid*.

solve a triangle (p. 591) To determine the measures of all the angles and sides of the triangle.

space (pp. 16, 65) The set of all points.

sphere (pp. 16, 154, 648) The set of all points in space that are a given distance from a given point. The given point is the *center* of the sphere, and the given distance is the *radius*.

square (pp. 6, 374, 391) A quadrilateral with four right angles and four congruent sides.

square numbers (p. 82) The terms of the sequence 1, 4, 9, 16, 25, ... The function f that relates the value of a square number to its order n in the sequence is $f(n) = n^2$.

square pyramid (pp. 16, 420) A pyramid with a square base.

standard form of a linear equation (p. 167) The form $Ax + By = C$, where A and B are real numbers and A and B are not both zero.

straight angle (p. 93) An angle whose measure is equal to 180°.

straightedge (p. 48) A tool used as a guide in drawing lines, rays, and segments.

strip pattern (p. 242) See *frieze pattern*.

supplementary angles (p. 93) Two angles whose measures have a sum of 180°. Each angle is called the *supplement* of the other.

surface (p. 731) The boundary of a three-dimensional figure that completely encloses a region of space.

surface area (p. 731) The area of the surface of a three-dimensional figure.

· · **T** · ·

tangent circles (p. 665) Circles in the same plane that are tangent to the same line at the same point. The circles are *externally tangent* if all points of one circle, except for the point of tangency, lie in the exterior of the other. The circles are *internally tangent* if all points of one circle, except for the point of tangency, lie in the interior of the other circle.

tangent of an angle (pp. 571, 573) The tangent of an acute angle of a right triangle is the ratio of the length of the side opposite the angle to the length of the side adjacent to it.

tangent to a circle (p. 663) A line in the plane of the circle that intersects the circle at exactly one point. The point is called the *point of tangency*.

taxicab geometry (p. 46) A system of geometry in which the distance between two points, called the *taxi distance*, is defined as the least number of horizontal and vertical units between the points on a coordinate plane.

theoretical probability (pp. 74, 634) The theoretical probability of an event, $P(E)$, is given by the ratio $\dfrac{\text{number of favorable outcomes}}{\text{total number of outcomes}}$.

three-dimensional coordinate system (p. 42) A method of locating points in space formed by three number lines that intersect at their origins, each line being perpendicular to the other two.

three-dimensional figure (p. 16) A figure that extends beyond a single plane into space.

traceable network (p. 78) A network in which it is possible to begin at one vertex and travel to all other vertices by tracing each edge exactly once.

transformation (pp. 206, 210) A correspondence between one figure, called a *preimage*, and a second figure, its *image*, such that each point of the preimage is paired with exactly one point of the image, and each point of the image is paired with exactly one point of the preimage. Also called a *mapping*.

translation (p. 218) A translation by a vector $\overrightarrow{AA}$ is a transformation that maps every point P to a point P' such that: (1) $PP' = AA'$; (2) $\overline{PP'} \parallel \overline{AA'}$ or $\overline{PP'}$ is collinear with $\overline{AA'}$. The vector $\overrightarrow{AA}$ is the *translation vector*, and P' is the *translation image* of P.

transversal (p. 139) A line that intersects two or more coplanar lines at different points.

trapezoid (pp. 374, 397) A quadrilateral with exactly one pair of parallel sides.

triangle (p. 196) A polygon with three sides.

triangular numbers (p. 81) The terms of the sequence 1, 3, 6, 10, 15, ... The function f that relates the value of a triangular number to its order n in the sequence is $f(n) = \dfrac{n(n+1)}{2}$.

trigonometric ratio (p. 571, 573) A ratio of the lengths of two sides of a right triangle.

two-dimensional figure (p. 16) See *plane figure*.

two-point perspective (p. 24) A drawing with two vanishing points has two-point perspective.

vanishing point (p. 23) A point at a distance from the observer where two or more lines of a drawing appear to intersect.

vector (p. 176) A quantity that has both a size, called its *magnitude*, and a direction.

vertex angle of an isosceles triangle (p. 324) The angle opposite the base.

vertex of an angle (pp. 31, 92) The common endpoint of the sides.

vertex of a cone (p. 647) See *cone*.

vertex of a network (p. 78) See *network*.

vertex of a polygon (pp. 24, 195) See *polygon*.

vertex of a polyhedron (p. 724) A point that is the intersection of three or more edges.

vertex of a pyramid (p. 420) See *pyramid*.

vertical angles (p. 129) Two angles whose sides form two pairs of opposite rays.

volume of a three-dimensional figure (p. 737) The number of nonoverlapping cubic units contained in the interior of the figure.

x-axis (p. 41) The horizontal number line in a coordinate plane.

x-coordinate (p. 41) The first number of an ordered pair representing a point on the coordinate plane, indicating the distance left or right of the y-axis.

x-intercept of a line (p. 168) The x-coordinate of the point where the line intersects the x-axis.

y-axis (p. 41) The vertical number line in a coordinate plane.

y-coordinate (p. 41) The second number of an ordered pair representing a point on the coordinate plane, indicating the distance above or below the x-axis.

y-intercept of a line (p. 168) The y-coordinate of the point where the line intersects the y-axis.

GLOSSARY

GLOSSARY/GLOSARIO

• • A • •

acute angle/ángulo agudo (p.93) Un ángulo cuya medida es mayor de 0° y menor de 90°.

acute triangle/triángulo acutángulo (p. 256) Un triángulo con tres ángulos agudos.

adjacent angles/ángulos adyacentes (p. 94) Dos ángulos que tienen un lado y el vértice en común pero no tienen puntos interiores en común.

adjacent arcs/arcos adyacentes (p. 669) Arcos de un mismo círculo que se intersecan en un solo punto.

alternate exterior angles/ángulos alternos externos (p. 140) Cuando dos líneas son intersecadas por una línea transversal determinan dos ángulos exteriores no adyacentes en los lados opuestos de la transversal.

alternate interior angles/ángulos alternos interiores (p. 140) Cuando dos líneas son intersecadas por una línea transversal determinan dos ángulos interiores no adyacentes en los lados opuestos de la transversal.

altitude of a cone/altura de un cono (p. 647) El segmento desde el vértice al plano de la base y que es perpendicular a ésta.

altitude of a cylinder/altura del un cilindro (p. 646) Un segmento de línea perpendicular a las bases con un extremo en cada base.

altitude of a parallelogram/altura de un paralelogramo (p. 441) Cualquier segmento perpendicular a la línea que contiene una base desde cualquier punto en el lado opuesto.

altitude of a prism/altura de un prisma (p. 468) Un segmento perpendicular a las bases con un extremo en cada base.

altitude of a pyramid/altura de una pirámide (p. 468) El segmento desde el vértice a la base y perpendicular a ésta.

altitude of a trapezoid/altura del trapezoide (p. 448) Cualquier segmento perpendicular al lado que contiene una base desde cualquier punto en la base opuesta.

altitude of a triangle/altura de un triángulo (pp. 333, 443) Un segmento perpendicular desde el vértice al lado opuesto.

angle/ángulo (pp. 31, 92) Figura formada por dos rayos con un punto extremo en común.

angle bisector/bisectriz de un ángulo (p. 95) Un rayo que divide el ángulo en dos ángulos congruentes.

angle of depression/ángulo de depresión (p. 575) Ángulo cuyo vértice está en el ojo de un observador y cuyos lados son una línea horizontal y la línea de visión del observador hacia un objeto más bajo en relación al observador.

angle of elevation/ángulo de elevación (p. 575) Un ángulo cuyo vértice está en el ojo de un observador y cuyos lados son una línea horizontal y la línea de visión del observador hacia un objeto más alto en relación al observador.

angle of rotation/ángulo de rotación (p. 224) Véase *rotación*.

apothem of a regular polygon/apotema de un poligono regular (p. 594) Un segmento de línea, o la longitud del segmento, cuyos extremos son el centro del polígono y el punto medio de un lado.

arc/arco (pp. 48, 614) Parte de un círculo determinada por dos puntos. Cualesquiera dos puntos dividen al círculo en dos arcos. Estos puntos son los extremos de cada arco.

area of a plane figure/área de una figura plana (p. 435) Número de unidades cuadradas en el interior de la figura.

auxiliary figure/figura auxiliar (pp. 212, 257) Un punto, una línea, un rayo o un segmento que se le añade a una figura para que se pueda probar.

axis of a cone/eje de simetría de un cono (p. 647) El segmento cuyos extremos son el vértice y el centro de la base.

axis of a cylinder/eje de simetría de un cilindro (p. 646) El segmento cuyos extremos son los centros de las bases.

axis of symmetry/eje de simetría (p. 18) Si se hace girar una figura tridimensional alrededor de una línea de manera que la figura coincida con su posición original dos o más veces durante una vuelta completa, la línea es su eje de simetría.

base angles of an isosceles triangle/ángulos de la base de un triángulo isósceles (p. 324) Los ángulos opuestos a sus lados.

base angles of a trapezoid/ángulos de la base de un trapezoide (p. 397) Dos ángulos consecutivos cuyos vértices son los extremos de una base.

base of a cone or of a cylinder/base de un cono o de un cilindro (pp. 646, 647) Véase *cono* y *cilindro*.

base of an isosceles triangle/base del triángulo isósceles (p. 324) Véase *triángulo isósceles*.

base of a parallelogram/base de un paralelogramo (p. 441) Cualquier lado de un paralelogramo puede considerarse como su base. La palabra base también se refiere a la longitud de ese lado.

base of a prism or pyramid/base de un prisma o de una pirámide (pp. 419, 420) Véase *prisma* y *pirámide*.

base of a triangle/base de un triángulo (p. 443) El lado del triángulo sobre el cual se traza su altura. La palabra base también se refiere a la longitud de ese lado.

base of a trapezoid/base de un trapezoide (p. 397) Los lados paralelos o las longitudes de los lados paralelos

bearing of an object/orientación de un objeto (p. 160) La medida, tomada en la dirección de las manecillas del reloj, del ángulo cuyo vértice está en el lugar del observador y cuyos lados son un rayo en la dirección de un *meridiano de referencia* y un rayo en la dirección de la línea de visión del observador en relación al objeto. Si el meridiano de referencia es la línea longitudinal que pasa a través del Polo Norte, la orientación es una *orientación verdadera*. Si el meridiano de referencia es la línea longitudinal que pasa a través del polo magnético, la orientación es una *orientación magnética*.

between/entre (p. 73) En una recta numérica, el punto *C* está entre los puntos *A* y *B* si la coordenada de *C* está entre las coordenadas de *A* y *B*.

biconditional statement/equivalencia lógica (p. 105) La proposición que se forma al combinar una proposición condicional y su recíproca cuando ambas son verdaderas, usando el conectivo *si* y *solamente si* para indicarla. También se *llama proposición bicondicional*.

bisector of an angle/bisectriz de un ángulo (p. 96) Véase *bisectriz de un ángulo*.

bisector of a segment/bisectriz de un segmento (pp. 49, 74,88) Véase *bisectriz de un segmento*.

center of a circle/centro del círculo (pp. 48, 609, 612) El punto en el plano del círculo que es equidistante de todos los puntos en el círculo.

center of gravity/centro de gravedad (p. 332) Véase *baricentro de un triángulo*.

center of a regular polygon/centro de un polígono regular (p. 594) El punto que es equidistante de todos los vértices del polígono.

center of rotation/centro de rotación (p. 224) Véase *rotación*.

center of a sphere/centro de la esfera (pp. 154, 648) El punto que es equidistante de todos los puntos de la esfera.

center of symmetry/centro de simetría (p. 10) Si una figura plana se puede girar alrededor de un punto de manera que coincida con su posición original dos o más veces durante una vuelta completa, el punto es su centro de simetría.

central angle of a circle/ángulo central de un círculo (p. 614) Un ángulo en el plano del círculo cuyo vértice es el centro del círculo.

central angle of a regular polygon/ángulo central de un polígono regular (p. 595) Un ángulo cuyo vértice es el centro del polígono y cuyos lados contienen dos vértices consecutivos.

centroid of a triangle/centroide de un triángulo (p. 332) El punto donde concurren (se intersecan) sus medianas. También se le llama el *centro de gravedad*.

chord of a circle/cuerda (de un círculo) (p. 612) Un segmento cuyos extremos son puntos en el círculo.

chord of a sphere/cuerda (de un esfera) (p. 648) Un segmento cuyos extremos son puntos en la esfera.

circle/círculo (pp. 48, 609, 612) El conjunto de todos los puntos en un plano que equidistan de un punto dado en el plano.

circumcenter of a triangle/circuncentro (de un triángulo) (p. 332) El punto de concurrencia o donde se intersecan las bisectrices perpendiculares de los lados.

GLOSSARY/GLOSARIO

circumference of a circle/circunferencia del círculo (longitud) (pp. 609, 613) El perímetro del círculo.

circumscribed circle/círculo circunscrito (pp. 332, 626) Un círculo está circunscrito a un polígono si cada vértice del polígono cae sobre el círculo.

circumscribed polygon/polígono circunscrito (p. 332) Un polígono está circunscrito a un círculo si cada lado del polígono interseca al círculo en exactamente un punto.

collinear points/puntos colineales (p. 66) Puntos que caen en la misma línea. Los puntos que no caen en la misma línea se llaman *puntos no colineales*.

common tangent/tangente común (p. 665) Una línea que es tangente a dos círculos en el mismo plano. Una *tangente común interna* interseca al segmento cuyos extremos son los centros de los círculos; una *tangente común externa* no lo interseca.

compass/compás (p. 48) Instrumento que se usa para trazar circunferencias y arcos.

compass-and-straightedge construction/ construcción con regla y compás (p. 48) Un trazado geométrico preciso con sólo dos instrumentos: un compás y una regla sin marcas.

complementary angles/ángulos complementarios (p. 93) Dos ángulos cuyas medidas suman 90°. Cada ángulo es el *complemento* del otro.

composition of transformations/composición de transformaciones (p. 231) El proceso por el cual se puede transformar una figura seguida de una segunda transformación en la imagen de la figura misma.

concave polygon/polígono cóncavo (p. 196) Un polígono en el cual una línea que contiene un lado del polígono también contiene un punto en su interior. Se le llama también *polígono no convexo*.

concave polyhedron/poliedro cóncavo (p. 725) Un poliedro en el cual el plano que contiene un lado del poliedro también contiene un punto en su interior.

concentric circles/círculos concéntricos (p. 612) Círculos en el mismo plano con el mismo centro.

conclusion/conclusión (p. 53) En una proposición condicional, la parte que le sigue a *entonces*.

concurrent lines/líneas incidentes (que se cruzan) (p. 332) Dos o más líneas que se intersecan en un solo punto. El punto de intersección se llama *punto de incidencia*.

conditional statement/proposición condicional (p. 53) Una proposición que se puede escribir en la forma si, entonces. También llamada una *implicación*.

cone/cono (pp. 16, 647) Una figura tridimensional que consiste de una cara circular, llamada la *base*, un punto llamado el vértice que no está en el plano de la base y una *superficie lateral* que conecta el vértice de cada punto en la frontera de la base.

congruent angles/ángulos congruentes (p. 95) Ángulos con la misma medida.

congruent arcs/arcos congruentes (p. 670) Arcos del mismo círculo o de círculos congruentes que tienen la misma medida.

congruent circles/círculos congruentes (p. 612) Círculos que tienen radios congruentes.

congruent figures/figuras congruentes (p. 33) Figuras que tienen exactamente la misma forma y tamaño.

congruent polygons/polígonos congruentes (p. 211) Polígonos cuyos lados se pueden colocar en una correspondencia tal que los lados correspondientes son congruentes y los ángulos correspondientes son congruentes.

congruent segments/segmentos congruentes (p. 74) Segmentos que tienen la misma longitud.

conic section/sección cónica (p. 651) La sección transversal que se forma cuando un plano interseca a un cono recto doble.

conjecture/conjetura (p. 85) Una generalización que resulta de un razonamiento inductivo.

contrapositive/contrapositiva (p. 106) La proposición que resulta cuando la hipótesis y la conclusión de una proposición condicional se niegan y luego se intercambian.

converse/recíproco (p. 104) La proposición que resulta cuando la hipótesis y la tesis de una porposición condicional se intercambian.

convex polygon/polígono convexo (p. 196) Un polígono en el cual ninguna línea que contiene un lado del polígono contiene un punto en su interior.

convex polyhedron/poliedro convexo (p. 725) Un poliedro en el cual ningún plano que contiene un lado del poliedro contiene un punto en su interior.

coordinate plane/plano coordenado (p. 41) Un plano numerado formado por dos rectas numéricas perpendiculares que se intersecan en el origen.

coordinate(s) of a point/coordenada(s) de un punto (pp. 40, 72) El número o números reales que corresponden al punto. En una recta numérica, la coordenada de cada punto es un solo número real. En un plano coordenado, cada punto tiene un par ordenado (x, y) de números reales. En un sistema coordenado tridimensional, las coordenadas de cada punto son triples ordenados (x, y, z) de números reales.

coplanar points/puntos coplanares (p. 66) Puntos que caen en el mismo plano. Puntos que no están en el mismo plano se llaman *puntos no coplanares*.

corollary/corolario (p. 261) Un teorema que se deriva fácilmente de otro teorema ya probado.

corresponding angles/ángulos correspondientes (p. 139) Cuando una transversal interseca dos líneas, los ángulos correspondientes son dos ángulos que están en posiciones similares en relación a las líneas y a la transversal.

cosine/coseno (pp. 572, 580) El coseno de un ángulo agudo de un triángulo rectángulo es la razón de la longitud entre el cateto adyacente al ángulo a la longitud de la hipotenusa.

counterexample/contraejemplo (p. 53) Un contraejemplo de una proposición condicional es un instante por el cual la hipótesis es verdadera y la conclusión es falsa. En general, un contraejemplo es un ejemplo que muestra que una conjetura es falsa.

cross products of a proportion/productos cruzados de una proporción (p. 494) El producto de los medios y el producto de los extremos. En la proporción $a : b = c : d$, los productos cruzados son ad y bc.

cross section/sección o corte transversal (p. 17) La intersección de una figura tridimensional y un plano.

cube/cubo (pp. 16, 724) Un poliedro regular con seis caras, cada una limitada por un cuadrado. También se le llama *hexaedro regular*.

cylinder/cilindro (pp. 16, 646) Una figura tridimensional que consiste de dos regiones circulares paralelas congruentes llamadas *bases* y una *superficie lateral* que conecta las fronteras de las bases.

• • **D** • •

decagon/decágono (p. 196) Un polígono con diez lados.

deductive reasoning/razonamiento deductivo (p. 116) El proceso de razonar lógicamente de una hipótesis aceptada para mostrar que una conclusion deseada tiene que ser verdadera.

definition/definición (p. 89) Un(a) (proposición) o enunciado del significado de una palabra, un término o una frase.

degree measure of an angle/medida de un ángulo (pp. 31, 92) Un número positivo menor de o igual a 180 que mide su apertura.

degree measure of an arc/medida de un arco (p. 669) La medida de un arco menor es la medida del ángulo central que lo interseca. La medida de una arco mayor es 360° menos la medida del arco menor que completa el resto de círculo. La medida de un semicírculo es 180°.

degree of a vertex/grado de un vértice (p. 78) En una red o grafo, el número de segmentos que tiene ese vértice como extremo. Un *vértice par* es el que tiene un grado par. Un *vértice impar* es el que tiene un grado impar.

diagonal of a polygon/diagonal de un polígono (p. 197) Un segmento cuyos extremos no son vértices consecutivos.

diameter of a circle/diámetro (pp. 609, 612) Una cuerda, o la longitud de la cuerda, que contiene el centro del círculo.

diameter of a sphere/diámetro de una esfera (p. 648) Un cuerda, o la longitud de la cuerda, que contiene el centro de la esfera.

dihedral angle/ángulo diedro (pp. 138, 321) La figura formada por dos semiplanos que tienen un borde (o arista) en común.

dilation/dilatación (p. 512) Una dilatación con centro C y un factor de escala positivo k es una transformación en un plano que lleva cada punto P del plano a un punto P' tal que: (1) Si el punto P no es el punto C, P' está en $\overrightarrow{AB}$ y $CP' = k(CP)$; (2) Si el punto P es el punto C, entonces P' es P.

direction of a vector/dirección de un vector (p. 578) La medida, tomada en la dirección en contra de las manecillas del reloj, del ángulo que forma el vector con el eje de x positivo.

dissection/disección (p. 334) El proceso de separar una figura geométrica en "pedazos" y arreglarlas de nuevo para formar una figura diferente.

distance between a point and a line (or plane)/ distancia entre un punto y una línea (o plano) (p. 159) La longitud del segmento perpendicular desde el punto a la línea (o plano).

distance between two parallel lines (or planes)/ distancia entre dos líneas paralelas (o planos) (p. 158) La distancia entre una línea (o plano) y un punto en la otra línea (o plano).

distance between two points (number line)/distancia entre dos puntos (recta numérica) (pp. 40, 72) El valor absoluto de la diferencia de las coordenadas de los dos puntos.

distance formula on a coordinate plane/fórmula de distancia en un plano coordenado (p. 41) La distancia d entre puntos con coordenadas (x_1, y_1) y (x_2, y_2) se da por
$$d = \sqrt{(x_2 - x_1)^2 + (y_2 - y_1)^2}.$$

distance formula in three dimensions/fórmula de distancia en tres dimensiones (p. 42) La distancia d entre puntos con coordenadas (x_1, y_1, z_1) y (x_2, y_2, z_2) se da por
$$d = \sqrt{(x_2 - x_1)^2 + (y_2 - y_1)^2 + (z_2 - z_1)^2}.$$

dodecahedron/dodecaedro (p.724) Un poliedro con doce caras.

dot product of vectors/producto escalar de vectores (p. 178) Para vectores $\vec{u} = (a_1, b_1)$ y $\vec{v} = (a_2, b_2)$, el producto escalar $\vec{u} \cdot \vec{v}$ es el número real $a_1 a_2 + b_1 b_2$.

• • E • •

Edge of a dihedral angle/arista de un ángulo diedro (p. 138) La arista común de las caras del ángulo.

edge of a half plane/borde de un semiplano (p. 138) Véase *semiplano*.

edge of a network/borde de una red o grafo (p. 78) Véase *red o grafo*.

edge of a polihedron/arista de un poliedro (p. 724) Un segmento que es la intersección de las dos caras.

endpoint/extremo (pp. 30, 67, 92, 614) Véase *rayo*, *segmento* y *arco*.

endpoint of a kite/extremo de una cometa o papalote (p. 404) El extremo común de un par de lados congruentes.

enlargement/expansión (p. 512) Una dilatación con factor de escala k de manera que $k > 1$.

equal vectors/vectores iguales (p. 176) Vectores que tienen la misma magnitud y la misma dirección.

equiangular polygon/polígono equiangular (p. 197) Un polígono cuyos ángulos son congruentes.

equilateral polygon/polígono equilátero (p. 1297) Un polígono cuyos lados son congruentes.

equilateral triangle/triángulo equilátero (p. 356) Un triángulo con tres lados congruentes.

exterior angle of a polygon/ángulo exterior de un polígono (p. 196) Un ángulo que forma un par lineal con un ángulo del polígono.

exterior angles/ángulos exteriores (p. 139) En la figura a la derecha, las líneas ℓ y m son intersecadas por una transversal t. Los ángulos exteriores son $\angle 1$, $\angle 2$, $\angle 7$ y $\angle 8$.

externally tangent circles/círculos tangentes exteriores (p. 665) Véase *círculos tangentes*.

extremes of a proportion/extremos de una proporción (p. 494) En la proporción $a : b = c : d$, a y d son los extremos.

• • F • •

face of a dihedral angle/cara de un ángulo diedro (p. 138) Uno de dos semiplanos que forman el ángulo.

face of a polyhedron/cara de un poliedro (p. 724) Una de las superficies planas.

figurate number/número figurado (p. 81) Un número que puede ser modelado por un patrón específico de puntos.

figure/figura (pp. 9, 65) Cualquier conjunto de puntos.

frieze pattern/patrón de friso (p. 242) Un patrón que se extiende infinitamente en dos direcciones opuestas de manera que el patrón se puede trazar en él mismo por una traslación en cualquier dirección.

frustrum of a pyramid or cone/tronco de una pirámide o cono (p. 754) La figura formada cuando una sección transversal paralela a la base se une a todos los puntos de la pirámide o el cono que está entre la sección transversal y la base.

function/función (p.80) Cuando una cantidad depende de otra, la primera cantidad se dice que es una función de la segunda.

• • G • •

geometric mean/media geométrica (p. 557) La media geométrica de dos números positivos a y b es el número x de manera que $\frac{a}{x} = \frac{x}{b}$ donde x es positivo.

geometric model/modelo geométrico (p. 7) Una figura geométrica que representa un objeto real.

glide reflection/refelxión deslizante (p. 232) Una composición de transformaciones que consiste de la reflexión de una figura a través de una línea seguida de una traslación de la imagen en la dirección de la línea.

golden ratio/razón áurea (p. 497) La razón $\frac{1 + \sqrt{5}}{2}$.

golden rectangle/rectángulo áureo (p. 498) Un rectángulo para el cual la razón $\frac{\text{longitud}}{\text{ancho}}$ es igual a la razón áurea.

graph/gráfica (p. 78) Véase *red o grafo*.

great circle/círculo máximo (pp. 154, 649) Una sección transversal de una esfera cuyo centro es el centro de la esfera.

• • H • •

half-plane/semiplano (p. 138) Una línea que está en un plano y separa los puntos del plano que no están en la línea en dos regiones distintas llamadas semiplanos. La línea es el *borde* de cada semiplano.

height of a figure/altura de una figura (p. 441) La longitud de un altura.

hemisphere/hemisferio (p. 649) El círculo máximo de la esfera divide a la esfera en dos partes congruentes cada una de las cuales se llama un hemisferio.

heptagon/heptágono (p 196) Un polígono con siete lados.

hexagon/hexágono (pp. 82, 196) Un polígono con seis lados.

hexahedron/hexaedro (p. 724) Un poliedro con seis caras.

horizon line/horizonte (p. 23) Una línea horizontal que contiene el punto en el horizonte de un objeto.

hypotenuse/hipotenusa (p. 290) El lado de un triángulo rectángulo opuesto al ángulo recto.

hypothesis/hipótesis (p. 53) En una proposición condicional, la parte que sigue a *si*.

• • I • •

icosahedron/icosaedro (p. 724) Un poliedro con veinte caras.

image/imagen (p. 210) Véase *transformación*.

incenter of a triangle/incentro de un triángulo (p. 332) El punto de concurrencia de las bisectrices de los ángulos.

indirect measurement/medición indirecta (p. 532) Determinar una medida desconocida usando relaciones matemáticas entre las medidas conocidas en lugar de usar instrumentos de medir como una regla o un transportador.

indirect reasoning/razonamiento indirecto (p. 347) Un tipo de razonamiento en el cual todas las conclusiones posibles excepto una se eliminan resultando que la conclusión que queda debe ser verdadera.

inductive reasoning/razonamiento inductivo (p. 85) El proceso de analizar patrones entre un conjunto de datos y usar los patrones para establecer generalidades.

infinite sequence/sucesión infinita (p. 628) Una sucesión que no tiene un término mayor.

initial point of a vector/origen de un vector (p. 176) El vector denominado $\overrightarrow{PQ}$ comienza en el punto P y termina en el punto Q. El punto P es el origen del vector.

inscribed angle/ángulo inscrito (p. 681) Un ángulo cuyo vértice está en el círculo y cuyos lados contienen cuerdas del círculo.

inscribed circle/círculo inscrito (p. 332) Un círculo está inscrito en un polígono si cada lado del polígono interseca el círculo exactamente en un punto.

inscribed polygon/polígono inscrito (pp. 332, 626) Un polígono está inscrito en un círculo si cada vértice del polígono está en el círculo.

intercepted arc/arco intersecado (p. 614) Un arco de un círculo está intersecado por un ángulo si sus extremos están en los lados del ángulo y todos sus otros puntos están en el interior del ángulo.

interior angles/ángulos interiores (p. 139) En la figura a la derecha, la transversal *t* interseca las líneas ℓ y *m*. Los ángulos interiores son ∠3, ∠4, ∠5 y ∠6.

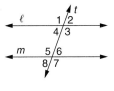

internally tangent circles/círculos tangentes interiormente (p. 665) Véase *círculos tangentes*.

intersection of geometry figures/intersección de figuras geométricas (p. 17, 65) El conjunto de puntos comunes a todas las figuras.

inverse/inverso (p. 106) La declaración que resulta al negar la hipótesis y la conclusión de una proposición condicional.

isometric drawing/dibujo isométrico (p. 25) Un dibujo de un objeto tridimensional en el cual las aristas paralelas del objeto se muestran como segmentos paralelos en el dibujo.

isometry/isometría (p. 211) Una transformación en la cual una figura y su imagen son congruentes. También se le llama *transformación rígida*.

isosceles trapezoid/trapezoide isósceles (pp. 374, 398) Un trapezoide con lados no paralelos que son congruentes.

isosceles triangle/triángulo isósceles (pp. 256, 324) Un triángulo con por lo menos dos lados congruentes. Los dos lados congruentes se llaman simplemente lados. El tercer lado se llama *base*.

• • **K** • •

kite/cometa o papalote (p. 404) Un cuadrilátero convexo en el cual dos pares distintivos de lados consecutivos son congruentes.

• • **L** • •

lateral area of a prism/área lateral del prisma (p. 732) La suma de la áreas de sus caras laterales.

lateral edge of a prism or pyramid/aristas laterales de un prisma o una pirámide (pp. 419, 420) La intersección de dos caras laterales.

lateral faces of a prism or pyramid/caras laterales de un prisma o una pirámide (pp. 419, 420) Véase *prisma* y *pirámide*.

lateral surface of a cone or cylinder/superficie lateral de un cono o un cilindro (pp. 646, 647) Véase *cono* y *cilindro*.

legs of an isosceles triangle/lados de un triángulo isósceles (p. 324) Véase *triángulo isósceles*.

legs of a right triangle/catetos de un triángulo rectángulo (p. 290) Los lados opuestos a los ángulos agudos.

legs of a trapezoid/lados de un trapezoide (p. 397) Los lados no paralelos.

length of a segment/longitud de un segmento (pp. 40, 72) La distancia entre los extremos de un segmento.

line/línea (pp. 9, 65) Un conjunto de puntos que se extiende infinitamente en dos direcciones opuestas. Éste es uno de los términos básicos indefinidos de la geometría.

linear pair/par lineal (p. 94) Dos ángulos adyacentes cuyos lados no comunes son rayos opuestos.

line of reflection/línea de reflexión (p. 210) Véase *reflexión*.

line of symmetry/línea de simetría (pp. 7, 10) Una línea que divide una figura plana en dos partes de tal manera que cada una de ellas es la imagen de la otra.

line segment/segmento de línea (pp. 30, 67) Véase *segmento*.

line symmetry/simetría axial (pp. 7, 10) Véase *simetría de reflexión*.

locus/lugar geométrico (p. 314) El conjunto de todos los puntos que satisfacen una condición o un conjunto de condiciones. En inglés, el plural de locus es loci.

· · **M** · ·

magnitude of a vector/magnitud de un vector (p. 176) El tamaño del vector. La magnitud de un vector $\overrightarrow{AB}$ es la longitud de $\overline{AB}$.

major arc/arco mayor (p. 614) Un arco que es más grande que un semicírculo.

mapping/planimetría (p. 210) Véase *transformación*.

means of a proportion/medias de una proporción (p. 494) En la proporción $a : b = c : d$, b y c son las medias.

measure of an angle/medida de un ángulo (pp. 31, 92) Véase *medida (grados) de un ángulo*.

measure of an arc/medida de un arco (p. 669) Véase *medida (grados) de un arco*.

median of a triangle/mediana de un triángulo (p. 332) Un segmento cuyos extremos son un vértice del triángulo y el punto medio del lado opuesto.

midpoint of a segment/punto medio de un segmento (pp. 49, 74) El punto que divide al segmento en dos segmentos congruentes.

midsegment of a trapezoid/medio segmento de un trapezoide (p. 397) El segmento cuyos extremos son los puntos medios de sus lados no paralelos.

midsegment of a triangle/medio segmento de un triángulo (p. 336) Un segmento cuyos extremos son los puntos medios de dos lados.

minor arc/arco menor (p. 614) Un arco que es más pequeño que un semicírculo.

· · **N** · ·

net/desarrollo (p. 18) Una figura de dos dimensiones que al doblarse forma la superficie de una figura tridimensional.

network/red o grafo (p. 78) Un conjunto finito de puntos llamados *vértices* o *nudos* del sistema, unidos por segmentos de arcos llamados *bordes* del sistema. También se llama un *grafo*.

n-gon/n-polígono (p. 196) Un polígono con n lados.

nonagon/nonágono (p. 196) Un polígono con nueve lados.

non-Euclidean geometry/geometría no euclidiana (pp. 47, 155) Una geometría en la cual uno o más postulados no son los mismos a los de la geometría euclidiana. Las geometrías clásicas no euclidianas asumen un postulado paralelo que es fundamentalmente diferente al de Euclides.

number line/recta numérica (p. 40) Una línea cuyos puntos se colocan en correspondencia con el conjunto de números reales.

· · **O** · ·

oblique cone/cono oblicuo (p. 647) Un cono que no es un cono recto.

oblique cylinder/cilindro oblicuo (p. 646) Un cilindro que no es un cilindro recto.

oblique line and plane/línea y plano oblicuos (p. 137) Una línea y un plano que no son ni paralelos ni perpendiculares.

oblique lines/líneas oblicuas (p 131) Dos líneas que se intersecan y no son perpendiculares.

oblique planes/planos oblicuos (p. 137) Dos planos que no son ni paralelos ni perpendiculares.

oblique prism/primas oblicuos (p. 419) Un prisma que no es un prisma recto.

oblique pyramid/pirámide oblicua (p. 420) Una pirámide que no es una pirámide recta.

oblong numbers/números oblongos (p. 81) Los términos de una sucesión 2, 6, 12, 20, 30, ... La función f que relaciona el valor de un número oblongo a su orden n en la sucesión es $f(n) = n(n + 1)$.

obtuse angle/ángulo obtuso (p. 93) Un ángulo cuya medida es mayor de 90° y menor de 180°.

obtuse triangle/triángulo obtusángulo (p. 256) Un triángulo con un ángulo obtuso.

octagon/octágono (p. 196) Un polígono con ocho lados.

octahedron/octaedro (p. 724) Un poliedro con ocho caras.

octant/octante (p. 42) Una de las ocho regiones en que los ejes de x, y, y z dividen un sistema de coordenadas tridimensionales.

one-point perspective/perspectiva de un punto
(p. 24) Un dibujo con un punto de desvanecimiento que tiene la perspectiva de un punto.

opposite of a vector/opuesto de un vector (p. 180)
Un vector que tiene la misma magnitud del vector dado pero dirección opuesta.

opposite rays/rayos opuestos (p. 93) $\overrightarrow{BA}$ y $\overrightarrow{BC}$ son rayos opuestos si el punto B está entre los puntos A y C.

ordered pair/par ordenado (p. 41) En un plano coordenado, un par ordenado (x, y) es el par de números reales que corresponde a un punto.

ordered triple/triple ordenado (p. 42) En un sistema coordenado tridimensional, un triple ordenado (x, y, z) son tres números reales que corresponde a un punto.

order of rotation symmetry/orden de rotación de simetría (p. 10) El número de veces que una figura plana coincide con su posición original durante una rotación completa alrededor de su centro de simetría.

orientation/orientación (p. 211) El orden en el cual los vértices de un polígono se nombran en dirección de las manecillas del reloj o en contra.

origin of a coordinate plane/origen del plano coordenado (p. 41) El punto donde se intersecan los ejes.

origin of a number line/origen de la recta numérica (p. 40) El punto que corresponde a cero.

orthocenter of a triangle/ortocentro de un triángulo (p. 333) El punto de concurrencia de las líneas que contienen las alturas.

orthogonal drawing/dibujo ortogonal (p. 24) Dibujo que representa un objeto tridimensional al detallar tres vistas del objeto. También se llama *dibujo ortográfico*.

• • • **P** • • •

paper-folding construction/construcción por dobleces de papel (p. 49) Un dibujo preciso de una figura geométrica que se hace por medio de dobleces de una hoja de papel.

parallel line and plane/línea y plano paralelos (p. 136) Una línea y un plano que no se intersecan.

parallel lines/líneas paralelas (pp. 23, 131) Líneas coplanares que no se intersecan.

parallelogram/paralelogramo (pp. 178, 373, 377) Un cuadrilátero con dos pares de lados paralelos.

parallel planes/planos paralelos (p. 136) Planos que no se intersecan.

parallel vectors/vectores paralelos (p. 176) Vectores que tienen la misma dirección o direcciones opuestas.

pentagon/pentágono (pp. 82, 196) Un polígono con cinco lados.

perimeter of a plane figure/perímetro de una figura plana (p. 435) La distancia alrededor de la figura. El perímetro de un polígono es la suma de las longitudes de sus lados.

perpendicular bisector of a segment/mediatriz de un segmento (p. 157) Una línea, rayo o segmento que es perpendicular al segmento dado en su punto medio.

perpendicular line and plane/línea y plano perpendiculares (p. 137) Una línea y un plano son perpendiculares si y solamente si la línea es perpendicular a dos líneas distintivas en el plano en su punto de intersección.

perpendicular lines/líneas perpendiculares (pp. 41, 130) Dos líneas que se intersecan para formar un ángulo recto.

perpendicular planes/planos perpendiculares (p. 137) Dos planos son perpendiculares si y solamente si un plano contiene una línea que es perpendicular al otro plano.

perpendicular vectors/vectores perpendiculares (p. 178) Vectores cuyas direcciones son perpendiculares.

perpespective drawing/dibujo en perspectiva (p. 24) Un dibujo hecho en una superficie de dos dimensiones de manera que objetos tridimensionales se parecen a la imagen real.

plane/plano (pp. 9, 65) Un conjunto de puntos que se extiende infinitamente en todas direcciones a lo largo de una superficie plana. Éste es uno de los términos básicos indefinibles de geometría.

plane figure/figura plana (p. 9) Un conjunto de puntos que está en el mismo plano. También se llama *figura bidimensional*.

plane of symmetry/plano de simetría (p. 16) Un plano que divide a una figura tridimensional en dos partes que son imágenes la una de la otra.

plane symmetry/simetría (p. 16) Véase *simetría de reflexión de una figura tridimensional.*

Platonic solid/sólido platónico (pp. 724, 725) Véase *poliedro regular.*

point/punto (pp. 9, 65) Un lugar. Éste es uno de los términos básicos indefinibles de geometría.

point of concurrency/punto de concurreencia (p. 332) Véase *líneas incidentes.*

point of tangency/punto de tangencia (p. 663) Véase *tangente a un círculo.*

polygon/polígono (p. 195) Una figura plana formada por tres o más segmentos de manera que cada segmento interseca exactamente a otros dos, uno a cada extremo y no dos segmentos con un extremo común son colineales. Cada segmento es un *lado* del polígono. El extremo común de dos lados es un *vértice*. El polígono encierra completamente una región del plano llamada su *interior.*

polyhedron/poliedro (p. 724) Una figura tridimensional formada por superficies planas limitadas por polígonos y unidas en pares a lo largo de sus lados. El poliedro encierra completamente una región de espacio llamada su *interior.*

polyomino/poliomino (pp. 216, 217) Una figura plana formada por un conjunto de unidades cuadradas arregladas de manera que cada cuadrado comparte un lado con por lo menos otro cuadrado.

postulate/postulado (p. 66) Una declaración que se acepta como verdadera sin prueba.

preimage/pre-imagen (p. 210) Véase *transformación.*

prism/prisma (p. 419) Una figura tridimensional que consiste de dos *bases* paralelas que son regiones poligonales congruentes y *caras laterales* que están limitadas por paralelogramos que conecta los lados correspondientes a las bases.

probability of an event/probabilidad de un suceso (p. 74) La posibilidad de que un suceso ocurra.

proof/prueba o demostración (p. 116) Un conjunto de declaraciones justificadas organizadas en orden lógico de manera que juntas lleguen a convencer al lector de que una declaración dada es verdadera. La justificación para una declaración puede ser una definición, una propiedad algebraica, un postulado o un teorema previamente probado.

proportion/proporción (pp. 34, 494) Una declaración en la que dos razones son iguales.

protactor/transportador (pp. 31, 92) Un instrumento que se usa para encontrar la medida de un ángulo.

pyramid/pirámide (p. 420) Una figura tridimensional que consiste de una cara poligonal, llamada *base*, un punto llamado el *vértice* que no está en el plano de la base y *caras laterales* triangulares que conectan el vértice con cada lado de la base.

Pythagorean triple/triple pitagórico (p. 470) Un conjunto de tres enteros positivos *a*, *b* y *c* que satisfacen la ecuación $a^2 + b^2 = c^2$.

• • **Q** • •

quadrant/cuadrante (p. 41) Una de las cuatro regiones en que los ejes de *x* e *y* dividen al plano coordenado.

quadrilateral/cuadrilátero (p. 196) Un polígono con cuatro lados.

• • **R** • •

radius of a circle/radio de un círculo (pp. 48, 609, 612) Un segmento, o la longitud del segmento, cuyos extremos son el centro del círculo y un punto del círculo.

radius of a regular polygon/radio de un polígono regular (p. 594) Un segmento, o la longitud del segmento, cuyos extremos son el centro y un vértice del polígono.

radius of a sphere/radio de una esfera (p. 648) Un segmento, o la longitud del segmento, cuyos extremos son el centro de la esfera y un punto en la esfera.

ratio/razón (pp. 34, 493) Una comparación de dos cantidades por la división.

ray/rayo (pp. 30, 92) Parte de una línea que empieza en un punto y se extiende infinitamente en una dirección. El punto se llama el *extremo* del rayo.

rectangle/rectángulo (pp. 6, 374, 391) Un cuadrilátero con cuatro ángulos rectos.

rectangular prism/prisma rectangular (pp. 16, 468) Un prisma con bases rectangulares.

reduction/reducción (p. 512) Una dilatación con un factor de escala *k* de manera que $0 < k < 1$.

reflection/reflexión (p. 210) Una reflexión en una línea ℓ es una transformación que traza cada punto P a un punto P' de manera que: (1) si P no están en ℓ, entonces ℓ es la mediatriz perpendicular de $\overline{PP'}$; (2) si P está en ℓ, entonces P' es P. La línea ℓ es la *línea de reflexión* y P' es la *imagen de reflexión* de P.

reflection symmetry of a plane figure/simetría de reflexión de una figura plana (pp. 7, 10, 212) Una figura plana tiene simetría de reflexión si se puede dividir en dos partes a lo largo de una línea siendo la una la imagen de la otra. También se llama *simetría axial*.

reflection symmetry of a three-dimensional figure/simetría de reflexión de una figura tridimensional (p. 16) Una figura tridimensional tiene simetría de reflexión si se puede dividir en dos partes a lo largo de un plano siendo la una la imagen de la otra. También se llama *simetría*.

regular polygon/polígono regular (p. 197) Un polígono que es equilátero y equiángulo.

regular polyhedron/poliedro regular (p. 725) Un poliedro convexo en el cual las caras están limitadas por polígonos congruentes regulares y el mismo número de aristas intersecan en cada vértice. También se llama *sólidos platónicos*.

regular prism/prisma regular (p. 419) Un prisma recto cuyas bases son regiones poligonales regulares.

regular pyramid/pirámide regular (p. 420) Una pirámide recta cuya base es una región poligonal regular.

regular tessellation of the plane/teselación regular del plano (p. 268) Una teselación en la cual cada forma es un polígono regular y todas las formas son congruentes.

rhombus/rombo (pp. 374, 391) Un cuadrilátero con cuatro lados congruentes.

right angle/ángulo recto (p. 93) Un ángulo cuya medida es 90°.

right cone/cono recto (p. 647) Un cono en el cual el eje y la altura son el mismo segmento.

right cylinder/cilindro recto (p. 646) Un cilindro en el cual el eje es también la altura.

right prism/prisma recto (p. 419) Un prisma cuyas aristas laterales son perpendiculares a las bases.

right pyramid/pirámide recta (p. 420) Una pirámide en la cual la base tiene simetría de rotación y el segmento desde el vértice al centro de simetría es perpendicular a la base.

right triangle/triángulo rectángulo (p. 256) Un triángulo con un ángulo recto.

rigid transformation/transformación rígida (p. 211) Véase *isometría*.

rotation/rotación (p. 224) Una rotación a través de $a°$ alrededor de un punto O es una transformación que traza cada punto P a un punto P' de manera que: (1) si el punto P es diferente del punto O, entonces $OP' = OP$ y $m\angle POP' = a°$; (2) si el punto P es el punto O, entonces P' es P. El punto O es el *centro de rotación*, $a°$ es el *ángulo de rotación* y P' es la *imagen de rotación* de P.

rotation symmetry of a plane figure/simetría de rotación de una figura plana (p. 10) Una figura plana tiene simetría de rotación si se puede girar alrededor de un punto de modo que coincida con su posición original dos o más veces durante una vuelta completa.

rotation symmetry of a three-dimensional figure/ simetría de rotación de una figura tridimensional (p. 18) Una figura tridimensional tiene simetría de rotación si se puede girar alrededor de una línea de modo que coincida con su posición original dos o más veces durante una vuelta completa.

ruler/regla (pp. 29, 72) Un instrumento que se usa para hallar la longitud de un segmento.

● ● ● **S** ● ● ●

same-side exterior angles/ángulos externos del mismo lado de la transversal (p. 145) Cuando dos líneas son intersecadas por una transversal, los ángulos externos son dos ángulos externos en el mismo lado de la transversal.

same-side interior angles/ángulos internos del mismo lado de la transversal (p. 141) Cuando dos líneas son intersecadas por una transversal, los ángulos internos son dos ángulos interiores en el mismo lado de la transversal.

scale drawing/dibujo a escala (p. 34) Un dibujo bidimensional que es similar al objeto que representa. La razón del tamaño del dibujo al tamaño real del objeto es la *escala* del dibujo.

scale factor of similar polygons/factor de escala de polígonos similares (p. 502) La razón de las longitudes de los lados correspondientes.

scale factor of similar solids/factor de escala de sólidos similares (p 770) La razón de las medidas lineales correspondientes.

scale model/modelo a escala (p. 35) Una figura tridimensional cuyas superficies son similares a las superficies correspondientes del objeto que representa. La razón del tamaño del modelo al tamaño real del objeto es la *escala* del modelo

scalene triangle/triángulo escaleno (p. 256) Un triángulo sin lados congruentes.

secant of a circle/secante de un círculo (p. 663) Una línea que interseca a un círculo en dos puntos.

sector/sector (p. 621) Una región limitada por dos radios de un círculo y su arco intersecado.

segment/segmento (pp. 30, 67) Parte de una línea que comienza en un punto y termina en otro. Los puntos se llaman *extremos* del segmento.

segment bisector/bisector de un segmento (pp. 49, 74) Cualquier segmento, rayo, línea o plano que interseca el segmento en su punto medio.

segment of a circle/segmento de un círculo (p. 622) Una región limitada por un arco y la cuerda que tiene los mismos extremos que el arco.

semicircle/semicírculo (p. 614) Un arco cuyos extremos son los extremos del diámetro.

sequence/sucesión (p. 628) Un conjunto de números, llamados *términos*, arreglados en un orden específico.

side of an angle/lado de un ángulo (pp. 31, 92) Cada uno de los dos rayos que forman el ángulo.

side of a polygon/lado de un polígono (p. 195) Véase *polígono*.

similar figures/figuras semejantes (p. 34) Figuras que tienen la misma forma pero no necesariamente el mismo tamaño.

similar polygons/polígonos semejantes (p. 501) Polígonos cuyos vértices se pueden aparear de tal manera que los ángulos correspondientes son congruentes y los lados correspondientes son proporcionales.

similar solids/sólidos semejantes (p. 770) Dos sólidos son semejantes si tienen la misma forma y todas las medidas lineales correspondientes, como la altura y los radios, son proporcionales.

sine/seno (pp. 572, 580) El seno de un ángulo agudo de un triángulo rectángulo es la razón de la longitud del cateto opuesto al ángulo a la longitud de la hipotenusa.

skew lines/líneas alabeadas (p 131) Líneas que no son coplanares.

slant height of a regular pyramid/apotema de una pirámide regular (p. 468) La altura de una cara lateral.

slant height of a right cone/generatriz de un cono regular (p 647) La distancia desde el vértice del cono a cualquier punto en el borde de la base.

slope-intercept form of a linear equation/forma de pendiente e intersección de una ecuación lineal (p. 167) La forma $y = mx + b$ donde m es la pendiente de la línea y b es su intersección en y.

slope of a line/pendiente de una línea (p. 161) En un plano coordenado, la razón del cambio de valor de las coordenadas y (el ascenso) al cambio correspondiente en las coordenadas x (el curso), como están medidas entre cualesquiera dos puntos en la línea. Para una línea no vertical que contiene los puntos (x_1, y_1) y (x_2, y_2) la pendiente m se da por la fórmula

$$m = \frac{\text{elevación}}{\text{curso}} = \frac{\text{cambio en y}}{\text{cambio en x}} = \frac{y_2 - y_1}{x_2 - x_1}$$

small circle of a sphere/círculo menor de una esfera (p 648) Una sección transversal de la esfera cuyo centro no es el centro de la esfera.

solid figure/figura sólida (p. 16) Una figura tridimensional que consiste de todos sus puntos en la superficie y todos sus puntos que la superficie encierra. También se le llama *sólido*.

solve a triangle/resolver un triángulo (p. 591) Determinar las medidas de todos los ángulos y lados de un triángulo.

space/espacio (pp. 16, 65) El conjunto de todos los puntos.

sphere/esfera (pp. 16, 154, 648) El conjunto de todos los puntos en el espacio que están a una distancia dada de un punto dado. El punto dado se llama el centro de la esfera y la distancia dada es el *radio*.

square/cuadrado (pp. 6, 374, 391) Un cuadrilátero con cuatro ángulos rectos y cuatro lados congruentes.

square numbers/números cuadrados (p. 82) Los términos de la sucesión 1, 4, 9, 16, 25, ... La función *f*, que relaciona el valor de un número cuadrado a su orden *n* en la sucesión, es $f(n) = n^2$.

square pyramid/pirámide cuadrada (pp. 16, 420) Una pirámide con una base cuadrada.

standard form of a linear equation/forma estándar de una ecuación lineal (p. 167) La forma $Ax + By = C$ donde *A* y *B* son números reales y ambos no son iguales a cero.

straight angle/ángulo plano (p. 93) Un ángulo cuya medida es igual a 180°.

straightedge/regla recta (p. 48) Un instrumento que se usa como guía para trazar líneas, rayos y segmentos.

strip pattern/patrón de franja (p. 242) Véase *patrón de friso*.

supplementary angles/ángulos suplementarios (p. 93) Dos ángulos cuyas medidas suman 180°. Cada ángulo se dice es el *suplemento* del otro.

surface/superficie (p. 731) La frontera de una figura tridimensional que encierra completamente una región del espacio.

surface area/área de una superficie (p. 731) El área de la superficie de una figura tridimensional.

· · **T** · ·

tangent circles/círculos tangentes (p. 665) Círculos en el mismo plano que son tangentes a la misma línea en el mismo punto. Los círculos son *tangentes exteriormente* si todos los puntos de un círculo, excepto el punto de tangencia, está en el exterior del otro círculo. Los círculos son tangentes *interiormente* si todos los puntos de un círculo, excepto el punto de tangencia, están en el interior del otro círculo.

tangent of an angle/tangente de un ángulo (pp. 571, 573) La tangente de un ángulo agudo de un triángulo rectángulo es la razón de la longitud del cateto opuesto al ángulo a la longitud del cateto adyacente.

tangent of a circle/tangente de un círculo (p. 663) Una línea en el plano del círculo que interseca el círculo en exactamente un punto. El punto se llama *punto de tangencia*.

taxicab geometry/geometría taxi (p. 46) Un sistema de geometría en el cual la distancia entre dos puntos, llamada *distancia taxi*, se define como el menor número de unidades horizontales y verticales entre los puntos en un plano coordenado.

terminal point of a vector/punto terminal de un vector (p. 176) El vector $\overrightarrow{PQ}$ comienza en el punto *P* y termina en el punto *Q*. El punto *Q* es el punto terminal.

terms of a sequence/términos de una sucesión (p. 628) Véase *sucesión*.

tessellation of the plane/teselación del plano (p. 267) Un patrón de formas que se repite y que cubre completamente el plano sin separación ni superposición.

tetrahedron/tetraedro (p. 724) Un poliedro con cuatro caras.

theorem/teorema (p. 117) Una declaración que se puede probar verdadera.

theoretical probability/probabilidad teórica (pp. 74, 634) La probabilidad teórica de un evento o suceso, $P(E)$, se da por la razón
$$\frac{\text{número de resultados favorables}}{\text{número total de resultados}}.$$

three-dimensional coordinate system/sistema de coordenadas en tres dimensiones (p. 42) Un método para localizar puntos en el espacio formado por rectas numéricas que se intersecan en sus orígenes, cada línea es perpendicular a las otras dos.

three-dimensional figure/figura tridimensional (p. 16) Una figura que se extiende más allá de un solo plano al espacio.

traceable network/red o grafo que se puede trazar (p. 78) Una red en la cual es posible comenzar en un vértice y moverse hacia los otros vértices trazando cada borde exactamente una vez.

transformation/transformación (pp. 206, 210) Una correspondencia entre una figura, llamada *pre-imagen*, y una segunda figura, su *imagen*, de tal manera que cada punto de la pre-imagen se aparea exactamente con un punto de la imagen y cada punto de la imagen se aparea exactamente con un punto de la pre-imagen. También se llama una *planimetría*.

translation/traslación (p. 218) Una traslación por un vector $\overrightarrow{AA}$ es una transformación que traza cada punto P a un punto P' de manera que: (1) $PP' = AA'$; (2) $\overline{PP'} \parallel \overline{AA'}$ o $\overline{PP'}$ es colineal con $\overline{AA'}$. El vector $\overrightarrow{AA}$ es el *vector de traslación* y P' es la *imagen de traslación* de P.

transversal/transversal o secante (p. 139) Una línea que interseca a dos o más líneas coplanares en diferentes puntos.

trapezoid/trapezoide (pp. 374, 397) Un cuadrilátero con exactamente un par de lados paralelos.

triangle/triángulo (p. 196) Un polígono con tres lados.

triangular numbers/números triangulares (p. 81) Los términos de la sucesión 1, 3, 6, 10, 15, ... La función f, que relaciona el valor de un número triangular con su orden n en la sucesión, es $f(n) = \frac{n(n+1)}{2}$.

trigonometric ratio/razón trigonométrica (p. 571, 573) Una razón de las longitudes de dos lados de un triángulo recto.

two-dimensional figure/figura bidimensional (p. 16) Véase *figura plana.*

two-point perspective/perspectiva de dos puntos (p. 24) Un dibujo con dos puntos de desvanecimiento tiene perspectiva de dos puntos.

vanishing point/punto de desvanecimiento (p. 23) Un punto a una distancia desde el observador donde dos o más líneas de un dibujo parecen que se intersecan.

vector/vector (p. 176) Una cantidad que tiene un tamaño, llamado su *magnitud*, y una dirección.

vertex angle of an isosceles triangle/vértice de un ángulo de un triángulo isósceles (p 324) El ángulo opuesto a la base.

vertex of an angle/vértice de un ángulo (pp. 31, 92) El extremo común de los lados.

vertex of a cone/vértice de un cono (p. 647) Véase *cono.*

vertex of a network/vértice de una red o grafo (p. 78) Véase *red o grafo.*

vertex of a polygon/vértice de un polígono (pp. 24, 195) Véase *polígono.*

vertex of a polyhedron/vértice de un poliedro (p. 724) Un punto que es la intersección de tres o más aristas.

vertex of a pyramid/vértice de una pirámide (p. 420) Véase *pirámide.*

vertical angles/ángulos opuestos por el vértice (p. 129) Dos ángulos cuyos lados forman dos pares de rayos opuestos.

volume of a three- dimensional figure/volumen de una figura tridimensional (p. 737) El número de unidades cúbicas no superpuestas en el interior de la figura.

x-axis/eje de x (abcisa) (p. 41) La recta numérica horizontal en un plano coordenado.

x-coordinate/coordenada de x (abcisa) (p. 41) El primer número en un par ordenado que representa un punto en el plano coordenado e indica la distancia a la izquierda o derecha del eje de y.

x-intercept of a line/intercepto de x de una línea (p. 168) La coordenada x (abcisa) del punto donde la línea interseca el eje de x.

y-axis/eje de y (ordenada) (p. 41) La recta numérica vertical en un plano coordenado.

y-coordinate/coordenada de y (ordenada) (p. 41) El segundo número de un par ordenado que representa un punto en el plano coordenado e indica la distancia por encima o por debajo del eje de x (abcisas).

y-intercept of a line/intercepto de y de una línea (p. 168) La coordenada de y (ordenada) del punto donde la línea interseca el eje de y.

••• SELECTED ANSWERS •••

Chapter 1 Exploring Geometry

Lesson 1.2, pages 9–15

TRY THESE
3.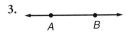

5. none   7.

9. none   11. 5-fold   13.

15. Each has a vertical and a horizontal line of symmetry and rotation symmetry. Explanations will vary.   17. Any line that passes through the center of the circle is a line of symmetry.

PRACTICE   19. plane   21.

23. none   25. 7-fold   27. none

EXTEND   33.   35.

37.   39.

41. A, H, I, M, O, T, U, V, W, X, Y; Answers will vary. Possible words: HUT, YOU.   43. It has a line of symmetry through the O. yes; possible words: TOT, WOW

THINK CRITICALLY   45. An ellipse has a horizontal and a vertical line of symmetry and rotation symmetry. An egg-shape has only one line of symmetry.
47. Answers will vary. Since the manhole cover rests on a lip and a circle has infinitely many symmetries, there is no way that it can be turned so that it will fall into the hole.

ALGEBRA AND GEOMETRY REVIEW   48. 3   49. −1.5
50. −6.5   51. −2   52. −1.5   53. C   54. 45¢
55. 77¢   56. 75¢   57. 75¢ and $1.00

Lesson 1.3, pages 16–22

TRY THESE   1. 3   3.   5. cylinder

7. one   9. cylinder   11. The figures all have infinitely many planes and one axis of symmetry.

PRACTICE   13. rectangular prism   15. a. 2   15. b. 1
15. c. 3   17. cone   19. 1   21.

23. IV   25. III   27. rectangle

EXTEND
29. There are 22 total three-dimensional symmetries.
3 planes through a face: 3 axes through faces:

6 planes through opposite   3 axes through opposite
edge;   vertices:

6 axes through opposite edge:

31. Answers will vary. Accept all answers students can justify.

THINK CRITICALLY   33. spiral

Algebra and Geometry Review
34. 6 lines of symmetry and 6-fold rotation symmetry.
35. point, line, and plane   36. $-\dfrac{5}{2}$ 37. $\dfrac{4}{3}$ 38. 7

39. −12 **40.** $-\frac{2}{5}$ **41.** 25 **42.** triangle **43.** 10.8
44. 120 **45.** $66\frac{2}{3}$% **46.** 150%

Lesson 1.4, pages 23–28

TRY THESE **1.**
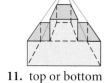

3.
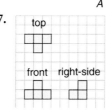

5.
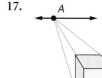

7.
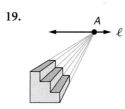

11. top or bottom **13.** back **15.**
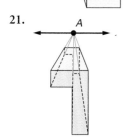

PRACTICE

17.
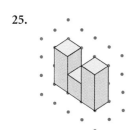

19.
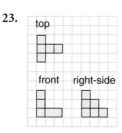

21.
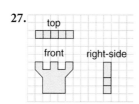

23.
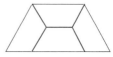

25.

27.

EXTEND
29.
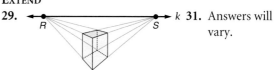
k **31.** Answers will vary.

THINK CRITICALLY
35. There are 8 four-cube polycubes.
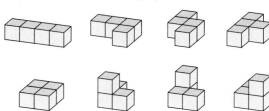
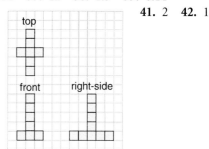

Algebra and Geometry Review
36. 14 **37.** 12 **38.** 1.5 **39.** 0.14
40.
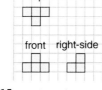
41. 2 **42.** 1

Lesson 1.6, pages 33–39

TRY THESE **1.** C **5.** 125 mi **7.** 468.75 mi
9. 1 in.:$\frac{1}{8}$ in. **11. a.** 3 in. **11. b.** 2.8 in.
11. c. 1.9 in.

PRACTICE **13.** C and E **15.** 4.5 in. **17.** 1.375 in.
19. 0.1875 in. **21.** 4 in. 120° **23.** $2\frac{2}{3}$ in. 120°

EXTEND **27.** Answers will vary. **29.** $\frac{1}{8}$ mi
31. 21.3 in. × 16.3 in.

THINK CRITICALLY
33. a.
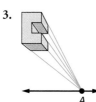
b.

c.

Algebra and Geometry Review
34. $(-4, 2)$ **35.** $(0, 3)$ **36.** $(2, 5)$ **37.** $(-3, 0)$
38. $(0, 0)$ **39.** $(3, 0)$ **40.** $(-4, -4)$ **41.** $(0, -4)$
42. $(3, -3)$ **43.** $(4, -5)$ **44.** E, G **45.** E, J

46. D, E, F **47.** B, E, H **48.** B **49.** D **50.** A, E
51. B, C, D
52.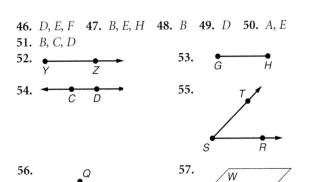

53.

54.

55.

56. $\bullet Q$ **57.** $\boxed{W}$

58. 3 in. wide, 5 in. long

Lesson 1.7, pages 40–45

TRY THESE **1.** 4 **3.** 6 **5.** 13 **7.** W **9.** R **11.** Q
13. S **15.** 17 **17.** 15.7 **19.** 14.2 km **21.** $(6, 0, 4)$
23. $(6, 4, 4)$ **25.** 5

PRACTICE **27.** 5 **29.** 5 **31.** 16 **33.** 5 **35.** 8.6
37. $(0, 0, 7)$ **39.** $(0, 5, 0)$

EXTEND **43.** inside **45.** on **47.** outside
49. $MA = \sqrt{(-3+1)^2 + (5+1)^2} = \sqrt{4+36} =$
$\sqrt{40} = 2\sqrt{10}$
$MB = \sqrt{(-3+5)^2 + (5-11)^2} = \sqrt{4+36} =$
$\sqrt{40} = 2\sqrt{10}$

THINK CRITICALLY **51.** Infinitely many; explanations
will vary. **53.** a vertical line at $x = 2$

ALGEBRA AND GEOMETRY REVIEW **55.** 12-fold rotation
symmetry **56.** 6 **57.** $-\sqrt{4}, -1\frac{7}{8}, -1.5, 2.3, 2\frac{2}{3}, \sqrt{8}$
58. $-\frac{16}{5}, -3, -\sqrt{8}, 2.8, \sqrt{13}, \sqrt{17}$ **59.** $962\frac{1}{2}$ mi
60. D

Lesson 1.8, pages 48–51

APPLY THE STRATEGY **15.** Student should first bisect
$\overline{PQ}$ and the bisect each of the two new segments
formed. **17.** First bisect $\overline{AZ}$ to obtain a segment that is
half as long as $\overline{AZ}$. Then construct this segment together
with $\overline{AZ}$ on a line so that the share a common endpoint.
19. Answers will vary. Using a compass is more
mechanical and using a ruler involves reading the marks
on it accurately.

Chapter 2 Geometry and Logic

Lesson 2.1, pages 65–71

TRY THESE **1.** Y, X; Postulate 1 **3.** line ℓ; Postulate 6
5. Postulate I; The sticks are used as points to make
sure the line, wall, is straight. **7.** K **9.** C, N, F

PRACTICE **11.** true **13.** true **15.** true **17.** true
19. 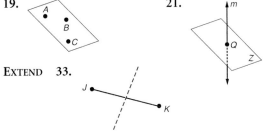 **21.**

EXTEND **33.**

THINK CRITICALLY **35.** Postulate I; You can find the
number of combinations of 8 points taken 2 at a time.
37. 28; 20

ALGEBRA AND GEOMETRY REVIEW **39.** 2.75 **40.** 0.44
41. 2 **42.** 8 **43.** 1 **44.** -1 **45.** 4 **46.** 3 **47.** 3
48. 7 **50.** R, S, T **51.** P, R, S, T or Q, R, S, T
52. P, Q, R, S **53.** line ℓ **54.** true **55.** $\frac{1}{2}$ **56.** 21
57. $\frac{4}{5}$ **58.** $\frac{1}{8}$

Lesson 2.2, pages 72–77

TRY THESE **1.** 6.5 cm **5.** 15.75 **7.** 2 **9.** 0
11. $(0, 5)$ **13.** $(1, 2)$ **15.** 0.3

PRACTICE **17.** $|2 - 5| = 3$ in.
19. $\left|1 - 2\frac{7}{8}\right| = 1\frac{7}{8}$ in. **21.** $\left|2\frac{7}{8} - 5\frac{1}{2}\right| = 2\frac{5}{8}$ in.
23. 56 **25.** false **27.** true **29.** false
31. $(3, 9)$ **33.** $\left(4\frac{3}{4}, 4\right)$

EXTEND **39.** $(13, -12)$ **41.** false **43.** $\frac{1}{6}$
THINK CRITICALLY **45.** $\left(2\frac{1}{2}, 2, 4\right)$ **47.** false

ALGEBRA AND GEOMETRY REVIEW **48.** $-4, -19,$
$4.25, -5\frac{1}{2}$ **49.** $y = x - 6$ **50.** $f(x) = -5x$
51. $-5.5, -\frac{5}{2}$ **52.** A **53.** false

Lesson 2.3, pages 80–83

APPLY THE STRATEGY **17.** 1, 4, 9, 16, 25, 36, 49, 64, 81,
100, 121, 144 **19.** 385
21.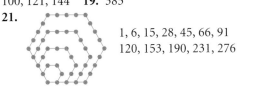

1, 6, 15, 28, 45, 66, 91
120, 153, 190, 231, 276

23. $p(n) = \frac{1}{2}n(3n - 1)$

Lesson 2.6, pages 92–99

TRY THESE **1.** 120°; obtuse **3.** 120°; obtuse
5. 180°; straight **7.** $\angle EZF$ and $\angle CZB$ **9.** $\angle AZE$ and
$\angle CZB$, $\angle BZE$ and $\angle CZA$ **11.** $\angle AZB$ and $\angle BZF$,

∠AZC and ∠CZF, ∠AZD and ∠DZF, ∠AZE and ∠EZF   **13.** 37°   **15.** 90°   **17.** 54°; 36°   **19.** angle bisector

PRACTICE   21. ∠PNT   **23.** ∠PNS, 125°; ∠QNT, 145°
27. ∠AGB and ∠AGE, ∠CGD and ∠EGD
29. $\overline{GB}$ and $\overline{GD}$ **31.** ∠FGA and ∠FGE   **33.** 83.5°
35. $(60 - c)°$ **37.** 18.6°   **39.** $(130 - n)°$ **41.** 26°
43. 47°   **45.** 60°   **47.** 72°

EXTEND   49.

51. $f(n) = \dfrac{n(n - 1)}{2}$

53. 115°

THINK CRITICALLY   55. false   **57.** false

ALGEBRA AND GEOMETRY REVIEW   59. 5   **60.** 3.2
61. 9.4   **62.** C   **63.** yes   **64.** no   **65.** no
66.

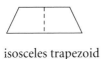

    kite   or   isosceles trapezoid

Lesson 2.8, pages 104–109

TRY THESE   1. If an angle has a measure greater than 0° and less than 90°, then it is acute.   statement: true   converse: true   **3.** Segments are of equal length if and only if they are congruent segments.   **5.** If the figure is a line, then it has at least two distinct points.
7. Inverse: If two angles are not both right angles, then they are not congruent. Contrapositive: If two angles are not congruent, then they are not both right angles. inverse: false   contrapositive: true

PRACTICE   11. If two angles form a linear pair, then the two angles are adjacent. true   **13.** If two angles do not form a linear pair, then the two angles are not adjacent. false   **15.** Two angles are complementary angles if and only if their measures have a sum of 90°.   **17.** If the angle is straight, then its sides are opposite rays.
19. Converse: If $AC + CB = AB$, then point C is between points A and B. Inverse: If point C is not between points A and B, then $AC + CB \neq AB$. Contrapositive: If $AC + CB \neq AB$, then point C is not between points A and B.   statement: true   converse: true   inverse: true   contrapositive: true
21. Converse: If something is part of a line, then it is a ray. Inverse: If something is not a ray, then it is not part of a line. Contrapositive: If something is not part of a line, then it is not a ray.   statement: true   converse: false   inverse: false   contrapositive: true   **23.** Just because winds of 100 mi/h or greater are hurricane force winds, does not mean that 100 mi/h has been shown to be the minimum velocity required to be considered hurricane force winds. Perhaps 90 mi/h and over are hurricane force winds. This new information would not render the given statement false, but it would show that Kanella had reasoned incorrectly.

EXTEND   27. Converse: If $r = -5$, then $|r| = 5$. Inverse: If $|r| \neq 5$, then $r \neq -5$. Contrapositive: If $r \neq -5$, then $|r| \neq 5$.   statement: false   converse: true   inverse: true   contrapositive: false   **29.** Converse: If $y = -3$, then $y^3 = -27$. Inverse: If $y^3 \neq -27$, then $y \neq -3$. Contrapositive: If $y \neq -3$, then $y^3 \neq -27$. statement: true   converse: true   inverse: true   contrapositive: true   **31.** conditional statement

THINK CRITICALLY   33. T, T, T, T   **35.** T, F, F, T
ALGEBRA AND GEOMETRY REVIEW   37.

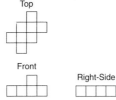

38. If three points are noncollinear, then they determine a plane.   **39.** Converse: If three points determine a plane, then they are collinear. Inverse: If three points are not noncollinear, then they don't determine a plane. Contrapositive: If three points don't determine a plane, then they are not noncollinear. statement: true   converse: true   inverse: true   contrapositive: true   **40.** $m^3 + 4m^2 + m$
41. $5v^3 - 14v^2 + 4v - 10$

Lesson 2.9, pages 110–114

TRY THESE   1. symmetric property   **3.** $GH = JK$; subtraction property   **5.** Definition of midpoint
7. transitive   **9.** none   **11.** symmetric

PRACTICE   15. Unique Plane Postulate
17. Definition of Right Angle   **19.** Division Property
21. ∠WAZ is a straight angle.   **23.** ∠WAX ≅ ∠XAY
25. Transitive Property

THINK CRITICALLY   31. false; Examples will vary. Possible example: Is a friend of   Reflexive: Bob is a friend of Bob.   Symmetric: Bob is a friend of Sue. Sue is a friend of Bob.   Transitive: If Bob is a friend of Sue and Sue is a friend of Mary, then Bob is not necessarily a friend of Mary.

ALGEBRA AND GEOMETRY REVIEW   33. 17; 7   **34.** $\dfrac{7}{24}$
35. Transitive Property   **36.** $10\sqrt{5}$ **37.** 3   **38.** 7

39. $\dfrac{\sqrt{15}}{5}$ **40.** $|y|$ **41.** $4|K|\sqrt{3}$

Chapter Review, pages 120–121

1. c **2.** a **3.** e **4.** b **5.** d **6.** P and Q; Postulate 1
7. Any 3 of the points P, Q, R, and S; Postulate 3
8. line n or $\overleftrightarrow{PQ}$; Postulate 6 **9.** plane $\mathcal{A}$; Postulate 5
10. 14 **11.** 25 **12.** 19.5
13. 7,776 regions: The pattern is

Figure 1	3
Figure 2	$3 \times 2 = 6$
Figure 3	$6 \times 3 = 18$
Figure 4	$18 \times 2 = 36$
Figure 5	$36 \times 3 = 108$
Figure 6	$108 \times 2 = 216$
Figure 7	$216 \times 3 = 648$
Figure 8	$648 \times 2 = 1,296$
Figure 9	$1,296 \times 3 = 3,888$
Figure 10	$3,888 \times 2 = 7,776$

14. It is not precise enough, "something" is too vague and "two parts" is incomplete. It should say "two equal parts." **15.** It is not concise. The second part, "and they can be an equal distance apart," is not necessary and should be omitted. **16.** $\angle APB$ (60°), $\angle BPC$ (30°), or $\angle DPE$ (60°) **17.** $\angle APE$ (120°), $\angle BPD$ (120°), or $\angle CPE$ (150°) **18.** $\angle APC$ (90°) or $\angle CPD$ (90°)
19. $\angle APD$ (60°) and $\angle BPC$ (30°) or $\angle DPE$ (60°) and $\angle BPC$ (30°) **20.** $\angle APD$ (60°) and $\angle BPD$ (120°), $\angle APC$ (90°) and $\angle CPD$ (90°), $\angle BPC$ (30°) and $\angle CPE$ (150°), $\angle BPD$ (120°) and $\angle DPE$ (60°), or $\angle DPE$ (60°) and $\angle EPA$ (120°). **22.** If a point is the midpoint of a segment, then the point divides the segment in two congruent parts. True Converse: If a point divides a segment in two congruent parts, then the point is the midpoint of the segment. True Inverse: If a point is not the midpoint of a segment, then the point does not divide the segment in two congruent parts. True Contrapositive. If a point does not divide a segment in two congruent parts, then the point is not the midpoint of the segment. True **23.** If two angles are a linear pair, then they are adjacent. True If two angles are adjacent, then they are a linear pair. False If two angles are not a linear pair, then they are not adjacent. False If two angles are not adjacent, then they are not a linear pair. True **24.** Two angles are congruent if and only if they have the same measure. **25.** A set of points is the intersection of two figures if and only if the set contains all points common to the two figures.
26. Given $\angle A$ is an acute angle
$\angle B$ is complementary to $\angle A$
$\angle C$ is supplementary to $\angle A$
Prove $m\angle B > m\angle C$
Since $\angle B$ is complementary to $\angle A$, $m\angle B + m\angle A =$

90°. Since $\angle C$ is supplementary to $\angle A$, $m\angle C + m\angle A$ = 180°. By the Subtraction Property $m\angle A = 90 - m\angle B$ and $m\angle A = 180 - m\angle C$. So, by substitution, $90 - m\angle B = 180 - m\angle C$ or $-m\angle B = 90 - m\angle C$, by the Subtraction Property. This means $m\angle B = m\angle C - 90$, by the Multiplication Property. Therefore, $m\angle B$ is 90° less than $m\angle C$ or $m\angle B > m\angle C$.
27. Given $\angle A \cong \angle B$
$\angle A$ and $\angle B$ are supplementary
Prove $\angle A$ and $\angle B$ are right angles

Statements	Reasons
1. $\angle A \cong \angle B$; $\angle A$ and $\angle B$ are supplementary	1. Given
2. $m\angle A = m\angle B$	2. Definition of congruent angles
3. $m\angle A + m\angle B = 180°$	3. Definition of supplementary
4. $m\angle A + m\angle A = 180°$ or $2\,m\angle A = 180°$	4. Substitution Property
5. $m\angle A = 90°$	5. Division Property
6. $m\angle B = 90°$	6. Substitution Property
7. $\angle A$ and $\angle B$ are right angles	7. Definition of right angle

Chapter 3
Lines, Planes, and Angles

Lesson 3.1, pages 129–134

TRY THESE **1.** $m\angle 3 = 25°$; $m\angle 2 = m\angle 4 = 155°$
3. $m\angle 2 = m\angle 4 = 72°$; other angles: 108°
5. perpendicular **7.** 28° **9a.** perpendicular
9b. parallel **9c.** oblique **9d.** skew

PRACTICE **11.** $m\angle 1 = 120°$; $m\angle 2 = m\angle 4 = 60°$
13. $m\angle 1 = m\angle 3 = 122°$; other angles: 58°
15. $m\angle 2 = 34°$; $m\angle 1 = m\angle 3 = 146°$
17. $m\angle 1 = m\angle 3 = 94°$; $m\angle 2 = m\angle 4 = 86°$
19. $m\angle 1 = m\angle 3 = 132°$; $m\angle 2 = m\angle 4 = 48°$
21. adjacent offices get smaller; opposite gets larger; right angles or perpendicular **23.** $\overrightarrow{AB}$ **25.** 55°

EXTEND **27.** b

THINK CRITICALLY **33.** $m\angle 1 = m\angle 3 = 175°$; $m\angle 2 = m\angle 4 = 5°$ **35.** $m\angle 1 = m\angle 3 = 137°$; $m\angle 2 = m\angle 4 = 43°$

ALGEBRA AND GEOMETRY REVIEW **36.** $\angle ABF$ and $\angle EBF$
37. Possible answers: $\angle BEF$ and $\angle CEF$ **38.** Possible answers: $\angle BEF$ and $\angle CEF$ **39.** 33° **41.** B
42. $x = -1$; $y = 3$ **43.** $x = 4$; $y = \dfrac{1}{2}$
44. $x = 1$; $y = 0$; $z = -2$

Lesson 3.3, pages 139–145

TRY THESE **1.** $m\angle 2 = m\angle 4 = m\angle 6 = 40°$ $m\angle 1 = m\angle 3 = m\angle 5 = m\angle 7 = 140°$ **3.** All angles measure 90°. **5.** $m\angle 3 = m\angle 1 = m\angle 5 = m\angle 7 = 75°$ $m\angle 2 = m\angle 4 = m\angle 6 = m\angle 8 = 105°$ **7.** $\angle 1, \angle 3$; $\angle 2, \angle 4$; $\angle 5, \angle 7$; $\angle 6, \angle 8$; $\angle 1, \angle 9$; $\angle 5,$ $\angle 11$, $\angle 2, \angle 10$; $\angle 6, \angle 12$; $\angle 4, \angle 12$; $\angle 3, \angle 10$; $\angle 8, \angle 11$; $\angle 7, \angle 9$ **9.** $\angle 6, \angle 3$; $\angle 2, \angle 7$; $\angle 6, \angle 9$; $\angle 5, \angle 10$; $\angle 7,$ $\angle 12$; $\angle 8, \angle 10$ **11.** $\angle 1, \angle 6$; $\angle 5, \angle 2$; $\angle 9, \angle 12$; $\angle 10,$ $\angle 11$; $\angle 3, \angle 8, \angle 7, \angle 4$ **13.** All angles measure 68° or 112°.

PRACTICE **17.** $m\angle 1 = m\angle 3 = m\angle 8 = m\angle 6 = 155°$ $m\angle 4 = m\angle 7 = m\angle 5 = 25°$ **19.** $m\angle 1 = m\angle 3 = m\angle 5 = m\angle 7 = 145°$ $m\angle 2 = m\angle 4 = m\angle 6 = m\angle 8 = 35°$ **21.** $\angle 1, \angle 3$; $\angle 2, \angle 4$; $\angle 5, \angle 7$; $\angle 6, \angle 8$ **23.** $\angle 6,$ $\angle 3$; $\angle 2, \angle 7$ **25.** $\angle 1, \angle 6$; $\angle 5, \angle 2$; $\angle 3, \angle 8$; $\angle 4, \angle 7$ **27.** $\angle 3, \angle 5, \angle 7$ **29.** all measure 110° or 70° **31.** all measure 43° or 137° **33.** All of the hip jack rafters must meet the hip rafter at an angle of 30°, because they are corresponding angles. **35.** $x = 30$; $y = 12$ **37.** $x = 9$; $y = 25$

EXTEND **41.** $\angle 3, \angle 4$; $\angle 3 \cong \angle 4$ **43.** $m\angle 3 = m\angle 4 = m\angle 6 = 45°$

ALGEBRA AND GEOMETRY REVIEW **47.** The sum of the measures of complementary angles is 90°; the sum of the measures of supplementary angles is 180°. **48a.** $2\sqrt{2}$ **48b.** $3\sqrt{3}$ **48c.** $5\sqrt{3}$ **48d.** $2\sqrt{10}$ **49.** $x = -8, 3$

Lesson 3.4, pages 147–153

TRY THESE **1.** Converse of Alternate Interior Angles Theorem. **3.** Converse of Corresponding Angles Postulate. **5.** $x = 9$ **7.** $a \parallel b$

PRACTICE **13.** Converse of Corresponding Angles Postulate. **15.** Converse of Same-Side Exterior Angles Theorem. **17.** $x = 7$ **19.** $\overline{AB} \parallel \overline{CD}$; Converse of Alternate Interior Angles Theorem.

ALGEBRA AND GEOMETRY REVIEW **36.** $2x^2 + 3x - 20$ **37.** $(x - 10)(x + 2)$ **38.**

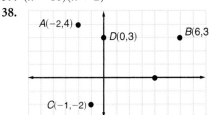

Lesson 3.6, pages 161–166

TRY THESE **1.** d **3.** c **5.** b **7.** perpendicular

9. parallel **11.** 0 **13.** undefined **19.** positive **21.** undefined **PRACTICE** **25.** $\frac{1}{2}$ **27.** $\frac{5}{3}$ **29.** $-\frac{4}{3}$ **31.** 0 **33.** 2 **41.** Yes, the stair ratio would exceed the minimum. Reasons will vary. Possible reason: A scale drawing shows that the stair ratio would form approximately a 30° angle. **43.** 2223 ft

THINK CRITICALLY **47.** 25,093 people/y; cannot have fractional people **49.** $\frac{6}{5}$

Lesson 3.7, pages 167–171

TRY THESE **1.** c **3.** b **5.** d **7.**

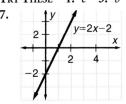

9.

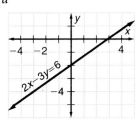

13. $y = -5x$ **15.** $y = -\frac{4}{3}x - 1$ **17.** $y = \frac{1}{2}x - 4$ **19.** $x = 7$

PRACTICE **21.** $3; -1$ **23.** $0; -2$ **25.** $-\frac{1}{4}; 1$ **27.** $-2; 4$ **29.** $y = \frac{1}{2}x$ **31.** $y = \frac{7}{2}x$ **33.** $y = -\frac{5}{2}x + 3$ **37.** $y = -x + 3$; $y = x - 7$ **39.** $y = -x + 9$; $y = x + 1$; $y = -x + 1$; $x = 4$; $y = -1$ **41.** $y = -3$; $x = -3$; $x = 6$; $y = 3x - 6$; $y = 3x + 21$

ALGEBRA AND GEOMETRY REVIEW **46.** 2; infinite **47.** $\sqrt{205}$ **48.** 22, 29

Lesson 3.9, pages 176–181

TRY THESE **1.** none **3.** $\vec{t}, \overline{w}$ **5.** $(7, 4)$; $\sqrt{65}$ **7.**

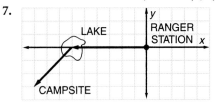

The hiker is west, southwest of the ranger station.

13. 0; perpendicular **15.** -8; not perpendicular **PRACTICE** **17.** $\sqrt{61}$ **19.** $(3, 1)$ **21.** $(4, 3)$; 5 **23.** $(-2, 0)$; 2 **25.** $(-2, 5)$; $\sqrt{29}$ **27.** $(-1, -4)$; $\sqrt{17}$

37. 20; not perpendicular   **39.** 0; perpendicular
41. 5; not perpendicular

EXTEND   45. $(-x, -y)$

THINK CRITICALLY   51. lowering the handle

ALGEBRA AND GEOMETRY REVIEW   52. No, the product of slopes $\neq -1$ **53.** No, the dot product is not 0
54. 51.8; 45; none; 74   **55.** 63°; 153°

Lesson 3.10, pages 182–185

APPLY THE STRATEGY   19. 306°; 142.5 mi/h   **21.** 23°; 570 km/h   **23.** 1.3 m/s

Chapter Review, pages 186–187

1. b   **2.** c   **3.** d   **4.** a
5. m∠1 = m∠3 = 77°; m∠2 = 103°
6. m∠1 = m∠3 = 155°; m∠2 = 25°
7. m∠1 = m∠2 = 35°
8. perpendicular   parallel   skew

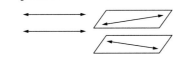

9. m∠3 = m∠5 = m∠7 = 128°;
m∠2 = m∠4 = m∠6 = m∠8 = 52°
10. m∠1 = m∠2 = m∠4 = m∠6 = 137°; m∠5 = m∠3 = m∠(5x + 3) = m∠(8x − 21) = 43°
11. m∠(10x + 8) = m∠1 = m∠3 = m∠5 = 118°; m∠6 = m∠4 = m∠(6x − 4) = m∠2 = 62°
12. Converse of the Corresponding Angles Postulate
13. Converse of the Same-Side Interior Angles Theorem   **14.** Converse of the Alternate Interior Angles Theorem   **1**
5.

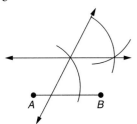

16.

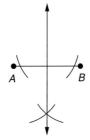

17. ∥ line, $m = 2$ **18.** ∥ line, $m = -\frac{3}{4}$
⊥ line, $m = -\frac{1}{2}$ ⊥ line, $m = \frac{4}{3}$
19. ∥ line, $m = -5$ ⊥ line, $m = \frac{1}{5}$
20. $m = -\frac{1}{5}$ **21.** $m = 2$ **22.** $m = -1$

23. $y = -3x + 2; m = -3, 6 = -2$

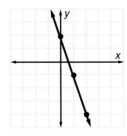

24. $y = \frac{1}{4}x$ **25.** $y = x - 5$
$m = \frac{1}{4}, b = 0$ $m = 1, b = -5$

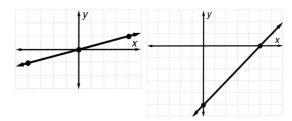

26. $y = 6x - 13$ **27.** $y = -\frac{4}{5}x + 1$ **28.** $x = 3$

29.

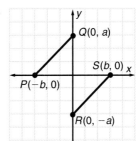

$\overrightarrow{PQ}, m = \dfrac{a - 0}{0 - (-b)} = \dfrac{a}{b}$
$\overline{RS}, m = \dfrac{0 - (-a)}{b - 0} = \dfrac{a}{b}$
Thus, $\overrightarrow{PQ} \parallel \overline{RS}$, because, if two lines have equal slope, then they are parallel.

30. $(-7, 3)$ **31.** $\sqrt{58}$ **32.** $\sqrt{5}$ **33.** $(-8, 1)$ **34.** 1
35.

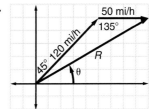

$R = 159.3$ mi/h
$\theta = 32.2°$ N of E

Chapter 4 Transformations in the Plane

Lesson 4.1, pages 195–201

TRY THESE   1. pentagon; concave   **3.** 16-gon; concave   **5.** 99°; 45°   **7.** polygon *RTVX*: regular; polygon *RSTUVWXY*: regular

PRACTICE **9.** 14-gon; concave **11.** pentagon; concave **13.** 100°; 80° **15.** $d(12) = \frac{12(12-3)}{2} = 54$

EXTEND **23.** *LN*: $y = x + 1$; *KM*: $y = -x + 2$; Yes, diagonals are $\perp$ since slopes are 1 and -1. (Negative reciprocals);

$P\left(\frac{1}{2}, \frac{3}{2}\right)$

25. $x = 25$; $y = 45$ **27.** 108° **29.** 1.92 **31.** 108°

ALGEBRA AND GEOMETRY REVIEW **38.** $x < 7.5$ **39.** $x \le -9$ **40.** $x \ge 4$ **41.** $x \ge -50$ **42.** $x > \frac{7}{4}$ **43.** $x \ge 3$ **44.** 8 units **45.** A regular quadrilateral; convex

Lesson 4.2, pages 202–205

APPLY THE STRATEGY **13.** 12:24 PM

Patsi's distance	Kristine's distance

Patsi's distance	Kristine's distance

15. 12 arrangements. Let *A* represent Rocky Raccoon, let *B* represent Joe Bear, let *C* represent Jerri Giraffe, *D* represent Tabbie Cat, and *O* represent the globe.
Arrangements possible:

ABOCD	*BAOCD*	*CAOBD*	*DBOCA*
ABODC	*BAODC*	*CAODB*	*DBOAC*
ADOBC	*BCODA*	*CBODA*	*DCOAB*
ADOCB	*BCOAD*	*CBOAD*	*DCOBA*
ACODB	*BDOAC*	*CDOAB*	*DAOCB*
ACOBD	*BDOCA*	*CDOBA*	*DAOBC*

17. 90 cars

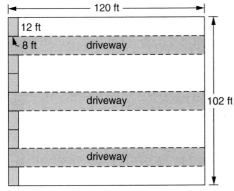

19. yes

Lesson 4.4, pages 210–215

TRY THESE **1.** $\overline{CB}$ **3.** $\overline{SA}$ **5.** $\overline{TZ}$
7. $A'(0, 4)$; $B'(-4, 4)$; $C'(-4, 1)$; $D'(0, 1)$

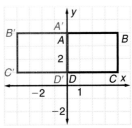

9. $W'(0, 2)$; $X'(-3, 4)$; $Y'(-4, 2)$; $Z'(-3, 0)$

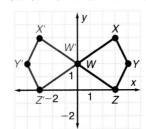

11.

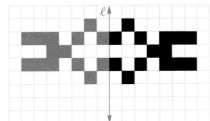

PRACTICE **13.** $\angle UHC$ **15.** $\angle UYQ$ **17.** $\angle HCT$

EXTEND
25.

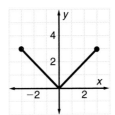

27.

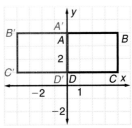

The *y*-axis is a line of symmetry.

31.

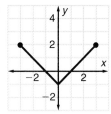

33.

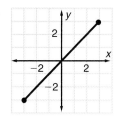

The *y*-axis is a line of symmetry.

The *y*-axis is not a line of symmetry.

ALGEBRA AND GEOMETRY REVIEW **41.** 576 **42.** −9 **43.** 25 **44.** 134 **45.** ≈ 8.5; (4, 4) **46.** 7; (2.5, 0) **47.** ≈ 3.2; (3.5, −2.5) **48.** $A'(-2, 3)$, $B'(-3, 5)$, and $C(-2, -2)$

Lesson 4.5, pages 218–223

TRY THESE

1.

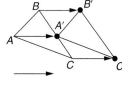

3.

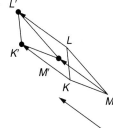

5. $A'(-4 + 3, -1 - 2) = A'(-1, -3)$
$B'(-4 + 3, 4 - 2) = B'(-1, 2)$
$C'(-3 + 3, 1 - 2) = C'(0, -1)$
$D'(-1 + 3, -1 - 2) = D'(2, -3)$
$E'(-1 + 3, -4 - 2) = E'(2, -6)$
$AE = \sqrt{(-4 - (-1))^2 + (-1 - (-4))^2} = \sqrt{9 + 9} = \sqrt{18};$
$A'E' = \sqrt{(-1 - 2)^2 + (-3 - (-6))^2} = \sqrt{9 + 9} = \sqrt{18}$

PRACTICE

9.

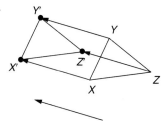

11.

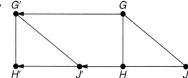

13. $A'(-4 - 2, -1 + 5) = A'(-6, 4)$

$B'(-4 - 2, 4 + 5) = B'(-6, 9)$
$C'(-3 - 2, 1 + 5) = C'(-5, 6)$
$D'(-1 - 2, -1 + 5) = D'(-3, 4)$
$E'(-1 - 2, -4 + 5) = E'(-3, 1)$
$CD = \sqrt{(-3 - (-1))^2 + (1 - (-1))^2} = \sqrt{4 + 4} = \sqrt{8} = 2\sqrt{2}; C'D' = \sqrt{(-5 - (-3))^2 + (6 - 4)^2} = \sqrt{4 + 4} = \sqrt{8} = 2\sqrt{2}; CD = C'D'$

17.

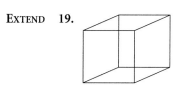

EXTEND **19.**

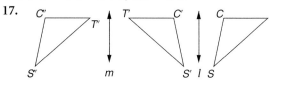

21.

THINK CRITICALLY **25.** By definition of reflection, $PA = AP'$ and $P'B = BP''$. By the Segment Addition Postulate, $AP' + P'B = AB = d$. Also, $PA + AP' + P'B + BP'' = PP''$. By substitution, $AP' + AP' + P'B + P'B = PP''$. So $2(AP' + P'B) = PP''$ and $2d = PP''$.

ALGEBRA AND GEOMETRY REVIEW **27.** $x = -5$ or $x = 15$ **28.** $t = -7$ or $t = 1$ **29.** no solution **30.** If a triangle has at least two angles that are congruent, then at least two sides of the triangle are congruent; True **31.** $K'(0, 2)$, $L'(2, 6)$, $M'(5, 4)$

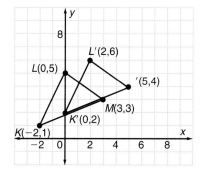

Lesson 4.6, pages 224–229

TRY THESE **1.** 90°, 180°, 270°, 360° **5.** $(2x)° = 55°$, $x = 27.5°$

7.

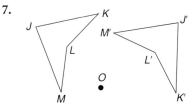

9. $(2x)° = 137°$, $x = 68.5°$ **11.** 90°

EXTEND **13.** counterclockwise **15.** 120°; clockwise
17. 144°

THINK CRITICALLY
19. $\vec{u} = (-2, 3)$ or $\vec{u}$ (2, −3); no; $\vec{u}$, and $\vec{v}$ are 90° or
−90° rotations. **23.** When n = positive odd number
$\overline{OP}$ is in Quadrant III, when n = positive even
number; $\overline{OP}$ is in Quadrant I; when n = negative odd
number, $\overline{OP}$ is in Quadrant III, when n = negative
even number, $\overline{OP}$ is in Quadrant I

ALGEBRA AND GEOMETRY REVIEW
24.

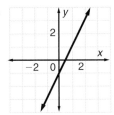

25.

26.

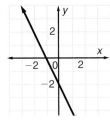

27. 123° **28.** 57° **29.** 57°
30. 123° **31.** $A(0, 0) \rightarrow$
$A'(0, 0)$; $B(0, -2) \rightarrow$
$B'(2, 0)$; $C(2, 1) \rightarrow$
$C'(-1, 2)$

31.

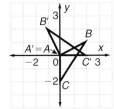

$A(0, 0) \rightarrow A'(0, 0)$
$B(0, -2) \rightarrow B'(2, 0)$
$C(2, 1) \rightarrow C'(-1, 2)$

Lesson 4.7, pages 231–236

TRY THESE **1.** $D''(-5, 0)$; $T''(-7, -5)$, $P''(-3, -2)$;
glide reflection **3.** $D''(0, -2)$, $T''(2, 3)$, $P''(-2, 0)$;
glide reflection **5a.** $P''(4, 1)$ **5b.** $P''(4, 1)$

PRACTICE **9.** $D''(3, 0)$, $T''(5, -5)$, $P''(1, -2)$; 180°
rotation **11.** $D''(0, 2)$, $T''(-2, 3)$, $P''(2, 0)$; glide

reflection **13a.** $P''(0, 3)$ **13b.** $P''(8, 3)$, $P''(3, -4)$

EXTEND **21.** B; 45° clockwise or 315° counterclockwise
23. reflection across y-axis; reflection across x-axis

THINK CRITICALLY **27a.** picking it up: reflected it over
his head; walking to vanglide; putting it on truck:
reflected it back down; pushing it back in the van: glide.
27b. pushing chair to back of truck: glide; picking it up:
reflection; up the sidewalk into front room; glide; down;
reflection.

ALGEBRA AND GEOMETRY REVIEW **30.** 7.4 **31.** 935
32. 383.25 **33.** if m∠1 + m∠2 ≠ 180°, then ∠1 and
∠2 do not form a linear pair; true **34.** $X''(-1, -3)$,
$Y''(3, 0)$, $Z''(6, -4)$

Chapter Review, pages 246–247

1. b **2.** d **3.** c **4.** a **5.** pentagon; concave; 145°
6. pentagon; convex; 100° **7.** hexagon; concave; 45°
8. hexagon; convex; 70° **9.** 15 **10.** $P'(1, 3)$, $Q'(0, 1)$,
$R'(4, -2)$ **11.** $A'(-1, -2)$, $B'(-2, 3)$, $C'(-5, 2)$;
$D((-4, -3)$ **12.** $W'(1, 3)$, $X'(-1, -2)$, $Y'(4, -3)$,
$Z'(3, 1)$ **13.** $P'(-4, 4)$, $Q'(-3, 2)$, $R'(-7, -1)$
14. $A'(3, -4)$, $B'(4, 1)$, $C'(4, 1)$, $D'(7, 0)$
15. $W'(-1, -1)$, $X'(1, -6$, $Y'(-4, -7)$, $Z'(-3, -3)$
16. $P'(-3, -1)$, $Q'(-1, 0)$, $R'(2, -4)$
17. $A'(-2, -1)$, $B'(3, -2)$, $C'(2, -5)$, $D'(-3, -4)$
18. $W'(1, -3)$, $X'(-1, 2)$, $Y'(4, 3)$, $Z'(3, -1)$
19. $P''(3, 3)$, $Q''(2, 1)$, $R''(6, -2)$; glide reflection across
the y-axis and 2 units right **20.** $A''(1, -2)$,
$B''(-3, -2)$, $C''(-2, -5)$, $D''(3, -4)$; reflection across
the line $y = -x$ **21.** $W''(1, -3)$, $X''(-1, 2)$, $Y''(4, 3)$,
$Z''(3, -1)$; 180° rotation about the origin
22. $P'(-4, -1)$ **23.** $P'(-2, -3)$ **24.** $P'(-5, 2)$
25. $P'(-2, 10)$ **26.** $P'(-1, 5)$ **27.** TR

Chapter 5 Triangles and Congruence

Lesson 5.1, pages 255–259

TRY THESE **1.** scalene; right **3.** scalene; obtuse
5. 123° **7.** m∠1 = 90°; m∠2 = 56°; m∠3 = 38°;
m∠4 = 128°

PRACTICE **9.** isosceles; acute **11.** isosceles; right
13. 102° **15.** m∠1 = 62°; m∠2 = 50°; m∠3 = 40°

EXTEND **17.** $x = 30$ **19.** $x = 22$ **21.** $x = 69°$
ALGEBRA AND GEOMETRY REVIEW **29.** 12 **30.** 45.9
31. 103.95 **32.** $t = \dfrac{d}{r}$ **33.** $p = \dfrac{l}{rt}$ **34.** $h = \dfrac{V}{lw}$
35. $x = \dfrac{c - b}{a}$ **36.** $b = y - mx$ **37.** $y = \dfrac{c - 9x}{b}$
38. In △KLM, m∠KLM = 42° and m∠LMK = 65°,
find m∠MKL = 73°

Lesson 5.2, pages 260–266

Try These **1.** 125° **3.** 235° **5.** 108°
7. m∠A = 38°; m∠Y = 38°; m∠Z = 104° **9.** Yes; a
12-sided regular polygon

Practice **11.** 138° **13.** 141° **17.** m∠P = 103°;
m∠L = 113°; m∠O = 34°; m∠W = 110°

Extend **19.** 540° **21.** 110° **23.** It makes no
difference where *P* is located in the interior of the
pentagon. If *P* is located on a side, between two
consecutive vertices, only four triangles are formed and
the sum of the angles with vertex *P* is only 180°.
Therefore, the sum of the measures of the angles of the
pentagon is 4(180) − 180 = 540°.

Think Critically **31.** 60, 90, 108, 120, 135, 140, 144,
150, 156, 160, 162, 165, 168, 170, 171, 172, 174, 175,
176, 177, 178, and 179

Algebra and Geometry Review **33.** ±7 **34.** ±14.1
35. ±20 **36.** ±12.2 **37.** ±12.6 **38.** no real
solution **39.** $m = 0$ **40.** $m = -2.3$ **41.** no slope
42. $m = \frac{2}{3}$ **43.** interior angle: 144°; exterior angle: 36°

Lesson 5.5, pages 275–285

Try These **1.** You are given in the figure that
∠DAC ≅ ∠BCA and ∠BAC ≅ ∠DCA. By the reflexive
property of congruence, $\overline{AC} \cong \overline{CA}$. Since the conditions
of the ASA Congruence Postulate are satisfied. △ABC ≅
△CDA.

Practice **7.** You are given in the figure that
$\overline{PQ} \cong \overline{ON}$, $\overline{QR} \cong \overline{OM}$, and that ∠PQR and ∠MON
are right angles. Since all right angles are congruent,
∠PQR ≅ ∠MON. Since the conditions of the SAS
Congruence Postulate are satisfied. △PQR ≅ △NOM.

Extend

17.

Statements	Reasons
1. $\overline{A'B} \parallel \overline{BA'}$; $\overline{AB} \parallel \overline{A'B'}$ $\overline{AB'} \cong \overline{A'B}$	1. Given
2. ∠B'AX ≅ ∠BA'X ∠AB'X ≅ ∠A'BX	2. Alternate Interior Angles Theorem
3. △AXB' ≅ △A'XB	3. ASA Congruence Postulate

19.

Statements	Reasons
1. $\overline{C'D'} \parallel \overline{CD}$; $\overline{C'D} \parallel \overline{D'C}$ $\overline{C'D} \cong \overline{CD'}$	1. Given
2. ∠CD'X ≅ ∠C'DX ∠D'CX ≅ ∠DC'X	2. Alternate Interior Angles Theorem
3. △D'CX ≅ △DC'X	3. ASA Congruence Postulate

Think Critically **23.** $R(4, 0)$

Algebra and Geometry Review **24.** −3, −2
25. −3, 4 **26.** −9, −2 **27.** −5, $\frac{3}{2}$ **28.** $\frac{-15 \pm \sqrt{193}}{2}$
29. $-\frac{2}{3}, \frac{1}{3}$ **30.** 1 **31.** $2\sqrt{5}$ **32.** $\sqrt{26}$
33.

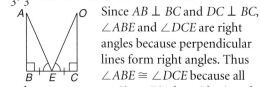

Since $AB \perp BC$ and $DC \perp BC$,
∠ABE and ∠DCE are right
angles because perpendicular
lines form right angles. Thus
∠ABE ≅ ∠DCE because all
right angles are congruent. Since *E* is the midpoint of
$\overline{BC}$, $\overline{BE} \cong \overline{EC}$ because of the definition of a midpoint.
You are also given ∠AEB ≅ ∠DEC. You have now
satisfied the hypothesis of the ASA Congruence
Postulate. Therefore, △ABE ≅ △DCE.

Lesson 5.6, pages 282–285

Apply the Strategy **11.** △ALO ≅ △GKN
13. △TSE ≅ △ARF; △DSE ≅ △NRF **15.** Reflect the
red overlay about line ℓ. **17. a.** yes **b.** 8 **c.** yes; 4
19. △GJE

Lesson 5.7, pages 288–293

Try These **1.** AAS **3.** HL **5.** AAS **7.** Since the
tower is perpendicular to the ground, ∠AXP and
∠DXP are right angles and are, therefore, congruent. It
is given that ∠XAP ≅ ∠XDP. By the reflexive property,
$\overline{PX} \cong \overline{PX}$. Therefore, △XAP ≅ △XDP, by AAS.

Practice **9.** AAS **11.** AAS **13.** ASA **15.** Since
quadrilateral *ABCD* is a square, ∠WDC and ∠YBA are
right angles and $\overline{CD} \cong \overline{AB}$. You are given that $\overline{CW} \cong$
$\overline{AY}$. Therefore, △WDC ≅ △YBA by HL.

Extend

19.

Statements	Reasons
1. $\overline{AE} \perp \overline{FC}$	1. Given
2. ∠FCA and ∠FCE are right angles	2. Perpendicular lines form right angles
3. $\overline{BF} \cong \overline{DF}$	3. Given
4. $\overline{CF} \cong \overline{CF}$	4. Reflexive property
5. △BFC ≅ △DFC	5. HL

25. Since the ladder doesn't change length, $AB \cong EF$.

Think Critically **27.** To assure that △EYW ≅
△FXZ, WX must equal YZ. If WX = YZ then WX + XY
= XZ + YZ by the addition property. By the Segment
Addition Postulate, WX + XY = WY and XY + YZ =
XZ. So, WY = XZ by substitution or $WY \cong XZ$. It is
given that ∠W and ∠Z are right angles and ∠E ≅ ∠F.
Therefore, △EYW ≅ △FXZ by ∠A.

Algebra and Geometry Review **29.** 49 **30.** 343

31. 128 **32.** $\frac{1}{125}$ **33.** $\frac{1}{4}$ **34.** $\frac{127}{4}$ **35.** If a line joining two distinct points lies in a plane, then the two points lie in that plane; true. **36.** If two angles are supplementary, then they form a linear pair; false.

37.

Statements	Reasons
1. $\angle X \cong \angle Y$	1. Given
2. $\overline{ML}$ bisects $\angle XLY$	2. Given
3. $\angle XLM \cong \angle YLM$	3. Definition of angle bisector
4. $\overline{LM} \cong \overline{LM}$	4. Reflexive property
5. $\triangle XLM \cong \triangle YLM$	5. AAS

38. $67\frac{1}{2}°$ **39.** $\angle 1$ and $\angle 3$, $\angle 2$ and $\angle 4$, $\angle 8$ and $\angle 6$, or $\angle 7$ and $\angle 5$ **40.** $\angle 2$ and $\angle 3$ or $\angle 7$ and $\angle 6$

Lesson 5.9, pages 297–303

Try These **1.** HL; $\overline{PQ} \cong \overline{NT}$, $\overline{QR} \cong \overline{TM}$, $\overline{PR} \cong \overline{NM}$, $\angle P \cong \angle N$, $\angle Q \cong \angle T$, $\angle R \cong \angle M$ **3.** Since $\angle LDO \cong \angle HOD$ and $\angle LDH \cong \angle HOL$, $m\angle LDO = m\angle HOD$ and $m\angle LDH = m\angle HOL$. By the addition property $m\angle LDO + m\angle LDH = m\angle HOD + m\angle HOL$. The Angle Addition Postulate tells you that $m\angle LDO + m\angle LDH = m\angle HDO$ and $m\angle HOD + m\angle HOL = m\angle LOD$. So, by substitution, $m\angle HDO = m\angle LOD$ or $\angle HDO \cong \angle LOD$. By reflexive property $\overline{DO} \cong \overline{DO}$. Therefore, $\triangle HDO \cong \triangle LOD$ by ASA. Thus $\overline{HD} \cong \overline{LO}$ by CPCTC.

7. 3.9 **9.** 70°

Practice **11.** SAS; $\overline{DC} \cong \overline{SW}$, $\overline{CE} \cong \overline{WT}$, $\overline{DE} \cong \overline{ST}$, $\angle C \cong \angle W$, $\angle D \cong \angle S$, $\angle E \cong \angle T$

17. 39° **19.** 51°; 51°

Think Critically

29. $AE = \sqrt{(5-3)^2 + (3-5)^2} = 2\sqrt{2}$
$DE = \sqrt{(5-1)^2 + (3-1)^2} = 2\sqrt{5}$
$AD = \sqrt{(3-1)^2 + (5-1)^2} = 2\sqrt{5}$
$BC = \sqrt{(9-7)^2 + (5-1)^2} = 2\sqrt{5}$
$EC = \sqrt{(7-5)^2 + (1-3)^2} = 2\sqrt{2}$
$BE = \sqrt{(9-5)^2 + (5-3)^2} = 2\sqrt{5}$

31. $\angle D \cong \angle B$; CPCTC

Algebra and Geometry Review **34.** $7\frac{1}{3}$ **35.** $-\frac{7}{3}$ **36.** $-\frac{2}{3}$ **37.** -6 **38.** 2 **39.** -1 **40.** perpendicular **41.** perpendicular **42.** parallel **43.** neither

Chapter Review, pages 304–305

1. e **2.** c **3.** d **4.** a **5.** 6 **6.** 108° **7.** 119° **8.** 61° **9.** 115° **10.** 30 **11.** 150° **12.** Yes; Justifications will vary **13.** Yes; ASA

14.

Statements	Reasons
1. $\overline{RT} \cong \overline{TV}$; $\angle S \cong \angle U$	1. Given
2. $\angle RTS \cong \angle VTU$	2. Vertical angles are congruent
3. $\triangle RST \cong \triangle VUT$	3. AAS

15.

Statements	Reasons
1. $\overline{SU} \cong \overline{TW}$	1. Given
2. $SU = TW$	2. Definition of congruent segments
3. $SU + UT = TW + UT$	3. Addition property
4. $SU + UT = ST$ $TW + UT = UW$	4. Segment Addition Postulate
5. $ST = UW$	5. Substitution property
6. $\overline{ST} \cong \overline{UW}$	6. Definition of congruent segments
7. $\overline{RS} \cong \overline{VW}$; $\angle S \cong \angle W$	7. Given
8. $\triangle RST \cong \triangle VWU$	8. SAS

16.

Statements	Reasons
1. $\angle EFK$ and $\angle HKF$ are right angles	1. Given
2. $\triangle KEF$ and $\triangle FHK$ are right triangles	2. Definition of right triangle
3. $\overline{EK} \cong \overline{HF}$	3. Given
4. $\overline{KF} \cong \overline{KF}$	4. Reflexive property
5. $\triangle KEF \cong \triangle FHK$	5. HL

17. $\triangle BCF \cong \triangle NFC$; SAS

18.

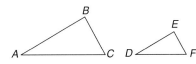

19.

Statements	Reasons
1. $\angle W$ and $\angle E$ are right angles	1. Given
2. $\angle W \cong \angle E$	2. All right angles are congruent
3. $\angle NOW \cong \angle TOE$	3. Given
4. $\overline{NW} \cong \overline{TE}$	4. Given
5. $\triangle NOW \cong \triangle TOE$	5. AAS
6. $\angle WNO \cong \angle ETO$	6. CPCTC

20.

Statements	Reasons
1. $\overline{FS} \cong \overline{YP}$; $\angle FSP \cong \angle YPS$	1. Given
2. $\overline{SP} \cong \overline{SP}$	2. Reflexive property
3. $\triangle FSP \cong \triangle YPS$	3. SAS
4. $\overline{FP} \cong \overline{SY}$	4. CPCTC

21.

Statements	Reasons
1. $\overline{HM} \cong \overline{HM}$	1. Reflexive property
2. $\angle O \cong \angle E$	2. Given
3. $\overline{HO} \parallel \overline{ME}$	3. Given
4. $\angle OHM \cong \angle EMH$	4. Alternate Interior Angles Theorem
5. $\triangle OHM \cong \triangle EMH$	5. AAS
6. $\overline{HE} \cong \overline{MO}$	6. CPCTC

Chapter 6 A Closer Look at Triangles

Lesson 6.2, pages 316–323

TRY THESE **1.** right angle **3.** Perpendicular Bisector Theorem **5.** yes **7.** no **9.** no **11.** E is four units from $\overrightarrow{BC}$ and four units from $\overrightarrow{BA}$. So, $\overrightarrow{BD}$ bisects $\angle ABC$ by the converse of the Angle Bisector Theorem.

PRACTICE

19. no **21.** yes **23.** yes
EXTEND **27.** $y = \frac{1}{3}x + 6$ **29.** 3 **31.** $(8, 0)$ and $(0, -12)$ **33.** $y = -\frac{2}{3}x - \frac{10}{3}$
THINK CRITICALLY **39.** $(-4, -4)$ **41.** yes
43. $\overline{AB} \parallel \overline{WX}$ or $\overline{AB}$ is collinear with $\overline{WX}$
ALGEBRA AND GEOMETRY REVIEW **45.** $(x - 4)(x - 3)$
46. $(2x - 5)(x + 3)$ **47.** $5(1 + 2x)(1 - 2x)$

Lesson 6.3, pages 324–330

TRY THESE **1.** yes; $\angle A$, $\angle C$ **3.** no **5.** no **7.** 40°, 100° **9.** 60°, 60° **11.** 39.5°, 39.5° **13.** 0; 5.5
15. $0; \frac{1}{2}j$ **17.** $6r; 0$ **19.** 14
PRACTICE **23.** no **25.** yes; $\angle X$, $\angle Z$ **27.** no
29. $-4; 0$ **31.** $j, 0$ **33.** $-8s; 0$ **35.** 45°, 90°
37. 45°, 45° **39.** 54, 54, 36; $\angle P$ and $\angle R$
ALGEBRA AND GEOMETRY REVIEW **58.** 65°; Alternate interior angles are congruent. **59.** 50°; Use corresponding angles to show m $\angle FED = 65°$ and vertical angles to show m $\angle FDE = 65°$. So m$\angle EFD$ is $180° - 130° = 50°$. **60.** $\angle FDE \cong \angle EAB$ because corresponding angles are congruent. Thus m$\angle FED = 65°$. $\angle FDE \cong \angle GDC$ because vertical angles are congruent. Thus, m$\angle FDE = 65°$. Since $\angle FED \cong \angle FDE$, $\overline{EF} \cong \overline{FD}$ by the converse of the Base Angles Theorem. **61.** $\frac{5x - 6y}{3}$ **62.** 84.8; 85; 85

Lesson 6.5, pages 336–341

TRY THESE **1.** $\overline{XY}$, $\overline{XZ}$, $\overline{YZ}$ **3.** $XZ = 2$, Triangle Midsegment Theorem **5.** $ZY = 2.5$, Triangle Midsegment Theorem **7.** $XY = 3.5$, Triangle

Midsegment Theorem **9.** 135° **11.** $AT = 6$; $PY = 12$ **13.** Use slope formula to show $\overline{EF}$ and $\overline{AC}$ have slope equal to $\frac{1}{2}$. Therefore, $\overline{EF} \parallel \overline{AC}$. Use distance formula to find $EF = 2\sqrt{5}$ and $AC = 4\sqrt{5}$. So, $EF = \frac{1}{2}AC$. **15.** Since $\overline{FD}$ and $\overline{AB}$ are both vertical, they are parallel. Since $FD = 4$ and $AB = 8$, $FD \frac{1}{2}AB$.
PRACTICE **19.** 12.5 **21.** 18 **23.** 12.5 **25.** 24°
27. 22° **29.** 158° **31.** $n = 21$; $PQ = 30$; $ST = 60$
33. $n = 16$; m$\angle D = $ m$\angle NMF = 82°$; m$\angle DMN = 98°$
35. 15; Triangle Midsegment Theorem **39.** Use distance formula to find $DE = t$ and $AC = 2t$. $DE = \frac{1}{2}AC$.
ALGEBRA AND GEOMETRY REVIEW **45.** $x = -2$, $x = 8$
46. $y = -10$, $y = 8$ **47.** $w = -2$, $w = 5$ **48.** no; Corresponding angles are not congruent.
49. m$\angle A = 96°$, m$\angle B = 74°$, m$\angle C = 10°$, obtuse, or m$\angle A = 76°$, m$\angle B = 84°$, m$\angle C = 20°$, acute.

Lesson 6.6, pages 342–345

APPLY THE STRATEGY **15.** The figures that are out-a-shapes are divided in such a way that the two shapes formed inside the main shape, are congruent to each other. In this figure, a trapezoid is divided into a square and triangle. Since these created shapes are not congruent, this figure is not an out-a-shape.

Lesson 6.9, pages 352–356

TRY THESE **1.** largest: $\angle C$, smallest: $\angle B$ **3.** largest: $\angle G$, smallest: $\angle I$ **5.** longest: $\overline{MO}$, shortest: $\overline{NO}$
7. longest: $\overline{TU}$, shortest: $\overline{ST}$ **9.** yes, $5 + 6 > 7$
11. yes, $15 + 15 > 15$ **13.** yes, $3.5 + 6.5 > 7.5$
15. $3 < AC < 21$ **17.** $1\frac{3}{4} < BC < 9\frac{1}{4}$
19. $9.5 - \sqrt{3} < AB < 9.5 + \sqrt{3}$
PRACTICE **23.** largest: $\angle E$, smallest: $\angle F$ **25.** largest: $\angle K$, smallest: $\angle J$ **27.** longest: $\overline{QR}$, shortest: $\overline{PR}$
29. longest: $\overline{WX}$, shortest: $\overline{VW}$ **31.** no, $1 + 2 = 3$
33. yes, $3\frac{3}{7} + 5 > 5\frac{5}{8}$ **35.** no, $5\sqrt{2} + 10\sqrt{2} < 18\sqrt{2}$
37. $5 < AB < 31$ **39.** $5\sqrt{13} - 5 < CA < 5\sqrt{13} + 5$
41. $|s - r| < BC < s + r$ **43.** $\angle S$, $\angle R$, $\angle Q$. The smallest angle lies opposite the shortest side and the largest angle lies opposite the longest side.
ALGEBRA AND GEOMETRY REVIEW **48.** 108°
49. acute **50.** obtuse **51.** straight **52.** $61\frac{1}{3}°$
53. $-25r + 10s - 2sr$ **54.** $y > 2$ or $y < -3$

Chapter Review, pages 364–365

1. d **2.** c **3.** a **4.** b **5.** a circle **6.** an angle bisector   **7.** a perpendicular bisector   **8.** R is equidistant from the two axes   **9.** $SD = SE = \sqrt{10}$ **11.** $m\angle X = m\angle Y = 71.5°$ **12.** $(5, -1); (1, 3)$ **13.** $(1, 3); (1, -1)$ **14.** $(0, 0)$ and $(3, -2); \overline{AD}$ and the midsegment both have slope $-\frac{2}{3}$.   **15.** $AS = 4$ units and midsegment $= 2$ units   **16.** Since m intersects $\overline{AC}$ at D, A and C must be different sides of m. Assume that m doesn't intersect $\overline{AB}$ or $\overline{BC}$. Since A and B are on the same side of m and B and C are on the same side of m then A and C must be on the same side of m which contradicts the given information. Therefore, m must intersect either $\overline{AB}$ or $\overline{BC}$.   **17.** longest side: $\overline{MN}$; shortest side: LM **18.** $4 < y < 28$ **19.** $m\angle A > m\angle D$ **20.** $BC > EF$

Chapter 7 Quadrilaterals

Lesson 7.2, pages 377–383

TRY THESE   1. $73°$ **3.** $73°$ **5.** 2.9 units   **9.** $(4, 7)$ **11.** $(2j - 2h, 2k)$

PRACTICE   13. 10 units   **15.** $111°$ **17.** false   **19.** false   **21.** true   **23.** false   **25.** true   **27.** not enough information   **29.** not enough information   **31.** $45°$ **33.** $(14, 0)$ **35.** $(b - 2a, 2c)$ assuming $2a < b$ **37.** $m\angle A = 106°$, $m\angle C = 106°$, $m\angle D = 74°$, $CD = 1.35$m, $AD = 0.96$ m   **39.** Opposite sides of a parallelogram are congruent.   **41.** Opposite angles of a parallelogram are congruent.   **43.** Consecutive angles of a parallelogram are supplementary.

EXTEND   45. $x = 50$, $y = 32$, $z = 98$ **47.** 5 **49.** 11 **53.** $AE = \sqrt{(r + 5 - 0)^2 + (t - 0)^2}$, $CE = \sqrt{[(2r + 2s) - (r + s)]^2 + (2t - t)^2}$

ALGEBRA AND GEOMETRY REVIEW   59. 17 **60.** 25 **61.** 3.75 **62.** k^{12} **63.** $-6a^7$ **64.** $-4a$ **65.** $\frac{9g^4}{4h^8}$ **66.** $z^2 + 10z + 25$ **67.** $144°$ **68.** 15

Lesson 7.3, pages 384–389

TRY THESE   1. Theorem 7.6   **3.** Theorem 7.8   **5.** Yes, Theorem 7.9   **7.** Yes, Theorem 7.6   **9.** true   **11.** false   **13.** yes

PRACTICE   17. Yes, Theorem 7.5   **19.** Yes, Theorem 7.6   **21.** no   **23.** The diagonals of $RUST$ bisect each other. Thus, by theorem 7.8, $RUST$ is a parallelogram. Since $RUST$ is a parellelogram, $\overline{RU} \parallel \overline{TS}$ by definition.   **29.** yes, Theorem 7.9   **31.** yes, definition of parallelogram

EXTEND   33. $x = 14$, $y = 35\frac{2}{3}$ **35.** no   **37.** yes, Theorem 7.6   **39.** no

ALGEBRA AND GEOMETRY REVIEW   50. $51°$ **51.** $51°$ **52.** $39°$ **53.** $39°$ **54.** $141°$ **55.** $141°$ **56.** $P'(5, 0)$, $Q'(8, -1)$, $R'(6, -5)$

Lesson 7.4, pages 391–396

TRY THESE   1. rectangle, $2,800$ ft   **3.** $58°, 6.9$ **5.** $2.15, 90°$ **7.** $29°, 29°, 8$ **9.** rhombus   **11.** square

PRACTICE   15. $28°, 4.41$ **17.** $17°, 35.2$ **19.** $90°, 90°$ **21.** square   **23.** rhombus

EXTEND   27. $(r + t, s)$ **29.** $C(0, -r), D(-r, 0)$ **31.** rectangle   **35.** $41°$; Since $URST$ is a rhombus, consecutive angles are supplementary and each diagonal bisects a pair of opposite angles. Therefore, $m\angle URS = 180 - 98 = 82°$ and $m\angle URZ = \frac{1}{2}(82) = 41°$.   **37.** 4.88; You can prove $ERTF$ is a parallelogram and use opposite sides of a parallelogram are congruent to prove $RT = EF = 4.88$.   **39.** $90°$; Since $AEFD$ is a square, $\angle AEF$ is a right angle. Since $\angle AEF$ and $\angle UER$ are a linear pair they are supplementary. Therefore, $\angle UER$ is a right angle.

ALGEBRA AND GEOMETRY REVIEW   45. $-\frac{1}{8}$ **46.** $\frac{7}{4}$ **47.** 0 **48.** 1 **49.** $y \quad 2x + 1.5$ **50.** $y = x^2 - 3$

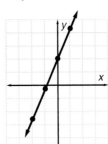
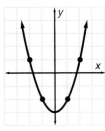

51. $y = 1.5x - 3$ **52.** $y = 2x^2 + 1$

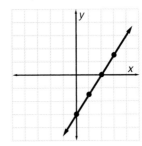
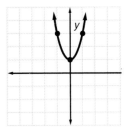

53. parallelogram   **54.** $70°$ **55.** $-4 < x < 41$

Lesson 7.5, pages 397–403

TRY THESE **1.** 6

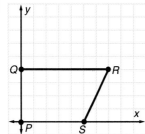

3. $5\frac{1}{2}$

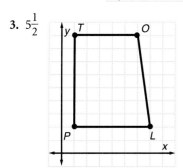

5. m∠A = 106°, m∠B = 106°, m∠C = 74°, EF = 33.2
7. In an isosceles trapezoid the legs are congruent and in a non-isosceles trapezoid the legs have different lengths. Student lists will vary. **9.** YA = 5.32, XZ = 6.77

PRACTICE **11.** 8 **13.** 8 **15.** m∠W = 109°, m∠H = 71°, m∠E = 71°, midsegment: 2.75 **17.** Yes, m∠H = 56°, Base angles are congruent. **19.** No, Base angles are not congruent. **21.** Yes, Base angles are congruent.

EXTEND **23.** 126° **25.** 54° **27.** 39° **29.** 39° **31.** 6.4 **33.** 6 **35.** 2.17 **37.** 6.29 **39.** 4.12 **41.** $(-t, 0), (-r, s)$ **43.** $AB = CD = \sqrt{(t-r)^2 + s^2}$

ALGEBRA AND GEOMETRY REVIEW **54.** 720° **55.** 1080°
56. 1800° **57.** 3240° **58.** $x = -2, y = 5$
59. $x = 0, y = -2$ **60.** no solution **61.** Yes, Theorem 7.18

Lesson 7.6, pages 404–409

TRY THESE **1.** $\overline{CO} \perp \overline{HW}$, Theorem 7–19
3. 5.95, 90° **5.** 3.1, 80° **9.** $AB = BC = 10$ and $CD = AD = 2\sqrt{5}$

PRACTICE **11.** 4.98, 62°
15. $RA = ER = \sqrt{65}$ and $AS = SE$ 5
17. $WO = OL = 7, LD = DW = \sqrt{85}$ **19.** $C(-a, 0)$, $D(0, n)$ where $n < 0$ **21.** (0, 0)

EXTEND **23.** 1704 ft **25.** 1347 ft **27.** 2464 ft
29. This proof is a generalization of the proof in Exercise 28. The proof is done in the same way with the

general expression $\frac{1}{n}$ replacing $\frac{1}{4}$. **31.** Since reflection preserves segment length, $\overline{UT'} \cong \overline{ST'}$. So, quadrilateral $RST'U$ has congruent sides $\overline{UT'}$ and $\overline{ST'}$ and congruent sides $\overline{UR}$ and $\overline{SR}$. Therefore quadrilateral $URST'$ is a kite. **33.** $AB \cong BC$ and $CD \cong DA$ or $\overline{AB} \cong \overline{DA}$ and $\overline{BC} \cong \overline{CD}$

ALGEBRA AND GEOMETRY REVIEW **36.** $-6, \frac{8}{3}$
37. $-3, 4$ **38.** $-\frac{4}{5}, \frac{12}{5}$ **39.** $-\frac{28}{3}, \frac{14}{3}$ **40.** $-1, \frac{3}{5}$
41. no solution
42.

Statements	Reasons
1. $\ell \parallel m$	1. Given
2. ∠CBE and ∠FEB are supplementary	2. Same-Side Interior Angles Theorem
3. m∠CBE + m∠FEB = 180°	3. Definition of supplementary
4. ∠ABW ≅ ∠CBE; ∠DEY ≅ ∠FEB	4. Vertical Angles Theorem
5. m∠ABW = m∠CBE; m∠DEY = m∠FEB	5. Definition of congruent angles
6. m∠ABW + m∠DEY = 180°	6. Substitution property
7. ∠ABW and ∠DEY are supplementary	7. Definition of supplementary

43.

Statements	Reasons
1. $\ell \parallel m$	1. Given
2. ∠WBC ≅ ∠DEY supplementary	2. Same-Side Interior Angles Theorem

44. 14.5 **45.** 42°

Lesson 7.7, pages 412–415

APPLY THE STRATEGY **16.** The angle measures suggest that the figure is a parallelogram, but the unequal measures of opposite sides contradicts this information.
17. The angle measures suggest that the figure is an equilateral triangle, but the sides are not of equal length.

Lesson 7.9, pages 419–425

TRY THESE **1.** right rectangular pyramid **3.** oblique hexagonal prism

17. 5 **19.** 4 **21.** 3

PRACTICE **25.** oblique triangular pyramid
27.

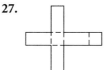

29.

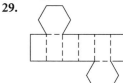

31.

hex

33. 9 **35.** 8 **37.** 11

EXTEND **41.** six

43.

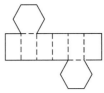

THINK CRITICALLY **49.** $S(n) = n$

ALGEBRA AND GEOMETRY REVIEW **51.** none

52. $1, -\frac{5}{2}$ **53.** none **54.** $7, -7$ **55.** none

56. $\frac{-7 \pm 61}{6}$ **57.** 121° **58.** 59° **59.** 59° **60.** 59°

CHAPTER REVIEW, PAGES 426–427 **1.** b **2.** a **3.** c
4. square **5.** trapezoid **7.** parallelogram
7. trapezoid **8.** 74° **9.** 106° **10.** 4.2 **11.** 7.6

12. slope of $\overline{PR}$ is $\frac{10 - (-7)}{1 - 7} = -\frac{17}{6}$

slope of $\overline{AL}$ is $\frac{9 - (-6)}{9 - (-1)} = \frac{3}{2}$

$PR = \sqrt{(1 - 7)^2 + (10 + 7)^2} = 5\sqrt{13}$

$AL = \sqrt{(9 + 1)^2 + (9 + 6)^2} = 5\sqrt{13}$

Since the diagonals of parallelogram $PARL$ are congruent but not perpendicular, parallelogram $PARL$ is a rectangle.

13. slope of $\overline{PR}$ is $\frac{8 - (-2)}{-5 - 5} = -1$

slope of $\overline{AL}$ is $\frac{-3 - 3}{0 - 6} = 1$

$PR = \sqrt{(-5 - 5)^2 + (8 + 2)^2} = 10\sqrt{2}$

$AL = \sqrt{(-3 - 3)^2 + (0 - 6)^2} = 6\sqrt{2}$

Since the diagonals of parallelogram $PARL$ are perpendicular but not congruent, parallelogram $PARL$ is a rhombus.

14. slope of $\overline{PR}$ is $\frac{0 - 8}{2 - 6} = 2$

slope of $\overline{AL}$ is $\frac{2 - 6}{8 - 0} = -\frac{1}{2}$

$PR = \sqrt{(2 - 6)^2 + (0 - 8)^2} = 4\sqrt{5}$

$AL = \sqrt{(8 - 0)^2 + (2 - 6)^2} = 4\sqrt{5}$

Since the diagonals of parallelogram $PARL$ are congruent and perpendicular, parallelogram $PARL$ is a square.

15. slope of $\overline{PR}$ is $\frac{4 - 4}{1 - (-7)} = 0$ (horizontal line)

slope of $\overline{AL}$ is $\frac{10 - (-2)}{-3 - (-3)} =$ Undefined (vertical line)

$PR = \sqrt{(1 + 7)^2 + (4 - 4)^2} = 8$

$AL + \sqrt{(-3 + 3)^2 + (10 + 2)^2} = 12$

Since the diagonals of parallelogram $PARL$ are perpendicular but not congruent, parallelogram $PARL$ is a rhombus.

16. yes **17.** yes **18.** no **19.** yes
20. $\sqrt{65.25} \approx 8.08$ **21.** no **22.** 62° **23.** 99°

24. 106° **25.** 31 **26.** no; Theorem 7.4 **27.** yes; defintion of parallelogram **28.** determines the figure
29. regular hexagon; rectangle
31.

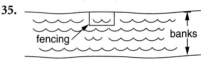

Chapter 8 Perimeters and Areas of Polygons

Lesson 8.1, pages 435–440

TRY THESE **1.** false; justifications will vary **3.** 43.4 cm; 98.8 cm² **5.** $4\sqrt{11}$ units; 11 units² **7.** 56 m; 160 m² **9.** 60 in; 125 in² **11.** $(x + 5)(x + 2) = x^2 + 7x + 10$ **13.** 70 ft **15.** $683.11

PRACTICE **17.** 40 in.; 64 in.² **19.** 168 m; 1764 m²
21. 58 cm; 168 cm² **23.** 48 in.; 44 in.² **25.** $x(x + 3) = x^2 + 3x$ **27.** $(3x + 8)(x + 3) = 3x^2 + 17x + 24$
29. $420 **31.** Divide the area by the known dimension

EXTEND **33.** $4(x + 3)$ or $4x + 12$

35.

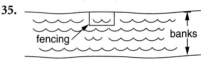

37.

width = $\frac{a}{2}$

a

THINK CRITICALLY **43.** 52″; explanations will vary

ALGEBRA AND GEOMETRY REVIEW **45.** 52 in.; 129 in.²
46. $-\frac{2}{5}, \frac{5}{2}, -\frac{5}{2}$ **47.** $\frac{5}{2}, 0.41, \frac{2}{5}, 0.3, -\frac{2}{5}, -\frac{5}{2}$

Lesson 8.2, pages 441–447

TRY THESE **1.** 44 ft² **3.** 244 in.² **5.** 27.5 cm²
9. 81 ft.²

PRACTICE **11.** 48 m² **13.** 21 in.² **15.** 28 cm²
17. 144 ft² **19.** 19 cm **21.** 31 m **25.** 220 ft²
27. 440 ft² **29.** $16\sqrt{2}$ units and $8\sqrt{2}$ units **31.** 103.5 ft² **33.** 4 in.; 16 in. **35.** 84; they are the same

ALGEBRA AND GEOMETRY REVIEW **38.** $y = -12x + 6$
39. 28° **40.** 60° **41.** sometimes **42.** never
43. 228 **44.** 60 in. **45.** 192 cm²

Lesson 8.3, pages 448–453

TRY THESE   **1.** 51 ft^2 **3.** 32 in.^{2} **5.** 28 cm^2
7. 29 units^2 **9.** 4 cm

PRACTICE   **11.** 90 cm^2 **13.** 55 ft^2 **15.** $1540.71
17. 12 units^2 **19.** 8 units^2
EXTEND   **23.** 6.5 cm and 15.5 m   **25.** $A = \frac{1}{2}h(b_1 + b_2)$
27. $A = \frac{1}{4}h(b_1 + m)$

ALGEBRA AND GEOMETRY REVIEW   **33.** 111°   **34.** −9, 2
35. rational   **36.** $-\frac{1}{2}$ **37.** 28 cm   **38.** 162 cm^2

Lesson 8.6, pages 462–465

APPLY THE STATEGY   **15.** 2880 feet^2 **17.d.** Using the
Trapezoidal Rule, the area is 84 units^2. The area
produced by the Monte Carlo method should be close
to 84 units^2.

Lesson 8.7, pages 466–471

TRY THESE

1. $x = 15$ cm
      perimeter = 36 cm
             area = 54 cm^2
3. $z = 8.9$ cm
      perimeter = 28.9 m
             area = 35.6 m^2
5. 10   **7.** 13.9
9. $x = 15$ cm
      $y = 17$ cm

PRACTICE

11. $x = 17$ cm
      perimeter = 40 cm
             area = 60 cm^2
13. $z = 3.3$ cm
      perimeter = 8.0 m
             area = 2.7 m^2
15. 8.2   **17.** 16   **19.** $x = 10$ in.;   **21.** 187 m $y = 18$ in.

EXTEND   **23.** When $p = 5$ and $q = 2$, the triple is 20,
21, 29. This is a primitive triple because 20, 21, and 29
have no common factors.
THINK CRITICALLY   **29.** $A = \frac{s^2\sqrt{3}}{4}$
ALGEBRA AND GEOMETRY REVIEW
31. 36°   **33.** $\frac{5}{2}$

Lesson 8.8, pages 474–480

TRY THESE
1. It is a triangle. It is not a right triangle.   **3.** It is not
a triangle.   **5.** It is a triangle. It is a right triangle.   **7.**
acute   **9.** right   **11.** obtuse   **13.** 127.3 ft   **15.** acute
PRACTICE

17. It is a triangle. It is a right triangle.   **19.** It is not a
triangle.   **21.** right   **23.** obtuse   **25.** acute   **27.** She
should move the tape closer together.

EXTEND
29. no   **31.** no   **33.** 90°

THINK CRITICALLY
35. always   **39.** no; A triangle that has three equal
sides is an equilateral triangle. All three angles of an
equilateral triangle are 60° and cannot be changed.

ALGEBRA AND GEOMETRY REVIEW
41. mean : 85.8; median : 87; mode : 87   **42.** 115°
43. 140°   **44.** 0.099999, 14%, 0.2, $\frac{1}{4}$, 1^8, 1.3, 28
45. $x = -5, 9$ **46.** 210 cm^2

CHAPTER REVIEW, PAGES 484–485
1. c   **2.** a   **3.** e   **4.** b   **5.** d   **6.** 144.5 m^2; 51 m
7. 196 ft^2; 56 ft   **8.** 78 in.^{2}; 44 in.   **9.** 59 cm^2; 42 cm
10. 82 ft^2 **11.** 2000 in.   **12.** 130 m^2 **13.** 75 cm^2
14. 116 m^2 **15.** 192.5 ft^2 **16.** 42.5 cm; 47.5 cm
17. 18.75 units^2 **18.** 70.5 units^2 **19.** 48.5 units^2
20. 48.5 units^2 **21.** 100 units^2 **22.** 26 units^2 **23.**
97 units   **24.** 5 units   **25.** right   **26.** no triangle
27. acute   **28.** obtuse   **29.** no triangle   **−30.** Fido
barks; Law of Detachment   **31.** If an animal is a dog,
then it has fur; Law of Syllogism.

Chapter 9 Similarity

Lesson 9.3, pages 501–509

TRY THESE   **1.** yes; $\angle F \cong \angle C$, $\angle D \cong \angle A$, $\angle E \cong \angle B$;
and $\frac{FD}{CA} = \frac{DE}{AB} = \frac{FE}{CB} = \frac{2}{1}$; $\triangle FDE \approx \triangle CAB$.
7. $\frac{0.75}{x \text{ mi}} = \frac{1 \text{ in.}}{140 \text{ mi}}$, 105 mi
PRACTICE   **9.** No; corresponding sides are not
proportional   **11.** scale factor: 5 : 4; $\frac{AB}{ML} = \frac{15}{12} = \frac{5}{4}$
13. scale factor: 4 : 1; $\frac{MP}{SR} = \frac{3+1}{1} = \frac{4}{1}$ **15.** yes; The
two bills are similar with a scale factor of approximately
12 : 5. The play money is clearly smaller than the actual
currency.   **17.** sometimes   **19.** always   **21.** always
23. sometimes   **25.** $BE = 4$ in; $BD = 10$ in;
m$\angle A = 53°$; m$\angle DBC =$ m$\angle ABE = 37°$ **27.** VZ
6 cm; $WY = 12$ cm; m$\angle W = 45°$; m$\angle Y =$ m$\angle Z = 26°$;
m$\angle V = 109°$ **29.** 96 ft^2

31. $x = 32$ in.   **33.** $x = 9$ cm
      $y = 30$ in. $y = 4.5$ cm

EXTEND   **35.** $x = 17$ **37.** No; only if the original is a
square   **39.** $x = 12$ m, $y = 9$ m

THINK CRITICALLY   **41.** yes, the A series rectangles are

similar. The angle measures of any rectangle are all 90°. Also, the ratio of the lengths of any two rectangles is equal to the ratio of the widths of those rectangles.
43. 16 cm

ALGEBRA AND GEOMETRY REVIEW

45. 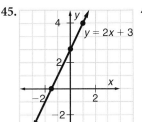 **46.**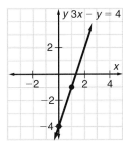

48. 9 **49.** $\frac{34}{25}$ **50.** 6 **52.** $\pm 4\sqrt{5}$ **53.** $\pm 7\sqrt{2}$
54. $x = 19$ or -11 **55.** $BC = 6$ cm; $LM = 3$ cm; $MN = 7.5$ cm; $NJ = 4.5$ cm; $m\angle B = m\angle K = 152°$; $m\angle L = 34°$; $m\angle D = 160°$; $m\angle N = 160°$; $m\angle A = 34°$.

Lesson 9.5, pages 515–523

TRY THESE

1. Scale factor: $\frac{1}{4}$ **5.** 3 : 2
   Perimeter ratio: $\frac{1}{4}$
   Area ratio: $\frac{1}{16}$

7. Students should draw a right triangle with a height of $9\sqrt{2}$ cm (≈ 12.7 cm) and a base of $15\sqrt{2}$ cm (≈ 21.2 cm).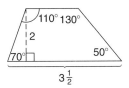

9. The student should draw a trapezoid with the following measurements:

PRACTICE

13. scale factor = 9 : 2
   ratio of perimeters = 9 : 2
   ratio of areas = 81 : 4
17. Approximately $340 **19.** The scale factor needed to produce a garden with 480 ft² is $\sqrt{2}$: 1. This is the square root of the ratio of the areas. Using this scale factor, the garden will measure 16.97 × 28.28 ft.
EXTEND **23.** 132 **29.** 44 cm², 99 cm² **31.** 5\2

ALGEBRA AND GEOMETRY REVIEW

32. $\frac{1}{4}$ **33.** $\frac{1}{2}$ **34.** 0 **35.** 36 cm **36.** 9 in.
37. 360° **38.** 360° **39.** 360°

Lesson 9.6, pages 524–531

TRY THESE

1. Yes; $\angle UVY \cong \angle UWY$ by the Corresponding Angles Postulate both triangles have the common $\angle C$. Thus, the triangles are similar by AA Similarity Postulate. $\triangle UVY \sim \triangle UWX$.
3. Yes; $\triangle BCA \sim \triangle DCE$ by the SAS Similarity Theorem. $\angle BCA \cong \angle DCE$ by the Vertical Angles Theorem. $\frac{DC}{CE} = \frac{BC}{CA} = \frac{5}{6}$, so the sides that include the congruent vertical angles are proportional.
5. $x = 2.5$ in.   **7.** They must be parallel.   **9.** 3.2m

PRACTICE

11. no; corresponding sides are not proportional.
19. $x = 3$ cm **21.** It is unknown which triangle contains the 6 cm median.   **23.** 700 ft **25.** 70°

ALGEBRA AND GEOMETRY REVIEW

35 $\angle 3$ and $\angle 5$, $\angle 4$ and $\angle 6$ **36.** $\angle 4$ and $\angle 5$, $\angle 3$ and $\angle 6$ **37.** $\angle 1$ and $\angle 5$, $\angle 2$ and $\angle 6$, $\angle 4$ and $\angle 8$, $\angle 3$ and $\angle 7$ **38.** $x = \frac{4}{7}$ **39.** $x = 3$ **41.** 45 **42.** 230 **43.** 53

Lesson 9.7, pages 532–535

APPLY THE STRATEGY

13. $\frac{h}{p} = \frac{\frac{1}{2}b + s}{r}$ **15.** 140 ft

17. no; The tree is 17.5m tall, too short for take off clearance **19.** Since TF and AP are both vertical segments, iTF | AP. So, $\angle DTF \cong \angle DAP$ by the Corresponding Angles Postulate. Since $DA \perp PB$, $\angle DAP \cong \angle BPA$ by the Alternate Interior Angles Theorem. Therefore, $\angle DTF \cong \angle BPA$ by the transitive property. Both $\angle ABP$ and $\angle TFD$ are right angles and they are congruent. Thus, $\triangle ABP \sim \triangle DFT$ by the AA Similarity Postulate.

Lesson 9.8, pages 536–543

TRY THESE

1. $x = 12$ cm; $y = 21.6$ cm **3.** $t = 10.5$ cm
5. $x = 7$ **7.** The studs would not be perpendicular to the floor support.

PRACTICE

9. $x = 3$ m
   $BC = 4$ m
13. They are parallel because of the Corresponding Angles Postulate. **15.** no; Section per section, comet 1 travels a distance that is two-thirds the distance that comet 2 travels. **17.** 20 cm

ALGEBRA AND GEOMETRY REVIEW

28. $x^2 + 8x + 15$ **29.** $4y^2 - 1$ **30.** $x^3 + 9x^2 + 19x - 4$ **31.** 226 **32.** 65 **33.** 34 **34.** 4m **35.** 74 in.

36. 35 cm **37.** $x = \pm 6$ **38.** $x = \pm 10$
39. $x = -8$ or 6

Chapter Review, pages 544–545

1. d **2.** a **3.** c **4.** b **5.** $x = 8$ **6.** $x = -\frac{11}{6}$ **7.**
$x = 6$ or $x = -6$ **8.** 4 **9.** yes; $1.5:1$ **10.** no **11.**
yes; $1:3$ **12.** $A'(-3, 9)$, $B'(6, 3)$, $C'(9, -6)$
13. $2:3; 2:10$ **14.** yes; AA; $\triangle ABD \sim \triangle ACE$
$1:4; 2:10$
$1:16; 4:9$
15. yes; SAS; $\triangle ABC \sim \triangle EDC$ **16.** yes; SSS; $\triangle ABC \sim$
$\triangle EFD$ **17.** 17.5 cm **18.** 27.5 ft **19.** $x = 14$ units
20. $x = 8\frac{1}{3}$ units **21.** no

Chapter 10 Applications of Similarity and Trigonometry

Lesson 10.2, pages 556–561

TRY THESE **1.** $YWZ; ZWX$ **3.** $WZ; XZ; XY$ **5.** $\sqrt{70}$
7. $2\sqrt{21}$ **9.** $x = 2\sqrt{3}$ **11.** $x = 2.5; y = 7.5$
13. 6.2 ft

PRACTICE **17.** $\triangle ADC; \triangle DBC$ **19.** 14 **21.** $3\sqrt{5}$
23. $x = 5$ **25.** $x = 4; Y = 2\sqrt{5}; Z = 3\sqrt{5}$

EXTEND **27. 1.** Given; **2.** Corollary 10.3; **3.** $cx = a^2$,
$cy = b^2$; **4.** Addition Property of Equality; **5.** $a^2 + b^2 =$
$c(x + y)$; **6.** Substitution **29.** $\sqrt[3]{60}$ **31.** $\frac{BC}{AB} = \frac{AB}{AC}$
ALGEBRA AND GEOMETRY REVIEW **37.** 18°, 72°, 90°,
22.5°, 67.5°, 90° **38.** 42° **39.** 10, 24, 26
40. octagon **41.** 7.90 **42.** 3.30 **43.** 3.45
44. 0.5333 **45.** 0.4286 **46.** 0.5556 **47.** 5 **48.** 60°
49. 120° **50.** ≈ 1.73 **51.** 2 **52.** ≈ 8.65

Lesson 10.3, pages 564–569

TRY THESE
1. $x = 3; y = 3\sqrt{2}$ **3.** $x = 5\sqrt{2}; y = 5\sqrt{2}$
5. short leg $= \frac{15}{2}$; long leg $= \frac{15\sqrt{3}}{2}$ **7.** $3\sqrt{2} \approx 4.24$ ft
PRACTICE **9.** $x = 10; y = 5\sqrt{3}$ **11.** $x = 3; y = 3$
13. $x = 8\sqrt{3}; y = 4\sqrt{3}$ **15.** $x = \frac{7\sqrt{6}}{2}; y = \frac{7\sqrt{2}}{2}$
17. 53 cm **19.** 127 ft
EXTEND **23.** $AD = 4$, $BF = 2\sqrt{3}$

Lesson 10.5, pages 573–579

TRY THESE **1.** $\angle Z$ is included in each triangle, and
$m\angle B = m\angle D = m\angle F = 90°$. The triangles are similar

by the AA Similarity Postulate. **3. 1.** $\angle Z$ is included
in each triangle, and $m\angle B = m\angle D = m\angle F = 90°$. The
triangles are similar by the AA Similarity Postulate.
2. a. $\frac{2}{3}$ **b.** $\frac{2}{3}$ **c.** $\frac{2}{3}$ **3.** 54 units
5. 0.4663 **7.** 2.1445 **9.** $x = 6.71$ **11.** $x = 14.93$
13. 23° **15.** 78° **17.** 37° **19.** 67° **23.** The
distance from Jaime to the building. **25.** The height of
the building minus the distance from the ground to
Jaime's eyes. **27.** $\tan 33° = \frac{x}{20}; x \approx 13$ m **29.** 947 ft
PRACTICE
31. $\tan C = \frac{5}{4} = 1.25$; $\tan D = \frac{4}{5} = 0.8$ **33.** 4.9
35. 53° **37.** 4.4 **39.** 36° **41.** 54° **43.** 3.4°
EXTEND **45.** $A = 45.7$ units^2 **47.** $A = 32.5$ units^2
49. 9°, $\sqrt{37}$ **51.** 74°, $\sqrt{53}$ **53.** $\sqrt{29}, 68°$
THINK CRITICALLY **55.** 34°
ALGEBRA AND GEOMETRY REVIEW **59.** $x \le 16$
60. no; one has slope $\frac{3}{2}$; and the other $-\frac{3}{2}$;
61. $A = 24\sqrt{3}$ cm^2; $P = 12 + 12\sqrt{3}$ cm **62.** $\frac{16}{121}$
63. $10\sqrt{2}$ m

Lesson 10.6, pages 580–586

TRY THESE **1.** $\frac{6}{10} = 0.6$ **3.** $\frac{6}{8} = 0.75$ **5.** $\frac{6}{10} = 0.6$
7. 0.6428 **9.** 0.7660 **11.** $x = 3.15$ **13.** $x = 7.33$
15. 33° **17.** 56° **19.** 66° **21.** 22° **23.** 132 ft
PRACTICE **25.** b, d **27.** f **29.** no match **31.** 39°
33. 11.9 **35.** 4.4 **37.** 54° **41.** 3431 ft
43. 30° $\frac{\sqrt{3}}{3}$ 0.5774 $\frac{1}{2}$ 0.5 $\frac{\sqrt{3}}{2}$ 0.8660
45. 60° $\sqrt{3}$ 1.7321 $\frac{\sqrt{3}}{2}$ 0.8660 $\frac{1}{2}$ 0.5
EXTEND **47.** $P = 22.4$ cm; $A = 18.2$ cm^2
THINK CRITICALLY **51.** Slant height $= 11.7$ in.; altitude
$= 11.5$ in. **53.** $45° < x < 90°$ **55.** $(90 - x)$
ALGEBRA AND GEOMETRY REVIEW
56. $\sin A = 0.6250$ **57.** $\sin A = 0.5151$
  $\cos A = 0.7806$ $\cos A = 0.8571$
  $\tan A = 0.8006$ $\tan A = 0.6009$
58. $\sin A = 0.8122$ **59.** $x = 8$
  $\cos A = 0.5833$
  $\tan A = 1.3924$
60. $x = 1.5$ **61.** $x = \frac{27}{4}$

Lesson 10.8, pages 590–593

APPLY THE STRATEGY **19.** 17 in. **21.** $FG = 8$; $FH =$
10.9; $GH = 7.5$; $m\angle F = 43°$, $m\angle G = 90°$, $m\angle H = 47°$

23. 15.7 ft; yes; At this height the angle measures 79°

Lesson 10.9, pages 594–599

Try These **1.** 60°, 8 in., 6.9 in. **3.** 51°, 2.3 ft., 2.1 ft
5. 166 in.^{2} **7.** 33 m^2 **9.** 755 in.^{2}

Practice **13.** $P = 35.3$ ft; $A = 86$ ft^2 **17.** 9.2 ft **19.**
152 m^2 **21.** 5.5 cm **23.** 179 ft^2 **25.** 7725 cm^2

Extend **27.** $P = 1780$ ft; $A = 218{,}050$ ft^2 **29.** no; It
would have considerably less space. **31.** Four times
larger.

Chapter Review, pages 600–601

1. c **2.** d **3.** F **4.** e **5.** a **7.** *JMK; KML*
8. $x = 6$, $y = 2\sqrt{13}$, $z = 3\sqrt{13}$ **9.** $x = 53$; $y = 10$
10. $x = 3$; $y = 3\sqrt{2}$ **11.** $x = 7$; $y = 73$ **12.** $x = 7$;
$y = 7$ **13.** 5 **14.** 4.7 **15.** 32° **16.** 2.3 **17.** 8.8
18. 50° **19.** $AC = 5.6$, $BC = 7.5$; m∠$B = 48°$
20. $EF = 8.1$; m∠$E = 68°$, m∠$F = 22°$
21. $GH = 3.8$; $HI = 9.8$; m∠$I = 23°$ **22.** 18 in.^{2}
23. 178 cm^2 **24.** 94 in.^{2}

Chapter 11 Geometry of the Circle

Lesson 11.2, pages 612–619

Try These **1.** Z **3.** $\overline{RT}$ **5.** L **9.** $\overarc{RAT}$, $\overarc{TBR}$
11. 10π m, 31.4 m **13.** 14.5π cm, 45.6 cm
15. radius: 100 in.; diameter: 200 in.
17. radius; $\dfrac{20.9}{\pi}$ m; 6.7 m **19.** 4πm, 12.6 m
21. 10π in., 31.4 in.

Practice **27.** 220π ft; 691.2 ft **29.** 18π in.; 56.5 in.
31. $7\frac{1}{8}\pi$ cm; 22.4 cm **33.** 21.6π yd; 67.9 yd
35. radius 200 in.; diameter 400 in. **37.** radius:
$\dfrac{49.3}{\pi}$ m; 15.7 m; diameter: $\dfrac{98.6}{\pi}$ m; 31.4 m
39. 2.5π m; 7.9 m **41.** $\dfrac{13}{3}\pi$ mm; 13.6 mm **43.** 9.0 in.

Extend **45.** 41.7 in. **47.** 71.4 ft. **49.** 44.4 cm

Think Critically **51.** Two circles that are equal in
circumference have the same radius. Circles that have
the same radius are congruent. Two rectangles are not
necessarily congruent if they have the same perimeter.
Examples vary. Two squares are congruent if they have
the same perimeters. **53.** sometimes **55.** 6.3 m

Algebra and Geometry Review **56.** A **57.** A
58. C **59.** B **60.** C **61.** A **62.** C

Lesson 11.3, pages 620–625

Try These **1.** 144π m^2; 452.4 m^2 **3.** 156.25π cm^2;
490.9 cm^2 **5.** 11 in. **7.** $\sqrt{\dfrac{89}{\pi}}$ m; 5.3 m **9.** 12.5π in.^{2}

11. 16π ft^2 **15.** $4\sqrt{3}$ in. **17.** $\dfrac{32}{3}\pi$ in.^{2}
19. $121\left(\dfrac{\pi}{4} - \dfrac{1}{2}\right)$m^2 **21.** 28π ft^2 **23.** One-eighth of the
large pizza has greater area. The total area of the small
pizza is 36π in.^{2}, so $\dfrac{1}{6}$ of the small pizza is 6π in.^{2}. The
total area of the large pizza is 49π in.^{2}, so $\dfrac{1}{8}$ of the large
pizza is $\dfrac{49}{8}\pi$ in.^{2} or 6.125π in.^{2}.

Practice **25.** 72.25π ft.^{2}; 227.0 ft^2 **27.** 515.29π in.^{2};
1618.8 in.^{2} **29.** $\dfrac{10}{7}$ m; 1.4 m **31.** $\sqrt{\dfrac{137}{3\pi}}$ in.; 3.8
33. 6.75π ft.^{2} **35.** 468π cm^2 **37.** $(6\pi - 9\sqrt{3})$cm^2
39. $225\left(\dfrac{\pi}{2} - 1\right)$ in.^{2} **41.** 264 in.^{2}

Extend
43. 0.6 m^2 **45.** 382.3 ft.^{2} **47.** 14π ft **49.** 24°

Think Critically **51.** True. The area of the respective
sectors and triangles are congruent. Therefore the
difference between their areas are equal. **53.** $a^2 : b^2$

Lesson 11.5, pages 630–633

Apply the Strategy **21.** A regular octagon with sides
of 50 cm has an area of approximately 12,071 cm^2.

Lesson 11.6, pages 634–639

Try These **1.** $\dfrac{16}{55}$ **3.** $\dfrac{1}{36}$ **5.** 0 **7.** $\dfrac{1}{9}$
11. radius + 5 in.
Practice **13.** $\dfrac{3}{40}$ **15.** $\dfrac{11}{5}$ **17.** $\dfrac{1}{12}$ **19.** 0 **21.** $\dfrac{9}{16}$
Extend **25.** $\dfrac{3}{8}$ **27.** $\dfrac{3}{4}$ **29.** 1 **31.** an infinite
number **33.** **a.** $\dfrac{1}{2}$ **b.** $\dfrac{1}{2}$ **c.** $\dfrac{1}{8}$ **d.** $\dfrac{7}{9}$ **e.** 0 **f.** 1
35. 5

Lesson 11.7, pages 640–645

Try These **1.** $x^2 + y^2 = 9$ **3.** $(x - 3)^2 + (y - 5)^2 = 4$
5. $C(0, 0)$, $r = 12$ **7.** $C(0, 0)$, $r = 5$
9. $C(-11, 5)$, $r = 2$
Practice **13.** $x^2 + y^2 = 324$ **15.** $x^2 + y^2 = 400$
17. $(x + 9)^2 + (y + 5)^2 = 81$ **19.** $C(0, 0)$, $r = 9\sqrt{2}$
21. $C(1, 10)$, $r = 9$
25. $(x - 3)^2 + (y - 6)^2 = 9$
$(x - 9)^2 + (y - 6)^2 = 9$
$(x - 6)^2 + (y - 3)^2 = 9$
$(x - 6)^2 + (y - 9)^2 = 9$
Extend **27.** $C = 6\pi\sqrt{2} = 26.7$; $A = 18\pi = 56.5$
29. $(x - 1)^2 + (y - 4)^2 = 64$
31. $(x - 4)^2 + (y - 8)^2 = 18$

THINK CRITICALLY   35. $y = \pm\sqrt{4 - (x - 2)^2} - 9$
37. $k = 0; k > 0$ **39.** $x^2 + y^2 = r^2$ **41.** $x^2 + y^2 = r^2$
ALGEBRA AND GEOMETRY REVIEW   43. $-2, 10$
45. 44 ft   **46.** 22   **47.** 3

Lesson 11.8, pages 646–653

TRY THESE

3. $r = 2$ in., $d = 4$ in., $c = 4\pi$ in. ≈ 12.6 in., $h = 12$ in., $s = 2\sqrt{37}$ in. ≈ 12.2 in.   **5.** rectangle, width $= 8\pi$ cm, height $= 12$ cm

PRACTICE

17. 15 in.   **19.** 9.2 ft or 221 ft   **21.** 120π ft

EXTEND   25. infinite number of planes, one axis
27. one plane, no axis   **29.** $(5, -2, 0)$, 8, 16, 16π
ALGEBRA AND GEOMETRY REVIEW   36. $-\dfrac{1}{2}$ **37.** $4\sqrt{5}$

39. scalene, obtuse   **40.** 60 ft^2, 40 ft.   **41.** sometimes
42. always

Chapter 12   Circles: Lines, Segments, and Angles

Lesson 12.1, pages 663–668

TRY THESE   1. yes   **3.** no   **5.** $3\sqrt{11}$

PRACTICE   9. no   **11.** yes   **13.** 15.2 in.   **17.** 5; $(1, 8)$
19. $\dfrac{4}{3}$; $-\dfrac{3}{4}$
EXTEND   21. 11.8 in.   **23.** 0.87   **25.** $CX = CP\sqrt{5}$
27. $4 - \pi$
31. $PX = PR, PY = PS, PZ = PT$

ALGEBRA AND GEOMETRY REVIEW   35. $(-2, -5)$ **36.** $(2, 5)$ **37.** $(-5, 2)$ **38.** $(5, -2)$ **39.** $(-10, -30)$
40. $\left(\dfrac{2}{3}, \dfrac{4}{3}\right)$ **41.** infinite solutions   **42.** $\left(\dfrac{4}{7}, \dfrac{1}{7}\right)$
43. no   **44.** $(0, 0)$ **45.** 16.0

Lesson 12.2, pages 669–674

TRY THESE   1. 90°   **3.** 101°   **5.** 79°

PRACTICE   9. 90°   **11.** 125°   **13.** 60°

EXTEND   17. 22.5°   **19.** $5^2 + 12^2 = 25 + 144 + 169$
21. 15.3   **27.** Since rotation preserves the measure of an angle, $\angle B'PA' \cong \angle BPA$. Therefore, arc length and arc measures are preserved.

THINK CRITICALLY
29. Given: In circle O, m $\overset{\frown}{AB}$ = m $\overset{\frown}{CD}$.
Prove: m$\angle AOB$ = m$\angle COD$.
Proof: let m $\overset{\frown}{AB}$ = $x°$. Then, by the definition of the degree measure of an arc. m$\angle AOB = x°$. It is given that m $\overset{\frown}{AB}$ = m $\overset{\frown}{CD}$, so m $\overset{\frown}{CD}$ = $x°$, and, by the definition of

the degree measure of an arc. m$\angle COD = x°$. Since m$\angle AOB = x°$ and m$\angle COD = x°$, by the Transitive Property it follows that m$\angle AOB$ = m$\angle COD$. Using similar reasoning, if $\overset{\frown}{AB}$ and $\overset{\frown}{CD}$ are minor arcs of circles O and P, respectively, it can be shown that m$\angle AOB$ = m$\angle CPD$.
31. $\dfrac{L}{2\pi r} = \dfrac{d}{360}$; $L = \dfrac{dr\pi}{180}$; the length of a minor arc of a circle with radius r is equal to the measure of the central angle d times r times π divided by 180.
ALGEBRA AND GEOMETRY REVIEW   33. $(x + 1)(x - 15)$
34. $(x + 8)(x + 8)$ **35.** $(x - 6)(x + 5)$ **36.** $(x + 11)(x - 11)$ **37.** $(2x - 3)(x + 7)$ **38.** $(x - 8)(x - 8)$ **39.** 71°   **40.** AC, AB, BC **41.** 45°, 107°, 118°

Lesson 12.3, pages 675–680

TRY THESE   1. 2.8   **1.6**   **5.** 3.2   **9.** 15 ft; 27 ft

PRACTICE   11. 2.0   **13.** 3.0   **15.** 3.3   **17.** 9.6 in.
19. 30.2 in.

EXTEND   21. 1.8 units   **23.** 244 m   **25.** $PC = 9.1$, $AB = 8.5$ **27.** 328 ft   **29.** 13.9   **31.** 15.5
THINK CRITICALLY   33. $PC = r\cos\dfrac{x}{2}$; $AB = 2r\sin\dfrac{x}{2}$
35. Given $AB = CD$ Prove: $QX = QY$
ALGEBRA AND GEOMETRY REVIEW   38. 108°   **39.** 140°
40. 150°   **41.** 35   **42.** ± 26 **43.** -30 **44.** 12.5 cm

Lesson 12.4, pages 681–687

TRY THESE   1. 50°   **3.** 40°   **5.** 68°   **9.** 90°; rectangle or square

PRACTICE   11. 90°   **13.** 61°   **15.** 58°   **19.** $XY = 7.43$ cm, $YZ = 6.69$ cm, perimeter $= 24.12$ cm

ALGEBRA AND GEOMETRY REVIEW   39. 4   **40.** 9
41. -3 **42.** $-\dfrac{2}{3}$ **43.** $x = -2$ **44.** $x = 5$
45. $x = 2$ **46.** 62.5° or 10.5°

Lesson 12.6, pagaes 694–699

TRY THESE   1. 102.5°   **3.** 69°   **7.** 50°; 53°

PRACTICE   9. 40.5°   **11.** 15°   **13.** 69°   **15.** 61°
17. 81°   **19.** 72°   **21.** 108°

EXTEND   23. 10.5°; m$\overset{\frown}{XY}$ = 169.5 $\overset{\frown}{XYZ}$ = 190.5°
25. m$\angle XPY$ = 18.5

ALGEBRA AND GEOMETRY REVIEW   33. 9; $(-4, 7)$
34. 11: $(3, 0)$ **35.** $\dfrac{15}{17}$ **36.** $\dfrac{15}{17}$ **37.** $\dfrac{8}{17}$ **38.** $\dfrac{8}{17}$
39. $\dfrac{15}{8}$ **40.** $\dfrac{5}{15}$ **41.** 89°

Lesson 12.7, pages 700–705

TRY THESE   1. 4.4

PRACTICE   7. 2.3   **13.** $BC \approx 13.4$ cm, $BD \approx 12$ cm

15. $x = 1$ **17.** $x = 5.7$

EXTEND **19.** 5.6 **21.** 6 **23.** $CR = 4$, $EP = 2$

ALGEBRA AND GEOMETRY REVIEW **29.** 100° **30.** 80°
31. 10° **32.** 100° **33.** 80° **34.** 90° **35.** 9 **36.** 72
37. 46.25 **38.** 6 **39.** 72.08 **40.** 24 **41.** 2.4

Lesson 12.9, pages 710–713

APPLY THE STRATEGY **15.** about 17.7 in. **17.** 224.9
19. $1; \frac{1}{2}; \frac{1}{3}; \frac{1}{4}; \frac{1}{3813}$ or 1: 05; $0.\overline{333}$; 0.25; 0.00026
CHAPTER REVIEW, PAGES 714–715 **1.** c **2.** d **3.** b
4. a **5.** no **6.** yes **7.** yes **8.** 5 **9.** 5 **10.** 3
11. 5 **12.** 44° **13.** 88° **14.** 44° **15.** 10° **17.** 40°
18. 100° **19.** 80° **20.** 10 in. **21.** 12 in. **22.** 18 in.
23. F, F, T; F, T, T; T, F, F; T, T, T **24.** about 12 mi

Chapter 13 Surface Area and Volume

Lesson 13.3, pages 731–736

TRY THESE

5. $LA = ph$
  $= 18 \cdot 12$
  $= 216$ in.^{2}
$SA = ph + 2B$
  $= 216 + 2(23.4)$
  $= 262.8$ in.^{2}

7. LA: 251 cm^2; SA: 352 cm^2

9. LA: 240 m^2; SA: 288 m^2 **11.** 39 in.^{2}

PRACTICE

15. $LA = ph$
  $= (5 + 12 + 13)8$
  $= 30 \cdot 8$
  $= 240$ cm^2
$SA = ph + 2B$
  $= 240 + 21\backslash2(5 \cdot 12)$
  $= 240 + 60$
  $= 300$ cm^2

17. LA: 22 in.^{2} SA: 33 in.^{2}

19. LA: 240 ft^2 SA: 295 ft^2 **21.** LA: 585 in.^{2} SA: 665
in.^{2} **25.** No, surface area is 486 in.^{2}

EXTEND **27.** 76,000

THINK CRITICALLY **33.** yes; a cube having a height
of 6 in.

ALGEBRA AND GEOMETRY REVIEW

35.

Statement	Reason
1. $\triangle ABC$, D is midpoint of $\overline{AC}$, $\angle ADB \cong \angle CDB$	1. Given
2. $\overline{AD} \cong \overline{DC}$	2. definition of midpoint
3. $\overline{DB} \cong \overline{DB}$	3. reflexive property

4. $\triangle ADB \cong \triangle CDB$ **4.** SAS
37. false **38.** 18.03 in. **39.** $\frac{1}{11}$
40. $x < 4$

41. $h^3 - h^2 - 5h - 3$ **42.** $-2 - \sqrt{3}$

Lesson 13.4, pages 737–743

TRY THESE **1.** 216 ft^3 **3.** 550 m^3 **5.** 368 cm^3
7. 52.5 cm^3 **9.** V = 8 units^3 SA = 30 units^2
11. V = 8 units^3 SA = 34 units^2

PRACTICE **13.** 707 ft^3 **15.** 6182 cm^3 **17.** 16,085 in.^{3}
19. 48 cm^3 **21.** 204 ft^3

EXTEND **25.** 22 in. **27.** 5316 in.^{3} **29.** 15.625 m^3

THINK CRITICALLY **31.** If $a > b$, the volume of the first
cylinder is greatest. If $b > a$, the volume of the second
cylinder is greatest. **33.** false

ALGEBRA AND GEOMETRY REVIEW **35.** 151 **36.** 1980°
37. 24 **38.** 4.67×10^{-9}

Lesson 13.5, pages 744–747

APPLY THE STRATEGY **15.** 8 in., 8 in., 2 in.
17. 375 in.^{3} **19.** 15.3 in., 6.6 in., 2.7 in.

Lesson 13.6, pages 748–755

TRY THESE **3.** $LA = 520$ cm^2 $SA = 779$ cm^2
5. $LA = 81$ m^2 $SA = 109$ m^2 **7.** 1265 cm^2 **9.** 9;
square or triangle **11.** 107.7 in.^{2}

PRACTICE **13.** 204 in.^{2} **15.** 56 cm^2 **17.** 758 cm^2
19. 145 ft^2 **21.** 902 cm^2 **23.** 217 in.^{2} **25.** 169 in.^{2}

EXTEND **27.** 806 cm^2 **29.** 121 cm^2 **31.** 6.4 cm

THINK CRITICALLY **33.** 103.7 cm^2 **35.** 1583 in.^{2}
37. doubled

ALGEBRA AND GEOMETRY REVIEW **40.** $9\sqrt{3}$ cm
41. 115° **42.** 17.5 **43.** $(x + 3)(y - 5)$ **44.** 2

Lesson 13.7, pages 756–762

TRY THESE **1.** 500 cm^3 **3.** 452 m^3 **5.** 1050 ft^3
7. 3 m $\times$ 3 m **9.** 106.3 in.^{3} **13.** 615.8 in.^{3}
15. 384 cm^3 **17.** 528 cm^3 **19.** 73 m^3 **21.** 445 m^3
23. 4 ft **27.** 62.2 in.^{3}

EXTEND **29.** 3024 m^3 **31.** 378 in.^{3}

THINK CRITICALLY **33.** 840 in.^{3} **35.** 1081 ft^3
37. doubled; quadrupled

ALGEBRA AND GEOMETRY REVIEW **39.** $\sqrt{265}$ **40.** 7
41. $(7, 6)$ **42.** $\frac{6 - 5x}{-3x - 3}$

Lesson 13.8, pages 763–769

TRY THESE **1.** SA: $400\pi \approx 1257$ cm^2; V: $1333.3\pi \approx$ 4189 cm^3 **3.** SA: $132.25\pi \approx 415$ in.^{2}; V: $253.479\overline{16}\pi \approx 796$ in.^{3} **5.** SA: $36\pi \approx 113$ yd^2; V: $36\pi \approx 113$ yd^3 **7.** 4189 in.^{3} **9.** about 661.8 in.^{3}

PRACTICE

11. SA: $324\pi \approx 1018$ cm^2 **13.** SA: $114.7\pi \approx 357$ in.^{2}
$\quad\quad V$: $972\pi \approx 3054$ cm^3 $\quad\quad V$: $202.2716\pi \approx 635$ in.^{3}
17. SA: 2.89×10^7 mi^2 **19.** SA: 1.97×10^8 mi^2
$\quad\quad V$: 1.46×10^{10} mi^3 $\quad\quad V$: 2.09×10^{12} mi^3
21. SA: 2.47×10^{10} mi^2 **23.** SA: 3.26×10^9 mi^2
$\quad\quad V$: 3.65×10^{14} mi^3 $\quad\quad V$: 1.75×10^{13} mi^3
25. SA: 6.36×10^6 mi^2
$\quad\quad V$: 1.51×10^9 mi^3

EXTEND **27.** 44.67 in.^{2} **29.** 10000π mm$^2 \approx 31416$ mm^2; 166666.6π mm$^3 \approx 523599$ mm^3 **31.** $17.78\pi \approx 223$ mm^2

THINK CRITICALLY **33.** 64π in.$^2 \approx 201$ in.^{2}

ALGEBRA AND GEOMETRY REVIEW **35.** 0 **36.** 0 **37.** undefined **38.** cube: $SA = 864$ cm^2, $V = 1728$ cm^3 **39.** 823 cm^3; sphere: $SA = 452$ cm^2, $V = 905$ cm^3 **40.** $-1 \le y \le \frac{7}{3}$

Lesson 13.9, pages 770–775

TRY THESE **1.** no **3a.** $\frac{2}{1}$ **3b.** $\frac{4}{1}$ **3c.** $\frac{8}{1}$ **5.** 711 cm^2

PRACTICE **9.** 2:7, 2:7, 4:49, 4:49, 8:343 **11.** 2:3, 2:3, 2:3, 4:9, 4:9 **13.** $2343.75

EXTEND **17.** 0.0625 in.^{2} **21.** 1.15 in.

THINK CRITICALLY **23.** always **25.** always **27.** always **29.** no **31.** no **33.** pqr times, no

ALGEBRA AND GEOMETRY REVIEW **34.** 72 **35.** D **36.** -3 **37.** -70 **38.** 660 ft

Chapter Review, pages 776–777

1. a **2.** d **3.** b **4.** c **5.** 9 **6.** 7 **7.** 21 **8.** All have triangular bases. **9.** The single figure in one group has an apex while the other figures do not. **10.** $SA = 408$ units^2, $LA = 360$ units^2, $V = 360$ units^3 **11.** $SA \approx 447.6$ cm^2, $LA \approx 277.7$ cm^2, $V \approx 722.1$ cm^3 **12.** $3\frac{1}{3}$ in. $\times$ $3\frac{1}{3}$ in. **13.** $SA = 96$ in.^{2}, $LA = 60$ in.^{2}, $V = 48$ in.^{3} **14.** $SA \approx 489.2$ cm^2, $LA \approx 344.2$ in.^{2}, $V \approx 718.4$ cm^3 **15.** $V \approx 74$ in.^{3} **16.** $SA \cong 1520.5$ cm^2, $V = 5575.3$ cm^3 **17.** 4050 in.^{3}

INDEX

INDEX

equal, 176
magnatude, 176
parallel, 176
perpendicular, 178
terminal point of, 585
Venn diagram, 106–108, 147, 727–730
Vertex
of angle, 92
of cone, 647
defined, 24, 31
of polygon, 195
of polyhedron, 724
of prism, 420
of pyramid, 420, 748
Vertical lines, slope-intercept form and, 168
Volume
of cones, 756–761
maximizing, 744–746
of prisms and cylinders, 737–742
of pyramids, 756–761
of spheres, 763–767
of three-dimensional figure, 723–775

Volume Addition Postulate, 737
Volume Congruence Postulate, 737
Volumetric weight, 762

• • **W** • •

Width, of rectangular prism, 468
Working Together, 3, 193, 253, 371, 491, 551, 607, 661, 721
Writing Mathematics, 8, 12, 19–20, 26–28, 32, 36–37, 43–44, 67–68, 75–76, 81–82, 91, 96–97, 103, 107, 113, 118–119, 132–134, 138, 142–143, 144–145, 151, 153, 155, 158, 164–165, 170–171, 175, 179–180, 184, 188, 190, 198–199, 203, 209, 213–214, 217, 221, 227, 234, 244, 245, 258, 263, 270, 274, 278, 283, 286, 291–292, 296, 300, 306, 315, 320–321, 328–329, 333, 339, 340–341, 343–344, 351, 355–356, 376, 380, 382, 387, 394–395, 400, 407, 413, 414, 422–423, 428, 430, 437–438, 440, 443–444, 451–453, 457, 460, 462, 463, 469, 471–473, 477, 479, 500, 504–505, 512, 518–519, 523, 528, 532–533, 539–540, 555, 559, 567–568, 572, 576–578, 583, 589, 591, 592, 597–599, 611, 615–616, 623, 628, 632, 636, 637, 643, 649, 650, 666, 671, 678, 684, 690–691, 693, 697, 703, 711, 712, 725, 726, 734–735, 740–741, 745–746, 751–752, 759–760, 766, 769, 773–774

• • **X** • •

x-axis, 41
x-coordinate, 41, 161

• • **Y** • •

y-axis, 41
y-coordinate, 41, 161, 167

Photo Credits

CONTENTS

p. vi: Roberto De Gugliemo/Science Photo Library/Photo Researchers, Inc.; p viii: Alan Goldsmith/The Stock Market; p. ix: Ryan J. Hulvat; p. x: Earl Glass/Stock, Boston; p. xi: PhotoDisc, Inc.; p. xii: Ryan J. Hulvat; p. xiii: Greg Grosse; p. xiv: Frank Siteman/Stock, Boston; p. xv: Ryan J. Hulvat; p. xvi: Ron Kimball

CHAPTER 1

p. 1: PhotoDisc, Inc. (top right); Joseph Higgins (bottom right); p. 2: PhotoDisc, Inc. (all); p. 3: Stephen Frisch/Stock, Boston; p. 4: Grant Heilman Photography, Inc.; p. 5: Grant Heilman Photography, Inc. (top left, top right, middle left); p. 5: Joseph Higgins (middle right); p. 5: Nuridsany et Perennou/Photo Researchers, Inc. (bottom right); p. 8: PhotoDisc, Inc. (all); p. 12: Grant Heilman Photography, Inc.; p. 13: Terry E. Eiler/Stock, Boston, Inc.; p. 15: PhotoDisc, Inc. (left); p. 15: Federal Bureau of Investigation (insets); p. 17: PhotoDisc, Inc.; p. 18: David Sutherland/Tony Stone Images; p. 19: Joseph Higgins; p. 20: Joseph Higgins; p. 21: David R. Frazier/Tony Stone Images; p. 22: Dennis O'Clair/Tony Stone Images; p. 23: Allen Russell/ Index Stock Photography; p. 25: Peter Beck/The Stock Market; p. 26: Dave Wendt/Wendt Worldwide Photography; p. 28: Bob Daemmrich/Stock, Boston; p. 29: Greg Grosse; p. 30: Aaron Haupt/Photo Researchers, Inc.; p. 31: PhotoDisc, Inc.; p. 33: Linc Cornell/Stock, Boston; p. 34: Ed Wheeler/The Stock Market; p. 36: Courtesy of International Business Machines Corp.; p. 39: Jeff Gnass Photography/ The Stock Market; p. 42: PhotoDisc, Inc.; p. 46: Jeff Greenberg; p. 49: Ryan J. Hulvat; p. 52: Ryan J. Hulvat; p. 53: Mark E. Gibson/ The Stock Market

CHAPTER 2

p. 62: John Madere/The Stock Market (top); Photonics, Graphics (bottom); p. 64: Courtesy of TransAmerica Corporation; p. 65: Ryan Hulvat; p. 67: Piet Mondrian, "Tableau No. IV; Lozenge Composition with Red, Gray, Blue, Yellow, and Black," 1924/1925, Gift of Herbert and Nannette Rothschild, c Board of Trustees, National Gallery of Art, Washington; p. 68: Joseph Higgins; p. 71: Lonnie Duka/Tony Stone Images; p. 73: Ryan J. Hulvat; p. 76: David Young Wolff/Tony Stone Images; p. 78: Anthony Edgeworth/The Stock Market; p. 80: PhotoDisc, Inc.; p. 84: Ryan J. Hulvat; p. 90: Focus on Sports; p. 93: PhotoDisc, Inc.; p. 97: PhotoDisc, Inc.; p. 98 Jim Pickerell/Tony Stone Images; p. 99: NASA; p. 100: Ryan J. Hulvat; p. 103: John Gilmore; p. 105: PhotoDisc, Inc.; p. 106: Don Smetzer/Tony Stone Images; p. 108: Churchill & Klehr/Tony Stone Images; p. 111: Greg Grosse; p. 116: Jeff Greenberg

CHAPTER 3

p. 126: David Tejada/Tony Stone Images; p. 127: Courtesy of the U.S. Coast Guard; p. 128: David Schultz/Tony Stone Images; p. 131: Ryan J. Hulvat; p. 132: Greg Grosse; p. 135: Joseph Higgins; p. 136: Ryan J. Hulvat; p. 141: Ryan J. Hulvat; p. 145: Fred J. Maroon/Photo Researchers, Inc.; p. 146: Yva Momatiuk/Photo Researchers, Inc.; p. 148: Wendt Worldwide Photography; p. 151: Wendt Worldwide Photography; p. 153: Carl Purcell/Photo Researchers, Inc.; p. 154: Benjamin Rondel/The Stock Market; p. 156: Jeff Zarube/The Stock Market; p. 157: Joseph Higgins; p. 160: Jeff Greenberg/Photo Researchers, Inc.; p. 161: Greg Grosse; p. 162: Michael Tamborrino/The Stock Market; p. 163: James Prince/Photo Researchers, Inc.; p. 165: Chris Jones/The Stock Market; p. 170: Bob Daemmrich/Stock, Boston; p. 171: Focus on Sports; p. 173: Ryan J. Hulvat; p. 177: Wendt Worldwide Photography; p. 179: Rose Hamilton/Tony Stone Images

CHAPTER 4

p. 192: Burlington Industries, Inc.; p. 192: Allis-Chalmers Corporation; p. 193: Chuck Savage/The Stock Market; p. 194: p. Erich Lessing/Art Resource, NY; p. 198: Joseph Higgins; p. 201: CORBIS-BETTMANN; p. 202: Ryan J. Hulvat; p. 204: Bob Daemmrich/Stock, Boston; p. 207: Ryan J. Hulvat; p. 209: PhotoDisc, Inc.; p. 210: Joseph Higgins; p. 213: Courtesy of International Business Machines, Inc.; p. 216: Joseph Higgins; p. 223: Robert Frerck/Tony Stone Images

CHAPTER 5

p. 252: Phyllis Picardi/Stock, Boston; p. 253: Michael Keller/The Stock Market; p. 254: The Stock Market; p. 256: ChromoSohm/Sohm/The Stock Market; p. 258: Churchill & Klehr; p. 262: Greg Grosse; p. 263: Robert Rathe/Stock, Boston; p. 266: Andy Levin/Photo Researchers; p. 270: M.C. Escher's "Symmetry Drawing E 99" Cordon Art-Baam-Holland. All rights reserved.; p. 274: Ryan J. Hulvat; p. 277: Joseph Higgins; p. 279: PhotoDisc, Inc.; p. 281: Peter Beck/The Stock Market; p. 284: Cary Wolinsky/ Stock, Boston; p. 286: PhotoDisc, Inc.; p. 289: Jeff Greenberg; p. 290: Bob Daemmrich/Stock, Boston; p. 294: Ryan J. Hulvat; p. 298: Tony Stone Images; p. 303: Mark Sparacio

CHAPTER 6

p. 310: John Lamb/Tony Stone Images; p. 310: Steve Elmore/The Stock Market; p. 311: Jon Feingersh/The Stock Market; p. 315: PhotoDisc, Inc. (all); p. 316: Ryan J. Hulvat; p. 318: PhotoDisc, Inc.; p. 320: Eric Simmons/Stock, Boston; p. 323: Jeffrey Muir Hamilton/Stock, Boston; p. 326: PhotoDisc, Inc.; p. 329: PhotoDisc, Inc. (all); p. 333: Peter Dazeley/Tony Stone Images; p. 334: Joseph Higgins; p. 339: Ilene Perlman/Stock, Boston; p. 341: Rob Crandall/Stock, Boston; p. 342: Peter Beck/The Sock Market; p. 343: Pal Hermansen/Tony Stone Images; p. 346: Ryan J. Hulvat; p. 348: Ed Kashi; p. 350: Ryan J. Hulvat; p. 351: Joseph Higgins; p. 353: PhotoDisc, Inc.; p. 355: Doris De Witt/Tony Stone Images; p. 357: Robert E. Daemmrich/Tony Stone Images; p. 358: Joseph Higgins; p. 361: David Barnes/ The Stock Market; p. 363: Gabe Palmer/The Stock Market

CHAPTER 7

p. 370: PhotoDisc, Inc. (all); p. 371: PhotoDisc, Inc.; p. 379: Greig Cranna/Stock, Boston; p. 380: Barbara Alper/Stock, Boston; p. 389: David Weintraub/Stock, Boston; p. 390: PhotoDisc, Inc.; p. 392: William Johnson/Stock, Boston; p 395: Grant Heilman Photography, Inc.; p. 398: Peter Menzel/Stock, Boston; p. 400: Grant Heilman Photography, Inc.; p. 403: Melissa Grimes-Guy/Photo Researchers; p. 404: Joseph Higgins; p. 406: Ryan J. Hulvat; p. 407: Joseph Higgins; p. 410: PhotoDisc, Inc.; p. 413: Frank Cezus/Tony Stone Images; p. 414: Joseph Nettis/Stock, Boston; p. 417: Jeff Greenberg; p. 419: David Parker/Science Photo Library/Photo Researchers; p. 423: Robert Huntzinger/The Stock Market; p. 425: John Henley/The Stock Market

CHAPTER 8

p. 431: Grant Heilman Photography, Inc.; p. 433: Charles Gupton/Stock, Boston, Inc.; p. 479: Jeff Greenberg; p. 436: Hans Reinhard/Photo Researchers, Inc.; p. 438: Joseph Higgins; p. 442: Ryan J. Hulvat; p. 445: Mark Burnett/Stock, Boston, Inc.; p. 447: Bob Daemmrich/Stock, Boston, Inc.; p. 448: Courtesy of Citicorp; p. 451: Ryan J. Hulvat; p. 452: The Image Bank; p. 457: Greg Grosse; p. 461: Bill Horsman/Stock, Boston, Inc.; p. 462: Grant Heilman Photography, Inc.; p. 464: Joe McBride/Tony Stone Images; p. 467: SPRINGER/CORBIS-BETTMANN; p. 472: High-Res Solutions/Phototake NYC; p. 475: Mimi Ostendorf-Smith/Photonics; p. 476: Henley and Savage/Tony Stone Images; p. 480: Steve Mason/The Stock Market; p. 481: Greg Grosse

CHAPTER 9

p. 490: Earth Imaging/Tony Stone Images; p. 491: Greg Mancuso/Tony Stone Images; p. 492: Joseph Higgins; p. 493: PhotoDisc, Inc.; p. 495: Peter Menzel/Stock, Boston; p. 497: Ryan J. Hulvat; p. 499: PhotoDisc, Inc.; p. 500: Charles D. Winters/Photo Researchers, Inc.; p. 503: PhotoDisc, Inc.; p. 507: Mimi Ostendorf-Smith/Photonics Graphics; p. 511: Ryan J. Hulvat; p. 517: Courtesy Frederick & Nelson Co.; p. 519: Peter Beck/ The Market; p. 520: Michael Dwyer/Stock, Boston; p. 521: Ryan J. Hulvat; p. 522: PhotoDisc, Inc.; p. 527: New York City Department of Sanitation; p. 529: Mark C. Burnett/Photo Researchers, Inc.; p. 532: R.J. Erwin/Photo Researchers, Inc.; p. 540: NASA; p. 542: Bill Gallery/Stock, Boston; p. 543: Joseph Pobereskin/Tony Stone Images (bottom left); PhotoDisc, Inc. (bottom middle)

CHAPTER 10

p. 550: Courtesy of Miles Homes, A division of Insilco Corporation; p. 550: Mark E. Gibson; p. 551: PhotoDisc, Inc.; p. 552: Charles Feil/Stock, Boston; p. 554: Courtesy of International Business Machines Corp.; p. 555: Springer-Verlag, New York; p. 562: Art Resource, NY; p. 568: Joseph Higgins